Lecture Notes in Computer Science 5978

Daniele Micciancio (Ed.)

Theory of Cryptography

7th Theory of Cryptography Conference, TCC 2010 Zurich, Switzerland, February 9-11, 2010 Proceedings

Volume Editor

Daniele Micciancio
University of California, San Diego
Computer Science & Engineering Department
9500 Gilman Drive, La Jolla, CA 92093-5004, USA
E-mail: daniele@cs.ucsd.edu

Library of Congress Control Number: 2009943559

CR Subject Classification (1998): B.3, C.2, D.4.6, K.6.5, G.2, I.1

LNCS Sublibrary: SL 4 – Security and Cryptology

ISSN 0302-9743
ISBN-10 3-642-11798-8 Springer Berlin Heidelberg New York
ISBN-13 978-3-642-11798-5 Springer Berlin Heidelberg New York

springer.com

Printed in Germany

Typesetting: Camera-ready by author, data conversion by Scientific Publishing Services, Chennai, India
Printed on acid-free paper SPIN: 12990504 06/3180 5 4 3 2 1 0

Preface

TCC 2010, the 7th Theory of Cryptography Conference, was held at ETH Zurich, Zurich, Switzerland, during February 9–11, 2010. TCC 2010 was sponsored by the International Association of Cryptologic Research (IACR) and was organized in cooperation with the Information Security and Cryptography group at ETH Zurich. The General Chairs of the conference were Martin Hirt and Ueli Maurer.

The conference received 100 submissions, of which the Program Committee selected 33 for presentation at the conference. The Best Student Paper Award was given to Kai-Min Chung and Feng-Hao Liu for their paper "Parallel Repetition Theorems for Interactive Arguments." These proceedings consist of revised versions of those 33 papers. The revisions were not reviewed, and the authors bear full responsibility for the contents of their papers. In addition to the regular papers, the conference featured two invited talks: "Secure Computation and Its Diverse Applications," given by Yuval Ishai and "Privacy-Enhancing Cryptography: From Theory Into Practice," given by Jan Camenisch. Abstracts of the invited talks are also included in this volume.

As in previous years, TCC received a steady stream of high-quality submissions. Consequently, the selection process was very rewarding, but also very challenging, as a number of good papers could not be accepted due to lack of space. I would like to thank the TCC Steering Committee, and its Chair Oded Goldreich, for entrusting me with the responsibility of selecting the conference program. Since its inception, TCC has been very successful in attracting some of the best work in theoretical cryptography every year and offering a compelling program to its audience. I am honored I had the opportunity to contribute to the continuation of the success of the conference.

I wish to thank all the people who contributed to the conference. Many thanks to the authors of all submitted papers, whose work is the very reason for TCC to exist. I am grateful for the dedication of the Program Committee members, who spent countless hours reviewing and discussing the submissions, to determine which papers to include in the program, and to provide the authors with useful feedback. I also thank the many external reviewers who assisted the Program Committee in its work. Thank you to past TCC Chairs Ran Canetti, Salil Vadhan and Omer Reingold, who promptly answered my questions each time I needed their advice. Special thanks to Shai Halevi, who wrote a wonderful software package to facilitate all aspects of the PC work. I am very grateful to TCC 2010 General Chairs Martin Hirt and Ueli Maurer without whom the conference would not have happened. Thanks also to our corporate Sponsors, Credit Suisse, Microsoft Research and Omnisec. Finally, thanks to the Springer LNCS editorial staff, including Ingrid Haas, Alfred Hofmann, Frank Holzwarth, and Anna Kramer, for their assistance in assembling these proceedings.

December 2009 Daniele Micciancio

Organization

TCC 2010

Theory of Cryptography Conference

February 9–11, 2010, Zurich, Switzerland

Sponsored by the
International Association for Cryptologic Research (IACR)

Organized in cooperation with the
Information Security and Cryptography group, ETH Zurich

With financial support from
Credit Suisse, Microsoft Research and Omnisec

General Chairs

Martin Hirt and Ueli Maurer	ETH, Switzerland

Program Committee

Michael Backes	Saarland University and MPI SWS, Germany
Amos Beimel	Ben-Gurion University, Israel
Alexandra Boldyreva	Georgia Institute of Technology, USA
Dario Catalano	University of Catania, Italy
Nelly Fazio	City University of New York, USA
Jens Groth	University College London, UK
Iftach Haitner	Microsoft Research New England, USA
Johan Håstad	KTH, Sweden
Susan Hohenberger	Johns Hopkins University, USA
Daniele Micciancio (Chair)	University of California, San Diego, USA
Phong Nguyen	INRIA and ENS, France
Rafail Ostrovsky	University of California, Los Angeles, USA
Alon Rosen	IDC Herzliya, Israel
Victor Shoup	New York University, USA
Adam Smith	Pennsylvania State University, USA
Vinod Vaikuntanathan	IBM T.J. Watson Research Center, USA
Bogdan Warinschi	University of Bristol, UK
Stefan Wolf	ETH Zurich, Switzerland

TCC Steering Committee

Mihir Bellare	University of California, San Diego, USA
Ivan Damgård	University of Aarhus, Denmark

Oded Goldreich (Chair)	Weizmann Institute of Science, Israel
Shafi Goldwasser	MIT and Weizmann Institute of Science, USA and Israel
Johan Håstad	KTH, Sweden
Russell Impagliazzo	UC San Diego and IAS, USA
Ueli Maurer	ETH Zurich, Switzerland
Silvio Micali	MIT, USA
Moni Naor	Weizmann Institute of Science, Israel
Tatsuaki Okamoto	NTT Laboratories, Japan

External reviewers

M. Abdalla
J.H. Ahn
J. Alwen
J. Bethencourt
C. Bosley
Z. Brakerski
E. Bresson
P. Bunn
R. Canetti
N. Chandran
M. Chase
C. Cho
S. Chow
K.-M. Chung
O. Ciobotaru
S. Coull
Y. Deng
Y. Dodis
M. Duermuth
O. Dunkelman
D. Fiore
M. Fischlin
M. Fitzi
P.-A. Fouque
S. Garg
P. Golle
P. Gopal
D. Gordon
R. Gradwohl
M. Green
K. Haralambiev
C. Hazay
B. Hemenway
J. Herranz
D. Hofheinz
T. Holenstein
Y. Ishai
A. Jain
R. Jaiswal
Y. Kalai
B. Kanukurthi
J. Katz
E. Kiltz
G. Kreitz
V. Kumar
E. Kushilevitz
F. Laguillaumie
H. Lin
Y. Lindell
M. Liskov
V. Lyubashevsky
M. Mahmoody-Ghidary
T. Moran
S. Myers
A. Nicolosi
K. Nissim
A. O'Neill
P. Paillier
O. Pandey
R. Pass
C. Peikert
K. Pietrzak
D. Pointcheval
T. Rabin
O. Regev
R. Reischuk
A. Sahai
G. Segev
Y. Seurin
H. Seyalioglu
E. Shi
H. Sibert
W.E. Skeith
I. Spharlinski
M. Stam
T. Tassa
M. Tennenholtz
A. Tentes
B. Terelius
S. Vadhan
M. Varia
M. Venkitasubramaniam
D. Vergnaud
I. Visconti
A. Wadia
H. Wee
D. Wichs
D. Wikström
J. Wullschleger
D. Xiao
Q. Yang
V. Zikas

Table of Contents

Parallel Repetition

Obfuscation

Invited Talk

Multiparty Computation

CCA Security

Threshold Cryptography and Secret Sharing

Symmetric Cryptography

Key-Leakage and Tamper-Resistance

Rationality and Privacy

Public-Key Encryption

Invited Talk

Zero-Knowledge

An Efficient Parallel Repetition Theorem

Johan Håstad[1], Rafael Pass[2], Douglas Wikström[3], and Krzysztof Pietrzak[4]

[1] KTH, Stockholm, supported by ERC grant 226-203
[2] Cornell University, Ithaca, supported in part by a Microsoft New Faculty Fellowship, NSF CAREER Award CCF-0746990, AFOSR Award FA9550-08-1-0197, and BSF Grant 2006317
[3] KTH, Stockholm
[4] CWI, Amsterdam

Abstract. We present a general parallel-repetition theorem with an efficient reduction. As a corollary of this theorem we establish that parallel repetition reduces the soundness error at an exponential rate in any public-coin argument, and more generally, any argument where the verifier's messages, but not necessarily its decision to accept or reject, can be efficiently simulated with noticeable probability.

1 Introduction

When the soundness error of an interactive proof [7] or interactive argument [3], or more generally computationally-sound interactive proofs, is too large for applications, one might hope to prove a direct-product theorem which applies to the protocol at hand. A direct-product theorem for some class of problems states that if an adversary has some probability of succeeding in a single instance, then his chance in solving many independent instances of the problem drops exponentially. Running several independent instances of a protocol can be done sequentially or in parallel. Sequential repetition means that the $(i+1)$st execution of the protocol is only started after finishing the ith execution. Parallel repetition means that all protocols are run simultaneously. It is well-known that sequential repetition reduces the soundness error at an exponential rate for both proofs and arguments. However, although parallel repetition is known to reduce the soundness error in interactive proofs [1,6], Bellare, Impagliazzo and Naor [2] demonstrate the existence of argument systems where parallel repetition does not reduce the soundness error, leaving open the following question:

For what computationally-sound proof systems does parallel repetition reduce the soundness error?

There have been several works addressing this question. Yao's [17] work on hardness amplification of one-way functions can be viewed as establishing that parallel repetition reduces the soundness error at an asymptotically optimal rate in every "publicly-verifiable" two-round argument—namely arguments where one can efficiently check if a transcript is accepting without knowing the verifier's internal randomness. Bellare, Impagliazzo and Naor [2] extended this result to show that parallel repetition reduces the error for general (not necessarily publicly-verifiable) arguments with at most three rounds. For two-round

D. Micciancio (Ed.): TCC 2010, LNCS 5978, pp. 1–18, 2010.

protocols, Canetti, Halevi and Steiner [4] obtain a quantitatively better bound (approaching Yao's original bound for publicly-verifiable arguments), and Impagliazzo, Jaswal and Kabanets [11] show a more fine-grained "Chernoff-type" theorem. Finally, Pass and Venkitasubramaniam [13] show that parallel repetition also reduces the error for any constant-round public-coin protocol.

On the negative side, as shown by Bellare et al [2] and Pietrzak and Wikström [14], parallel repetition does not decrease the error for general (non public-coin) protocols with eight rounds; furthermore, black-box reductions cannot be used to establish such a result even for general four round protocols.

Thus, given the current state of the art, it is unknown whether parallel-repetition reduces the soundness error even in public-coin protocols with a super-constant number of rounds, or any general classes of non public-coin protocols with more than 3 rounds. The former of these questions was stated as an open problem by Bellare et al [2]. In this work we identify a general class of computationally sound protocols for which parallel repetition reduces the soundness error. This class encompasses—and significantly extends—all earlier classes of computationally sounds protocols for which parallel repetition had been established; in particular, it includes *all* public-coin protocols but also natural classes of private-coin protocols.

1.1 Our Results

We say that a verifier is *δ-simulatable* if, roughly speaking, given the prover's view of any partial interaction, with probability δ, the next-message function of the verifier (excluding its verdict to accept or reject) can be simulated for all remaining rounds (with a small statistical error). In other words, it is possible to efficiently simulate a δ-fraction of the verifier's continuations without knowing the verifier's internal randomness.

Note that any public-coin or three-round protocol trivially is 1-simulatable, but this notion captures many other protocols. For instance, public-coin protocols in the public-key model—where the verifier has a secret key and might determine whether to accept or reject based on this key—are also 1-simulatable.

Our main result is an efficient parallel repetition theorem (i.e., a parallel repetition theorem with an efficient reduction) for any $\frac{1}{poly}$-simulatable verifier. More precisely, our main theorem says that for any protocol where the verifier is δ-simulatable, we can turn an arbitrary parallel prover $\mathcal{P}^{(k)}$ for the k-fold repetition of $\mathcal{V}$ with success probability ϵ into a single instance prover $\tilde{\mathcal{P}}$ with success probability $1 - O(\frac{m}{\delta}\sqrt{-\log(\epsilon)/k} + \sqrt{m}\log(mk)/\sqrt{k})$ where $2m+1$ is the number of rounds. Note that this implies that the error probability decreases exponentially down to some negligible function when the number of repetitions is sufficiently larger than the number of rounds. Following Impagliazzo et al. [11] we can actually prove a more general "Chernoff-type" theorem, where one only assumes that the parallel prover convinces a certain fraction (and not all) of the individual verifiers.

As any public-coin protocol or three-round protocol satisfies 1-simulatability, we get as corollaries parallel repetition theorems for three-round protocols [2] and for public-coin protocols [13]. Note that whereas [13] only shows a parallel repetition theorem for *constant-round* protocols, our theorem applies to protocols with an arbitrary polynomial number of rounds. Our parameters are, however, worse that those of [13], which establishes an essentially optimal error reduction for the case of constant-round protocols.

As can be seen from the expression above, the success probability of the single-instance prover decreases linearly with the number of rounds in the protocol. If we restrict our attention to public-coin verifiers, or more generally, 1-simulatable verifiers *with verdict*—i.e., verifiers where the next messages function and its verdict to accept or reject—can be simulated with a small statistical error—then we can show a stronger parallel repetition theorem, where the decrease in error probability is independent of the number of rounds.

Finally, we show using a simple argument that our results hold also for *concurrent* provers, which may schedule their interaction with the individual verifiers arbitrarily.

1.2 Some History and Related Papers

An earlier version of this paper [9], where we established a parallel repetition only for interactive arguments with $(1 - \frac{1}{poly})$-simulatable verifiers (and some generalizations thereof), dates back to April 2008. Recent works extend it.

Most notably, Haitner [8] gave a modification of any interactive protocol by introducing a "random-termination verifier" where the verifier decides to stop and accept immediately with small but noticeable probability at each round. Haitner proved that any interactive protocol modified in this way, satisfies a parallel repetition theorem.

His construction is the main motivation of our study of δ-simulatable verifiers as it is easy to simulate a verifier that has halted. As a consequence our results gives a new proof of Haitner's theorem which is, in our eyes, simpler and which gives better parameters.

In an even more recent paper Chung and Liu [5] improves the analysis of our reduction. They manage to avoid the use of any lemma of the type obtained by Raz getting optimal reduction of the error rate for the public-coin case and almost optimal result in the case of 1-simulatable verifiers. It does not seem that their result extends to the case of δ-simulatable verifiers.

In a different direction, Pass, Tseng and Wikström [12] rely on our techniques to show that parallel repetition of public-coin protocols also gives a qualitative (rather than quantitative) improvement in soundness: any public-coin argument, when sufficiently repeated in parallel, becomes sound also against a "resetting"-attack if the verifier uses a pseudo-random function to pick its messages. As a corollary of this result, they establish impossibility of public-coin black-box zero-knowledge protocols (for non-trivial languages) that remain secure under parallel repetition. Interestingly, [12] show that the dependence on m in our security reduction for the main theorem is inherent in their setting; this stands in contrast with our sharper reduction for the case of public-coin protocols.

1.3 Our Techniques

We show how to turn any parallel-prover $\mathcal{P}^{(k)}$ into a single-instance prover $\tilde{\mathcal{P}}$; furthermore, we require that $\tilde{\mathcal{P}}$'s success probability is significantly higher that of $\mathcal{P}^{(k)}$. Traditionally, $\tilde{\mathcal{P}}$ achieves this by internally incorporating $\mathcal{P}^{(k)}$, appropriately feeding it messages, while at the same time picking one of the parallel executions that it feeds to an external verifier. In other words, out of the k instances that $\mathcal{P}^{(k)}$ believes it is participating in, $\tilde{\mathcal{P}}$ controls $k-1$ of them, while one of them is externally forwarded.

The crux of this approach is how to determine the $k-1$ messages sampled in some particular round are good. In the public-coin case, in the work of Pass and Venkitasubramaniam [13], the "goodness" of a message is determined by estimating (using sampling) the probability with which $\tilde{\mathcal{P}}$ would be able to complete the partial interaction if this message was fixed; and $\tilde{\mathcal{P}}$ selects the message which leads to the highest success probability. This procedure requires recursively sampling $\tilde{\mathcal{P}}$ and results in a blow-up of the running-time as a function of the number of rounds and thus only a constant number of rounds can be handled. In the case of private-coin protocols, another problem arises already for the case of three-round protocols: we might not be able to determine if the verifier $\mathcal{V}_i$ accepts in a particular transcript as we do not know its random tape. Bellare et al. [2] overcome this problem by "guessing" that $\mathcal{V}_i$ accepts, if, intuitively, "many" other verifiers accept; as we are internally running all the other verifiers we know their random tapes and thus their decision.

A-priori, it would seem that a combination of these approaches would at least give a parallel-repetition theorem for constant-round private-coin protocols as long as it is possible to appropriately sample the next messages of the verifier. The problem is that when selecting "good" messages, we might be biasing the distribution of continued executions. It is, thus, no longer clear that the procedure of "guessing" that $\mathcal{V}_i$ accepts if many other verifiers accept, yields a good estimate of whether $\mathcal{V}_i$ actually accepts.

The key technique introduced in this paper is a method for selecting "good" messages without biasing the distribution too much. We essentially choose the first continuation that can be seen to lead to a good outcome. The fact that this procedure does not bias the distribution of interactions by too much follows from a powerful lemma of Raz [15] which was used in an essential way in the proof of the parallel repetition theorem for two-prover interactive proofs. Additionally, this approach does not lead to a blow-up in running-time and can be applied to any polynomial number of rounds.

Let us first outline the idea for the case of public-coin protocols. Instead of trying to recursively estimate how good a message is, we use the following simple procedure to pick messages to forward to $\mathcal{V}_i$. Given a partial interaction, repeatedly sample random completions of this transcripts, until a successful transcript is reached, i.e., one where all verifiers accept. When this happens, select the next message to forward to the external verifier based on what that message was in the sampled accepting transcript. In other words, sample a random message conditioned on it leading to a successful transcript. To analyze why this works, consider the following mental experiment, where messages from $\tilde{\mathcal{P}}$ are determined in the same way, but now also $\mathcal{V}_i$'s messages are selected conditioned on them leading to an accepting execution. Clearly, in this mental experiment $\tilde{\mathcal{P}}$ succeeds in convincing $\mathcal{V}_i$ with probability 1. It is also not hard to see that the expected number of samples required by $\tilde{\mathcal{P}}$ is not too high and that its running-time is still polynomial. The problem is that the real external verifier does not pick its messages conditioned on them leading to an accepting execution; rather, it picks them uniformly at random. However, by relying on Raz' lemma, we can show, provided that i is picked uniformly at random from $[k]$, that the distribution of messages actually sent by the real external verifier are statistically close to those sent in the mental experiment, where we condition on them leading to an accepting execution. By applying the union bound over each round in the protocol, we conclude that also in the real execution, $\tilde{\mathcal{P}}$ succeeds which high probability.

Note that the above argument directly applies also to 1-simulatable verifiers with verdict; we only require it to be possible to 1) emulate continuations of partial interactions with the external verifier, and 2) determine if the external verifier would have accepted in those executions. To extend this analysis to 1-simulatable verifiers without verdict, we augment the argument by first showing that in the mental experiment it is sufficient to guess the decision of $\mathcal{V}_i$ based on the decisions of the other verifiers, in analogy with [2]. Now we can no longer claim that the success probability in the mental experiment is 1, but it will still be sufficiently high; the rest of the proof remains the same, and we conclude that also in the real execution $\tilde{\mathcal{P}}$ succeeds with high probability. We mention that to simplify the analysis, and to generalize it to handle "Chernoff-type" bounds, we generalize the "guessing" procedure of [2].

Finally, consider the case of $\frac{1}{poly}$-simulatable verifiers. Here we can only emulate continuations of the external verifier for a small, but noticeable, fraction of its true continuations. Nevertheless, by another application of Raz's lemma, we can show that the distribution of messages sent to the internal prover does not change by too much even if we condition the ith execution on a noticeable subset of continuations, and thus $\frac{1}{poly}$-simulatability suffices. More precisely, by Raz's lemma, it follows that the probability the external verifier chooses a continuation that we can simulate is not affected much if we condition on getting an accepting interaction; this, in particular means that (on average) the probability that a partial transcript leads to an accepting transcript does not change much even if we condition on only continuations that we can simulate.

Note that in the above proof sketch we lose a factor of m, i.e., the number of rounds in the protocol, by the application of the union bound. For the special case of 1-simulatable verifiers with verdict, we go back to the underlying tool of *relative entropy* used to prove Raz's lemma, and use it to prove a generalization that considers multiple rounds at once, without losing the factor of m.

1.4 Outline of Paper

We first introduce some basic definitions in Section 2. Then we give a definition of δ-simulatable verifiers in Section 3. In Section 4 we state the parallel repetition theorem. Then in Section 5 we prove the general parallel repetition theorem, leaving the sharper theorem for the full version. Finally, we explain in Section 6 how to generalize our results to *concurrent* provers.

2 Notation and Basic Definitions

We denote the set $\{1, \ldots, m\}$ by $[m]$. We use n to denote the security parameter. All random variables are written in uppercase and usually we use the corresponding lower case for outcomes of the variable. When we say that a random variable X over a set $\mathcal{X}$ is chosen *randomly*, we mean that it is uniformly and independently distributed of all other variables. We use $\log a$ to denote the logarithm of a in base 2. We write $x \leftarrow_R X$ when x is chosen randomly from the set X.

If X is a random variable we write $\mathsf{P}_X(x) = \Pr[X = x]$ to denote the probability that it assumes the value x, and we denote its support by $[X]$. If X and Y are random variables we denote the conditional distributions of Y given X by $\mathsf{P}_{Y|X}$, and when we condition on a fixed value $x \in [X]$ we denote the corresponding probability function by $\mathsf{P}_{Y|X}(\cdot\,|x)$. Thus, $\mathsf{P}_{Y|X}(y\,|x) = \mathsf{P}_{XY}(x, y) / \mathsf{P}_X(x)$. When W is an event, we write $\mathsf{P}_{X|W}(x) = \Pr[X = x\,|W]$.

We often use the chain-rule for distributions and we use dots, when we are interested in a specific conditional distribution, e.g., we write $\mathsf{P}_{XY} = \mathsf{P}_Y \mathsf{P}_{X|Y}$ and $\mathsf{P}_{X|Y}(\cdot\,|y)$.

Definition 1. *The statistical distance between two distributions* P_X *and* P_Y *over a set* $\mathcal{X}$ *is*

$$\|\mathsf{P}_X - \mathsf{P}_Y\| = \frac{1}{2} \sum_{x \in \mathcal{X}} |\mathsf{P}_X(x) - \mathsf{P}_Y(x)| \ .$$

In a computationally sound protocol, soundness only holds against *efficient* (i.e., polynomial-time) provers. In general, a computationally sound protocol accepts a joint parameter λ that may, or may not, contain an instance of a language. We use $\mathcal{P}$ and $\mathcal{V}$ to denote the prover and verifier of a protocol, and we write $\langle \mathcal{P}, \mathcal{V} \rangle(\lambda)$ for the output of $\mathcal{V}$ after an interaction with $\mathcal{P}$ on common input λ. For notational convenience, we consider the security parameter n and any additional advice to the prover as encoded into λ. We denote the k-wise parallel repetition of a verifier $\mathcal{V}$ by $\mathcal{V}^k$. The repeated verifier simulates the individual verifiers independently, except that their message rounds are synchronized. It accepts if all the individual verifiers accept. The ith verifier is denoted by $\mathcal{V}_i$, but all verifiers run the same program $\mathcal{V}$. We are also interested in repeated threshold verifiers, denoted by $\mathcal{V}^k_\gamma$, that accept if at least $(1 - \gamma)k$ of the individual verifiers accept.

The number of exchanges in the protocol is denoted by m, where one exchange consists of two rounds, and the very first message of the prover is considered part of the 0th exchange.

We denote the lth message of the ith verifier $\mathcal{V}_i$ by $C_{l,i}$ and its state after the lth message has been computed by $T_{l,i}$. We denote the lth message sent by the prover to the ith verifier $\mathcal{V}_i$ by $A_{l,i}$, and we denote the state of the prover after it has computed its lth message by S_l. The decision of $\mathcal{V}_i$ is denoted by D_i, i.e., 1 for accept and 0 for reject. We define $C_l = (C_{l,1}, \ldots, C_{l,k})$ and $A_l = (A_{l,1}, \ldots, A_{l,k})$. The variables are then related as follows given a random joint parameter Λ

$$
\begin{aligned}
T_{0,i} &= \Lambda && (1)\\
(S_0, A_0) &= \mathcal{P}^{(k)}(\Lambda) &&\\
(T_{l+1,i}, C_{l+1,i}) &= \mathcal{V}_{R_{l,i}}(T_{l,i}, A_{l,i}) && \text{for } 0 \leq l < m\\
(S_l, A_l) &= \mathcal{P}^{(k)}(S_{l-1}, C_l) && \text{for } 0 < l \leq m\\
D_i &= \mathcal{V}(T_{m,i}, A_{m,i}) \ , &&
\end{aligned}
$$

where we think of both the prover and verifier as deterministic algorithms and denote the random tape used by $\mathcal{V}_i$ in round l by $R_{l,i}$. The verifier may of course "store" randomness from one round to be used in later rounds.

To collect random variables belonging to different exchanges we write, e.g., $C_{[l],i} = (C_{1,i}, \ldots, C_{l,i})$ and $C_{[l]} = (C_1, \ldots, C_l)$. Sometimes we wish to exclude only a single index i. Then we write $C_{l,\langle i \rangle} = (C_{l,1}, \ldots, C_{l,i-1}, C_{l,i+1}, \ldots, C_{l,k})$. We mostly view $\mathcal{V}$ and $\mathcal{P}^{(k)}$ as deterministic functions, but when convenient and clear from the context we drop the the random tape from our notation.

3 Simulatable Verifiers

Our parallel repetition theorem is applicable to δ-*simulatable* verifiers. Roughly speaking, we say that a verifier is δ-simulatable if given only the prover's view of any partial interaction (which thus excludes the verifier's internal state), we can efficiently simulate a δ fraction of the verifier's actual continuations.

Recall that given a prover $\mathcal{P}$ and a verifier $\mathcal{V}$, a partial transcript of length l is denoted $(\lambda, a_{[l]}, c_{[l]})$, the lth states of $\mathcal{P}$ and $\mathcal{V}$ are denoted s_l and t_l respectively, and that these values are defined formally by Equation (1) in Section 2. Thus, the prover's view after producing its lth message a_l is given by $(s_{[l]}, \lambda, a_{[l]}, c_{[l]})$.

Definition 2 (δ-Simulatable Verifier). *A verifier $\mathcal{V}$ is said to be δ-*simulatable *if there exists a PPT simulator $\mathcal{S}$ such that for every prover strategy $\mathcal{P}$ and every partial history $(s_{[l]}, t_{[l]}, \lambda, a_{[l]}, c_{[l]})$, there is a subset Δ of $\mathcal{V}$'s random tapes, compatible with the history so far, of density δ such that the output of $\mathcal{S}$ on input $(s_{[l]}, \lambda, a_{[l]}, c_{[l]})$ is statistically close to the prover's view of a continued interaction between $\mathcal{P}$ and $\mathcal{V}$, including $\mathcal{V}$'s verdict, when $\mathcal{V}$'s random tape is chosen uniformly from Δ. When the verdict of $\mathcal{V}$ is removed from consideration, we say that $\mathcal{V}$ is δ-simulatable without verdict, or simply δ-simulatable.*

Remark 1. Note that the definition requires the simulator to simulate a probability distribution that is allowed to be dependent on the state of the verifier and that this state is unknown. This seems like an impossible task in general unless we minimize the information contained in the state. In the early version of this paper [9] this state was not included in the probability distribution but instead we required that the next message of the internal, fully simulated verifiers could be efficiently generated based on the conversation up to this point. If this is indeed possible then we can instead let the state be given by the messages already sent and then use this generation process to replace the original verifier. With the current definition we need no condition on the internal verifiers and hence it is, in our eyes, preferable.

Remark 2. The property that we only demand the two distributions to be statistically close and not identical is only a technicality. In fact, when using the definition in this abstract we assume that the two distributions are the same, to avoid cumbersome notation to take care of the error terms given by a small statistical distance.

Remark 3. A careful reading of the proof reveals that we can let the probability δ of successful sampling depend on the round but not on the partial history it extends.

Remark 4. We can allow a weaker definition of simulatability where the ability to simulate $\mathcal{V}$ also depends on the $\mathcal{P}$'s messages. This leads to a more complicated proof of Lemma 4 that either loses a factor of m in the error bounds or uses the methods of [16] to get the same bounds. In order to keep this extended abstract self-contained we use the weaker definition here.

Clearly, any public-coin protocol is 1-simulatable with verdict. It is also easy to see that the "random-termination verifiers" of Haitner are $\frac{1}{4m}$-simulatable with verdict: the simulator simply aborts (accepting) with probability $\frac{1}{4m}$. Furthermore, public-coin protocols in a public-key model (where the verifier only sends random messages, but bases is decision on its secret key), as well as three-round protocols, are 1-simulatable without verdict.

4 The Parallel Repetition Theorem

We prove a parallel repetition theorem for any verifier that is δ-simulatable without verdict. The theorem implies that a $(2m+1)$-round protocol when repeated $k = \Omega(\frac{m^2}{\delta^2}t)$ times in parallel reduces the error probability from 1/2 to $2^{-t} + \mathsf{negl}(n)$ if we require that all parallel verifiers accept. In the general statement we consider also repeated threshold verifiers $\mathcal{V}_\gamma^k$ that accept if at least $(1-\gamma)k$ of the k parallel verifiers accept.

Theorem 1. *Assume $\epsilon \leq 1/2$, let $\mathcal{V} \in \mathsf{PPT}$ be a verifier that is δ-simulatable without verdict, and let $\mathcal{P}^{(k)}$ be a polynomial-time parallel prover. Then there exists a prover $\tilde{\mathcal{P}}$ running in time $\mathsf{Poly}(n, k, m, 1/\epsilon)$ such that for every $\lambda \in \{0,1\}^*$ where $\Pr[\langle \mathcal{P}^{(k)}, \mathcal{V}_\gamma^k \rangle(\lambda) = 1] \geq \epsilon$, for some threshold $0 \leq \gamma < 1$,*

$$\Pr\left[\langle \tilde{\mathcal{P}}, \mathcal{V} \rangle(\lambda) = 1\right] \geq 1 - \gamma - O\left(\frac{m}{\delta}\sqrt{-\log(\epsilon)/k} + \sqrt{m}\log(mk)/\sqrt{k}\right) ,$$

where n is the security parameter, m is the number of messages sent by $\mathcal{V}$, and k is the number of verifiers interacting with the parallel prover.

The constants hidden in the $O(\cdot)$-notation in Theorem 1 are small and given explicitly in our proof. It turns out that in the case of 1-simulatable verifiers we can get a stronger theorem.

Theorem 2. *Assume $\epsilon \leq 1/2$, let $\mathcal{V} \in \mathsf{PPT}$ be a verifier, and let $\mathcal{P}^{(k)}$ be a polynomial-time parallel prover. Then there exists a prover $\tilde{\mathcal{P}}$ running in time* $\mathsf{Poly}(n, k, m, 1/\epsilon)$ *such that for every $\lambda \in \{0,1\}^*$ where $\Pr[\langle \mathcal{P}^{(k)}, \mathcal{V}_\gamma^k \rangle(\lambda) = 1] \geq \epsilon$, for some threshold $0 \leq \gamma < 1$,*

1. ***if $\mathcal{V}$ is 1-simulatable with verdict**, then*

$$\Pr\left[\langle \tilde{\mathcal{P}}, \mathcal{V} \rangle(\lambda) = 1\right] \geq 1 - \gamma - 2\sqrt{-\log(\epsilon)/k} - \sqrt{1/k} \ , \ \textit{and}$$

2. ***if $\mathcal{V}$ is 1-simulatable without verdict**, then*

$$\Pr\left[\langle \tilde{\mathcal{P}}, \mathcal{V} \rangle(\lambda) = 1\right] \geq 1 - \gamma - O\left(\sqrt{m}\sqrt{-\log(\epsilon)/k} + \sqrt{m}\log(mk)/\sqrt{k}\right) \ ,$$

where n is the security parameter, m is the number of messages sent by $\mathcal{V}$, and k is the number of verifiers interacting with the parallel prover.

Due to the lack of space the proof of Theorem 2 is omitted but can be found in [16]. It relies on the notion of *relative entropy (Kullback-Leibler distance)* and uses a lemma extending Lemma 1 below to treat multiple rounds.

Readers familiar with the recent result of Pass et al. [12], may find Case 1 of Theorem 2 surprising, since superficially it seems the same technique should be applicable to remove the dependence on the number of rounds in [12], which would contradict their results. The reason this is not the case is that in [12], the reduction samples messages in a given round conditioned on *two* events: (1) that "all verifiers accept", and (2) that the "right" message is output by the embedded "resetting" attacker. Thus, in each round a *distinct* event is considered. Another way to say this is that the probability that the "right" messages are output in all rounds in a straight-line execution of the resetting attacker is $\mathsf{Poly}(n)^{-m}$. Thus, we could apply this technique to simplify the proof in [12], but the dependence on m would not disappear.

5 Proof of Theorem 1

We prove Theorem 1 in three steps. First we prove the theorem for public-coin verifiers in the case where $\gamma = 0$. This immediately generalizes to 1-simulatable verifiers with verdict. Then we show how to generalize the proof to verifiers that are only δ-simulatable with verdict. Finally, we prove that the result can be generalized to $\gamma > 0$ and verifiers that are δ-simulatable without verdict.

5.1 Proof of Theorem 1 in the Public-Coin Case

It is quite natural to simulate an interaction between the parallel prover $\mathcal{P}^{(k)}$ and the repeated verifier $\mathcal{V}^k$ and let the external verifier play the role of $\mathcal{V}_i$ for some i. In other words any message to $\mathcal{V}_i$ would instead be forwarded to the external verifier and its reply is taken as the reply of $\mathcal{V}_i$. The question is how to choose the index i and how the other verifiers should be simulated. We solve this in a simple way by picking a uniformly random i, simulating the other verifiers by picking random messages and then taking the first answers that can be seen to lead to making all verifiers accept. Let us discuss the intuition behind this approach.

Consider the tree of all possible interactions between $\mathcal{P}^{(k)}$ and $\mathcal{V}^k$, where each leaf encodes which verifiers accept and the edges on level l are labeled with the random choices of the verifiers in exchange l. If we could sample a random leaf such that all verifiers accept, then clearly $\mathcal{V}_i$ also accepts for a any choice of i. If the success probability of $\mathcal{P}^{(k)}$ is ϵ we can efficiently sample from this distribution in time polynomial in $1/\epsilon$ and the security parameter n as follows. In exchange l we repeatedly choose the messages $c_l = (c_{l,1}, \ldots, c_{l,k})$ of all verifiers randomly and simulate a completion conditioned on the interaction so far and our choice of messages in exchange l. If the completion gives a leaf where all verifiers accept, then we take c_l to be the messages of the verifiers in exchange l. Clearly, if a suitable c_l is found for each l, then all verifiers accept.

Suppose now that we pick a random index i and in exchange l pick the message $c_{l,i}$ of $\mathcal{V}_i$ only once. The messages $c_{l,\langle i \rangle} = (c_{l,1}, \ldots, c_{l,i-1}, c_{l,i+1}, \ldots, c_{l,k})$ of all other verifiers are still repeatedly sampled, but now conditioned on $c_{l,i}$ in addition to the interaction so far. The key observation is that this modified distribution is quite close to the original one, and that we may view $\mathcal{V}_i$ as the external verifier. Thus, we avoid sampling too much to stay close to the original distribution on the leaves where all verifiers accept.

More Details. Denote by Complete the probabilistic algorithm that given a partial interaction between $\mathcal{P}^{(k)}$ and $\mathcal{V}^k$ returns a random sample from the distribution of the decisions of the verifiers, conditioned on the partial interaction given as input. The detailed reduction is given by Algorithm 1 below. The parameter

Algorithm 1 (b). $\tilde{\mathcal{P}}_u(x)$

```
if x is a joint parameter λ then                          // Read joint parameter
   (s_0, a_0) ← P^(k) (λ)                                 // Compute prover's first message
   i ←_R [k]                                              // Choose random index
   return ([i, s_0, λ, ∅, a_[0]] , a_{0,i})               // Output state and first message
else
   Interpret x as ([i, s_{l-1}, λ, c_[l-1], a_[l-1]] , c_{l,i})  // Read state & verifier's message
   for v = 1, ..., u do
      c_{l,⟨i⟩} ←_R {0,1}^{p(n)×(k-1)}                    // Sample verifiers' messages
      (s_l, a_l) ← P^(k) (s_{l-1}, c_l)                    // Compute prover's reply
      if Complete(λ, c_[l], a_[l]) = 1̄ then               // If messages are good,
         return ([i, s_l, λ, c_[l], a_[l]] , a_{l,i})      // then output reply
      done
   done
   return (fail, fail)                                    // Give up
end
```

u denotes the maximal number of samples generated by the prover in each round to find a suitable reply from the parallel prover. For simplicity we assume that the message of the verifier in each exchange is drawn from $\{0,1\}^{p(n)}$ for some polynomial p.

Note that the prover keeps as its state the index i corresponding to the external verifier, the state of the simulated parallel prover, and a partial interaction. We now consider the error probability of the constructed prover.

The sampling lemma of Raz [15] says that given independently distributed random variables $U_1, \ldots, U_k$, the distribution of U_i does not change much on average over the index i by conditioning on an event E, provided that the probability of E is not too small. (We mention that the sampling lemma was previously used by Impagliazzo et al. [11] in the context of parallel-repetition of 2-round arguments). We make use of the following variant that appears as Corollary 6 in Holenstein's simplified proof of Raz' theorem [10].

Lemma 1. *[10] Let* $\mathsf{P}_{YU^kV} = \mathsf{P}_Y\big(\prod_{i=1}^k \mathsf{P}_{U_i|Y}\big)\mathsf{P}_{V|YU^k}$ *be a probability distribution and* E *an event. Then*

$$\frac{1}{k}\sum_{i=1}^{k} \left\|\mathsf{P}_{YU_iV|E} - \mathsf{P}_{YV|E}\,\mathsf{P}_{U_i|Y}\right\| \leq k^{-1/2}\sqrt{\log|V^*| - \log\Pr[E]}\;.$$

where V^* *is the set of values of* v *that can occur conditioned on* E *occurring.*

In our application, the variable Y represents the interaction so far and U_i are the messages of the verifiers in the current round. We let V be a binary variable such that $\mathsf{P}_{V|YU^k}(1\,|y,u)$ is the probability that all verifiers accept in a random completion, for every partial interaction $(y,u) \in [Y, U^k]$. The lemma then implies for a random Y that most U_i are, even if we condition on extending Y to an accepting interaction, distributed very closely to their unconditional distribution which in this case is the uniform distribution.

Thus, we can conclude that in any single round, if we have chosen Y up to this point with the conditional distribution of a partial interaction leading to an accepting leaf then if we, in this round, pick a random i, the distribution of U_i is likely to be close to uniform. A problem to be taken care of is that i is chosen once and remains fixed for all rounds.

Let us consider a modified process where the external verifier $\mathcal{V}_i$ instead of choosing $c_{l,i}$ with the uniform distribution does a process similar to that of $\tilde{\mathcal{P}}_u$. It samples complete interactions that extend the current interaction of all verifiers until it finds a complete interaction where all verifiers accept and then chooses the value of $c_{l,i}$ in this interaction as its response. Furthermore, let us remove the restriction that $\tilde{\mathcal{P}}_u$ only makes u attempts to find a complete interaction where all verifiers accept and let it sample until it finds a completion. Let D_{real} be the distribution on interactions produced by $\tilde{\mathcal{P}}_u$ interacting with $\mathcal{V}_i$ and let D_{ideal} be the distribution on interactions in this modified process.

Clearly, D_{ideal} outputs a uniformly selected interaction in which all verifiers accept. Thus, in this modified process $\mathcal{V}_i$ always accepts. Below we estimate the statistical distance between this process and the original process. This statistical distance is an upper bound on the probability that $\mathcal{V}_i$ rejects. Let us first see that it is unlikely that the modified process ever needs to sample a large number of times. This is intuitively not surprising. For the sampling to take a long time we need to choose a partial interaction that is very unlikely to lead to a complete accepting interaction. But as we are choosing partial interactions as part of an accepting interaction we are very unlikely to choose such a partial interaction. This is made formal by the following easy lemma, a proof of which is given below.

Lemma 2. *Let Y be a random variable and let $X_0, X_1, X_2, \ldots$ be identically distributed binary random variables which are only dependent through Y, i.e., $\mathsf{P}_{Y,X_0,\ldots X_j} = \mathsf{P}_Y \prod_{i=0}^{j} \mathsf{P}_{X_i|Y}$ and $\mathsf{P}_{X_i|Y} = \mathsf{P}_{X_j|Y}$ for any i, j. Let J be the random variable denoting the smallest nonzero index such that $X_J = 1$. Then $\mathrm{E}\left[J \,|X_0 = 1\right] \leq \frac{1}{\Pr[X_0=1]}$.*

Let us see how this lemma proves that the expected number of samples needed to find an accepting completion is small. We let Y be a random partial interaction which is chosen by picking a complete accepting interaction, i.e., Y is $C_{[l-1]}$ for some l, and we let X_i be one if a particular random completion of Y makes all verifiers accept. Then $\mathrm{E}\left[J \,|X_0 = 1\right]$ is exactly the expected number of attempts to complete the interaction Y to make all verifiers accept given that Y was picked by first picking a complete interaction which makes all verifiers accept and then truncating to the appropriate length.

Let $\delta = \sqrt{-\log(\epsilon)/k} + (\epsilon u)^{-1}$. We claim that the statistical difference between D_{real} and D_{ideal} when truncated to t rounds is bounded by $t\delta$. This is clearly true for $t = 0$ and we proceed by induction using the following lemma.

Lemma 3. *Let X_0 and X_1 be two random variables over $\mathcal{X}$, and let Z_x and Z'_x be two families of random variables parameterized by $x \in \mathcal{X}$ such that*

$$\|\mathsf{P}_{X_0} - \mathsf{P}_{X_1}\| = \delta_1 \quad \textit{and} \quad \mathrm{E}_x\left[\|\mathsf{P}_{Z_x} - \mathsf{P}_{Z'_x}\|\right] = \delta_2 \ ,$$

where x is distributed according to P_{X_0}. Then

$$\left\|\mathsf{P}_{X_0,Z_{X_0}} - \mathsf{P}_{X_1,Z'_{X_1}}\right\| \leq \delta_1 + \delta_2 \ .$$

Before we prove Lemma 3, let us see how it enables us to complete the induction step. We let X_0 be a $(t-1)$-round interaction chosen according to D_{ideal}, X_1 a $(t-1)$ round interaction chosen according to D_{real}, Z_{X_0} the next round message chosen by the verifiers according to D_{ideal} and Z'_{X_0} the next round message chosen from D_{real}. We need to estimate the expected statistical distance between Z'_{X_0} and Z_{X_0} over X_0.

We have two differences between the two distributions, how $\mathcal{V}_i$'s message is chosen and the limited sampling. The latter is, by Lemma 2 and Markov's inequality, bounded by $(\epsilon u)^{-1}$ and we claim that former difference is bounded by $\sqrt{-\log(\epsilon)/k}$. Let us see how this follows from Lemma 1.

As stated before, we let Y be the interaction up to the $(t-1)$st round and U_i the message of $\mathcal{V}_i$ in round t and V a bit which is one with the probability that a random completion of the given interaction accepts. The event E is that "$V = 1$". Then D_{ideal} picks messages with the distribution given by $\mathsf{P}_{U_i|YVE}$ while $\mathcal{V}_i$ picks messages with the uniform distribution which in this case is $\mathsf{P}_{U_i|Y}$. Lemma 1 now tells us exactly that for a random Y and i the statistical distance between these two distributions is at most $\sqrt{-\log(\epsilon)/k}$.

Finally, setting $u = \epsilon^{-1}m\sqrt{k}$ completes the proof of Theorem 1 in the public-coin case as claimed. The missing proofs of Lemma 3 and Lemma 2 are given below.

Proof (Lemma 2). We can consider only values y such that $\Pr[X_0=1 \,|Y=y] > 0$ and summing over those we have

$$\begin{aligned}
\mathrm{E}[J\,|X_0 = 1] &= \sum_y \Pr[Y = y\,|X_0 = 1]\,\mathrm{E}[J\,|Y = y \wedge X_0 = 1] \\
&= \sum_y \Pr[Y = y\,|X_0 = 1] / \Pr[X_1 = 1\,|Y = y \wedge X_0 = 1] \\
&= \sum_y \Pr[Y = y\,|X_0 = 1] / \Pr[X_1 = 1\,|Y = y] \\
&= \sum_y \frac{\Pr[Y = y \wedge X_1 = 1]}{\Pr[X_0 = 1]} \cdot \frac{\Pr[Y = y]}{\Pr[X_1 = 1 \wedge Y = y]} \le \frac{1}{\Pr[X_0 = 1]},
\end{aligned}$$

where the third equality follows from the conditional independence of the X_i's and the fourth equality follows since the X_i's are also identically distributed. □

Proof (Lemma 3). We use the characterization that two distributions are at statistical distance δ if and only if there is a coupled way of choosing elements from the two distributions such that the two samples are equal with probability $1 - \delta$. We need to choose coupled pairs (x, z) and (x', z') from the given distributions. First choose a coupled pair (x, x') distributed according to P_{X_0} and P_{X_1}, respectively. If they are unequal, which happens with probability δ_1, we give up. If they are are equal we choose a coupled pair (z, z') according to the distributions P_{Z_x} and $\mathsf{P}_{Z'_x}$. The probability that these are unequal (over the choice of x and the second choice) is upper bounded by δ_2. This completes the proof. □

5.2 Proof of Theorem 1 for δ-Simulatable Verifiers with Verdict

When the verifier is no longer public-coin and only δ-simulatable for some $\delta \geq 1/\mathsf{Poly}(n)$, it may keep its state hidden from the prover inbetween exchanges. To deal with this, we replace each call to $\mathsf{Complete}$ in Algorithm 1 by a call to the δ-simulator on input $(i, s_{[l]}, t_{[l],\langle i\rangle}, \lambda, a_{[l]}, a_{[l]})$.

We consider a fixed round l and all variables below depend on the value of l but, for notational convenience, we omit this dependence. Let us define $X_i = (T_{l-1}, C_{[l-1]}, C_{l,i})$ and $Y_i = (T_{l,\langle i\rangle}, C_{l,\langle i\rangle})$. Recall that $C_{l,\langle i\rangle}$ denotes the array $(C_{l,1}, \ldots, C_{l,i-1}, C_{l,i+1}, \ldots, C_{l,k})$ and similarly for $T_{l,\langle i\rangle}$.

By δ-simulatability, there is a subset, Δ of the external verifiers possible random tapes for which we can simulate $\mathcal{V}_i$.

Let W be an indicator variable of the event $D = \overline{1}$ (that all verifiers accept). Then define $\delta^i_{x_i,y_i}$ as the probability that the prover's view of a random completion of (x_i, y_i), conditioned on the event $W = 1$, is an output from the simulator. Furthermore, let $\delta^i_{x_i}$ be the expected value of $\delta^i_{x_i,y_i}$ over y_i, where y_i is chosen according to the distribution $\mathsf{P}_{Y_i|X_i,W}(\cdot\,|x_i,1)$. Due to the conditioning on $W = 1$, δ-simulatability does not immediately say anything about these quantities, but for any fixed x_i the distribution of Y_i conditioned on both $W = 1$ and the event that the output is from simulator is given by the probability function

$$\mathsf{P}_{Y_i|X_i,W}(y_i\,|x_i,1)\,\frac{\delta^i_{x_i,y_i}}{\delta^i_{x_i}}\ .$$

We want to prove that this, for a uniformly random i, is statistically close to the distribution $\mathsf{P}_{Y_i|X_i,W}(\cdot\,|x_i,1)$ and thus we should estimate

$$\frac{1}{k}\sum_{i=1}^{k}\sum_{x_i,y_i}\mathsf{P}_{X_i,Y_i|W}(x_i,y_i\,|1)\left|1-\frac{\delta^i_{x_i,y_i}}{\delta^i_{x_i}}\right|\ . \tag{2}$$

The following lemma is the key to estimating this distance.

Lemma 4

$$\frac{1}{k}\sum_{i=1}^{k}\sum_{x_i,y_i}\mathsf{P}_{X_i,Y_i|W}(x_i,y_i\,|1)\,|\delta^i_{x_i,y_i}-\delta| \le O(\sqrt{-\log(\epsilon)/k})\ . \tag{3}$$

We postpone the proof of the lemma until we have seen how it is used. Fix i and x_i and consider the contribution to the sums in (2) and (3) over a random Y_i conditioned on $W = 1$. Define a random variable Z which takes the value $\delta^i_{x_i,y_i}/\delta^i_{x_i}$ with probability $\mathsf{P}_{Y_i|X_i,W}(y_i\,|x_i,1)$. Then the contribution to Equation (3) is at most $\delta\,\mathrm{E}[|1-sZ|]$ with $s = \delta^i_{x_i}/\delta$ while the contribution to Equation (2) is $\mathrm{E}[|1-Z|]$. Now consider the following lemma.

Lemma 5. *Assume that Z is a positive random variable with $\mathrm{E}[Z] = 1$. Then for any $s > 0$ we have $\mathrm{E}[|1-Z|] \le 2\,\mathrm{E}[|1-sZ|]$.*

Again, we postpone the proof until we have completed the argument. Since $\mathrm{E}[Z] = 1$, we see that Equation (2) is bounded by $O(\delta^{-1}\sqrt{-\log(\epsilon)/k})$. Thus the additional statistical distance between the ideal distribution and that obtained by our parallel prover introduced in round l is bounded by this quantity.

Using coupling and the union bound as in Section 5.1, we conclude that replacing the 1-simulator by a δ-simulator introduces an additional error of at most $O(\frac{m}{\delta}\sqrt{-\log(\epsilon)/k})$.

Finally, let us prove the two lemmas above, completing the proof of Theorem 1 in this case.

Proof (Lemma 4). We apply Lemma 1 with U_i representing $\mathcal{V}_i$'s random tape compatible with the interaction up to this point. We need to analyze the probability that we can simulate $\mathcal{V}_i$ conditioned upon all verifiers accepting. Without conditioning this probability is statistically close to δ by the definition of δ-simulatability (for notational convenience we assume here that this probability equals δ). The deviation from this is bounded by the statistical distance of the conditioned distribution from the uniform distribution. The lemma now follows from Lemma 1. □

Proof (Lemma 5). Note that $\sum_{z \leq 1} \mathsf{P}_Z(z)(1-z) = \frac{1}{2}\mathrm{E}[|1-Z|]$, since $\mathrm{E}[Z] = 1$ and $|1-z|$ is symmetric around 1. If $s \leq 1$, then $|1-z| < |1-sz|$ for every $z \leq 1$ and the claim follows. If $s > 1$, then we instead consider the partial sum for $z > 1$ and apply the corresponding argument. □

5.3 Proof of Theorem 1 for δ-Simulatable Verifiers without Verdict

First we note that it is easy to generalize the above result to the case with a repeated threshold verifier that accepts if at least $(1-\gamma)k$ verifiers accept. Replace the definition of the indicator variable W such that it is one if and only if $\sum_{i=1}^{k} D_i \geq (1-\gamma)k$. Then in the corresponding "modified process" discussed in Section 5.1 the probability that $\mathcal{V}_i$ accepts is at least $1-\gamma$, since i is chosen uniformly in $[k]$ and independently of the "modified process". A trivial modification of the analysis above then gives the same additional statistical error due to having an external verifier, the use of limited sampling, and a δ-simulatable verifier.

To generalize the theorem to δ-simulatable verifiers *without verdict*, starting from the result established in Section 5.2 for δ-simulatable verifiers *with* verdict, we modify the reduction by redefining W using "soft" decisions as was already done in [2]. Suppose that instead of accepting only samples where at least $(1-\gamma)k$ verifiers accept, we define a binary random variable W that is one with probability $\min(1, 2^{\nu(\gamma k - z)})$, where z is the number of rejecting verifiers, and accept a sample if $W = 1$. Then it turns out that, provided that, we choose ν small enough, this acceptance criteria can be approximated well even if we do not know the verdict of the external verifier $\mathcal{V}_i$. Let us start with the key lemma, of which the proof is postponed to the end of this section.

Lemma 6 (Soft Decision). *Let $D_1, \ldots, D_k$ be binary random variables such that* $\Pr[\sum_{i=1}^{k} D_i \geq (1-\gamma)k] \geq \epsilon$*, let* $Z = k - \sum_{i=1}^{k} D_i$*, let* $\gamma > 0$*,* $\nu > 0$*, and* $m \geq 1$*, and let W be a binary random variable such that* $\Pr[W = 1 \,|\, Z = z] = \min(1, 2^{\nu(\gamma k - z)})$*. Then*

$$\frac{1}{k}\sum_{i=1}^{k} \Pr[D_i = 0 \,|\, W = 1] \leq \gamma + \frac{1}{k\nu}(\log m + \log k - \log \epsilon) + \frac{4}{\nu^2 m k^2} \quad .$$

Remark 5. Although setting $\nu = 1$ and $\gamma = 0$ recovers the decision procedure in [2], our analysis differs from theirs. They implicitly use Raz's lemma to argue that the variables W_i and W are close in distribution on *average* over i. We need the stronger statement that these variables are close in distribution for *any* i. This is why we need the additional parameter ν.

Now set $\nu = \frac{1}{\sqrt{m}}\sqrt{-\log(\epsilon)/k}$ and suppose now at first that we did know the verdict of $\mathcal{V}_i$. The old argument carries over and we end up at a random point where $W = 1$. Before we could conclude that $\mathcal{V}_i$ accepted while currently by applying Lemma 6, we see that the probability that $\mathcal{V}_i$ rejects is at most

$$\gamma + \frac{1}{k\nu}(\log m + \log k - \log \epsilon) + \frac{4}{\nu^2 mk^2}$$
$$= \gamma + \frac{\sqrt{m}}{\sqrt{-\log(\epsilon)k}}(\log(mk) - \log(\epsilon)) + \frac{1}{-\log(\epsilon)k}$$
$$\leq \gamma + \sqrt{m}\log(mk)/\sqrt{k} + \sqrt{m}\sqrt{-\log(\epsilon)/k} + 1/k \ ,$$

and this is enough to prove Theorem 1.

The key to case the case when we do not know the verdict of $\mathcal{V}_i$ is that if ν is small then the decision of an individual verifier is does not affect the behavior very much. In fact, let us simply approximate Z by assuming than $D_i = 1$ and let us run our parallel prover using this approximation. Compare a run of this modified prover and a run of an ideal prover that uses the correct value of Z using the same randomness.

These two provers only behave differently when the the modified prover accepts a history that the ideal prover would have rejected. To be precise, each time the modified prover accepts a history the probability that the ideal prover would have rejected the same history is $1 - 2^{-\nu} \leq \nu$.

As the modified prover only accepts m histories over the course of a run, the statistical difference between the behavior of modified prover and the ideal prover is bounded by νm.

This gives a total additional error from using soft decisions when sampling of $(m+1)\nu \leq (\sqrt{m}+1)\sqrt{-\log(\epsilon)/k}$. Combined with the proof of Lemma 6 below, this concludes the proof of Theorem 1 in its full generality.

Proof (Lemma 6). Let $p_j = \Pr[Z = j]$. We know by assumption that

$$\sum_{j=0}^{k\gamma} p_j \geq \epsilon \ . \tag{4}$$

We know that $\Pr[Z = j \,|W = 1]$ is proportional to $p_j 2^{-\min(0,\nu(j-\gamma k))}$. This implies that the expected number of D_i's equal to zero is

$$\mathrm{E}\left[Z \,|W = 1\right] = \frac{\sum_{j=0}^{k} j p_j 2^{-\min(0,\nu(j-\gamma k))}}{\sum_{j=0}^{k} p_j 2^{-\min(0,\nu(j-\gamma k))}} \ . \tag{5}$$

The denominator is lower bounded by $\sum_{j=0}^{\gamma k} p_j 2^{-\min(0,\nu(j-\gamma k))} = \sum_{j=0}^{\gamma k} p_j$ and is thus, by Equation (4), at least ϵ. Let t be a parameter to be determined, then the numerator is bounded by

$$\sum_{j=1}^{k} \max(\gamma k + t, j) p_j 2^{-\min(0,\nu(j-\gamma k))}$$

$$\leq (\gamma k + t) \sum_{j=1}^{k} p_j 2^{-\min(0,\nu(j-\gamma k))} + \sum_{j=1}^{k-(\gamma k+t)} j p_{\gamma k+t+j} 2^{-\nu(t+j)} \ . \quad (6)$$

It is not difficult to see that $\sum_{j=1}^{\infty} j2^{-\nu j} \leq \frac{4}{\nu^2}$ and thus the upper bound in Equation (6) is at most

$$(\gamma k + t) \sum_{j=1}^{k} p_j 2^{-\min(0,\nu(j-\gamma k))} + \frac{4}{\nu^2} 2^{-\nu t} \ .$$

Setting $t = \frac{1}{\nu}(\log m + \log k - \log \epsilon)$ and using that the denominator of Equation (5) is at least ϵ we see that

$$\mathrm{E}\left[Z \,|W = 1\right] \leq \gamma k + \frac{1}{\nu}(\log m + \log k - \log \epsilon) + \frac{4}{\nu^2 mk} \ .$$

The proof is concluded by remembering that i is chosen uniformly at random from $[k]$. □

6 Concurrent Repetition

Although verifiers repeated in parallel perform their computations independently and use independently generated randomness, their communication is *synchronized.* It is natural to consider a more general form of repetition where this restriction is removed, i.e., the prover may *arbitrarily schedule* its interaction with the individual verifiers.

Only minor modifications are needed to generalize Theorem 1 and Theorem 2, with the same parameters, to the setting where a concurrent prover interacting with the k-wise *concurrent* repetition of $\mathcal{V}$ is converted into a prover $\tilde{\mathcal{P}}$ interacting with $\mathcal{V}$. The key observation for this extension is that a concurrent prover only sends $m+1$ messages to $\mathcal{V}_i$. Thus, $\tilde{\mathcal{P}}$ need only sample completions at m points during an interaction with $\mathcal{V}$, and Lemma 1 is only applied m times. Furthermore, the δ-simulator and soft decisions are only used at each point where $\tilde{\mathcal{P}}$ samples completions, i.e., exactly m times. More details will be given in the full version of this paper.

References

1. Babai, L.: Trading group theory for randomness. In: 17th ACM Symposium on the Theory of Computing (STOC), pp. 421–429. ACM Press, New York (1985)
2. Bellare, M., Impagliazzo, R., Naor, M.: Does parallel repetition lower the error in computationally sound protocols? In: 38th IEEE Symposium on Foundations of Computer Science (FOCS), pp. 374–383. IEEE Computer Society Press, Los Alamitos (1997)

3. Brassard, G., Chaum, D., Crépeau, C.: Minimum disclosure proofs of knowledge. Journal of Computer and System Sciences 37(2), 156–189 (1988)
4. Canetti, R., Halevi, S., Steiner, M.: Hardness amplification of weakly verifiable puzzles. In: Kilian, J. (ed.) TCC 2005. LNCS, vol. 3378, pp. 17–33. Springer, Heidelberg (2005)
5. Chung, K.-M., Liu, F.-H.: Parallel repetition theorems for interactive arguments. In: Micciancio, D. (ed.) TCC 2010. LNCS, vol. 5978, pp. 19–36. Springer, Heidelberg (2010)
6. Goldreich, O.: Modern Cryptography, Probabilistic Proofs and Pseudorandomness. Algorithms and Combinatorics. Springer, Heidelberg (1998)
7. Goldwasser, S., Micali, S., Rackoff, C.: The knowledge complexity of interactive proof systems. SIAM Journal on Computing 18(1), 186–208 (1989)
8. Haitner, I.: A parallel repetition theorem for any interactive argument. In: 50th IEEE Symposium on Foundations of Computer Science (FOCS). IEEE Computer Society Press, Los Alamitos (2009)
9. Håstad, J., Pass, R., Pietrzak, Wikström, D.: An efficient parallel repetition theorem (April 2008) (manuscript)
10. Holenstein, T.: Parallel repetition: simplifications and the no-signaling case. In: 39th ACM Symposium on the Theory of Computing (STOC), pp. 411–419. ACM, New York (2007)
11. Impagliazzo, R., Jaiswal, R., Kabanets, V.: Chernoff-type direct product theorems. In: Menezes, A. (ed.) CRYPTO 2007. LNCS, vol. 4622, pp. 500–516. Springer, Heidelberg (2007)
12. Pass, R., Tseng, D., Wikström, D.: On the composition of public-coin zero-knowledge protocols. In: Halevi, S. (ed.) CRYPTO 2009. LNCS, vol. 5677, pp. 160–176. Springer, Heidelberg (2009)
13. Pass, R., Venkitasubramaniam, M.: An efficient parallel repetition theorem for arthur-merlin games. In: 39th ACM Symposium on the Theory of Computing (STOC), pp. 420–429. ACM, New York (2007)
14. Pietrzak, K., Wikström, D.: Parallel repetition of computationally sound protocols revisited. In: Vadhan, S.P. (ed.) TCC 2007. LNCS, vol. 4392, pp. 86–102. Springer, Heidelberg (2007)
15. Raz, R.: A parallel repetition theorem. SIAM Journal on Computing 27(3), 763–803 (1998)
16. Wikström, D.: An efficient concurrent repetition theorem (2009), `http://eprint.iacr.org/`
17. Yao, A.C.: Theory and application of trapdoor functions. In: 23rd IEEE Symposium on Foundations of Computer Science (FOCS), pp. 80–91. IEEE Computer Society Press, Los Alamitos (1982)

Parallel Repetition Theorems for Interactive Arguments*

Kai-Min Chung[1,**] and Feng-Hao Liu[2]

[1] School of Engineering & Applied Sciences, Harvard University, Cambridge MA, USA
kmchung@fas.harvard.edu

[2] Department of Computer Science, Brown University, Providence RI, USA
fenghao@cs.brown.edu

Abstract. We study efficient parallel repetition theorems for several classes of interactive arguments and obtain the following results:

1. We show a *tight* parallel repetition theorem for public-coin interactive arguments by giving a tight analysis for a reduction algorithm of Håstad et al. [HPPW08]. That is, n-fold parallel repetition decreases the soundness error from δ to δ^n. The crux of our improvement is a new analysis that avoid using Raz's Sampling Lemma, which is the key ingredient to the previous results.
2. We give a new security analysis to strengthen a parallel repetition theorem of Håstad et al. for a more general class of arguments. We show that n-fold parallel repetition decreases the soundness error from δ to $\delta^{n/2}$, which is almost tight. In particular, we remove the dependency on the number of rounds in the bound, and as a consequence, extend the "concurrent" repetition theorem of Wikström [Wik09] to this model.
3. We obtain a way to turn *any* interactive argument to one in the class above using fully homomorphic encryption schemes. This gives a way to amplify the soundness of any interactive argument without increasing the round complexity.
4. We give a simple and generic transformation which shows that tight direct product theorems imply almost-tight Chernoff-type theorems. This extends our results to Chernoff-type theorems, and gives an alternative proof to the Chernoff-type theorem of Impagliazzo et al. [IJK09] for weakly-verifiable puzzles.

Keywords: Parallel repetition, interactive argument, public-coin, Arthur-Merlin, direct product theorem.

1 Introduction

In an interactive protocol $\langle \mathsf{P}, \mathsf{V} \rangle$, the prover P wants to convince the verifier V of the validity of some statement (e.g., $x \in L$ for some language L). Two desired

* A full version of this paper can be found on [CL09].

** Supported by US-Israel BSF grant 2006060 and NSF grant CNS-0831289.

D. Micciancio (Ed.): TCC 2010, LNCS 5978, pp. 19–36, 2010.

properties are *completeness*: for a valid statement, the honest prover can always convince the honest verifier; and *soundness*: for an invalid statement, an honest verifier, even when interacting with an adversarial prover, should accept with bounded probability, namely at most some δ, where δ is called the *soundness error* or error probability of the protocol. A protocol is called an interactive *proof* if the soundness holds against computationally unbounded provers, and an interactive *argument* if the soundness only holds against efficient provers.

When the soundness error of a protocol is too high, a natural way to decrease it is by repetition. That is, a prover and a verifier run n copies of the protocol, and the verifier decides whether to accept or not based on the outcomes of the n executions. For example, a *direct product verifier* $\mathsf{V}^{n,n}$ accepts if all constituent verifiers accept, and more generally the *threshold verifier* $\mathsf{V}^{n,k}$ accepts if at least k constituent verifiers accept. Repetitions can be either sequential or parallel. Sequential repetition decreases soundness error for all known settings, but increases the round complexity, which is usually undesirable. Parallel repetition does not increase the number of rounds and decreases soundness error for interactive proofs. However, for interactive arguments, whether parallel repetition decreases soundness error is a subtle question.

For three-message arguments, a sequence of works [BIN97, CHS05, IJK09, CLLY09, HS09] shows that parallel repetition decreases the soundness error for the threshold verifier $\mathsf{V}^{n,k}$ at the optimal, information-theoretic rate, namely, the probability that n independent Bernoulli random variables with expectation δ have sum at least k. In contrast, Bellare, Impagliazzo, and Naor [BIN97], and Pietrzak and Wikström [PW07] construct some protocols where the soundness error does not decrease *at all* under parallel repetition. Thus, parallel repetition theorems for general arguments have been ruled out. (However, Haitner [Hai09] recently showed that any interactive arguments can be slightly modified so that parallel repetition decreases the error.) On the other hand, for public-coin arguments, recent study shows that the soundness error decreases even for protocols with an arbitrary (polynomial) number of messages.

1.1 Parallel Repetition for Public-Coin Arguments

The first parallel repetition theorem for public-coin arguments is by Pass and Venkitasubramaniam [PV07] for constant-round protocols. They give an efficient transformation that converts a (cheating) parallel prover P^{n*} who interacts with a direct product verifier $\mathsf{V}^{n,n}$ with success probability δ^n to a (cheating) prover P^* who interacts with V with success probability essentially[1] δ, where the success probability refers to the probability that P^{n*} (resp., P^*) successfully convinces the verifier $\mathsf{V}^{n,n}$ (resp., V). This is essentially optimal since one can easily turn a single-copy prover strategy P^* with success probability δ to a parallel prover strategy P^{n*} with success probability δ^n by applying P^* independently to each copy. However, their analysis is only efficient for constant-round protocols.

[1] Throughout the introduction, we ignore the required negligible slackness for such reductions in the discussion.

Håstad, Pass, Pietrzak, and Wikström [HPPW08] give a more efficient reduction algorithm that allows them to prove parallel repetition theorem for public-coin arguments with an arbitrary number of rounds. They actually proves a more general threshold theorem which says that a (cheating) prover P^{n*} interacting with a threshold verifier[2] $\mathsf{V}^{n,(1-\rho)n}$ with success probability ε can be converted to a (cheating) prover P^* interacting with V with success probability $1-\rho-O(m\sqrt{\log(1/\varepsilon)/n})$, where $\rho \in [0,1)$ and m is the number of rounds. In the literature (e.g., [IJK09]), this type of theorems is often referred as *Chernoff-type* theorems. In particular, when $\rho = 0$ (i.e., the direct product case), the success probability is $1 - O(m\sqrt{\log(1/\varepsilon)/n})$, which is suboptimal in comparison to $\varepsilon^{1/n} \approx 1 - O(\log(1/\varepsilon)/n)$. Their analysis uses Raz's Sampling Lemma [Raz98] in every round, which is the reason for the factor $O(m\sqrt{\log(1/\varepsilon)/n})$ in the bound.[3] An immediate question is whether the sub-optimality is inherent for super-constant round protocols.

Recently, Wikström [Wik09] strengthened the bound of Håstad et al. [HPPW08] by generalizing Raz's Lemma and applying it only once instead in every round. He improves the analysis of [HPPW08] and shows that the construction in [HPPW08] actually achieves success probability $1 - \rho - O(\sqrt{\log(1/\varepsilon)/n})$ for Chernoff-type case, and $1 - O(\sqrt{\log(1/\varepsilon)/n})$ for direct product case. Removing the dependency on m allows him to prove a more general "concurrent repetition" theorem. The previous works give bounds on the rate at which the soundness error decreases, but it remains open whether the bounds are tight for the parallel repetition of public-coin arguments.

Our Result. In this paper, we prove a *tight* parallel repetition theorem for public-coin interactive arguments. We show that n-fold parallel repetition decreases the soundness error of public-coin arguments from δ to δ^n. We use the same reduction algorithm as [HPPW08], and the crux of our improvement is a way to avoid using Raz's Sampling Lemma.

Techniques. The constructions of P^* from P^{n*} mentioned above share the following structure. Without loss of generality, let P^{n*} be a deterministic parallel prover. The constructed prover P^* simulates internally an interaction between P^{n*} (given as a black-box) and n verifiers $\mathsf{V}_1, \ldots, \mathsf{V}_n$, where one coordinate V_i for some $i \in [n]$ chosen by P^* is played by the external verifier V. That is, throughout the interaction, P^* forwards the message that P^{n*} sends to V_i to the external V, and forwards V's message to P^{n*} as V_i's message. Since P^{n*} is deterministic, the interaction of P^{n*} and $\mathsf{V}^{n,n}$ is determined by the verifiers' messages. In each round, V selects a uniformly random message for V_i, and the task of P^* is to select good messages for the rest verifiers (denoted by V_{-i}) that maximize the probability of $\mathsf{V} = \mathsf{V}_i$ accepting at the end of interaction.

For example, the prover P^* of Pass and Venkitasubramaniam [PV07] uses *recursive sampling* to select a good coordinate $i \in [n]$ and good messages for V_{-i}

[2] Recall that a threshold verifier $\mathsf{V}^{n,(1-\rho)n}$ accepts iff at least $(1-\rho)n$ constituent verifiers accept.

[3] We elaborate more detail in the Techniques paragraph below.

such that P^{n*} could convince V_i with the highest probability among the samples he sees. However, since P^* recursively takes many samples in each round, the number of samples grows exponentially in the number of rounds. Thus, this is only efficient for constant-round protocols.

To cope with the inefficiency, the prover P^* of Håstad et al. [HPPW08] selects coordinate $i \in [n]$ uniformly at random, and uses *rejection sampling* to select good messages for V_{-i}. More precisely, let $(\boldsymbol{v}_1, \boldsymbol{p}_1, \ldots, \boldsymbol{v}_m, \boldsymbol{p}_m)$ denote the messages of $\langle \mathsf{P}^{n*}, \mathsf{V}^{n,n} \rangle$, where $\boldsymbol{v}_j = (v_{j,1}, \ldots, v_{j,n})$ and $\boldsymbol{p}_j = (p_{j,1}, \ldots, p_{j,n})$ are messages of $\mathsf{V}^{n,n}$ and P^{n*} in round $j \in [m]$, respectively. In the j-th round, when P^* receives V's message, P^* considers the message as $v_{j,i}$, and repeatedly samples *random continuations* from the current partial interaction of P^{n*} and $\mathsf{V}^{n,n}$ for a polynomial number of times. That is, P^* samples messages $\boldsymbol{v}_{j,-i} = (v_{j,1}, \ldots, v_{j,i-1}, v_{j,i+1}, \ldots, v_{j,n})$, and $\boldsymbol{v}_{j+1}, \ldots, \boldsymbol{v}_m$ uniformly at random to complete the interaction. Once the continuation is *successful*, i.e., $\mathsf{V}^{n,n}$ accepts, P^* selects the $\boldsymbol{v}_{j,-i}$ of this continuation as V_{-i}'s messages, and forwards P^{n*}'s response $p_{j,i}$ to the external verifier V. If no successful continuations are found in polynomially many samples, P^* simply aborts.

To analyze the success probability, Håstad et al. [HPPW08] consider an "ideal" version of the procedure, where there is no external verifier, and the prover $\tilde{\mathsf{P}}^*$ simulates the interaction of P^{n*} and $\mathsf{V}^{n,n}$ alone by selecting each round of *all* internal verifiers' messages by rejection sampling, i.e., conditioning on a successful random continuation. Since successful continuation always exists by construction, $\tilde{\mathsf{P}}^*$ can always complete a successful interaction (i.e., $\mathsf{V}^{n,n}$ accepts) with probability 1. They then apply Raz's Lemma [Raz98] for every round to upper bound the statistical distance between the two experiments. Each application of Raz's Lemma incurs statistical distance $O(\sqrt{\log(1/\varepsilon)/n})$. Thus, the constructed prover P^* can succeed with probability at least $1 - O(m\sqrt{\log(1/\varepsilon)/n})$, where m is the number of the round. The analysis of Wikström [Wik09] follows the same structure as Håstad et al. [HPPW08]. He generalizes Raz's Lemma to a "multi-round" setting which allows him to bound the statistical distance by one application of the generalized lemma, and hence remove the dependency on m. However, to get a tight direct product theorem, we cannot afford the $O(\sqrt{\log(1/\varepsilon)/n})$ loss of applying the Raz's Lemma. It is also not clear whether the bound on the statistical distance of two experiments can be improved to $1 - \varepsilon^{1/n}$.

We instead analyze the construction directly, avoiding the use of any form of Raz's Lemma. We lower bound the success probability of the constructed prover P^* by induction. Let η_i be the success probability of P^* (i.e., the probability that P^* convinces V) when the external verifier V is embedded in the i-th coordinate, and γ the success probability of P^{n*} (i.e., the probability that P^{n*} convinces $\mathsf{V}^{n,n}$). We essentially[4] show by induction on the round $j \in [m]$ that

[4] Technically, this is for a stronger prover who can sample random continuation for unbounded number of times. For the real prover, we need to modify the inductive hypothesis to take into account the fact that the prover may fail to find a successful continuation and abort.

$$\prod_{i}^{n} \eta_i \geq \gamma, \text{ when conditioning on any partial interaction } (\boldsymbol{v}_1, \boldsymbol{p}_1, \ldots, \boldsymbol{v}_j, \boldsymbol{p}_j).$$

The base case where $j = m$ is trivial. The inductive step is proved by two applications of Hölder's Inequality. It follows that the success probability of P^* when $j = 0$ is

$$\frac{1}{n} \cdot \sum_{i=1}^{n} \eta_i \geq \left(\prod_{i=1}^{n} \eta_i \right)^{1/n} \geq \gamma^{1/n},$$

which is at least $\varepsilon^{1/n}$ by assumption.

1.2 Extension to Arguments with Simulatable Verifiers without Verdict

The results of Håstad et al. [HPPW08, HPWP10][5] extend to arguments with *simulatable* verifiers *without verdict* defined in [HPWP10]. The model generalizes both three-message arguments and public-coin arguments, and contains other natural protocols. Roughly speaking, simulatability of a verifier means that given only the prover's view of any partial interaction (which thus excludes the verifier's internal state) one can efficiently simulate the verifier in the rest of the interaction. However, since the verifier's coins are not given, one may not know the decision of the verifier in the end of the interaction. In such cases, it is referred as *simulatable verifiers without verdict*.

The argument of Håstad et al. [HPPW08] extends to this model, and gives parallel repetition theorems with the same parameters. That is, the constructed prover P^* achieves success probability $1-\rho-O(m\sqrt{\log(1/\varepsilon)/n})$ for Chernoff-type case, and $1 - O(m\sqrt{\log(1/\varepsilon)/n})$ for direct product case, where m is the number of rounds. The bounds are further improved to $1 - \rho - O(\sqrt{m}\sqrt{\log(1/\varepsilon)/n})$ and $1 - O(\sqrt{m}\sqrt{\log(1/\varepsilon)/n})$, respectively in the new version of Håstad et al. [HPWP10], which remain dependent on m.

Our Result. We give a new reduction algorithm that converts a parallel prover P^{n*} for $\mathsf{V}^{n,n}$ with success probability ε to a prover P^* for V with success probability $\varepsilon^{2/n} \approx 1 - O(\log(1/\varepsilon)/n)$, which is almost tight.

Techniques. Recall that the prover P^* of Håstad et al. [HPPW08] selects good messages of V_{-i} by sampling and selecting a "successful" random continuation. However, since the decision of the external verifier is not known, P^* needs to select a "successful" random continuation based only on the decisions of the internal verifiers. A naive approach for P^* is to choose a continuation where all the internal verifiers accept. However, such naive P^* cannot succeed with good probability if the "success pattern" has certain *bad correlations*, as illustrated by the following example.

[5] [HPWP10] is a new version of [HPPW08] that merges the paper [Wik09] and contains additional results.

Consider a two-message protocol $\langle \mathsf{P}, \mathsf{V} \rangle$, and a (deterministic) parallel prover P^{n*} such that when interacting with $\mathsf{V}^{n,n}$, (i) P^{n*} can convince the parallel verifier $\mathsf{V}^{n,n}$ with probability ε, and (ii) for every $i \in [n]$, P^{n*} can convince all except the i-th verifier with probability $(1-\varepsilon)/n$. As there are only two messages, the naive prover P^* receives a message v from the external verifier V, and selects a response as follows. P^* randomly selects $i \in [n]$, views v as v_i, selects v_{-i} such that P^{n*} convinces V_{-i}, and forwards the corresponding p_i to V. Observe that P^* will select continuations in both cases but can only successfully convince the external verifier V in case (i). It is possible that P^* may succeed with probability (i)/((i)+(ii)) $= \varepsilon/(\varepsilon + (1-\varepsilon)/n)$ for every external verifier V's message v. Thus, the success probability of P^* may be only (i)/((i)+(ii)) $\approx n\varepsilon \ll \varepsilon^{1/n}$.

Two techniques have been developed to handle this bad correlation issue since the study of three-message arguments. Bellare et al. [BIN97] use the idea of *soft decision*, namely, the more the number of accepting internal verifiers, the higher the probability that the prover selects a random continuation. This approach is taken in both Impagliazzo et al. [IJK09] and Håstad et al. [HPWP10]. All these results used Raz's Sampling Lemma in their analysis.

To avoid the use of Raz's Sampling Lemma, we adopt another technique developed by Canetti et al. [CHS05] who prove a tight parallel repetition theorem for three-message arguments. The key observation is that one can exploit such a bad correlation to decrease the problem size: they present a transformation that turns a badly correlated parallel prover P^{n*} (interacting with $\mathsf{V}^{n,n}$) to a parallel prover $\mathsf{P}^{(n-1)*}$ (interacting with $\mathsf{V}^{n-1,n-1}$) that still has good success probability. To illustrate the idea using the above example, a such $\mathsf{P}^{(n-1)*}$ can simply interact with $\mathsf{V}^{n-1,n-1}$ by simulating the interaction of P^{n*} and $\mathsf{V}^{n,n}$ with the first coordinate played by an internal verifier and the rest coordinates played by $\mathsf{V}^{n-1,n-1}$. It is not hard to see that such $\mathsf{P}^{(n-1)*}$ can succeed with probability $\varepsilon + (1-\varepsilon)/n$.

It follows that one can iteratively apply the transformation until either (i) $n = 1$ or (ii) no such bad correlations exist. In case (i), we trivially obtain a single-copy prover P^* with good success probability, while in case (ii), Canetti et al. manage to show that the naive approach works for three-message arguments. We observe that this idea is generic and applicable to our setting, where our analysis technique described in Section 1.1 can be generalized to prove that, in out setting, the naive approach works in case (ii) as well, which leads to an almost tight bound.

1.3 Reducing Soundness Error for *Any* Interactive Arguments

We obtain a way to turn *any* interactive argument $\langle \mathsf{P}, \mathsf{V} \rangle$ to an interactive argument $\langle \mathsf{P}', \mathsf{V}' \rangle$ with simulatable verifier without verdict that preserves the completeness and soundness of the original protocol. As a consequence of the above section, parallel repetition decreases soundness error of the modified protocol $\langle \mathsf{P}', \mathsf{V}' \rangle$ in a nearly optimal rate.

The idea is to run the protocol $\langle \mathsf{P}, \mathsf{V} \rangle$ with all messages under the encryption of a *fully homomorphic encryption scheme* [Gen09] using verifier's public key, which still allows the prover to simulate the original protocol with messages under encryption. Intuitively, completeness and soundness are preserved since the two parties effectively run the same protocol. Furthermore, since all the messages are encrypted under the verifier's key, they look random to the prover. Therefore, the verifier is easy to simulate – simply generate the verifier's message by encrypting some junks.

We remark that our result in this section is incomparable to the result of Haitner [Hai09] (subsequently improved by Håstad et al. [HPWP10]), who also gives a simple transformation that turns any interactive argument to one with comparable soundness where parallel repetition decreases the soundness error. Our result achieves nearly optimal rate, while the result of Haitner [Hai09] and Håstad et al. [HPWP10] has the undesirable dependency on the number of rounds m. In particular, the number of repetition is required to be at least $\Omega(m^4)$ for the soundness error to decrease. On the other hand, we use a relatively strong cryptographic assumption of fully homomorphic encryption schemes while their result holds unconditionally.

1.4 Extension to Chernoff-Type Theorems

We give a simple and generic transformation which shows that tight direct product theorems imply almost tight Chernoff-type theorems, and thus extend our results to Chernoff-type theorems. Our transformation applies to various models such as weakly-verifiable puzzles, and gives an alternative proof to the Chernoff-type theorem of Impagliazzo et al. [IJK09] as a consequence of the tight direct product theorem of Canetti et al. [CHS05].

The transformation converts a parallel prover P^{n*} for $\mathsf{V}^{n,k}$ to a parallel prover P^{t*} for $\mathsf{V}^{t,t}$ for any given $t \leq k$. The prover P^{t*} simply samples a random set of coordinate $S \subset [n]$ of size t, and interacts with $\mathsf{V}^{t,t}$ by simulating the interaction of P^{n*} and $\mathsf{V}^{n,k}$ with coordinates S played by $\mathsf{V}^{t,t}$ and the remaining coordinates played by internal verifiers. Clearly, P^{t*} convinces $\mathsf{V}^{t,t}$ if and only if P^{n*} convinces verifiers V_i's for $i \in S$ of $\mathsf{V}^{n,k}$. Let ε be the success probability of P^{n*}. It is not hard to show that P^{t*} has success probability at least $\varepsilon \cdot \binom{k}{t}/\binom{n}{t}$ by an averaging argument. Let $k = (1-\rho)n$, and suppose a tight direct theorem holds, then applying the reduction on P^{t*} with properly chosen t gives a prover P^* with success probability $(\varepsilon \cdot \binom{k}{t}/\binom{n}{t})^{1/t} \approx 1 - \rho - O(\sqrt{\log(1/\varepsilon)/n})$.[6]

For public-coin arguments, the transformation extends our direct product theorem to a Chernoff-type theorem with similar parameter to [Wik09]. For arguments with simulatable verifiers without verdict, the transformation and our improved direct product theorem yield a prover P^* with success probability $(1-\rho)^2 - O(\sqrt{\log(1/\varepsilon)/n})$. This bound is incomparable to the bound $1 - \rho - O(\sqrt{m}\sqrt{\log(1/\varepsilon)/n})$ of [HPWP10] in that our bound does not depend on m, but has a slightly worse dependency on ρ.

[6] Technically, for the reduction to be efficient, we cannot set the parameter t to be too large. Thus, the reduction P^* can only success with probability $1 - \rho - \max\{\alpha, O(\sqrt{\log(1/\varepsilon)/n})\}$ for an arbitrarily small constant α, which suffices for most applications.

As an additional contribution, we also prove that the reduction algorithm of Pass and Venkitasubramaniam [PV07] for *constant-round* public-coin arguments gives tight parallel repetition theorems for any threshold verifiers, i.e., if V has soundness error δ, then $\mathsf{V}^{n,k}$ has soundness error essentially $P(n,k,\delta)$, where $P(n,k,\delta) = \Pr[\sum_{i=1}^{n} X_i \geq k]$ with X_i's being i.i.d. binary random variables and $\Pr[X_i = 1] = \delta$.

2 Preliminary and Notation

We introduce the following notation for an interactive protocol $\langle \mathsf{P}, \mathsf{V} \rangle$. Let x denote the common input. We assume the verifier speaks first. One round contains two message exchanges – from the verifier to the prover and back. Let m denote the number of rounds. A transcript of an interaction is denoted by $(v_1, p_1, \ldots, v_m, p_m) = \langle \mathsf{P}, \mathsf{V} \rangle(x)$. When V is public-coin, verifier's messages $v_1, \ldots, v_m$ are independent uniformly random strings.

Consider parallel execution of a protocol. We use $\langle \mathsf{P}^n, \mathsf{V}^{n,k} \rangle$ to denote a n-fold parallel repetition of $\langle \mathsf{P}, \mathsf{V} \rangle$, where n copies of verifiers are denoted by $\mathsf{V}_1, \ldots, \mathsf{V}_n$, and $\mathsf{V}^{n,k}$ accepts iff at least k copies of V_i's accept. A transcript of an interaction is denoted by $(\boldsymbol{v}_1, \boldsymbol{p}_1, \ldots, \boldsymbol{v}_m, \boldsymbol{p}_m) = \langle \mathsf{P}^n, \mathsf{V}^{n,k} \rangle(x)$, where $\boldsymbol{v}_j = (v_{j,1}, \ldots, v_{j,n})$ and $\boldsymbol{p}_j = (p_{j,1}, \ldots, p_{j,n})$.

When a parallel prover P^{n*} is deterministic, the interaction $\langle \mathsf{P}^{n*}, \mathsf{V}^{n,k} \rangle$ is determined by the verifier's messages $(\boldsymbol{v}_1, \ldots, \boldsymbol{v}_m)$. Thus, we can skip prover's messages and describe an interaction by $(\boldsymbol{v}_1, \ldots, \boldsymbol{v}_m)$. We refers to a partial transcript as a *history* $\bar{h} = (\boldsymbol{v}_1, \ldots, \boldsymbol{v}_j)$.

The main tool used in our analysis is Hölder's Inequality.

Lemma 1 (Hölder's Inequality [Dur04])

– *Let F, G be two non-negative functions from Ω to $\mathbb{R}$, and $a, b > 0$ satisfying $1/a + 1/b = 1$. Let q be a uniformly random variable over Ω. We have*

$$\mathop{\mathrm{E}}_q[F(q) \cdot G(q)] \leq \mathop{\mathrm{E}}_q[F(q)^a]^{1/a} \cdot \mathop{\mathrm{E}}_q[G(q)^b]^{1/b}.$$

– *In general, let $F_1, \ldots, F_n$ be non-negative functions from Ω to $\mathbb{R}$, and $a_1, \ldots a_n > 0$ satisfying $1/a_1 + \ldots 1/a_n = 1$. We have*

$$\mathop{\mathrm{E}}_q[F_1(q) \cdots F_n(q)] \leq \mathop{\mathrm{E}}_q[F_1(q)^{a_1}]^{1/a_1} \cdots \mathop{\mathrm{E}}_q[F_n(q)^{a_n}]^{1/a_n}.$$

3 Tight Direct Product Theorem for Public-Coin Arguments

In this section, we prove a tight direct product theorem for public-coin interactive arguments.

Theorem 1. *Let $\mathsf{V} \in$ PPT be public-coin. There exists a prover strategy P^* such that for every common input x, every $n \in \mathbb{N}$, every $\varepsilon, \xi \in (0,1)$, and every parallel prover strategy P^{n*},*

1. $\mathsf{P}^*(x, n, \varepsilon, \xi)$ *runs in time* $\mathrm{poly}(|x|, n, \varepsilon^{-1}, \xi^{-1})$ *given oracle access to* $\mathsf{P}^{n*}(x)$.
2. $\Pr[\langle \mathsf{P}^{n*}, \mathsf{V}^{n,n}\rangle(x) = 1] \geq \varepsilon \Rightarrow$

$$\Pr[\langle \mathsf{P}^*(n, \varepsilon, \xi), \mathsf{V}\rangle(x) = 1] \geq \varepsilon^{1/n} \cdot (1 - \xi).$$

We remark that the theorem also holds for interactive arguments with simulatable verifier with verdict defined in Section 4.

Without loss of generality we assume that P^{n*} is deterministic, since by sampling, we can find a fixing of the coin tosses of P^{n*} with only a small loss in the success probability. Let us first recall the common approach of such a reduction algorithm. On input x, the constructed prover P^* simulates the interaction of $\langle \mathsf{P}^{n*}, \mathsf{V}^{n,n}\rangle(x)$ internally, where P^* simulates $n-1$ internal verifiers by himself, and lets the external verifier V play V_i for some coordinate $i \in [n]$ by forwarding the messages accordingly. Since P^{n*} is deterministic, the interaction is determined by $\mathsf{V}^{n,n}$'s message $(\boldsymbol{v}_1, \ldots, \boldsymbol{v}_m)$. Let $T_i(\cdot)$ denote whether V_i accepts. That is, $T_i(\boldsymbol{v}_1, \ldots, \boldsymbol{v}_m) = 1$ iff V_i accepts in $\langle \mathsf{P}^{n*}, \mathsf{V}^{n,n}\rangle(x)$ with history $(\boldsymbol{v}_1, \ldots, \boldsymbol{v}_m)$.

This can be viewed as a *game* $\mathcal{G}(\mathsf{P}^{n*}, x)$ played between P^* and V as follows. At beginning, P^* plays a *move* $i \in [n]$. Then for each round $j \in [m]$, V plays a random move $v_{j,i}$, and P^* plays a (carefully chosen) move $\boldsymbol{v}_{j,-i} = (v_{j,1}, \ldots, v_{j,i-1}, v_{j,i+1}, \ldots, v_{j,n})$ alternately. At the end, P^* *succeeds* if $T_i(\boldsymbol{v}_1, \ldots, \boldsymbol{v}_m) = 1$. Note that a node of the game tree is of the form either $(i; \boldsymbol{v}_1, \ldots, \boldsymbol{v}_j)$, in which case it is V's turn to move, or of the form $(i; \boldsymbol{v}_1, \ldots, \boldsymbol{v}_{j-1}, v_{j,i})$, in which case it if P^*'s turn to move. Phrased in this way, the task is to design a strategy for P^* such that if $\langle \mathsf{P}^{n*}, \mathsf{V}^{n,n}\rangle(x)$ accepts with probability at least ε, then P^* can succeed with probability close to $\varepsilon^{1/n}$ in game $\mathcal{G}(\mathsf{P}^{n*}, x)$. We present the "rejection sampling" reduction algorithm of Hastad et al. [HPPW08] as a strategy of P^* in this game:

Definition 1 (Strategy P^*_{rej}). *We define strategy P^*_{rej} as follows. Let P^{n*} be a deterministic parallel prover, x a common input, and $\mathcal{G}(\mathsf{P}^{n*}, x)$ the corresponding game defined as above.*

- *In the first P^*-move, P^*_{rej} selects a coordinate $i \in_R [n]$ uniformly at random.*
- *On P^*-move node $u = (i; \boldsymbol{v}_1, \ldots, \boldsymbol{v}_{j-1}, v_{j,i})$, P^*_{rej} simulates a* random continuation *of $\mathcal{G}(\mathsf{P}^{n*}, x)$ (i.e., the interaction of $\langle \mathsf{P}^{n*}, \mathsf{V}^{n,n}\rangle(x)$) at most $M \stackrel{\text{def}}{=} O(mn/\varepsilon\xi)$ times. That is, P^*_{rej} simulates the game from u with both parties playing random moves $\boldsymbol{v}_{j,-i}, \ldots, \boldsymbol{v}_{m,i}, \boldsymbol{v}_{m,-i}$. A continuation is* successful *if all verifiers accept, i.e., $T_\ell(\boldsymbol{v}_1, \ldots, \boldsymbol{v}_m) = 1$ for all $\ell \in [n]$. The first time a successful continuation is found, P^*_{rej} plays the corresponding move $\boldsymbol{v}_{j,-i}$. If no successful continuations are found, P^*_{rej} aborts.*

Note that if P^*_{rej} does not abort, P^*_{rej} plays move $\boldsymbol{v}_{j,-i}$ with the probability proportional to the conditional success probability of P^{n*} given on the history $(\boldsymbol{v}_1, \ldots, \boldsymbol{v}_j)$.

Clearly, strategy P^*_{rej} can be implemented in time $\mathrm{poly}(|x|, n, \varepsilon^{-1}, \xi^{-1})$. We next analyze the success probability of P^*_{rej} by induction on the round $j \in [m]$. For the sake of clarity, below we first present the analysis of an ideal version P^*_{ideal} of P^*_{rej}, where P^*_{ideal} can simulate random continuations for unbounded number of times. The analysis of P^*_{rej} is presented in the full version of this paper [CL09].

3.1 Analysis of P^*_{ideal}

In this subsection, we analyze the success probability of an ideal version P^*_{ideal} of strategy P^*_{rej}, which is the same as P^*_{rej} except that P^*_{ideal} can simulate the random continuations an unbounded number of times. Thus, P^*_{ideal} will never abort whenever there is a successful continuation from the current P^*-move node. We will show that if $\Pr[\langle \mathsf{P}^{n*}, \mathsf{V}^{n,n}\rangle(x) = 1] \geq \varepsilon$, then P^*_{ideal} can succeed with probability at least $\varepsilon^{1/n}$ in game $\mathcal{G}(\mathsf{P}^{n*}, x)$.

We first introduce the following notation to express the success probability of P^*_{ideal}. We define $\gamma(\bar{h}) \overset{\text{def}}{=} \Pr[\langle \mathsf{P}^{n*}, \mathsf{V}^n\rangle(x) = 1|\bar{h}]$, where $\bar{h}$ is a history of the form either $(\boldsymbol{v}_1, \ldots, \boldsymbol{v}_j)$ or $(\boldsymbol{v}_1, \ldots, \boldsymbol{v}_{j-1}, v_{j,i})$. That is, $\gamma(\bar{h})$ is the accepting probability of $\langle \mathsf{P}^{n*}, \mathsf{V}^n\rangle$ conditioning on the history $\bar{h}$. Note that $\gamma = \Pr[\langle \mathsf{P}^{n*}, \mathsf{V}^n\rangle(x) = 1] \geq \varepsilon$ by assumption. Next, for every $i \in [n]$, we define $\eta_i(\bar{h}) \overset{\text{def}}{=} \Pr[\mathsf{P}^*_{ideal} \text{ succeeds } |u = (i; \bar{h})]$ to be the success probability of P^*_{ideal} conditioning on node $u = (i; \bar{h})$ of the game tree. Note that the success probability of P^*_{ideal} is $(1/n) \cdot \sum_{i=1}^n \eta_i$.

Claim. For every $i \in [n]$ and full history $\bar{h} = (\boldsymbol{v}_1, \ldots, \boldsymbol{v}_m)$, we have $\eta_i(\bar{h}) = T_i(\bar{h})$. For every $i \in [n]$, $j \in [m]$, and history $\bar{h} = (\boldsymbol{v}_1, \ldots, \boldsymbol{v}_{j-1})$, we have[7]

$$\eta_i(\bar{h}) = \mathop{\mathrm{E}}_{\boldsymbol{v}_j}\left[\frac{\gamma(\bar{h}, \boldsymbol{v}_j) \cdot \eta_i(\bar{h}, \boldsymbol{v}_j)}{\gamma(\bar{h}, v_{j,i})}\right].$$

Proof. The first part follows by definition. For the second part, recall that V plays the random strategy and P^*_{ideal} plays the rejection sampling strategy. V plays each $v_{j,i}$ with probability $\Pr[v_{j,i}]$, which corresponds to the expectation operator over $v_{j,i}$. P^*_{ideal} plays each $\boldsymbol{v}_{j,-i}$ with probability $\Pr[\boldsymbol{v}_{j,-i}] \cdot (\gamma(\bar{h}, \boldsymbol{v}_j)/\gamma(\bar{h}, v_{j,i}))$, which corresponds to the expectation operator over $\boldsymbol{v}_{j,-i}$ with factor $\gamma(\bar{h}, \boldsymbol{v}_j)/\gamma(\bar{h}, v_{j,i})$ in the expectation.

We now prove that the success probability of P^*_{ideal} is at least $\varepsilon^{1/n}$ by induction. In fact, we induct on a slightly stronger inductive hypothesis: for every $j \in \{0, \ldots, m\}$ and history $\bar{h} = (\boldsymbol{v}_1, \ldots, \boldsymbol{v}_j)$, $\prod_{i=1}^n \eta_i(\bar{h}) \geq \gamma(\bar{h})$.

The base case $j = m$ is trivial. For every full history $\bar{h} = (\boldsymbol{v}_1, \ldots, \boldsymbol{v}_m)$, $\gamma(\bar{h}) = 1$ iff $\eta_i(\bar{h}) = T_i(\bar{h}) = 1$ for every $i \in [n]$. Assuming that the inductive hypothesis holds for j and every $\bar{h} = (\boldsymbol{v}_1, \ldots, \boldsymbol{v}_j)$, we want to prove the inductive hypothesis for $j - 1$ and every $\bar{h} = (\boldsymbol{v}_1, \ldots, \boldsymbol{v}_{j-1})$. More precisely, for every $\bar{h} = (\boldsymbol{v}_1, \ldots, \boldsymbol{v}_{j-1})$, we want to show that

[7] We use the convention that if $\gamma(\bar{h}, v_{j,i}) = 0$ (which implies $\gamma(\bar{h}, \boldsymbol{v}_j) = 0$), then the ratio is 0.

$$\prod_{i=1}^{n} \eta_i(\bar{h}) = \prod_{i=1}^{n} \mathop{\mathrm{E}}_{\boldsymbol{v}_j} \left[\frac{\gamma(\bar{h}, \boldsymbol{v}_j) \cdot \eta_i(\bar{h}, \boldsymbol{v}_j)}{\gamma(\bar{h}, v_{j,i})} \right] \geq \gamma(\bar{h}),$$

provided that for every $\boldsymbol{v}_j$, $\prod_{i=1}^{n} \eta_i(\bar{h}, \boldsymbol{v}_j) \geq \gamma(\bar{h}, \boldsymbol{v}_j)$. For notational simplicity, we abstract the above statement as the following lemma.

Lemma 2. *Let $\gamma, \eta_1, \ldots, \eta_n : \Omega^n \to [0,1]$ be $[0,1]$-valued functions over a product space Ω^n such that $\prod_i \eta_i(\boldsymbol{q}) \geq \gamma(\boldsymbol{q})$ for every $\boldsymbol{q} = (q_1, \ldots, q_n) \in \Omega^n$. Let $\gamma = \mathrm{E}_{\boldsymbol{q}}[\gamma(\boldsymbol{q})]$. For every $i \in [n]$, let*

$$\gamma(q_i) = \mathop{\mathrm{E}}_{\boldsymbol{q}_{-i}} [\gamma(\boldsymbol{q})] \quad \text{and} \quad \eta_i = \mathop{\mathrm{E}}_{\boldsymbol{q}} \left[\frac{\gamma(\boldsymbol{q}) \cdot \eta_i(\boldsymbol{q})}{\gamma(q_i)} \right],$$

where the above expectation is over uniform distribution over Ω^n. We have

$$\prod_{i=1}^{n} \eta_i = \prod_{i=1}^{n} \mathop{\mathrm{E}}_{\boldsymbol{q}} \left[\left(\frac{\gamma(\boldsymbol{q}) \cdot \eta_i(\boldsymbol{q})}{\gamma(q_i)} \right) \right] \geq \gamma.$$

Proof. The trick is to apply Hölder's Inequality to "swap the operators". We present the whole computation first, and then explain how Hölder's Inequality is applied.

$$\begin{aligned}
&\prod_{i=1}^{n} \mathop{\mathrm{E}}_{\boldsymbol{q}} \left[\left(\frac{\gamma(\boldsymbol{q}) \cdot \eta_i(\boldsymbol{q})}{\gamma(q_i)} \right) \right] \\
&\geq \mathop{\mathrm{E}}_{\boldsymbol{q}} \left[\left(\frac{\gamma(\boldsymbol{q})^n \cdot \prod_{i=1}^{n} \eta_i(\boldsymbol{q})}{\prod_{i=1}^{n} \gamma(q_i)} \right)^{1/n} \right]^n && \text{(by Hölder's Inequality)} \\
&\geq \mathop{\mathrm{E}}_{\boldsymbol{q}} \left[\left(\frac{\gamma(\boldsymbol{q})^{n+1}}{\prod_{i=1}^{n} \gamma(q_i)} \right)^{1/n} \right]^n && \text{(by inductive hypothesis)} \\
&\geq \left[\left(\frac{\mathrm{E}_{\boldsymbol{q}}[\gamma(\boldsymbol{q})]^{n+1}}{\mathrm{E}_{\boldsymbol{q}}[\prod_{i=1}^{n} \gamma(q_i)]} \right)^{1/n} \right]^n && \text{(by Hölder's Inequality)} \\
&= (\gamma^{n+1}/\gamma^n) = \gamma.
\end{aligned}$$

We now explain the application of Hölder's Inequalities.

- The first inequality uses $\mathrm{E}[X_1^n]^{1/n} \cdot \cdots \cdot \mathrm{E}[X_n^n]^{1/n} \geq \mathrm{E}[X_1 \cdot \cdots \cdot X_n]$ with

$$X_i = \left(\frac{\gamma(\boldsymbol{q}) \cdot \eta_i(\boldsymbol{q})}{\gamma(q_i)} \right)^{1/n}.$$

- The third inequality uses $\mathrm{E}\left[B^{n+1}\right]^{1/(n+1)} \cdot \mathrm{E}\left[(A/B)^{(n+1)/n}\right]^{n/(n+1)} \geq \mathrm{E}[A]$, or equivalently,

$$\mathrm{E}\left[\left(\frac{A^{n+1}}{B^{n+1}} \right)^{1/n} \right] \geq \left(\frac{\mathrm{E}[A]^{n+1}}{\mathrm{E}[B^{n+1}]} \right)^{1/n}$$

with

$$\begin{cases} A = \gamma(\boldsymbol{q}), \\ B^{n+1} = \prod_{i=1}^{n} \gamma(q_i). \end{cases}$$

Remark 1. One might worry about the legitimacy of the manipulation when the denominators are zeros. One way to justify it is by adding some μ in the denominators before the manipulation. Formally, we have

$$\prod_{i=1}^{n} \mathop{\mathrm{E}}_{\boldsymbol{q}}\left[\left(\frac{\gamma(\boldsymbol{q}) \cdot \eta_i(\boldsymbol{q})}{\gamma(q_i)}\right)\right] \geq \prod_{i=1}^{n} \mathop{\mathrm{E}}_{\boldsymbol{q}}\left[\left(\frac{\gamma(\boldsymbol{q}) \cdot \eta_i(\boldsymbol{q})}{\gamma(q_i) + \mu}\right)\right] \geq \cdots \geq (\gamma^{n+1}/(\gamma + \mu)^n),$$

which is valid for arbitrary $\mu > 0$. Taking $\mu \to 0$, we obtain the desired result.

Applying the above lemma directly completes the proof of the induction. It follows that the success probability of P^*_{ideal} is

$$\frac{1}{n} \cdot \sum_{i=1}^{n} \eta_i \geq \left(\prod_{i=1}^{n} \eta_i\right)^{1/n} \geq \gamma^{1/n} \geq \varepsilon^{1/n}.$$

The next step is to analyze P^*_{rej} in a similar way as above. The challenge is that P^*_{rej} may abort due to the failure of finding a successful continuation in M trials, which makes the success probability a more complicated formula. Details of the analysis of P^*_{rej} can be found in the full version of this paper [CL09].

4 Arguments with Simulatable Verifier without Verdict

In this section, we present a new reduction algorithm that extends our results to interactive arguments with simulatable verifiers defined by Håstad et al. [HPWP10]. Roughly speaking, a verifier is simulatable if given only the prover's view of any partial interaction (which thus excludes the verifier's internal state), one can efficiently simulate verifier in the rest of the interaction. In terms of the terminology in [HPWP10], our results holds for arguments with "1-simulatable verifiers without verdict," which we refer to as just simulatable verifiers below for simplicity. For the sake of completeness, we repeat their definition in this special case. For a more general definition of simulatability, we refer the reader to [HPWP10].

The definition requires the following notation. Recall that we use p_j and v_j to denote the prover and verifier's j-th messages, respectively. We let s_j and t_j be the states of the prover and verifier after computing the j-th messages, respectively. We think of the verifier as using independent random tape R_j for computing j-th message. Namely, V computes message v_j from its previous state t_{j-1}, prover's message p_j, and fresh randomness R_j. Note that the verifier's state t_j implicitly contains the content of the random tapes $r_1, \ldots, r_j$ (generated in the previous rounds) of V. For convenience, we use $p_{[j]}$ to denote $p_1, \ldots, p_j$, and the same rule applies to other variables.

Definition 2 (Simulatable Verifier [HPWP10]). *A verifier* V *is said to be* simulatable without verdict, *or just* simulatable, *if for every* PPT *prover strategy* P^* *there exists a* PPT *simulator* S *such that for every partial interaction* $(s_{[j]}, t_{[j]}, x, p_{[j]}, v_{[j]})$, *the distribution of* P^**'s view of an interaction with* V *(not including the decision bit of* V*), starting from states* s_j *and* t_j *and message* p_j, *is* computationally indistinguishable *to the distribution of* P^**'s view of an interaction with* S *starting from states* s_j *and* $[s_{[j]}, x, p_{[j]}, v_{[j]}]$ *and message* p_j. *When the decision bit of* V *is included in the consideration, we say that* V *is* simulatable with verdict.

Remark 2. In the above definition, we only require the distributions to be computational indistinguishable, as opposed to the statistical closeness defined in [HPWP10]. Håstad et al. requires statistical closeness since they need to handle a general notion of "δ-simulatability." On the other hand, for the case of 1-simulatability, it can be shown (e.g., in the old version of Håstad el al. [HPPW08]) that the requirement can be relaxed to computational indistinguishability. The relaxation to computational indistinguishability is essential to our application of fully-homomorphic encryption in Section 5.

Remark 3. Another deviation from [HPWP10] is that, in the above definition, our simulator S interacts with P^*, as opposed to generate the view by himself in [HPWP10]. This difference is not essential. We adopt to the above definition since it makes the simulation of the random continuation described below more intuitive.

We observe that for arguments with simulatable verifier, in the corresponding game $\mathcal{G}(\mathsf{P}^{n*}, x)$, P^* can still simulate a random continuation from any P^*-move node u. Each internal verifier's next message is easy to generate since the message depends only on the verifier's state, the prover's message, and fresh randomness. For the external verifier V, although P^* does not know V's state, P^* can invoke the simulator to generate the verifier's message. However, P^* is not able to know the decision of the external verifier. Thus, P^* needs to select a "successful" random continuation based only on the decisions of the internal verifiers. As illustrated in the example in Section 1.2, there is an issue of "bad correlations." We resolve this issue in the spirit of Canetti et al. [CHS05], where we iteratively exploit bad correlations to decrease the problem size in a preprocssessing stage, and use a modified rejection sampling strategy when no such bad correlations exist. Our reduction turns a parallel prover P^{n*} for $\mathsf{V}^{n,n}$ with success probability $\delta^n \stackrel{\text{def}}{=} \varepsilon$ to a prover P^* for a single simulatable verifier V with success probability $\delta^2 = \varepsilon^{2/n} \approx 1 - O(\log(1/\varepsilon)/n)$.[8] Formally, we obtain the following theorem.

Theorem 2. *Let* $\mathsf{V} \in$ PPT *be simulatable without verdict. There exists a prover strategy* P^* *such that for every common input* x, *every* $n \in \mathbb{N}$, *every* $\varepsilon, \xi \in (0, 1)$, *and every parallel prover strategy* P^{n*},

[8] It is more convenient to present our proof using parameter δ^n instead of ε in this section.

1. $\mathsf{P}^*(x, n, \varepsilon, \xi)$ *runs in time* $\mathrm{poly}(|x|, n, \varepsilon^{-1}, \xi^{-1})$ *given oracle access to* $\mathsf{P}^{n*}(x)$.
2. $\Pr[\langle \mathsf{P}^{n*}, \mathsf{V}^{n,n}\rangle(x) = 1] \geq \varepsilon \Rightarrow$

$$\Pr[\langle \mathsf{P}^*, \mathsf{V}\rangle(x) = 1] \geq \varepsilon^{2/n} \cdot (1 - \xi).$$

Detailed description of the reduction algorithms and analysis can be found in the full version of this paper [CL09].

5 Reducing Soundness Error for Any Arguments

(The ideas in this section were obtained in discussions with Boaz Barak, Yael Tauman Kalai, and Salil Vadhan.)

In this section, we present a way to turn *any* interactive argument $\langle \mathsf{P}, \mathsf{V}\rangle$ to an interactive argument $\langle \mathsf{P}', \mathsf{V}'\rangle$ with simulatable verifier that preserves the completeness and soundness of the original protocol. It follows that parallel repetition reduces soundness error of the modified protocol $\langle \mathsf{P}', \mathsf{V}'\rangle$ in a nearly optimal rate by Theorem 2.

Recall that the idea is to run the protocol $\langle \mathsf{P}, \mathsf{V}\rangle$ with all messages under the encryption of a *fully homomorphic encryption scheme.* Roughly speaking, a fully homomorphic encryption scheme is a public key encryption scheme with the additional property that given a public key pk, and an encryption $\mathsf{Enc}_{\mathsf{pk}}(m)$, one can homomorphically evaluate any function f (described by a poly-size circuit C) on the underlying message to obtain an encrypted function value $\mathsf{Enc}_{\mathsf{pk}}(f(m))$ without knowing the message m. That is, in addition to the standard functions $(\mathsf{KeyGen}, \mathsf{Enc}, \mathsf{Dec})$ in public key encryption schemes, a fully homomorphic encryption scheme has an additional efficient function Eval that on inputs a public key pk, a description of a poly-size circuit $C(\cdot)$, and a cipher text c that is a valid encryption of m, outputs a cipher text c' which is a valid encryption of $C(m)$.

Recently in a breakthrough, Gentry [Gen09] showed the first construction of a fully homomorphic encryption scheme under reasonable hardness assumptions on ideal lattice problems and sparse subset sum problems. We refer the reader to [Gen09] for the formal definitions and constructions.

Let $\langle \mathsf{P}, \mathsf{V}\rangle$ be any interactive argument. Recall our notation, P and V receive some common input x and alternately send to each other messages denoted as $(v_1, p_1, v_2, p_2, \ldots, v_m, p_m)$ where m is the number of the rounds. We define a modified protocol $\langle \mathsf{P}', \mathsf{V}'\rangle$ that executes the protocol $\langle \mathsf{P}, \mathsf{V}\rangle$ under a fully homomorphic encryption of the verifier's key as follows. For simplicity, we assume that V always makes his decision in the end of the protocol, and all messages of $\langle \mathsf{P}, \mathsf{V}\rangle$ have some fixed length. We also assume that the encryption scheme has perfect correctness and the decryption algorithm Dec always outputs some messages (perhaps junks).

– In the first round, the verifier V' generates $(\mathsf{pk}, \mathsf{sk}) \leftarrow \mathsf{KeyGen}()$,[9] prepares V's first message v_1, and sends the public key pk and the encrypted message $v_1' = \mathsf{Enc}_{\mathsf{pk}}(v_1)$ to P'.

[9] For simplicity, we omit the security parameter throughout this section.

- The prover P' on the received message v_1', homomorphically computes p_1', a valid encryption of the first message p_1 of P. Namely, let $C_1(x, v_1)$ be the next-message function of P. The prover P' uses Eval to compute $p_1' = \mathsf{Eval}_{\mathsf{pk}}(v_1', C_1(x, \cdot))$.
- In general, in the ℓ-th round, the verifier V' receives message $p_{\ell-1}'$. V' first decrypts the message $p_{\ell-1}'$ to obtain $p_{\ell-1} = \mathsf{Dec}_{\mathsf{sk}}(p_{\ell-1}')$. V' simulates V to generate the next message v_ℓ, and sends the encrypted message $v_\ell' = \mathsf{Enc}_{\mathsf{pk}}(v_\ell)$ to P'.
- The prover P' on the received message v_ℓ', homomorphically computes p_ℓ', a valid encryption of the first message p_ℓ of P. Namely, let $C_\ell(x, v_{[\ell]}, p_{[\ell-1]})$ be the next-message function of P. The prover P' uses Eval to compute $p_\ell' = \mathsf{Eval}_{\mathsf{pk}}((v_{[\ell]}', p_{[\ell-1]}'), C_\ell(x, \cdot))$.
- At the end, V' decrypts the last message p_m'. V' accepts iff V accepts.

We first observe that $\langle \mathsf{P}', \mathsf{V}' \rangle$ has exactly the same completeness and soundness as $\langle \mathsf{P}, \mathsf{V} \rangle$ suppose the homomorphic encryption scheme has perfect correctness. The completeness is trivially the same, since $\langle \mathsf{P}', \mathsf{V}' \rangle$ simply simulates $\langle \mathsf{P}, \mathsf{V} \rangle$ under a fully homomorphic encryption. For the soundness, note that for every (cheating) prover strategy P'^* for $\langle \mathsf{P}', \mathsf{V}' \rangle$, we can construct a (cheating) prover strategy P^* that interacts with V by simulating the interaction of P'^* and V' as follows. P^* first generates $(\mathsf{pk}, \mathsf{sk})$ by himself and forwards pk to P'^*. P^* then simulates the interaction of P'^* and V' by (i) encrypting the messages of V and forwarding them to P'^*, and (ii) decrypting the messages of P'^* and forwarding them to V. It follows that P^* can convince V with the same probability as P'^* convincing V'. Similarly, for every P^*, there is a P'^* that applies the same strategy as P^* (homomorphically) and convinces V' with the same probability as P^* convincing V.

It remains to show that V' is simulatable. To argue this, we need to specify the random tape used by V' in each round, since this affects the states t_ℓ's of the verifier. For convenience, we define $m+1$ random tapes $R_0, R_1, \ldots, R_m$ for V', where both R_0 and R_1 are generated in the first round. We let R_0 be the random tape that contains all the randomness used in V. For $\ell \in [m]$, we let R_ℓ be the randomness that V' uses to encrypt the ℓ-th round message. Note that defined in this way, given the state $t_{\ell-1}'$ of V' and prover P''s message p_ℓ', the underlying verifier V's message v_i is *deterministic*, and the randomness of V''s message v_i' comes only from the encryption. Now it is trivial to simulate V'. A simulator S simply ignores the prover's message, and sends a fresh encryption of junks in each round. By the semantic security of the encryption scheme, the prover's view when interacting with V' is computationally indistinguishable from that when interacting with S.

We summarize the above discussion in the following theorem.

Theorem 3. *Let $\langle \mathsf{P}, \mathsf{V} \rangle$ be any interactive argument with soundness error δ. Suppose there exists a fully homomorphic encryption scheme with perfect correctness, then the modified interactive argument $\langle \mathsf{P}', \mathsf{V}' \rangle$ defined above satisfies the following properties.*

– $\langle \mathsf{P}', \mathsf{V}' \rangle$ *has exactly the same completeness and soundness as* $\langle \mathsf{P}, \mathsf{V} \rangle$.
– V' *is simulatable without verdict, and thus n-fold parallel repetition reduces soundness error from* δ *to* $\delta^{n/2} + \mathsf{ngl}$.

6 Extension to Chernoff-Type Theorems

In this section, we present a generic transformation that converts a parallel prover P^{n*} that has good success probability against a threshold verifier to a parallel prover P^{t*} that has good success probability against a direct product verifier for some $t \leq n$. The transformation can be used to show that tight direct product theorems implies Chernoff-type theorems. For example, using our transformation with the direct product theorem of Canetti et al. [CHS05] yields an alternative proof of the Chernoff-type theorem of Impagliazzo et al. [IJK09] for weakly-verifiable puzzles. The transformation also extends our direct product theorems to Chernoff-type theorems.

The transformation is defined as follows. P^{t*} first selects a set $S \subset [n]$ of size t uniformly at random, and then interacts with $\mathsf{V}^{t,t}$ by simulating the interaction of $\langle \mathsf{P}^{n*}, \mathsf{V}^{n,k} \rangle$ with $\mathsf{V}^{t,t}$ playing the coordinates of $\mathsf{V}^{n,k}$ in S and the remaining $n-t$ coordinates played by internal verifiers. The following simple lemma easily follows by the definition.

Lemma 3. *Let* $\langle \mathsf{P}, \mathsf{V} \rangle$ *be an interactive protocol, and* $t, k, n \in \mathbb{N}$ *such that* $1 \leq t \leq k \leq n$. *Let* P^{n*} *be a parallel prover strategy, and* P^{t*} *the induced parallel prover strategy defined as above. For every common input* x, *we have*

$$\Pr[\langle \mathsf{P}^{t*}, \mathsf{V}^{t,t} \rangle(x) = 1] \geq \Pr[\langle \mathsf{P}^{n*}, \mathsf{V}^{n,k} \rangle(x) = 1] \cdot \frac{\binom{k}{t}}{\binom{n}{t}}.$$

When V is public-coin, the above lemma and Theorem 1 implies that for every parallel prover P^{n*}, every $t \leq k$ and $\xi \in (0,1)$, there exists a prover P^* such that for every x with $\Pr[\langle \mathsf{P}^{n*}, \mathsf{V}^{n,k} \rangle(x) = 1] \geq \varepsilon$, we have $\Pr[\langle \mathsf{P}^*, \mathsf{V} \rangle(x) = 1] \geq \left(\varepsilon \cdot \binom{k}{t} / \binom{n}{t}\right)^{1/t} \cdot (1-\xi)$. However, P^* runs in time $\mathrm{poly}(|x|, n, \binom{n}{t}/\binom{k}{t}, \varepsilon^{-1}, \xi^{-1})$, which may not be efficient[10] for large t. Nevertheless, we can obtain the following Chernoff-type theorem by setting the parameters properly. We state the theorem in a similar form to [HPPW08] and [Wik09].

Theorem 4. *Let* $\alpha, \rho \in (0,1)$ *be any* constants *such that* $\alpha + \rho < 1$. *Let* $\mathsf{V} \in \mathrm{PPT}$ *be public-coin. There exists a prover strategy* P^* *such that for every common input* x, *every* $n \in \mathbb{N}$, *every* $\varepsilon, \xi \in (0,1)$ *with* $n \geq 4\log(1/\varepsilon)/\alpha^2$, *and every parallel prover strategy* P^{n*},

1. $\mathsf{P}^*(x, n, \varepsilon, \xi)$ *runs in time* $\mathrm{poly}(|x|, n, \varepsilon^{-1}, \xi^{-1})$ *given oracle access to* $\mathsf{P}^{n*}(x)$.
2. $\Pr[\langle \mathsf{P}^{n*}, \mathsf{V}^{n,(1-\rho)n} \rangle(x) = 1] \geq \varepsilon \Rightarrow$

$$\Pr[\langle \mathsf{P}^*(n, \varepsilon, \xi), \mathsf{V} \rangle(x) = 1] \geq 1 - \rho - \alpha.$$

[10] Here, by efficient we mean the running time is polynomial in $|x|, n, \varepsilon^{-1}, \xi^{-1}$.

In comparison, the simple reduction and tight direct product theorem yields a Chernoff-type theorem with a slightly restricted parameter range where α and ρ are constants. Nevertheless, it suffices for conceivable applications and achieves almost tight bound $1 - \rho - 2\sqrt{\log(1/\varepsilon)/n}$ in this regime.

Similarly, when V is simulatable, we can extend Theorem 2 to the following Chernoff-type theorem.

Theorem 5. *Let $\alpha, \rho \in (0,1)$ be any constants such that $\alpha+\rho < 1$. Let $\mathsf{V} \in$ PPT be exteandable and simulatable. There exists a prover strategy P^* such that for every common input x, every $n \in \mathbb{N}$, every $\varepsilon, \xi \in (0,1)$ with $n \geq 16\log(1/\varepsilon)/\alpha^2$, and every parallel prover strategy P^{n*},*

1. *$\mathsf{P}^*(x, n, \varepsilon, \xi)$ runs in time $\mathrm{poly}(|x|, n, \varepsilon^{-1}, \xi^{-1})$ given oracle access to $\mathsf{P}^{n*}(x)$.*
2. *$\Pr[\langle \mathsf{P}^{n*}, \mathsf{V}^{n,(1-\rho)n}\rangle(x) = 1] \geq \varepsilon \Rightarrow$*

$$\Pr[\langle \mathsf{P}^*(n, \varepsilon, \xi), \mathsf{V}\rangle(x) = 1] \geq (1-\rho)^2 - \alpha.$$

Detailed proofs of Lemma 3, Theorem 4, 5 can be found in the full version of this paper [CL09].

7 Constant-Round AM Arguments Systems

In this section, we prove a tight parallel repetition theorem for threshold verifiers $\mathsf{V}^{n,k}$ for *constant-round* public-coin arguments, which generalizes the direct product theorem of Pass and Venkitasubramaniam [PV07]. We state the theorem and further details of the proofs can be found in the full version of this paper [CL09].

Theorem 6. *Let $m \in \mathbb{N}$ be an arbitrary constant, and $\mathsf{V} \in$ PPT be m-round and public coin. There exists a prover strategy P^* such that for every common input x, every $n, k \in \mathbb{N}$ with $k \in [n]$, every $\delta, \xi \in (0,1)$, and every parallel prover strategy P^{n*},*

1. *$\mathsf{P}^*(x, n, k, \delta, \xi)$ runs in time $\mathrm{poly}(|x|, n, \delta^{-m}, P(n,k,\delta)^{-m}, \xi^{-m})$ given oracle access to $\mathsf{P}^{n*}(x)$.*
2. *$\Pr[\langle \mathsf{P}^{n*}, \mathsf{V}^{n,k}\rangle(x) = 1] \geq P(n,k,\delta) \Rightarrow$*

$$\Pr[\langle \mathsf{P}^*(n, k, \delta, \xi), \mathsf{V}\rangle(x) = 1] \geq \delta \cdot (1-\xi).$$

Acknowledgments

We thank Boaz Barak, Yael Tauman Kalai, and Salil Vadhan for the useful discussion that leads to the results in Section 5. We also thank Salil Vadhan for very helpful discussions throughout this work.

References

[BIN97] Bellare, M., Impagliazzo, R., Naor, M.: Does parallel repetition lower the error in computationally sound protocols? In: FOCS, pp. 374–383 (1997)

[CHS05] Canetti, R., Halevi, S., Steiner, M.: Hardness amplification of weakly verifiable puzzles. In: Kilian, J. (ed.) TCC 2005. LNCS, vol. 3378, pp. 17–33. Springer, Heidelberg (2005)

[CL09] Chung, K.-M., Liu, F.-H.: Parallel repetition theorems for interactive arguments. Electronic Colloquium on Computational Complexity (ECCC) (109) (2009), `http://eccc.uni-trier.de/report/2009/109/`

[CLLY09] Chung, K.-M., Liu, F.-H., Lu, C.-J., Yang, B.-Y.: Efficient string-commitment from weak bit-commitment and full-spectrum theorem for puzzles (2009) (unpublished manuscript)

[Dur04] Durrett, R.: Probability: Theorey and Examples, 3rd edn. Duxbury (2004)

[Gen09] Gentry, C.: Fully homomorphic encryption using ideal lattices. In: STOC, pp. 169–178 (2009)

[Hai09] Haitner, I.: A parallel repetition theorem for any interactive argument. In: FOCS (2009)

[HPPW08] Håstad, J., Pass, R., Pietrzak, K., Wikström, D.: An efficient parallel repetition theorem (2008) (unpublished manuscript)

[HPWP10] Håstad, J., Pass, R., Wikström, D., Pietrzak, K.: An efficient parallel repetition theorem. In: Micciancio, D. (ed.) TCC 2010. LNCS, vol. 5978, pp. 1–18. Springer, Heidelberg (2010)

[HS09] Holenstein, T., Schoenebeck, G.: General hardness amplification of predicates and puzzles (2009) (unpublished manuscript)

[IJK09] Impagliazzo, R., Jaiswal, R., Kabanets, V.: Chernoff-type direct product theorems. J. Cryptology 22(1), 75–92 (2009)

[PV07] Pass, R., Venkitasubramaniam, M.: An efficient parallel repetition theorem for arthur-merlin games. In: STOC, pp. 420–429 (2007)

[PW07] Pietrzak, K., Wikström, D.: Parallel repetition of computationally sound protocols revisited. In: Vadhan, S.P. (ed.) TCC 2007. LNCS, vol. 4392, pp. 86–102. Springer, Heidelberg (2007)

[Raz98] Raz, R.: A parallel repetition theorem. SIAM J. Comput. 27(3), 763–803 (1998)

[Wik09] Wikström, D.: An efficient concurrent repetition theorem. Cryptology ePrint Archive, Report 2009/347 (2009)

Almost Optimal Bounds for Direct Product Threshold Theorem

Charanjit S. Jutla

IBM T. J. Watson Research Center
Yorktown Heights, NY 10598

Abstract. We consider weakly-verifiable puzzles which are challenge-response puzzles such that the responder may not be able to verify for itself whether it answered the challenge correctly. We consider k-wise direct product of such puzzles, where now the responder has to solve k puzzles chosen independently in parallel. Canetti et al have earlier shown that such direct product puzzles have a hardness which rises exponentially with k. In the threshold case addressed in Impagliazzo et al, the responder is required to answer correctly a fraction of challenges above a threshold. The bound on hardness of this threshold parallel version was shown to be similar to Chernoff bound, but the constants in the exponent are rather weak. Namely, Impagliazzo et al show that for a puzzle for which probability of failure is δ, the probability of failing on less than $(1-\gamma)\delta k$ out of k puzzles, for any parallel strategy, is at most $e^{-\gamma^2 \delta k/40}$.

In this paper, we develop new techniques to bound this probability, and show that it is arbitrarily close to Chernoff bound. To be precise, the bound is $e^{-\gamma^2(1-\gamma)\delta k/2}$. We show that given any responder that solves k parallel puzzles with a good threshold, there is a uniformized parallel solver who has the same threshold of solving k parallel puzzles, while being oblivious to the permutation of the puzzles. This enhances the analysis considerably, and may be of independent interest.

1 Introduction

Consider challenge-response puzzles where the responder may not be able to determine if its answer is a correct response or not, either because the challenge may have multiple correct responses (and the challenger seeks a particular one of those), or because the responder is computationally constrained, e.g. in CAPTCHA puzzles [8]. Such puzzles are called weakly-verifiable puzzles [2].

In cryptography, and other applications, the challenge-response puzzles are often used to distinguish between a real and fake responder, where the differentiation is obtained by the probability of their solving a randomly chosen challenge. For example, the authentic party may have a probability α of solving the challenge correctly, whereas non-authentic parties may have a probability only β ($< \alpha$) of solving the puzzles correctly. However, if the gap is not large then direct product, or (parallel) repetition of such puzzles may be sought. Ideally, one would like that if k puzzles are chosen independently in parallel, then the

D. Micciancio (Ed.): TCC 2010, LNCS 5978, pp. 37–51, 2010.

probability of the non-authentic party solving all puzzles correctly is at most β^k [2]. Unfortunately, this also makes the success probability of the authentic party go down (if $\alpha < 1$).

In [4], the authors observe that the authentic party is on average expected to solve αk puzzles, and if a Chernoff-like bound holds, then the probability of fake parties solving αk puzzles may go down exponentially. They show that their intuition is correct, and indeed give an exponential bound. However, the bound they obtain has a weak constant in the exponent. In particular they show that (setting $\delta = 1 - \beta$) the probability of the non-authentic party responding *incorrectly* to less than $(1 - \gamma)\delta k$ puzzles (out of k parallel puzzles) is at most $e^{-\gamma^2 \delta k/40}$. For real problems like CAPTCHA, the 1/40 factor in the exponent is debilitating, and the authors mention it as an open problem to improve this constant.

As is to be expected, the result in [4] is proved by reducing a single puzzle instance to a (simulated) direct product puzzle instance. However, multiple simulations are required to get a good reduction. The complication in analyzing the reduction then stems from the fact that the given single puzzle instance must be embedded in all simulated direct product puzzle instances, and hence they are not independent. In [4], the authors use a nice duality property of good (bi-partite graph based) samplers to analyze the dependent simulations.

In this paper we develop further new techniques to analyze this probability and show that one can indeed bound the probability arbitrarily close to as in Chernoff bound. In particular we upper bound the above probability by about

$$e^{-\gamma^2(1-\gamma)\delta k/2}$$

Since γ is usually tiny, the above is almost as good as can be expected. Further, the techniques developed have potential to improve the bound further, e.g. replacing $(1 - \gamma)$ by $(1 - \gamma^2)$.

We show that a uniformized parallel solver, which first permutes its given k-puzzles randomly, solves them as before, and permutes the results back, has the same probability of success as before. However, this uniformized solver is much easier to analyze. While this in itself, when plugged into the "trust reduction" strategy of [4] gives better bounds than before, to get the bounds similar to Chernoff bound we need further new techniques. In particular, while a count of other simulated puzzles being answered incorrectly gives a good guess of whether the given puzzle may be answered incorrectly, a linearly weighted metric we consider leads to more optimal bounds.

While the idea of uniformized parallel solver also applies to Raz's Theorem [7], in particular because of Holenstein's observation that the two provers can use shared randomness [3], it is to be seen if it leads to improved analysis.

The rest of the paper is organized as follows. In Section 2 we describe a result about samplers which we employ, as well as give definitions of threshold weakly-verifiable puzzles. In section 3 we consider uniformized parallel solvers and give the main technical lemmas. In section 4 we give the main theorem and its proof. In section 5 we describe the pre-processing phase.

2 Preliminaries

2.1 Basics

Lemma 1. *[Chernoff Bound [1]] Let $X = (X_1 + X_2 + ... + X_n)/n$, where the X_i are mutually independent indicator random variables, each with mean μ. Then, for $\beta \geq 0$,*

$$\begin{aligned} \Pr[X \geq (1+\beta)\mu] &< (e^{\beta}(1+\beta)^{-1-\beta})^{\mu n} \\ \Pr[X < (1-\beta)\mu] &< e^{-\beta^2 \mu n/2} \end{aligned}$$

2.2 Samplers

Consider bipartite graphs $F = G(L \cup R, E)$. We allow graphs with multiple edges. For a vertex v of G, we denote by $N_G(v)$ the multi-set of its neighbours in G. When the graph G is clear from context, we will drop the subscript G, and simply write $N(v)$. We say that G is *bi-regular* if the degree of each vertex in L is same, and the degree of each vertex in R is same.

Let $G = G(L \cup R, E)$ be any bi-regular bipartite graph. For a function $\lambda : [0,1] \times [0,1] \rightarrow [0,1]$, we say that G is a *$\lambda$-sampler* [4] if, for every function $F : L \rightarrow [0,1]$ with the average value $\mathbf{E}_{x \in L}[F(x)] \geq \mu$ and any $0 < \nu < 1$, there are at most $\lambda(\mu, \nu) \cdot |R|$ vertices $r \in R$ such that $\mathbf{E}_{y \in N(r)}[F(y)] \leq (1-\nu)\mu$.

We will employ the following lemma from [5,4]. It says that for any two large vertex subsets W and F of a sampler, the fraction of edges between W and F is close to the product of the densities of W and F.

Lemma 2. *[5,4] Suppose $G = G(L \cup R, E)$ is a λ-sampler. Let $W \subseteq R$ be any set of measure at least τ, and let $V \subseteq L$ be any set of measure at least β. Then, for all $0 < \nu < 1$ and $\lambda_0 = \lambda(\beta, \nu)$, we have*

$$\Pr_{x \in L, y \in N(x)}[x \in V \,\&\, y \in W] \geq \beta(1-\nu)(\tau - \lambda_0)$$

where the probability is over first picking x uniformly from L, and then picking y uniformly from $N(x)$.

We will also need the following observation from [4], which shows that the direct product is an extremely good sampler. Consider the following bipartite graph $G = G(L \cup R, L)$: the set of left vertices is the set of n-bit strings $\{0,1\}^n$; the right vertices R are a pair $\langle r, c\rangle$, where r range over all k-tuples of n-bit strings $\{0,1\}^{nk}$, and c range over m-bit strings $\{0,1\}^m$; for every $y = \langle (r_1, r_2, ..., r_k), c\rangle \in R$, there are k edges (y, r_1), (y, r_2),...,(y, r_k) in E.

Lemma 3. *[4] The graph G defined above is a λ-sampler for $\lambda(\mu, \nu) = e^{-\nu^2 \mu k/2}$.*

2.3 Weakly-Verifiable Puzzles

Definition 1. [2] A weakly-verifiable puzzle $\mathcal{P} = (C, R, d(n))$, with security parameter n, consists of a polynomial time computable function C, a polynomial time computable predicate R, and a polynomial $d(n)$. For any functions $t(n)$ and $c(n)$, the $(t(n), c(n))$-value (*failure value*) of the puzzle is

$$\mathrm{val}(\mathcal{P}, t, c) := \min_X \Pr_{r \in \mathcal{U}_{d(n)}, s \in \mathcal{U}_{c(n)}} [\neg R(r, X(s, C(r)))]$$

where the minimization is over $t(n)$-computable randomized algorithms X using $c(n)$ bits of randomness.

For a parameter δ, $0 \leq \delta \leq 1$, we say that a puzzle $\mathcal{P}$ is $(\delta, t(n), c(n))$-hard if the $(t(n), c(n))$-value of $\mathcal{P}$ is at most δ. In other words, every algorithm X running in time $t(n)$, and using $c(n)$ bits of randomness, has probability at least δ of answering the puzzle wrong.

Definition 2. The k-wise direct product $\mathcal{P}^k$ of a weakly-verifiable puzzle $\mathcal{P} = (C, R, d(n))$ is the weakly-verifiable puzzle $(C^k, R^k, kd(n))$, where $C^k(\langle r_1, ..., r_k\rangle)$ is defined to be $(C(r_1),..., C(r_k))$, and

$$R^k(\langle r_1, ..., r_k\rangle, \langle x_1, ..., x_k\rangle) := \bigwedge_{i=1}^{k} R(r_i, x_i)$$

For any parameters ν and δ, $0 \leq \nu, \delta \leq 1$, and any functions $t(n), c(n)$, the puzzle $\mathcal{P}^k$ is said to be ν-approximate $(\delta, t(n), c(n))$-hard if the following minimum probability

$$\min_X \Pr_{r \in \mathcal{U}^k_{d(n)}, s \in \mathcal{U}_{c(n)}} \left[|\{i \in [1..k] \ : \ \neg R(r_i, X_i(s, C^k(\mathbf{r})))\}| \ > \ \nu k \right]$$

is at least δ, where the minimization is over all randomized algorithms X running in time $t(n)$ and using $c(n)$ bits of randomness. Note that X here takes k puzzles and returns k answers, $\langle X_1, ..., X_k\rangle$. Such an algorithm X will be referred to as a **k-parallel solver**.

3 Uniformized Parallel Solvers

Given a k-parallel solver X, we consider its *uniformized* version $\overline{X}$, which first randomly permutes its given k puzzles, solves them using X, and permutes back the results. In other words, for all $i = 1..k$,

$$\overline{X}_i(\langle s, \pi\rangle, \langle C(r_1), ..., C(r_k)\rangle) := X_{\pi^{-1}(i)}(s, \langle C(r_{\pi(1)}), ..., C(r_{\pi(k)})\rangle)$$

where π is any permutation of $[1..k]$.

It is easy to see that the "failure value" of the uniformized parallel solver remains the same, as the following shows.

$$\Pr_{r_1,...,r_k} \Pr_{s,\pi} \left[|\{i \in [1..k] \ : \ \neg R(r_i, \overline{X}_i(\langle s,\pi\rangle, \langle C(r_1),...,C(r_k)\rangle))\}| \ > \ \nu k \right]$$
$$= \Pr_{r_1,...,r_k} \Pr_{s,\pi} \left[|\{i : \ \neg R(r_i, X_{\pi^{-1}(i)}(s, \langle C(r_{\pi(1)}),...,C(r_{\pi(k)})\rangle))\}| \ > \ \nu k \right]$$
$$= \Pr_{r_1,...,r_k} \Pr_{s,\pi} \left[|\{j = \pi^{-1}(i) : \ \neg R(r_{\pi(j)}, X_j(s, \langle C(r_{\pi(1)}),...,C(r_{\pi(k)})\rangle))\}| \ > \nu k \right]$$
$$= \Pr_{r_1,...,r_k} \Pr_{s} \left[|\{j : \ \neg R(r_j, X_j(s, \langle C(r_1),...,C(r_k)\rangle))\}| \ > \ \nu k \right]$$

where the last equality follows because r_1,...,r_k are chosen independently and identically. Thus, without loss of generality, we can consider only uniformized parallel solvers.

Notation
Let us fix a parallel solver X, and its uniformized parallel solver $\overline{X}$. We will use the following shorthands to denote some useful quantities and predicates. Let $C^k(\boldsymbol{r},\pi)$ denote $\langle C(r_{\pi(1)}),...,C(r_{\pi(k)})\rangle$. Thus, $C^k(\boldsymbol{r},\mathbf{1})$ (where $\mathbf{1}$ is the identity permutation) just denotes $\langle C(r_1),...,C(r_k)\rangle$. Given the randomness $r_1,...r_k$ to generate the k puzzles, and the randomness $\langle s,\pi\rangle$ used by $\overline{X}$, **define random variables**.

- total($\overline{X}$) := $|\{i \in [1..k] \ : \ \neg R(r_i, \overline{X}_i(\langle s,\pi\rangle, C^k(\boldsymbol{r},\mathbf{1})))\}|$
- F($\overline{X}$) (short for first) := $\neg R(r_1, \overline{X}_1(\langle s,\pi\rangle, C^k(\boldsymbol{r},\mathbf{1})))$
- others($\overline{X}$) := $|\{i \in [2..k] \ : \ \neg R(r_i, \overline{X}_i(\langle s,\pi\rangle, C^k(\boldsymbol{r},\mathbf{1})))\}|$
- others($\overline{X}, j$) := $|\{i \in [1..j-1, j+1..k] \ : \ \neg R(r_i, \overline{X}_i(\langle s,\pi\rangle, C^k(\boldsymbol{r},\mathbf{1})))\}|$
- for $\Gamma \subseteq [1..k]$, (failure-) pattern($\overline{X}, \Gamma$) denotes

$$\bigwedge_{i\in\Gamma} R(r_i, \overline{X}_i(\langle s,\pi\rangle, C^k(\boldsymbol{r},\mathbf{1}))) \wedge \bigwedge_{i\notin\Gamma} \neg R(r_i, \overline{X}_i(\langle s,\pi\rangle, C^k(\boldsymbol{r},\mathbf{1})))$$

From now on, unless otherwise stated, all probabilities will be over $r_1,...,r_k$ each chosen uniformly and independently from $\mathcal{U}_{d(n)}$, s chosen uniformly (and independently) from $\mathcal{U}_{c(n)}$, and π chosen uniformly (and independently) from all permutations of $[1..k]$. Further, define

- For any t, $0 \le t \le k$, let p_t denote Pr[total($\overline{X}$) = t].
- Let $\tau = (1-\gamma)\delta k$.
- Define $P = \sum_{t\le\tau} p_t$.
- For $j \ge 0$, let $\psi_j = \gamma\delta(1-\gamma) + j\cdot(\gamma/k)$. Let $\alpha = 1/(1-\tau/k-\psi_0)$.

Lemma 4. *For any integer t, $0 \le t \le k$,*

$$\Pr[F(\overline{X}) \mid total(\overline{X}) = t] \ = \ \frac{t}{k}$$

Proof. We first show that for any t, $0 \le t \le k$, and any subset Γ of $[1..k]$ of size t, the probability of pattern($\overline{X}$, Γ) is a function only of t, and is independent of the subset Γ.

Indeed, consider Γ, and another subset Γ' of size t, and let σ be any permutation of $[1..k]$, such that $\Gamma' = \sigma(\Gamma)$ (a permutation applied to a subset Γ just yields the set which is the range of the permutation with domain Γ). It is clear that such a permutation exists. Then,

$$\begin{aligned}
&\Pr[\text{pattern}(\overline{X}, \Gamma')] \\
&= \Pr\Big[\bigwedge_{i\in\Gamma'} R(r_i, X_{\pi^{-1}(i)}(s, C^k(\boldsymbol{r},\pi))) \bigwedge_{i\notin\Gamma'} \neg R(r_i, X_{\pi^{-1}(i)}(s, C^k(\boldsymbol{r},\pi)))\Big] \\
&= \Pr\Big[\bigwedge_{i\in\sigma(\Gamma)} R(r_i, X_{\pi^{-1}(i)}(s, C^k(\boldsymbol{r},\pi))) \bigwedge_{i\notin\sigma(\Gamma)} \neg R(r_i, X_{\pi^{-1}(i)}(s, C^k(\boldsymbol{r},\pi)))\Big] \\
&= \Pr\Big[\bigwedge_{j\in\Gamma} R(r_{\sigma(j)}, X_{\pi^{-1}(\sigma(j))}(s, C^k(\boldsymbol{r},\pi))) \wedge \\
&\qquad \bigwedge_{j\notin\Gamma} \neg R(r_{\sigma(j)}, X_{\pi^{-1}(\sigma(j))}(s, C^k(\boldsymbol{r},\pi)))\Big]
\end{aligned}$$

Now, denote $r_{\sigma(j)}$ by w_j. Then, the above becomes (with probability now over $w_{\sigma^{-1}(1)},\ldots, w_{\sigma^{-1}(k)}, s, \pi$)

$$\begin{aligned}
\Pr\Big[&\bigwedge_{j\in\Gamma} R(w_j, X_{\pi^{-1}(\sigma(j))}(s, C^k(\boldsymbol{w},\sigma^{-1}\pi))) \wedge \\
&\bigwedge_{j\notin\Gamma} \neg R(w_j, X_{\pi^{-1}(\sigma(j))}(s, C^k(\boldsymbol{w},\sigma^{-1}\pi)))\Big]
\end{aligned}$$

Now, $\pi^{-1}\sigma = (\sigma^{-1}\pi)^{-1}$. Denote $\sigma^{-1}\pi$ by $\hat{\pi}$. Since permutations form a group, $\hat{\pi}$ is independent of σ, with π chosen uniformly and independently of σ. Then, the above probability can be written as (with probability now over $w_{\sigma^{-1}(1)}$, ...,$w_{\sigma^{-1}(k)}$, s, $\hat{\pi}$)

$$\Pr\Big[\bigwedge_{j\in\Gamma} R(w_j, X_{\hat{\pi}^{-1}(j)}(s, C^k(\boldsymbol{w},\hat{\pi}))) \bigwedge_{j\notin\Gamma} \neg R(w_j, X_{\hat{\pi}^{-1}((j)}(s, C^k(\boldsymbol{w},\hat{\pi})))\Big]$$

Since, $w_1, ..., w_k$ are chosen identically and independently, the above remains same even when the probability is considered over $w_1, ...w_k, s, \hat{\pi}$. This proves that the above probability is a function only of t, and independent of the particular subset Γ. Now,

$$\begin{aligned}
&\Pr[\mathrm{F}(\overline{X}) \mid \text{total}(\overline{X}) = t] \\
&= \sum_{\Gamma:|\Gamma|=t} \Pr[\mathrm{F}(\overline{X}) \wedge \text{pattern}(\overline{X}, \Gamma) \mid \text{total}(\overline{X}) = t] \\
&= \sum_{\Gamma:|\Gamma|=t, 1\in\Gamma} \Pr[\text{pattern}(\overline{X}, \Gamma) \mid \text{total}(\overline{X}) = t] \\
&= \frac{\sum_{\Gamma:|\Gamma|=t,1\in\Gamma} \Pr[\text{pattern}(\overline{X}, \Gamma)]}{\Pr[\text{total}(\overline{X}) = t]}
\end{aligned}$$

$$= \frac{\sum_{\Gamma:|\Gamma|=t,1\in\Gamma} \Pr[\text{pattern}(\overline{X},\Gamma)]}{\sum_{\Gamma:|\Gamma|=t} \Pr[\text{pattern}(\overline{X},\Gamma)]}$$
$$= \binom{k-1}{t-1} / \binom{k}{t}$$
$$= t/k$$
□

Lemma 5. *For $t < k$,*

$$\Pr[F(\overline{X}) \mid \mathit{others}(\overline{X}) \le t] = \frac{\sum_{t' \le t+1} (t'/k) p_{t'}}{\sum_{t' \le t} p_{t'} + ((t+1)/k) p_{t+1}}$$

Proof. First note that, for $t < k$

$$\begin{aligned} &\Pr[\text{others}(\overline{X}) = t] \\ &= \Pr[\text{F}(\overline{X}) \wedge \text{others}(\overline{X}) = t] + \Pr[\neg\text{F}(\overline{X}) \wedge \text{others}(\overline{X}) = t] \\ &= \Pr[\text{F}(\overline{X}) \wedge \text{total}(\overline{X}) = t+1] + \Pr[\neg\text{F}(\overline{X}) \wedge \text{total}(\overline{X}) = t] \\ &= \frac{t+1}{k} p_{t+1} + \frac{k-t}{k} p_t \quad \text{(by Lemma 4)} \end{aligned}$$

The lemma follows easily from this observation. □

Lemma 6. *For any $i > 0$, suppose for all j, $0 \le j < i$*

$$\Pr[F(\overline{X}) \mid \mathit{others}(\overline{X}) \le \tau + j] > \frac{\tau}{k} + \psi_j$$

then

$$p_{\tau+i} > \psi_0 P \cdot \frac{k\alpha}{\tau + i} \cdot \prod_{0<j<i} (1 + (\psi_j - \frac{j}{k})(\frac{k\alpha}{\tau + j})) \tag{1}$$

Proof. For any j, $j < i$, we first note that Lemma 5, along with the hypothesis of the lemma for j, yields (by simple manipulation)

$$p_{\tau+j+1} > \frac{k\alpha}{\tau + j + 1} \cdot (\psi_j P + \sum_{0<j'\le j} p_{\tau+j'}(\psi_j - \frac{j'}{k})) \tag{2}$$

The base case, i.e. $i = 1$, follows immediately from this by considering $j = 0$. Now suppose the induction hypothesis holds for i, and we will prove the lemma for $i+1$. The antecedent for $i+1$ completely yields the antecedent for $i' < i+1$. Thus, inequality (1) holds for all such i'.

Let $\Psi(j)$ stand for $(\psi_j - \frac{j}{k})(\frac{k\alpha}{\tau+j})$.

Then by inequality (2), and plugging in inequality (1) for each $p_{\tau+j}$ $(j < i+1)$, while noting that ψ_j is an increasing function of j, we get that $p_{\tau+i+1}$ is greater than

$$\frac{\psi_0 Pk\alpha}{\tau+i+1}\cdot\Big(1+\sum_{0<j\le i}\Psi(j)\prod_{0<j'<j}(1+\Psi(j'))\Big)$$

$$=\frac{\psi_0 Pk\alpha}{\tau+i+1}\cdot\Big(1+\sum_{0<j\le i}(1+\Psi(j)-1)\prod_{0<j'<j}(1+\Psi(j'))\Big)$$

$$=\frac{\psi_0 Pk\alpha}{\tau+i+1}\cdot\Big(1+\sum_{0<j\le i}\prod_{0<j'\le j}(1+\Psi(j')) - \sum_{0<j\le i}\prod_{0<j'<j}(1+\Psi(j'))\Big)$$

$$=\frac{\psi_0 Pk\alpha}{\tau+i+1}\cdot\Big(1+\prod_{0<j'\le i}(1+\Psi(j'))-\prod_{0<j'<1}(1+\Psi(j'))\Big)$$

$$=\frac{\psi_0 Pk\alpha}{\tau+i+1}\cdot\prod_{0<j'<i+1}(1+\Psi(j'))$$

□

Lemma 7. *For $\gamma<1$, and for any positive integer $t<\gamma\delta k\,(=M)$,*

$$\prod_{j=1}^{t}\frac{\tau+\psi_j k}{\tau+j} > (1-\gamma^2)\cdot e^{\gamma(1-\gamma)(1-\frac{t}{2M})t\,-\,O(1/(\delta k))}$$

Proof. From the definition of ψ_j, the product can be written as

$$\prod_{j=1}^{t}\frac{\delta k-\gamma^2\delta k+\gamma j}{\delta k-\gamma\delta k+j} = \gamma^t\prod_{j=1}^{t}\frac{(\delta/\gamma)k-\gamma\delta k+j}{\delta k-\gamma\delta k+j}$$

Using the gamma function, which for $x>0$ satisfies $\Gamma(x+1)=x\Gamma(x)$, the above can be written as

$$\gamma^t\cdot\frac{\Gamma((\delta/\gamma)k-\gamma\delta k+t+1)\,\Gamma(\delta k-\gamma\delta k+1)}{\Gamma((\delta/\gamma)k-\gamma\delta k+1)\,\Gamma(\delta k-\gamma\delta k+t+1)}$$

Now, using Stirling's approximation for gamma function [6]

$$\Gamma(z+1)=\sqrt{\frac{2\pi}{z}}\left(\frac{z}{e}\right)^z e^{O(1/z)}$$

the above is greater than

$$\gamma^t(1-\gamma^2)e^{-O(1/\delta k)}\cdot\frac{((\delta/\gamma)k-\gamma\delta k+t)^{(\delta/\gamma)k-\gamma\delta k+t}\cdot(\delta k-\gamma\delta k)^{\delta k-\gamma\delta k}}{((\delta/\gamma)k-\gamma\delta k)^{(\delta/\gamma)k-\gamma\delta k}\cdot(\delta k-\gamma\delta k+t)^{\delta k-\gamma\delta k+t}}$$

Taking just the product of γ^t and the big fraction, and factoring out δk from all terms, we get

$$\frac{(1-\gamma^2+\gamma t/(\delta k))^{(\delta/\gamma)k-\gamma\delta k+t}\cdot(1-\gamma)^{\delta k-\gamma\delta k}}{(1-\gamma^2)^{(\delta/\gamma)k-\gamma\delta k}\cdot(1-\gamma+t/(\delta k))^{\delta k-\gamma\delta k+t}} \tag{3}$$

Now, we use the following series expansion (convergent for $z<1$)

$$-\ln(1-z)=\sum_{i\ge 1}\frac{z^i}{i}$$

Recalling that $M = \gamma\delta k$, and denoting $(1 - t/M)$ by θ, the log of the above fraction (3) is sum of four terms I_1, I_2, I_3 and I_4, where

$$I_1 = -M(1/\gamma^2 - 1 + t/M)\sum \frac{(\gamma^2\theta)^i}{i}$$
$$I_2 = -M(1/\gamma - 1)\sum \frac{\gamma^i}{i}$$
$$I_3 = M(1/\gamma^2 - 1)\sum \frac{\gamma^{2i}}{i}$$
$$I_4 = M(1/\gamma - 1 + t/M)\sum \frac{(\gamma\theta)^i}{i}$$

where all the sums have i ranging from 1 to infinity. Now, a little manipulation shows that

$$I_2 + I_3 \;=\; M/\gamma \cdot \sum_{i\geq 2} \left(\frac{1}{i-1} - \frac{1}{i}\right)(\gamma^i - \gamma^{2i-1})$$

Similarly, $I_1 + I_4$ is

$$-M/\gamma \cdot \sum_{i\geq 2} \left(\frac{\theta^{i-1}}{i-1} - \frac{\theta^i}{i}\right)(\gamma^i - \gamma^{2i-1}) + t \cdot \sum_{i\geq 1} \left(\frac{(\gamma\theta)^i}{i} - \frac{(\gamma^2\theta)^i}{i}\right)$$

Thus, all four terms together sum up to

$$M/\gamma \cdot \sum_{i\geq 2} \left(\frac{1}{i-1}(1-\theta^{i-1}) - \frac{1}{i}(1-\theta^i)\right)(\gamma^i - \gamma^{2i-1}) + t \cdot \sum_{i\geq 1} \left(\frac{(\gamma\theta)^i}{i} - \frac{(\gamma^2\theta)^i}{i}\right)$$

Now, $(\frac{1}{i-1}(1-\theta^{i-1}) - \frac{1}{i}(1-\theta^i))$ is non-negative for all i, as long as $\theta \leq 1$: it is positive at $\theta = 0$, is non-negative at $\theta = 1$, and the derivative (w.r.t. θ) is non-zero everywhere except at $\theta = 1$.

Thus, we will only take the term corresponding to $i = 2$ from the first sum, and the term corresponding to $i = 1$ from the second sum. This leads to a lower bound of

$$t(\gamma\theta - \gamma^2\theta) + M(\gamma - \gamma^2)(1/2 - \theta + \theta^2/2)$$

Since $\theta = 1 - t/M$, the above simplifies to $\gamma(1-\gamma)(1 - \frac{t}{2M})t$. □

Finally, we need the following simple calculation. Define

$$q_{\tau+i} = \psi_0 \cdot \frac{k}{\tau + i} \cdot \prod_{j=1}^{i-1} \frac{\tau + \psi_j k}{\tau + j}$$

Lemma 8. *Let $M = \lceil \gamma\delta k \rceil$, and suppose $\delta k \geq 1$.*

1. *For any i, $0 \leq i < M$, and for any $\chi \geq 1$,*

$$q_{\tau+i} \cdot \chi \cdot \frac{2}{\gamma(1-\gamma)^2} \cdot e^{-\gamma^2(1-\gamma)\delta k/2} - e^{-(1-\gamma)(\delta - \tau/k - \psi_i)k/2} > \frac{\chi}{2} \cdot e^{-\gamma^2(1-\gamma)\delta k/2}$$

2. $q_{\tau+M} \cdot \frac{2}{\gamma(1-\gamma)^2} \cdot e^{-\gamma^2(1-\gamma)\delta k/2} > 1$

The detailed calculations can be found in Appendix A.

4 The Main Theorem

Theorem 9. *Let $\mathcal{P} = (C, R, d(n))$ be a weakly-verifiable puzzle that is $(\delta,\ t(n),\ c(n))$-hard. Let k be any positive integer such that $\delta k \geq 1$, and γ $(1 > \gamma > 0)$ be arbitrary. Further, let ϵ_0 be any arbitrary positive real, and let*

$$\epsilon \geq \frac{2}{\gamma(1-\gamma)^2} \cdot e^{(1-\gamma)(-\gamma^2\delta+\epsilon_0)k/2}.$$

Then the direct product puzzle $\mathcal{P}^k$ is $(1-\gamma)\delta$-approximate $((1-\epsilon), t'(n), c'(n))$-hard with $t'(n) = t(n) \cdot poly(\epsilon, 1/n, 1/(\gamma\delta k), 1/\ln(1/\epsilon_0))$, and $c'(n) = c(n) \cdot poly(\epsilon, 1/(\gamma\delta k), 1/\ln(1/\epsilon_0))$.

In the following let $\epsilon_1 = \epsilon_2 = \epsilon_3 = \epsilon_0/6$. Recall the definitions of τ, P, and ψ_j from Section 3.

Consider, for contradiction sake, a k-parallel solver X which for the k-wise direct product $\mathcal{P}^k$ has $(1-\gamma)\delta$-approximate (failure) value less than $1-\epsilon$, i.e

$$P = \Pr_{r \in \mathcal{U}^k_{d(n)}, s \in \mathcal{U}_{c(n)}} \left[|\{i \in [1..k] \ : \ \neg R(r_i, X_i(s, C^k(r)))\}| \ \leq \ \tau \right] \ > \ \epsilon$$

As explained earlier in Section 3, we consider its uniformized version $\overline{X}$, which has the same failure value $(1-P)$. Using $\overline{X}$ as an oracle, we will give an algorithm Y to solve the underlying puzzle $\mathcal{P}$ with failure value less than δ, leading to a contradiction.

The algorithm Y will have a pre-processing phase (i.e. independent of the given target puzzle instance x, and function of security parameter n), where it runs some statistical tests using X to determine the appropriate algorithm $\mathcal{C}[i]$ to run, where $\mathcal{C}[0], ..., \mathcal{C}[M-1]$ are M $(= \lceil \gamma\delta k \rceil)$ algorithms as follows:

$\mathcal{C}[i]$: On input x, run $\mathcal{C}'[i]$ below on x. If the value returned is different from $\perp$, then return that value; otherwise repeat by calling $\mathcal{C}'[i]$ on x again, for a total of at most T iterations ($T = \frac{8}{\epsilon\gamma^2(1-\gamma)^2} \ln(1/\epsilon_1)$). If no output is produced in these T iterations, return $\perp$.

$\mathcal{C}'[i]$: On input x, choose $k-1$ random tapes $\alpha_2, ..., \alpha_k$ uniformly and independently from $\{0,1\}^{d(n)}$. Let $x_2, ..., x_k$ be the corresponding puzzles, i.e. $x_l = C(\alpha_l)$, for $l = 2..k$. Set $\bar{x} = \langle x, x_2, ..., x_k \rangle$. Run $\overline{X}$ on $\bar{x}$. Check if **others** $\leq \tau + i$, and if so return $\overline{X}_1(\bar{x})$; otherwise return $\perp$.

The pre-processing phase η returns $\eta(\overline{X}, n, \delta, \gamma, k)$, a value between 0 and $M-1$. When it is clear from context, we just call the value η. Thus, $0 \leq \eta \leq M-1$. As mentioned above, Y runs $\mathcal{C}[\eta]$ on x.
The event **valid** stands for the following being satisfied by the returned η:

1. For all $i < \eta$, $\Pr[\mathrm{F}(\overline{X}) \mid \mathrm{others}(\overline{X}) \ \leq \tau + i] \ > \ \frac{\tau}{k} + \psi_i$, and
2. $\Pr[\mathrm{F}(\overline{X}) \mid \mathrm{others}(\overline{X}) \ \leq \tau + \eta] \ \leq \ \frac{\tau}{k} + \psi_\eta + \epsilon_2$.

We will later bound the probability of *valid* not happening by ϵ_3 (lemma 12); i.e. after we describe how the pre-processing works. In rest of this section, we condition on the event *valid* being true, and we will not mention it explicitly in the probabilities.

We first need to bound the probability of $\mathcal{C}[\eta]$ timing out, i.e. returning $\perp$. Note that, $\mathcal{C}'[\eta]$ returns something other than $\perp$ if (others $\leq \tau + \eta$). As in Lemma 5, it is easy to see that the probability of this happening is at least P which is at least ϵ (by hypothesis of the theorem). However, multiple calls to $\mathcal{C}'[\eta]$ are *not* independent, as they all include the query x. However, as shown in [4], the corresponding graph is a good sampler, and that helps us analyze the probability of $\mathcal{C}[\eta]$ timing out. Of course, we require Lemma 6, and the idea therein of a linearly increasing ψ_j, to obtain better bounds.

To this end, we consider a (k-colored) bipartite graph $G = G(L \cup R, E)$; the set of left vertices is the set of $d(n)$-bit strings $\{0,1\}^{d(n)}$; the right vertices are triples $\langle \bar{\alpha}, s, \pi \rangle$, where $\bar{\alpha}$ ranges over all k-tuples of $d(n)$-bit strings, and s ranges over $c(n)$ bit strings, and π ranges over permutations of $[k]$; for every $y = \langle (\alpha_1, ..., \alpha_k), s, \pi \rangle \in R$ there are k edges (y, α_1), ..., (y, α_k) in E, *colored* 1..k respectively.

By lemma 3, this graph is a λ-sampler for $\lambda(\mu, \nu) = e^{-\nu^2 \mu k/2}$.

Corresponding to each $(\alpha_1, ..., \alpha_k)$ are puzzles $(x_1, ..., x_k)$. Now, *define* Good_η to be the subset of R (the right vertices) such that $\overline{X}$ when run on input $(x_1, ..., x_k)$, with randomness s and π, has the following property

$$|\{i \in [1..k] \;:\; \neg R(\alpha_i, \overline{X}_i(\langle s, \pi \rangle, \langle x_1, ..., x_k \rangle))\}| \;\leq \tau + \eta$$

In other words, $\text{total}(\overline{X}) \leq \tau + \eta$. Let the density of Good_η in R be g_η. We now *define* $H_\eta \subseteq L$ to be all those vertices α such that α has less than $(\epsilon \cdot \frac{\gamma^2(1-\gamma)^2}{8})$ fraction of its neighbours in the set Good_η. We will later see in Lemma 11 how H_η is relevant, even though $\mathcal{C}'[\eta]$ embeds α (or it's x) only in the first position. We can bound the size of H_η, just as in [4], by employing Lemma 2.

Lemma 10. *H_η has density at most $\delta - \tau/k - \psi_\eta - \epsilon_0$.*

Proof. Suppose to the contrary, the density of H_η is greater than $\beta = \delta - \tau/k - \psi_\eta - \epsilon_0$. Let $H' \subseteq H_\eta$ be any subset of density exactly β. Now, by definition of H_η, we have $\Pr_{\alpha \in L, w \in N(\alpha)}[\alpha \in H' \,\&\, w \in \text{Good}_\eta] < \beta\epsilon\gamma^2(1-\gamma)^2/8$. On the other hand, by Lemma 2, we get that the same probability is at least $\beta(g_\eta - \lambda_0)(1 - \bar{\nu})$ for $\lambda_0 = \lambda(\beta, \bar{\nu})$, for any $0 \leq \bar{\nu} \leq 1$. We set $\bar{\nu} = \sqrt{1-\gamma}$.

Now, note that $g_\eta = \Pr[\text{total}(\overline{X}) \leq \tau + \eta]$. If $\eta = 0$, then $g_\eta = P > \epsilon$. Otherwise, since event *valid* is true, we can use Lemma 6 to lower bound $p_{\tau+\eta}$. Next, noting that in Lemma 6, α is greater than one, we can use Lemma 7 to get an explicit lower bound for $p_{\tau+\eta}$, and hence for g_η.

Then, using Lemma 8.1, and noting that $1 - \bar{\nu} > \gamma/2$, it can be seen by a simple calculation that $\beta(g_\eta - \lambda_0)(1 - \bar{\nu})$ is more than $\beta\epsilon\gamma^2(1-\gamma)^2/8$, a contradiction. □

Lemma 11. *For every $\alpha \notin H_\eta$ and the puzzle x corresponding to that random α, we have $\Pr[\mathcal{C}[\eta](x) = \perp] \leq \epsilon_1$, where the probability is over the random coins of $\mathcal{C}[\eta]$ (including those of $\overline{X}$ and X).*

Proof. We consider a variation of $\mathcal{C}[i]$, where instead of calling $C'[i]$, it calls the following $\mathcal{C}''[i]$ instead.

$\mathcal{C}''[i]$: On input x, choose $k-1$ random tapes $\alpha_1, ..., \alpha_{k-1}$ uniformly and independently from $\{0,1\}^{d(n)}$. Let $x_1, ..., x_{k-1}$ be the corresponding puzzles, i.e. $x_l = C(\alpha_l)$, for $l = 1..k-1$. Pick $j \in [1..k]$ at random and set $\bar{x} = \langle x_1, ..., x_{j-1}, x, x_j, ..., x_{k-1}\rangle$. Run $\overline{X}$ on $\bar{x}$. Check if **others**$(\overline{X}, j) \leq \tau + i$, and if so return $\overline{X}_j(\bar{x})$; otherwise return $\perp$.

For each fixed α, the behaviour of $\mathcal{C}'[i]$ and $\mathcal{C}''[i]$ is statistically identical, because placing x in the random j-th place is just a permutation of placing x in the first place, and that the permutations form a group.

Further, picking a color $j \in [1..k]$ at random, and then picking $\alpha_1, ..., \alpha_{k-1}$ at random and placing α in the j-th place to form $\bar{\alpha}$ is same as picking a random neighbour of α (random element of $N_G(\alpha)$, and note that $N_G(\alpha)$ is defined to be a multi-set)[1].

Now, $\mathcal{C}''[\eta]$ returns something other than $\perp$ if the neighbour satisfies **others** $(\overline{X}, j) \leq \tau + \eta$, which is implied by total $\leq \tau + \eta$. But, for $\alpha \notin H_\eta$, the density of neighbours satisfying total $\leq \tau + \eta$ is more than $\epsilon\gamma^2(1-\gamma)^2/8$. Hence for such α, the probability of $\mathcal{C}''[\eta]$ returning something other than $\perp$ is more than $\epsilon\gamma^2(1-\gamma)^2/8$.

But, the probability of $\mathcal{C}[\eta](x)$, using $\mathcal{C}'$, returning $\perp$ is same as probability of $\mathcal{C}$, using $\mathcal{C}''$, returning $\perp$, which is at most $(1-(\epsilon \cdot \frac{\gamma^2(1-\gamma)^2}{8}))^T = \epsilon_1$. □

Proof of Main Theorem. Now we are ready to prove the main theorem. Since there are a potential T attempts by $\mathcal{C}[\eta]$ on x, we call the values returned in the q-th attempt by $\mathcal{C}'[\eta]^q$ $(1 \leq q \leq T)$. Now, the input x was set by choosing α uniformly from $\{0,1\}^{d(n)}$. Thus,

$$\Pr_\alpha[\,\mathcal{C}[\eta](x) \text{ is wrong}] \leq \Pr_\alpha[\,\alpha \in H_\eta] + \Pr_\alpha[\,\mathcal{C}[\eta](x) \text{ is wrong } \& \ \alpha \notin H_\eta]$$

The first term on the right-hand side is at most $\delta - \tau/k - \psi_\eta - \epsilon_0$ by Lemma 10. We now focus on the second term.

$$\begin{aligned}
&\Pr[\,\mathcal{C}[\eta](x) \text{ is wrong } \& \ \alpha \notin H_\eta] \\
&\leq \Pr[\,\mathcal{C}[\eta](x) = \perp \& \ \alpha \notin H_\eta] + \Pr[\,\mathcal{C}[\eta](x) \text{ is wrong } \& \ \mathcal{C}[\eta](x) \neq \perp \& \ \alpha \notin H_\eta] \\
&\leq \epsilon_1 + \Pr[\,\mathcal{C}[\eta](x) \text{ is wrong } \& \ \mathcal{C}[\eta](x) \neq \perp] \qquad \text{(by Lemma 11)} \\
&\leq \epsilon_1 + \Pr[\,\mathcal{C}[\eta](x) \text{ is wrong } \mid \mathcal{C}[\eta](x) \neq \perp] \\
&= \epsilon_1 + \Pr[\,\mathcal{C}'[\eta]^q(x) \text{ is wrong } \mid \exists q : \mathcal{C}'[\eta]^q(x) \neq \perp] \\
&= \epsilon_1 + \Pr[F(\overline{X}) \mid \mathit{others}(\overline{X}) \leq \tau + \eta] \\
&\leq \epsilon_1 + \frac{\tau}{k} + \psi_\eta + \epsilon_2 \qquad \text{(by event } \textbf{valid}\text{)}
\end{aligned}$$

Thus,

$$\Pr_\alpha[\,\mathcal{C}[\eta](x) \text{ is wrong}] = \delta + \epsilon_1 - \epsilon_0 + \epsilon_2.$$

[1] This follows formally by noting that $\sum_{j=1}^{k} j\binom{k}{j}(A-1)^{k-j} = kA^{k-1}$, for any A.

Finally, by Lemma 12 (of the following section), the probability of η being not valid is at most ϵ_3, and this leads to a contradiction as $\epsilon_0 > \epsilon_1 + \epsilon_2 + \epsilon_3$. □

5 Pre-processing and Hypothesis Testing

As mentioned in Section 4, the algorithm Y first does some pre-processing using algorithm η, and using $\overline{X}$ as an oracle. The inputs to η are the security parameter n, k, as well as γ and δ. It returns a value η, $0 \leq \eta \leq M-1$ ($M = \lceil \gamma\delta k \rceil$).

Before we describe this pre-processing algorithm, we remark that it is intended to compute the smallest $j < M$, such that $\Pr[\mathrm{F}(\overline{X}) \mid \mathrm{others}(\overline{X}) \ \leq \tau + j] \leq \frac{\tau}{k} + \psi_j$. Now, we had assumed that $P > \epsilon$, and the hypothesis of Theorem 9 assumes a lower bound on ϵ. Then, by Lemmas 6, 7 and 8.2, it follows that if such a j does not exist, then $p_{\tau+M} > 1$, an impossibility. So, let $\bar{\eta}$ be that smallest $0 \leq j < M$.

The algorithm $\eta(\overline{X}, n, \delta, \gamma, k)$ does the following:

η : For each $i = 1..M-1$, compute the following statistics

$$t_i = \frac{\#(F(\overline{X})\ \&\ \mathrm{others}(\overline{X})\ \leq \tau + i)}{1 + \#(\mathrm{others}(\overline{X})\ \leq \tau + i)}$$

where the count is over running $\overline{X}$ on random and independent $\bar{\alpha}$ (each in $\{0,1\}^{d(n)k}$), for a total of N times (N to be determined below). Set η to be the smallest i such that $t_i < \tau/k + \psi_i + \epsilon_2/2$. If no such η exists then set $\eta = M - 1$. Return η.

Lemma 12. *There is a polynomial ϕ, independent of n, such that with $N = \phi(\gamma\delta k, \ln(1/\epsilon_2), \ln(1/\epsilon_3))$,*

$$\Pr[\ \textit{not}\ \textbf{valid}\ ] < \epsilon_3$$

Proof. Clearly, for $i = \bar{\eta}$, the actual conditional probability of F is is no more than $\tau/k + \psi_i$. Hence, $t_i > \tau/k + \psi_i + \epsilon_2/2$ is an exponentially low probability event by Chernoff bound. Now, for some smaller i, if conditional probability of F is greater than $\tau/k + \psi_i + \epsilon_2$, then again t_i being less than $\tau/k + \psi_i + \epsilon_2/2$ is an exponentially low probability event. □

References

1. Alon, N., Spencer, J.: The Probabilistic Method. John Wiley and Sons, Chichester (1992)
2. Canetti, R., Halevi, S., Steiner, M.: Hardness Amplification of Weakly-Verifiable Puzzles. In: Kilian, J. (ed.) TCC 2005. LNCS, vol. 3378, pp. 17–33. Springer, Heidelberg (2005)
3. Holenstein, T.: Parallel Repetition: Simplifications and the No-Signalling Case. In: Proc. ACM STOC (2007)

4. Impagliazzo, R., Jaiswal, R., Kabanets, V.: Chernoff-Type Direct Product Theorem. J. Cryptology 22, 75–93 (2009)
5. Impagliazzo, R., Jaiswal, R., Kabanets, V., Wigderson, A.: Uniform direct-product theorems: Simplified, optimized, and de-randomized. In: Proc. ACM STOC (2008)
6. Knuth, D.: Art of Computer Programming, vol. 1. Addison Wesley, Reading (1973)
7. Raz, R.: A Parallel Repetition Theorem. SIAM J. of Computing 27(3), 763–803
8. von Ahn, L., Blum, M., Hopper, N.J., Langford, J.: CAPTCHA: Using hard AI problems for security. In: Biham, E. (ed.) EUROCRYPT 2003. LNCS, vol. 2656, pp. 294–311. Springer, Heidelberg (2003)

Appendix A

Recall,

$$q_{\tau+i} = \psi_0 \cdot \frac{k}{\tau+i} \cdot \prod_{j=1}^{i-1} \frac{\tau+\psi_j k}{\tau+j}$$

Lemma 8. Let $M = \lceil \gamma\delta k \rceil$, and suppose $\delta k \geq 1$.

1. For any i, $0 \leq i < M$, and for any $\chi \geq 1$,

$$q_{\tau+i} \cdot \chi \cdot \frac{2}{\gamma(1-\gamma)^2} \cdot e^{-\gamma^2(1-\gamma)\delta k/2} - e^{-(1-\gamma)(\delta-\tau/k-\psi_i)k/2} > \frac{\chi}{2} \cdot e^{-\gamma^2(1-\gamma)\delta k/2}$$

2. $q_{\tau+M} \cdot \frac{2}{\gamma(1-\gamma)^2} \cdot e^{-\gamma^2(1-\gamma)\delta k/2} > 1$

Proof. For the first item in the lemma, we have

$$\begin{aligned} \delta - \tau/k - \psi_i &= \delta - (1-\gamma)\delta - \gamma(1-\gamma)\delta - i \cdot (\gamma/k) \\ &= \gamma^2\delta - \gamma \cdot (i/k) \end{aligned}$$

Now, by Lemma 7,

$$q_{\tau+i} > \frac{\gamma\delta(1-\gamma)k}{\tau+i} \cdot (1-\gamma^2) \cdot e^{\gamma(1-\gamma)(1-\frac{i-1}{2M})(i-1) - O(1/(\delta k))}$$

But, $\tau + i < \delta k$, and $(1-\gamma^2)(1-\gamma) > (1-\gamma)^2$. Further, $e^{-\gamma(1-\gamma)} > 1 - \gamma(1-\gamma) \geq 3/4$. Thus,

$$\begin{aligned} & q_{\tau+i} \cdot \chi \cdot \frac{2}{\gamma(1-\gamma)^2} \cdot e^{-\gamma^2(1-\gamma)\delta k/2} \; - \; e^{-(1-\gamma)(\delta-\tau/k-\psi_i)k/2} \\ & > \frac{3}{2} \cdot \chi \cdot e^{-\gamma^2(1-\gamma)\delta k/2+\gamma(1-\gamma)(1-\frac{i-1}{2M})i} - e^{\gamma(1-\gamma)(i/2)-\gamma^2(1-\gamma)\delta k/2} \\ & = e^{-\gamma^2(1-\gamma)\delta k/2+\gamma(1-\gamma)(i/2)} \cdot (\frac{3}{2} \cdot \chi \cdot e^{\gamma(1-\gamma)(\frac{1}{2}-\frac{i-1}{2M})i} - 1) \\ & > e^{-\gamma^2(1-\gamma)\delta k/2} \cdot \frac{\chi}{2} \end{aligned}$$

We have ignored the $e^{-O(1/\delta k)}$ factor, since the constant in the exponent is known to be a small fraction (i.e. in Sterling's formula), and hence this factor is more than compensated by $\frac{\delta k}{\tau+i}$ which we ignored.

For the second item in the lemma, again using Lemma 7 we have

$$
\begin{aligned}
& q_{\tau+M} \cdot \frac{2}{\gamma(1-\gamma)^2} \cdot e^{-\gamma^2(1-\gamma)\delta k/2} \\
& > \frac{3}{2} \cdot e^{-\gamma^2(1-\gamma)\delta k/2+\gamma(1-\gamma)(1-\frac{M-1}{2M})M} \\
& > \frac{3}{2} \cdot e^{-\gamma^2(1-\gamma)\delta k/2+\gamma(1-\gamma)\gamma\delta k/2}
\end{aligned}
$$

□

On Symmetric Encryption and Point Obfuscation

Ran Canetti[1,⋆], Yael Tauman Kalai[2], Mayank Varia[3,⋆⋆], and Daniel Wichs[4,⋆⋆⋆]

[1] School of Computer Science, Tel Aviv University
canetti@post.tau.ac.il
[2] Microsoft Research New England
yael@microsoft.com
[3] Massachusetts Institute of Technology
varia@csail.mit.edu
[4] New York University
wichs@cs.nyu.edu

Abstract. We show tight connections between several cryptographic primitives, namely encryption with weakly random keys, encryption with key-dependent messages (KDM), and obfuscation of point functions with multi-bit output (which we call multi-bit point functions, or MBPFs, for short). These primitives, which have been studied mostly separately in recent works, bear some apparent similarities, both in the flavor of their security requirements and in the flavor of their constructions and assumptions. Still, rigorous connections have not been drawn.

Our results can be interpreted as indicating that MBPF obfuscators imply a very strong form of encryption that *simultaneously* achieves security for weakly-random keys and key-dependent messages as special cases. Similarly, each one of the other primitives implies a certain restricted form of MBPF obfuscation. Our results carry both constructions and impossibility results from one primitive to others. In particular:

- The recent impossibility result for KDM security of Haitner and Holenstein (TCC '09) carries over to MBPF obfuscators.
- The Canetti-Dakdouk construction of MBPF obfuscators based on a strong variant of the DDH assumption (EC '08) gives an encryption scheme which is secure w.r.t. *any* weak key distribution of super-logarithmic min-entropy (and in particular, also has very strong leakage resilient properties).
- All the recent constructions of encryption schemes that are secure w.r.t. weak keys imply a weak form of MBPF obfuscators.

1 Introduction

Symmetric encryption is an algorithmic tool that allows a pair of parties to communicate secret information over open communication media that are accessible to eavesdroppers. In order to achieve this goal, the communicating parties need to have some

⋆ Supported by the Check Point Institute for Information Security, An ISF Grant, A Marie Curie Grant, A US-Israel BSF grant.

⋆⋆ Supported by the Department of Defense through the NDSEG Program.

⋆⋆⋆ Work completed while the author was visiting Microsoft Research, New England.

D. Micciancio (Ed.): TCC 2010, LNCS 5978, pp. 52–71, 2010.

shared secret randomness (a *key*). The classic view of symmetric encryption allows the encryption scheme to determine the distribution of the key precisely (typically it is a uniformly random string). It also assumes that the encryption and decryption algorithms are executed in a completely sealed way, so no information about the key is leaked to the eavesdroppers. Finally, the classic model assumes that the parties only use the key in the encryption and decryption routines and not for any other purpose. In particular, their messages are never related to the key.

In recent years, much research has been done to investigate various relaxations of this classic (and somewhat naive) model. One relaxation is to consider the case where the key is chosen using a "defective" source of randomness that does not generate uniform and independent random bits. (See e.g. [1,14,21,2,25] and the references therein). Namely, the key is assumed to be taken from a distribution that is adversarially chosen under some restriction. Typically the restriction is that the min-entropy of the distribution of the secret key is at least α, for some value of α. In this case the scheme is said to be secure w.r.t. α-weak keys.

A different relaxation of the classic model considers the case where the key is chosen uniformly but some *arbitrary* information on the key is leaked to the adversary (see e.g. [1,25]). This models both direct attacks where the adversary gains access to the internal storage of the parties, such as the freezing attack of [18], and indirect information leakage that occurs when the shared key is derived from the communication between the parties, such as the information exchange used to agree on the key. Of course, all security is lost of the adversary learns the key in its entirety, and therefore some restriction needs to be imposed on the *amount* of information that the adversary can get. One possibility is to require that the key has some significant statistical entropy left, even given the leakage. We call this the *entropic* setting. Another, stronger, security notion only insists that it is *computationally* infeasible to compute the secret key from the leaked information, but allows the leakage to completely determine the key statistically. We call this the *computational* setting.[1] It turns out that encryption resilient to weak keys is also resilient to a comparable amount of leakage in the entropic setting. Conversely, in some settings there is a simple transformation from leakage resilient encryption to one that withstands comparably weak keys.[2]

Yet another relaxation of the classic model considers the case where the messages may depend on the shared key. Security in this more demanding setting was termed *key-dependent message security* (KDM security) by Black, Rogaway and Shrimpton in [7]. In the last few years, the notion of KDM security has been extensively studied [19,5,9,4,20,17,8,3], and several positive results emerged, most notably the results of [8,3] who showed how to obtain KDM security w.r.t. the class of affine functions (the former under the DDH assumption and the latter under the LWE assumption). In

[1] Many other models of leakage-resilience, such as the "only computation leaks information" model [23,15], place further restrictions on the type of information that may be leaked, and are not considered in this work.

[2] In the case of semantic security for symmetric-key encryption (without chosen-plaintext attacks), we can use the following transformation: Given a scheme (Enc, Dec) that's secure against key leakage, construct the weak-key scheme $(Enc'_k(m) = (r, Enc_{k+r}(m))$ for a random $|k|$-bit r, $Dec'_k(r,c) = Dec_{k+r}(c))$.

contrast, [17] show that there exist no black-box reductions from the KDM security of any encryption scheme w.r.t. all efficient functions to *"any standard cryptographic assumption"*.

While the constructions for KDM-secure schemes and the constructions of schemes that are secure w.r.t. α-weak keys bear significant similarities to each other (eg., see [8,25], [14,3], and [1,3]), no formal connections between the problems have been made so far.

Another recently studied primitive, which may seem unrelated at a cursory look, is obfuscation of point functions (programs) with multi-bit output. Obfuscation is the task of constructing an algorithm, called an *obfuscator* $\mathcal{O}$, that takes as input a program p from a family P of programs and outputs a program $q = \mathcal{O}(p)$ that has essentially the same functionality as p, but where the code of q gives no information (or, rather, no computational ability) that cannot be determined given only oracle access to p. A central point here is that $\mathcal{O}$ should work correctly and securely for *any* program in P.

A point function with multi-bit output (or a MBPF) is a function $I_{(k,m)}$ which, on input x, outputs m if $x = k$ and $\perp$ otherwise. In the special case of point functions, the value m is fixed to some constant, say 1. Obfuscators for point functions are constructed in [10,26] under strong assumptions (and in [22] in the random oracle model). Obfuscators for MBPF are only known based on very strong and specific assumptions (specifically, the existence of fully-composable point function obfuscators) [11]. Different constructions exist for restricted settings, such as the case where m is shorter than k, or the case where m and k are distributed independently from each other [11,14]. In all of these constructions the obfuscator is given the values k and m explicitly.

The applicability of MBPF obfuscation to symmetric encryption has been pointed out in [11], who proposed to encrypt a message m with key k by letting $O(I_{(k,m)})$ be the ciphertext. The fact that security holds for any k was used to suggest that m remains hidden even when k is taken from a distribution which is not uniform, as long as it has sufficient min-entropy (i.e., it cannot be guessed in polynomial time.) Also, [14] show that their construction of leakage resilient encryption can be used as a restricted variant of MBPF obfuscation.

1.1 Our Results

We show tight relations between the above primitives. Specifically, we show that weak key resilience, leakage resilience, and KDM security, each with its own variants, can all be viewed as natural special cases of the MBPF obfuscation problem. In fact, a generalized version of KDM security, which also withstands the case where the key is taken from a weakly random distribution, is also a special case of MBPF obfuscation. In addition to providing some insight and intuition to these primitives, the drawn connections provide new results — both constructions and hardness results — for the primitives considered.

The remainder of the introduction overviews our results. We first present the general connections between obfuscation and symmetric encryption; next we sketch some conclusions and corollaries.

As a preliminary step towards drawing general connections, we set up a framework for relaxing the standard notion of security of MBPF obfuscation. This notion, called virtual black-box (VBB) security [6], essentially requires that for any adversary with binary output there exists a simulator such that, for *any* k, m, the output of the adversary given $\mathcal{O}(I_{(k,m)})$ is indistinguishable from the output of the simulator given oracle access to $I_{(k,m)}$. We wish to consider the relaxed case where k and m are taken from an unknown *distribution* from a given class. We capture this relaxation by replacing the "for any k, m" requirement in the VBB definition with "for any distribution on k, m from a given class of distributions". Note that here the simulator knows the class of distributions, but not the distribution itself. This relaxation allows us to relate the different classes of strong encryption to MBPF obfuscators for different classes of distributions. Specifically:

Obfuscation vs. Weak-Key and Leakage Resilient Encryption: We say that an MBPF obfuscator is α-entropic with independent messages if it is an *MBPF obfuscator* for product distributions on k, m, where the distribution of k has min-entropy at least α, and m is drawn *independently* of k, but need not have any entropy. We say that the obfuscator is a fully-entropic IM MBPF if it has α-entropic security for all super-logarithmic α. We show:

From IM MBPF obfuscators to encryption. Any α-entropic IM MBPF obfuscator with independent messages allows us to construct semantically secure encryption scheme with security for α-weak keys, via the transformation $\mathsf{Enc}_k(m) = \mathcal{O}(I_{(k,m)})$.

From encryption to IM MBPF obfuscators. Conversely, any encryption scheme with semantic security for α-weak keys allows us to construct α-entropic IM MBPF obfuscators. The transformation is simple: To obfuscate a pair k, m, simply encrypt m with key k to obtain a ciphertext c; then, the obfuscated program simply has a hard-coded ciphertext c, and on input x, runs the decryption algorithm on c with the key x. Here, for the correctness of obfuscation, we require that the encryption scheme can *detect* if it is decrypting a ciphertext with an incorrect secret key. We show that this property can be added generically to any semantically secure encryption scheme.

CPA security vs. self-composability. If we start with a *CPA secure* encryption for α-weak keys, then the resulting IM MBPF obfuscator $\mathcal{O}$ is *self-composable*, in the sense that security is preserved even if $\mathcal{O}$ is run multiple times on MBPFs with the *same input* k and (possibly) different outputs m_i. As was shown by [11], this property is not, in general, implied by obfuscation alone. Conversely, if we start with a self-composable IM MBPF obfuscator then we derive an encryption scheme which is CPA secure for α-weak keys.

Fully-entropic obfuscation and fully-weak key security. If we start with an IM MBPF obfuscator that has full-entropic security (i.e., it works for any distribution where k is independent from m and has some super-logarithmic min-entropy) then we obtain an encryption scheme with semantic-security for fully-weak keys. (i.e. security for any key-distribution with super-logarithmic entropy).

Computational leakage vs. auxiliary information. If we start from a computational leakage resilient encryption then the resulting MBPF obfuscator is secure with respect to dependent auxiliary input, as defined in [16]. Similarly, if we start from a MBPF obfuscator that's secure with dependent auxiliary input then the resulting encryption scheme is computationally leakage resilient.

KDM security: All of the above equivalence results in the preceding paragraph were stated with respect to the restricted notion of obfuscation to *independent messages*. Interestingly, the standard notion of MBPF obfuscation provides the additional (and very powerful) security guarantee for encryption with *key-dependent messages (KDM)*.

We say that $\mathcal{O}$ is a *α-entropic (dependent) MBPF obfuscator* if it withstands any *joint* distribution on k, m where the projection distribution on k has min-entropy at least α (and m may depend on k). We say that $\mathcal{O}$ is a fully-entropic (dependent) MBPF obfuscator if the above holds for all super-logarithmic α .

We also define α-KDM encryption schemes which provide security *even* when the key is taken from any distribution of entropy α, *and* the message can be an arbitrary function of the secret key. We show:

Obfuscation vs. encryption. Any α-entropic (dependent) MBPF obfuscator provides, via the same transformation as before, an α-KDM semantically secure encryption scheme.

Multi message resilience vs. self composability. If the encryption scheme we start with is *multi-message* α-KDM secure, in the sense that it withstands the case where the adversary obtains encryptions of any polynomial number of functions of the secret key, then the resulting (dependent) MBPF obfuscator has α-entropic security and is *self composable*. The converse implication holds as well.

To connect our new α-entropic definition to previous works, we show that any MBPF obfuscator that is α-entropic for any super-logarithmic α also satisfies the virtual black-box property, i.e., it works for *any* k, m. (We note that the proof of this result is trickier than it might seem, the main difficulty being that in the case of α-entropic security the simulator has the bound α, whereas in the VBB case no such bound exists.)

1.2 Implications

We show some implications of the above correspondence results:

Secure encryption w.r.t. (fully) weak keys. Known constructions of encryption schemes that are secure w.r.t. weak keys are parameterized by the min-entropy α tolerated. That is, a bound α must be chosen in advance, and then a scheme is constructed based on α. Using our transformations, we get that, under the strong DDH assumption in [10], the [10,11] MBPF obfuscator provides an encryption scheme that is secure w.r.t. α-weak keys, for *any* super-logarithmic function α. The main advantage is that the min-entropy α does not need to be chosen in advance. More specifically, we obtain a single encryption scheme, parameterized only by the security parameter n (and *not* by α), which simultaneously achieves security for all $\alpha(n) \in \omega(\log n)$.

We remark that the hardness assumption we use has a similar flavor - it explicitly makes an assumption for every distribution with super logarithmic min-entropy. The crucial point is however that the construction does *not* depend on α and so it provides a tradeoff between the strength of the assumption and the strength of the obtained guarantee. See Section 6.1 for further details.

Impossibility for MBPF Obfuscators and fully composable point function obfuscators. Using our transformations, the negative result due to Haitner and Holenstein [17] implies that there are no constructions of MBPF obfuscators that can be proven secure via a "black box reduction to standard cryptographic primitives." Since full MBPF obfuscators can be constructed in a black-box way from fully composable point function obfuscators [11], the impossibility carries over to this primitive as well. See Section 6.2 for further details.

Constructing self-composable MBPF obfuscators with independent messages. Using our transformations, we can use constructions of encryption schemes that are secure w.r.t. α-weak keys, to get self composable MBPF obfuscators with independent messages. More specifically, we construct self composable obfuscators for MBPFs $\{I_{(k,m)}\}$ as long as the distribution of m is independent of the distribution of k, both distributions are efficiently sampleable, and the distribution of k has min-entropy α. See Section 6.2 for further details.

Organization. Section 2 contains some basic definitions for obfuscation and encryption. Section 3 draws connections between obfuscation and weak key and leakage resilient encryption. Section 4 draws connections between obfuscation and encryption resilient to key dependent messages. Section 6 states the corollaries that we draw from the general connections. Many proofs are left out and appear only in the full version [12].

2 Definitions

2.1 Obfuscation of Point Functions with Multi-bit Output

Let $I_{(k,m)} : \{0,1\}^* \cup \{\bot\} \to \{0,1\}^* \cup \bot$ denote the function

$$I_{(k,m)}(x) = \begin{cases} m & \text{if } x = k \\ \bot & \text{otherwise} \end{cases}$$

which outputs the *message* m given the *key* k, and $\bot$ otherwise. Let $\mathcal{I} = \{I_{(k,m)} \mid k, m \in \{0,1\}^*\}$ be the family of all such functions, which we call the family of *point functions with multi-bit output* or just *multi-bit point functions (MBPF)* for short.

Definition 1 (Obfuscation of Point Functions with Multi-bit Output). *A multi-bit point function (MBPF) obfuscator is a PPT algorithm $\mathcal{O}$ which takes as input values (k, m) describing a function $I_{(k,m)} \in \mathcal{I}$ and outputs a circuit C. We will abuse notation and write $\mathcal{O}(I_{(k,m)})$, but will always assume that $\mathcal{O}$ gets k and m as clearly delineated inputs.*

Correctness: *For all $(k, m) \in \{0,1\}^*$ with $|k| = n$, $|m| = \text{poly}(n)$, all $x \in \{0,1\}^n$,*

$$\Pr[C(x) \neq I_{(k,m)}(x) \mid C \leftarrow \mathcal{O}(I_{(k,m)})] \leq \text{negl}(n)$$

where the probability is taken over the randomness of the obfuscator algorithm.

Polynomial Slowdown: *For any k, m, the size of the circuit $C = \mathcal{O}(I_{(k,m)})$ is polynomial in $|k| + |m|$.*

Entropic Security: *We say that the scheme has $\alpha(n)$-**entropic security** if for any PPT adversary $\mathcal{A}$ with 1 bit output, any polynomial $\ell(\cdot)$, there exists a PPT simulator $\mathcal{S}$ such* *that for all jointly-distributed* $\{X_n, Y_n\}_{n\in\mathbb{N}}$ *where* X_n *takes values in* $\{0,1\}^n$, Y_n *takes values in* $\{0,1\}^{\ell(n)}$ *and* $H_\infty(X_n) \geq \alpha(n)$, *we have:*

$$\left|\Pr\left[\mathcal{A}(\mathcal{O}(I_{(k,m)})) = 1\right] - \Pr\left[\mathcal{S}^{I_{(k,m)}(\cdot)}(1^n) = 1\right]\right| \leq \mathrm{negl}(n)$$

where the probability is taken over the randomness of $(k,m) \leftarrow (X_n, Y_n)$, *the randomness of the obfuscator* $\mathcal{O}$ *and the randomness of* $\mathcal{A}, \mathcal{S}$. *We say that a scheme has* ***fully-entropic security*** *if it has* $\alpha(n)$*-entropic security for all* $\alpha(n) \in \omega(\log(n))$.

We relate the notion of fully-entropic security, defined above, to the standard security guarantee provided by obfuscation called the *virtual black-box property*:

Definition 2 (Virtual black-box property [10,6,26]). *For any PPT adversary* $\mathcal{A}$ *with 1 bit output and any polynomials* $p(\cdot)$, $\ell(\cdot)$, *there exists a PPT simulator* $\mathcal{S}$ *such that for all distributions* $\{X_n, Y_n\}_{n\in\mathbb{N}}$ *with* X_n *taking values in* $\{0,1\}^n$ *and* Y_n *taking values in* $\{0,1\}^{\ell(n)}$, *we have:*

$$\left|\Pr\left[\mathcal{A}(\mathcal{O}(I_{(k,m)})) = 1\right] - \Pr\left[\mathcal{S}^{\mathcal{I}_{(k,m)}}(1^n) = 1\right]\right| \leq \frac{1}{p(n)}.$$

The probability is taken over the randomness of $(k,m) \leftarrow (X_n, Y_n)$, $\mathcal{A}$, $\mathcal{S}$, *and* $\mathcal{O}$.

Note the difference between the fully-entropic definition and the VBB definition: the former allows a different simulator for each entropy threshold $\alpha(\cdot)$, but requires a negligible error in simulation, while the latter allows a different simulator for each simulation-error $p(\cdot)$, but requires the simulator to work for all distributions regardless of entropy. Interestingly, we show that the fully-entropic definition implies VBB (but don't know whether the converse holds as well).

Theorem 1. *If* $\mathcal{O}$ *is a MBPF obfuscator that satisfies* fully-entropic security *(as in Definition 1) then* $\mathcal{O}$ *also satisfies* virtual black-box obfuscation *(as in Definition 2).*

The proof of this theorem appears in the full version of this paper [12]. The idea is to extend the technique used in [10] to show that a distribution-based definition implies the virtual black box property in the case of point functions. At a high level, the distributional definition there says that if a user chooses a key from a well-spread distribution, then an adversary cannot learn anything from an obfuscated point function beyond the fact that the key is from this distribution, so in particular the key is hard to determine. We show how to extend the distributional definition to the MBPF setting and use this to prove that fully-entropic security provides this distributional requirement, and therefore the virtual black-box property as well.

Fully entropic security, as well as virtual black box security, are quite strong, and difficult to satisfy. The notion of $\alpha(n)$-entropic security, for some particular $\alpha(n) \in \omega(\log(n))$, corresponds to a meaningful weakening of that notion where security is only provided when the input comes from a reasonably random source. A similar weakening of obfuscation, in the special case of point functions, was also considered by Canetti, Micciancio and Reingold [13] in the context of perfectly one-way hash functions.

Instead of restricting attention to distribution with $\alpha(n)$ min-entropy, one might instead give the simulator the ability to ask its oracle more queries, by a factor of $2^{\alpha(n)}$ (i.e. the simulator is no longer polynomial time). In the full version [12], we show that this alternative relaxed notion is actually implied by α-entropic security.

We consider several additional variants of obfuscation throughout the paper. First, we propose an additional weakening of the definition, which we call security for *independent messages*, and where we require that the distribution on the output m is independent from that of the input k for a point function $I_{(k,m)}$.

Definition 3 (Independent Messages). *We say that an obfuscator $\mathcal{O}$ is $\alpha(n)$-entropically secure* for independent messages *if we restrict the definition of $\alpha(n)$-entropic security only to distributions $\{X_n, Y_n\}$ where X_n and Y_n are* independently *distributed. We define the notion of* fully-entropic security for independent messages *analogously.*

We also define a stronger variant of plain obfuscation, which provides some *composability* guarantees. There are two variants: For *full composition* we require that the security of obfuscation is preserved even if the adversary gets (freshly and independently) obfuscated circuits for many functions, where the various obfuscated functions are related in arbitrary ways (i.e., both the keys and the messages may differ). For *self composition* we require that all the obfuscated functions have the same value of the key k. That is, one should obfuscate the functions $I_{(k,m_1)}, \ldots, I_{(k,m_t)}$ with the *same key* k but *potentially different messages* $m_1, \ldots, m_t$. (For point functions, self composition boils down to the case of many obfuscated versions of the same function.)

Definition 4 (Composability). *A multi-bit point function obfuscator $\mathcal{O}$ with $\alpha(n)$-entropic security is said to be* **fully-composable** *if for any PPT adversary $\mathcal{A}$ with 1 bit output, any polynomials $t(\cdot), \ell(\cdot)$, there exists a PPT simulator $\mathcal{S}$ such that for all distributions $\{(X_n, Y_n)\}_{n\in\mathbb{N}}$, where $X_n = X_n^{(1)}, \ldots, X_n^{(t)}$, $Y_n = Y_n^{(1)}, \ldots, Y_n^{(t)}$, and $X_n^{(i)}$ taking values in $\{0,1\}^n$, $Y_n^{(i)}$ taking values in $\{0,1\}^{\ell(n)}$ and $H_\infty(X_n) \geq \alpha(n)$, we have:*

$$|\Pr[\mathcal{A}(\mathcal{O}(I_{k_1,m_1}), \ldots, \mathcal{O}(I_{k_t,m_t}))=1] - \Pr[\mathcal{S}^{I_{(k_1,m_1)},\ldots,I_{(k_t,m_t)}}(1^n)=1]| \leq \mathrm{negl}(n),$$

where the probabilities are over $(k_1, \ldots, k_t, m_1, \ldots, m_t) \leftarrow (X_n, Y_n)$ and over the randomness of $\mathcal{A}, \mathcal{S}, \mathcal{O}$.

If the above holds only for the distributions X_n where $\Pr[k_1 = k_2 \ldots = k_t] = 1$, then we say that $\mathcal{O}$ is **self-composable.**

The notions of composability extend naturally to obfuscators with fully-entropic security, where we require that the above definition holds for all $\alpha(n) \in \omega(\log(n))$. It also extends to obfuscators for independent messages, where we restrict the definition to the case where X_n and Y_n are independent. (It is stressed that there is no independence assumption among the coordinates within X_n or Y_n.)

2.2 Definitions for Encryption with Weak Keys

A symmetric encryption scheme consists of efficient algorithms (Enc, Dec).[3] We say that the encryption scheme is semantically secure *for $\alpha(n)$-weak keys* if the usual notion of semantic security holds even when the key comes from any weak-source of entropy $\alpha(n)$. We propose the following definition of symmetric key encryption with weak keys.

Definition 5 (Symmetric Encryption with Weak Keys). *We say that an encryption scheme has* **CPA security** *for $\alpha(n)$-weak keys if there exists an efficient algorithm $D(n,\ell)$ running in time $\mathrm{poly}(n,\ell)$, such that, for all PPT adversaries $\mathcal{A}$ and all distribution-ensembles $\{X_n\}_{n\in\mathbb{N}}$ with $H_\infty(X_n) \geq \alpha(n)$, we have:*

$$|\Pr[\mathrm{CPA}_0^{X,D}(\mathcal{A},n)=1] - \Pr[\mathrm{CPA}_1^{X,D}(\mathcal{A},n)=1]| \leq \mathrm{negl}(n)$$

where the games $\mathrm{CPA}_b^{X,D}(\mathcal{A},n)$ for $b=0,1$ are defined via the following experiment:

1. *$k \leftarrow X_n$*
2. *Repeat: $\mathcal{A}$ submits a query m and receives a ciphertext c where:*
   In game $\mathrm{CPA}_0^{X,D}$, the challenger sets $c \leftarrow \mathsf{Enc}_k(m)$.
   In game $\mathrm{CPA}_1^{X,D}$, the challenger sets $c \leftarrow D(n,|m|)$.
3. *The output of the game is the output of $\mathcal{A}$.*

The algorithm $D(n,\ell)$ can keep persistent state during stage 2. We define **semantic security** *with $\alpha(n)$-weak keys via the games $\mathrm{SEM}_0^{X,D}$, $\mathrm{SEM}_1^{X,D}$, which are equivalent to the CPA games* except *that step (2) is performed only once.*

We say that an encryption scheme is CPA-secure (resp. semantically-secure) for **fully weak** *keys if it is CPA-secure (resp. semantically-secure) secure for $\alpha(n)$-weak keys for all $\alpha(n) \in \omega(log(n))$.*

Note that, in case of $\alpha(n) = n$ (i.e. uniformly random secret keys), the above definition is equivalent to the standard notion of CPA/semantic security, since we can always simply define $D(n,\ell)$ to always output fresh encryptions $\mathsf{Enc}_k(0^\ell)$, where k is initially chosen uniformly at random and re-used for all queries. On the other hand, when considering $\alpha(n)$-weak keys, the above definition is somewhat stronger than just requiring that the adversary cannot distinguish between an encryption of m and that of some set message, such as 0^ℓ. In particular, it requires that there is a single *universal* distribution D on ciphertexts, which is indistinguishable from encryption with *any* key distribution X_n of sufficient entropy. For example, consider an encryption scheme which, along with the ciphertext, always outputs the first bit of the secret key. Although such scheme might satisfy a natural definition where encryption of m_0 and m_1 are indistinguishable, it could never satisfy the above definition, even for $\alpha(n) = n - 1$. The reason is that the ciphertext distribution is now different depending on whether the keys come from a distribution that fixes the first bit at 0 versus one which fixes the first bit at 1. Although our definition is stronger than one may need, we will show that it is necessary and sufficient for our equivalence with obfuscation to hold. Moreover, all natural constructions of encryption schemes with weak-keys that we know of achieve the above definition.

---

[3] That is, the key generation algorithm is implicit and is assumed to always generate a uniform n-bit string.

We also define a "wrong-key detection" property, which will be needed to achieve correctness in obfuscation.

Definition 6 (Wrong-Key Detection). *We say that an encryption scheme satisfies the* wrong-key detection *property if for all* $k \neq k' \in \{0,1\}^n$, *all* $m \in \{0,1\}^{\text{poly}(n)}$, $\Pr[\mathsf{Dec}_{k'}(\mathsf{Enc}_k(m)) \neq \perp] \leq \text{negl}(n)$.

We note that a similar, but weaker, property called confusion freeness, was defined in [24]. For confusion freeness, the keys k, k' are random and independent, while we consider a worst-case choice of k, k' and the probability above is only over the randomness of the encryption scheme.

Lemma 1 (see the full version [12] for proof) shows that, in the case of semantic security, wrong-key detection can always be achieved via a simple transformation. We note, however, that this transformation no longer works in the case of CPA security.

Lemma 1. *Let* $(\mathsf{Enc}, \mathsf{Dec})$ *be a semantically-secure encryption scheme for* $\alpha(n)$*-weak keys and let* $\mathcal{H}$ *be a pairwise-independent permutation family. Define an encryption scheme* $(\mathsf{Enc}', \mathsf{Dec}')$ *by:*

$$\mathsf{Enc}'_k(m) \triangleq \begin{cases} \textit{Choose: } h \leftarrow \mathcal{H}, r \leftarrow U_n \\ \textit{Output: } \langle r, h, c = \mathsf{Enc}_{h(k)}(r||m) \rangle \end{cases}$$

$$\mathsf{Dec}'_k(\langle r, h, c \rangle) \triangleq \begin{cases} \textit{Compute: } (r'||m') = \mathsf{Dec}_{h(k)}(c) \\ \textit{Output: } m' \textit{ if } r' = r \textit{ and } \perp \textit{ otherwise} \end{cases}$$

Then $(\mathsf{Enc}', \mathsf{Dec}')$ *is a semantically-secure encryption scheme for* $\alpha(n)$*-weak keys, with* wrong-key detection. *The above also holds if we replace "*$\alpha(n)$*" with "fully".*

3 Encryption with Weak Keys and MBPF Obfuscation

3.1 Sem. Sec. Encryption and Obfuscation with Independent Messages

In this section, we show equivalence between semantically secure encryption with weak keys and MBPF obfuscators for *independent messages.*

Theorem 2. *Let* $\alpha(n) \in \omega(\log(n))$. *There exist MBPF obfuscators with* $\alpha(n)$*-entropic security for independent messages if and only if there exist semantically secure encryption schemes with wrong key detection for* $\alpha(n)$*-weak keys. Furthermore, the above also holds if we replace "*$\alpha(n)$*" with "fully".*

We prove the "if" and "only if" directions in Lemmas Lemma 2 and Lemma 3, respectively.

Lemma 2. *Let* $\alpha(n) \in \omega(\log(n))$ *and let* $\mathcal{O}$ *be a MBPF obfuscator with* $\alpha(n)$*-entropic security for independent messages. Let* $\mathsf{Enc}_k(m) \triangleq \mathcal{O}(I_{(k,m)})$, $\mathsf{Dec}_k(C) \triangleq C(k)$ *where the ciphertext* C *is interpreted as a circuit. Then the encryption scheme* $(\mathsf{Enc}, \mathsf{Dec})$ *is semantically secure with* $\alpha(n)$*-weak keys and has the wrong-key detection property.*

Proof. The correctness of decryption follows from the correctness of obfuscation. For the security of the encryption scheme with $\alpha(n)$-weak keys. Fix any adversary $\mathcal{A}$ and any distribution $\{X_n\}_{n\in\mathbb{N}}$ with $H_\infty(X_n) \geq \alpha(n)$. The distribution $\{Y_n\}$ is defined by running $\mathcal{A}(1^n)$ and outputting the message m that $\mathcal{A}$ gives to its challenger. Define the distribution $D(n,\ell) = \mathcal{O}(I_{(k,m)})$ where $(k,m) \leftarrow (U_n, U_\ell)$. Then, by the $\alpha(n)$-entropic security of obfuscation, there must be a simulator $\mathcal{S}$ such that

$$\begin{aligned}
&\left|\Pr[\mathrm{SEM}_0^{X,D}(\mathcal{A},n)=1]-\Pr[\mathrm{SEM}_1^{X,D}(\mathcal{A},n)=1]\right| \\
&= \left|\Pr_{(k,m)\leftarrow(X_n,Y_n)}[\mathcal{A}(\mathcal{O}(I_{(k,m)}))=1] - \Pr_{(k,m)\leftarrow(U_n,U_\ell)}[\mathcal{A}(\mathcal{O}(I_{(k,m)}))=1]\right| \\
&\leq \left|\Pr_{(k,m)\leftarrow(X_n,Y_n)}[\mathcal{A}(\mathcal{O}(I_{(k,m)}))=1] - \Pr_{(k,m)\leftarrow(X_n,Y_n)}\left[\mathcal{S}^{I_{(k,m)}(\cdot)}(1^n)=1\right]\right| \quad (1) \\
&+ \left|\Pr_{(k,m)\leftarrow(X_n,Y_n)}[\mathcal{S}^{I_{(k,m)}}(1^n)=1] - \Pr_{(k,m)\leftarrow(U_n,U_\ell)}[\mathcal{S}^{I_{(k,m)}}(1^n)=1]\right| \quad (2) \\
&+ \left|\Pr_{(k,m)\leftarrow(U_n,U_\ell)}[\mathcal{S}^{I_{(k,m)}}(1^n)=1] - \Pr_{(k,m)\leftarrow(U_n,U_\ell)}[\mathcal{A}(\mathcal{O}(I_{(k,m)}))=1]\right| \quad (3) \\
&\leq \mathrm{negl}(n)
\end{aligned}$$

where (1),(3) follow by the definition of entropic security of obfuscation, and (2) follows since the only way that a PPT simulator can get anything from its oracle is by querying it on the input k, which happens with negligible probability when k comes from a source of super-logarithmic entropy $\alpha(n)$. □

Lemma 3. *Let* (Enc, Dec) *be an encryption scheme with semantic security for $\alpha(n)$-weak keys and with the* wrong-key detection *property. We define the obfuscator $\mathcal{O}$ which, on input $I_{(k,m)}$, computes a ciphertext $c = \mathsf{Enc}_k(m)$ and outputs the circuit $C_c(\cdot)$ defined by $C_c(x) \triangleq \mathsf{Dec}_x(c)$. Then the obfuscator $\mathcal{O}$ has $\alpha(n)$-entropic security for independent messages.*

Proof. First, we show the correctness property of the obfuscator. Fix $k, x \in \{0,1\}^n$ and $m \in \{0,1\}^{\mathrm{poly}(n)}$. If $k = x$ then

$$\Pr[C(x) \neq I_{(k,m)}(x) \mid C \leftarrow \mathcal{O}(I_{(k,m)})] = \Pr[\mathsf{Dec}_k(\mathsf{Enc}_k(m)) \neq m] \leq \mathrm{negl}(n)$$

by the correctness of encryption. On the other hand, if $k \neq x$ then

$$\Pr[C(x) \neq I_{(k,m)}(x) \mid C \leftarrow \mathcal{O}(I_{(k,m)})] = \Pr[\mathsf{Dec}_x(\mathsf{Enc}_k(m)) \neq \bot] \leq \mathrm{negl}(n)$$

by the *wrong-key detection* of encryption.

The polynomial slowdown property of the obfuscator follows from the fact that the size of the circuit is only proportional to the ciphertext size and the size of the decryption circuit, which are polynomial in $|k|, |m|$.

Lastly, we show $\alpha(n)$-entropic security for independent messages. Let $D(n,\ell)$ be the distribution defined by the semantic-security of the encryption scheme. For any polynomial $\ell(n)$ any PPT adversary $\mathcal{A}$ which attacks the obfuscation scheme, we define the simulator $\mathcal{S}$ which chooses a random ciphertext c from the distribution $D(n, \ell(n))$ and runs $\mathcal{A}$ on a circuit C_c constructed using the ciphertext c. Then

$$\left| \Pr_{(k,m)\leftarrow(X_n,Y_n)} [\mathcal{A}(\mathcal{O}(I_{(k,m)})) = 1] - \Pr_{(k,m)\leftarrow(X_n,Y_n)} [\mathcal{S}^{I_{(k,m)}}(1^n, 1^\ell) = 1] \right| \quad (4)$$

$$= \left| \Pr\left[\mathcal{A}(C_c) = 1 \;\middle|\; \begin{array}{c}(k,m) \leftarrow (X_n, Y_n) \\ c \leftarrow \mathsf{Enc}_k(m)\end{array}\right] - \Pr\left[\mathcal{A}(C_c) = 1 \mid c \leftarrow D(n,\ell)\right] \right|$$

$$\leq \text{negl}(n) \quad (5)$$

Where (5) follows by semantic-security. □

3.2 CPA Encryption and Composable Obfuscation for Indep. Messages

In this section, we show equivalence between CPA secure encryption with weak keys and self-composable MBPF obfuscators for *independent messages.*

Theorem 3. *Let* $\alpha(n) \in \omega(\log(n))$. *There exist* **self-composable** *MBPF obfuscators with* $\alpha(n)$*-entropic security for independent messages if and only if there exist* **CPA** *secure encryption schemes for* $\alpha(n)$*-weak keys and the wrong-key detection property. The above also holds if we replace "*$\alpha(n)$*" with "fully".*

We prove the two sides of the "if and only if" separately. First we show that composable obfuscation implies encryption (Lemma 4) and then we show that encryption implies obfuscation (Lemma 5).

In the next lemma, going from obfuscation to encryption, it would be natural to define $\mathsf{Enc}_k(m) = \mathcal{O}(I_{(k,m)})$. However, we instead define $\mathsf{Enc}_k(m) = (\mathcal{O}(I_{(k,r)}), m \oplus r)$ for a uniform r. The reason for this is that the messages m chosen by the adversary in the CPA game can depend adaptively on prior ciphertexts. However, for composable obfuscation, the distributions Y_i of the messages m_i are independent of prior obfuscated circuits. We get around this by making sure that the obfuscation is applied to a random value.

Lemma 4. *Let* $\alpha(n) \in \omega(\log(n))$ *be an arbitrary function. Let* $\mathcal{O}$ *be a* **self-composable** *MBPF obfuscator with* $\alpha(n)$*-entropic security for independent messages. We define* (Enc, Dec) *by*

$$\mathsf{Enc}_k(m) \triangleq (\mathcal{O}(I_{(k,r)}), m \oplus r) \quad , \quad \mathsf{Dec}_k(C, y) \triangleq C(k) \oplus y$$

where r *is uniformly random, and* C *is interpreted as a circuit. The resulting encryption scheme is* **CPA** *secure with* $\alpha(n)$*-weak keys.*

The other direction is shown via the same construction as in the case of semantic security:

Lemma 5. *Let* (Enc, Dec) *be an encryption scheme with* **CPA** *security for* $\alpha(n)$*-weak keys and having the* wrong-key detection *property. We define the obfuscator* $\mathcal{O}$ *which, on input* $I_{(k,m)}$, *computes a ciphertext* $c = \mathsf{Enc}_k(m)$ *and outputs the circuit* $C_c(\cdot)$ *defined by* $C_c(x) = \mathsf{Dec}_x(c)$. *Then,* $\mathcal{O}$ *is a* **self-composable** *MBPF obfuscator with* $\alpha(n)$*-entropic security for independent messages.*

4 KDM Encryption and MBPF Obfuscation

First, we define the notion of semantically-secure encryption with *key dependent messages* (KDM) and $\alpha(n)$-weak keys.

Definition 7 (Semantic KDM Encryption with Weak Keys)
A symmetric encryption scheme (Enc, Dec) *is semantically secure for* **key dependent messages (KDM)** *and* $\alpha(n)$-weak keys *if there exists a distribution* $D(n, \ell)$*, which is efficiently sampleable in time* $\text{poly}(n, \ell)$*, such that for all functions* f*, all PPT adversaries* $\mathcal{A}$*, and all distribution-ensembles* $\{X_n\}_{n\in\mathbb{N}}$ *with* $H_\infty(X_n) \geq \alpha(n)$*, we have:*

$$|\Pr[\text{KDM}_0^{X,D}(\mathcal{A}, n) = 1] - \Pr[\text{KDM}_1^{X,D}(\mathcal{A}, n) = 1]| \leq \text{negl}(n), \quad (6)$$

where $\text{KDM}_b^{X,D}(\mathcal{A}, n)$ *is defined via the following experiment:*

$k \leftarrow X_n$

$c_0 \leftarrow \mathsf{Enc}_k(f(k)), c_1 \leftarrow D(n, \ell)$ *where* ℓ *is the output size of* f.

Output: $\mathcal{A}(c_b)$

We now show that semantically secure encryption with KDM and security for weak keys is equivalent to MBPF obfuscation.

Theorem 4. *Let* $\alpha(n) \in \omega(\log(n))$*. There exist MBPF obfuscators with* $\alpha(n)$*-entropic security for the* **standard notion of dependent messages** *if and only if there exist semantically-secure* **KDM** *encryption schemes with* $\alpha(n)$*-weak keys and the "wrong-key detection" property. In particular, the above also holds if we replace "*$\alpha(n)$*" with "fully".*

The proof of the above theorems follows from essentially the same arguments as in Lemma 2 and Lemma 3. We simply observe that allowing the adversary to get encryption of a value $f(k)$ in the proofs of those lemmas, corresponds to having a distribution Y_n that depends on X_n, that it $Y_n = f(X_n)$. Conversely, for any joint distribution $\{X_n, Y_n\}$, we can define some (probabilistic, and possibly inefficient) function f so that $Y_n = f(X_n)$.

In the full version of this paper [12], we also explore a notion of CPA security with KDM and weak-keys. We essentially show results analogous to those in Section 3.2 connecting CPA encryption (without KDM) to obfuscation with independent messages, but only if we restrict ourselves to a non-adaptive attacker who chooses the function f of the secret key prior to seeing any ciphertexts.

5 Encryption/Obfuscation with Auxiliary Input

In the full version of this work [12] we also define encryption with semantic/CPA security *with auxiliary input family* $\mathcal{F}$, where the adversary gets to learn $f(k)$ for any $f \in \mathcal{F}$.[4] Similarly, we define (self-composable) MBPF obfuscation with auxiliary input family $\mathcal{F}$, where the adversary and simulator both get $f(k)$ for some $f \in \mathcal{F}$ and

[4] This is only interesting for families $\mathcal{F}$ where each $f \in \mathcal{F}$ is *hard* to invert, as otherwise $f(k)$ completely reveals k and no security is possible. Often, it makes sense to restrict $\mathcal{F}$ much further, such as requiring that $f(k)$ is exponentially-hard to invert ...

the obfuscated point k (we only consider this notion for obfuscation with independent messages). Both notions can be defines for $\alpha(n)$-weak keys as well as fully weak keys.

We show that all of the results of Section 3 extend naturally to the auxiliary input setting. That is:

- We extend Theorem 2, to show an equivalence between semantically secure encryption with auxiliary-input family $\mathcal{F}$ and wrong-key detection, and obfuscation of MBPF with auxiliary-input family $\mathcal{F}$ and independent messages. The equivalence holds for $\alpha(n)$-weak keys or "fully weak" keys. The constructions are the same as those of Lemma 2 and Lemma 3.
- We similarly extend Theorem 3 showing a similar equivalence for CPA secure encryption and self-composable onfuscation with auxiliary input. The constructions are the same as those of Lemma 4 and Lemma 5.

6 Implications

We now show how to use the above equivalence results between encryption with weak keys and obfuscation of multi-bit point functions to derive new results in both directions.

6.1 Encryption with Fully Weak Keys

Encryption with $\alpha(n)$-weak keys vs. fully-weak keys. Prior work on leakage-resilient encryption and encryption with weak-keys has given results of the following form:

1. Fix any constant $\varepsilon > 0$ and let $\alpha(n) = n^{\varepsilon}$.
2. Construct an encryption scheme, which depends on ε, and achieves security for $\alpha(n)$-weak keys.

We note that there are several issues with the above two-step approach. Firstly, we may not know the exact level of key-entropy, or correspondingly the value of ε, at design time. Therefore, in practice, it may be difficult to decide on what ε to use when choosing the encryption scheme. A scheme which is designed for some specific ε does not provide any security guarantees for key-distributions whose entropy is strictly less than n^{ε}, and so we may be tempted to be conservative with the choice of ε at design time. On the other hand, when taking an excessively small value of ε in the above constructions, we are forced to reduce the exact-security of the system (e.g. working in a group of description-length n^{ε}) or reduce the efficiency of the system proportionally with $n^{1/\varepsilon}$, leading to poorer security or performance even if the system is later only used with uniformly random keys! Secondly, none of the prior results generalize to allow for specific super-logarithmic entropy thresholds such as $\alpha(n) = \log^{1+\varepsilon}(n)$, even if ε is specified a-priori.

In contrast, an encryption scheme with security for fully-weak keys provides the corresponding advantages. More specifically, the order of quantifiers now requires that there is a *single encryption scheme*, parameterized only by the security parameter n (but *not* by ε), which simultaneously achieves security for all $\alpha(n) \in \omega(\log(n))$. The exact-security of the scheme may depend on $\alpha(n)$ (since there is always a way to break the scheme in time $2^{\alpha(n)}$), but this relationship is now more fluid, with the exact-security gracefully degrading for smaller $\alpha(n)$. In particular, the security guarantees are meaningful even for $\alpha(n) = \log^{1+\varepsilon}(n)$, and there is no single threshold above which the scheme is secure and below which it is insecure. This is a significant advantage, as it does not require one to decide at design time on the tradeoff between allowed entropy levels and achieved security/efficiency.

New construction of encryption with fully-weak keys. We now describe the point-function obfuscation scheme of Canetti [10], and notice that it yields a self-composable MBPF obfuscator with *fully-entropic security* for independent messages. It is based on a strengthened version of the DDH assumption, which we describe shortly. Using this simple observation and our connection between obfuscation and encryption (Lemma 4), we get the first symmetric-key encryption scheme with CPA security for *fully-weak* keys (albeit under a strong assumption). We begin by defining the *strengthened DDH* assumption for a prime-ordered group $\mathbb{G}$.

Definition 8 (Strengthened DDH Assumption [10]). *Let $\mathbb{G}$ be a group of prime order $p = 2^{\text{poly}(n)}$ and let g be a random generator of $\mathbb{G}$. The* strengthened DDH assumption *states that, for any distribution $\{X_n\}$ over $\mathbb{Z}_p$ with entropy $H_\infty(X_n) \geq \omega(\log(n))$, we have $\langle g^a, g^b, g^{ab}\rangle \approx \langle g^a, g^b, g^c\rangle$ where $a \leftarrow_R X_n$, and $b, c \leftarrow_R \mathbb{Z}_p$.*

We now define the function $F : \mathbb{Z}_p \to \mathbb{G} \times \mathbb{G}$ by $F(k) = \langle r, r^k\rangle$ where $r \leftarrow_R \mathbb{G}$. In [10], this was shown to be a secure *point*-function obfuscator (with fully-entropic security) under the strengthened DDH assumption. In addition, this point-function obfuscator is self-composable since, given a (random) obfuscation $\langle g_1, g_2\rangle$ of some point x, it is easy to generate freshly random (and independent) new obfuscation of x by taking $\langle g_1^u, g_2^u\rangle$ for a random $u \in \mathbb{Z}_p$. We use the construction of Canetti and Dakdouk [11] to turn a point-function obfuscator into a multi-bit point-function obfuscator. Define the function:

$$\mathcal{O}(I_{(k,m)}) = \begin{cases} \text{Sample } r_0, r_1, \ldots, r_\ell \leftarrow_R \mathbb{G} \text{ for } \ell = |m|. \\ \text{Set } g_0 = r_0^k \\ \text{For each } i \in \{1, \ldots, \ell\} : \text{ if } m_i = 1 \text{ set } g_i = r_i^k \text{ else } g_i \leftarrow_R \mathbb{G}. \\ \text{Output: } c = (\langle r_0, g_0\rangle, \ldots, \langle r_\ell, g_\ell\rangle). \end{cases}$$

Using the techniques of [11], it is easy to show that $\mathcal{O}$ is a *self-composable* obfuscator with fully-entropic security for *independent messages* under the strengthened DDH assumption. Combining this with Lemma 4, we get the following theorem.

Theorem 5. *Under the strengthened DDH assumption, there exists a CPA-secure symmetric encryption scheme with security against* fully-weak keys. *In particular, this means that there is a* single *scheme, parameterized only by the security parameter n, such that security of the scheme is maintained when the key is chosen from* any *distribution of entropy $\alpha(n) \in \omega(\log(n))$.*

The strengthened DDH assumption is indeed a strong one. A potentially weaker formulation would be to limit the min-entropy of X_n to be at least some specific super-logarithmic function $\alpha(n)$. This way, we would obtain a parameterized version of Theorem 5 that relates the strength of the security guarantee to the strength of the assumption.

It is important to note that the construction itself is independent of the parameter α. That is, we obtain a single encryption scheme that provides a range of security guarantees, depending on the strength of the assumption.

6.2 Obfuscation

Entropically Secure Obfuscation for Independent Messages: It is fairly simple to construct $\alpha(n)$-entropically secure obfuscation for independent messages, when $\alpha(n) = n^{\varepsilon}$ for some constant $\varepsilon \geq 0$. First we construct a semantically secure encryption scheme with $\alpha(n)$-weak keys. This can be done by simply extracting a sufficient amount of uniform randomness from the key k, using a strong randomness extractor Ext, and then using the result as a one time pad to encrypt the message. For variable-length messages, we also need to expand the extracted randomness to an appropriate size, using a pseudo-random generator PRG. In particular, we define

$$\mathsf{Enc}_k(m) = \langle r, \mathsf{PRG}(\mathsf{Ext}(k; r)) \oplus m \rangle$$

where r is a uniformly random seed for the extractor. The output length of Ext and the input length of PRG are set to some value v which is sufficiently small that the outputs of the extractor is (statistically) close to uniform, and sufficiently large that the output of the PRG is pseudo-random.[5]

One can use this encryption scheme to construct one which also has the wrong-key detection property using Lemma 1. Such a scheme yields an multi-bit point function obfuscator with $\alpha(n)$-entropic security for independent messages, by Lemma 3.

Self-Composable *Entropically Secure Obfuscation for Independent Messages:* One problem with the above construction of semantically-secure encryption using extractors, is that it does not generalize to CPA security. In fact, achieving CPA secure encryption with weak keys seems to be a much harder problem, which has received much attention in recent works [1,14,25]. We now show how to use these results to achieve self-composable entropically secure obfuscation for independent messages. On a high level, we would simply like to just apply our result connecting such encryption and obfuscation (Lemma 5) "out of the box". However, there are several issues that we must deal with first.

- *Efficiently-Sampleable Distributions:* The works of [1,14,25] are concerned with "key leakage", where the adversary gets to learn some (short) function of the secret key, whose output length is bounded by λ bits. Conditioned on such leakage, the key can be thought of as being derived from a (special type) of weak source with entropy $\alpha(n) \approx n - \lambda$. It turns out that the constructions are also secure when the key is chosen from an *arbitrary*, but *efficiently-sampleable* weak source of entropy $\alpha(n)$ [25]. Therefore, our results for obfuscation will only translate to the case where the distribution obfuscated program is efficiently sampleable.

[5] For example, if we choose $v = n^{\varepsilon}/2$, then an extractor based on universal-hash functions will produce an output which is $2^{-v/2} = \text{negl}(n)$-close to uniform, and the output of the PRG is $\text{negl}(n^{\varepsilon}/2) = \text{negl}(n)$-pseudorandom. However, this does not generalize to smaller values of α such as, $\alpha(n) = \log^2(n)$.

- *Public Keys/Parameters:* Only the scheme of [14] is explicitly designed for the symmetric key setting. The schemes of [1,25] are public-key encryption schemes. As noted, such schemes are secure when the key-generation procedure uses randomness that comes from a weak source. Therefore such schemes naturally translate to the symmetric key setting, where the randomness of the key-generation algorithm is the shared secret key. Unfortunately, these schemes also rely on *public parameters* which are chosen uniformly at random, and are available to the key-generation algorithm. Therefore, we will only get an obfuscator in the presence of public parameters. Note that in the context of standard obfuscation, public parameters are never needed since the obfuscator $\mathcal{O}$ could always sample fresh parameters each time it runs. However, when considering *composable obfuscation*, this equivalence does not hold since future uses of the obfuscator might compromise security of prior uses. Therefore, having randomness in the form of public parameters, which are re-used for all invocations of the obfuscator, can be useful in this context.
- *Uniform Ciphertexts:* Recall that our definition of CPA security is slightly different than the standard (we require that the ciphertexts of any message are indistinguishable from some universally specified distribution) and has not been explicitly analyzed by these schemes. However, in all of these schemes explicitly show in their proofs that the ciphertexts are indistinguishable from uniform, which satisfies our definition.
- *Wrong-Key Detection:* The wrong-key detection property is explicitly analyzed in [14]. For the schemes of [1,25], we get the property that, given the public parameters it is computationally difficult to find k, k' such that $\mathsf{Dec}_{k'}(\mathsf{Enc}_k(m)) \neq \perp$. This translates to a *computational-correctness* property for the obfuscator where, given the public parameters, it is computationally difficult to find k, m, x such that $\mathcal{O}(I_{(k,m)})(x) \neq I_{(k,m)}(x)$.

Using our connection between CPA-secure symmetric key encryption and self-composable obfuscation with independent messages, we get the following new constructions of obfuscators as a corollary of Lemma 5, using the schemes of [1,14,25].

Theorem 6. *For any constant $\varepsilon > 0$, there exists a* self-composable *MBPF obfuscator with independent messages under* any *of the following assumptions:*

1. *Decisional Diffie-Hellman (DDH) with n^ε-entropic security, based on [25].* $^{(*,\dagger)}$.
2. *Learning With Errors (LWE) with n^ε-entropic security, based on [1].* $^{(*,\dagger)}$.
3. *Learning Subspaces with Noise (LSN) with εn-entropic security, based on [14].* $^{(*)}$.

where the restrictions are:

* *Only works for* efficiently *sampleable key-distributions.*
† *Requires public parameters and only achieves computational-correctness.*

Difficulty of Achieving Obfuscation with Dependent Messages. The connection between encryption and obfuscation also yields new negative results for the more standard notion of obfuscation that allows for *dependent* messages, and in particular for the standard VBB notion. We rely on a recent result of Haitner and Holenstein [17], which shows that there can be *no* black-box reduction from a semantically secure encryption scheme with security against key-dependent messages to, essentially, *any standard cryptographic assumption*. The notion of "cryptographic assumption" is formalized in [17] as (essentially) any game between an attacker and a challenger in which we assume that all PPT attackers have a negligible success probability. In particular, this includes all standard assumptions such as existence of Trapdoor One-Way Permutations or Claw-Free Permutations, as well as specific algebraic assumptions like the hardness of factoring, DDH, Learning with Errors and many others.[6] Since, by Theorem 4, we have a reduction from a semantically secure encryption schemes with security against key-dependent messages to obfuscation of multi-bit point functions with n-entropic security (i.e. even uniformly random keys), we see that this latter notion of obfuscation cannot be realized from essentially any cryptographic assumption under black-box reductions.

Theorem 7. *No construction of an MBPF obfuscator with $\alpha(n)$-entropic security* for dependent messages *can be proven secure via a black-box reduction to* any *"standard cryptographic assumption", even for $\alpha(n) = n$ (i.e. even uniformly random keys).*

We note that Canetti and Dakdouk [11] showed that *composable obfuscation of point functions (with no output)* (i.e. functions $I_k(x)$ which output 1 when $x = k$ and $\perp$ otherwise) implies multi-bit point function obfuscators *with dependent messages*. Thus we get the following as a corollary.

Corollary 1. *No construction of a* composable *obfuscator for* single-value point functions *with $\alpha(n)$-entropic security can be proven secure via a black-box reduction to* any *"standard cryptographic assumption", for any $\alpha()$ (even for $\alpha(n) = n$, namely uniformly random keys).*

We note that the impossibility result of [17] only considers semantically secure encryption with *variable length messages* and does not rule out KDM security when the message size is shorter than the key. Correspondingly, the work of [11] constructs MBPF obfuscators with $\alpha(n)$-entropic security (for some $\alpha(n) \ll n$) and *for dependent messages* in this special case, where the message size is (significantly) smaller than the key size (i.e. functions $I_{(k,m)}$ where $|m| < |k|$). These constructions only relied on *standard cryptographic assumptions* such as collision-resistant hash functions. The above theorem shows that such constructions do not generalize to variable-length messages, where the message size can exceed the key size. Alternatively, in this work we show how to leverage prior results on leakage-resilient cryptography to construct self-composable MBPF obfuscators with $\alpha(n)$-entropic security (for some $\alpha(n) \ll n$), under standard assumptions, in the special case of (variable-length) *independent messages*. It seems that there is little hope in generalizing this approach to the standard notion of obfuscation, which also allows key-dependent messages.

[6] On the other hand, the impossibility result does not exclude proofs of security in the Random Oracle model, reductions to non-standard assumptions (which cannot be formulated as a game between an adversary and a challenger) such as "Knowledge of Exponent", or non-black-box reductions.

References

1. Akavia, A., Goldwasser, S., Vaikuntanathan, V.: Simultaneous hardcore bits and cryptography against memory attacks. In: Reingold, O. (ed.) TCC 2009. LNCS, vol. 5444, pp. 474–495. Springer, Heidelberg (2009)
2. Alwen, J., Dodis, Y., Wichs, D.: Leakage-resilient public-key cryptography in the bounded-retrieval model. In: Halevi, S. (ed.) CRYPTO 2009. LNCS, vol. 5677, pp. 36–54. Springer, Heidelberg (2009)
3. Applebaum, B., Cash, D., Peikert, C., Sahai, A.: Fast cryptographic primitives and circular-secure encryption based on hard learning problems. In: Halevi, S. (ed.) CRYPTO 2009. LNCS, vol. 5677, pp. 595–618. Springer, Heidelberg (2009)
4. Backes, M., Dürmuth, M., Unruh, D.: Oaep is secure under key-dependent messages. In: Pieprzyk, J. (ed.) ASIACRYPT 2008. LNCS, vol. 5350, pp. 506–523. Springer, Heidelberg (2008)
5. Backes, M., Pfitzmann, B., Scedrov, A.: Key-dependent message security under active attacks - brsim/uc-soundness of symbolic encryption with key cycles. In: CSF, pp. 112–124 (2007)
6. Barak, B., Goldreich, O., Impagliazzo, R., Rudich, S., Sahai, A., Vadhan, S., Yang, K.: On the (Im)possibility of obfuscating programs. In: Kilian, J. (ed.) CRYPTO 2001. LNCS, vol. 2139, pp. 1–18. Springer, Heidelberg (2001)
7. Black, J., Rogaway, P., Shrimpton, T.: Encryption-scheme security in the presence of key-dependent messages. In: Nyberg, K., Heys, H.M. (eds.) SAC 2002. LNCS, vol. 2595, pp. 62–75. Springer, Heidelberg (2003)
8. Boneh, D., Halevi, S., Hamburg, M., Ostrovsky, R.: Circular-secure encryption from decision diffie-hellman. In: Wagner, D. (ed.) CRYPTO 2008. LNCS, vol. 5157, pp. 108–125. Springer, Heidelberg (2008)
9. Camenisch, J., Chandran, N., Shoup, V.: A public key encryption scheme secure against key dependent chosen plaintext and adaptive chosen ciphertext attacks. In: Joux, A. (ed.) EUROCRYPT 2009. LNCS, vol. 5479, pp. 351–368. Springer, Heidelberg (2009)
10. Canetti, R.: Towards realizing random oracles: Hash functions that hide all partial information. In: Kaliski Jr., B.S. (ed.) CRYPTO 1997. LNCS, vol. 1294, pp. 455–469. Springer, Heidelberg (1997)
11. Canetti, R., Dakdouk, R.R.: Obfuscating point functions with multibit output. In: Smart, N.P. (ed.) EUROCRYPT 2008. LNCS, vol. 4965, pp. 489–508. Springer, Heidelberg (2008)
12. Canetti, R., Kalai, Y.T., Varia, M., Wichs, D.: On symmetric encryption and point obfuscation (2010); Full Version Cryptology ePrint Archive
13. Canetti, R., Micciancio, D., Reingold, O.: Perfectly one-way probabilistic hash functions. In: Proceedings of the 30th ACM Symposium on Theory of Computing, pp. 131–140 (1998)
14. Dodis, Y., Kalai, Y.T., Lovett, S.: On cryptography with auxiliary input. In: STOC, pp. 621–630 (2009)
15. Dziembowski, S., Pietrzak, K.: Leakage-resilient cryptography. In: FOCS, pp. 293–302 (2008)
16. Goldwasser, S., Kalai, Y.T.: On the impossibility of obfuscation with auxiliary input. In: FOCS, pp. 553–562 (2005)
17. Haitner, I., Holenstein, T.: On the (im)possibility of key dependent encryption. In: Reingold, O. (ed.) TCC 2009. LNCS, vol. 5444, pp. 202–219. Springer, Heidelberg (2009)
18. Halderman, J.A., Schoen, S.D., Heninger, N., Clarkson, W., Paul, W., Calandrino, J.A., Feldman, A.J., Appelbaum, J., Felten, E.W.: Lest we remember: cold-boot attacks on encryption keys. Commun. ACM 52(5), 91–98 (2009)
19. Halevi, S., Krawczyk, H.: Security under key-dependent inputs. In: ACM Conference on Computer and Communications Security, pp. 466–475 (2007)

20. Hofheinz, D., Unruh, D.: Towards key-dependent message security in the standard model. In: Smart, N.P. (ed.) EUROCRYPT 2008. LNCS, vol. 4965, pp. 108–126. Springer, Heidelberg (2008)
21. Katz, J., Vaikuntanathan, V.: Signature schemes with bounded leakage resilience. In: Matsui, M. (ed.) ASIACRYPT 2009. LNCS, vol. 5912, pp. 703–720. Springer, Heidelberg (2009), http://www.mit.edu/~vinodv/papers/asiacrypt09/KV-Sigs.pdf
22. Lynn, B., Prabhakaran, M., Sahai, A.: Positive results and techniques for obfuscation. In: Cachin, C., Camenisch, J.L. (eds.) EUROCRYPT 2004. LNCS, vol. 3027, pp. 20–39. Springer, Heidelberg (2004)
23. Micali, S., Reyzin, L.: Physically observable cryptography (extended abstract). In: Naor, M. (ed.) TCC 2004. LNCS, vol. 2951, pp. 278–296. Springer, Heidelberg (2004)
24. Micciancio, D., Warinschi, B.: Completeness theorems for the abadi-rogaway language of encrypted expressions. Journal of Computer Security 12(1), 99–130 (2004)
25. Naor, M., Segev, G.: Public-key cryptosystems resilient to key leakage. In: Halevi, S. (ed.) CRYPTO 2009. LNCS, vol. 5677, pp. 18–35. Springer, Heidelberg (2009)
26. Wee, H.: On obfuscating point functions. In: Proceedings of the 37th ACM Symposium on Theory of Computing, pp. 523–532 (2005)

Obfuscation of Hyperplane Membership

Ran Canetti[1,*], Guy N. Rothblum[2,**], and Mayank Varia[3,***]

[1] School of Computer Science, Tel Aviv University
canetti@cs.tau.ac.il
[2] Princeton University
rothblum@princeton.edu
[3] Massachusetts Institute of Technology
varia@csail.mit.edu

Abstract. Previous work on program obfuscation gives strong negative results for general-purpose obfuscators, and positive results for obfuscating simple functions such as equality testing (point functions). In this work, we construct an obfuscator for a more complex algebraic functionality: testing for membership in a hyperplane (of constant dimension). We prove the security of the obfuscator under a new strong variant of the Decisional Diffie-Hellman assumption. Finally, we show a cryptographic application of the new obfuscator to digital signatures.

1 Introduction

The problem of program obfuscation has been of long-standing interest to practitioners, and has recently been an active topic of research in theoretical cryptography. The high-level goal of program obfuscation is to compile a computer program in such a way that an adversary cannot learn anything from seeing the program beyond could be learned by running the program and observing its input-output behavior.

Barak *et al.* [1] formalized the notion of obfuscation using simulation-based definitions. Over the past decade, the theory community has found a few positive obfuscation results for specific families of programs. In this paper, we provide an obfuscator for a new family of programs.

Virtual black-box obfuscation. The procedure of "obfuscating" a computer program should garble the program's code and make it unintelligible. The extent of the garbling is limited by the fact that the program's functionality should be preserved. As a result, both honest and adversarial users of the obfuscated program can learn some information by observing the program's input-output functionality, and we do not wish to prevent users from learning information this

* Supported by the Check Point Institute for Information Security, an ISF grant, an EU Marie Curie grant, and an Israel-US BSF grant.

** Supported by NSF Grants CCF-0635297, CCF-0832797 and by a Computing Innovation Fellowship.

*** Supported by the Department of Defense through the NDSEG Program.

D. Micciancio (Ed.): TCC 2010, LNCS 5978, pp. 72–89, 2010.

way. Instead, obfuscation ensures that this is the only way that an adversary can learn information from the obfuscated program.

There are several ways to formalize this intuitive notion [2,3,4]. This paper uses the virtual black-box formalization of [1]. A significant obstacle to obtaining positive results with respect to this definition is that the security notion must hold for *all* programs in a given family, and not just a random instance. This is one reason why standard cryptographic tools and analytical techniques (which usually deal with randomly chosen instances) are not always helpful for obfuscation.

Previous results. Several works have disproved the existence of a "general-purpose obfuscator" that can simultaneously obfuscate every program [1,2,5]. In fact, these papers demonstrate specific programs that cannot be obfuscated, and these programs come from a relatively low complexity class. While these negative results are disheartening, they focus on specific (often contrived) functionalities. Obfuscation remains possible for many programs of interest.

Still, very few positive results are known even for specific, simple programs (or boolean circuits). One family of programs for which positive results are known is the family of "point circuits": password-checking programs that accept a single input string and reject all other inputs. This family can be obfuscated under a variety of cryptographic assumptions [6,7,8,9]. Some of these constructions can be generalized in two ways. First, we can obfuscate "multi-point circuits," which accept a polynomially-sized list of input strings, and second, we can obfuscate "point circuits with multi-bit output," which store a hidden output string that is revealed only for a single input value [10]. Other formalizations of program obfuscation [3,4] allow for the obfuscation of cryptographic tasks such as checking proximity to a hidden point [11], vote mixing [12], and re-encryption [4]. The latter two applications use a different security guarantee that only holds over a random choice of the circuit from the family.

Our result. In this paper, we obfuscate programs that perform hyperplane membership testing. Let P be a hyperplane in a vector space, and let H_P be a program that tests whether its input is a point on the hyperplane. An obfuscation of H_P allows a user to determine whether her input point is on the hyperplane, but reveals no additional information such as the distance from her point to the hyperplane or any other points that are on the hyperplane.

More precisely, given a prime p and positive integer d, consider the family of hyperplanes through the origin in the vector space $(\frac{\mathbb{Z}}{p\mathbb{Z}})^d$ over the finite field $\frac{\mathbb{Z}}{p\mathbb{Z}}$. In this setting, a hyperplane can be defined by a vector that is orthogonal to every point in the plane. Let $\boldsymbol{a}$ be a vector in $(\frac{\mathbb{Z}}{p\mathbb{Z}})^d$ and consider the program

$$H_{\boldsymbol{a}}(\boldsymbol{x}) = \begin{cases} 1 & \text{if } \langle \boldsymbol{a}, \boldsymbol{x} \rangle = 0 \\ 0 & \text{otherwise.} \end{cases}$$

We construct an obfuscator for this family of programs.

This primitive subsumes many of the previously-known results. In the $d = 2$ case, these "hyperplanes" turn out to be equivalent to point circuits, and our specific construction and assumption reduce to those in [6]. Furthermore, the technique of [10] can be applied to our primitive as well, so we can obfuscate circuits that output a hidden message when its input is on the hyperplane.

We also note that hyperplane membership testing has been considered in the context of private predicate encryption schemes by Shen, Shi, and Waters [13], although our results are incomparable to theirs.

Application to digital signatures. As an example of the proposed primitive's usefulness, we demonstrate an application of our obfuscator to leakage-resilient one-time signatures. We emphasize, though, that the main motivation for this work is the new obfuscator, rather than any single application.

The signature scheme is constructed as follows: the secret key is a randomly chosen plane in 3-dimensional space, and the public key is the obfuscated membership program. To sign a message m, find a point on the plane that is related to m. This signature can be verified by running the public obfuscated program.

This signature scheme satisfies a weaker form of the unforgeability game, where the adversary is required to submit a message m to be signed before receiving the public key. Techniques from [14] allows us to transform the weak scheme into an ordinary one-time signature scheme. Additionally, this one-time signature scheme remains unforgeable even when a function of the secret key is leaked whose output length is up to half as long as the secret key. For schemes that do not use general zero-knowledge proofs, this matches the leakage bound of [15] (albeit under much stronger assumptions).

Construction. The construction is as follows. Let G be a group of order p that satisfies a strengthened version of the Decisional Diffie-Hellman assumption, which we describe in more detail below. When the obfuscator is given a hyperplane $H_{\boldsymbol{a}}$ to obfuscate, where $\boldsymbol{a} = (a_1, \ldots, a_d)$, it chooses a random generator $g \overset{U}{\leftarrow} G$ and outputs g^{a_i} for all i. This allows the user to compute whether a given point $\boldsymbol{x} = (x_1, \ldots, x_d)$ is on the hyperplane by computing

$$(g^{a_1})^{x_1} \times \cdots \times (g^{a_d})^{x_d} = g^{\langle \boldsymbol{a}, \boldsymbol{x} \rangle} ,$$

and checking whether this equals G's identity element (i.e. whether $\langle \boldsymbol{a}, \boldsymbol{x} \rangle = 0$).

Our assumption. We are not able to prove the security of our construction based on the standard Decisional Diffie-Hellman assumption, which states that g^{ab} is indistinguishable from uniform, given g, g^a, and g^b for uniformly-chosen exponents a and b. We describe the difficulty with basing our scheme on DDH, as it motivates and clarifies our new assumption.

For our construction, it is crucial that the adversary not be able to compute any polynomial relationships in the exponent (not just quadratic ones). Consider, for instance, whether it is possible to compute g^{a^3} given just g and g^a. What if we wish to compute g^{abc} from g, g^a, g^b, and g^c? Can elements of the form g^{a^3} or g^{abc} even be distinguished from uniform? No efficient algorithms for running these computations are known (e.g. in groups where DDH is hard), but standard assumptions such as DDH do not seem to rule out the existence of such algorithms. In general, we wish to understand when $g^{p(a,b)}$ is distinguishable from uniform, given a polynomial p and group elements g, g^a, and g^b. Clearly, this is true when p is linear, or closely resembles a line. Our new assumption states that these are the only such polynomials for which $g^{p(a,b)}$ can be distinguished from uniform.

We also consider the effect of choosing exponents from weak entropy distributions. This setting has been previously considered by Canetti [6], who forms a modified DDH assumption in which g^{ab} is considered to be indistinguishable from uniform, even given g, g^a, and g^b, where a is chosen from the uniform distribution but b is chosen from any distribution of super-logarithmic min-entropy. Our assumption expands upon this idea and considers many exponents that are not only chosen from weak entropy distributions, but which may also be related.

Specifically, given a tuple of group elements $\langle g^{a_1}, g^{a_2}, \ldots, g^{a_d} \rangle$ where the a_i's are chosen from some joint distribution, we ask for which polynomials p is $g^{p(a_1,\ldots,a_d)}$ still indistinguishable from uniform? If the polynomial p looks linear when restricted to the support of the joint distribution, then of course $g^{p(a_1,\ldots,a_d)}$ can be distinguished from uniform. Our new assumption states that indistinguishability holds in all other cases.

This new assumption is stronger than the standard DDH assumption, or even the modified DDH assumption of [6], but we provide evidence of its feasibility by proving that it holds in the generic group model. Furthermore, we believe that resolving the status of this new assumption would be interesting either way. If it holds, then we obtain an obfuscator for a new and interesting family of functions. Showing that the assumption does not hold would shed new light on which computations can be run efficiently in the exponent of DDH groups.

Organization. Section 2 defines virtual black-box obfuscation and the hyperplane membership testing programs. Section 3 describes our assumption in detail and compares it to previous assumptions. Section 4 presents our obfuscator for the family of hyperplanes and proves its security. Section 5 extends our construction to the multi-bit setting. Section 6 presents our one-time signature scheme. Some of the proofs are relegated to the full version of this paper [16].

2 Definitions

2.1 Virtual Black-Box Obfuscation

In [1], [5], and other works, an obfuscator is defined as a compiler that takes a circuit as input and returns another circuit. The output circuit should be a "garbled" version of the input circuit, in the sense that the circuits should have the same functionality, but it should be difficult for an adversary to learn information from the output circuit.

Consider an imaginary world in which people can give others access to oracles at will. In this imaginary world, we can easily perform perfect obfuscation by giving users oracle access to a computer program. The oracle allows them to learn the program's input-output functionality, but any other aspect of the program's behavior is hidden from the users. Unfortunately, in the real world we cannot hand out oracles to other people. Instead, we want obfuscators to be able to replicate the power of oracles in the imaginary world. The formalization of obfuscation provided by Barak *et al.* [1], called the *virtual black-box* property, achieves this goal.

The virtual black-box property considers two different worlds. In the real world, an efficient adversary has access to the obfuscated program code and attempts to learn a one bit predicate about the underlying program. The definition ensures that there exists a simulator in imaginary world that only interacts with an oracle to the program but can still can learn the same predicate that the adversary learns in the real world. Hence, the virtual black-box property ensures that access to the code of an obfuscated program is *no more* useful than access to the oracle.

We only require that the obfuscator operate over a specified family of circuits. Throughout this work, all circuits are assumed to be non-uniform.

Definition 1 (Obfuscation). *Let $\mathcal{C} = \{\mathcal{C}_n\}_{n \in \mathbb{N}}$ be a family of polynomial-size circuits, where $\mathcal{C}_n$ denotes all circuits of input length n. A probabilistic polynomial time (PPT) algorithm $\mathcal{O}$ is an* obfuscator *for the family $\mathcal{C}$ if the following three conditions are met.*

1. Approximate functionality: *There exists a negligible function ε such that for every n, every circuit $C \in \mathcal{C}_n$ and every x in the input space of C,*
$$\Pr[\mathcal{O}(C)(x) = C(x)] > 1 - \varepsilon(n),$$
where the probability is over the randomness of $\mathcal{O}$. If this probability always equals 1, then we say that $\mathcal{O}$ has exact functionality.
2. Polynomial slowdown: *There exists a polynomial q such that for every n, every circuit $C \in \mathcal{C}_n$, and every possible sequence of coin tosses for $\mathcal{O}$, the circuit $\mathcal{O}(C)$ runs in time at most $q(|C|)$.*
3. Virtual black-box: *For every PPT adversary A and polynomial δ, there exists a PPT simulator S such that for all sufficiently large n, and for all $C \in \mathcal{C}_n$,*
$$|\Pr[A(\mathcal{O}(C)) = 1] - \Pr[S^C(1^n) = 1]| < \frac{1}{\delta(n)},$$
where the first probability is taken over the coin tosses of A and $\mathcal{O}$, and the second probability is taken over the coin tosses of S.

2.2 Vector Spaces

In this section, we define the vector spaces over which our constructions operate. Let $d \in \mathbb{N}$, p be a prime number, and $\mathbb{F}_p = \frac{\mathbb{Z}}{p\mathbb{Z}}$. Then, $\mathbb{F}_p$ is a field and $\mathbb{F}_p^d$ is a vector space over $\mathbb{F}_p$. We denote a vector in the vector space by $\boldsymbol{x} = (x_1, \ldots, x_d)$, and we have an inner product-style operation given by $\langle \boldsymbol{x}, \boldsymbol{y} \rangle = \sum_{i=1}^d x_i y_i$.

Definition 2. *Let $S \subseteq \mathbb{F}_p^d$ be a set.*

1. *Two vectors $\boldsymbol{x}, \boldsymbol{y} \in \mathbb{F}_p^d$ are said to be* orthogonal *if their inner product is zero, so $\langle \boldsymbol{x}, \boldsymbol{y} \rangle = 0$. Note that the set of all vectors orthogonal to $\boldsymbol{x}$ forms a $(d-1)$ dimensional hyperplane.*
2. *The* closure *of S, written $\bar{S}$, is the subspace of all linear combinations of vectors in S.*
3. *The* orthogonal complement *of S, written $S^\perp$, is the subspace of all vectors that are orthogonal to every vector in S. That is,*

$$S^\perp = \{\boldsymbol{x} \in \mathbb{F}_p^d : \langle \boldsymbol{x}, \boldsymbol{s} \rangle = 0 \, \forall \boldsymbol{s} \in S\} .$$

We caution that $\mathbb{F}_p^d$ does not satisfy all of the axioms of an inner product space. Nevertheless, the following theorem about inner product spaces, which we need in the proof of our main theorem, does hold over $\mathbb{F}_p^d$.

Theorem 3. *Let $S \subseteq \mathbb{F}_p^d$ be a set. Then, $(S^\perp)^\perp = \bar{S}$.*

Proof (sketch). First, $S^\perp$ and $S^{\perp\perp}$ are subspaces of $\mathbb{F}_p^d$ because the conditions imposed on them are linear. Second, $\bar{S} \subseteq S^{\perp\perp}$ because the vectors in $\bar{S}$ are orthogonal to those in $S^\perp$, so they are in $S^{\perp\perp}$. Third, $\dim(\bar{S}) = \dim(S^{\perp\perp})$ because both of them are equal to $d - \dim(S^\perp)$.

Therefore, $\bar{S}$ and $S^{\perp\perp}$ are subspaces of $\mathbb{F}_p^d$ of the same dimension such that one is included in the other, so they are equal. □

We also note that the vector space $\mathbb{F}_p^d$ is a bit redundant for our needs. We wish to identify a hyperplane P with a vector $\boldsymbol{x}$ that is orthogonal to every vector in the hyperplane. However, the vector $\boldsymbol{x}$ is not unique: indeed, for any $c \in \mathbb{F}_p \setminus 0$, the vector $c\boldsymbol{x}$ is also orthogonal to every vector in P, so the normal vector to the hyperplane is only unique up to scalar multiplication. As a result, we note that there are only $d-1$ degrees of freedom when choosing a normal vector, which is why the $d = 2$ case corresponds to point functions. One canonical representation of the normal vector, which we will use when convenient throughout the paper, is to consider all of the vectors in $\mathbb{F}_p^d$ whose first non-zero coordinate equals 1.[1]

3 Our Assumption

In this section, we define the main assumption. Then, we relate our assumption to a DDH variant found in [6] and consider the assumption in the generic group model.

Our assumption uses groups of increasing prime order. We use the following definition to encapsulate the order requirement.

[1] In fact, the appropriate ambient space from which to consider the normal vectors is the projective space $\mathbb{P}^{d-1}(\mathbb{F}_p)$, and we are using its embedding in $\mathbb{F}_p^d$ as a concrete instantiation of the projective space.

Definition 4. *A function $\rho(n)$ is called a* prime sequence *if for every $n \in \mathbb{N}$, $\rho(n)$ is a prime number in the range $(2^{n-1}, 2^n]$.*

Our assumption is parametrized by $d \in \mathbb{N}$. We abuse notation a bit and denote $\mathbb{F}_\rho^d = \{\mathbb{F}_{\rho(n)}^d\}_{n \in \mathbb{N}}$.

Assumption 5. *Given $d \in \mathbb{N}$, there exists a family of groups $\mathcal{G} = \{G_n\}_{n \in \mathbb{N}}$ (written multiplicatively) such that the following three conditions hold:*

1. *There are efficient algorithms to perform the group operation, to test for equality with the identity element, and to sample uniformly from $\mathcal{G}$.*
2. *The orders of the groups form a prime sequence $\rho(n) = |G_n|$.*
3. *For every PPT adversary A and for all families of distributions $\mathcal{L} = \{\mathcal{L}_n\}_{n \in \mathbb{N}}$ and $\mathcal{R} = \{\mathcal{R}_n\}_{n \in \mathbb{N}}$ over $\mathbb{F}_\rho^d$, there exists a polynomial q such that for all n,*

$$\begin{aligned} &|\Pr[\boldsymbol{l} \leftarrow \mathcal{L}_n, g \stackrel{U}{\leftarrow} G_n : A(g^{l_1}, \ldots, g^{l_d}) = 1] \\ &\qquad - \Pr[\boldsymbol{r} \leftarrow \mathcal{R}_n, g \stackrel{U}{\leftarrow} G_n : A(g^{r_1}, \ldots, g^{r_d}) = 1]| \\ &\quad \leq q(n) \cdot \max_{\boldsymbol{x} \in \mathbb{F}_{\rho(n)}^d} |\Pr[\boldsymbol{l} \leftarrow \mathcal{L}_n : \langle \boldsymbol{l}, \boldsymbol{x} \rangle = 0] - \Pr[\boldsymbol{r} \leftarrow \mathcal{R}_n : \langle \boldsymbol{r}, \boldsymbol{x} \rangle = 0]| . \end{aligned} \tag{1}$$

In words, this assumption states that an adversary can distinguish two distributions of vectors if and only if linear tests can do so as well.

3.1 Discussion

We make several remarks:

1. The right-hand side of (1) depends on ρ but not on any other property of $\mathcal{G}$.
2. Note that the adversary is allowed to distinguish $\mathcal{L}$ and $\mathcal{R}$ better than any single linear test does. For example, the adversary may try many linear tests. The assumption merely states that the left-hand side of (1) is negligible whenever the right-hand side is.
3. For a given adversary A, we denote $A_{\boldsymbol{l}} = \Pr[g \stackrel{U}{\leftarrow} G_n : A(g^{l_1}, \ldots, g^{l_d}) = 1]$ and $A_{\mathcal{L}} = \Pr[\boldsymbol{l} \leftarrow \mathcal{L}_n : A_{\boldsymbol{l}}]$ for simplicity. We will say that $\mathcal{L}$ and $\mathcal{R}$ are *indistinguishable by linear tests* if

$$\varepsilon(n) = \max_{\boldsymbol{x} \in \mathbb{F}_{\rho(n)}^d} |\Pr[\boldsymbol{l} \leftarrow \mathcal{L} : \langle \boldsymbol{l}, \boldsymbol{x} \rangle = 0] - \Pr[\boldsymbol{r} \leftarrow \mathcal{R} : \langle \boldsymbol{r}, \boldsymbol{x} \rangle = 0]|$$

   is a negligible function of n. Thus, the assumption states that for all $\mathcal{L}$ and $\mathcal{R}$ that are indistinguishable by linear tests, $|A_{\mathcal{L}} - A_{\mathcal{R}}|$ is negligible as well for all PPT adversaries A.
4. This assumption is computationally falsifiable, though perhaps inefficiently. There are two possible obstructions to efficiency. First, the descriptions of $\mathcal{L}$ and $\mathcal{R}$ may be inefficient, although this is not a problem for the distributions constructed in our proof. Second, it may not be efficient to determine which linear test performs the best. An interesting question is whether this computation can be performed efficiently, leading to an efficient falsification procedure.

3.2 On the Assumption's Hardness

In this section, we categorize the hardness of our assumption. To begin with, we present a DDH-based assumption due to Canetti [6].

Assumption 6. *Let n be a security parameter and let $p = 2q+1$ be a randomly chosen n-bit safe prime. Consider the group Q of squares in $\mathbb{F}_p^*$. For any well-spread distribution ensemble $\{X_q\}$ where the domain of X_q is $\mathbb{F}_q$, for $g \overset{U}{\leftarrow} Q$, $a \leftarrow X_q$, $b, c \overset{U}{\leftarrow} \mathbb{F}_q$, the ensembles $\langle g, g^a, g^b, g^{ab} \rangle$ and $\langle g, g^a, g^b, g^c \rangle$ are computationally indistinguishable.*

In this assumption, a "well-spread ensemble" means that the min-entropy $H_\infty(X_q)$ is a super-logarithmic function of n.

The following theorem exemplifies the strength of our assumption by relating it to Assumption 6, which is already considered by the cryptographic community to be quite strong.

Theorem 7. *Assumption 6 implies Assumption 5 for dimension 2. For higher dimensions, our assumption may be stronger because Assumption 5 for dimension $d+1$ implies Assumption 5 for dimension d.*

On the other hand, we provide evidence that our assumption is feasible by showing that it holds in the generic group model.

Theorem 8. *For all $d \in \mathbb{N}$, Assumption 5 for dimension d holds in the generic group model.*

Finally, we note that Assumption 5 for dimension 1 trivially holds, even by groups that do not satisfy DDH. A precise definition of the generic group model and a proof of Theorem 8 can be found in [16]. The rest of this section is devoted to a proof of Theorem 7, which we break into the following two lemmas.

Lemma 9. *Assumption 5 for dimension $d+1$ implies Assumption 5 for dimension d.*

Lemma 10. *Assumption 6 implies Assumption 5 for dimension 2.*

Proof (Lemma 9). Assume that Assumption 5 for dimension d is false, so for every prime sequence ρ and every set of groups $\mathcal{G} = \{G_n\}_{n \in \mathbb{N}}$, there exists a PPT adversary A and two distributions $\mathcal{L}, \mathcal{R}$ over vectors in $\mathbb{F}_\rho^d$ that are indistinguishable by linear tests but such that $|A_\mathcal{L} - A_\mathcal{R}|$ is noticeable.

Now construct distributions $\mathcal{L}'$ and $\mathcal{R}'$ over vectors in $\mathbb{F}_\rho^{d+1}$ that sample $\mathcal{L}$ and $\mathcal{R}$, respectively, to obtain the first d components of the vector, and then sample the final component uniformly over $\mathbb{F}_\rho$. We claim that linear tests do not distinguish $\mathcal{L}'$ from $\mathcal{R}'$.

Any linear test $\boldsymbol{x}_n' \in \mathbb{F}_{\rho(n)}^{d+1}$ that has a non-zero final component will not distinguish $\mathcal{L}_n'$ from $\mathcal{R}_n'$ because the final component of these two distributions is uniform, so the inner product will have the uniform distribution in both cases as well. Furthermore, if there exists a sequence of linear tests $\{\boldsymbol{x}_n'\} \in \mathbb{F}_\rho^{d+1}$ that have zero for the final component and distinguish $\mathcal{L}'$ from $\mathcal{R}'$, then the sequence $\{\boldsymbol{x}_n\} \in \mathbb{F}_\rho^d$ formed by deleting the final component from $\boldsymbol{x}'_n$ distinguishes $\mathcal{L}$ and $\mathcal{R}$, contradicting our assumption that $\mathcal{L}$ and $\mathcal{R}$ are indistinguishable by linear tests.

Finally, let A' be the adversary that drops its final component and feeds the rest to A. It is clear that $A'_{\mathcal{L}'} = A_{\mathcal{L}}$ and $A'_{\mathcal{R}'} = A_{\mathcal{R}}$, so $|A'_{\mathcal{L}'} - A'_{\mathcal{R}'}|$ is noticeable. Therefore, Assumption 5 for dimension $d + 1$ is false as well. □

Proof (Lemma 10). Suppose that Assumption 6 holds. For every n, the assumption holds for a randomly chosen safe prime p, and thus for every n there exists some safe prime $p_n = 2q_n + 1$ for which it holds. Let G_n be the subgroup of quadratic residues in $\mathbb{F}^*_{p_n}$, and let $\mathcal{G} = \{G_n\}_{n\in\mathbb{N}}$. We claim that Assumption 5 for dimension 2 holds for the family $\mathcal{G}$ and prime sequence $\rho(n) = q_n$.

It is clear that the first two properties of Assumption 5 for dimension 2 hold. Also, using our convention that the first non-zero coordinate of a vector is fixed to be 1, we may assume without loss of generality that every vector in $\mathbb{F}_\rho^2$ has the form $(1, x)$ for $x \in \mathbb{F}_{q_n}$ except for the vector $(0, 1)$, which is easy to test for. Thus, a "vector" is really just a group element. Furthermore, a "linear test" is just an equality check because the vector $(y, -1)$ has an inner product of zero with the vector $(1, x)$ if and only if $y = x$.

Hence, it remains to prove the following: for every PPT adversary A and for all families of distributions $\mathcal{L}$ and $\mathcal{R}$ over $\{\mathbb{F}_{q_n}\}_{n\in\mathbb{N}}$ such that

$$\max_{x\in\mathbb{F}_{q_n}} |\Pr[l \leftarrow \mathcal{L} : l = x] - \Pr[r \leftarrow \mathcal{R} : r = x]|$$

is negligible, the quantity

$$|\Pr[l \leftarrow \mathcal{L}_n, g \overset{U}{\leftarrow} G_n : A(g, g^l) = 1] - \Pr[r \leftarrow \mathcal{R}_n, g \overset{U}{\leftarrow} G_n : A(g, g^r) = 1]|$$

is negligible as well.

First, we prove that the statement holds when $\mathcal{L}$ is well-spread and $\mathcal{R}$ is the uniform distribution. Hence, we wish to show that for all PPT A,

$$|\Pr[l \leftarrow \mathcal{L}_n, g \overset{U}{\leftarrow} G_n : A(g, g^l) = 1] - \Pr[r \overset{U}{\leftarrow} \mathbb{F}_{q_n}, g \overset{U}{\leftarrow} G_n : A(g, g^r) = 1]|$$

is negligible. The proof of this statement closely follows the proofs in [6], so we only sketch the details here. If this statement is not true, then the probability $P_x = A_{(1,x)} = \Pr[A(g, g^x) = 1]$ is noticeably different from the mean value $\bar{P} = A_{\mathcal{R}}$ for super-polynomially many values of x. Without loss of generality, there exist super-polynomially many values a for which P_a is noticeably larger than $\bar{P}$. Let X_{q_n} be the uniform distribution over all such a. Then, the ensembles $\langle g, g^a, g^b, g^c\rangle$ and $\langle g, g^a, g^b, g^{ab}\rangle$ are distinguishable when $a \leftarrow X_{q_n}$ by running A on the final two components of the ensemble. In the first case, A outputs 1 with probability $\bar{P}$, and in the second case, A outputs 1 with noticeably higher probability. This contradicts Assumption 6.

Next, we note that the statement immediately extends to the setting where both $\mathcal{L}$ and $\mathcal{R}$ are well-spread by a simple hybrid argument.

Finally, we consider arbitrarily distributions $\mathcal{L}$ and $\mathcal{R}$ such that

$$\max_{x \in \mathbb{F}_{q_n}} |\Pr[\boldsymbol{l} \leftarrow \mathcal{L} : l = x] - \Pr[r \leftarrow \mathcal{R} : r = x]|$$

is negligible. In words, this equation means that for every x that occurs with noticeable probability in $\mathcal{L}$, it occurs with the same probability in $\mathcal{R}$ as well up to a negligible difference. Thus, the distributions $\mathcal{L}$ and $\mathcal{R}$ can only differ on outcomes that occur with negligible probability. Therefore, it suffices to consider $\mathcal{L}$ and $\mathcal{R}$ that are well-spread, and in this case we showed that for every PPT A,

$$|\Pr[l \leftarrow \mathcal{L}_n, g \overset{U}{\leftarrow} G_n : A(g, g^l) = 1] - \Pr[r \leftarrow \mathcal{R}_n, g \overset{U}{\leftarrow} G_n : A(g, g^r) = 1]|$$

is negligible, so Assumption 5 for dimension 2 holds as desired. □

We note that a literal converse to this lemma does not quite make sense because Assumption 6 is specific to the group of quadratic residues modulo $\mathbb{F}_p^*$ for a safe prime p, whereas Assumption 5 makes the more general claim that there exists some family of groups that satisfy a certain condition (potentially quite different from the groups used in Assumption 6).

4 Construction

In this section, we define the family of programs that we obfuscate, present the obfuscator, and prove its security under Assumption 5.

Let d be an integer and ρ be a prime sequence. Given a vector $\boldsymbol{a} \in \mathbb{F}_{\rho(n)}^d$, let $H_{\boldsymbol{a}}$ be the circuit that has $\boldsymbol{a}$ hardwired, and on input $\boldsymbol{x} \in \mathbb{F}_{\rho(n)}^d$, computes $\langle \boldsymbol{a}, \boldsymbol{x} \rangle$ in the obvious way and accepts if and only if the inner product equals 0. Let $\mathcal{F}_\rho^d = \{H_{\boldsymbol{a}} : n \in \mathbb{N}, \boldsymbol{a} \in \mathbb{F}_{\rho(n)}^d\}$ be the family of all such circuits.

We show how to obfuscate the family $\mathcal{F}_\rho^d$ for any $d \in \mathbb{N}$, prime sequence ρ, and set of groups $\mathcal{G}$ (written multiplicatively) that satisfy Assumption 5 for dimension d. The obfuscator $\mathcal{O}_{\mathcal{G},d}$ operates as follows.

Algorithm 1. Obfuscator $\mathcal{O}_{\mathcal{G},d}$ for the family of hyperplanes $\mathcal{F}_\rho^d$

Input: vector $\boldsymbol{a} = (a_1, \ldots, a_d)$ in $\mathbb{F}_{\rho(n)}^d$

1: choose a generator $g \overset{U}{\leftarrow} G_n \setminus \{1_{G_n}\}$ uniformly at random
2: compute $g_i \leftarrow g^{a_i}$ for $i = 1, \ldots, d$

Output: circuit that has $g_1, \ldots, g_d$ hardwired, and on input a vector $\boldsymbol{x}$, accepts if and only if $\prod_{i=1}^d g_i^{x_i} = 1_{G_n}$

We stress that the generator g is not made public in addition to the g_i. However, recall that the vector $\boldsymbol{a}$ is only defined uniquely up to scalar multiplication, and that one way to enforce this requirement is to assume that the first non-zero coordinate of $\boldsymbol{a}$ equals 1. With this convention, the generator g is revealed.

This convention makes it clear that in the $d = 2$ case, this construction is the same as the one in [6], and it can be based off the same DDH assumption by Theorem 7. Hence, our construction subsumes the one in [6].

We note that in the work of Shen, Shi and Waters [13] on *private* inner-product predicate encryption schemes, their construction also tests whether an inner product is 0 by running it in the exponent of a group where CDH is hard. Otherwise the settings, constructions, and assumptions are quite different. In particular, a user who wants to check whether a vector $\boldsymbol{x}$ has inner product 0 with a hidden vector $\boldsymbol{v}$ needs to first encrypt $\boldsymbol{v}$ using a secret key (so their predicate encryption scheme does not directly yield an obfuscation).

We now show that $\mathcal{O}_{\mathcal{G},d}$ is an obfuscator, based on Assumption 5.

Theorem 11. *Let $d \in \mathbb{N}$ and $\mathcal{G}$ be a set of groups satisfying Assumption 5. Then, the algorithm $\mathcal{O}_{\mathcal{G},d}$ is an obfuscator for the family $\mathcal{F}_\rho^d$ with exact functionality.*

It is clear that $\mathcal{O}_{\mathcal{G},d}$ satisfies the exact functionality and polynomial slowdown properties required of an obfuscator, so it remains to prove the virtual black-box property. Before doing so, we present a definition that will be useful throughout the proof and an intermediate lemma.

Definition 12. *Let $d \in \mathbb{N}$ and p be a prime number. We say that the set $V \subseteq \mathbb{F}_p^d$* distinguishes *two vectors $\boldsymbol{l}, \boldsymbol{r} \in \mathbb{F}_p^d$ if there exists $\boldsymbol{x} \in V$ such that exactly one of the inner products $\langle \boldsymbol{l}, \boldsymbol{x} \rangle$ and $\langle \boldsymbol{r}, \boldsymbol{x} \rangle$ equals 0. Otherwise, we say that $\boldsymbol{l}$* and *$\boldsymbol{r}$* are indistinguishable by *V, which means that for all $\boldsymbol{x} \in V$, $\langle \boldsymbol{l}, \boldsymbol{x} \rangle = 0$ if and only if $\langle \boldsymbol{r}, \boldsymbol{x} \rangle = 0$.*

At a high level, this lemma states that for every adversary A, there exists a set V that can distinguish vectors in $\mathbb{F}_{\rho(n)}^d$ as well as A can.

Lemma 13. *Suppose $(\mathcal{G}, d)$ satisfy Assumption 5. For every PPT adversary A and polynomial ε, there exists a polynomial s (that can depend on A) such that for every $n \in \mathbb{N}$, there exists a set $V \subseteq \mathbb{F}_{\rho(n)}^d$ of size at most $s(n)$, such that for every pair of vectors $\boldsymbol{l}, \boldsymbol{r} \in \mathbb{F}_{\rho(n)}^d$ that are indistinguishable by V, $|A_{\boldsymbol{l}} - A_{\boldsymbol{r}}| < \frac{1}{\varepsilon(n)}$.*

Using standard techniques found in [6] and other papers, we can show that the lemma implies that $\mathcal{O}_{\mathcal{G},d}$ is an obfuscator.

Proof (Theorem 11 from Lemma 13). Let A be an adversary and ε be a polynomial, and we must construct a simulator S such that for every $n \in \mathbb{N}$ and every vector $\boldsymbol{r} \in \mathbb{F}_{\rho(n)}^d$,

$$|\Pr[A(\mathcal{O}_{\mathcal{G},d}(H_{\boldsymbol{r}})) = 1] - \Pr[S^{H_{\boldsymbol{r}}}(1^n) = 1]| < \frac{1}{\varepsilon(n)} .$$

By Lemma 13, there exists a polynomial s such that for every $n \in \mathbb{N}$, there exists a set $V \subseteq \mathbb{F}_{\rho(n)}^d$ of size at most $s(n)$ such that the property in the lemma holds. Let $S^{H_{\boldsymbol{r}}}(1^n)$ be the nonuniform circuit that receives V as advice and does the following:

1: **for all** $\boldsymbol{x} \in V$ **do**
2:     query the oracle on input $\boldsymbol{x}$ and record the response

3: **end for**
4: choose a vector $\boldsymbol{l} \in \mathbb{F}^d_{\rho(n)}$ such that $\forall \boldsymbol{x} \in V$, $\langle \boldsymbol{l}, \boldsymbol{x} \rangle = 0$ iff $H(\boldsymbol{x})$ accepts
5: **output** $A(\mathcal{O}_{\mathcal{G},d}(H_{\boldsymbol{l}}))$

Finally, since $\Pr[A(\mathcal{O}_{\mathcal{G},d}(H_{\boldsymbol{r}})) = 1] = A_{\boldsymbol{r}}$ by definition and $\Pr[S^{H_{\boldsymbol{r}}}(1^n) = 1] = A_{\boldsymbol{l}}$ by construction, Lemma 13 ensures that S satisfies the virtual black-box condition. □

Next, we provide some high-level intuition about why the lemma is true. Suppose there is an adversary A that breaks the obfuscation (and thus the lemma as well). We build a new adversary A^* that runs A many times. Also, we construct two distributions $\mathcal{L}$ and $\mathcal{R}$. These distributions will be uniform over their support, so we can really just think of them as sets. The construction of $\mathcal{L}$ and $\mathcal{R}$ proceeds iteratively, subject to two invariant conditions: first, A^* must be able to distinguish these distributions, and second, no linear test should do so. These constraints together violate Assumption 5.

We achieve the first invariant using the negation of Lemma 13, which continually gives us a pair of vectors $(\boldsymbol{l_i}, \boldsymbol{r_i})$ that A (and thus A^*) can distinguish. We add $\boldsymbol{l_i}$ to the support of $\mathcal{L}$ and $\boldsymbol{r_i}$ to the support of $\mathcal{R}$. The second invariant is achieved by continually monitoring $\mathcal{L}$ and $\mathcal{R}$ as they grow. We "trap" any linear test $\boldsymbol{x}$ once it is able to distinguish d of the pairs $(\boldsymbol{l_i}, \boldsymbol{r_i})$. Once we have identified such a linear test, we ensure that subsequent pairs of vectors that we add to $\mathcal{L}$ and $\mathcal{R}$ are indistinguishable by $\boldsymbol{x}$. Hence, any linear test only distinguishes a constant number of the pairs, so by making the distributions $\mathcal{L}$ and $\mathcal{R}$ well-spread, we ensure that all linear tests succeed with only negligible probability. The only downside to the proof is that the "trapping" procedure requires a simulator whose runtime is exponential in d, so the proof only holds for constant dimension.

The rest of this section is devoted to a formal proof of the lemma, which uses some techniques from the proofs in [6], some novel proof concepts, and some linear algebra. We use the set notation $[k] = \{1, 2, \ldots, k\}$ in this proof.

Proof (Lemma 13). Given $\mathcal{G}$ and d, assume for the sake of contradiction that the obfuscator $\mathcal{O}_{\mathcal{G},d}$ does not satisfy Lemma 13. Hence, there exists an adversary A and polynomial ε such that for all polynomials s, there exist infinitely many $n \in \mathbb{N}$ such that for every set $V \subseteq \mathbb{F}^d_{\rho(n)}$ of size at most $s(n)$, there exist vectors $\boldsymbol{l}, \boldsymbol{r} \in \mathbb{F}^d_{\rho(n)}$ with the property that $\langle \boldsymbol{l}, \boldsymbol{x} \rangle = 0$ if and only if $\langle \boldsymbol{r}, \boldsymbol{x} \rangle = 0$ for all $\boldsymbol{x} \in V$, such that $|A_{\boldsymbol{r}} - A_{\boldsymbol{l}}| \geq \frac{1}{\varepsilon(n)}$.

Because these probabilities are separated by a noticeable amount, an efficient algorithm is able to determine which of $A_{\boldsymbol{l}}$ and $A_{\boldsymbol{r}}$ is larger by taking n samples of each one (using independent randomness for A and the choice of $g \xleftarrow{U} G_n$ each time) and observing which sample probability is greater. By a Chernoff bound, this algorithm succeeds with overwhelming probability. Thus, from now on we assume without loss of generality that $A_{\boldsymbol{r}} > A_{\boldsymbol{l}}$, which allows us to drop the absolute value.

Given a constant c, apply this statement to the polynomial $s_c(n) = n^c$ and the resulting $n \in \mathbb{N}$ in order to build two large sets $\hat{L}_n^c$ and $\hat{R}_n^c$ iteratively as follows.

1: initialize $V \leftarrow \emptyset$ and $i \leftarrow 1$
2: **while** $|V| \leq n^c$ **do**
3: given the set V, let $\boldsymbol{l_i}$ and $\boldsymbol{r_i}$ be vectors that violate Lemma 13
4: insert $\boldsymbol{l_i} \in \hat{L}_n^c$ and $\boldsymbol{r_i} \in \hat{R}_n^c$
5: **for all** subsets $T \subseteq \hat{L}_n^c \cup \hat{R}_n^c$ of size at most $d-2$ **do**
6: add to V random bases of $(T \cup \{\boldsymbol{l_i}\})^\perp$ and $(T \cup \{\boldsymbol{r_i}\})^\perp$
7: **end for**
8: increment $i \leftarrow i+1$
9: **end while**

This algorithm iteratively finds pairs of vectors that the adversary A can distinguish but the set V cannot. Then, it adds many points to V. We now describe in detail how these additional points affect future iterations of the loop.

When $T = \emptyset$ in the for loop, the algorithm adds to V a basis of vectors orthogonal to $\boldsymbol{l_i}$. Since $\boldsymbol{l_i}$ is the only vector (up to scalar multiplication) that is orthogonal to every vector in this basis, it follows that in all future iterations $i' > i$ of the loop, $\boldsymbol{l_{i'}}$ and $\boldsymbol{r_{i'}}$ are linearly independent from $\boldsymbol{l_i}$, because $\boldsymbol{l_{i'}}$ and $\boldsymbol{r_{i'}}$ must be indistinguishable by V. The same is true for $\boldsymbol{r_i}$, so the sets $\hat{L}_n^c$ and $\hat{R}_n^c$ are continually increasing in size.

When T is not equal to the empty set, the additional points added to V ensure that linear tests cannot distinguish $\hat{L}_n^c$ from $\hat{R}_n^c$. Specifically, we claim that for every vector $\boldsymbol{x} \in \mathbb{F}_{\rho(n)}^d$, there are at most d indices such that $\langle \boldsymbol{x}, \boldsymbol{l_i} \rangle = 0$ but $\langle \boldsymbol{x}, \boldsymbol{r_i} \rangle \neq 0$, or vice-versa.

To see this, suppose without loss of generality that there exists a vector $\boldsymbol{x} \in \mathbb{F}_{\rho(n)}^d$ and J indices $i_1 < i_2 < \cdots < i_J$ such that $\langle \boldsymbol{x}, \boldsymbol{l_{i_j}} \rangle = 0$ but $\langle \boldsymbol{x}, \boldsymbol{r_{i_j}} \rangle \neq 0 \; \forall j \in [J]$. We show by induction that the vectors $\boldsymbol{l_{i_1}}, \ldots, \boldsymbol{l_{i_J}}$ are linearly independent. As the base case, we showed above that any two vectors from $\hat{L}_n^c \cup \hat{R}_n^c$ are linearly independent. Now, for $j \geq 2$ suppose that $S_j = \{\boldsymbol{l_{i_1}}, \ldots, \boldsymbol{l_{i_j}}\}$ contains linearly independent vectors. At iteration i_j of the loop, a basis $\{\boldsymbol{b_1}, \ldots, \boldsymbol{b_k}\}$ of the space $S_j^\perp$ is added to V. By definition, the basis vectors are linearly independent. If $\boldsymbol{l_{i_{j+1}}}$ were linearly dependent on S_j, say $\boldsymbol{l_{i_{j+1}}} = \alpha_1 \boldsymbol{l_{i_1}} + \cdots + \alpha_j \boldsymbol{l_{i_j}}$, then

$$\langle \boldsymbol{l_{i_{j+1}}}, \boldsymbol{b_{i'}} \rangle = \langle \alpha_1 \boldsymbol{l_{i_1}} + \cdots + \alpha_j \boldsymbol{l_{i_j}}, \boldsymbol{b_{i'}} \rangle = \alpha_1 \langle \boldsymbol{l_{i_1}}, \boldsymbol{b_{i'}} \rangle + \cdots + \alpha_j \langle \boldsymbol{l_{i_j}}, \boldsymbol{b_{i'}} \rangle = 0$$

for all $i' \in [k]$. Because $\boldsymbol{l_{i_{j+1}}}$ and $\boldsymbol{r_{i_{j+1}}}$ are indistinguishable by V, it follows that $\langle \boldsymbol{r_{i_{j+1}}}, \boldsymbol{b_{i'}} \rangle = 0$ for all $i' \in [k]$ as well, so

$$\boldsymbol{r_{i_{j+1}}} \in \overline{\{\boldsymbol{b_1}, \ldots, \boldsymbol{b_k}\}}^\perp = (S_j^\perp)^\perp = \overline{S_j}$$

by Theorem 3, which means that $\boldsymbol{r_{i_{j+1}}}$ is linearly dependent on the vectors in S_j so $\langle \boldsymbol{x}, \boldsymbol{r_{i_{j+1}}} \rangle = 0$. This contradicts the assumption that $\boldsymbol{x}$ distinguishes $\boldsymbol{l_{i_{j+1}}}$ from $\boldsymbol{r_{i_{j+1}}}$, so the vectors $\boldsymbol{l_{i_1}}, \ldots, \boldsymbol{l_{i_{j+1}}}$ must be linearly independent, which completes the induction. The vectors come from a space with dimension d, so there can only be d linearly independent vectors, so $J \leq d$ as desired.

Next, we find a lower bound on the size of the sets $\hat{L}_n^c$ and $\hat{R}_n^c$. The loop condition is to stop when $|V| > n^c$. On each iteration of the loop, $|\hat{L}_n^c|$ and $|\hat{R}_n^c|$ each increase by 1 and $|V|$ increases by at most

$$2d \times \left[\sum_{k=0}^{d-2} \binom{|\hat{L}_n^c \cup \hat{R}_n^c|}{k}\right] \leq 2d^2 \binom{|\hat{L}_n^c \cup \hat{R}_n^c|}{d-2} \leq \mathrm{O}(|\hat{L}_n^c \cup \hat{R}_n^c|^{d-2}),$$

which means that the size of V is

$$|V| = \mathrm{O}(2^{d-2}) + \mathrm{O}(4^{d-2}) + \cdots + \mathrm{O}(|\hat{L}_n^c \cup \hat{R}_n^c|^{d-2}) = \mathrm{O}(|\hat{L}_n^c \cup \hat{R}_n^c|^{d-1}).$$

We also know that $|V| \leq n^c$, so it follows that $|\hat{L}_n^c|$ and $|\hat{R}_n^c|$ are $\Omega(n^{c/d})$.

Consider the $2\varepsilon(n)$ intervals $[0, \frac{1}{2\varepsilon(n)}], [\frac{1}{2\varepsilon(n)}, \frac{1}{\varepsilon(n)}], \ldots, [1 - \frac{1}{2\varepsilon(n)}, 1]$ that partition the unit interval. We say that an interval $[\alpha, \beta]$ "separates" an $\boldsymbol{l_i}$, $\boldsymbol{r_i}$ pair if $A_{\boldsymbol{l_i}} < \alpha$ and $A_{\boldsymbol{r_i}} > \beta$. Since $A_{\boldsymbol{r_i}} - A_{\boldsymbol{l_i}} > \frac{1}{\varepsilon(n)}$, each pair is separated by at least one of the $2\varepsilon(n)$ intervals. Hence, by the pigeonhole principle, there exists one interval that separates a $\frac{1}{2\varepsilon(n)}$ fraction of the pairs. Call this interval $[\alpha_c^*, \beta_c^*]$. Let L_n^c and R_n^c be subsets of $\hat{L}_n^c$ and $\hat{R}_n^c$, respectively, consisting only of the $\boldsymbol{l_i}$, $\boldsymbol{r_i}$ pairs that are separated by $[\alpha_c^*, \beta_c^*]$. Note that $|L_n^c|$ and $|R_n^c|$ are $\Omega(\frac{n^{c/d}}{\varepsilon(n)})$.

Furthermore, there is an algorithm A_c^* that distinguishes L_n^c from R_n^c. It is nonuniformly hardcoded with the value $\mu_c^* = \frac{\alpha_c^* + \beta_c^*}{2}$, and operates as follows.

Input: a vector $\boldsymbol{v} \in \mathbb{F}_{\rho(n)}^d$

1: run $A(\mathcal{O}_{\mathcal{G},d}(H_{\boldsymbol{v}}))$ a total of $32n \cdot \varepsilon(n)^2$ times using fresh randomness for A and $\mathcal{O}$ each time

2: let τ denote the fraction of iterations that A accepts

Output: "L_n^c" if $\tau \leq \mu_c^*$ and "R_n^c" otherwise

If the input to this algorithm is a vector $\boldsymbol{l} \in L_n^c$, then we know that $A_{\boldsymbol{l}} \leq \alpha_c^*$. By a Chernoff bound, the probability that the empirical acceptance rate τ is greater than $\mu_c^* = \alpha_c^* + \frac{1}{4\varepsilon(n)}$ is at most e^{-n}. The same is true for vectors in R_n^c, so this algorithm succeeds with probability $1 - e^{-n}$. On the other hand, we argued above that linear tests distinguish L_n^c from R_n^c with probability at most $\frac{d}{|L_n^c|} = \mathrm{O}(\frac{\varepsilon(n)}{n^{c/d}})$.

Finally, we construct the distributions $\mathcal{L}$ and $\mathcal{R}$ that break Assumption 5. Recall that the negation of Lemma 13 yields a function $n(c)$ as follows: for every poly s_c the lemma provides some value n of the security parameter where there is a counterexample to the lemma. Furthermore, we note that as $c \to \infty$, the sequence $\{n(c)\}_{c\in\mathbb{N}} \to \infty$ as well. This is due to the fact that if $c > nd$, then the lemma considers sets V of size up to $n^{nd} > \rho(n)^d$, so the entire collection of vectors in $\mathbb{F}_{\rho(n)}^d$ can fit in V and the lemma is obviously true in this case.

We form a sort of inverse to this function as follows: given n, let c_n be the biggest value of c such that the counterexample with c applies to n. Note that c_n is not well-defined for all values of n, but it is defined for infinitely large set of values which we will denote by $N \subseteq \mathbb{N}$. It follows from the above argument that as $n \to \infty$, the sequence $\{c_n\}_{n \in N} \to \infty$ as well. Hence, there exists an infinitely large subset $N' \subseteq N$ such that $\{c_n\}_{n \in N'}$ is monotonically increasing. We form the families of distributions $\mathcal{L}$ and $\mathcal{R}$ such that $\mathcal{L}_n$ and $\mathcal{R}_n$ are uniform over the sets $L_n^{c_n}$ and $R_n^{c_n}$, respectively, for all $n \in N'$. We set $\mathcal{L}_n = \mathcal{R}_n$ arbitrarily for all $n \notin N'$.

Consider the following unified adversary A^* that is nonuniformly hardcoded with the values $\mu^*_{c_n} = \frac{1}{2}(\alpha^*_{c_n} + \beta^*_{c_n})$ for all $n \in N'$ (and arbitrarily values of $\mu^*_{c_n}$ for $n \notin N'$).

Input: a vector $\boldsymbol{v} \in \mathbb{F}^d_{\rho(n)}$

1: run $A(\mathcal{O}_{\mathcal{G},d}(H_{\boldsymbol{v}}))$ a total of $32n \cdot \varepsilon(n)^2$ times using fresh randomness for A and $\mathcal{O}$ each time

2: let ξ denote the fraction of iterations that A accepts

Output: "L_n" if $\xi \leq \mu^*_{c_n}$ and "R_n" otherwise

This adversary will succeed at distinguishing $\mathcal{L}$ from $\mathcal{R}$ with overwhelming probability $1 - e^{-n}$ for all $n \in N'$ (and of course the adversary will fail on all $n \notin N'$). On the other hand, any sequence of linear tests only succeeds with probability $O(\frac{\varepsilon(n)}{n^{c_n/d}})$ which is negligible since $c_n \to \infty$ as $n \to \infty$. Hence, there is no polynomial $q(n)$ that bounds the ratio of success probabilities for the infinitely many $n \in N'$, so Assumption 5 is false as desired. □

5 Obfuscation of Hyperplanes with Multi-bit Output

Given an obfuscator for the family of point functions, the work of [10] shows how to construct an obfuscator for the family of point functions with multi-bit output. This family also accepts a single point, but instead of just having a yes or no output, it returns a hidden message on the correct input value. Such an obfuscator can be used to create a strong symmetric-key encryption scheme that satisfies leakage resilience and circular security [17]. Their construction applies in our case too, so we can obfuscate the family of "hyperplanes with multi-bit output," with the nice property that the message is not revealed when the input is the zero vector (the one vector that is known to be in every hyperplane).

Formally, let $H_{\boldsymbol{a},m}$ be the circuit that has the vector $\boldsymbol{a} \in \mathbb{F}^d_{\rho(n)}$ hardwired, and on input a vector $\boldsymbol{x} \in \mathbb{F}^d_{\rho(n)}$, outputs m if $\langle \boldsymbol{a}, \boldsymbol{x} \rangle = 0$ but $\boldsymbol{x} \neq \boldsymbol{0}$, and outputs $\perp$ otherwise. Let $\mathcal{M}^d_{\rho,l} = \{H_{\boldsymbol{a},m} : \boldsymbol{a} \in \mathbb{F}^d_{\rho(n)}, m \in \{0,1\}^{l(n)}\}$ be the family of all such circuits. In particular, we can think of the hyperplanes family $\mathcal{F}^d_\rho$ as a special case of this family where $l = 0$ (i.e. there is only one possible message).

We show how to obfuscate the family $\mathcal{M}^d_{\rho,l}$ given any $d \in \mathbb{N}$ and obfuscator for hyperplanes $\mathcal{O}_{\mathcal{G},d}$ that is $(l+1)$-composable.

Definition 14 (t-composable obfuscation [10]). *A PPT $\mathcal{O}$ is a t-composable obfuscator for the family $\mathcal{C}$ if functionality and polynomial slowdown hold as before, and the virtual black-box property holds whenever the adversary and simulator are given up to t circuits in $\mathcal{C}$. Formally, for every PPT adversary A and* *polynomial δ, there exists a PPT simulator S such that for all sufficiently large n, and for all $C_1, \ldots, C_t \in \mathcal{C}_n$,*

$$| \Pr[A(\mathcal{O}(C_1), \ldots, \mathcal{O}(C_t)) = 1] - \Pr[S^{C_1, \ldots, C_t}(1^n) = 1]| < \frac{1}{\delta(n)} ,$$

where the first probability is taken over the coin tosses of A and $\mathcal{O}$, and the second probability is taken over the coin tosses of S.

Unfortunately, we do not know how to prove from Assumption 5 that $\mathcal{O}_{\mathcal{G},d}$ is even 2-composable. All we can show is that the composability of $\mathcal{O}_{\mathcal{G},d}$ is related to the length of messages that we can obfuscate. Let $\tilde{\mathcal{O}}_{\mathcal{G},l,d}$ be an obfuscator for the family $\mathcal{M}^d_{\rho,l}$ that operates as follows.

Algorithm 2. Obfuscator $\tilde{\mathcal{O}}_{\mathcal{G},l,d}$ for the family $\mathcal{M}^d_{\rho,l}$

Input: vector $\boldsymbol{a} \in \mathbb{F}^d_{\rho(n)}$
1: set $C_0 = \mathcal{O}_{\mathcal{G},d}(\boldsymbol{a})$
2: **for** $i = 1$ to l **do**
3: **if** $m_i = 1$ **then**
4: set $C_i = \mathcal{O}_{\mathcal{G},d}(\boldsymbol{a})$
5: **else**
6: choose $\boldsymbol{a}' \xleftarrow{U} \mathbb{F}^d_{\rho(n)}$ and set $C_i = \mathcal{O}_{\mathcal{G},d}(\boldsymbol{a}')$
7: **end if**
8: **end for**
Output: circuit that hardwires $C_0, \ldots, C_l$ and operates as follows on input $\boldsymbol{x} \in \mathbb{F}^d_{\rho(n)}$: output $\perp$ if $C_0(\boldsymbol{x})$ rejects or if $\boldsymbol{x} = \boldsymbol{0}$, otherwise output the string s formed by $s_i = C_i(\boldsymbol{x})$ for $i = 1, \ldots, l$

Theorem 15. *Suppose that the obfuscator $\mathcal{O}_{\mathcal{G},d}$ is $(l+1)$-composable for some $l = \text{poly}(n)$, and let $\rho(n) = |G_n|$. Then, $\tilde{\mathcal{O}}_{\mathcal{G},l,d}$ is an obfuscator for the family of hyperplanes with multi-bit output $\mathcal{M}^d_{\rho,l}$ with approximate functionality.*

The proof of this theorem is similar to the one in [10].

Proof. The approximate functionality and polynomial slowdown of $\tilde{\mathcal{O}}_{\mathcal{G},l,d}$ are clear from the construction and the corresponding properties of $\mathcal{O}_{\mathcal{G},d}$. For $i = 1$ to l, let $\boldsymbol{a_i}$ be the vector such that $\boldsymbol{a_i} = \boldsymbol{a}$ if $m_i = 1$ or $\boldsymbol{a_i}$ is uniformly chosen otherwise. By the $(l+1)$-composable virtual black-box property, we know that there exists a simulator S such that the output of

$$A(\tilde{\mathcal{O}}_{\mathcal{G},l,d}(\boldsymbol{a})) = A(\mathcal{O}_{\mathcal{G},d}(\boldsymbol{a_1}), \ldots, \mathcal{O}_{\mathcal{G},d}(\boldsymbol{a_d}))$$

can be simulated by $S^{H_{\boldsymbol{a_1}}, \ldots, H_{\boldsymbol{a_d}}}$. Furthermore, the oracles $H_{\boldsymbol{a_1}}, \ldots, H_{\boldsymbol{a_d}}$ can be simulated by the oracle $H_{\boldsymbol{a},m}$ up to a negligible simulation error in the following manner: if $H_{\boldsymbol{a},m}(\boldsymbol{x}) = \perp$, then we say that $H_{\boldsymbol{a_i}}(\boldsymbol{x}) = 0$ for all i. Otherwise $H_{\boldsymbol{a},m}(\boldsymbol{x}) = m$, in which case we say that $H_{\boldsymbol{a_i}}(\boldsymbol{x}) = m_i$. Hence, the simulator $T^{H_{\boldsymbol{a},m}}$ that runs $S^{H_{\boldsymbol{a_1}}, \ldots, H_{\boldsymbol{a_d}}}$ and emulates the oracle queries in this manner satisfies the virtual black-box property for $\tilde{\mathcal{O}}_{\mathcal{G},l,d}$. □

6 One-Time Signature Schemes

We can use an obfuscator for the family of planes in three-dimensional space to form a one-time signature scheme. Informally, the secret and public keys are a hidden plane and an obfuscation of the plane membership testing program, respectively. A signature of a message is a point on the hyperplane that is related to the message, and the verification procedure runs the obfuscated hyperplane testing circuit to verify signatures.

More formally, let ρ be a prime sequence, and $\mathcal{O}$ be an obfuscator for the family of hyperplanes over $\mathbb{F}_\rho^3$ (such as the obfuscator $\mathcal{O}_{\mathcal{G},d}$ constructed in Section 4). Consider the following three algorithms.

KeyGen(1^n): Choose field elements $sk_1, sk_2, c \xleftarrow{U} \mathbb{F}_{\rho(n)} \setminus 0$. Form the vector $\boldsymbol{sk} = (sk_1, sk_2, 1)$ in $\mathbb{F}^3_{\rho(n)}$ and the obfuscated plane $P = \mathcal{O}(\boldsymbol{sk})$. The secret key is (sk_1, sk_2), and the public key is (P, c).

Sign($m \in \mathbb{F}_{\rho(n)}$): Let σ_2 be the unique field element such that the inner product $\langle \boldsymbol{sk}, (cm, \sigma_2, 1)\rangle = 0$. The signature is (cm, σ_2).

Verify($m, (\sigma_1, \sigma_2)$): Accept if and only if $\sigma_1 = cm$ and $P(\sigma_1, \sigma_2, 1)$ accepts.

This signature scheme is unforgeable in a weak sense, described in [14] and other works, in which the forger must choose the message on which she requests a signature before being shown the public key. The techniques in [14] allow us to transform this scheme into one that is existentially unforgeable under chosen message attacks (the standard security notion for signature schemes). The transformation requires a chameleon hash function whose seed can be chosen with public coins. It is known how to construct such a hash function under the DDH assumption [18].

Furthermore, our one-time signature scheme is resilient to any leakage function whose output length is less than half as long as the secret key.

Theorem 16. *Let ρ be a prime sequence and $\mathcal{O}$ be an obfuscator for the family of hyperplanes over the vector space $\mathbb{F}_\rho^3$. Then, the above algorithm leads to an existentially unforgeable one-time signature scheme that is resilient to any leakage function whose output length is bounded by*

$$l(n) = n - \omega(\log(n)) .$$

In particular, leakage of $l(n) = \gamma n$ bits for any $\gamma < 1$ is permitted.

This theorem is proved in [16]. We note that the leakage bound in the theorem is tight. Consider the following leakage function that has a message m hardcoded: use the secret key to form a signature associated to m, and output σ_2. This leakage function has n bits of output, and permits a forgery of the message m by the signature (cm, σ_2).

The secret key consists of two elements of $\mathbb{F}_{\rho(n)}$, so it is $2n$ bits long. Thus, our signature scheme permits leakage of up to half of the length of the secret key. This matches the leakage bound attained in [15] for schemes that do not use general non-interactive zero-knowledge proofs (albeit under a much stronger assumptions).

Acknowledgments. The authors would like to thank the anonymous referees for their helpful feedback.

References

1. Barak, B., Goldreich, O., Impagliazzo, R., Rudich, S., Sahai, A., Vadhan, S., Yang, K.: On the (im)possibility of obfuscating programs. In: Kilian, J. (ed.) CRYPTO 2001. LNCS, vol. 2139, pp. 1–18. Springer, Heidelberg (2001)
2. Goldwasser, S., Rothblum, G.N.: On best-possible obfuscation. In: Vadhan, S.P. (ed.) TCC 2007. LNCS, vol. 4392, pp. 194–213. Springer, Heidelberg (2007)
3. Hofheinz, D., Malone-Lee, J., Stam, M.: Obfuscation for cryptographic purposes. In: Vadhan, S.P. (ed.) TCC 2007. LNCS, vol. 4392, pp. 214–232. Springer, Heidelberg (2007)
4. Hohenberger, S., Rothblum, G.N., Shelat, A., Vaikuntanathan, V.: Securely obfuscating re-encryption. In: Vadhan, S.P. (ed.) TCC 2007. LNCS, vol. 4392, pp. 233–252. Springer, Heidelberg (2007)
5. Goldwasser, S., Kalai, Y.T.: On the impossibility of obfuscation with auxiliary input. In: FOCS, pp. 553–562. IEEE Computer Society, Los Alamitos (2005)
6. Canetti, R.: Towards realizing random oracles: Hash functions that hide all partial information. In: Kaliski Jr., B.S. (ed.) CRYPTO 1997. LNCS, vol. 1294, pp. 455–469. Springer, Heidelberg (1997)
7. Canetti, R., Micciancio, D., Reingold, O.: Perfectly one-way probabilistic hash functions. In: Proceedings of the 30th ACM Symposium on Theory of Computing, pp. 131–140 (1998)
8. Lynn, B.Y.S., Prabhakaran, M., Sahai, A.: Positive results and techniques for obfuscation. In: Cachin, C., Camenisch, J.L. (eds.) EUROCRYPT 2004. LNCS, vol. 3027, pp. 20–39. Springer, Heidelberg (2004)
9. Wee, H.: On obfuscating point functions. In: Proceedings of the 37th ACM Symposium on Theory of Computing, pp. 523–532 (2005)
10. Canetti, R., Dakdouk, R.R.: Obfuscating point functions with multibit output. In: Smart, N.P. (ed.) EUROCRYPT 2008. LNCS, vol. 4965, pp. 489–508. Springer, Heidelberg (2008)
11. Dodis, Y., Smith, A.: Correcting errors without leaking partial information. In: STOC, pp. 654–663 (2005)
12. Adida, B., Wikström, D.: How to shuffle in public. In: Vadhan, S.P. (ed.) TCC 2007. LNCS, vol. 4392, pp. 555–574. Springer, Heidelberg (2007)
13. Shen, E., Shi, E., Waters, B.: Predicate privacy in encryption systems. In: Reingold, O. (ed.) TCC 2009. LNCS, vol. 5444, pp. 457–473. Springer, Heidelberg (2009)
14. Hohenberger, S., Waters, B.: Short and stateless signatures from the rsa assumption. In: Halevi, S. (ed.) CRYPTO 2009. LNCS, vol. 5677, pp. 654–670. Springer, Heidelberg (2009)
15. Katz, J., Vaikuntanathan, V.: Signature schemes with bounded leakage resilience. In: Matsui, M. (ed.) ASIACRYPT 2009. LNCS, vol. 5912, pp. 703–720. Springer, Heidelberg (2009)
16. Canetti, R., Rothblum, G., Varia, M.: Obfuscation of hyperplane membership. Cryptology ePrint Archive (2009), `http://eprint.iacr.org/`
17. Canetti, R., Kalai, Y., Varia, M., Wichs, D.: On symmetric encryption and point obfuscation. In: Micciancio, D. (ed.) TCC 2010. LNCS, vol. 5978, pp. 52–71. Springer, Heidelberg (2010)
18. Krawczyk, H., Rabin, T.: Chameleon signatures. In: NDSS. The Internet Society (2000)

Secure Computation and Its Diverse Applications

Yuval Ishai

Technion and UCLA
yuvali@cs.technion.il

Abstract. Secure multiparty computation (MPC) allows two or more parties to perform a joint distributed computation without revealing their secrets to each other. While MPC has traditionally been viewed as an ends rather than a means, in recent years we have seen a growing number of unexpected applications of MPC and connections with problems from other domains.

In this talk we will survey several of these connections and highlight some research directions which they motivate. In particular, we will discuss the following connections:

- MPC and locally decodable codes. How can the secrecy property of MPC protocols be useful for reliable and efficient access to data?
- MPC and the parallel complexity of cryptography. How can progress on the round complexity of MPC lead to better parallel implementations of one-way functions and other cryptographic primitives?
- MPC and private circuits. How can MPC be used to protect cryptographic hardware against side-channel attacks?
- MPC in the head. How can MPC protocols which assume an honest majority be useful in the context of two-party cryptography?

D. Micciancio (Ed.): TCC 2010, LNCS 5978, p. 90, 2010.

On Complete Primitives for Fairness

Dov Gordon[1,⋆], Yuval Ishai[2,3,⋆⋆], Tal Moran[4], Rafail Ostrovsky[3,⋆⋆⋆], and Amit Sahai[3,†]

[1] University of Maryland
gordon@cs.umd.edu
[2] Technion, Israel
yuvali@cs.technion.il
[3] University of California, Los Angeles
{rafail,sahai}@cs.ucla.edu
[4] Harvard SEAS
talm@seas.harvard.edu

Abstract. For secure two-party and multi-party computation with abort, classification of which primitives are *complete* has been extensively studied in the literature. However, for *fair* secure computation, where (roughly speaking) either all parties learn the output or none do, the question of complete primitives has remained largely unstudied. In this work, we initiate a rigorous study of completeness for primitives that allow fair computation. We show the following results:

- **No "short" primitive is complete for fairness.** In surprising contrast to other notions of security for secure two-party computation, we show that for fair secure computation, no primitive of size $O(\log k)$ is complete, where k is a security parameter. This is the case even if we can enforce parallelism in calls to the primitives (i.e., the adversary does not get output from any primitive in a parallel call until it sends input to all of them). This negative result holds regardless of any computational assumptions.
- **A fairness hierarchy.** We clarify the fairness landscape further by exhibiting the existence of a "fairness hierarchy". We show that for every "short" $\ell = O(\log k)$, no protocol making (serial) access to any ℓ-bit primitive can be used to construct even a $(\ell + 1)$-bit simultaneous broadcast.
- **Positive results.** To complement the negative results, we exhibit a k-bit primitive that *is* complete for two-party fair secure computation. We show how to generalize this result to the multi-party setting.
- **Fairness combiners.** We also introduce the question of constructing a protocol for fair secure computation from primitives that may be faulty. We show

⋆ Work supported by US-Israel Binational Science Foundation grant 2004240 and NSF grant 0830464.

⋆⋆ Supported in part by ISF grant 1310/06, BSF grants 2008411, and NSF grants 0627781, 0716835, 0716389, 0830803, 0916574.

⋆⋆⋆ Supported in part by IBM Faculty Award, Xerox Innovation Group Award, the Okawa Foundation Award, Intel, Teradata, BSF grant 2008411, NSF grants 0716835, 0716389, 0830803, 0916574 and U.C. MICRO grant.

† Supported in part from BSF grant 2008411, NSF grants 0916574, 0830803, 0627781 and 0716389, an equipment grant from Intel and an Okawa Foundation Research Grant.

D. Micciancio (Ed.): TCC 2010, LNCS 5978, pp. 91–108, 2010.

that this is possible when a majority of the instances are honest. On the flip side, we show that this result is tight: no functionality is complete for fairness if half (or more) of the instances can be malicious.

1 Introduction

In the setting of secure multi-party computation, n participants wish to jointly compute a function while maintaining several security properties, such as the privacy of their inputs, correctness of the outputs, and others. The security of their computation is defined by comparing their view in the protocol to an ideal world — one that embodies complete security — and proving that they are indistinguishable to an outside observer. In this ideal world there is an additional trusted party that privately receives the inputs from all participants, performs the computation on their behalf, and returns the outputs.

In the real world, of course, relying on an outside party is undesirable. In some of the most fundamental results in modern cryptography, many beautiful techniques have been developed in order to remove the trusted party while retaining most of the security properties that he affords. In fact, in the two-party setting there is only one security property for which the trusted third party has remained essential. Informally, we say a computation is *fair* if either all players receive their output or none of them do. It is easy to see that in the ideal world, where there is an additional trusted party, computations are always fair. This is a difficult property to achieve in the real world, as a malicious player may abort the protocol prematurely.

In 1986, Richard Cleve [20] proved that, for general functionalities, fairness is impossible to achieve unless the majority of parties is honest. Specifically, he showed that even the very basic functionality of coin-tossing cannot be fairly computed by a two-party protocol. Recalling that fairness is immediate with the help of a third party, in this paper we address a very natural question.

> What is the *minimum* amount of help required to be able to compute all functions fairly?

We think of this helper as a naive black box, or a *primitive*, with no knowledge of the function being computed. It is charged with a fixed task: it takes inputs from each player, and then simultaneously outputs some fixed function of the inputs to all players. Clearly we can compute any function fairly if this primitive is sufficiently complex: we can simply define its input to be a description of the function being computed, along with the inputs to that function. (Indeed, this was demonstrated by Fitzi et al. [26], as discussed below.) However, our interest is in reducing the complexity of these primitives. In particular, we study the minimum input size to such primitives that will enable the fair computation of any function.

Interestingly, there has been extensive research on very similar questions in the context of *unfair* secure computation. When there is no honest majority among the players, it is known that oblivious transfer is both necessary and sufficient for computational security (without fairness) [51,31,42]. Similarly, in addition to the impossibility of attaining fairness without an honest majority, it is also known that we cannot achieve information-theoretic security in this setting [19], and there is a long line of research identifying the minimum primitives that enable information theoretic security (without fairness). Surprisingly, very little work has addressed the parallel questions with respect to fairness.

1.1 Our Results

The main theme that recurs throughout our results is that when looking at primitives for fair computation, input size matters. We classify primitives according to the length of the inputs provided by the two parties: a k-bit primitive accepts inputs of size k from both parties.

No "Short" Primitive is Complete for Fairness. For many other notions of security which are unrealizable in the plain model (such as unconditional security and UC security) there exist *finite* functionalities which are complete for that setting (e.g., [32,39,41,42]). In other words, to enable (unfair) secure computation in these models, it suffices to give access to a trusted implementation of some simple function with short inputs. Surprisingly, we show that no primitive of size $O(\log \kappa)$ is complete for fair computation (where κ is the security parameter). Our impossibility result holds even if we allow parallel calls to the primitives (where the adversary does not get output from any primitive in a parallel call until it sends input to all of them), and even if the adversary may only deviate from the protocol by aborting early.

Coin-Flipping and Simultaneous Broadcast are not Complete for Fairness. Coin-flipping is perhaps the simplest fair functionality that is known to be unrealizable in the plain model [20]. Simultaneous broadcast is important because it is one of the most natural candidates for a complete primitive for fairness. It is often the first thing that comes to mind when thinking about how to construct a fair protocol. Lending weight to this intuition, Katz proved that simultaneous broadcast is complete for *partial fairness* [40]. Finally, although we know that is also unrealizable in the plain model (even a 1-bit simultaneous broadcast implies fair coin-flipping), it can be constructed from conceivable non-standard assumptions such as timed commitments [12] or physical limits on signal propagation. Surprisingly, our results imply that simultaneous broadcast of any size (as well as coin-flipping) is not complete for fair computation (this is a direct corollary of our "No Short Primitive" result).

A Fairness Hierarchy. We clarify the fairness landscape further by exhibiting the existence of a "fairness hierarchy." We show that for every "short" k (when the adversary can run in time poly(2^k)), no k-bit primitive can be used to construct even a $k+1$-bit simultaneous broadcast. This result is almost tight: given a "long" k'-bit simultaneous broadcast and a semantically-secure encryption scheme with keys of length k', we can construct a simultaneous broadcast of any length by first exchanging encrypted inputs and then exchanging keys. The fairness hierarchy complements the first proof of impossibility for achieving fair computation through short primitives (described above). The latter demonstrates that there exists *some* function that cannot be fairly computed, even with (parallel) access to a short primitive. In the hierarchy result, we only consider serial access to a short, k-bit primitive, but we demonstrate that it does not enable fair computation for even the most simple $k+1$ bit primitives.

Positive Results. Previous work by Fitzi et al. [26] proposed the *Universal Black Box* (UBB) primitive and showed that is is complete for fair secure computation in both the two-party and multi-party settings. However, the UBB primitive requires as input a description of functionality to be computed (a more detailed description appears in Sec. 3.2). Thus, its input size and running time depend on the complexity of the target functionality.

In this paper, we show that there exists a much simpler primitive that is complete for two-party fair computation. This primitive implements a "Fair Reconstruction" procedure for a secret sharing scheme. Before calling the primitive, the parties first run an unfair secure computation that outputs *non-malleable secret shares* of the desired function's output. They then call the primitive using these secret shares as their input. The input to the primitive depends on a security parameter and on the output size of the functionality being computed (but not on its description). With the addition of computational assumptions (the existence of one-way functions), the input size can be made to depend only on the security parameter.

We also show how this result can be generalized to the multi-party setting. This generalization is straightforward if we are satisfied by fairness without robustness (i.e., if a single malicious party can cause the entire computation to abort). In the full-security case, however, our results exhibit a trade-off between the input size and the number of primitive invocations that may be required to complete the protocol: for n parties, we describe a primitive that has input size $O(n^2)$ but requires $O(n)$ invocations in the worst case, and a more complex primitive that has input size $O(2^n)$, but only requires a single invocation. (In contrast, the input size of the UBB primitive grows with the number of parties, the functionality's input length and description size — however, it requires only a single invocation.)

Fairness Combiners. We next consider the orthogonal question of constructing a protocol for fair computation from primitives which may be faulty. Questions of this nature have been studied in the context of many other primitives (e.g., [37,38]). We show a functionality that is complete for two-party fair computation when the majority of its instances are honest. On the flip side, we show that this result is tight: no functionality is complete for fairness if half (or more) of the instances can be malicious.

1.2 Related Work

The issue of fairness has not been neglected; there has been a lot of diverging work on achieving fairness through relaxations in the security definition [10,21,23,29,40,34], relaxations in the communication model [43,40] and by enabling dispute resolution through a trusted third party [2,3,13,47]. The recent counter-intuitive results of Gordon, Hazay, Katz and Lindell [35,36], showing that some non-trivial functions *can* be fairly computed in the plain model, have caused a surge of renewed interest in the subject.

Fair Exchange and Contract Signing. One of the earliest related problems in the cryptographic literature is that of fairly exchanging items or information. In a fair exchange, either both parties receive the item, or neither party does. Although we are interested in fair computation of arbitrary functions, the problem of fair exchange turns out to be the crux of the matter. Exemplifying this, both the security definitions and the solutions for the more general setting were usually used in this setting first.

In the plain model, Cleve showed that perfectly fair exchange is impossible [20]. Boneh and Naor gave a similar lower bound for fair contract signing [12]. However, relaxed definitions of fairness are still possible to achieve.

One very common relaxation is that each party would have to perform a similar amount of computation to compute its output. This definition is weak enough to be satisfied by protocols in the plain model (using standard cryptographic assumptions), and is usually accomplished by releasing the information "gradually" ([10,25,28,21,24,11,48,29]). In a similar approach, the output of the parties is masked with noise that decreases over the time, allowing their confidence in the output to increase as the protocol proceeds ([45,6,33]).

More recently, the fair exchange problem has been studied in the *optimistic* model: this model, introduced by Asokan [2], uses a trusted third party (TTP) but requires that the TTP be involved in the protocol only if one of the parties is malicious.

A third model was proposed by Chen, Kudla and Paterson [16] and extended by Lindell [44]. In this model fairness is *legally* rather than technically enforceable: the guarantee is the honest party will either receive her output, or a "check" from the other party (for a pre-agreed amount). In order to invalidate the check in court, the paying party will have to reveal information which will allow the honest party to compute her output.

Fairness in General Secure Computation. Positive results in all three models have been extended to the general two-party computation setting [51,13,44]. Although a complete primitive for fairness is implied in these works, the construction in each is specific to the model. In contrast, our positive result (Thm. 1) is generic: in any of the fairness models above, it is sufficient to implement our simple complete primitive for fairness in order to get generic fair computation in that model.

A fourth relaxation of the fairness definition, *partial fairness*, was proposed by Katz [40]. This definition is phrased in the language of secure computation: informally, a protocol realizes a functionality with ε-partial fairness if there exists an ideal-world simulator whose output cannot be distinguished from the real-world adversary with advantage greater than ε. Katz showed that simultaneous broadcast (SB) is a complete primitive for ε-partial fairness (for any fixed ε). Our positive result uses techniques similar to his.

Gordon and Katz study partial fairness in the plain model [34], and show that even partial fairness is impossible to achieve in general in the plain model. Their proof gives a specific functionality that is complete for (perfect) fairness, and our proof of Thm. 2 has a similar flavor.

In the multi-party computation setting with an honest majority and a broadcast channel, completely fair computation is possible for any functionality, even without computational assumptions [9,15,49]. When there is no honest majority, Cleve's lower-bound applies and general fair computation cannot be achieved at all in the plain model. Lepinski, Micali, Peikert, and shelat devised a protocol for completely fair multi-party computation with any number of malicious parties by relying on "envelope" primitives; communication primitives with special physical properties [43]. Our complete primitives for the multi-party case are possibly less amenable to implementation via simple physical means (such as envelopes), but allow us to separate the unfair computation from the calls to the complete primitive (whereas the two are intertwined in [43]).

We note that we have mentioned only a small fraction of the related work on fair-exchange and fair computation. See [13,47,48,44] for more works in this area.

Classifying Primitives in Secure Computation. Our result in Sec. 5 defines a fairness hierarchy, based on the input size of the primitives. While classification based on input size seems less useful in other contexts of secure computation, other hierarchies have been studied. For example, classification according to the number of calls to a primitive [5,7], classification by privacy level [17] and by reductions to other primitives [46]. While these works may share some of our goals, namely to further understand the theoretical underpinnings of secure computation, their methods are quite different, and they do not address fairness.

2 Definitions

Definition 1 (Non-Malleable Secret Sharing). *A* 2-out-of-2 non-malleable secret sharing scheme *(NMSS scheme) is defined by a pair of algorithms* (Share, Rec) *with the following properties:*

- Share(s, r) *returns* 2 *shares,* (s_0, s_1) *(where* s_i *is the share of the i-th party) such that if* r *is picked at random, then a single share reveals no information about* s.
- Rec(Share(s, r)) $= (s, 0)$ *for every* s, r*. The second output of* Rec *serves as a flag which is set to 0 if the secret has been successfully reconstructed.*
- *Any attempt by a player to modify their share (independently of the remaining share) is detected with overwhelming probability. Formally, we say that* (Share, Rec) *is* ε*-non-malleable if for every secret* s*, every (computationally unbounded) adversary* $\mathcal{A}$ *can win the following game with at most* ε *probability:*
  - $\mathcal{A}$ *corrupts one of the parties.*
  - *Random shares* (s_0, s_1) *from* Share(s, r) *are given to the* 2 *parties.*
  - *Based on the share* $s_{\mathcal{A}}$ *it observed,* $\mathcal{A}$ *computes a new share* $s^*_{\mathcal{A}}$.
  - $\mathcal{A}$ *wins if* $s^*_{\mathcal{A}} \neq s_{\mathcal{A}}$ *and* Rec$(s^*_{\mathcal{A}}, s_H) = (s', 0)$ *for some secret* s'*, where* s_H *is the share received by the uncorrupted party.*

We note that similar notions of robust secret sharing from the literature (cf. [22] and references therein) are weaker in that they define $\mathcal{A}$ to win the above game only if $s' \neq s$. While this weaker notion does not suffice for our purposes, previous constructions (including the ones from [22]) in fact also satisfy our stronger requirement. We include a construction of an NMSS scheme in the full version.

The following functionality will play a role in several of our results:

Definition 2 (Fair Reconstruction). FairRec$_\ell(x, y)$ *is defined:*

$$\mathsf{FairRec}_\ell(x, y) = \begin{cases} (s, s) & \text{if } \mathsf{Rec}(x, y) = (s, 0) \\ (\bot, \bot) & \text{otherwise} \end{cases}$$

where $x, y \in \{0, 1\}^\ell$.

Intuitively, the FairRec functionality is just a fair implementation of Rec: it takes a non-malleable secret share from each player, and outputs the result of Rec to both players if and only if the secret was successfully reconstructed. We will prove that it is complete for fairness in Section 3.1. Interestingly, it will also play a key role in our proofs of impossibility in Section 4.

Secure Function Evaluation (SFE). We use the standard definitions for secure function evaluation in the standalone model. Due to limited space, we do not repeat them here, but refer the reader to [14] for a complete definition.

The definition of SFE given in [14] guarantees output delivery, as do all of the protocols that we present. Our impossibility results hold even if we only consider a slightly weaker definition of SFE where the adversary is allowed to abort before giving input to the trusted party. In the two-party setting these notions are equivalent. We note that any (polynomial-time computable) functionality can be computed according to a relaxed notion of security in which the adversary receives his output first and may choose to abort immediately afterwards. Since our interest in this work is in *fair* secure-computation, we will always refer to the stronger notion of security described above, except when explicitly stating a computation is "secure-with-abort". We sometimes (informally) refer to a protocol as *fair* when we actually mean that it is *secure* according to this stricter notion (which includes fairness).

Definition 3 **(k-bit Primitives).** *We say a protocol Π implementing some functionality $\mathcal{F}$ has access to a k-bit primitive g, if in every round of the protocol, the players may submit k-bit inputs to a trusted computation that securely implements g. We write Π^g to make explicit the fact that Π has access to g.*

We often consider k-bit primitives where $k = k(\kappa)$ is a function of the security parameter. In this case, when we say g is a k-bit primitive we mean that g is an infinite sequence of primitives, such that for every $\kappa \in \mathbb{N}$ there is defined a $k(\kappa)$-bit primitive g_κ in that sequence.

We sometimes informally refer to primitives as "short" or "long". A k-bit primitive is considered "short" if $k = O(\log \kappa)$, where κ is the security parameter. A k bit primitive is considered "long" if $k = \Omega(\kappa)$.

Definition 4 **(Complete Primitive for Fairness).** *For a functionality $\mathcal{F}$ and a primitive g, we say the* fairness of $\mathcal{F}$ reduces to g *if there exists a protocol Π^g that securely computes $\mathcal{F}$. Let C be a class of functionalities. We say that* g is C-complete for fairness *if, for all $\mathcal{F} \in C$, the fairness of $\mathcal{F}$ reduces to g.*

When g is C-complete for fairness and C is the class of all functions, we may omit it and say that g is *complete for fairness*.

Definition 5 **(Parallel Primitives).** *For a primitive g, we denote $\mathsf{par}_k(g)$ the primitive that consists of k independent copies of g with* enforced parallelism. *The parallelism is enforced in that none of the copies of g in $\mathsf{par}_k(g)$ send output to any party until all k copies have received input from all parties. We use Π^g_p to denote that protocol Π has access to $\mathsf{par}_k(g)$.*

Definition 6 (Simultaneous Broadcast). *This primitive was originally defined by Chor, Goldwasser, Micali and Awerbuch [18], in the context of multi-party computation. In the two-party setting, we define the primitive* Simultaneous Broadcast (SB) *that takes one input value from each player and (fairly) swaps them. Formally,* SB $(x, y) = (y, x)$. *We refer to a k-bit* SB *when the input sizes are at most k-bits long.*

3 Fairness-Complete Primitives

3.1 Fairness-Complete Primitives for Two-Party Computation

In this section we demonstrate an ideal function that is complete for two-party fairness. In order to compute some function $\mathcal{F}(x, y) = \{F_0(x, y), F_1(x, y)\}$ fairly, the parties will first compute a related function $\mathcal{F}'(x, y)$ that provides player i with an *encryption* of $F_i(x, y)$, along with a share of the corresponding decryption key (generated using a 2-out-of-2 NMSS scheme). This reduces the problem to a simple exchange of the secret shares. Of course, if the players exchanged these on their own, one player might abort just at the point of exchange, recovering the decryption key (and thus his output) all alone. Instead, the ideal functionality FairRec takes the shares from each player and performs the reconstruction fairly; the non-malleability property of the secret sharing scheme enables the functionality to verify that both players have provided correct shares. The details follow:

Theorem 1. *Any two-party functionality $\mathcal{F}$ with output length m can be fairly computed in the OT-hybrid model by using a single call to* $\mathsf{FairRec}_{O(m)}$. *If one-way functions exist, then $\mathcal{F}$ can be fairly computed in the OT-hybrid model with a single call to* $\mathsf{FairRec}_{O(\kappa)}$.

We begin by defining a function $\mathcal{F}'$ related to $\mathcal{F}$ in the way described above. Specifically, let $(\mathsf{Enc}, \mathsf{Dec})$ be the encryption and decryption functions for a semantically secure symmetric encryption scheme. When using $\mathsf{FairRec}_{O(m)}$, as in the first part of the theorem, the encryption scheme is a one-time pad (with key length m). When using $\mathsf{FairRec}_{O(\kappa)}$, any semantically-secure symmetric encryption scheme may be used (with key length $O(\kappa)$). Then we define:

$$\mathcal{F}'(x, y) = \left\{F'_0(x, y), F'_1(x, y)\right\} = \{(\mathsf{Enc}_{k_{\mathsf{Enc}}}(F_0(x, y)), s_0), (\mathsf{Enc}_{k_{\mathsf{Enc}}}(F_1(x, y)), s_1)\}$$

where $(s_0, s_1) = \mathsf{Share}(k_{\mathsf{Enc}}, r)$, and r is chosen uniformly at random.

The size of the input to FairRec is the size of the share of one decryption key. The fair computation of $\mathcal{F}(x, y)$ follows easily:

1. Execute a secure-with-abort protocol to compute $\mathcal{F}'(x, y)$ in the OT-hybrid model (e.g., [41]).
2. Player $i \in \{0, 1\}$ parses the output $F'_i(x, y)$ as
$$z_i = (z_{\mathsf{Enc}}, z_{\mathsf{FairRec}}) = (\mathsf{Enc}_{k_{\mathsf{Enc}}}(F_i(x, y)), \mathsf{Share}(k_{\mathsf{Enc}}, r)_i)$$
and submits z_{FairRec} to the ideal function FairRec
3. Let K_i denote the output that player i receives from FairRec. If $K_i = \perp$, output $\perp$. Otherwise, output $\mathsf{Dec}_{K_i}(z_{\mathsf{Enc}})$.

Due to space limitations, we defer the proof of security for this protocol to the full version of the paper.

Standalone vs. Composable Security. Note that the SFE security definitions we use in this paper are in the *standalone* setting, and in particular allow the ideal-world simulator to rewind the adversary. In the first part of Thm. 3.1, using $\mathsf{FairRec}_{O(m)}$ and a one-time pad, this is not a problem: we can use a *straight-line* simulator (that does not rewind the adversary) and prove security even under parallel composition. In the second part of the theorem, however (with a small FairRec functionality), we use rewinding in an essential way. Thus, $\mathsf{FairRec}_{O(\kappa)}$ may not be complete for fairness under parallel composition. The crux of the problem is that the adversary's actions may depend on the encryption used to reduce the input size to FairRec (this causes a *selective decommitment* problem). We note that given access to $\mathsf{par}_{O(m)}(\mathsf{FairRec}_{O(\kappa)})$ ($O(m)$ parallel calls to a small fair reconstruction functionality) we can construct a protocol that is provably secure under parallel composition: we can replace the encryption with an all-or-nothing transform of the output of $\mathcal{F}$, each share of which is then split between the parties using a 2-out-of-2 NMSS scheme. These shares are fairly reconstructed in parallel using $\mathsf{par}_{O(m)}(\mathsf{FairRec}_{O(\kappa)})$.

3.2 Fairness-Complete Primitives for Multi-party Computation

If we only cared about achieving fairness, and not about guaranteeing output delivery, then it is possible to extend the protocol of the previous section to multi-party setting in a straightforward way by using n-out-of-n NMSS. However, guaranteeing robustness as well is a bit more subtle (note that in the two-party case, robustness and fairness are equivalent). Below, we describe three different primitives that are complete for *robust* secure computation. Recall that in a robust, ideal-world computation, the trusted third party will never abort; instead it replaces the inputs of aborting players with a default value, guaranteeing that the honest parties always receive output. Each of the three primitives has a different trade-off between the call complexity (how many times the functionality must be invoked) and the input size to the primitive (depending on the number of participating parties and the description of the function to be evaluated).

Universal Black Box (UBB). This primitive was originally proposed by Fitzi et al. [26]: The UBB receives as input from each party both a circuit (specifying the function to be computed) and an argument to that function. It then partitions the parties according to the circuit that they provide. For each set of parties that gave the same circuit as input, the UBB primitive outputs to that set of parties the evaluation of that circuit on the arguments given by those parties, using default arguments for the remaining parties. It is easy to see that a single call to this primitive can be used to securely compute any function.

Fair Consistent Reconstruction (FCR). Before using the FCR primitive to compute a functionality F, the n parties first perform a secure computation with abort for the related function F'. The function $F'(x_1, \ldots, x_n)$ chooses random keys $k_1, \ldots, k_n$ and computes $k = \sum_{i=1}^{n} k_i$ and $o = \mathsf{Enc}_k(F(x_1, \ldots, x_n))$. It also generates a key pair (sk, vk) for some public-key signature scheme, and computes $\sigma_i = \mathsf{Sign}(sk, k_i)$ for every $i \in [n]$. The output to party i is (o, vk, k_i, σ_i). We note that if any player does abort during this computation[1], the remaining players simply substitute some default input for him and begin again.

The parties then invoke FCR primitive on the values (vk, k_i, σ_i) as received from the execution of F'. The output of FCR is as follows:

Case 1: If not all parties submit the same value for vk, then FCR partitions the players according to the values of vk they submitted and returns these partitions.
Case 2: If any player submits (vk, k_i, σ_i) such that $\mathsf{Vrfy}(vk, k_i, \sigma_i) = 0$ (i.e. they submitted an invalid signature on key k_i), FCR returns a message identifying player i as a cheater.
Case 3: If all parties submit valid signatures under the same key vk, then FCR returns $k = \sum_{i=1}^{n} k_i$.

If players receive output indicating that case 1 occurred, then they begin the entire computation again using default values for anyone outside their own partition. (Note that all honest players are guaranteed to be in the same partition.) If they receive output indicating that case 2 occurred, then they begin the computation again excluding the cheating party and using a default value for his input. Finally, if they receive output indicating that case 3 occurred, they simply use k to decrypt o and recover the output of F. This process may continue for a total of $n-1$ iterations until the protocol terminates. We defer the proof of security to the full version of the paper.

All-Subsets Reconstruction (ASR). The ASR primitive is, essentially, a version of the FCR primitive that can be used to fairly compute any functionality with a single invocation (rather than n). The price is that its input size is exponential in the number of parties (although still independent of the complexity of the target functionality). The input from party i is a set of inputs to the FCR primitive: one input to FCR for every subset $S \subseteq [n]$ of the parties such that $\{i\} \subset S$. The ASR primitive internally runs the entire reconstruction protocol using FCR:

1. ASR begins by running FCR using the inputs corresponding to the set of all parties.
2. If reconstruction of k fails because FCR ends in either case 1 or case 2, ASR recurses, either after partitioning the players, or after dropping a cheater, and calls FCR with the appropriate corresponding inputs.
3. If the reconstruction succeeds, ASR simply outputs k to all players still "active" in the successful call to FCR.

In order to use this primitive to compute a functionality F, the n parties first perform a secure computation with abort for the related function F'', described below. Let F' be the function computed in the protocol that uses FCR, and let out_i be the output of player i from F'. Then for every subset of parties S, $F''(x_1, \ldots, x_n)$ computes the outputs $(\mathsf{out}_1^{(S)}, \ldots, \mathsf{out}_n^{(S)}) = F'(x_1^{(S)}, \ldots, x_n^{(S)})$ where

$$x_i^{(S)} = \begin{cases} x_i & \text{if } i \in S \\ \perp & \text{if } i \notin S \end{cases}$$

[1] There exist protocols in which all players are correctly informed of the identity of the aborting player [31,30].

(i.e., $\mathsf{out}_i^{(S)}$ is the output of the function F' to party i assuming only the parties in S are honest and the remaining parties abort). Note that by the definition of F', $\mathsf{out}_i^{(S)} = (o^{(S)}, vk^{(S)}, k_i^{(S)}, \sigma_i^{(S)})$. F'' outputs $\mathsf{out}_i^{(S)}$ to player i, for each set S such that $\{i\} \subset S$.

The parties invoke ASR with these outputs. If an honest party i receives the output $\perp$, it assumes it is the only honest party and computes the functionality F on its own. Otherwise, it uses the output k to decrypt $o^{(S)}$ (for any S such that $i \in S$). We again defer the proof of security to the full version.

4 Limits on Fairness-Completeness

In this section we show that there does not exist a finite (i.e. "short") primitive that is complete for fairness. More specifically, we prove that the $\mathsf{FairRec}_\kappa$ function cannot be fairly computed even if the players are given parallel access to a primitive of size $O(\log\kappa)$. There are two main ideas behind the proof. For simplicity, imagine for now that the entire protocol consisted of a single call to this short primitive. Our first observation is that because the primitive is short, the adversary can locally simulate it, computing its output for each possible input of the other party. This will play a crucial role in our proof, but it does not itself suffice: so far the adversary has no way of knowing *which* of these outputs are correct. However, because the primitive is supposed to be complete for fairness, it allows us to compute the $\mathsf{FairRec}$ functionality, which has a very useful property: its output is *verifiable*. That is, when two parties are given inputs generated by Share, then the correct output of $\mathsf{FairRec}$ is $(s, 0)$, where the flag 0 indicates that s is the correct output. Furthermore, for incorrect inputs, with overwhelming probability the output of $\mathsf{FairRec}$ is $(\perp, \perp)$. The adversary simply computes the primitive for every possible input of the other player, and outputs s when he recovers it.

When we consider a protocol with many calls to the primitive (including parallel calls), we combine the above ideas using a standard hybrid argument. If the adversary aborts before any invocations of the primitive, he cannot learn anything about the output s. On the other hand, if he behaves honestly in all invocations, he should always recover s. We prove below that there is some specific invocation for which the adversary can gain a non-negligible advantage over the honest party by aborting and simulating the input to that invocation as described above. Finally, he can guess which invocation will allow this advantage with significant probability. Formally, we prove:

Theorem 2. *Let g be an $O(\log\kappa)$-bit primitive. Then for any polynomial p, $\mathsf{par}_{p(\kappa)}(g)$ is not complete for fairness.*

Note that for any $k \geq 1$, $\mathsf{par}_k(g)$ is a more powerful primitive than g (i.e., if the fairness of $\mathcal{F}$ reduces to g then the fairness of $\mathcal{F}$ also reduces to $\mathsf{par}_k(g)$). We are proving an impossibility, so starting with a more powerful primitive strengthens our results. Our proof will hold even if we restrict the adversarial behavior to aborting early.

Proof. Suppose there exists such a primitive g and polynomial p. Consider the $r = r(\kappa)$ round protocol Π_p^g that fairly computes $\mathsf{FairRec}_\kappa(x, y)$ while making a call to $\mathsf{par}_{p(\kappa)}(g)$ in each round. We can think of this call as $p(\kappa)$ parallel calls to g. Without loss of generality, we assume that these calls to g constitute the only communication between the players[2]. Let $q = p \cdot r$ be the total number of calls to g. For each round $i \in r$, we define some arbitrary ordering σ_i on the parallel calls to g that occur in that round. This induces a natural ordering over all q calls to g, where for $i < j$, calls in round i are ordered before calls in round j. We let g_k denote the kth call to g according to this ordering.

Consider an execution of Π_p^g in which $\mathsf{Share}(s, r) = (s_0, s_1)$ is used to generate shares, for a random s and r. Player j gets the share s_j. We let the value a_i denote the output of player 0 when player 1 acts honestly for the first i calls to g (according to the ordering previously described) and then aborts. We define b_i in the symmetric way. Note that by correctness of Π_p^g, and the definition of $\mathsf{FairRec}_\kappa$, for all i

$$\Pr[a_i \neq s \wedge a_i \neq \bot] = \mathsf{negl}(\kappa) = \Pr[b_i \neq s \wedge b_i \neq \bot]$$

and

$$\Pr[a_q = s] = \Pr[b_q = s] = 1 - \mathsf{negl}(\kappa).$$

where the probability is over the random tapes of the players. Furthermore, by the definition of FairRec and the properties of a NMSS scheme,

$$\Pr[a_0 \neq \bot] = \mathsf{negl}(\kappa) = \Pr[b_0 \neq \bot].$$

It follows that for every large enough κ, there exists a polynomial $p'(\kappa)$ and a round i such that either

$$\Pr[a_i = s] - \Pr[b_{i-1} = s] \geq \frac{1}{p'(\kappa)}.$$

or

$$\Pr[b_i = s] - \Pr[a_{i-1} = s] \geq \frac{1}{p'(\kappa)}.$$

Without loss of generality, we will assume the former, and we demonstrate an adversary $\mathcal{A}$ that breaks the security of Π_p^g with probability at least $1/(q \cdot p'(\kappa))$.

$\mathcal{A}$ begins by choosing a random value $i^* \in [1, \ldots, q]$, and plays honestly for the first $i^* - 1$ calls to g (i.e., submits correct values to g) and then aborts. Note that the resulting output of player 1 is b_{i^*-1}. The adversary now attempts to compute the value of a_{i^*} by simulating the side of player 1. Note, however, that by definition, the value of a_{i^*} depends on honest input to g_{i^*} from both players, and $\mathcal{A}$ may not know (anything) about player 1's input to g_{i^*}. Here we use the fact that g has short inputs, and that FairRec is verifiable. $\mathcal{A}$ goes through all possible inputs $\beta \in \{0, 1\}^{O(\log \kappa)}$ that player 1 might have sent to g_{i^*}, and for each such value he simulates g internally, using as input his own (honest) value that he *would* have sent if he had not aborted, and β. He computes a_{i^*} from his view in the (real) interaction with player 1, and the simulated output of g_{i^*}. Since one of these values of β is the value used by player 1 in the actual execution, it follows that the correct value of a_{i^*} is among this set of outputs. Furthermore, if some simulated $a_{i^*} = s' \neq \bot$ then $s' = s$ with overwhelming probability. $\mathcal{A}$ outputs

[2] This is without loss of generality because we can always modify g to do message transmission, in addition to its original functionality. Note also that if less than $p(\kappa)$ calls are needed in a particular round, the players can make extra calls with random inputs, ignoring the outputs, to make the total number of calls $p(\kappa)$.

$s' \neq \perp$ if this occurs, and $\perp$ otherwise. By our assumption, there exists an i such that $\Pr[a_i = s \wedge b_{i-1} = \perp] \geq \frac{1}{p'(\kappa)}$. Hence, $\mathcal{A}$ recovers s without the honest party receiving output with probability $1/(q \cdot p'(\kappa))$, contradicting the fairness of protocol.

Theorem 3. *Simultaneous broadcast is not complete for fairness.*

The fact that short simultaneous broadcast (cf. Section 2) is not complete for fairness follows from Theorem 2. We give two different proofs that a large simultaneous broadcast is not complete for fairness.

Proof (First proof of Thm. 3.). The first proof follows as a corollary of Theorem 2, and Lemma 1 (below). This is true because if long-SB were complete, then by Lemma 1, short-SB would also be complete, contradicting Theorem 2.

Lemma 1. *Let g denote the k-bit* SB *primitive. For any $p \in \mathbb{N}$, there exists a protocol Π_p^g that implements kp-bit* SB *with perfect security.*

Proof. The protocol is the (trivial) one round protocol in which both parties split their inputs into p blocks of size k, submit block i to instance g_i, and output the concatenation of the p outputs (maintaining the order). The proof of the lemma is straightforward, so we omit a formal exposition. We simply note that the decision of an adversary to change its input to any instance(s) of g (including the decision to abort in some instance(s)) is entirely independent of the actions or input of the honest party. The simulator simply recovers the p values that the adversary intended for g (recall that an abort is treated as input 0^k), concatenates them, and forwards them to the trusted party. After receiving output, the simulator rewinds the adversary, parses the output into blocks, and uses block i as the honest player's input to g_i.

In our full version we provide a second proof that is more direct and is of independent interest. Due to space constraints we omit the proof here.

5 A Fairness Hierarchy

In this section we show that for small k, fairness cannot be "amplified" at all (with regards to input size). Specifically, for small values of k we show that no k-bit functionality can be used to build $(k + 1)$-SB, even if standard cryptographic assumptions are allowed. Unlike in Section 4, here we assume that the players do not have parallel access to the primitive. More formally:

Theorem 4. *For $k = O(\log \kappa)$, the fairness of $(k + 1)$-*SB *does not reduce to any k-bit primitive.*

To gain some intuition for how we prove the theorem, consider that in an ideal world execution of simultaneous broadcast, if the players inputs are chosen independently, then (by definition) each player's output is independent of their own input. However, we show below that in any real world protocol constructing $(k + 1)$-SB from k-bit functionalities, this property cannot be guaranteed. We demonstrate that there always exists some round in which an adversary can gain information about the other party's input, as well as some later round in which it can still affect the other party's output by choosing whether or not to abort. By choosing whether to abort in the later round based on what is learned in the earlier round, the adversary can correlate the output of the honest player with his input.

There are two main ideas behind the proof. The first idea is the one used in the proof of Theorem 2: because the $k = O(\log \kappa)$-bit inputs to the primitive are small, the adversary can gain an "information lead" of one round by testing all $2^k = \text{poly}(\kappa)$ inputs that the honest party might send to the k-bit primitive in the next round. For each of these possible outputs from the primitive, the adversary computes the $k+1$-bit value that he *would* have output in the protocol if this were the last thing he received in the protocol (i.e. if the honest player aborted immediately afterwards). In this way the adversary computes the set of all "potential outputs" that he could possibly output if the honest party sends a single additional message and then aborts.

Unfortunately, unlike in the proof of Theorem 2, we cannot argue here that the adversary recognizes the correct output of the protocol among this set: in Theorem 2 the output was verifiable, while the output of SB is not. Instead, we rely on a different observation: there are twice as many possible input values to the $k+1$-SB protocol as there are potential outputs from the k-bit primitive used in any particular round. Thus, for every round, at least half of the possible $k+1$ bit inputs to SB will not appear in the set of potential outputs. We use this fact to show that there exist two inputs, y, y', and a round i such that if the honest player has input y, the potential output set in round i contains y but not y', while if his input is y', it contains y' but not y. Furthermore, we will prove that at least one of the parties can affect the other's output by sending a random message in some later round, $i' > i$, and then aborting. The adversarial strategy is as follows: he runs the protocol honestly until round i, and then determines which of the two inputs the honest party has. Depending on what he learns, the then chooses whether to complete the protocol honestly, or to later abort after round i'. By making this decision, he creates a correlation between the honest party's output and his input, violating the security of the simultaneous broadcast.

Due to space considerations, we defer the formal proof to the full version of the paper.

6 Fairness Combiners

We have demonstrated that one possible approach for achieving fair secure computation is to rely on a trusted third party to implement the FairRec functionality. A natural question that arises is, how much can we distribute that trust? Instead of trusting a single party, can we use multiple parties, guaranteeing both fairness and security so long as some number of them act honestly?

This can be thought of as a fairness analogue of *combiners* [37,38]; the two computing parties (clients) have access to n "fairness providers" (servers), of which at least $n - t$ are guaranteed to be honest. Two questions arise: what are the values of t and n for which this is possible, and what are the minimal requirements of the honest fairness providers? Below, we show that combining is possible if and only if $n > 2t$.

6.1 Combining with an Honest Majority

If the clients and servers had a broadcast channel and a complete point to point network, the problem would simply be a restricted case of secure multiparty computation where only two parties have input and output. In this case we could use the protocol of Rabin and Ben-Or to fairly compute any functionality [49], since a strict majority of the servers are honest, and at least one client is honest. However, we are interested in using *independent* servers, where each provider communicates only with the two clients and is not expected to know anything about the other primitives.

Instead we model the problem as a secure multiparty computation over an *incomplete* point-to-point network, in which there only exist communication channels between the two computing players, and from each player to each server. The adversary is allowed to corrupt at most one of the computing players, and at most t of the fairness providers. Finally, we assume the existence of oblivious transfer, although we discuss how to make our results unconditional in the full version. Viewing the problem this way, one solution is to emulate the missing channels (including a broadcast channel), enabling the use of the general result of Rabin and Ben-Or. We outline this solution below.

The two clients begin by executing an *unfair* secure computation (using oblivious transfer) to establish correlated randomness for each pair of servers. This randomness will include shared secret keys that enable any two servers to authenticate and encrypt messages to one another, as well as additional correlated randomness that will enable broadcast for any $n > 2t$ [4,27]. In order to prevent the clients from learning the correlated randomness, the computation will actually output NMSS shares of the output, one share to each client, which they then relay to the appropriate servers. If the protocol to establish these keys ends unfairly, then the honest client simply aborts; no information is leaked and no harm is done, since this execution is independent of their inputs. Otherwise, the servers inform both clients that they have successfully reconstructed their keys and randomness. If anyone indicates otherwise, all players immediately abort the protocol. With the communication channels in place, the players can now execute the protocol of Rabin and Ben-Or [49] to compute the desired functionality.[3]

In the full version of the paper, we show how to construct a more efficient protocol for this task and discuss in greater detail the security properties we can guarantee.

Impossibility of Combining with a Faulty Majority. The previous result is tight; if the majority of the fairness providers are corrupt, they do not help us to achieve fairness. Specifically, we consider the case where the players have access to two fairness providers, one of which is corrupt, and show that any function that can be computed in this model can be computed in the plain model. Since there exist functions that cannot be computed in the plain model [20], it follows that the same functions cannot be computed in our setting. Below, when a protocol Π permits calls to two instances of some primitive g, we denote this by Π^{g_1,g_2}.

[3] We note that the protocol of Rabin and Ben-Or is *robust*, which means that even if some servers abort, the clients can continue the computation with the remaining servers. This enables even two honest clients to complete their computation correctly, so long as less than half of the servers are corrupt.

Theorem 5. *Let g_1 and g_2 be two instances of some arbitrary functionality g. If Π^{g_1,g_2} securely computes function $\mathcal{F}$, even when an adversary corrupts one of the instances of g, then there exists a protocol Π' that securely computes $\mathcal{F}$ without access to any primitives.*

Proof (Sketch). The proof is a standard partitioning argument. Π' simply follows the description of Π^{g_1,g_2}, delegating the responsibilities for computing g_1 and g_2 to players 0 and 1 respectively. By our assumption, so long as one of the two instances is executed fairly and securely, Π' fairly and securely computes $\mathcal{F}$. Since one of the players is honest, the primitive that they control will always be executed honestly.

References

1. Proceedings of the Twentieth Annual ACM Symposium on Theory of Computing, Chicago, Illinois, USA, May 2-4. ACM, New York (1988)
2. Asokan, N., Schunter, M., Waidner, M.: Optimistic protocols for fair exchange. In: CCS, pp. 7–17. ACM, New York (1997)
3. Asokan, N., Shoup, V., Waidner, M.: Optimistic fair exchange of digital signatures. IEEE Journal on Selected Areas in Communications 18(4), 593–610 (2000)
4. Baum-Waidner, B., Pfitzmann, B., Waidner, M.: Unconditional byzantine agreement with good majority. In: Jantzen, M., Choffrut, C. (eds.) STACS 1991. LNCS, vol. 480, pp. 285–295. Springer, Heidelberg (1991)
5. Beaver, D.: Correlated pseudorandomness and the complexity of private computations. In: STOC, pp. 479–488. ACM, New York (1996)
6. Beaver, D., Goldwasser, S.: Multiparty computation with faulty majority, pp. 468–473 (1989)
7. Beimel, A., Malkin, T.: A quantitative approach to reductions in secure computation. In: Naor, M. (ed.) TCC 2004. LNCS, vol. 2951, pp. 238–257. Springer, Heidelberg (2004)
8. Bellare, M. (ed.): CRYPTO 2000. LNCS, vol. 1880. Springer, Heidelberg (2000)
9. Ben-Or, M., Goldwasser, S., Wigderson, A.: Completeness theorems for non-cryptographic fault-tolerant distributed computation (extended abstract). In: STOC [1], pp. 1–10
10. Blum, M.: How to exchange (secret) keys. ACM Transactions on Computer Systems 1(2), 175–193 (1983); Previously published in ACM STOC 1983 proceedings, pp. 440–447
11. Boneh, D., Naor, M.: Timed commitments, pp. 236–254
12. Boneh, D., Naor, M.: Timed commitments. In: Bellare [8], pp. 236–254
13. Cachin, C., Camenisch, J.: Optimistic fair secure computation. In: Bellare [8], pp. 93–111
14. Canetti, R.: Security and composition of multiparty cryptographic protocols. J. Cryptology 13(1), 143–202 (2000)
15. Chaum, D., Crépeau, C., Damgård, I.: Multiparty unconditionally secure protocols (extended abstract). In: STOC [1], pp. 11–19
16. Chen, L., Kudla, C., Paterson, K.G.: Concurrent signatures. In: Cachin, C., Camenisch, J.L. (eds.) EUROCRYPT 2004. LNCS, vol. 3027, pp. 287–305. Springer, Heidelberg (2004)
17. Chor, B., Geréb-Graus, M., Kushilevitz, E.: On the structure of the privacy hierarchy. J. Cryptology 7(1), 53–60 (1994)
18. Chor, B., Goldwasser, S., Micali, S., Awerbuch, B.: Verifiable secret sharing and achieving simultaneity in the presence of faults (extended abstract). In: FOCS, pp. 383–395. IEEE, Los Alamitos (1985)
19. Chor, B., Kushilevitz, E.: A zero-one law for boolean privacy. SIAM J. Discrete Math. 4(1), 36–47 (1991)

20. Cleve, R.: Limits on the security of coin flips when half the processors are faulty (extended abstract). In: STOC, pp. 364–369. ACM, New York (1986)
21. Cleve, R.: Controlled gradual disclosure schemes for random bits and their applications. In: Brassard, G. (ed.) CRYPTO 1989. LNCS, vol. 435, pp. 573–588. Springer, Heidelberg (1990)
22. Cramer, R., Dodis, Y., Fehr, S., Padró, C., Wichs, D.: Detection of algebraic manipulation with applications to robust secret sharing and fuzzy extractors. In: Smart, N.P. (ed.) EUROCRYPT 2008. LNCS, vol. 4965, pp. 471–488. Springer, Heidelberg (2008)
23. Damgård, I.: Practical and provably secure release of a secret and exchange of signatures. J. Cryptology 8(4), 201–222 (1995)
24. Damgård, I.: Practical and provably secure release of a secret and exchange of signatures. J. Cryptology 8(4), 201–222 (1995)
25. Even, S., Goldreich, O., Lempel, A.: A randomized protocol for signing contracts. Commun. ACM 28(6), 637–647 (1985)
26. Fitzi, M., Garay, J.A., Maurer, U.M., Ostrovsky, R.: Minimal complete primitives for secure multi-party computation. J. Cryptology 18(1), 37–61 (2005)
27. Fitzi, M., Gisin, N., Maurer, U.M., von Rotz, O.: Unconditional byzantine agreement and multi-party computation secure against dishonest minorities from scratch. In: Knudsen, L.R. (ed.) EUROCRYPT 2002. LNCS, vol. 2332, pp. 482–501. Springer, Heidelberg (2002)
28. Galil, Z., Haber, S., Yung, M.: Cryptographic computation: Secure fault-tolerant protocols and the public-key model, pp. 135–155.
29. Garay, J.A., MacKenzie, P.D., Prabhakaran, M., Yang, K.: Resource fairness and composability of cryptographic protocols. In: Halevi, S., Rabin, T. (eds.) TCC 2006. LNCS, vol. 3876, pp. 404–428. Springer, Heidelberg (2006)
30. Goldreich, O.: Foundations of Cryptography: Volume 2, Basic Applications. Cambridge University Press, New York (2004)
31. Goldreich, O., Micali, S., Wigderson, A.: How to play any mental game or a completeness theorem for protocols with honest majority. In: STOC, pp. 218–229. ACM, New York (1987)
32. Goldreich, O., Vainish, R.: How to solve any protocol problem - an efficiency improvement. In: Pomerance, C. (ed.) CRYPTO 1987. LNCS, vol. 293, pp. 73–86. Springer, Heidelberg (1988)
33. Goldwasser, S., Levin, L.: Fair computation and general functions in the presence of immoral majority
34. Gordon, D., Katz, J.: Partial fairness in secure two-party computation. Cryptology ePrint Archive, Report 2008/206 (2008), http://eprint.iacr.org/2008/206
35. Gordon, S.D., Hazay, C., Katz, J., Lindell, Y.: Complete fairness in secure two-party computation. In: Ladner, R.E., Dwork, C. (eds.) STOC, pp. 413–422. ACM, New York (2008)
36. Gordon, S.D., Katz, J.: Complete fairness in multi-party computation without an honest majority. In: Reingold [50], pp. 19–35
37. Harnik, D., Kilian, J., Naor, M., Reingold, O., Rosen, A.: On robust combiners for oblivious transfer and other primitives. In: Cramer, R. (ed.) EUROCRYPT 2005. LNCS, vol. 3494, pp. 96–113. Springer, Heidelberg (2005)
38. Herzberg, A.: Folklore, practice and theory of robust combiners. Journal of Computer Security 17(2), 159–189 (2009)
39. Ishai, Y., Prabhakaran, M., Sahai, A.: Founding cryptography on oblivious transfer - efficiently. In: Wagner, D. (ed.) CRYPTO 2008. LNCS, vol. 5157, pp. 572–591. Springer, Heidelberg (2008)
40. Katz, J.: On achieving the best of both worlds in secure multiparty computation. In: Johnson, D.S., Feige, U. (eds.) STOC, pp. 11–20. ACM, New York (2007)
41. Kilian, J.: Founding cryptography on oblivious transfer. In: STOC [1], pp. 20–31
42. Kilian, J., Kushilevitz, E., Micali, S., Ostrovsky, R.: Reducibility and completeness in private computations. SIAM Journal on Computing 29(4), 1189–1208 (2000)

43. Lepinski, M., Micali, S., Peikert, C., Shelat, A.: Completely fair sfe and coalition-safe cheap talk. In: Chaudhuri, S., Kutten, S. (eds.) PODC, pp. 1–10. ACM, New York (2004)
44. Lindell, A.Y.: Legally-enforceable fairness in secure two-party computation. In: Malkin, T.G. (ed.) CT-RSA 2008. LNCS, vol. 4964, pp. 121–137. Springer, Heidelberg (2008)
45. Luby, M., Micali, S., Rackoff, C.: How to simultaneously exchange a secret bit by flipping a symmetrically-biased coin. In: FOCS, pp. 11–21. IEEE, Los Alamitos (1983)
46. Maji, H.K., Prabhakaran, M., Rosulek, M.: Complexity of multi-party computation problems: The case of 2-party symmetric secure function evaluation. In: Reingold [50], pp. 256–273
47. Micali, S.: Simple and fast optimistic protocols for fair electronic exchange. In: PODC, pp. 12–19. ACM, New York (2003)
48. Pinkas, B.: Fair secure two-party computation. In: Biham, E. (ed.) EUROCRYPT 2003. LNCS, vol. 2656, pp. 87–105. Springer, Heidelberg (2003)
49. Rabin, T., Ben-Or, M.: Verifiable secret sharing and multiparty protocols with honest majority (extended abstract). In: STOC, pp. 73–85. ACM, New York (1989)
50. Reingold, O. (ed.): TCC 2009. LNCS, vol. 5444. Springer, Heidelberg (2009)
51. Yao, A.C.-C.: How to generate and exchange secrets (extended abstract). In: FOCS, pp. 162–167. IEEE, Los Alamitos (1986)

On the Necessary and Sufficient Assumptions for UC Computation*

Ivan Damgård, Jesper Buus Nielsen, and Claudio Orlandi

Department of Computer Science, Aarhus University
{ivan,jbn,claudio}@cs.au.dk

Abstract. We study the necessary and sufficient assumptions for universally composable (UC) computation, both in terms of setup and computational assumptions. We look at the common reference string model, the uniform random string model and the key-registration authority model (KRA), and provide new results for all of them. Perhaps most interestingly we show that:

- For even the minimal meaningful KRA, where we only assume that the secret key is a value which is hard to compute from the public key, one can UC securely compute any poly-time functionality if there exists a passive secure oblivious-transfer protocol for the stand-alone model. Since a KRA where the secret keys can be computed from the public keys is useless, and some setup assumption *is* needed for UC secure computation, this establishes the best we could hope for the KRA model: any non-trivial KRA is sufficient for UC computation.
- We show that in the KRA model one-way functions are sufficient for UC commitment and UC zero-knowledge. These are the first examples of UC secure protocols for non-trivial tasks which do not assume the existence of public-key primitives. In particular, the protocols show that non-trivial UC computation is possible in Minicrypt.

1 Introduction

We study the necessary and sufficient assumptions for universally composable (UC) computation [Can01], both in terms of which setup models are needed and how strong assumptions on the setup are needed, and in terms of necessary and sufficient computational assumptions.

One of the motivation is to study the minimal setup required for UC computation. It is known that some kind of setup *is* required, which makes it a theoretically interesting question exactly how strong an assumption must be made on the setup. We study both the common reference string model (CRS) and the key registration authority model (KRA), and some variations.

The goal of the study is also to determine the relationships between one way functions (OWF), passive secure stand-alone oblivious transfer (SA-OT)[1], UC

* A full version of this work can be found at http://eprint.iacr.org/2009/247

[1] If a passive secure SA-OT exists then also an active secure SA-OT exists via standard compilation techniques.

D. Micciancio (Ed.): TCC 2010, LNCS 5978, pp. 109–127, 2010.

commitments (UC-Com) and UC oblivious transfer (UC-OT) in different set-up models[2]: for stand-alone security, we know that OWF are sufficient for other cryptographic tasks such as commitments and zero-knowledge proofs. Are OWF equivalent to any of these tasks when it comes to UC security? For the CRS models, Damgård and Groth [DG03] answered the question negatively showing that UC-Com implies key agreement in the CRS models and therefore, given the black-box separation between OWF and key agreement [IR89, RTV04], OWF are not sufficient to realize UC-Com. We find it interesting to study if this is inherent or associated to the particular setup model. We include SA-OT because it is complete for stand alone computation [Kil88], and therefore a natural question is whether SA-OT is sufficient also for UC computation. On the other hand it is interesting to know whether this assumption is minimal, i.e. whether SA-OT is also necessary to implement UC-Com and UC-OT. The motivation for including UC-OT is that it is complete for general UC computation: it is possible to implement any well-formed ideal functionality given the UC-OT functionality, see [CLOS02, IPS08]. Finally the motivation for including UC-Com is that it is potentially weaker than UC-OT but still implies a number of non-trivial tasks like coin-flip and zero-knowledge.

Highlights. We highlight some of the new findings which we find particularly interesting: In the KRA model, we provide the first construction of UC commitment from one-way functions—all previous constructions, to the best of our knowledge, used special assumptions or assumed at least public-key encryption. A consequence of it is that zero-knowledge and coin-flip can be UC securely implemented in Minicrypt. Until now it was not known if any non-trivial UC computation was possible in Minicrypt.

Remembering that UC-Com implies SA-OT in the CRS models we get another new result: The choice of the setup model can make a difference in which ideal functionalities can be implemented under a given computational assumption. In the CRS model we need SA-OT for UC-Com, but in the KRA model we can do with just OWF. This seems to be the first such separation of the setup models.

It turns out that SA-OT is sufficient for UC-OT in any setup model we considered, in particular in the minimal version of both the CRS and KRA model. Since some setup assumption is needed for general UC computation, this seems a very positive addition to the UC theory: Some setup is needed, *but even the most trivial setup will allow to implement any well-formed functionality.*

Finally, we show how to implement authenticated channels given a minimal meaningful KRA: Implementing authentication in a public-key setting is of course trivial if one can choose the structure of the public keys—one includes a verification key in the public key and signs all messages. It is by far trivial in our relaxed KRA model as we make no assumption on the public keys except that they are in the range of some one-way function, which might itself be maliciously chosen. Standard constructions of signature schemes from one-way functions use verification keys with much more structure than this. This seems

[2] When writing UC-Com or UC-OT we mean the multi-party, multi-session version of the protocols.

to be the first result which shows that *any value you could hope for to act as a public key can actually be used to implement an authenticated channel.*

Setup models: We look at five setup models:

- In the *uniform common random string (U-CRS)* model we assume that a single uniformly random ℓ-bit string crs is chosen by a trusted party and made public. Here ℓ is chosen by the protocol.
- In the *chosen common reference string (C-CRS)* model the trusted party samples crs using a poly-time one-way[3] distribution $D : \{0,1\}^\kappa \rightarrow \{0,1\}^\ell$, which allows crs to have a particular form. We assume that the trusted party samples a single $crs = D(r)$ for uniformly random $r \in \{0,1\}^\kappa$ and makes crs public. The function D might be given by the protocol π.
- The *any common reference string (A-CRS)* model is like the C-CRS model, except that we let the adversary pick D, under the only restriction that D is one-way. The trusted party samples $crs = D(r)$ and makes (D, crs) public. [4]
- In the *chosen key registration authority (C-KRA)* model we assume that the protocol contains a poly-time function f_i for each party P_i. A trusted party will sample $pk_i = f_i(s_i)$ for each P_i and give s_i to P_i and pk_i to all other parties. This models a key-registration authority with public keys pk_i, secret keys s_i and where the parties are guaranteed to know their secret keys. [5]
- The *any key registration authority (A-KRA)* model is like the C-KRA model, except that we allow the adversary to specify each f_i, under the only restriction that f_i is a one-way function when P_i is honest.

In the CRS models we in addition assume the presence of authenticated channels, as the existence of a CRS clearly does not allow authentication: All parties know the CRS and nothing else, so nothing distinguishes an honest party from the adversary. In the KRA models we start from the unauthenticated channels model, as the existence of a public-key infrastructure has the potential to allow authentication. In the A-CRS model the protocol π does not choose D, and the security of π should hold for any one-way function D. Another way of phrasing the model is to say that the protocol π, parametrized by D, should be secure in the C-CRS model for any one-way function D. In some sense this models the minimal meaningful common random string: we do not make assumptions on how random it is, but the parties can agree on the fact that there is something about the string which neither of them knows.

The A-KRA model in some sense is the minimal meaningful assumption on a key registration authority: Each party has a publicly known public key pk_i and

[3] If D is invertible then the C-CRS and the U-CRS model are trivially equivalent.

[4] Note that the A-CRS model generalizes both the U-CRS and the C-CRS, for computational security: the U-CRS is computationally indistinguishable from a setup where the trusted party picks a random seed and expands it using a pseudo random generator.

[5] This is essentially the key registration with knowledge (KRK) model from [BCNP04]. We cannot start from the KR model from [BCNP04] as we need that parties know their secret keys to implemented authenticated channels.

there is some secret about pk_i which only P_i knows. A protocol for the A-KRA can therefore be run given any meaningful key registration authority, with no assumption whatsoever about the form of pk_i or the exact hardness of finding s_i. Note that one can think of simpler public key models, as the bare public key model (BPK) introduced in [CGGM00], that does not require corrupted parties to know their secret keys. However, even if the BPK model has been successfully used to break some impossibility results about concurrent ZK (see [OPV08] and reference therein), it was shown in [KL07] that UC computation is impossible even in the BPK model or any other public key model "without knowledge".

Results: Let $C_1, C_2 \in \{$ OWF, SA-OT, UC-Com, UC-OT $\}$ and M one of the described setup model. Then we can ask ourselves questions of the form *"is C_1 necessary/sufficient for C_2 in the M model?"* Some of these are of course trivial, like *is OWF necessary for UC-OT in the U-CRS model?*, but many are non-trivial and theoretically interesting, like *is SA-OT sufficient for UC-OT in the A-CRS model?*. The answers to these questions are pictorially illustrated in Figures 1 and 2. The main results are proved through the text while the implications that were already known, or that easily follow from our results and known facts are presented in Sec. 8.

When SA-OT appears on the left side of the implication, the assumption is that there exist a protocol that implement SA-OT in the stand alone mode. When UC-Com and UC-OT appear on the left side of the implication, the assumption is that there exist a protocol that implement that functionality *in the model in question*, i.e., we are not referring to an hybrid world where the functionality is given to the party: We assume that the parties know a protocol to implement the functionality.

Note that the figure states that UC-Com is not sufficient for UC-OT in the C-KRA model, while the answer is yes in the A-KRA model. This may seem surprising because the C-KRA model (unlike A-KRA) allows the protocol to

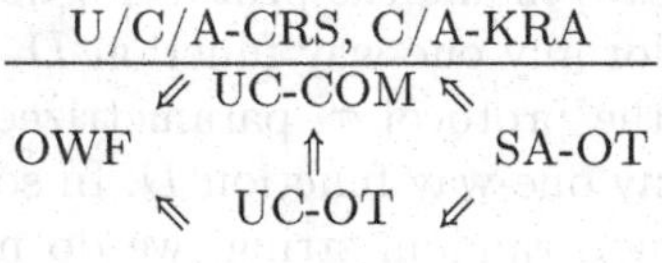

U/C/A-CRS, C/A-KRA

⇙ UC-COM ⇖
OWF ⇑ SA-OT
⇖ UC-OT ⇙

Fig. 1. One direction of the relationship between the primitives holds in any setup model

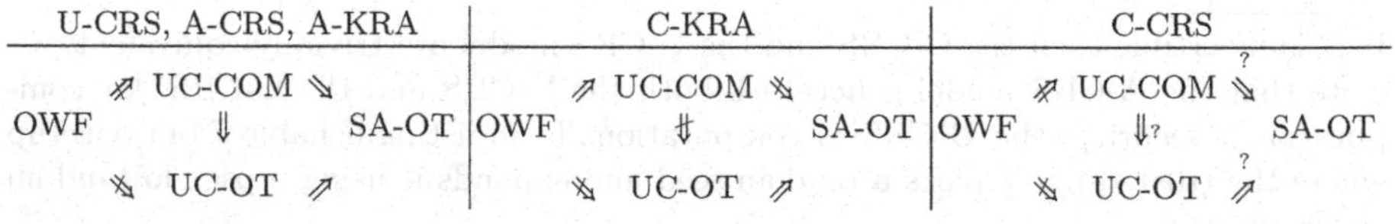

U-CRS, A-CRS, A-KRA	C-KRA	C-CRS
⇗̸ UC-COM ⇘ OWF ⇓ SA-OT ⇘̸ UC-OT ⇗	⇗ UC-COM ⇘̸ OWF ⇓̸ SA-OT ⇘̸ UC-OT ⇗	⇗̸ UC-COM ⇘? OWF ⇓? SA-OT ⇘̸ UC-OT ⇗?

Fig. 2. The other direction differs in different setup models

choose how public keys are computed and so it seems that anything that is possible in the A-KRA model should also possible in C-KRA. The catch is that the UC-Com assumption is *not* the same in the two models, in particular, having a UC-commitment scheme that works in the A-KRA model is a much stronger tool than one that needs C-KRA.

For all our negative answers to *sufficient* questions and positive answers to *necessary* questions, we mean that there is no black-box construction. We cannot answer whether OWF is sufficient for UC-Com with our current understanding of complexity theory: It might be that one-way functions do not exists, in which case the assumption OWF is false, and then OWF $\Rightarrow$ UC-Com is true. The result UC - Com $\overset{\text{(C-KRA)}}{\nRightarrow}$ UC - OT therefore means that there is no black-box construction which takes an implementation of UC commitment for the C-KRA model and gives an implementation of UC OT for the C-KRA model. For some of our positive answers to *sufficient* questions and negative answers to *necessary* questions, we appeal to non-black box constructions. As an example, the result OWF $\overset{\text{(C-KRA)}}{\Rightarrow}$ UC - Com uses a description of the circuit for the OWF.[6]

Related work. In [CLOS02] the main feasibility result for UC computation in the CRS model can be found. [CLOS02] needs to assume enhanced trapdoor permutation in order to achieve their results, while we use the strictly weaker assumption SA-OT. On the other hand this comparison is not quite fair as [CLOS02] tries to achieve adaptive security, and consider static security just as a special case, while our focus is entirely on static security. In [LPV09] a general framework for UC feasibility results is presented, showing how different setup assumptions (including timing model, tamper proof hardware, etc.) can be seen as different implementations of what the authors call *UC puzzles*. While in [LPV09] the results are proved assuming the existence of enhanced trapdoor permutation, we look at strictly weaker assumptions as OWF and SA-OT.

In a recent series of papers [PR08, MPR09], the classification of the cryptographic complexity of UC functionalities is studied. Perhaps most interestingly with respect to our work, in [MPR09] it is shown the SA-OT assumption is equivalent to any UC functionality being either trivial or complete. There is a clear overlap between these results and some of ours, however we focus on *setup* functionalities - that are invoked just once at the beginning of the protocol, while the constructions in [MPR09] use the ideal functionalities during the protocol in an on-line fashion.

2 The KRA Model

In this section we give our model of minimal public-key setup, where each party knows a secret which is not known by the other parties. We associate these

[6] It might be possible that $A \nRightarrow B$ and $A \Rightarrow B$ at the same time. However, if for any of the $\nRightarrow$ separations in Figure 1, 2 this is the case, then one would have a non-black-box construction of SA-OT using OWF. Such a construction is unlikely to exist - or at least requires completely new cryptographic techniques - see also [RTV04].

secrets to public values which we distribute via a key generator G. When sampled it outputs $((R_1, s_1), \dots, (R_n, s_n))$, where each R_i is a description of a PPT set and s_i is the *secret of* P_i, and $s_i \in R_i$ for $i \in [n] = \{1, \dots, n\}$. We call $R = (R_1, R_2, \dots, R_n)$. For a "normal" KRA we would have that the description of R contains the parties' public keys $pk_1, \dots, pk_n$ and that $s_i \in R_i$ if s_i is the secret key associated to pk_i; in this case we will write $(pk_i, s_i) \in R_i$. To model that a party's secret s_i is hard to find for the other parties we require that it is hard to find any s'_i such that $s'_i \in R_i$. This should hold even if one is given $R \cup \{s_j\}_{j \in [n] \setminus \{i\}}$.

To allow corrupted parties to use secrets different from those of the honest parties (maybe fixed instead of random or even of another form), we let G depend on the set C of corrupted parties, and we let the adversary $\mathcal{A}$ influence the key generation as follows: Both G and $\mathcal{A}$ are ITMs. First G is given input n and C, where n defines the number of parties and $C \subset [n]$ defines the set of corrupted parties. Then G and $\mathcal{A}$ interact and at some point G outputs $(R, s_1, \dots, s_n)$; we write $(R, s_1, \dots, s_n) \leftarrow (G(n, C), \mathcal{A})$. For G to be meaningful we require that $s_j \in R_j$ for *all* parties $P_{j \in [n]}$. We only require that the secrets of *honest* parties, $P_{i \in [n] \setminus C}$, are hard to find, as it is not necessarily meaningful to require that corrupted parties keep their secrets hidden.

We introduce some convenient notation for the case where all public keys are generated using the same function f. For a function $f : \{0,1\}^\kappa \rightarrow \{0,1\}^\ell$ we define the key generator G^f as follows: For each honest party P_i it samples $s_i \in \{0,1\}^\kappa$ and computes $pk_i = f(s_i)$. Then it outputs $\{pk_i\}_{i \in [n] \setminus C}$ to the adviser. It interprets the next message from the adviser as a set $\{s_i\}_{i \in C}$ and computes $pk_i = f(s_i)$ for $i \in C$. It defines R^f by $(pk, s) \in R^f$ iff $pk = f(s)$ and then outputs $((pk_1, s_1, R^f), \dots, (pk_n, s_n, R^f))$.

- Let n be the number of parties and C the set of corrupted parties, and run $G(n, C)$ with the UC adversary $\mathcal{A}$ as adviser.
- When $G(n, C)$ outputs $(R, s_1, \dots, s_n)$, then send $(R, \{s_i\}_{i \in C})$ to $\mathcal{A}$ and send (R, s_i) to each P_i, letting $\mathcal{A}$ determine the delivery time.

Fig. 3. The KRA ideal functionality $\mathcal{F}^G_{\text{KRA}}$ for a generator G

Definition 1. *We call G* meaningful *if $\forall$ $\mathcal{A}$, $\mathcal{A}$ wins the following game with negligible probability: Run $\mathcal{A}$ to get (n, C), with n polynomially bounded and $C \subset [n]$. Then sample $(R, s_1, \dots, s_n) \leftarrow (G(n, C), \mathcal{A})$. At this point $\mathcal{A}$* wins *if $\exists j \in [n]$ $s_j \notin R_j$. If $\mathcal{A}$ did not win here, run $\mathcal{A}$ on R to get $i \in [n] \setminus C$, and run $\mathcal{A}$ on $s_{-i} = \{s_j\}_{j \in [n] \setminus \{i\}}$ to get an output (i, s'_i). If $s'_i \in R_i$, then $\mathcal{A}$* wins.

Definition 2. *Let $\mathcal{G}$ be a set of key generators, let π be a protocol and $\mathcal{F}$ an ideal functionality. We say that π is a UC secure implementation of $\mathcal{F}$ with a $\mathcal{G}$ KRA if π is a UC secure implementation of $\mathcal{F}$ in the $\mathcal{F}^G_{\text{KRA}}$-hybrid model (Fig. 3) for all $G \in \mathcal{G}$. We say that π is a UC secure implementation of $\mathcal{F}$ with any meaningful KRA (A-KRA) if the above holds for $\mathcal{G}$ being the set of all meaningful key generators.*

3 Authentication in the A-KRA Model given OWF

We show here how to implement authentication with any meaningful KRA. We first construct a system for identification secure under concurrent composition, using Σ-protocols in a more or less standard manner. Then we extend this identification system to a UC secure authentication system in a novel manner.

Implementing authentication in a public-key setting is of course trivial if one can choose the structure of the public keys—one includes a verification key in the public key and signs all messages. It is by far trivial in our relaxed KRA model as we make no assumption on the public keys except that they are in the range of some one-way function, which might itself be maliciously chosen. Standard constructions of signature schemes from one-way functions use verification keys with much more structure than this.

3.1 Σ-Protocols

For details on the following brief introduction see [CDS94]. Let $R \subseteq \{0,1\}^* \times \{0,1\}^*$ be a binary relation. A Σ-protocol for R consists of (A, E, Z, J, W, S), where A is a poly-time algorithm which for all $(x, w) \in R$ and sufficiently long randomness r outputs a *commit message* $a = A(x, w, r)$; $E = \{0,1\}^\ell$ is a set of *challenges*; Z is a poly-time algorithm which given $(x, w) \in R$ and $e \in E$ and randomness r outputs a *reply* $z = Z(x, w, e, r)$; J is a poly-time algorithm, called the *judgment*, which given any (x, a, e, z) outputs $J(x, a, e, z) \in \{\texttt{accept}, \texttt{reject}\}$; and W is a poly-time algorithm called the *witness extractor* and S is a PPT algorithm called the *simulator*. Furthermore:

completeness: For all $(x, w) \in R$, all randomness r and $a = A(x, w, r)$ and $z = Z(x, w, e, r)$ it holds that $J(x, a, e, z) = \texttt{accept}$.

special soundness: For all (x, a, e, z) and (x, a, e', z') with $e \neq e'$, $V(x, a, e, z) = \texttt{accept}$ and $J(x, a, e', z') = \texttt{accept}$ it holds that $(x, W(x, a, e, z, e', z')) \in R$.

honest verifier zero-knowledge: For all $(x, w) \in R$ and all $e \in E$ the simulator outputs $(a, z) \leftarrow S(x, e)$ such that $J(x, a, e, z) = \texttt{accept}$ and such that the distribution of (x, a, e, z) is computationally indistinguishable from $(x, A(x, w, r), e, Z(x, w, e, r))$ for a uniformly random r. This holds even when the distinguisher is given w.

One round of the standard zero-knowledge protocol for Hamiltonian Cycle using a statistically binding commitment scheme is a Σ-protocol for Hamiltonian Cycle with $E = \{0,1\}$. Since Σ-protocols are closed under parallel composition, this gives a Σ-protocol for any NP relation R based on one-way function, with $E = \{0,1\}^\ell$ for any polynomial ℓ.

Let R_0 and R_1 be binary relations and define $R = R_0 \vee R_1$ by $((x_0, x_1), w) \in R$ iff $(x_0, w) \in R_0$ or $(x_1, w) \in R_1$. Then given two Σ–protocols Σ_0, Σ_1 for R_0, R_1 respectively, we can use the *OR-construction* to construct a Σ–protocol, $\Sigma = \Sigma_0 \vee \Sigma_1$ for the relation $R = R_0 \vee R_1$. Let $x = (x_0, x_1)$ be an instance for which there exists w_0 and w_1 such that $(x_0, w_0) \in R_0$ and $(x_1, w_1) \in R_1$. Then the OR-construction is *witness indistinguishable* in the following sense: Consider a

PPT adversary $\mathcal{A}$. Give it $(x,(w_0,w_1))$ and give it access to a proof oracle $\mathcal{O}_b$, which on input prv picks a fresh identifier I, samples $a^{(I)} \leftarrow A(x,w_b,r)$, stores the prover intermediate state $P^{(I)} = (I,b,r)$ and returns $a^{(I)}$ to $\mathcal{A}$. On an input $(\texttt{chl},I,e^{(I)})$ when some $P^{(I)} = (I,b,r)$ is stored, it deletes $P^{(I)}$, computes $z^{(I)} = Z(x,w_b,e^{(I)},r)$ and returns $z^{(I)}$ to $\mathcal{A}$. At the end $\mathcal{A}$ outputs a guess at b. Then $|\Pr[\mathcal{A}^{\mathcal{O}_0}(x,(w_0,w_1)) = 0] - \Pr[\mathcal{A}^{\mathcal{O}_1}(x,(w_0,w_1)) = 0]|$ is negligible.

We call G a *hard double-witness generator* for $R = R_0 \vee R_1$ if it is PPT and a random sample $(x,w_0,w_1) \leftarrow G$ has the property that $(x,w_0) \in R$ and $(x,w_1) \in R$ and that it is hard to compute w_0 from (x,w_1) and hard to compute w_1 from (x,w_0), i.e., a PPT algorithm given a random (x,w_b) outputs (x,w_{1-b}) with negligible probability. If G is a hard double-witness generator for $R = R_0 \vee R_1$, then $\Sigma = \Sigma_0 \vee \Sigma_1$ is witness hiding for G, i.e., an adversary cannot compute a witness after seeing any number of proofs. Since Σ-protocol are proofs of knowledge, the adversary cannot give a proof without knowing the witness. Putting these two observations together we get that the adversary cannot give a proof for a statement x even after seeing any number of proofs for x, in the following sense: We say that $\mathcal{A}$ *wins the reprove game* in Fig. 4 if at the end of the game there is a stored value (a,e,z) (from **reply verifier**), where $J(x,a,e,z) = \texttt{accept}$ and where $\mathcal{A}$ did not challenge a prover (in **reply verifier**) between receiving e and returning z. I.e., $\mathcal{A}$ did not challenge a prover while it had to compute its own challenge.

initialize Let $I = 0$. Sample $(x,w_0,w_1) \leftarrow G$ and give x to $\mathcal{A}$.
start prover Whenever $\mathcal{A}$ inputs $(\texttt{prv},b)$, let $I = I+1$, sample $a^{(I)} \leftarrow A(x,w_b,r)$, store the prover intermediate state $P^{(I)} = (I,b,r)$ and return $a^{(I)}$ to $\mathcal{A}$.
challenge prover Whenever $\mathcal{A}$ inputs $(\texttt{chl},I,e^{(I)})$ and some $P^{(I)} = (I,b,r)$ is stored, delete $P^{(I)}$, compute $z^{(I)} = Z(x,w_b,e^{(I)},r)$ and return $z^{(I)}$ to $\mathcal{A}$.
start verifier On input $(\texttt{verify},a)$ from $\mathcal{A}$, sample a uniformly random $e \in_R E$, store (a,e) and return e to $\mathcal{A}$.
reply verifier On input $(\texttt{reply},a,e,z)$ from $\mathcal{A}$, where (a,e) is stored, delete (a,e) and store (a,e,z).

Fig. 4. The reprove game for $\mathcal{A}$, Σ and G

Theorem 1. *Let Σ_0 be a Σ-protocol for R_0, Σ_1 be a Σ-protocol for R_1 and G be a hard double-witness generator for $R = R_0 \vee R_1$. Then for all $\mathcal{A}$ PPT verifiers, $\mathcal{A}$ wins the reprove game with $\Sigma = \Sigma_0 \vee \Sigma_1$ and G with negligible probability.*

The intuition behind the proof is that an adversary $\mathcal{A}$ which wins the game can be used to extract a witness by rewinding the winning conversation and sending a new challenge, to get two valid conversations. Since $\mathcal{A}$ did not challenge a prover between getting e and sending its reply, the rewinding does not give problems. Which of the two witnesses w_0 and w_1 is extracted by $\mathcal{A}$ does not change significantly if we give all the proofs to $\mathcal{A}$ using a random fixed witness w_b instead of letting $\mathcal{A}$ choose b from proof to proof: If it did, it would clearly allow us to break witness indistinguishability. So, with a non-negligible probability $\mathcal{A}$ computes the witness not used to give the proofs. This allows to break G.

3.2 Authentication

We now turn our focus to authentication. Given that we are in the A-KRA model, the sender S knows a secret s_S for his public key pk_S s.t. $(pk_S, s_S) \in R_S$ for some poly-time relation R_S. In the same way, the receiver R knows s_R s.t. $(pk_R, s_R) \in R_R$. Construct a Σ-protocol $\Sigma = \Sigma_0 \vee \Sigma_1$ for the relation $R = R_R \vee R_S$, i.e. the verifier V accepts if the prover P knows a secret key for pk_S or for pk_R. Now the parties can identify to each other using this Σ-protocol.

The way we build an authenticated channel from this identification protocol is as follows: S wants to send R a message $m \in \{0,1\}^\ell$, where ℓ is a fixed message length. We essentially let the receiver simulate a clock by identifying towards the sender ℓ times. In each "time period" the sender will then either identify itself or not. This defines the ℓ bits of the message. At the end the sender does a number of identifications to bring up the total number of identifications given by the sender to ℓ. The receiver will accept only if it sees a total of ℓ identifications. This is done to make it impossible for an adversary to drop identifications from the sender to the receiver. At the end, we add two last rounds where S identify to R and then R identifies to S. This is to inform the other party that the message was accepted.

For $m \in \{0,1\}^\ell$ define $\sigma(m) \in \{\mathtt{R}, \mathtt{S}\}^{2\ell+2}$ to be $\mathtt{S}^{m_1} \| \mathtt{R} \| \mathtt{S}^{m_2} \| \mathtt{R} \| \cdots \| \mathtt{S}^{m_\ell} \| \mathtt{R} \| \mathtt{S}^{\ell - \sum_{i=1}^{\ell} m_i} \| \mathtt{S} \| \mathtt{R}$, where m_i is the i-th bit of m. Note that $m \neq m' \Rightarrow \sigma(m) \neq \sigma(m')$, that $\sigma(m)$ contains exactly $\ell + 1$ symbols of each type, and that the last symbols are always $\mathtt{S} \| \mathtt{R}$. These are sufficient properties for the protocol to be secure. The protocol is given in Fig. 5.

Theorem 2. *If the public keys are set up as in Fig. 3, then the following holds except with negligible probability: If S outputs* `accept` *at the end of π_{au} then R outputs* (`accept`, m)*, where m was the message input by S.*

The intuition is as follows: For $I = 1, \ldots, 2\ell + 2$ we match the i'th instance run by S to the i'th instance run by R. If S and R open a prover respectively a verifier, or a verifier respectively a prover, then they might both continue to $I + 1$ without rejecting. If S and R both open a verifier, then one of them will be terminated without a prover were running. Therefore this verifier will reject (by Thm. 1), which makes the party running that verifier reject. If S and R both instantiate a prover, then one of these provers will close without a verifier having been running at the other party.

Wlog, say that a prover was running at S while no verifier was running at R (one can repeat the argument switching the role of R and S). This prover will not make any verifier accept at R, therefore S will run more provers than the number of accepting verifiers that R runs. Since S starts $\ell + 1$ provers, by construction of σ, it follows that R sees at most ℓ accepting verifiers. Therefore R will not output `accept`. It follows that if S and R have different σ, then one

setup: Sender S knows s_S, pk_R, R_S and R_R and receiver R knows pk_S, s_R, R_S and R_R such that $(pk_S, s_S) \in R_S$ and $(pk_R, s_R) \in R_R$.

sender: The first time S gets an input $m \in \{0,1\}^\ell$ it computes $\sigma = \sigma(m)$, sends m to R and runs the following:

1. Let $I = 1$.
2. If $\sigma_I = \mathtt{S}$, then instantiate a prover $P = P((pk_S, pk_R), s_S)$ and let it interact with R.
3. If $\sigma_I = \mathtt{R}$, then instantiate a verifier $V = V(pk_S, pk_R)$ and let it interact with R. If V rejects, then terminate the protocol with output `reject`.
4. When the above instance closes (either P or V), then let $I = I + 1$. If $I \leq 2\ell + 2$, then go to Step 2. If $I > 2\ell + 2$, then output `accept`.

receiver: The first time R receives $m \in \{0,1\}^\ell$ from S it computes $\sigma = \sigma(m)$ and runs the following:

1. Let $I = 1$.
2. If $\sigma_I = \mathtt{R}$, then instantiate a prover $P = P((pk_S, pk_R), s_R)$ and let it interact with S.
3. If $\sigma_I = \mathtt{S}$, then instantiate a verifier $V = V(pk_S, pk_R)$ and let it interact with S. If V rejects, then terminate the protocol with output `reject`.
4. When the above instance closes (either P or V), then let $I = I + 1$. If $I \leq 2\ell + 2$, then go to Step 2. If $I > 2\ell + 2$, then output $(\mathtt{accept}, m)$.

Fig. 5. The authentication protocol $\pi_{\mathsf{au}}((pk_S, pk_R), s_S, s_R)$

of them does not output `accept`. In other words, if both parties output accept, then they saw the same message m, as σ is a unique encoding of m.

Second, assume that R did not accept. This implies that R rejected when $I < 2\ell + 2$ or at least R never reached $I = 2\ell + 2$, as $\sigma_{2\ell+2} = \mathtt{R}$ implies that R cannot reject while $I = 2\ell + 2$. Therefore R ran at most ℓ provers and thus S saw at most ℓ verifiers accept. Therefore S did not accept either. In other words, if S accepts, then R accepts.

Putting these two observations together, we conclude that if S accepts, then both parties accept, and then S and R saw the same message m, as desired. This symmetric guarantee makes the protocol suitable also to authenticate messages from R to S, and we will use this property in Thm. 3.

3.3 Multiparty Authentication

Our ideal functionality for authenticated transmission is given in Fig. 6. We have it do a key setup as $\mathcal{F}^G_{\text{KRA}}$ and output the generated keys before the authenticated transfer phase begins. This is for compositional reasons—it allows outer protocols to use the same secrets, which we exploit in later sections. Here we focus on the phase after the keys are generated: The functionality allows to deliver only messages which were actually sent, which models authentication. It can deliver a message several times and reorder them. This can be handled outside $\mathcal{F}_{\text{MAU}}$ using e.g. sequence numbers. Any message sent is leaked to the adversary to model that the transmission is only authenticated, not private.

init: First it lets $\text{init}_i = 1$ for all P_i, and then it runs $\mathcal{F}^G_{\text{KRA}}$ with adversary $\mathcal{A}$, to generate $(R_1, pk_1, s_1), \ldots, (R_n, pk_n, s_n)$.
init done: If the adversary inputs $(\texttt{done}, i)$ at a point where $\text{init}_i = 1$ and after $\mathcal{F}^G_{\text{KRA}}$ terminated, then output (pki, s_i) to P_i, where $pki = ((R_1, pk_1), \ldots, (R_n, pk_n))$, and set $\text{init}_i = 0$.
authenticated transfer, send: On input (j, m) from P_i where $\text{init}_i = 0$, store (i, j, m) and output (i, j, m) to the adversary.
authenticated transfer, receive: On input (i, j, m) from the adversary, if (i, j, m) was previously stored, wait until $\text{init}_j = 0$ and then output (i, m) to P_j.

Fig. 6. The ideal functionality $\mathcal{F}^G_{\text{MAU}}$ for multiparty authenticated communication

Our implementation of $\mathcal{F}^G_{\text{MAU}}$ runs in the $\mathcal{F}^G_{\text{KRA}}$ hybrid model, see Fig. 7.

setup: When party P_i receives (pki, s_i) from $\mathcal{F}^G_{\text{KRA}}$, it parses pki as $((R_1, pk_1), \ldots, (R_n, pk_n))$ and sets init = 1.
key generation: On its first activation P_i generates a random verification key vk_i for a digital signature scheme, along with the corresponding signing key sk_i and stores $(\texttt{keys}, vk_i, sk_i)$. Then P_i sends vk_i to all other parties.
key authentication: After **key generation** each ordered pair of parties (P_i, P_j) with $i < j$ runs the following in parallel:
- The parties P_i and P_j run the protocol $\pi_{i,j} = \pi_{\text{au}}((pk_i, pk_j), s_i, s_j)$ from Fig. 5.
- Party P_i uses the input $m = (vk_i, vk'_j)$, where vk'_j is the value it received from P_j in **key generation**. Party P_j uses the input $m = (vk'_i, vk_j)$, where vk'_i is the value it received from P_i in **key generation**.
- If P_i accepts in $\pi_{i,j}$, then it stores $(\texttt{vk}, j, vk'_j)$. If P_j accepts in $\pi_{i,j}$ then it stores $(\texttt{vk}, i, vk'_i)$.

When P_i stored $(\texttt{keys}, vk_i, sk_i)$ and $(\texttt{vk}, j, vk'_j)$ for all P_j with $j \neq i$, then P_i outputs (pki, s_i) and sets init = 0.

KRA propagatioauthenticated transfer, send: When P_i gets input (j, m), where init = 0, P_i computes $S = \text{sig}_{sk_i}(i \| j \| m)$ and sends (i, j, m, S) to P_j.
authenticated transfer, receive: On a message (i, j, m, S) the party P_j waits until init = 0. Then it looks up $(\texttt{vk}, i, vk_i)$ and outputs (i, m) if $\text{ver}_{vk_i}(i \| j \| m, S) = \texttt{accept}$.

Fig. 7. The protocol π^G_{MAU} for multiparty authenticated communication

Theorem 3. *If G is a meaningful key generator, then π^G_{MAU} UC securely implements $\mathcal{F}^G_{\text{MAU}}$ against a static, active adversary.*

The proof is essentially a reduction to Thm. 2. If π^G_{MAU} is not secure it is possible to make an honest P_j output (i, m) for an honest P_i without giving input (j, m)

to P_i. We can reduce that to an attack on the protocol in Fig. 5. First of all, we can assume that all other parties than P_i and P_j are corrupted, as this can only help the adversary. Then, whenever P_i or P_j have to interact with any $P_k \notin \{P_i, P_j\}$, they run the protocol honestly, but use the secret s_k of P_k as witness. By witness indistinguishability (WI) this changes the probability that P_j outputs (i, m) without P_i having received input (j, m) at most negligibly. But now all interaction involving other parties than P_i and P_j can be run by the adversary in its head, as it knows s_k for all corrupted parties—whatever messages P_i would send to P_k can be computed using s_k. But this modified adversary is carrying out an attack on Fig. 7 with $n = 2$. This is essentially an attack on Fig. 5. The only difference is that in Fig. 7, during **KRA propagation**, the environment gets s_i and s_j from P_i, P_j. This happens after the protocol $\pi_{i,j}$ was run, and therefore it is not needed to run the adversary against Fig. 5.

4 UC-OT in the A-KRA Model given SA-OT

Suppose we are given an UC commitment functionality, $\mathcal{F}_{\text{MCOM}}$ as defined in [CF01]: then we can implement UC zero-knowledge, $\mathcal{F}_{\text{MZK}}$, for all NP relations, which in turn allows us to implement a static, active UC secure OT from the passive secure OT. We can therefore focus on implementing $\mathcal{F}_{\text{MCOM}}$ using SA-OT.

The main idea of the protocol in Fig. 8 is to "compile" the SA-OT into a UC-OT using the WI proof for statements of the kind "I followed to protocol or I know your secret key".

Theorem 4. *The protocol π_{MCOM} is a UC secure implementation of $\mathcal{F}_{\text{MCOM}}$ in the $\mathcal{F}^G_{\text{MAU}}$ hybrid model secure against a static, active adversary.*

The simulator extracts a commitment from a corrupted sender S^* to an honest receiver R by using selection bit $c = 1$ to learn m from the SA-OT. If the sender manages to send $m' \neq m$ in the opening phase for some commitment, we can extract the proofs in the SA-OT for this commitment and learn a secret s'_R for R's public key pk_R. Since R never uses s_R in the protocol, this contradicts the hardness of computing a witness for pk_R. To be able to use selection bit $c = 1$, the simulator gives the proof in the run of the SA-OT using the secret s'_S of the sender. This goes unnoticed by the computational hiding of the commitment scheme, the computational hiding of the SA-OT and the WI of the OR-construction. To trapdoor open a commitment to some m' the simulator simply sends m' and simulates the proof that this is the correct message, by using the secret of the receiver as witness. This goes unnoticed as for $c = 1$.

Corollary 1. *If there exists a passive secure OT protocol, then any well-formed functionality $\mathcal{F}$ can be UC implemented in the A-KRA model, against a static, active adversary.*

Proof: By Thm. 4 we can implement $\mathcal{F}_{\text{MCOM}}$ in the $\mathcal{F}^G_{\text{MAU}}$-hybrid model, which implies that we can implement any well-formed $\mathcal{F}$ in the $\mathcal{F}^G_{\text{MAU}}$-hybrid model, if there exists a passive secure SA-OT protocol. By Thm. 3 we can implement $\mathcal{F}^G_{\text{MAU}}$

The following describes a commitment to m from party S to party R. In the full protocol different instances use session identifiers to separate executions. Here commit$(\cdot)$ is a statistically binding commitment.

1. All communication is authenticated using $\mathcal{F}^G_{\text{MAU}}$. Use sequence numbers to guarantee that no identical messages are ever sent, and thus never accept the same message twice from any party.
2. R samples a uniformly random string u and sends $U \leftarrow \text{commit}(u)$ to S.
3. S samples a uniformly random string v, sets $m_0 = 1^{|m|}$, sets $m_1 = m$ and sends $V \leftarrow \text{commit}(v)$, $M_0 \leftarrow \text{commit}(m_0)$ and $M_1 \leftarrow \text{commit}(m_1)$ to R.
4. Then S and R run the SA-OT, where S takes inputs m_0 and m_1 and uses randomness v while R gives input $c = 0$ and uses randomness u. After sending each message in the SA-OT R shows that it knows an opening of U to u such that the message it sent is consistent with having run the SA-OT with randomness u, selection bit $c = 0$ and the messages received from R so far. After sending each message in the SA-OT S shows that it knows an opening of V, M_0 and M_1 to v, m_0 respectively m_1 such that the messages it sent are consistent with the execution of the SA-OT with randomness v, inputs m_0, m_1 and the messages received from S. The proofs are given via a Σ-protocol for NP and use the OR-construction to prove either knowledge of the openings mentioned above or the secret of the other party.
5. To open S sends m to R and shows that m is the message inside M_1 or that S knows s_R such that $(pk_R, s_R) \in R_R$. The proof is given using two Σ-protocols and the OR-construction.

Fig. 8. The protocol π_{MCOM} for UC commitments using SA-OT

in the unauthenticated $\mathcal{F}^G_{\text{KRA}}$-hybrid model for any meaningful G. It then follows from the UC composition theorem that we can implement any well-formed $\mathcal{F}$ in the unauthenticated $\mathcal{F}^G_{\text{KRA}}$-hybrid model for any meaningful G, if there exists a passive secure SA-OT protocol. □

5 UC Commitment in the C-KRA Model given OWFs

In a nutshell, to construct UC-Com in the C-KRA model, we let the public keys to be commitments of the secret keys. Then to commit the sender send an encryption of the message under his secret key. To open, he sends the message m together with a WI proof for a statement "m is the committed message or I know your secret key".

Theorem 5. *If one-way functions exist, then there exists a UC commitment scheme for the C-KRA model secure against a static, active adversary.*

Proof: The public key is an unconditionally binding commitment $pk_i = \text{commit}(K_i; r_i)$ to a uniformly random value $K_i \in_R \{0,1\}^\kappa$. Let $F_{\{0,1\}^\kappa} : \{0,1\}^{2\kappa} \to \{0,1\}^{2\kappa}$ be a pseudo-random permutation (PRP). Both can be instantiated using one-way functions.

To commit to $m \in \{0,1\}^\kappa$ with session identifier $sid \in \{0,1\}^\kappa$ towards P_j, P_i sends $M = F_{K_i}(sid\|m)$. To open the commitment to P_j, the sender sends m and gives a proof that it knows K and r such that "$pk_j = \text{commit}(K;r) \vee (pk_i = \text{commit}(K;r) \wedge M = F_K(sid\|m))$". The proof is given using two Σ-protocols combined with the OR-construction.

To extract, the simulator computes $m = F^{-1}_{K_i}(M)$, where K_i is found as part of the secret $s_i = (K_i, r_i)$ of the sender P_i. By pk_i binding the sender to K_i unconditionally and the soundness of the proof and the fact that the sender cannot open the commitment pk_j, this will yield the only m that the sender can open the commitment to later.

To equivocate the simulator sends a uniformly random M. When given m it sends m and gives the proof using the secret s_j of the receiver as witness. By computational hiding of the commitment scheme, pseudo-randomness of F and WI of the proof, this will go unnoticed. □

6 UC OT in the A-CRS Model given SA-OT

Here we implement UC OT from SA-OT in any CRS model. We prove it for the A-CRS model, and hence for the U-CRS and C-CRS models too. We already know how to do UC OT in the A-KRA model given SA-OT, so it is sufficient to implement $\mathcal{F}^G_{\text{KRA}}$ in the $\mathcal{F}^D_{\text{CRS}}$ for any meaningful G and all one-way D.

Theorem 6. *If D is OWF, then G^D is meaningful, and if the used OT protocol is a SA-OT, then π^D_{KRA} in Fig. 9 is a UC secure implementation of $\mathcal{F}^{G^D}_{\text{KRA}}$ in the $\mathcal{F}^D_{\text{CRS}}$-hybrid model against a static, active adversary.*

The protocol runs in the $\mathcal{F}^D_{\text{CRS}}$-hybrid model.

- All parties P_i receive (D, crs) from $\mathcal{F}^D_{\text{CRS}}$.
- Each P_i samples $pk_i = D(s_i)$ for a uniformly random s_i and sends pk_i to all parties. All parties resend the received value pk_i to all parties.
- Then in round-robin, for $i = 1, \ldots, n$, each P_i proves knowledge of s_i to all other parties. It does this in round robin, for $j = 1, \ldots, n$. With each P_j it runs the proof as in Fig. 8: It inputs $m_0 = 0^{|s_i|}$ and $m_1 = s_i$ to the SA-OT and P_j inputs $c = 0$. During the run of the SA-OT, P_i proves that either 1) its messages are consistent with a run of the SA-OT protocol and $pk_i = D(m_1)$ or 2) its messages are consistent with a run of the SA-OT protocol and $crs = D(m_1)$. Party P_j proves that either 1) its messages are consistent with a run of the SA-OT protocol with $c = 0$ or 2) it knows s such that $crs = D(s)$. The proofs are given via a Σ-protocol for NP and the OR-construction. When P_i and P_j are done, they both send `done` to the other parties. Parties only begin their proof when they received `done` from all previous pairs.
- If and when a party P_k received crs from $\mathcal{F}^D_{\text{CRS}}$, a value pk_i from each P_i and a resent value pk'_i from all other parties P_j with $pk'_i = pk_i$, and saw an accepting proof from each P_i, it outputs $((pk_1, R^D), \ldots, (pk_n, R^D)), s_k$.

Fig. 9. The protocol π^D_{KRA} that implements a KRA in the A-CRS model

The proof is very similar to the proof of Thm. 4. The simulator extracts the secret of corrupted parties using selection bit $c = 1$. It simulates proofs using the secret s of crs. We run the proofs in round-robin to ensure that the simulator will not give a simulated proof (using s) while a corrupted party has to give a proof. If it did so, we could not show that a corrupted party cannot give an acceptable proof unless it used m_1 such that $pk_i = D(m_1)$ in the SA-OT. When the proofs are run in round-robin, we can.

7 UC-Com in the A-KRA Model Implies SA-OT

Theorem 7. *SA-OT is necessary for UC-Com in the A-KRA model.*

Proof: We show how a UC secure commitment scheme for the A-KRA model can be turned into a SA-OT. Note that this UC-Com protocol needs to work for any KRA, so we can choose a special KRA that it's possible to "simulate" in some sense without using any setup assumptions.

To simplify the proof, consider the AND primitive, where A inputs $a \in \{0,1\}$ and B inputs a bit $b \in \{0,1\}$ and where A has no output and B gets output $c = ab$. It is well-known that if there exists passive, stand-alone secure AND (SA-AND), then there also exist SA-OT. Then it is sufficient to show how to implement SA-AND from UC-Com in the A-KRA model.

The existence of UC-Com clearly implies OWFs, so we can assume that we have a PRG $g : \{0,1\}^\kappa \to \{0,1\}^{\kappa+1}$. Consider the key generator G^f, where $f : \{0,1\} \times \{0,1\}^\kappa \times \{0,1\}^{\kappa+1} \to \{0,1\}^{\kappa+1} \times \{0,1\}^{\kappa+1}$ and $f(b, r^b, pk^{1-b}) = (pk^0, pk^1)$ for $pk^b = g(r^b)$. This is clearly a meaningful generator, as a PRG $g : \{0,1\}^\kappa \to \{0,1\}^{\kappa+1}$ is one-way.

From the assumption that there exist UC-Com in the A-KRA model, we have a protocol π which UC implements $\mathcal{F}_{\text{MCOM}}$ in the $\mathcal{F}_{\text{KRA}}^{G}$-hybrid model with sender S and receiver R. The sender gets key material (pk_S, s_S) and pk_R and the receiver gets key material pk_S and (pk_R, s_R). Here $pk_i = f(s_i)$ for $i = S, R$.

Consider the following adversary $\mathcal{A}$ against π for the case when the sender is corrupted: It samples uniformly random $c \in \{0,1\}$, $r_S^c \in \{0,1\}^\kappa$ and $r_S^{1-c} \in \{0,1\}^\kappa$ and lets $pk_S^c = g(r_S^c)$ and $pk_S^{1-c} = g(r_S^{1-c})$. Then it inputs $s'_S = (1-c, r_S^{1-c}, pk_S^c)$ to $\mathcal{F}_{\text{KRA}}^{G^f}$. Then it commits to some $m \in \{0,1\}$ by honestly running the commitment phase of the protocol π with key material (pk_S, s_S) and pk_R, where $s_S = (c, r_S^c, pk_S^{1-c})$. It's clear here that $s_S \neq s'_S$ as the first bit is different, and $f(s_S) = f(s'_S)$. Later it decommits by honestly running the opening phase of the protocol π.

Lemma 1. *When running with $\mathcal{A}$, the honest receiver R will accept the commitment and will later accept the opening to m, except with negligible probability.*

The proof follows from the fact that R cannot distinguish $\mathcal{A}$ from the honest sender S . By π being UC secure, and the above lemma, it follows that there exists a UC simulator $\mathcal{S}$ which can extract m from the conversation with $\mathcal{A}$ already in the commitment phase. Since $\mathcal{S}$ is simulating $\mathcal{F}_{\text{KRA}}^{G^f}$ to $\mathcal{A}$, it follows that $\mathcal{S}$ learns s'_S and chooses the value of pk_R.

1. First B samples $c \in \{0,1\}$ uniformly at random. Then, if $b = 1$, it uses $\mathcal{S}$ to sample pk_R, samples $r_S^{1-c} \in \{0,1\}^\kappa$ uniformly at random and lets $pk_S^{1-c} = g(r_S^{1-c})$. If $b = 0$, then B samples $pk_R = f(s_R)$ for uniformly random s_R and samples uniformly random $pk_S^{1-c} \in_R \{0,1\}^{\kappa+1}$. In both cases it sends (c, pk_S^{1-c}) and pk_R to A.
   At the same time A samples uniformly random $r_S^c \in_R \{0,1\}^\kappa$, lets $pk_S^c = g(r_S^c)$ and sends pk_S^c to B.
2. Both parties let $pk_S = (pk_0^S, pk_1^S)$. A sets $s_S = (c, r_S^c, pk_S^{1-c})$. If $b = 1$ then B lets $s'_S = (1-c, r_S^{1-c}, pk_S^c)$. *Note that in this case* $f(s_S) = pk_S = f(s'_S)$.
3. A inputs a by committing to $m = a$ by honestly running the commitment phase of π, playing the role of the sender S with key material (pk_S, s_S) and pk_R.
   If $b = 1$, then B runs $\mathcal{S}$ to extract a from the conversation with A, and outputs a. If $b = 0$, then B honestly runs the commitment phase of π, playing the role of the receiver R with key material (pk_R, s_R) and pk_S, and outputs 0.

Fig. 10. SA-AND protocol

Consider then the SA-AND in Fig. 10. If $b = 1$, then all values are distributed as in the simulation of π with $\mathcal{A}$ and $\mathcal{S}$, so B computes a, except with negligible probability. This established the correctness, hence it only remains to prove the following lemma.

Lemma 2. *1) When A and B are honest, then the view of A when $b = 0$ and $b = 1$ are computationally indistinguishable. 2) When A and B are honest and $b = 0$, then the views of B when $a = 0$ and $a = 1$ are computationally indistinguishable.*

Part 1) follows readily from the fact that by UC security R and $\mathcal{S}$ cannot be distinguished by $\mathcal{A}$. Part 2) follows readily from the fact that a commitment hides the message when both parties are honest. □

8 Conclusions

Combining our findings with some previous results it is possible to fill the rows of Table 1. We will make use of the following:

Theorem 8. *[IR89] There is no black-box construction of SA-OT from OWF.*

Theorem 9. *[IR89] There is no black-box construction of key-agreement (KA) from OWF.*

Theorem 10. *[DG03] UC-Com in the U/A-CRS model implies SA-OT.*

Theorem 11. *[DG03] UC-Com in the C-CRS model implies KA.*

The answer to (a) follows directly from Thm. 11 and Thm. 9 for the CRS models; in the same way it follows from (j) and Thm. 8 for the A-KRA model; (b) is shown in Thm. 5; (c) follows from Thm. 9 and the fact that UC-OT in any model

Table 1. Questions and answers: If a cell contains more than one element it means that the answer, Y(es) or N(o), in the row is true for all elements in the cell. As an example, row (g) says that the answer to the question *is SA-OT sufficient for UC-OT in the A-CRS model?* is yes.

	Assumption		Functionality	Model	Answer
(a)	OWF	suf.	UC-Com	U/C/A-CRS, A-KRA	N
(b)	OWF	suf.	UC-Com	C-KRA	Y
(c)	OWF	suf.	UC-OT	U/C/A-CRS, C/A-KRA	N
(d)	UC-Com	suf.	UC-OT	U/A-CRS, A-KRA	Y
(e)	UC-Com	suf.	UC-OT	C-CRS	open
(f)	UC-Com	suf.	UC-OT	C-KRA	N
(g)	SA-OT	suf.	UC-Com, UC-OT	U/C/A-CRS, C/A-KRA	Y
(h)	OWF	nec.	UC-Com, UC-OT	U/C/A-CRS, C/A-KRA	Y
(i)	UC-Com	nec.	UC-Com, UC-OT	U/C/A-CRS, C/A-KRA	Y
(j)	SA-OT	nec.	UC-Com	U/A-CRS, A-KRA	Y
(k)	SA-OT	nec.	UC-Com	C-CRS	open
(l)	SA-OT	nec.	UC-Com	C-KRA	N
(m)	SA-OT	nec.	UC-OT	U/A/C-CRS, A-KRA	Y
(n)	SA-OT	nec.	UC-OT	C-CRS	open

implies KA; the answer to (d) is built from the fact that UC-Com in those models implies SA-OT (see (j)), and that we can compile this into a UC-OT using the UC-Com, as it implies UC-ZK; the answer to (f) goes as follows: (m) tells us that UC-OT in the C-KRA model implies SA-OT while (b) tells us that OWF are sufficient for UC-OT in the C-KRA model. Therefore UC-Com is not sufficient for UC-OT, or we will get a contradiction with Thm. 8; (g) is proved in Thm. 4 and 6; (h) is trivial as OWF are minimal for cryptography, and (i) is trivial as UC-OT is complete for UC computation; (j) is proved in Thm. 7 and 10; (l) follows from (b) and Thm. 8; finally (m) follows from the following observation: semi-honest parties can efficiently simulate the U-CRS (or the A-CRS) setup model by letting one party pick a random string without learning the trapdoor and make the *crs* public. Then the parties will run the UC-OT protocol using this string as the CRS, therefore achieving a SA-OT. As for the C-KRA (or the A-KRA) models, they can be efficiently simulated by letting every party generate his own public/secret key pair and sending the public key to all other parties. Now the parties can run the UC-OT using those public keys, and they'll achieve a SA-OT.

8.1 The C-CRS Setup Assumption

In this section we discuss the C-CRS model, and the open questions (e), (k) and (n) left in the table. Consider (n): is SA-OT necessary for UC-OT in the C-CRS model? The way we positively answered the question for the other setup models is by letting one party honestly pick a random CRS and publish it, therefore simulating the setup model. We don't know how to do it in the C-CRS model: in fact, we don't know whether it is possible, for any chosen OWF f, to sample an image $y = f(x)$ *without* learning the pre-image x. For instance, if $x \in \mathbb{Z}_q$ and $f(x) = (g^x, h^x)$ for g, h elements in group of large prime order q, then it is believed that one *cannot* sample from the image of f without learning x, to the extent that people construct protocols based on this belief (the so-called knowledge of exponent assumption [Dam91]). This suggests very strongly that the open questions cannot be solved using the techniques we have used here. It could of course be possible to approach (n) in some other way. It seems counter-intuitive to think that it would be possible to implement UC-OT in a world where SA-OT does not exist: how much power does a symmetric setup as the C-CRS give to the parties? However, if it turns out that the answer to (n) is affirmative, then we could use (g) to turn any UC-OT in the C-CRS model into a UC-OT in the U/A-CRS model, and this would also be a surprising result. Similar considerations can be made for (e) and (k).

Acknowledgements. We would like to thank Yuval Ishai and the anonymous reviewers for many valuable comments.

References

[BCNP04] Barak, B., Canetti, R., Nielsen, J.B., Pass, R.: Universally composable protocols with relaxed set-up assumptions. In: FOCS, pp. 186–195 (2004)

[Can01] Canetti, R.: Universally composable security: A new paradigm for cryptographic protocols. In: FOCS, pp. 136–145 (2001)

[CDS94] Cramer, R., Damgård, I., Schoenmakers, B.: Proofs of partial knowledge and simplified design of witness hiding protocols. In: Desmedt, Y.G. (ed.) CRYPTO 1994. LNCS, vol. 839, pp. 174–187. Springer, Heidelberg (1994)

[CF01] Canetti, R., Fischlin, M.: Universally composable commitments. In: Kilian, J. (ed.) CRYPTO 2001. LNCS, vol. 2139, pp. 19–40. Springer, Heidelberg (2001)

[CGGM00] Canetti, R., Goldreich, O., Goldwasser, S., Micali, S.: Resettable zero-knowledge (extended abstract). In: STOC, pp. 235–244 (2000)

[CLOS02] Canetti, R., Lindell, Y., Ostrovsky, R., Sahai, A.: Universally composable two-party and multi-party secure computation. In: STOC, pp. 494–503 (2002)

[Dam91] Damgård, I.: Towards practical public key systems secure against chosen ciphertext attacks. In: Feigenbaum, J. (ed.) CRYPTO 1991. LNCS, vol. 576, pp. 445–456. Springer, Heidelberg (1992)

[DG03] Damgård, I., Groth, J.: Non-interactive and reusable non-malleable commitment schemes. In: STOC, pp. 426–437 (2003)

[IPS08] Ishai, Y., Prabhakaran, M., Sahai, A.: Founding cryptography on oblivious transfer - efficiently. In: Wagner, D. (ed.) CRYPTO 2008. LNCS, vol. 5157, pp. 572–591. Springer, Heidelberg (2008)

[IR89] Impagliazzo, R., Rudich, S.: Limits on the provable consequences of one-way permutations. In: STOC, pp. 44–61 (1989)

[Kil88] Kilian, J.: Founding cryptography on oblivious transfer. In: STOC, pp. 20–31 (1988)

[KL07] Kidron, D., Lindell, Y.: Impossibility results for universal composability in public-key models and with fixed inputs. Cryptology ePrint Archive, Report 2007/478 (2007)

[LPV09] Lin, H., Pass, R., Venkitasubramaniam, M.: A unified framework for concurrent security: universal composability from stand-alone non-malleability. In: STOC, pp. 179–188 (2009)

[MPR09] Maji, H.K., Prabhakaran, M., Rosulek, M.: A zero-one law for deterministic 2-party secure computation (2009) (manuscript)

[OPV08] Ostrovsky, R., Persiano, G., Visconti, I.: Constant-round concurrent non-malleable zero knowledge in the bare public-key model. In: Aceto, L., Damgård, I., Goldberg, L.A., Halldórsson, M.M., Ingólfsdóttir, A., Walukiewicz, I. (eds.) ICALP 2008, Part II. LNCS, vol. 5126, pp. 548–559. Springer, Heidelberg (2008)

[PR08] Prabhakaran, M., Rosulek, M.: Cryptographic complexity of multi-party computation problems: Classifications and separations. In: Wagner, D. (ed.) CRYPTO 2008. LNCS, vol. 5157, pp. 262–279. Springer, Heidelberg (2008)

[RTV04] Reingold, O., Trevisan, L., Vadhan, S.P.: Notions of reducibility between cryptographic primitives. In: Naor, M. (ed.) TCC 2004. LNCS, vol. 2951, pp. 1–20. Springer, Heidelberg (2004)

From Passive to Covert Security at Low Cost

Ivan Damgård, Martin Geisler, and Jesper Buus Nielsen

Dept. of Computer Science, University of Aarhus, Denmark
{ivan,mg,jbn}@cs.au.dk

Abstract. Aumann and Lindell defined security against *covert attacks*, where the adversary is malicious, but is only caught cheating with a certain probability. The idea is that in many real-world cases, a large probability of being caught is sufficient to prevent the adversary from trying to cheat. In this paper, we show how to compile a passively secure protocol for honest majority into one that is secure against covert attacks, again for honest majority and catches cheating with probability 1/4. The cost of the modified protocol is essentially twice that of the original plus an overhead that only depends on the number of inputs.

1 Introduction

When studying cryptographic protocols, the behavior of the adversary has traditionally been categorized as being either *semi-honest* (passive) or *malicious* (active). A semi-honest adversary will only listen in on the network communication and spy passively on the internal state of the corrupted protocol participants. At the other end of the spectrum, a malicious adversary can make corrupted parties behave arbitrarily and will try to actively disrupt the computation in order to gain extra information and/or cause incorrect results.

Aumann and Lindell [2] introduce a third type of adversary called a *covert* adversary. This is intuitively an adversary which is able to do an active attack, but will behave correctly if the risk of being caught is sufficiently large—even if that probability is not essentially 1. The argument for studying covert adversaries is that there are many real world situations where the consequences of being caught out-weights the benefit of cheating—even a small but non-negligible risk of being caught is a deterrent. An example could be companies that agree to conduct an auction using secure multiparty computation. If a company is found to be cheating it may be subject to fines and it will hurt its long-term relationships with customers and other companies.

In the standard simulation-based definition of secure multiparty computation a protocol is said to *securely evaluate a function* f if no attack against the protocol can do better than an attack on an ideal process where an ideal functionality evaluates f and hands the result to the parties. Aumann and Lindell [2] give three different models of what a covert adversary can do by defining two different ideal functionalities that may compute f as usual, but may also act differently, depending on what the adversary does. They also define what it

D. Micciancio (Ed.): TCC 2010, LNCS 5978, pp. 128–145, 2010.

means for a protocol to implement an ideal functionality securely, this is a fairly standard simulation-based definition for sequentially composable protocols.

Thus, the special ingredient in the model that allows to accommodate covert attacks is only in the definition of the functionalities, which correspond to different levels of security, which are called *Explicit Cheat Formulation* (ECF) and *Strong Explicit Cheat Formulation* (SECF).[1] The basic idea in both cases is that the adversary may decide to try to cheat and must inform the functionality about this. The functionality then decides if the cheating is detected which happens with probability ε, where ε is called the *deterrence factor*. In this case all parties are informed that some specific corrupt party cheated. Otherwise, with probability $1-\varepsilon$, the cheating is undetected, and there is no security guarantee anymore: the functionality gives all inputs to the adversary and lets him decide the outputs. The difference between the two variants is that for ECF, the adversary gets the inputs of honest parties and decides their outputs immediately when he decides to cheat. For SECF, this only happens if the cheat is not detected.

Thus, with ECF, the adversary is caught with probability ε, but will learn the honest parties' inputs even if he is caught. With SECF, he must try to cheat *and succeed* to learn anything he was not supposed to.

1.1 Our Contribution

In this paper we propose a new construction that "compiles" a passively secure protocol into a new protocol with covert security. The approach is generic, but for concreteness we describe the idea starting from the classical BGW protocol [6] for evaluating arithmetic circuits, and only give the full compiler in the full version of this paper [13].

We assume honest majority and synchronous communication with secure point-to-point channels. We also assume a poly-time adversary, as we use cryptographic tools.

The basic idea is to first use a protocol with full active security to do a small amount of computation. Here, we will prepare two sets of (secret-shared) inputs to the passively secure protocol. However, only one set of sharings contains the actual inputs, while the other—the *dummy* shares—contain only zeros. Initially, it is unknown which set is the dummy one. We then run the passively secure protocol on both sets of inputs until parties hold shares of the outputs, which they must commit to. Now we reveal which sharings contained dummy values, and everything concerning the dummy execution can be then made available to check that no cheating occurred here. If no cheating was detected, we open the outputs of the real execution.

The intuition is that the adversary has to decide whether to cheat without knowing which execution is the dummy one, and therefore we can catch him with probability $\frac{1}{2}$ if he cheats at all, so one would expect this to give a deterrence factor of $\frac{1}{2}$.

[1] They also have a so called Failed Simulation definition which is weaker and which we do not use here.

However, while the intuition is straightforward, there are several non-trivial technicalities to take care of to make this work. We need parties to be able to prove that they really sent/received a given message earlier, and we have to do the final check without introducing too much overhead. After solving these problems, we obtain a protocol with deterrence factor $\frac{1}{4}$ whose complexity is essentially twice that of the passive protocol plus the overhead involved in preparing the inputs (which does not depend on the size of the computation).

It should be noted that there is an overhead involved in proving what messages were sent in the past. For this, players need to sign the messages they send. However, unless the arithmetic circuit we compute has very large depth and small breath, the cost of signing can be amortized over several operations requiring communication, and so is not significant. For the most advanced version of our construction, players also need to UC commit at the end to the set of messages they sent to each player. Our solution to this in the standard model is based on Paillier encryption and is quite elaborate, but for a practical implementation one can use the random oracle model, in which case commitment reduces essentially to hashing the messages, and is not a major cost.

We note that we focus on the complexity we get when there is no deviation from the protocol. In our construction, the adversary can slow things down by a factor linear in the number of parties by deviating, but the protocol is still secure, the adversary can only make it fail if he runs the risk of actually cheating and hence of being caught. Now, the spirit of covert security is that the adversary is to some extent rational, he does not cheat because it does not pay off to do so. It seems to us that there is little benefit in practice for the adversary in only slowing things down, while he cannot learn extra information or influence the result. We therefore believe that the complexity in practice can be expected to be what we get when there is no deviation.

We show our protocol is secure by showing that it implements Aumann and Lindell's functionality *in the UC model* [9], i.e., we do not use their simulation notion. The only difference this makes is that we get a stronger composition property for our protocol.

We show that the classical passively secure protocol by Ben-Or et al. [6] can be compiled to give a protocol with SECF security. Our approach can be used in a more general way, to compile any passively secure protocols into a covert protocol, if the original protocol satisfies certain reasonable conditions. The conditions are essentially as follows. The protocol should be based on secret sharing and consist of a computation phase and a reconstruction phase.

Computation phase: The computation phase starts from sharings of the inputs and produces sharings of the outputs, where the view of $t < n/2$ passively corrupted parties is independent of the inputs being computed on.

Reconstruction phase: The reconstruction phase consists of a single message from each party to each other party—i.e., it is non-interactive.

Passive security: Suppose uniformly random sharings of the inputs are dealt by an ideal functionality. Consider the protocol that executes the computation phase on these sharings followed by the reconstruction phase. This protocol should be passively secure against $t < n/2$ statically corrupted parties.

The approach to obtaining covert security is basically the same as described above. The details are described in the full version [13]. If the computation phase leaks no information, even under active attacks (as is the case for the BGW protocol), we get SECF security, otherwise ECF security is obtained.

1.2 Related Work and Discussion

Goyal et al. [15] improve Aumann and Lindell's 2-party protocol and also give a general multiparty computation protocol with covert security for the case of dishonest majority.

Our work focuses instead on honest majority. The skeptical reader may ask whether this is really interesting: the motivation for covert security is to settle for less than full robustness in return for more efficient protocols, and it may seem that we already know how to have great efficiency with honest majority and full active security. For instance, in [5, 10], it is shown that unconditionally secure evaluation of a circuit $\mathcal{C}$ for n parties and $t < n/3$ corruptions can be done in complexity $O(|\mathcal{C}|n)$ plus an overhead that only depends on the depth of the circuit, and in [12], it is shown under a computational assumption that this can be reduced to $O(|\mathcal{C}|)$ except for logarithmic factors plus an overhead that is independent of the circuit. Here, the security threshold can be selected arbitrarily close to $\frac{1}{2}$.

How could we hope to be better than that? There are two answers to this: First, the previous protocols are not as efficient as it may seem: the result from [12] only works asymptotically for a large number of parties and very large computations, it makes non black-box use of a pseudo-random function and is, in fact, very far from being practical. The protocols in [5, 10] use only cheap information theoretic primitives, but the security threshold in non-optimal and there is an overhead implying that deep circuits are expensive.

However, these protocols can all become much simpler and more practical if we assume the adversary is passive. For instance, when the adversary is passive the protocols from [5, 10] can tolerate $t < n/2$ and no longer have an overhead that depends on the circuit depth. Our compiler works for any "reasonable" protocol that is based on secret sharing, so we can use it on these simpler passively secure protocols and get a protocol with covert security, but with efficiency and security threshold similar to the passively secure solutions.

The second answer is that general circuit evaluation is not the only application. There are many special purpose protocols that are designed for a passive adversary but where obtaining active security comes at a significant cost. One example is the protocol by Algesheimer et al. [1] for distributed RSA key generation. Another is the auction application described in [8]. In both cases the protocols do not go via evaluation of a circuit for the desired function, but gets significant optimizations by taking other approaches. We can use our construction here to get covert security at a cost essentially a factor of two.

2 Preliminaries

Aumann and Lindell [2] present three successively stronger notions of security in the presence of covert adversaries, of which we consider the two strongest ones. There the adversary is forced to decide whether to cheat without knowledge of the honest parties' inputs. As mentioned, these are called ECF and SECF and are defined by specifying two (very similar) ideal functionalities.

For convenience, we give the ECF and SECF functionalities here. The only differences from [2] is that we do not include an option for the adversary to abort the protocol, and also, if no cheating is detected, the adversary cannot stop the functionality from giving outputs to the honest parties. This gives a stronger notion of security, and we can obtain it as we assume an honest majority.

Another difference is that we relax the requirements on the detection mechanism slightly. In [2] it is required that only one corrupted party is detected and that the honest parties agree on that party. We allow that several corrupted parties are detected and allow that different honest parties detect different sets of corrupted parties. The only requirement is that there is at least one corrupted party which is detected by all honest parties. In the presence of an honest majority, the stronger detection requirement in [2] can then be implemented using a Byzantine agreement at the end of the protocol on who should take the blame. We prefer to see this negotiation as external to the protocol and thus allow the more relaxed detection. See Fig. 1.

The functionality $\mathcal{F}^f_{\text{SECF}}$ is defined exactly as $\mathcal{F}^f_{\text{ECF}}$, except that when the adversary sends a cheat message, the functionality does not send the inputs of

Let f be a function with n inputs and n outputs, where n is the number of parties. The ECF functionality $\mathcal{F}^f_{\text{ECF}}$ for function f with deterrence factor ε works as follows:

Inputs: Any honest party P_i sends input x_i to $\mathcal{F}^f_{\text{ECF}}$, while the adversary $\mathcal{A}$ sends input on behalf of the corrupted parties.

Cheat detection: Let $C \subset \{1, \ldots, n\}$ denote the indices of the corrupted parties and let $H = \{1, \ldots, n\} \setminus C$ be the honest parties. The adversary can at any time instruct $\mathcal{F}^f_{\text{ECF}}$ to give outputs of the form $(\texttt{corrupt}, j)$ for $j \in C$ to P_i with $i \in H$. For $i \in H$, let $J_i \subset C$ be the set of j for which P_i output $(\texttt{corrupt}, j)$.

Attempted cheat: If $\mathcal{F}^f_{\text{ECF}}$ receives $(\texttt{cheat})$ from $\mathcal{A}$, it will send $(x_1, \ldots, x_n)$ to $\mathcal{A}$. It then decides randomly if the cheating was detected or not:

Undetected: With probability $1 - \varepsilon$, $\mathcal{F}^f_{\text{ECF}}$ sends $(\texttt{undetected})$ to the adversary. Then $\mathcal{A}$ specifies for each $i \in H$ an output y_i and $\mathcal{F}^f_{\text{ECF}}$ outputs y_i to P_i for $i \in H$.

Detected: With probability ε, $\mathcal{F}^f_{\text{ECF}}$ sends $(\texttt{detected})$ to $\mathcal{A}$. In this case $\mathcal{A}$ also gets to decide the output y_i for $i \in H$, but must ensure that $\cap_{i \in H} J_i \neq \emptyset$ at the end of the execution.

Output generation: If $\mathcal{A}$ did not attempt to cheat, $\mathcal{F}^f_{\text{ECF}}$ computes outputs $(y_1, \ldots, y_n) = f(x_1, \ldots, x_n)$ and gives y_i to P_i.

Fig. 1. Functionality $\mathcal{F}^f_{\text{ECF}}$

honest parties to the adversary. This only happens if the cheating is undetected. We can now define security:

Definition 1. *Protocol π computes f with ε-ECF (SECF) security and threshold t if it implements $\mathcal{F}^f_{\text{ECF}}$ ($\mathcal{F}^f_{\text{SECF}}$) in the UC model, securely against poly-time adversaries corrupting at most t parties.*

This definition naturally extends to a hybrid UC model where certain functionalities are assumed to be available. By the UC composition theorem and given implementations of the auxiliary functionalities, a protocol follows that satisfy the above definition without auxiliary functionalities.

In the following, we will consider secure evaluation of an arithmetic circuit $\mathcal{C}$ over some finite field K. We assume that each input and output of $\mathcal{C}$ is assigned to some party, whence $\mathcal{C}$ induces in a natural way a function $f_{\mathcal{C}}$ of the form considered above. In the following, "computing $\mathcal{C}$ securely" will mean computing $f_{\mathcal{C}}$ securely in the sense of the above definition.

We will denote the participants in the protocol by $P_1, \ldots, P_n$ for a total of n parties. Shamir secret sharing of $a \in K$ with threshold t results in a set of shares denoted by $[a]_t$ or simply $[a]$ when the threshold is clear from the context. The share held by P_i is denoted a_i.

3 Auxiliary Functionalities

We define some ideal functionalities to make the presentation clearer. We show how to implement them Section 5.

Message Transmission Functionality. Functionality $\mathcal{F}_{\text{TRANSMIT}}$ is an enhancement of the standard model for secure point-to-point channels. It essentially allows to prove to third parties which messages one received during the protocol, and to further transfer such revealed messages. It does not commit the corrupted parties to what they sent to each other. See Fig. 2 for details.

The ideal functionality $\mathcal{F}_{\text{TRANSMIT}}$ works with message identifiers mid encoding a sender $s(mid) \in \{1, \ldots, n\}$ and a receiver $r(mid) \in \{1, \ldots, n\}$. We assume that no mid is used twice. The functionality works as follows:

Secure transmit: When receiving (`transmit`, mid, m) from $P_{s(mid)}$ and receiving (`transmit`, mid) from all (other) honest parties, store (mid, m), mark it as *undelivered*, and output $(mid, |m|)$ to the adversary. If P_s does not input a (`transmit`, mid, m) message, then output (`corrupt`, $s(mid)$) to all parties.

Synchronous delivery: At the end of each round, deliver each undelivered (mid, m) to $P_{r(mid)}$ and mark (mid, m) as delivered.

Reveal received message: On input (`reveal`, mid, i) from a party P_j which at any point received the output (mid, m), output (mid, m) to P_i.

Do not commit corrupt to corrupt: If both P_j and P_s are corrupt, then the adversary can ask $\mathcal{F}_{\text{TRANSMIT}}$ to output (mid, m') to any honest P_i for any m' and any mid with $s(mid) = s$.

Fig. 2. Ideal Functionality $\mathcal{F}_{\text{TRANSMIT}}$

This functionality will be used for all private communication in the following, and provides a way to reliably show what was received at any earlier point in the protocol. This is used when the dummy execution is checked for consistency.

Input Functionality. For notational convenience we assume that each P_i has one input $x_i \in K$. The input functionality is given in Fig. 3. Note that we let the adversary pick the dummy inputs, which is done simply not to decide at this abstract level on any specific set of dummy inputs. We also let the adversary pick the shares the functionality should produce for corrupt players. This is necessary to be able to implement the functionality with a real-life protocol.

The ideal functionality $\mathcal{F}_{\text{INPUT}}$ is parametrized by a secret sharing scheme, sss, and works as follows.

1. Receive an input x_i from each P_i and an input $(d_1, \ldots, d_n)$ from the adversary. The adversary also inputs x_i for $i \in C$.
2. Flip a uniformly random bit $d \in_{\mathsf{R}} \{0, 1\}$.
3. Let $e = 1 - d$. Let $x^{(i,d)} = d_i$ be the *dummy* inputs and let $x^{(i,e)} = x_i$ be the *enriched* inputs.
4. For every $x^{i,d}$ and $x^{i,e}$, the adversary inputs sets of shares $X^{i,d}$ and $X^{i,e}$. They each contain a share for every player in C, and we think of $X^{i,d}$ as the set of shares of $x^{i,d}$ that the adversary wants the functionality to produce for corrupt players.
5. For $j = 1, \ldots, n$ and $c = 0, 1$, sample $[x^{(j,c)}] \leftarrow \text{sss}(x^{(j,c)} | X^{j,c})$, by which we mean that shares of $x^{i,c}$ are sampled, conditioned on players in C receiving shares $X^{i,c}$.
6. Output $(x_i^{(j,0)})_{j=1}^n$ and $(x_i^{(j,1)})_{j=1}^n$ to P_i.
7. On a later input $(\texttt{reveal}, i, k)$, output d and $(x_i^{(j,d)})_{j=1}^n$ to P_k.

Fig. 3. Ideal Functionality $\mathcal{F}_{\text{INPUT}}$

Commitment Functionality. We use a flavor of commitment where the committer cannot avoid that a commitment is revealed. Details are in Fig. 4.

The functionality $\mathcal{F}_{\text{COMMIT}}$ uses commitment identifiers encoding the sender $s(cid)$ of the commitment. It works as follows:

Commit: On input $(\texttt{commit}, cid, m)$ from $P_{s(cid)}$ and input $(\texttt{commit}, cid)$ from all (other) honest parties, store (cid, m) and output $(\texttt{commit}, cid, |m|)$ to the adversary.

Reveal: On input $(\texttt{reveal}, cid, r)$ from all honest parties, where (cid, m) is stored, give (cid, m) to P_r.

Fig. 4. Ideal Functionality $\mathcal{F}_{\text{COMMIT}}$

Coin-Flip Functionality. We use the coin-flip functionality given in Fig. 5.

The functionality $\mathcal{F}_{\text{FLIP}}^{B}$ is parametrized by a positive integer B and works as follows:

1. Sample a uniformly random $k \in_{\text{R}} \{0, \dots, B-1\}$.
2. When the first honest party inputs (`flip`), output k to the adversary.
3. If in the round where the first honest party inputs (`flip`) there is some party P_i which does not input (`flip`), then output (`corrupt`, i) to all parties.

Fig. 5. Ideal Functionality $\mathcal{F}_{\text{FLIP}}^{B}$

4 Protocol

Having defined the necessary ideal functionalities, we will now describe how we use them to compile the classical passively secure protocol by Ben-Or et al. [6] based on Shamir secret-sharing into one with covert security. This protocol computes an arithmetic circuit $\mathcal{C}$ with passive security. Assuming the inputs to the arithmetic circuit have been secret shared, the protocol does addition by having parties add their shares locally, and multiplication by local multiplication of shares followed by a re-sharing by each parties of the local products. Due to space constraints, we assume the details are known to the reader.

The protocols in this section use the auxiliary functionalities we defined. Thus the actual complexity of our construction depends on the implementation of those auxiliary functionalities. It turns out that the overhead incurred includes a contribution coming from the cryptographic primitives we use, this overhead does not depend on the communication complexity of the protocol we compile. In addition, the adversary can choose to slow down $\mathcal{F}_{\text{TRANSMIT}}$ by a factor of n, but since he cannot make it fail, a covert adversary is unlikely to make such a choice as discussed in the introduction.

We begin with a simple construction which has a rather poor computational complexity. Following that, we show how the simple protocol can be adapted to yield a better complexity.

Theorem 1. *The protocol in Fig. 6 computes $\mathcal{C}$ with $\frac{1}{2}$-SECF security and threshold $t < n/2$ in the $(\mathcal{F}_{\text{TRANSMIT}}, \mathcal{F}_{\text{INPUT}}, \mathcal{F}_{\text{COMMIT}}, \mathcal{F}_{\text{FLIP}})$-hybrid world against a static adversary.*

Proof. Initially $\mathcal{S}$ is given the inputs of the corrupt parties. It passes them on to $\mathcal{A}$ and simulates the protocol execution up until the point where the bit d is revealed and it is determined which of the two executions were the dummy execution. $\mathcal{S}$ does this by inventing random shares whenever $\mathcal{A}$ would expect to see a share from an honest party. $\mathcal{A}$ will always see only t shares and any subset of size t look completely random in the real protocol execution. $\mathcal{S}$ can therefore simulate them perfectly by giving $\mathcal{A}$ random values.

During the protocol, $\mathcal{A}$ is observed by $\mathcal{S}$ and it can thus be determined if $\mathcal{A}$ ever sends an incorrect intermediate result to one of the honest parties.

– If $\mathcal{A}$ did not cheat at all, or if $\mathcal{A}$ cheated in both executions, then $\mathcal{S}$ simply follows the protocol. In the first case $\mathcal{F}_{\text{SECF}}^{fc}$ will give $\mathcal{S}$ the outputs for the

In general, if any of the ideal functionalities output $(\texttt{corrupt}, j)$ to P_i, then P_i also outputs $(\texttt{corrupt}, j)$. Not mentioning this further, the protocol proceeds in five steps:

1. All parties provide input to $\mathcal{F}_{\text{INPUT}}$. In return they obtain shares of secret sharings $[x^{(j,0)}]$ and $[x^{(j,1)}]$ for $j = 1, \ldots, n$. Nobody knows which sharings are dummy at this point.
2. Each party P_i generates random keys K_i^0 and K_i^1 and commit to them using $\mathcal{F}_{\text{COMMIT}}$ twice.
3. The passively secure protocol is run on both input sets $\{[x^{(j,0)}]\}_{j=1}^n$ and $\{[x^{(j,1)}]\}_{j=1}^n$. This evaluates the circuit $\mathcal{C}$ twice. The parties use $\mathcal{F}_{\text{COMMIT}}$ to commit to their shares of the output. All randomness used in the first and second protocol run come from pseudo-random generators seeded by K_i^0 and K_i^1, respectively.
4. The parties query $\mathcal{F}_{\text{INPUT}}$ for the random bit d and the shares of $\{[x^{(j,d)}]\}_{j=1}^n$. They then use $\mathcal{F}_{\text{COMMIT}}$ to reveal the key K_i^d used for the pseudo-random generator for all P_i. Knowing the initial inputs and the seed for the pseudo-random generator used, the entire message trace of all parties is fixed. The parties also open the commitments to the dummy output shares.
5. Each party locally simulates the entire dummy execution to determine if any cheating took place. This amounts to checking for each party whether his input shares of $[x^{(j,d)}]$ (revealed by $\mathcal{F}_{\text{INPUT}}$) and seed K_i^d (revealed by $\mathcal{F}_{\text{COMMIT}}$) together lead to the shares he claims to have obtained of the output (revealed by $\mathcal{F}_{\text{COMMIT}}$) if he follows the passively secure protocol on the messages that other parties would have sent if they followed the protocol on their shares and expanded randomness. If no discrepancies are found, the output shares of the real execution are opened.
   Otherwise, the honest parties must determine who cheated.[a]
   The parties have already locally simulated the dummy execution so they know the correct message trace. It is therefore simple to match this against the actual message trace revealed by $\mathcal{F}_{\text{TRANSMIT}}$ and pinpoint the first deviation. If P_j made the first mistake, the honest parties output $(\texttt{corrupt}, j)$ and halt.

[a] Note that it is possible for a corrupt party to "frame" an honest party by sending him wrong intermediate results. The honest party cannot tell the difference and will produce incorrect output. $\mathcal{F}_{\text{TRANSMIT}}$ is there to safeguard honest parties against this form of attack. The parties call it to reveal all messages that were received in the dummy execution.

Fig. 6. Simple version

corrupt parties, which $\mathcal{S}$ can pass along to $\mathcal{A}$ unchanged. In the second case, $\mathcal{A}$ will be caught with certainty before seeing anything which depend on the honest parties inputs. $\mathcal{S}$ can therefore simulate the protocol execution towards $\mathcal{A}$ using random shares only.

– If $\mathcal{A}$ cheats in execution d' (first or second execution), $\mathcal{S}$ will send $(\texttt{cheat})$ to $\mathcal{F}$. The functionality then determines if the cheat was successful:

Detected: The simulator must now ensure that $\mathcal{A}$ believes he cheated in the dummy execution.

$\mathcal{A}$ will want to query $\mathcal{F}_{\text{INPUT}}$ for the value of d and the shares of the dummy inputs. In response, $\mathcal{S}$ sends a response with $d = d'$, which means that $\mathcal{A}$ cheated in the dummy execution. $\mathcal{S}$ must also send back shares of the inputs $\{x^{(j,d)} = d_j\}_{j=1}^n$ consistent with the shares $\mathcal{A}$ has already seen. At this point $\mathcal{A}$ has only seen the shares it chose for the (*non-qualified*) subset of corrupt parties when $\mathcal{F}_{\text{INPUT}}$ was called initially. $\mathcal{S}$ can therefore choose polynomials that agree with these values and correspond to a sharing of the inputs d_j, and finally compute consistent shares of the honest parties using these polynomials.

If P_j were the first corrupt party who send an incorrect message to an honest party, $\mathcal{S}$ will send $(\texttt{corrupt}, j)$ to $\mathcal{F}_{\text{SECF}}^{f_C}$.

Undetected: In this case the functionality responded with $(\texttt{undetected})$ together with the honest parties' inputs. The simulator must therefore make it look as if $\mathcal{A}$ cheated in the execution that was not opened, i.e., the real execution. As above, $\mathcal{S}$ can compute polynomials that will give a correct sharing of inputs based on what $\mathcal{A}$ already knows and with $d = 1 - d'$.

Using these inputs together with the corrupt parties' inputs and outputs, $\mathcal{S}$ can now compute the consequence of $\mathcal{A}$'s cheating, i.e., the altered outputs of the honest parties. It passes these outputs to $\mathcal{F}_{\text{SECF}}^{f_C}$ as the honest parties' outputs.

It is clear that the above simulation matches the output of $\mathcal{A}$ in the hybrid world perfectly when $\mathcal{A}$ did not cheat and when $\mathcal{A}$ was foolish enough to cheat in both executions.

When $\mathcal{A}$ cheats in just one execution, $\mathcal{S}$ will make the honest parties output $(\texttt{corrupt}, j)$ for some corrupt P_j (if $\mathcal{A}$ was detected) or output normal outputs (if $\mathcal{A}$ was undetected). Each of these two cases are picked with probability exactly $\frac{1}{2}$ by the random choice made by $\mathcal{F}_{\text{SECF}}^{f_C}$. We get the same probability distribution in the hybrid world where $\mathcal{F}_{\text{INPUT}}$ picks the bit d uniformly at random.

In total, we can now conclude that the protocol in Fig. 6 computes f_C with $\frac{1}{2}$-SECF security.

The above protocol has each party execute the passively secure protocol twice after which each party simulates the actions of all other parties in the dummy execution. In the standard BGW protocol [6], each party has a computational complexity of $\mathcal{O}(n)$ per gate. By asking every party to simulate every other party, we increase the computational complexity to $\mathcal{O}(n^2)$ per gate.

The communication complexity is doubled by running the passively secure protocol twice. In the normal case where the dummy execution is found to contain no errors, the communication complexity is increased no further. When errors *are* detected, every party is sent the messages communicated by every other party. This will again introduce a quadratic blowup, now in the communication complexity. We argued in the introduction that even a small fixed probability of catching misbehavior is enough to deter the parties. Because of that, we expect to find no discrepancies most of the time, and thus obtain the same *communication* complexity as the original protocol within a constant factor. We still have a quadratic blowup in the *computational* complexity. However, local computations are normally considered free compared to the communication, i.e., the network is expected to be the bottleneck. So for a moderate number of parties, this simple protocol can still be quite efficient.

Still, we would like to lower the complexity when errors are detected. Below we propose a slightly more complex protocol which has only a constant overhead in both computation and communication both when no errors are detected and when the parties are forced to do a more careful verification.

This is a modification of the protocol in Fig. 6. After running Step 1–3 unchanged, it continues with:

1. All P_i use $\mathcal{F}_{\text{COMMIT}}$ to commit to their view of the protocol, i.e., all messages exchanged between P_i and P_j for all j. This results in commitments $\text{comm}_{\{i,j\}}^{(i,0)}$ for the first execution and $\text{comm}_{\{i,j\}}^{(i,1)}$ for the second, where $\text{comm}_{\{i,j\}}^{(m,c)}$ is the view of P_m of what was sent between P_i and P_j in execution number c.
2. The parties query $\mathcal{F}_{\text{INPUT}}$ for the random bit d and the shares of the dummy inputs. They then use $\mathcal{F}_{\text{FLIP}}^{n-1}$ to flip a uniformly random $k \in \{1, \ldots, n-1\}$ that will be used when checking. $\mathcal{F}_{\text{COMMIT}}$ is used by all parties to reveal the key K_i^d used for the pseudo-random generator for all P_i. Finally, the commitments to shares in the output from the dummy execution are opened.
3. Each party P_i checks P_l, where $l = (i - 1 + k \bmod n) + 1$, i.e., he checks P_{i+k} with wraparound from P_n back to P_1.
   The commitments $\text{comm}_{\{l,j\}}^{(j,d)}$ and $\text{comm}_{\{l,j\}}^{(l,d)}$ are opened to P_i, i.e., the committed views of P_l and P_j of what was exchanged between them. If there is a disagreement, then P_i broadcasts a complaint and P_l and P_j must decommit to all parties and use $\mathcal{F}_{\text{TRANSMIT}}$ to show which messages they received from the other. This will clearly detect at least one corrupt party among P_l and P_j if P_i was honest, or reveal P_i as corrupt if the commitments were equal after all, i.e., if P_i made a false accusation.
   If all committed views agree, then P_i simulates the local computations done by P_l and checks whether this leads to the shares of the dummy output opened by P_l and the messages sent according to $\text{comm}_{\{l,j\}}^{(l,d)}$. If a deviation is found, P_i broadcasts an accusation against P_l, and all parties check P_l as P_i did. If they verify the deviation they output $(\texttt{corrupt}, l)$, otherwise they output $(\texttt{corrupt}, i)$.
4. If no accusations were made, the output of the real execution is opened.

Fig. 7. Efficient version

If no errors are detected, each party does two protocol executions followed by a check of the input/output behavior of one other party. This is clearly a constant factor overhead compared to the passively secure protocol. When a party is accused, all other parties must check this party. This adds only a linear overhead to the overall protocol, and thus the protocol in Fig. 7 has a linear overhead compared to the passively secure protocol.

It might seem as an overkill in the protocol in Fig. 7 to use $\mathcal{F}_{\text{TRANSMIT}}$ for communication and then also have the parties commit to their communication using $\mathcal{F}_{\text{COMMIT}}$. The reason for the commitments is to commit the corrupted parties to what they sent among each other before it is revealed which parties check which parties. If we do not do that, they might decide on which of them was the deviator after the revelation of d and k and thus always pick the deviator to be one which is checked by a corrupted party. For an example of what can go wrong without the commitments the interested reader can refer to the CHINESE-WHISPERS protocol in the full version of this paper [13].

Theorem 2. *The protocol in Fig. 7 computes* $\mathcal{C}$ *with* $\frac{1}{4}$*-SECF security and threshold* $t < n/2$ *in the* $(\mathcal{F}_{\text{TRANSMIT}}, \mathcal{F}_{\text{INPUT}}, \mathcal{F}_{\text{COMMIT}}, \mathcal{F}_{\text{FLIP}})$*-hybrid world.*

Proof. The simulator for the protocol in Fig. 7 runs like the simulator for the protocol in Fig. 6, except that it must now only output $(\texttt{corrupt}, i)$ to $\mathcal{F}$ if it determines that a message trace for a corrupt party P_i was checked by an honest party, and it must do while maintaining the same probability distribution as in the hybrid world.

As before, $\mathcal{S}$ will simulate $\mathcal{A}$ and observe the messages sent to honest parties. As soon as an incorrect message is observed in execution d' and all parties committed to their communication with the other parties, we know there exists an offset $k' \in \{1, \ldots, n-1\}$ for which an honest P_i would catch a corrupt P_l, where $l = (i - 1 + k' \bmod n) + 1$ in execution d':

- If two parties P_l and P_j committed to $\text{comm}_{\{l,j\}}^{(l,d')} \neq \text{comm}_{\{l,j\}}^{(j,d')}$, then one of them is corrupted, P_l say, and we pick k' such that P_l is checked by an honest P_i.
- If $\text{comm}_{\{l,j\}}^{(l,d')} = \text{comm}_{\{l,j\}}^{(j,d')}$ for all pairs of parties, then the wrong message sent to an honest party in execution d' implies that some party P_l is committed to values which are not consistent with an execution of the protocol, and we pick k' to ensure that P_l is checked by an honest party.[2]

The simulator sends $(\texttt{cheat})$ to $\mathcal{F}_{\text{SECF}}^{fc}$. We have two outcomes:

Detected: Set $d = d'$ and sample k at random such that P_l is checked by an honest party.

Undetected: Set $d = d'$ with probability $\frac{1}{3}$, and $d = 1 - d'$ otherwise. Sample $k \in \{1, \ldots, n-1\}$ such that P_l is checked by an honest party with probability $\alpha = \frac{4}{3}(\frac{n-t}{n-1} - \frac{1}{4})$.

If $\mathcal{A}$ did not cheat, $\mathcal{S}$ selects d and k as in the hybrid protocol. The simulation continues as in the hybrid world with these choices for d and k. The ideal world output clearly match the hybrid world.

---

[2] Note that P_l need not be the one who sent the incorrect message to the honest party—P_l may have behaved locally consistent given its inputs—but $\mathcal{S}$ will be able to find a first deviator, and it will clearly not be one of the honest parties.

When $\mathcal{A}$ did cheat, we will show that d and k are picked with the correct distribution. First note that $\mathcal{S}$ pick $d = d'$ with probability $\frac{1}{4} \cdot 1 + \frac{3}{4} \cdot \frac{1}{3} = \frac{1}{2}$, as in the hybrid world.

For the selection of k, note that a cheating party will always have a unique distance to every honest party. These distances make up a subset of $\{1, \ldots, n-1\}$ of size $n - t$. The cheater is caught exactly when the offset is picked within this subset. This happens with probability $\frac{n-t}{n-1}$ in the hybrid world. The simulator picks k among the indices of honest parties with the same probability: $\frac{1}{4} + \frac{3}{4}\alpha = \frac{n-t}{n-1}$. We conclude that $\mathcal{S}$ will simulate the hybrid world.

5 Implementation of Sub-protocols

In this section we sketch how to implement the sub-protocols described above.

Detection. In all sub-protocols we will need a tool for stopping the protocol "gracefully" when corruption is detected This is done by all parties running the following rules in parallel.

1. If a party P_i sees that a party P_d deviates from the protocol, then P_i signs $(\texttt{corrupt}, d)$ to get signature γ_i and sends the signature to all parties. Then P_i outputs $(\texttt{corrupt}, d)$.
2. If P_k received a signature γ_i on $(\texttt{corrupt}, d)$ from $t + 1$ distinct parties P_i, it considers these as a *proof* that P_d is corrupted, sends this proof to all parties, outputs $(\texttt{corrupt}, d)$, waits for one round and then terminates all protocols.
3. If P_k receives a proof that P_d is corrupt from any party, it relays this proof to all parties, outputs $(\texttt{corrupt}, d)$, waits for one round and then terminates all protocols.

If the signature scheme are unforgeable and only corrupted parties deviate from the protocol, then the protocol has the following two properties, except with negligible probability.

Detection soundness: If an honest party outputs $(\texttt{corrupt}, d)$, then P_d is corrupt.

Common detection: If an honest party terminates the protocol prematurely, then there exists P_d such that *all* honest parties have output $(\texttt{corrupt}, d)$.

The reason why the relayer P_r waits for one round before terminating is that P_r wants all other parties to have seen a proof that P_i is corrupt before it terminates itself. Otherwise the termination of P_r would be considered a deviation and an honest P_r could be falsely detected. In the following we do not always mention explicitly that the detection sub-protocol is run as part of all protocols.

Transmission Functionality. The transmission protocol can run in two modes. In *cheap mode* $\mathcal{F}_{\text{TRANSMIT}}$ is implemented as follows.

1. On input ($\texttt{transmit}, mid, m$) party $P_{s(mid)}$ signs (mid, m) to obtain signature σ_s and sends (mid, m, σ_s) to $P_{r(mid)}$.
2. On input ($\texttt{transmit}, mid$) party $P_{r(mid)}$ waits for one round and then expects a message (mid, m, σ_s) from $P_{s(mid)}$, where σ_s is a valid signature from P_s on (mid, m). If it receives it, it outputs (mid, m).
3. On input ($\texttt{reveal}, mid, i$) party P_j, if it at some point output (mid, m), sends (mid, m, σ_s) to P_i, which outputs (mid, m) if σ_s is valid.

It is easy to check that this is a UC secure implementation under the following restrictions:

Synchronized input from honest parties: If some honest party receives input ($\texttt{transmit}, mid$), then all honest parties $P_i \neq P_{s(mid)}$ receives the same input ($\texttt{transmit}, mid$). Furthermore, if $P_{s(mid)}$ is honest, it receives input ($\texttt{transmit}, mid, m$) for some m.

Signatures: Even corrupted P_s send along the signatures σ_s.

The restriction *synchronized input from honest* can be enforced by the way the ideal functionality is used by an outer protocol, i.e., by ensuring that the honest parties agree on which message identifiers are used for which message in which rounds. This is the case for the way we use $\mathcal{F}_{\text{TRANSMIT}}$. The restriction *signatures* is unreasonable, and we show how to get rid of it below. We need the rule **Do not commit corrupt to corrupt** in $\mathcal{F}_{\text{TRANSMIT}}$ as we cannot prevent a corrupt P_s from providing a corrupt P_i with signatures on arbitrary messages, i.e., we cannot commit the corrupted parties to what they have sent among themselves.

As mentioned, the above implementation only works if all senders honestly send the needed signatures. If at some point some P_r does not receive a valid signature from P_s, it publicly accuses P_s of being corrupted and the parties switch to the below *expensive mode* for transmissions from P_s to P_r.

1. On input ($\texttt{transmit}, mid, m$) party $P_{s(mid)}$ signs (mid, m) to obtain signature σ_s and sends (mid, m, σ_s) to all $P_i \neq P_{s(mid)}$.
2. On input ($\texttt{transmit}, mid$) parties $P_i \neq P_{s(mid)}$ wait for one round and then expects a message (mid, m, σ_s) from $P_{r(mid)}$, where σ_s is a valid signature of $P_{s(mid)}$ on (mid, m). If P_i receives it, it sends (mid, m, σ_s) to $P_{r(mid)}$. Otherwise, it sends a signature γ_i on ($\texttt{corrupt}, i$) to all parties.
3. On input ($\texttt{transmit}, mid$) party $P_{r(mid)}$ waits for two rounds and then expects a message (mid, m, σ_s) from each P_i, where σ_s is a valid signature of $P_{s(mid)}$ on (mid, m). If it arrives from some P_i, then P_r outputs (mid, m).

Note that now each round of communication on $\mathcal{F}_{\text{TRANSMIT}}$ takes two rounds on the underlying network. Between two parties where there have been no accusations, messages are sent as before (Step 1 in the above protocol) and the extra round is used for silence it is necessary that also non-accusing parties use two rounds to not lose synchronization.

If P_s sends a valid signature to just one honest party, then P_r gets its signature and can proceed as in optimistic mode. If P_s does not send a valid signature to any honest party, then all $n - t$ honest P_i send γ_i to all parties and hence all honest parties output (`corrupt`, s) in the following round, meaning that P_s was detected. Using these observations it can easily be shown that the above protocol is a UC implementation of $\mathcal{F}_{\text{TRANSMIT}}$ against covert adversaries with deterrence factor 1. Note that it is not a problem that we send m in cleartext through all parties, as an accusation of P_s by P_r means that P_s or P_r is corrupt, and hence m need not be kept secret.

We skipped the details of how the accusations are handled. We could in principle handle accusations by using one round of broadcast after each round of communication to check if any party wants to make an accusation. After broadcasting the accusations, the appropriate parties can then switch to expensive model. To avoid using a Byzantine agreement primitive in each round, we use a slightly more involved, but much cheaper technique which communicates less than n^2 bits in each round and which only uses a BA primitive when there are actually some accusations to be dealt with. The details are given in the next section.

In cheap mode, using $\mathcal{F}_{\text{TRANSMIT}}$ adds an overhead $N\kappa$ bits compared to plain transmission, where κ is the length of a signature and N is the number of messages sent. In expensive mode this overhead is a factor n larger.

Cheap Exception Handling. Consider a protocol consisting of two protocols π_{MAIN} and π_{EXCEPT}, both for the authenticated, synchronous point-to-point model. Initially the parties run π_{MAIN}. The goal is to allow any party to raise a flag, which stops π_{MAIN} and starts π_{EXCEPT}. With some details left out for now, this is handled as follows.

- If a party P_i wants to stop the main protocol, it sends (`stop`) to all parties and stops the execution of π_{MAIN}. It records the round R_i in which it stopped running π_{MAIN}.
- If a party P_i receives (`stop`) from any party while running π_{MAIN}, it sends (`stop`) to all parties and stops the execution of π_{MAIN}. It records the round R_i in which it stopped running π_{MAIN}.
- After all parties stopped they resynchronize and then run π_{EXCEPT}.
- After having run π_{EXCEPT}, the parties agree on a round C of π_{MAIN} which was executed completely, i.e., $R_i > C$ for all honest P_i, and then they rerun from round $C+1$. If a party P_r already received a message from P_s for one of the rounds that are now rerun, then P_r ignores any new message sent by P_s for that round. This is to avoid that corrupted parties can change their mind on what they sent in a previous round.

The resynchronization is needed as honest parties might stop in different rounds—though at most with a staggering of one round.

The resynchronization uses a sub-protocol where the input of P_i is the round R_i in which it stopped. The output is some common R such that it is guaranteed that $R_i = R$ for some honest P_i, i.e., at least one honest party stopped in round R. Since the honest parties stop within one round of each other, it follows that all honest parties stopped in round $R-1$, R or $R+1$. In particular, no honest party stopped in round $R-2$. The parties can therefore safely set $C = R-2$, i.e., rerun from round $R-1$.

The protocol used to agree on the round R proceeds as follows:

1. Each P_i has input $R_i \in \mathbb{N}$ and it is guaranteed that $|R_i - R_j| \leq 1$ for all honest P_i and P_j.
2. Let $r_i = R_i \bmod 4$ and make 4 calls to the BA functionality—name the calls BA_0, BA_1, BA_2 and BA_3. The input to BA_c is 1 if $c = r_i$ or $c = r_i - 1 \bmod 4$ and the input to BA_c is 0 if $c = r_i + 1 \bmod 4$ or $c = r_i + 2 \bmod 4$.
3. Let $o_c \in \{0,1\}$ for $c = 0,1,2,3$ denote the outcome of BA_c. Now P_i finds the largest $R \in \{R_i - 1, R_i, R_i + 1\}$ for which $o_{R \bmod 4} = 1$ and outputs R.

It is fairly straight forward to see that the honest parties output the same R and that R was always the input of some honest party. Look at two cases.

- If there exists ρ such that $R_i = \rho$ for all honest P_i, then all honest parties input the same to the BA functionalities, and then trivially $o_{\rho-1 \bmod 4} = 1$, $o_{\rho \bmod 4} = 1$, $o_{\rho+1 \bmod 4} = 0$ and $o_{\rho+2 \bmod 4} = 0$. Consequently, at honest parties outputs $R = \rho$.
- If there exists ρ such that $R_i = \rho$ for some honest P_i and $R_j = \rho + 1$ for some honest P_j, then $R_k \in \{\rho, \rho+1\}$ for all honest P_k, and thus all honest P_k input 1 to $BA_{\rho \bmod 4}$, and so $o_{\rho \bmod 4} = 1$. Furthermore, all honest parties input 0 to $BA_{\rho+2 \bmod 4}$, so $o_{\rho+2 \bmod 4} = 0$. It follows that all honest parties output $R = \rho$ if $o_{\rho+1 \bmod 4} = 0$ and that all honest parties output $R = \rho + 1$ if $o_{\rho+1 \bmod 4} = 1$. Both outputs are valid.

The above protocol is an improved version of a protocol by Bar-Noy et al. [3], which in turn uses techniques from Berman et al. [7]). The protocol in [3] uses $\log(B)$ calls to the BA functionality, where B is an upper bound on the input of the parties. We use just 4.

Note that at the point where the four BAs are run, the honest parties might still be desynchronized by one round. We handle this using a technique from [16] which simulates each round in the BA protocols by three synchronous rounds in the authenticated channel model.

Commitment Functionality. The protocol uses a one-round UC commitment scheme with a constant overhead (commit to κ bits using $\mathcal{O}(\kappa)$ bits), which can be realized with static security in the PKI model [4] given any mixed commitment scheme [11] with a constant overhead. Concretely we can instantiate such a scheme under Paillier's DCR assumption. Note that opposed to Barak et al. [4] we do not need a setup assumption: We assume honest majority and can thus, once and for all, use an active secure MPC to generate the needed setup [14]. The protocol also uses an error-correcting code (ECC) for n parties which allows to compute the message from any $n - t$ correct shares.

If one is willing to use the random oracle model, UC commitment can instead be done by calling the oracle on input the message to commit to, followed by some randomness. In practice, this translates to a very efficient solution based on a hash function.

The protocol proceeds as follows.

1. On input $(\texttt{commit}, cid, m)$, $P_{s(cid)}$ computes an ECC $(m_1, \ldots, m_n)$ of m. The sender then computes $c_i \leftarrow \text{commit}_{pk_i}(m_i)$ and sends c_i to P_i via $\mathcal{F}_{\text{TRANSMIT}}$.
2. On input $(\texttt{reveal}, cid, r)$, P_i opens each c_i to P_i. The opening is sent via $\mathcal{F}_{\text{TRANSMIT}}$. If any P_i receives an invalid opening, it transfers c_i and m_i to all parties and P_s is detected as a cheater. Otherwise, P_i transfers c_i and the opening to P_r.
3. Then P_r collects validly opened c_i. Let I be the index of these and let m_i be the opening of c_i for $i \in C$. If $|I| < n - t$, then P_r waits for one round and terminates.[3] If $(m_i)_{i \in I}$ is not consistent with a codeword in the ECC, then P_r transfers $(c_i)_{i \in I}$ and the valid openings to the other parties which detect P_s as corrupted. Otherwise, P_r uses $(m_i)_{i \in I}$ to determine m and outputs (cid, m).

Assuming that a commitment to ℓ bits have bit-length $\mathcal{O}(\max(\kappa, \ell))$, where κ is the security parameter, the complexity of a commitment to ℓ bits followed by an opening is $\mathcal{O}(n \max(\kappa, \ell/n)) = \mathcal{O}(n(\kappa + \ell/n)) = \mathcal{O}(\ell + n\kappa)$. This is assuming that there are no active corruptions, such that $\mathcal{F}_{\text{TRANSMIT}}$ has constant overhead.

Flip Functionality. To implement $\mathcal{F}^B_{\text{FLIP}}$ the parties proceed as follows.

1. On input $(\texttt{flip})$, all P_i commit to a uniformly random $k_i \in \{0, \ldots, B-1\}$.
2. In the next round all P_i reveal k_i to all parties.
3. All parties output $k = \sum_{i=1}^{n} k_i \bmod B$.

Under the condition that the protocol is used by the honest parties in a way that guarantees that they input $(\texttt{flip})$ in the same round, the argument that the protocol implements the functionality against a covert adversary (with deterrence 1) is straight forward.

Input Functionality. The input functionality can be implemented using a VSS with a multiplication protocol active secure against $t < n/2$ corruptions. The VSS should have the property that it is possible to verifiable reconstruct the secret and the share of all parties given the shares of the honest parties—standard bivariate sharing has this property. We sketch the protocol.

1. Each P_i deals a VSS $[\![x_i]\!]$ of its input x_i.
2. The parties use standard techniques to compute a VSS $[\![d]\!]$ of a uniformly random $d \in_{\mathsf{R}} \{0,1\} \subset K$.
3. For each input $[\![x_i]\!]$ the parties use an actively secure multiplication protocol to compute $[\![x^{(i,0)}]\!] = [\![d_i \cdot x_i]\!]$ and $[\![x^{(i,1)}]\!] = [\![(1 - d_i) \cdot x_i]\!]$.
   Each P_i takes its output to be $(x_i^{(j,0)})_{j=1}^n$ and $(x_i^{(j,1)})_{j=1}^n$, where $x_i^{(j,c)}$ is its point on the polynomial used by the sharing $[\![x^{(j,c)}]\!]$. The other values of the VSS are internal to the implementation of $\mathcal{F}_{\text{INPUT}}$ and only used for the below command.
4. On input $(\texttt{reveal}, i, k)$ the parties reconstruct $[\![d]\!]$ and all $[\![x^{i,d}]\!]$ towards P_k and P_k computes the points $x^{(j,d)}$ of P_j in all sharings and output $(x_i^{(j,d)})_{j=1}^n$.

[3] Since we assume that at most t parties are corrupted, we can assume that either P_s is detected or P_r receives $n-t$ commitments with corresponding valid decommitments.

References

1. Algesheimer, J., Camenisch, J., Shoup, V.: Efficient computation modulo a shared secret with application to the generation of shared safe-prime products. In: Yung, M. (ed.) CRYPTO 2002. LNCS, vol. 2442, pp. 417–432. Springer, Heidelberg (2002)
2. Aumann, Y., Lindell, Y.: Security against covert adversaries: Efficient protocols for realistic adversaries. In: Vadhan, S.P. (ed.) TCC 2007. LNCS, vol. 4392, pp. 137–156. Springer, Heidelberg (2007)
3. Bar-Noy, A., Deng, X., Garay, J.A., Kameda, T.: Optimal amortized distributed consensus. Information and Computation 120(1), 93–100 (1995)
4. Barak, B., Canetti, R., Nielsen, J.B., Pass, R.: Universally composable protocols with relaxed set-up assumptions. In: FOCS, pp. 186–195. IEEE Computer Society, Los Alamitos (2004)
5. Beerliová-Trubíniová, Z., Hirt, M.: Perfectly-secure MPC with linear communication complexity. In: Canetti, R. (ed.) TCC 2008. LNCS, vol. 4948, pp. 213–230. Springer, Heidelberg (2008)
6. Ben-Or, M., Goldwasser, S., Wigderson, A.: Completeness theorems for non-cryptographic fault-tolerant distributed computation (extended abstract). In: STOC, pp. 1–10 (1988)
7. Berman, P., Garay, J.A., Perry, K.J.: Optimal early stopping in distributed consensus. In: Segall, A., Zaks, S. (eds.) WDAG 1992. LNCS, vol. 647, pp. 221–237. Springer, Heidelberg (1992)
8. Bogetoft, P., Christensen, D.L., Damgård, I., Geisler, M., Jakobsen, T., Krøigaard, M., Nielsen, J.D., Nielsen, J.B., Nielsen, K., Pagter, J., Schwartzbach, M.I., Toft, T.: Secure multiparty computation goes live. In: Dingledine, R., Golle, P. (eds.) FC 2009. LNCS, vol. 5628, pp. 325–343. Springer, Heidelberg (2009)
9. Canetti, R.: Universally composable security: A new paradigm for cryptographic protocols. In: FOCS, pp. 136–145. IEEE, Los Alamitos (2001)
10. Damgård, I., Nielsen, J.B.: Scalable and unconditionally secure multiparty computation. In: Menezes, A. (ed.) CRYPTO 2007. LNCS, vol. 4622, pp. 572–590. Springer, Heidelberg (2007)
11. Damgård, I.B., Nielsen, J.B.: Perfect hiding and perfect binding universally composable commitment schemes with constant expansion factor. In: Yung, M. (ed.) CRYPTO 2002. LNCS, vol. 2442, pp. 581–596. Springer, Heidelberg (2002)
12. Damgård, I., Ishai, Y., Krøigaard, M., Nielsen, J.B., Smith, A.: Scalable multiparty computation with nearly optimal work and resilience. In: Wagner, D. (ed.) CRYPTO 2008. LNCS, vol. 5157, pp. 241–261. Springer, Heidelberg (2008)
13. Damgård, I., Geisler, M., Nielsen, J.B.: From passive to covert security at low cost. Cryptology ePrint Archive, Report 2009/592 (2009), `http://eprint.iacr.org/`
14. Goldreich, O., Micali, S., Wigderson, A.: How to play any mental game – a completeness theorem for protocols with honest majority. In: STOC, pp. 218–229. ACM, New York (1987)
15. Goyal, V., Mohassel, P., Smith, A.: Efficient two party and multi party computation against covert adversaries. In: Smart, N.P. (ed.) EUROCRYPT 2008. LNCS, vol. 4965, pp. 289–306. Springer, Heidelberg (2008)
16. Lindell, Y., Lysyanskaya, A., Rabin, T.: Sequential composition of protocols without simultaneous termination. In: Proceedings of the twenty-first annual symposium on Principles of distributed computing, pp. 203–212. ACM Press, New York (2002)

A Twist on the Naor-Yung Paradigm and Its Application to Efficient CCA-Secure Encryption from Hard Search Problems

Ronald Cramer[1], Dennis Hofheinz[2,*], and Eike Kiltz[3,**]

[1] CWI, Amsterdam, and Leiden University
cramer@cwi.nl
[2] Karlsruhe Institute of Technology
Dennis.Hofheinz@kit.edu
[3] CWI, Amsterdam
kiltz@cwi.nl

Abstract. The Naor-Yung (NY) paradigm shows how to build a chosen-ciphertext secure encryption scheme from three conceptual ingredients:

– a weakly (i.e., IND-CPA) secure encryption scheme,
– a "replication strategy" that specifies how to use the weakly secure encryption scheme; concretely, a NY-encryption contains several weak encryptions of the same plaintext,
– a non-interactive zero-knowledge (NIZK) proof system to show that a given ciphertext is consistent, i.e., contains weak encryptions of the *same* plaintext.

The NY paradigm served both as a breakthrough proof-of-concept, and as an inspiration to subsequent constructions. However, the NY construction leads to impractical encryption schemes, due to the usually prohibitively expensive NIZK proof.

In this contribution, we give a variant of the NY paradigm that leads to practical, fully IND-CCA secure encryption schemes whose security can be based on a generic class of algebraic complexity assumptions. Our approach refines NY's approach as follows:

– Our sole computational assumption is that of a Diffie-Hellman (DH) type two-move key exchange protocol, interpreted as a weakly secure key encapsulation mechanism (KEM).
– Our "replication strategy" is as follows. Key generation consists of replicating the KEM several times, but *only the first pass*. Encryption then consists of performing the second pass with respect to all of these, *but with the same random coins* in each instance.
– For proving consistency of a given ciphertext, we employ a practical universal hash proof system, case-tailored to our KEM and replication strategy.

* Work performed while the author was with CWI. Supported by NWO.
** Supported by the research program Sentinels.

D. Micciancio (Ed.): TCC 2010, LNCS 5978, pp. 146–164, 2010.

We instantiate our paradigm both from *computational* Diffie-Hellman (CDH) and from RSA type assumptions. This way, practical IND-CCA secure encryption schemes based on *search* problems can be built and explained in a generic, NY-like fashion.

We would like to stress that while we generalize universal hash proof systems *as a proof system*, we do *not* follow or generalize the approach of Cramer and Shoup to build IND-CCA secure encryption. Their approach uses specific hash proof systems that feature, on top of a NIZK property, a computational indistinguishability property. Hence they necessarily build upon *decisional* assumptions, whereas we show how to implement our approach with *search* assumptions. Our approach uses hash proof systems in the NY way, namely solely as a device to prove consistency. In our case, secrecy is provided by the "weak encryption" component, which allows us to embed search problems.

Keywords: Public-key encryption, chosen-ciphertext security.

1 Introduction

One of the main fields of interest in cryptography is the design and the analysis of the security of encryption schemes in the public-key setting (PKE schemes). The notion of security against chosen-ciphertext attack (IND-CCA security) is due to Rackoff and Simon [23] and is now widely accepted as the standard security notion for public-key encryption schemes. In contrast to security against passive adversaries (security against chosen-plaintext attacks aka semantic security), in a chosen-ciphertext attack the adversary plays an active role by obtaining the decryptions of ciphertexts (or even arbitrary bit-strings) of his choosing. The practical significance of such attacks was demonstrated by Bleichenbacher [1] by means of an IND-CCA attack against schemes following the encryption standard PKCS #1.

The Naor-Yung paradigm. Historically, the first scheme that was provably secure against a weaker variant of IND-CCA attacks (namely, "lunch-time attacks") is due to Naor and Yung (NY) [21]. Dolev, Dwork, and Naor [9] later showed how to modify the paradigm of NY to achieve full IND-CCA security. We briefly outline the general idea in the following. To start, let's assume a weakly (i.e., IND-CPA) secure encryption scheme. (IND-CPA secure encryption schemes are a well-understood primitive, and can be constructed from various search or decisional computational problems, see, e.g., [11].) Now a ciphertext contains two weakly secure encryptions of *the same message* (under different public keys of the weakly secure scheme), along with a non-interactive zero-knowledge (NIZK) proof that indeed the same messages were encrypted. During the security proof, a simulator will know precisely one of the two secret keys for the weakly secure encryption schemes. (Note that in order to implement the decryption oracle, the simulator only needs to decrypt one ciphertext component and rely on the soundness of the NIZK proof.) Hence, we can carve out three conceptual ingredients for the NY paradigm:

- a weakly (i.e., IND-CPA) secure encryption scheme,
- a "replication strategy" that specifies how to use the weakly secure encryption scheme, and
- a NIZK proof system to show that a given ciphertext is consistent.

Of course, these ingredients are not independent (e.g., the statement to be proven by the NIZK proof depends both on the weakly secure encryption scheme and on the replication strategy). The system of Dolev, Dwork and Naor uses a similar overall approach, the main difference the used replication strategy. (Additionally, they also require a more special type of "non-malleable" NIZK proof systems.)

Hash proof systems. The first *practical* schemes provably IND-CCA secure under standard cryptographic hardness assumptions were due to Cramer and Shoup [8,8]. They later generalized their initial scheme to the paradigm of "hash proof systems" (HPSs) [7], thereby yielding new practical schemes from a number of alternative intractability assumptions. The approach of Cramer and Shoup is inspired by the NY paradigm. And indeed, a HPS, as used by Cramer and Shoup, combines the encryption and the proof part of the NY paradigm in one primitive. However, even though the concept of HPSs is generic, its use in [7] to build encryption schemes inherently relies on *decisional assumptions*, such as the assumed hardness of deciding if a given integer has a square root modulo a composite number with unknown factorization, or if deciding if a given tuple is a Diffie-Hellman tuple or not (DDH assumption).

The theory of hash proof systems has since been developed further (e.g., [10,25,18]), and particular instances of the HPS-based schemes could be optimized, leading to a number of even more efficient schemes (e.g., [10,19,14,18]). However, all of these schemes are based on decisional assumptions (such as the DDH assumption).

Lossy trapdoor functions and the approach of Rosen and Segev. An alternative generic framework of constructing IND-CCA secure encryption schemes is given by the recent concept of lossy trapdoor functions [22] that led to the first construction based on a (decisional) assumption related to finding shortest vectors on lattices. However, also lossy trapdoor functions inherently rely on decisional assumptions rather than computational assumptions.[1]

Recently, Rosen and Segev [24] proposed a refinement of the NY approach that does *not* require an explicit NIZK part. (In fact, consistency of a ciphertext can be checked deterministically be the decryptor, since they employ weak encryption schemes that are actually *functions*.) Unfortunately, their approach requires a nonstandard computational assumption, namely one-way security of several independent functions under *correlated inputs*. They show that security under correlated inputs is implied by lossy trapdoor functions; however, they do not show security under a standard *search* assumption.

[1] Unless, of course, the decisional assumption can be proved equivalent to a computational assumption, as it is the case with cryptosystems based on the problem of "learning with error" [22].

IND-CCA security from identity-based encryption. Boneh et al [3] describe a completely generic transformation of a selective-ID secure identitity-based encryption scheme into an IND-CCA secure PKE scheme. Their assumption (a selective-ID secure IBE scheme) cannot be directly counted as a search or decisional assumption. However, as it is based upon indistinguishability of adversarial views, it is arguably closer to a decisional assumption.

Decisional vs. search assumptions. We conclude that all known *generic paradigms* of constructing practical IND-CCA secure encryption seem to rely on *decisional* assumptions, as opposed to *search* assumptions. No generic paradigm is known under which practical IND-CCA secure schemes based on, say, the CDH problem could be constructed or explained.

In most known cases related to cryptography, decisional assumptions form a much stronger class of assumptions than the corresponding search assumptions. For example, deciding if a given integer has a modular square root or not may be much easier than actually computing a square root (or, equivalently, factoring the modulus). Only recently were practical schemes proposed whose IND-CCA security does not rely on decisional assumptions (e.g., [3,5,13,16]). In particular, the first practical encryption scheme IND-CCA secure under the *Computational* Diffie-Hellman (CDH) assumption was proposed by Cash, Kiltz, and Shoup [5] in 2008, and improved by Hanaoka and Kurosawa [13] later that year. In 2009, Hofheinz and Kiltz proposed a very efficient IND-CCA secure encryption scheme under the factoring assumption [16].

However, there seems to be no overarching concept that explains these schemes. Each of these schemes relies on different techniques to achieve security, and in particular to conduct a reduction in the security proof.

Our contribution. In this work, we refine the abstract NY paradigm in a way that allows to construct and explain *practical* IND-CCA secure encryption schemes whose security is based on general widely believed *search* assumptions. Concretely, we modify the NY paradigm as follows:

- Our sole computational assumption is that of a Diffie-Hellman type two-move key exchange protocol. This assumption is implied, e.g., under the CDH assumption in cyclic groups, or under the RSA assumptions. We interpret the key exchange protocol as a weakly secure key encapsulation mechanism. (That is, the first KE message is the KEM public key, and the second KE message is the KEM ciphertext.)
- Our "replication strategy" uses the above KEM several times, but with the same *encryption random coins* (and not with the same key or plaintext as in the original NY paradigm). Our replication strategy comes with a special simulation setup, such that the simulator in the security proof can decrypt *all* consistent ciphertexts, except for *one* predetermined target/challenge ciphertext.
- For proving consistency of a given ciphertext, we employ a generic but practical universal hash proof system. This forms a special type of designated-verifier NIZK proof system, case-tailored to our KEM and replication strategy. We stress that we do *not* use the HPS in the same way that Cramer and Shoup do to achieve IND-CCA security. In fact, [7] require a *NIZK proof property and a*

computational property. We only use a proof property of our HPS, and obtain secrecy from our assumption on the KEM, much like in the original NY paradigm. Hence we do *not* generalize the Cramer-Shoup approach to achieving IND-CCA secure encryption.

The technical assumption we use to capture a Diffie-Hellman type key exchange protocol is that of a *hard algebraic set system.* Roughly, an algebraic set system consists of a finite Abelian group S, together with a commutative, unitary subring Φ of group endomorphisms over S that fulfil a number of natural algebraic properties. It is a hard algebraic set system if a Diffie-Hellman style computational problem is intractable. Examples of hard algebraic set systems can be obtained from standard computational assumptions such as the CDH and the RSA assumptions (using hardcore bit extraction). Our main result is an efficient transformation from any hard algebraic set system into a practical IND-CCA secure encryption scheme. With respect to the results our construction can be seen as a generalization of the recent specific constructions from computational problems [3,5,13,16].

1.1 Technical Details

We now give some technical details of our transformation.

An IND-CPA secure construction. We start by describing a simple IND-CPA secure construction from any hard algebraic set system (S, Φ). It is actually a key encapsulation mechanism [8] (KEM) that can be viewed as a natural abstraction of the Diffie-Hellman key-exchange protocol. The scheme's secret key consists of a random $\chi \in \Phi$ and the public-key is a random $g \in S$ and $u = \chi(g) \in S$. Encryption picks random $\psi \in \Phi$, computes the ciphertext $c = \psi(g) \in S$ and uses the encapsulated key $K = \mathsf{Ext}(\psi(u))$ to blind the message. (Here Ext is an extractor function that is part of the underlying hard computational problem of the algebraic set system.) Decryption reconstructs the key by computing $K = \mathsf{Ext}(\chi(c))$. In our construction, we will have that χ and ψ commute. This directly implies correctness of the scheme, since then $\chi(c) = \chi(\psi(g)) = \psi(\chi(g)) = \psi(u)$.

Our IND-CCA secure construction. We augment the above IND-CPA secure construction in a clean and modular way (much like Naor and Yung) by adding a "replication part" and a "NIZK part" to the scheme. The two new parts require no computational assumptions, and so the resulting scheme is IND-CCA secure if the old scheme is IND-CPA secure. More concretely, ciphertexts are now tuples of the form $(c, \mathbf{d}, \pi)$, where c is from the IND-CPA construction, $\mathbf{d}$ is the "trapdoor element", and π is the "NIZK element" that proves consistency of the ciphertext. We now explain our construction by showing how the different parts affect the ability to perform decryption.

The idea behind the trapdoor element $\mathbf{d}$ in the ciphertext is that can be set up by a simulator such that it is possible to decrypt (without the knowledge of the scheme's secret key χ) all *consistent ciphertexts* $(c, \mathbf{d})$ except the ciphertext that is used to challenge the adversary (in the security reduction to the IND-CPA secure scheme). This "all-but-one" simulation technique can be traced back at least to [20], where it was used in the context of pseudorandom functions.[2] In the encryption context, "all-but-one" simulations have been used in identity-based encryption [2] and were already applied to several encryption schemes in [3,4,5,14,17,22,16].

The above all-but-one simulation technique allows to correctly simulate decryption of arbitrary for *consistent ciphertexts* $(c, \mathbf{d})$ but consistency can only be checked using the secret key which is not available during simulation. To provide an alternative consistency check we add the NIZK element π to the ciphertext. Actually, the NIZK element is generated using a hash proof system [7] and proves that $(c, \mathbf{d})$ is contained in the *trapdoor language* consisting of all consistent ciphertexts. However, we stress that we use hash proof system techniques here without relying on a (computational or decisional) assumption. Instead, we use a hash proof system only as a NIZK proof, in which case the hash proof system's soundness is information-theoretic.

We also note that the "trapdoor part" of a consistent ciphertext, along with the NIZK proof that the ciphertext is consistent, can be seen as a variant of an (extractable) NIZK proof of knowledge. However, in our case, the challenge ciphertext plays a special role: we need to construct the trapdoor language *from* a given challenge ciphertext. (Hence, extraction is—naturally—not possible for the challenge ciphertext.)

Our technical contribution (that may be of independent interest) is to bootstrap the trapdoor part and the the NIZK part (i.e., the hash proof system for the trapdoor language) generically from the abstract algebraic properties of algebraic set systems. In contrast to the generic NIZK-based constructions from [9,21] our constructions are relatively efficient: the key-size and ciphertexts of the obtained IND-CCA secure scheme contain $O(k)$ elements in S, where k is the security parameter. In many cases the ciphertexts can be "compactified" into a constant number of elements in S, giving truly practical schemes.

2 Preliminaries

2.1 Notation

Generic notation. A probabilistic polynomial-time (PPT) algorithm is a randomized algorithm which runs in strict polynomial time. By k we denote the security parameter, which indicates the "amount of security" we desire. A function $f : \mathbb{N} \rightarrow \mathbb{R}$ is negligible if for all $c \in \mathbb{N}$, there exists $k_0 \in \mathbb{N}$ such that $|f(k)| < k^{-c}$ for all $k > k_0$. Furthermore, f is overwhelming if $1 - f$ is negligible. For random variables X and Y, we write $X \overset{c}{\approx} Y$ if X and Y are

[2] We stress that our use of the term "all-but-one" refers to the ability to generate a secret key that can be used to decrypt all consistent ciphertexts except for an *externally given* ciphertext. This is very different from the techniques of, e.g., [8]: in this latter framework, the first step in the proof consists in *making the challenge ciphertext inconsistent*, and then constructing a secret key that can be used to construct *all* consistent ciphertexts. Hence, "all-but-one" really refers to an "artificially punctured" secret key.

computationally indistinguishable, i.e., if for all PPT algorithms D, we have that $\Pr[\mathsf{D}(X)=1]-\Pr[\mathsf{D}(Y)=1]$ is negligible. Similarly, we write $X \overset{s}{\approx} Y$ if the statistical distance between X and Y is negligible. For a vector $\mathbf{h}=(h_1,\ldots,h_\ell)$ and a nonempty set $J \subseteq \{1,\ldots,\ell\}$, we write $\mathbf{h}_J$ for the restricted vector $(h_i)_{i\in J}$. Furthermore, if ϕ is a function, then $\phi(\mathbf{h})$ denotes the component-wise application of ϕ, i.e., $\phi(\mathbf{h})=(\phi(h_i))_i$.

Group endomorphisms. For an abelian group, we denote its group operation additively. If S is an abelian group, then $\mathrm{End}(S)$ consists of all group-homomorphisms $\chi : S \to S$. It has a ring-structure, where point-wise-addition is ring-addition (denoted "+") and functional composition is ring-multiplication (denoted "$\circ$"). Suppose Φ is an additive sub-group of $\mathrm{End}(R)$. Then $\mathrm{Ann}(\Phi) \subset S$, consists of all $g \in S$ for which $\chi(g)=0$ for all $\chi \in \Phi$, and it is a sub-group of S. A sub-ring of $\mathrm{End}(R)$ is unitary if it contains the identity endomorphism.

2.2 Key Encapsulation Mechanisms

Instead of a public-key encryption scheme we consider the conceptually simpler KEM framework. It is well-known that an IND-CCA secure KEM combined with a (one-time-)IND-CCA secure symmetric cipher (DEM) yields a IND-CCA secure public-key encryption scheme [8]. Efficient one-time IND-CCA secure DEMs can be constructed even without computational assumptions, using an encrypt-then-MAC paradigm [8], or using strong pseudorandom permutations.

Syntactics. A *key encapsulation mechanism (KEM)* $\mathsf{KEM}=(\mathsf{Gen},\mathsf{Enc},\mathsf{Dec})$ consists of three PPT algorithms. Via $(pk,sk) \leftarrow \mathsf{Gen}(1^k)$, the key generation algorithm produces public/secret keys for security parameter $k \in \mathbb{N}$; via $(K,C) \leftarrow \mathsf{Enc}(pk)$, the encapsulation algorithm creates a symmetric key[3] $K \in \{0,1\}^k$ together with a ciphertext C; via $K \leftarrow \mathsf{Dec}(sk,C)$, the possessor of secret key sk decrypts ciphertext C to get back a key K which is an element in $\{0,1\}^k$ or a special reject symbol $\perp$. For correctness, we require that for all possible $k \in \mathbb{N}$, and all $(K,C) \leftarrow \mathsf{Enc}(pk)$, we have $\Pr[\mathsf{Dec}(sk,C)=K]=1$, where the probability is taken over the choice of $(pk,sk) \leftarrow \mathsf{Gen}(1^k)$, and the coins of all the algorithms in the expression above.

Security. The common requirement for a KEM is indistinguishability against chosen-ciphertext attacks (IND-CCA) [8], where an adversary is allowed to adaptively query a decapsulation oracle with ciphertexts to obtain the corresponding key. Formally:

Definition 1 (IND-CCA security of a KEM). *Let* $\mathsf{KEM}=(\mathsf{Gen},\mathsf{Enc},\mathsf{Dec})$ *be a KEM. For any PPT algorithm* A, *we define the following experiments* $\mathsf{Exp}^{\mathsf{CCA\text{-}real}}_{\mathsf{KEM},\mathsf{A}}$ *and* $\mathsf{Exp}^{\mathsf{CCA\text{-}rand}}_{\mathsf{KEM},\mathsf{A}}$*:*

[3] For simplicity we assume that the KEM's keyspace are bitstrings of length k.

Experiment $\mathsf{Exp}^{\mathsf{CCA\text{-}real}}_{\mathsf{KEM,A}}(k)$	**Experiment** $\mathsf{Exp}^{\mathsf{CCA\text{-}rand}}_{\mathsf{KEM,A}}(k)$
$(pk, sk) \leftarrow \mathsf{Gen}(1^k)$	$(pk, sk) \leftarrow \mathsf{Gen}(1^k)$
	$R \leftarrow \{0,1\}^k$
$(K^*, C^*) \leftarrow \mathsf{Enc}(pk)$	$(K^*, C^*) \leftarrow \mathsf{Enc}(pk)$
Return $\mathsf{A}^{\mathsf{Dec}(sk,\cdot)}(pk, K^*, C^*)$	*Return* $\mathsf{A}^{\mathsf{Dec}(sk,\cdot)}(pk, R, C^*)$

In the above experiments, the decryption oracle $\mathsf{Dec}(sk, C)$ *returns* $K \leftarrow \mathsf{Dec}(sk, C)$*, for all* $C \neq C^*$*. We define* A*'s advantage in breaking* KEM*'s* IND-CCA *security as*

$$\mathsf{Adv}^{\mathsf{CCA}}_{\mathsf{KEM,A}}(k) := \frac{1}{2}\left|\Pr\left[\mathsf{Exp}^{\mathsf{CCA\text{-}real}}_{\mathsf{KEM,A}}(k) = 1\right] - \Pr\left[\mathsf{Exp}^{\mathsf{CCA\text{-}rand}}_{\mathsf{KEM,A}}(k) = 1\right]\right|.$$

We say that KEM is IND-CCA secure *if* $\mathsf{Adv}^{\mathsf{CCA}}_{\mathsf{KEM,A}}$ *is negligible for all PPT* A.

As a stepping stone, we will also consider the weaker requirement of IND-CPA security of a KEM. The IND-CPA security experiment is very similar to the IND-CCA security experiment, only without a decryption oracle for the adversary:

Definition 2 **(IND-CPA security of a KEM).** *Let* $\mathsf{KEM} = (\mathsf{Gen}, \mathsf{Enc}, \mathsf{Dec})$ *be a KEM. For any PPT algorithm* A*, we define the following experiments* $\mathsf{Exp}^{\mathsf{CPA\text{-}real}}_{\mathsf{KEM,A}}$ *and* $\mathsf{Exp}^{\mathsf{CPA\text{-}rand}}_{\mathsf{KEM,A}}$ *as identical to the experiments* $\mathsf{Exp}^{\mathsf{CCA\text{-}real}}_{\mathsf{KEM,A}}$ *and* $\mathsf{Exp}^{\mathsf{CCA\text{-}rand}}_{\mathsf{KEM,A}}$ *from Definition 1, only that* A *does not get access to a decryption oracle* Dec. *Let*

$$\mathsf{Adv}^{\mathsf{CCA}}_{\mathsf{KEM,A}}(k) := \frac{1}{2}\left|\Pr\left[\mathsf{Exp}^{\mathsf{CPA\text{-}real}}_{\mathsf{KEM,A}}(k) = 1\right] - \Pr\left[\mathsf{Exp}^{\mathsf{CPA\text{-}rand}}_{\mathsf{KEM,A}}(k) = 1\right]\right|.$$

We say that KEM is IND-CPA secure *if* $\mathsf{Adv}^{\mathsf{CCA}}_{\mathsf{KEM,A}}$ *is negligible for all PPT* A.

3 Set Systems

3.1 Basic Definition

Definition 3 **(Set system).** *A* set system $\mathcal{SS} = (S, \Phi)$ *consists of the following*

- *A finite, non-empty set* S*.*
- *A non-empty set* Φ *of functions* $\chi : S \to S$*.*

Furthermore, we require that efficient algorithms exist for the following tasks:

- *Sampling* with the uniform distribution from* S*.*
- *Sampling* with the uniform distribution from* Φ*.*
- *Evaluating* $\chi(g)$ *when given* $\chi \in \Phi$ *and* $g \in S$*.*

*Here, * means that it is sufficient if sampling can be performed* approximatively uniform *(that is, if a distribution can be sampled which is statistically close to uniform).*

We stress that while our definitions are typically asymptotic, an explicit security parameter is sometimes suppressed for ease of exposition.

Definition 4 (Commutative set system). *A set system* (S, Φ) *is* commutative *if the functions in* Φ commute pairwise, *i.e., for all* $\chi, \psi \in \Phi$, *we have* $\chi \circ \psi = \psi \circ \chi$.

3.2 Hard Set Systems

The following definition encapsulates the computational hardness assumption associated with set systems.

Definition 5 (Hard set system). *Let* (S, Φ) *be a commutative set system, and let* $\mathsf{Ext} : S \to \{0,1\}^n$ *be efficiently computable. We say that* (S, Φ) *is a* hard set system with randomness extractor Ext *if*

$$(g, \chi(g), \psi(g), E) \quad \overset{c}{\approx} \quad (g, \chi(g), \psi(g), R),$$

where $g \in S$, $\chi, \psi \in \Phi$, *and* $R \in \{0,1\}^n$ *are uniformly chosen, and* $E = \mathsf{Ext}(\chi(\psi(g))) \in \{0,1\}^n$.

3.3 Algebraic Set Systems

We now set abstract algebraic conditions that are sufficient for the existence of a quite efficient transformation that we will use to achieve CCA security.

Definition 6 (Algebraic set system). *A set system* (S, Φ) *is an* algebraic set system *if the following algebraic conditions are fulfilled.*

Group structure. *S is a finite Abelian group.*

Recognizability. *S is efficiently recognizable.*

Commutative endomorphisms. *Φ is a commutative, unitary sub-ring of* $\mathrm{End}(S)$.

Almost-transitivity. *A* $g \in S$ *is called* normal *if*

$$\forall h \in S \ \exists \phi \in \Phi : \ h = \phi(g) \ .$$

We require that a uniformly chosen $g \in S$ *is normal with overwhelming probability.*

Uniformity. *For uniformly chosen* $g, u \in S$ *and* $\chi \in \Phi$, *we have* $(g, \chi(g)) \overset{s}{\approx} (g, u)$.

Remark 1. If Φ consists of all multiplications by non-negative integers then Φ is a commutative, unitary sub-ring of $\mathrm{End}(S)$.

3.4 Examples

Diffie-Hellman. Let S be a cyclic group $\mathbb{G} = \langle g \rangle$ of prime order p. (For this and the next example, we stick to the more common notation and write the group multiplicatively.) We define Φ as

$$\Phi := \{\chi(g) = g^x \ : \ x \in \mathbb{Z}_p\}.$$

If we require that S is efficiently recognizable, this makes (S, Φ) a set system. Let $\chi, \psi \in \Phi$, i.e., $\chi(g) = g^x$ and $\psi(g) = g^y$, for some $x, y \in \mathbb{Z}_p$. Now $\chi(\psi(g)) = (g^y)^x = g^{xy} = (g^x)^y = \psi(\chi(g))$ and therefore (S, Φ) is commutative. Since S is efficiently recognizable, it is easy to see that (S, Φ) is also algebraic.

If the DDH assumption holds in $\mathbb{G}$, then (S, Φ) is a hard set system with randomness extractor $\mathsf{Ext} : \mathbb{G} \to \{0,1\}^n$, where Ext is an arbitrary pseudorandom generator. If the CDH assumption holds in $\mathbb{G}$, then (S, Φ) is a hard set system with randomness extractor $\mathsf{Ext}_s : \mathbb{G} \to \{0,1\}$. Here, Ext_s maps $g \in \mathbb{G}$ to the Goldreich-Levin bit $\sum_{i=1}^{|g|} g_i s_i$, where $|g|$ denotes the bit length and g_i the i-th bit of g in some canonical bit representation, and $s = (s_1, \ldots, s_{|g|}) \in \{0,1\}^{|g|}$.

We stress that knowledge of the *order* of $\mathbb{G}$ is not required. (Only one must be able to approximatively sample uniform exponents.) In particular, $\mathbb{G}$ could be instantiated over a higher-genus curve.

RSA. We use the group of *signed quadratic residues* [12,15]. Fix a Blum integer $N = PQ$ for safe primes $P, Q \equiv 3 \bmod 4$ (such that $P = 2p + 1$ and $Q = 2q + 1$ for primes p, q). Let $\mathbb{J}_N \subseteq \mathbb{Z}_N^*$ denote the set of elements with Jacobi symbol 1 modulo N and let $\mathbb{QR}_N \subset \mathbb{J}_N$ denote the set of quadratic residues modulo N. Consider the quotient group $S := \mathbb{QR}_N^+ := \mathbb{QR}_N / \pm 1$. Together with the group operation $a \circ b := |a \cdot b \bmod N|$ this forms a finite Abelian group of order pq. Furthermore, since $\mathbb{QR}_N^+ = \mathbb{J}_N^+ := \mathbb{J}_N / \pm 1 = \{|x| \,:\, x \in \mathbb{J}_N\}$, S is efficiently recognizable. Define

$$\Phi := \{\chi(g) = |g^x| \,:\, x \in \mathbb{Z}_{\lfloor N/4 \rfloor}\}.$$

Observe that we can sample uniformly from S and Φ. Furthermore, $(g, \chi(g))$ is statistically close to (g, u) for uniform $g, u \in S$ and $\chi \in \Phi$, since $\lfloor N/4 \rfloor$ approximates pq, the order of S, suitably well. This makes (S, Φ) a set system.

Finally, if the RSA assumption holds in $\mathbb{Z}_N$, then (S, Φ) is also hard with randomness extractor $\mathsf{Ext} : S \to \{0,1\}$, where Ext maps $g \in S$ to the least significant bit $\mathsf{LSB}(g)$ of g.

4 IND-CPA Secure KEMs from Commutative Set Systems

Construction 7 (Semantically secure KEM). *Assume that (S, Φ) is a hard commutative set system with randomness extractor $\mathsf{Ext} : S \to \{0,1\}^n$. Then, our basic key encapsulation scheme $\mathsf{KEM} = (\mathsf{Gen}, \mathsf{Enc}, \mathsf{Dec})$, which is an obvious abstraction of the Diffie-Hellman scheme, is defined as follows.*

Key Generation. *$\mathsf{Gen}(1^k)$ chooses $g \in S$ and $\chi \in \Phi$ uniformly, and computes $u = \chi(g) \in S$. Public key is $pk = (g, u) \in S \times S$, and secret key is $sk = \chi \in \Phi$.*

Encapsulation. *Given $pk = (g, u) \in S \times S$, Enc chooses $\psi \in \Phi$ uniformly and computes the ciphertext $c = \psi(g) \in S$. Next, Enc derives the encapsulated key*

$$K = \mathsf{Ext}(\psi(u)) \in \{0,1\}^n. \quad (1)$$

Decapsulation. *Given $sk = \chi \in \Phi$ and $c \in S$,* Dec *computes*

$$\chi(c) = (\chi \circ \psi)(g) = (\psi \circ \chi)(g) = \psi(u)$$

to derive the encapsulated key $K \in \{0,1\}^n$ as in (1). Note that here it is exploited that the functions in Φ commute.

Theorem 1 (Definition 7 is an IND-CPA secure KEM). *If (S, Φ) is a hard commutative set system, then the KEM from Definition 7 is* IND-CPA *secure in the sense of Definition 2.*

Proof. This follows directly from Definition 5.

5 Hash Proof Systems

5.1 Definitions

We will use hash proof systems for a language L, as defined in [7, Section 5 of full version]. However, we stress that we will neither define nor use the concept of a subset membership problem (which essentially would require that elements in the language are computationally indistinguishable from elements outside of the language, see [7, Section 4 of full version]). For our purposes, only the proof system itself (whose security is defined information-theoretically) is relevant.

Definition 8 (Hash proof system). *Let L be a language and let ϵ be a real number with $0 \leq \epsilon < 1$. A hash proof system* with error probability *ϵ consists of the following.*

- *A finite non-empty set $\mathcal{V}$: this is where the verifier samples a secret verification-key from, to enable him to check proofs.*
- *A finite non-empty set $\mathcal{P}$ and a function $\alpha : \mathcal{V} \to \mathcal{P}$: this maps a verification key to its projection, which is an auxiliary input for the prover to construct a proof.*
- *A non-empty finite set Π: this is where proof strings will be sampled from.*

Furthermore, efficient algorithms for the following tasks exist.

- *Sampling with the uniform distribution from $\mathcal{V}$.*
- *Computing $\alpha(\kappa) \in \mathcal{P}$ when given $\kappa \in \mathcal{V}$.*
- *Computing the proof $\pi \in \Pi$ when given the statement $x \in L$, along with either the projection $\alpha(\kappa)$ and a witness $\phi \in \Phi$ (that $x \in L$), or, alternatively, the verification key κ itself.*

The following security properties hold, even in the presence of an unbounded adversary.

Completeness. *If indeed $x \in L$, a proof $\pi \in \Pi$ thus computed is* accepted *when verified using the secret verification key κ. This verification is performed efficiently by the verifier.*

Soundness. *For every $x \notin L$, every projection $P \in \mathcal{P}$, and every purported proof $\tilde{\pi} \in \Pi$: the probability (over uniform $V \in \mathcal{V}$ with $\alpha(V) = P$) that $\tilde{\pi}$ will be accepted is at most ϵ.*

Uniqueness. *The proof $\pi \in \Pi$ is unique. In the verification procedure referred to above, the verifier actually first computes π' from $x \in L$ and the verification key κ. The decision is then made by checking whether $\pi' = \pi$. In other words, the verifier can compute the proof himself from seeing the statement, using his secret verification key.*

Note that the uniqueness property implies a non-interactive zero-knowledge property, in the following sense. In the zero-knowledge setting, the verification key can be set up by a simulator, who then can generate the *unique* proofs π as $\pi = \kappa(x)$ for arbitrary statements x and without witness.

We make a number of remarks and comments concerning our definitions:

- The error probability ϵ can be decreased exponentially by running copies based on independently selected keys in parallel.
- Such a hash proof system will be "global" in the sense that it does not essentially depend on the length ℓ or on the choice of the base vectors $g, \mathbf{h}$. Furthermore, is assumed that the generation of the secret verification key does not depend on the choice of base vectors.
- Obviously, however, several technical details in the definition above will typically "scale with ℓ." Also, all algorithms involved may take ℓ and $g, \mathbf{h}$ as input (except secret key generation, which may not depend on $g, \mathbf{h}$, see above). But this dependence is suppressed in the notation.

5.2 Our Trapdoor Language

We define a natural language derived from set systems that simply "singles out" sequences of elements obtained by applying the same function to (a subset of) some fixed sequence elements. We note that [24] use the related but dual concept of "correlated products" to obtain chosen-ciphertext security. Namely, they apply several trapdoor functions to the same preimage, while in our approach, we apply one function to several preimages. We also note that in their work, it is crucial that the functions can be inverted (using a trapdoor). We do not have this requirement.

Definition 9 (Trapdoor language). *Let (S, Φ) be a set system, let ℓ be a positive integer, and let*

$$g \in S, \quad \mathbf{h} = (h_1, \ldots, h_\ell) \in S^\ell,$$

be base vectors*. Then the* trapdoor language L associated to (S, Φ) and $g, \mathbf{h}$ *is defined as*

$$L = \{(c, \mathbf{d}, J) \in S \times S^J \times \mathcal{J} \mid \exists \chi \in \Phi \text{ such that } c = \chi(g) \wedge \mathbf{d} = \chi(\mathbf{h}_J)\},$$

where $\mathcal{J}$ consists of all non-empty subsets of $\{1, \ldots, \ell\}$. (Recall our abbreviation $\chi(\mathbf{h}_J) = (\chi(h_i))_{i \in J}$.) Such a function $\chi \in \Phi$ (not necessarily unique) is called a witness.

In the remaining part of this section we show the following theorem.

Theorem 2 (HPS for our trapdoor language). *Let (S, Φ) be a an algebraic set system and let $g \in S$, $\mathbf{h} \in S^\ell$ be randomly chosen base vectors. If g is normal (in the sense of Definition 6) and $h_i \neq 0$ for all i, then there exists a hash proof system for the language L. The error probability is at most ℓ/p, where p is the smallest prime divisor of $|S|$.*

The proof proceeds in two steps. First we prove the case $\ell = 1$ and then we show how the general case follows from that by induction.

Let $g \in S$ be normal, and let $h \in S$. Since g is normal, $h = \rho(g)$ for some $\rho \in \Phi$. We now construct a hash proof system for the trapdoor language L. The hash proof system is defined as follows.

$$Z = \{(\chi(g), \psi(h)) \ : \ \chi, \psi \in \Phi\} \subset S \times S, \tag{2}$$

$$L = \{(\chi(g), \chi(h)) \ : \ \chi \in \Phi\} \subset Z. \tag{3}$$

(For simplicity and ease of presentation, we omit the J component of Z and L, since in case $\ell = 1$ this component is trivial.)

Setup. The verifier chooses a random secret verification key $(\delta, \rho) \in \Phi \times \Phi$, and computes its projection

$$\alpha = \delta(g) + \rho(h).$$

Proof phase. The prover holds $(c, d) \in L$ and a witness $\chi \in \Phi$ such that

$$(c, d) = (\chi(g), \chi(h)).$$

He computes the proof

$$\pi = \chi(\alpha).$$

Verification. The verifier checks whether

$$\pi = \delta(c) + \rho(d).$$

Note that if the prover is honest, then indeed by commutativity

$$\pi = \chi(\alpha) = \chi(\delta(g) + \rho(h)) = \delta(\chi(g)) + \rho(\chi(h)) = \delta(c) + \rho(d) \ .$$

We sketch why the above hash proof system satisfies the conditions of Definition 8. The full proof (as well as the case of general ℓ) is contained in the full version of this paper [6]. Let $(c, d) \in Z$. Suppose the prover falsely claims that $(c, d) \in L$. The pair (δ, ρ) is randomly distributed on $\Phi \times \Phi$ conditioned on the projection being equal to α. Then, by a technical lemma, each solution z of the two equations $\alpha = \delta(g) + \rho(h)$ and $z = \delta(c) + \rho(d)$ is equally likely to be the "correct proof." Since there are at least p such solutions Theorem 2 now follows (for $\ell = 1$).

6 IND-CCA Secure KEMs from Algebraic Set Systems

Construction 10 (Chosen-ciphertext secure KEM). *Let (S, Φ) a hard algebraic set system with randomness extractor $\mathsf{Ext} : S \to \{0,1\}^n$. Further, assume a target collision resistant hash function T on S (whose formal definition can be looked in [6]). For $c \in S$, $\mathsf{T}(c)$ is encoded as a subset of $\{1, \ldots, 2k\}$, with $|\mathsf{T}(c)| = k$. Note that if $\mathsf{T}(c) \neq \mathsf{T}(c')$, then these two sets are incomparable by inclusion.*

Key generation. *Let (S, Φ) be an algebraic set system. Choose*

$$g \in S, \quad \mathbf{h} = (h_1, \ldots, h_{2k}) \in S^{2k}.$$

Using Theorem 2, set up an instance of the hash proof system from Section 5 (with negligible error probability ϵ) for the trapdoor language L, resulting in a verification key $\kappa \in \mathcal{V}$. Note that proofs for membership in L are from a set $\Pi \subseteq S^m$ for some m. Next, compute the projection value $\alpha = \alpha(\kappa) \in \mathcal{P}$. Finally, choose a function $\chi \in \Phi$ uniformly and compute

$$u = \chi(g) \in S.$$

The public/secret key pair is

$$pk = (g, u, \mathbf{h}, \alpha) \in S \times S \times S^{2k} \times \mathcal{P}, \quad sk = (\chi, \kappa) \in \Phi \times \mathcal{V}.$$

Encapsulation. *Given $pk = (g, u, \mathbf{h}, \alpha)$, choose a function $\psi \in \Phi$ at random, and compute*

$$c = \psi(g)$$

Next, compute $J = \mathsf{T}(c) \subset \{1, \ldots, 2k\}$ and

$$\mathbf{d} = \psi(\mathbf{h}_J) \in S^k$$

Using $\psi \in \Phi$, $\alpha \in \mathcal{P}$ and $\mathbf{d} \in L$, compute the proof $\pi \in \Pi \subseteq S^m$ that $(c, \mathbf{d}, J) \in L$. The ciphertext consists of the pair

$$(c, \mathbf{d}, \pi) \in S \times S^k \times S^m,$$

and the session key is computed as

$$K = \mathsf{Ext}(\psi(u)) \in \{0,1\}^n. \tag{4}$$

Decapsulation. *Given $sk = (\chi, \kappa)$ and a ciphertext $(c, \mathbf{d}, \pi) \in S^{1+k+m}$, compute $J = \mathsf{T}(c) \subset \{1, \ldots, 2n\}$ and verify that $\pi \in S^m$ proves $(c, \mathbf{d}, J) \in L$. If the proof is invalid, reject. Otherwise, compute the session key as*

$$K = \mathsf{Ext}(\chi(c)) \in \{0,1\}^n.$$

Correctness. *We argue that the above KEM satisfies correctness. Note that for correctly generated ciphertexts, we have that*

$$(c, \mathbf{d}, \pi) = (\psi(g), \psi(\mathbf{h}_J), \pi),$$

where π is a proof that $(c, \mathbf{d}, J) \in L$. Hence, correctly generated ciphertexts are not rejected. Furthermore,

$$\chi(c) = (\chi \circ \psi)(g) = \psi(u),$$

which implies that decapsulation extracts the same key as encapsulation.

Theorem 3. *If (S, Φ) is a hard algebraic set system, then the above KEM is* IND-CCA *secure in the sense of Definition 1.*

Proof. We give a simulation of the IND-CCA experiment for an arbitrary PPT adversary A. It suffices to construct a simulator S such that the following holds. On input

$$(g, \chi(g), \psi(g), E^*)$$

(with g, χ, ψ, E^* as in Definition 5), S simulates the real IND-CCA experiment $\mathsf{Exp}^{\mathsf{CCA\text{-}real}}_{\mathsf{KEM,A}}$, and on input

$$(g, \chi(g), \psi(g), R^*),$$

S simulates the random IND-CCA experiment $\mathsf{Exp}^{\mathsf{CCA\text{-}rand}}_{\mathsf{KEM,A}}$.

Setup. So say that S is invoked on input (g, u, c^*, P), for $c^* = \psi(g)$, $u = \chi(g)$, and unknown $\chi, \psi \in \Phi$. Furthermore, $P \in \{0,1\}^n$ is either equal to the extraction E^* or random.

First, S sets up a substitute decapsulation key that can be used to decrypt all ciphertexts except the challenge ciphertext, which will be constructed around $\psi(g)$. Concretely, S computes from its own challenge (g, u, c^*, P) the value $J^* = \mathsf{T}(c^*) \subset \{1, \dots, 2k\}$. Then, S chooses uniformly $\eta = (\eta_1, \dots, \eta_{2k}) \in \Phi$ and defines

$$h_i = \eta_i(g) \qquad \text{for } i \in J^*, \tag{5}$$

$$h_i = \eta_i(g) \cdot u \qquad \text{for } i \notin J^*. \tag{6}$$

Finally, S sets up a hash proof system for the trapdoor language L induced by g and $\mathbf{h}$ (see Definition 9). Let κ and α be the corresponding verification key and its projection. Then, S defines a public key pk along with a substitute secret key sk' as follows:

$$pk = (g, u, \mathbf{h}, \alpha) \in S \times S^\ell \times S^{2k} \times \mathcal{P} \qquad sk' = (\eta, \kappa) \in \Phi^\ell \times \mathcal{V}.$$

Note that by the uniformity of (S, Φ) (see Definition 6), the public keys prepared by S are statistically close to authentic public keys as produced by the key generation from Definition 10.

Challenge ciphertext and key. Next, S prepares a challenge ciphertext $(c^*, \mathbf{d}^*, \pi^*) \in S \times S^{J^*} \times S^m$. We have already defined c^* above, so it remains to define $\mathbf{d}^* = (d_i^*)_{i \in J^*}$ and π. Namely, S sets $d_i^* = \eta_i(c^*)$ for $i \in J^*$. Since

$$d_i^* = \eta_i(c^*) = \eta_i(\psi(g)) = \psi(\eta_i(g)) \overset{i \in J^*}{=} \psi(h_i),$$

this gives a $(c^*, \mathbf{d}^*)$ exactly as produced by the encapsulation algorithm of Definition 10. Because $(c^*, \mathbf{d}^*, J^*) \in L$, a proof π for that statement can be produced using the verification key κ. This yields a challenge ciphertext $(c^*, \mathbf{d}^*, \pi^*)$ exactly as produced by the encapsulation algorithm.

Note that if S's challenge P satisfies

$$P = E^* = \mathsf{Ext}((\chi \circ \psi)(g)),$$

then P equals the real key K as the encapsulation algorithm would have computed in (4), and hence P is distributed as the challenge key K in the real IND-CCA experiment $\mathsf{Exp}^{\mathsf{CCA\text{-}real}}_{\mathsf{KEM},\mathsf{A}}$. On the other hand, if P is random, then clearly P is distributed as a random challenge key in the IND-CCA experiment $\mathsf{Exp}^{\mathsf{CCA\text{-}rand}}_{\mathsf{KEM},\mathsf{A}}$.

Decapsulation queries. S then invokes adversary A with public key pk', challenge ciphertext $(c^*, \mathbf{d}^*, \pi^*)$, and challenge key P. By the above, this yields a view for A as in the real, resp. random IND-CCA experiment, depending on whether $P = E^*$ or P is random.

It remains to implement a decapsulation oracle for A. To this end, assume that A makes a decapsulation query $(c, \mathbf{d}, \pi)$. First, we may assume $c \in S$, $\mathbf{d} \in S^J$ (for $J = \mathsf{T}(c)$), and $\pi \in S^m$, since S is efficiently recognizable. If π is not a correct proof of $(c, \mathbf{d}, J) \in L$ according to κ, then S rejects, exactly as the authentic decapsulation algorithm would have done. In the following, we hence may further assume that π is a valid (with respect to verification key κ) proof that $(c, \mathbf{d}, J) \in L$. By the soundness of the hash proof system,[4] this in particular implies that, with overwhelming probability, there exists $\tilde{\psi} \in \Phi$ with $\tilde{\psi}(g) = c$ and $\tilde{\psi}(h_i) = d_i$ for all $i \in J$.

Observe that $c = c^*$ would imply $J^* = J$, so that for all $i \in J^* = J$,

$$d_i = \tilde{\psi}(h_i) = \eta_i(\tilde{\psi}(g)) = \eta_i(c) = \eta_i(c^*) = d_i^*.$$

By the uniqueness of valid proofs, this would hence imply $(c, \mathbf{d}, \pi) = (c^*, \mathbf{d}^*, \pi^*)$, which is a forbidden decapsulation query for A. Thus, we may even assume that $c \neq c^*$.

Without loss of generality, from $c \neq c^*$ it follows that $J = \mathsf{T}(c) \neq \mathsf{T}(c^*) = J^*$. (Otherwise, A has found a T-collision.) But $J \neq J^*$ implies that there exists an $i \in J \setminus J^*$, i.e., an $i \in J$ for which $h_i = \eta_i(g) \cdot u$. This allows S to derive $\chi(c)$ using

$$d_i = \tilde{\psi}(h_i) = \tilde{\psi}(\eta_i(g)) \cdot \tilde{\psi}(u) = \eta_i(\tilde{\psi}(g)) \cdot \chi(\tilde{\psi}(g)) = \eta_i(c) \cdot \chi(c)$$

and its knowledge about η_i. On the other hand, $\chi(c)$ allows to compute $K = \mathsf{Ext}(\chi(c))$ exactly as the decapsulation algorithm. Hence, the prepared substitute secret key $sk' = (\eta, \kappa)$ can be used to answer A's decapsulation queries.

Summarizing, the prepared simulation shows Theorem 3.

[4] We stress that A only gets to see a proof π^* of a *valid* statement, which could have already been derived from the projected key α. Hence π^* does not disturb a reduction to the soundness of the hash proof system. This distinguishes our use of hash proof systems from the one in [8]. (In [8], the challenge ciphertext contains a proof of an *invalid* statement, which reveals information about the verification key κ beyond what is known from its projection α).

7 Discussion and Variants

Global parameters. Note that the set system (S, Φ) employed in our encryption scheme can be re-used in many instances of the scheme. (In other words, there is no trapdoor related directly to the definition of (S, Φ) itself.) In particular, in the RSA set system from Section 3.4, no knowledge about the factorization of the modulus N is required. That means that N can be used as a *global* parameter for many parties.

Parallelization. In some of our examples from Section 3.4, the extracted values are only bits. This means that when implementing our generic CCA-secure encryption scheme with these examples, the corresponding KEM keys are only bits. However, it is possible to get larger keys by running several instances of the encryption scheme at once, without damaging the chosen-ciphertext security. Concretely, instead of publishing $u = \chi(g)$ in the public key, one can publish $u_i = \chi_i(g)$ for independently chosen $\chi_i \in \Phi$ $(i = 1, \ldots, n)$. The sender still only uses one witness $\psi \in \Phi$ to compute $c = \psi(g)$, but now can extract from n separate values $\psi(u_1), \ldots, \psi(u_n)$. The adaptation of hash proof system and trapdoor language are straightforward. (However, we stress that in order to decrypt, there must be $2k$ elements $h_{i,1}, \ldots, h_{i,\ell}$ *for each* $i = 1, \ldots, n$. Hence, not only the public key size, but also the ciphertext size grows linearly in n.)

Compact ciphertexts. For concrete set system platforms, we can substantially reduce the size of ciphertexts (from $O(k)$ group elements to $O(1)$). To see how, recall that in the IND-CCA secure encryption scheme, the ciphertext contains (the projection of) a vector $\mathbf{d} = \psi(\mathbf{h}_J)$, where $\mathbf{h}$ is part of the public key. The setup of $\mathbf{h}$ during the security proof (see (5)) has been chosen such that $\mathbf{d}$ allows to recover $\chi(c)$ as $\chi(c) = d_i / \eta_i(c)$ for any $i \in J \setminus J^*$. Now consider what happens if we substitute the vector $\mathbf{d}$ in the ciphertext with a single element

$$D := \psi\left(\prod_{i \in J} h_i\right) = \left(\prod_{i \in J} \eta_i(c)\right) \cdot \left(\prod_{i \in J \setminus J^*} \chi(c)\right).$$

Then, the simulation in the security proof can still derive $\prod_{i \in J \setminus J^*} \chi(c) = \chi(c)^{\Delta}$ for $\Delta := |J \setminus J^*|$. (Note that $0 < \Delta \leq 2k$.) If we set $L := \mathrm{lcm}(1, \ldots, 2k)$, then Δ divides L, so that the simulation can *always* compute $\psi(u)^L$. We can then modify the randomness extraction into $\mathsf{Ext}'(z) := \mathsf{Ext}(z^L)$, such that the decapsulation can be computed from $\chi(c)^L$ (instead of $\chi(c)$). Note that this automatically allows to compress the proof part π of the ciphertext down to one element. In particular, the ciphertext size (in group elements) is now *constant.* However, our modifications require that

$$(g, \chi(g), \psi(g), E') \overset{c}{\approx} (g, \chi(g), \psi(g), R), \tag{7}$$

where $g \in S$, $\chi, \psi \in \Phi$, and $R \in \{0,1\}^n$ are uniformly chosen, and $E' = \mathsf{Ext}'(\chi(\psi(g)) = \mathsf{Ext}(\chi(\psi(g))^L) \in \{0,1\}^n$. Note that (7) holds in the case of the Diffie-Hellman- and RSA-based set systems from Section 3.4 (since L and the order of S are coprime).

References

1. Bleichenbacher, D.: Chosen ciphertext attacks against protocols based on the RSA encryption standard PKCS #1. In: Krawczyk, H. (ed.) CRYPTO 1998. LNCS, vol. 1462, pp. 1–12. Springer, Heidelberg (1998)
2. Boneh, D., Boyen, X.: Efficient selective-ID secure identity based encryption without random oracles. In: Cachin, C., Camenisch, J.L. (eds.) EUROCRYPT 2004. LNCS, vol. 3027, pp. 223–238. Springer, Heidelberg (2004)
3. Boneh, D., Canetti, R., Halevi, S., Katz, J.: Chosen-ciphertext security from identity-based encryption. SIAM Journal on Computing 36(5), 915–942 (2006)
4. Boyen, X., Mei, Q., Waters, B.: Direct chosen ciphertext security from identity-based techniques. In: ACM CCS 2005, pp. 320–329. ACM Press, New York (2005)
5. Cash, D., Kiltz, E., Shoup, V.: The twin Diffe-Hellman problem and applications. In: Smart, N.P. (ed.) EUROCRYPT 2008. LNCS, vol. 4965, pp. 127–145. Springer, Heidelberg (2008)
6. Cramer, R., Hofheinz, D., Kiltz, E.: Chosen-ciphertext Secure Encryption from Hard Algebraic Set Systems. Cryptology ePrint Archive, Report 2009/142
7. Cramer, R., Shoup, V.: Universal hash proofs and a paradigm for adaptive chosen ciphertext secure public-key encryption. In: Knudsen, L.R. (ed.) EUROCRYPT 2002. LNCS, vol. 2332, pp. 45–64. Springer, Heidelberg (2002)
8. Cramer, R., Shoup, V.: Design and analysis of practical public-key encryption schemes secure against adaptive chosen ciphertext attack. SIAM Journal on Computing 33(1), 167–226 (2003)
9. Dolev, D., Dwork, C., Naor, M.: Nonmalleable cryptography. SIAM Journal on Computing 30(2), 391–437 (2000)
10. Gennaro, R., Lindell, Y.: A framework for password-based authenticated key exchange. ACM Transactions on Information and System Security 9(2), 181–234 (2006)
11. Goldwasser, S., Micali, S.: Probabilistic encryption. Journal of Computer and System Sciences 28(2), 270–299 (1984)
12. Goldwasser, S., Micali, S., Rivest, R.L.: A digital signature scheme secure against adaptive chosen-message attacks. SIAM Journal on Computing 17(2), 281–308 (1988)
13. Hanaoka, G., Kurosawa, K.: Efficient chosen ciphertext secure public key encryption under the computational Diffie-Hellman assumption. In: Pieprzyk, J. (ed.) ASIACRYPT 2008. LNCS, vol. 5350, pp. 308–325. Springer, Heidelberg (2008)
14. Hofheinz, D., Kiltz, E.: Secure hybrid encryption from weakened key encapsulation. In: Menezes, A. (ed.) CRYPTO 2007. LNCS, vol. 4622, pp. 553–571. Springer, Heidelberg (2007)
15. Hofheinz, D., Kiltz, E.: The group of signed quadratic residues and applications. In: Halevi, S. (ed.) CRYPTO 2009. LNCS, vol. 5677, pp. 637–653. Springer, Heidelberg (2009)
16. Hofheinz, D., Kiltz, E.: Practical chosen ciphertext secure encryption from factoring. In: Joux, A. (ed.) EUROCRYPT 2009. LNCS, vol. 5479, pp. 313–332. Springer, Heidelberg (2009)
17. Kiltz, E.: Chosen-ciphertext security from tag-based encryption. In: Halevi, S., Rabin, T. (eds.) TCC 2006. LNCS, vol. 3876, pp. 581–600. Springer, Heidelberg (2006)
18. Kiltz, E., Pietrzak, K., Stam, M., Yung, M.: A new randomness extraction paradigm for hybrid encryption. In: Joux, A. (ed.) EUROCRYPT 2009. LNCS, vol. 5479, pp. 590–609. Springer, Heidelberg (2009)

19. Kurosawa, K., Desmedt, Y.: A new paradigm of hybrid encryption scheme. In: Franklin, M. (ed.) CRYPTO 2004. LNCS, vol. 3152, pp. 426–442. Springer, Heidelberg (2004)
20. Naor, M., Reingold, O., Rosen, A.: Pseudo-random functions and factoring. SIAM Journal on Computing 31(5), 1383–1404 (2002)
21. Naor, M., Yung, M.: Public-key cryptosystems provably secure against chosen ciphertext attacks. In: 22nd ACM STOC. ACM Press, New York (1990)
22. Peikert, C., Waters, B.: Lossy trapdoor functions and their applications. In: 40th ACM STOC, pp. 187–196. ACM Press, New York (2008)
23. Rackoff, C., Simon, D.R.: Non-interactive zero-knowledge proof of knowledge and chosen ciphertext attack. In: Feigenbaum, J. (ed.) CRYPTO 1991. LNCS, vol. 576, pp. 433–444. Springer, Heidelberg (1992)
24. Rosen, A., Segev, G.: Chosen-ciphertext security via correlated products. In: Reingold, O. (ed.) TCC 2009. LNCS, vol. 5444, pp. 419–436. Springer, Heidelberg (2009)
25. Gonzalez-Vasco, M.I., Villar, J.: In search of mathematical primitives for deriving universal projective hash families. Applicable Algebra in Engineering, communication and Computing 19(2), 161–173 (2008)

Two Is a Crowd? A Black-Box Separation of One-Wayness and Security under Correlated Inputs

Yevgeniy Vahlis*

University of Toronto
Canada
evahlis@cs.toronto.edu

Abstract. A family of trapdoor functions is one-way under correlated inputs if no efficient adversary can invert it even when given the value of the function on multiple correlated inputs. This powerful primitive was introduced at TCC 2009 by Rosen and Segev, who use it in an elegant black box construction of a chosen ciphertext secure public key encryption. In this work we continue the study of security under correlated inputs, and prove that there is no black box construction of correlation secure injective trapdoor functions from classic trapdoor permutations, even if the latter is assumed to be one-way for inputs from high entropy, rather than uniform distributions. Our negative result holds for all input distributions where each x_i is determined by the remaining $n-1$ coordinates. The techniques we employ for proving lower bounds about trapdoor permutations are new and quite general, and we believe that they will find other applications in the area of black-box separations.

1 Introduction

In this paper we study the following question: can classic trapdoor permutations be used to construct trapdoor functions that remain one way even when the adversary is given $F_{pub_1}(x_1), \ldots, F_{pub_n}(x_n)$ for independently chosen keys pub_i, but where the inputs x_i are correlated. In [17] Rosen and Segev introduce this problem, and highlight its importance by using such "correlation secure" injective trapdoor functions in a black box construction of chosen ciphertext secure public key encryption. Although this important type of public key encryption can be constructed from classic trapdoor permutations (see e.g., the seminal work of Dolev *et al* [8,9]), the constructions that achieve this goal make use of non-black-box techniques, which tend to be quite inefficient. In recent years there has been renewed effort to obtain constructions that use simpler primitives in a black box manner, yet so far no such constructions have been based on either semantically secure public key encryption, or even trapdoor permutations.

More generally, trapdoor permutations are a powerful public key primitive that is sufficient for many difficult applications in cryptography. Yet certain

* Supported by the Natural Sciences and Engineering Research Council of Canada.

D. Micciancio (Ed.): TCC 2010, LNCS 5978, pp. 165–182, 2010.

tasks, such as Identity Based Encryption [4,5], have so far resisted attempts to be solved using this tool. Indeed, the limits of what can be constructed from trapdoor permutations are not well understood. In this paper we show that trapdoor permutations do not permit a black box construction of correlation secure injective trapdoor functions, even if the inputs are chosen from a distribution with very little correlation.

On a parallel line of research, our work is a step in the study of resettable security, a notion introduced by Canetti *et al* [6] in the context of Zero-Knowledge Proofs, and recently extended to general secure computation by Goyal and Sahai [14]. Informally, in resettable security the adversary is allowed to restart his security experiment while forcing the target primitive to reuse some of its previous randomness. The study of correlation security can be seen as another form of resettability: the adversary is allowed to selectively restart the experiment by preserving the random input to the functions, but forcing the regeneration of the function keys. In light of the positive results on resettability it is quite interesting that trapdoor permutations cannot be easily made resettably secure. Hence, our result can be seen as a step towards identifying which functionalities can be made resettably secure, and what is the required amount of interaction for achieving that level of security.

We now describe our problem and results in more detail, followed by an overview of related work, and an exposition of our technical approach.

Black-Box Cryptography. A common approach for constructing cryptographic primitives is to base them on some other, simpler, primitives that are believed to be secure. A construction of primitive A from primitive B is black box if the algorithms of A use the algorithms of B as oracles. A security reduction from A to B is black-box if there exists an adversary S such that for every adversary T that breaks B, S breaks A. Furthermore, S uses T as an oracle. In recent years, several breakthrough papers provide non-black-box solutions to some cryptographic tasks. Nevertheless, black-box constructions remain the most common approach. In this paper, all our results concern black-box constructions with a black-box security reduction. Such constructions are called "Fully Black-Box" in the taxonomy of Reingold *et al* [16].

Our Contributions. One-way trapdoor functions were introduced by Diffie and Hellman in [7]. Informally, a family of functions is one-way if given a description of randomly chosen function f_{pub} of that family, and its image $f_{pub}(x)$ on a randomly chosen input x, no efficient adversary can output x. A family of functions is "trapdoor" if there is a key generation algorithm that outputs a pair of strings (pub, pri) such that it is easy to invert $F_{pub}(\cdot)$ given *pri*. The notion of correlation security, introduced by Rosen and Segev in [17], extends the above experiment by giving the adversary $(F_{pub_1}(x_1), \ldots, F_{pub_n}(x_n))$ where the pub_i are independently generated public keys, and $(x_i)_{i\in[n]}$ are sampled from some distribution $\mathcal{C}$. The family of functions is considered $\mathcal{C}$-correlation secure if no efficient adversary can invert the function on one of the coordinates. Of particular interest are distributions where the entire tuple (x_i) is reconstructible given some subset of the coordinates. Correlation security under such distributions was shown in [17] to be sufficient to obtain chosen ciphertext security. In this paper we prove the following black-box separations:

- Let $\mathcal{C}_1$ be the uniform 2-repetition distribution: pairs of the form (x, x) where x is chosen uniformly at random. We show that there does not exist a black box construction of a $\mathcal{C}_1$-correlation secure family of injective trapdoor functions from classic trapdoor permutations.
- We then extend the above result to all input distributions that are $(n-1)$-reconstructible. That is, distributions of the form $(x_1, \ldots, x_n)$ where each x_i is determined by $(x_1, \ldots, x_{i-1}, x_{i+1}, \ldots, x_n)$. This includes distributions with very weak correlation among the coordinates, such as $(n-1)$-wise independent distributions that are reconstructible in the above sense.

The base primitive in our separation actually has much stronger security properties than mere one-wayness. Indeed, our proofs show that even if one assumes a trapdoor permutation that is one-way for non-uniform (but high entropy) inputs then correlation security still cannot be achieved. Trapdoor permutations that are one-way for high entropy inputs have been shown by Bellare *et al* [1] to be sufficient to obtain deterministic public key encryption – a type of injective trapdoor functions, introduced by Bellare *et al* in [2], that hide almost all information about their input.

Related Work. In [15] Impagliazzo and Rudich introduced an approach for proving black-box separation results. In that seminal paper they prove a separation between one-way permutations and secure key-agreement. Since then a large body of research has followed their basic methodology. We provide a survey of the most relevant results, and recommend reading [16] for a more complete overview.

In this paper we study limits of public key primitives. In [11] Gertner *et al* show that public key encryption and Oblivious Transfer are incomparable under black-box reductions. They also show that trapdoor permutations cannot be constructed in a black-box way from public key encryption, and from trapdoor functions (functions which are not necessarily permutations, but allow sampling from the pre-image given trapdoor information). In [12] Gertner *et al* show that there is no black-box reduction from poly-to-one trapdoor functions to semantically secure public key encryption. Intuitively, [12] show that public key encryption is weaker than trapdoor functions because the latter allows the recovery of the complete input of the encryption algorithm, including the randomness. In contrast, a public key decryption algorithm recovers only the encrypted message, but not the randomness that was used by the encryptor. In [10] Gennaro *et al* show limits on the efficiency of cryptographic primitives constructed in a black-box way from basic tools such as one-way permutations, and trapdoor permutations. In particular they show bounds on the number of times that a trapdoor permutation needs to be invoked in order to construct a semantically secure public key encryption. Their lower bound is a function of the number of bits that the public key encryption scheme encrypts. Towards obtaining their result, Gennaro *et al* define an oracle model which provides all algorithms access to a random trapdoor permutation family. We adopt this model partially in our work.

In [13] Gertner *et al* prove that chosen ciphertext secure public key encryption cannot be constructed in a black-box way from semantically secure public key encryption, provided that the decryption algorithm does not query the encryption oracle of the underlying primitive. In light of previous results that separate trapdoor permutations from semantically secure public key encryption the [13] result leaves open the possibility of achieving chosen ciphertext security using trapdoor permutations. Interestingly, the decryption algorithm in the construction of [17] *does* query the "encryption" algorithm of the underlying trapdoor function. In [5] Boneh *et al* show that Identity Based Encryption cannot be constructed in a black-box way from trapdoor permutations. In the context of the transformation by Boneh *et al* [3] of any Identity Based Encryption to a chosen ciphertext secure public key encryption, the work of [5] rules out one possible method of getting CCA public key encryption from trapdoor permutations. Our work rules out another such approach.

Overview of Techniques. The basic approach of most black-box separation results can be described as follows. Given a target primitive A and a base primitive B first define an "idealized" version of B. The idealized B is usually a distribution on functions that satisfy B's correctness requirements. Then, show that an adversary that is given oracle access to the ideal B cannot break its security, even if that adversary is computationally unbounded[1]. Then, describe an adversary that, by making a small number of queries to the ideal B, breaks *any* construction of A. Note that the fact that the adversaries are computationally unbounded requires any non-trivial A to make essential use of the ideal B oracle (otherwise that A is trivially broken). A common final step is to "project" the result into the realm of polynomial time computation by adding a **PSPACE** complete oracle. This oracle makes a polytime adversary effectively unbounded, but it does not help break the ideal B. For more details about this general approach we direct the reader to [15,16,18].

We use the work of [10] and [5] as a basis for defining our ideal trapdoor permutation oracle. In their work, Gennaro *et al* define a distribution on triples of functions (g, e, d) where $g(\cdot)$ is a random function that maps trapdoors to public keys, $e(pub, \cdot)$ is an independent random permutation for every public key pub, and $d(pri, \cdot)$ inverts the permutation $e(pub, \cdot)$ if pri is a trapdoor for pub. Although this model captures nicely the concept of an ideal trapdoor permutation, we cannot adopt it directly. The problem is that the triple (g, e, d) is in fact correlation secure: for each public key pub the permutation $e(pub, \cdot)$ is chosen independently at random. So, seeing $e(pub_1, x)$ and $e(pub_2, x)$ does not provide any additional information about x over just seeing $e(pub_1, x)$. Our solution is to introduce an additional oracle which we call Break so that given access to $(g, e, d, \mathsf{Break})$ it is no longer possible to obtain correlation security, but

[1] Note that this necessarily implies that there is no polynomial time implementation of the idealized B.

one-wayness is preserved with respect to independently random inputs. It is the main technical contribution of this paper to find the delicate balance that leaves the entire oracle just strong enough to preserve one-wayness, but weak enough to obtain the negative result.

On a high level, the oracle Break can be described as follows: Break takes as input a triple of circuits G, E, D which are a candidate correlation secure trapdoor function. The other inputs are, two public keys PUB_1 and PUB_2, and the values $E_{PUB_1}(x)$ and $E_{PUB_2}(x)$. The naive solution would be to return x. However, this would easily allow an adversary to invert any function simply by setting $pub_1 = pub_2$. Ideally we would like to restrict Break to return x only when pub_1 and pub_2 are independently generated public keys of the provided trapdoor permutation. This, however, seems hard to check. Indeed, how can we verify that the public keys were properly generated? Moreover, even performing a simpler test: that the public keys are valid (that is, they are outputs of the key generation algorithm), may give too much power to the adversary. In particular, an adversary trying to invert $f(x)$, where f is any function, may design a trapdoor permutation scheme where pub_1 is a valid public key if and only if the first bit of x is 0. To prevent the adversary from performing the above attack, we require her to provide a triple of functions $\mathcal{O}' = (g', e', d')$ that is defined on a small part of the domain of (g, e, d) but such that relative to $\mathcal{O}'$, pub_1 and pub_2 are valid public keys. The partial oracle $\mathcal{O}'$ is then superimposed on $\mathcal{O}$ to obtain a new complete oracle $\mathcal{O}''$ relative to which pub_1 and pub_2 are valid public keys. The oracle Break then performs its computation relative to the new oracle $\mathcal{O}''$.

This modification successfully deals with the issue of the validity of public keys, but we are still left with no way of knowing that the public keys were generated independently. This causes a problem because an adversary trying to break the random trapdoor permutation (g, e, d) may simply set pub_1 to be the public key that she is trying to break, and set $pub_2 = pub_1$. Thus, some kind of additional check seems necessary, yet testing for independence of pub_1 and pub_2 seems too much to hope for. As it turns out, we do not need the two public keys to be completely independent. As illustrated by the above example, we run into a problem when our Break oracle allows the adversary to use the same public key of (g, e, d) in both public keys of (G, E, D). But, if we require that the sets of public keys of (g, e, d) that are used to generate PUB_1 and PUB_2 are *disjoint*, then it becomes difficult to invert $y = e(pub, x)$. To do so the adversary would have to find $e(pub', x)$ for some pub' different from pub. However, this is as hard as finding x since the permutations $e(pub, \cdot)$ and $e(pub', \cdot)$ are random and independent. We formalize the above ideas in Section 3.

2 Preliminaries

2.1 Notation

Denote by $\mathbb{N}$ the set of natural numbers. For $n \in \mathbb{N}$, $[n]$ denotes the set $\{1, \ldots, n\}$. For a set S we denote by $x \leftarrow_{\mathrm{R}} S$ the procedure of uniformly sampling an element from S and assigning the value to x. We write $x \in_{\mathrm{R}} S$ to denote the fact that x

is a uniformly sampled element of S. Although the distinction between families of functions and functions is very important, we occasionally write "trapdoor permutation" and "trapdoor function" instead of "family of trapdoor permutations" and "family of trapdoor functions". We do so to improve exposition, and only when the meaning is clear from context.

2.2 Probabilistic Lemmas

We will make use of the following simple fact:

Lemma 1. *Let $X_1, \ldots, X_{n+1}$ be independent Bernoulli random variables, where $\Pr[X_i = 1] = p$ and $\Pr[X_i = 0] = 1 - p$ for $i = 1, \ldots, n+1$ and some $p \in [0,1]$. Let $\mathcal{E}$ be the event that the first n variables are sampled at 1, but X_{n+1} is sampled at 0. Then, $\Pr[\mathcal{E}] \leq \frac{1}{e \cdot n}$. Note that the bound is independent of p.*

2.3 Non-uniform Trapdoor Permutations in the Presence of Oracles

We prove our theorems in a non-uniform computational model. Thus, a collection of Trapdoor Permutations is specified, for each value of the security parameter m, by a triple of deterministic PT algorithms (G, E, D), and the following additional parameters:

- $\lambda(m)$ is the length parameter of the trapdoor permutation oracle (g, e, d) that is used by (G, E, D).
- $q = q(m)$ is the maximum number of oracle queries that any of the algorithms make in a single execution. For convenience we assume that the algorithms always make exactly q queries.

The functionality of each of the algorithms is as follows: $G(\cdot)$ takes as input a trapdoor $PRI \in \{0,1\}^m$, and outputs a function public key $PUB \in \{0,1\}^m$. $E_{PUB}(\cdot)$ is a permutation on strings of length m. Finally, $D_{PRI}(\cdot)$ computes the inverse function of $E_{G(PRI)}(\cdot)$.

We now define two security conditions: one-wayness, and correlation-security (or one-wayness in the presence of correlated products). Let A be an algorithm with access to the same oracle as (G, E, D). We define the advantage of A in the one-wayness experiment with respect to an input distribution D over $\{0,1\}^m$ as follows:

$$\delta_{OW}(A, D) \stackrel{def}{=} \Pr[A(PUB, E_{PUB}(x)) = x;\ PUB \leftarrow G(PRI)]$$

where PRI is chosen uniformly at random from $\{0,1\}^m$ and x is sampled according to D. For convenience, we denote $\delta_{OW}(A) \stackrel{def}{=} \delta_{OW}(A, U_m)$, where U_m is the uniform distribution over $\{0,1\}^m$.

For a distribution $\mathcal{C}$ on $(\{0,1\}^m)^n$ for some $n \in \mathbb{N}$, we define the advantage of A in the $\mathcal{C}$-*correlation security* experiment as follows:

$$\delta_{CS}(A) \stackrel{def}{=} \Pr[A((PUB_i, E_{PUB_i}(x_i))_{i\in[n]}) \in \{x_i\}_{i\in[n]}\ ;\ PUB_i \leftarrow G(PRI_i)]$$

where $(x_i)_{i\in[n]} \in_R \mathcal{C}$, and PRI_i for $i \in [n]$ are chosen uniformly at random from $\{0,1\}^m$. As a convention, we will frequently omit the lengths of strings when discussing trapdoor permutations, when the length is clear from context.

We measure the complexity of algorithms in the oracle model only by the number of oracle queries that they make. Using a standard technique of adding a **PSPACE** oracle we obtain the fully black-box separation for probabilistic polynomial time Turing Machines (see [16,18] for a detailed exposition of the approach).

3 Our Oracles

We prove our theorem in a relativized model where all algorithms have access to three random oracles (g, e, d) that roughly correspond to the algorithms G, E, and D of a Trapdoor Permutation. For every $\lambda \in \mathbb{N}$, the oracles (g, e, d) are sampled uniformly at random from the set of all functions satisfying the following conditions:

- $g : \{0,1\}^\lambda \to \{0,1\}^\lambda$. We view g as taking a secret key *pri* as input and outputting a public key.
- $e : \{0,1\}^\lambda \times \{0,1\}^\lambda \to \{0,1\}^\lambda$ is a function that on input $pub \in \{0,1\}^\lambda$ and $x \in \{0,1\}^\lambda$ outputs $e(pub, x) \in \{0,1\}^\lambda$. We require that for every $pub \in \{0,1\}^\lambda$ the function $e(pub, \cdot)$ be a permutation of $\{0,1\}^\lambda$.
- $d : \{0,1\}^\lambda \times \{0,1\}^\lambda \to \{0,1\}^\lambda$ is a function that on input $pri \in \{0,1\}^\lambda$ and $y \in \{0,1\}^\lambda$ outputs an $x \in \{0,1\}^\lambda$ that is the (unique) pre-image of y under the permutation defined by the function $e(g(pri), \cdot)$.

Redundancy of d. The function d is completely determined by g, e. Thus, when discussing a description of a trapdoor permutation oracle $\mathcal{O} = (g, e, d)$ we will assume that d is deduced from g, e rather than being part of the description.

Partial Oracles. In our proofs we will occasionally need to refer to trapdoor permutation oracles that are defined on a subset of the domain. We call a function $\mathcal{O}' = (g', e')$ which is defined on a subset of the domain of $\mathcal{O}$, a valid *partial oracle* if for every *pub*, $e'(pub, \cdot)$ is 1-1. Again, d is not part of the description of $\mathcal{O}'$. Instead, it is determined from (g', e') as follows: for strings *pri*, and y, $d(pri, y)$ is defined and equal to x if and only if $g'(pri) = pub$ and $e'(pub, x) = y$.

Conventions. Without loss of generality we assume that whenever an algorithm makes an oracle query of the form $d(pri, y)$, it first queries $g(pri)$. This assumption is useful for a cleaner presentation of our proofs.

The Oracle Break. In addition to the oracles $\mathcal{O} = (g, e, d)$, our adversary will have access to an oracle Break that weakens the above random trapdoor permutation. Before we can describe Break we must introduce some additional notation.

3.1 Oracle Notation

Before proceeding with the description of the oracle Break, let us introduce additional notation that we use when discussing various aspects of the trapdoor permutation oracle.

Oracle algorithms. For a function $\mathcal{O}$ and algorithm A we denote by $A^{\mathcal{O}}$ the fact that A may make queries to $\mathcal{O}$.

Functions as sets. It will occasionally be convenient to view the trapdoor permutation oracle $\mathcal{O} = (g, e, d)$ as sets of input-output pairs. We will use square brackets to denote the symbolic form of a mapping. For instance: to denote that $e(pub, x) = y$ we may write $[e(pub, x) = y] \in \mathcal{O}$. We will occasionally use a wild card form of this notation. For instance: we write $[e(pub, \cdot) = y] \in \mathcal{O}$ to denote that there exists x such that $[e(pub, x) = y] \in \mathcal{O}$.
When discussing queries we write $[e(pub, x)]$ to denote a query to $e(\cdot, \cdot)$ with inputs (pub, x). This is to differentiate the query from the actual value $e(pub, x)$ which is the image of (pub, x) under the function e. Given a query q in symbolic form we write $\mathcal{O}(q)$ to denote the mapping under $\mathcal{O}$ of that query. For example: if $q = [e(pub, x)]$ and $e(pub, x) = y$ then $O(q) = [e(pub, x) = y]$.

Adding answers. Given $\mathcal{O}$ and a set of queries Q we define $\mathcal{O}(Q)$ to be the set of queries in Q with their answers according to $\mathcal{O}$. Namely, $\mathcal{O}(Q) = \{[\alpha = \beta] | \alpha \in Q, \mathcal{O}(\alpha) = \beta\}$.

Public keys that are used in a query. Given a trapdoor permutation oracle $\mathcal{O} = (g, e, d)$, and a set Q of (g, e, d) queries we define $PK_e(Q)$ to be the set of all pub such that $[e(pub, \cdot)] \in Q$. Similarly, we define $PK_g(Q)$ to be the set of all pub such that $[g(\cdot) = pub] \in \mathcal{O}(Q)$.

Combining trapdoor permutation oracles. Let $\mathcal{O}_1 = (g_1, e_1)$ and $\mathcal{O}_2 = (g_2, e_2)$ be two (possibly partial) trapdoor permutation oracles. We write $\mathcal{O}_1 \diamond \mathcal{O}_2$ to denote the oracle which answers according to $\mathcal{O}_2$ if possible, and according to $\mathcal{O}_1$ otherwise. More precisely, $(\mathcal{O}_1 \diamond \mathcal{O}_2)(\alpha)$ returns $\mathcal{O}_2(\alpha)$ if it is defined, and $\mathcal{O}_1(\alpha)$ otherwise.
We also define a "corrected" version of the $\diamond$ operator. We write

$$\mathcal{O}_1 \diamond_c \mathcal{O}_2 \stackrel{def}{=} (g_1 \diamond g_2, e_1 \diamond_c e_2)$$

The oracle $e = e_1 \diamond_c e_2$ is defined as follows: let pub, x, y such that $e_2(pub, x) = y$. We set $e(pub, x) = y$. Furthermore, if there exists $x' \neq x$ such that $e_1(pub, x') = y$ then let $y' = e_1(pub, x)$ (note that y' may be equal to $\perp$ if $e_1(pub, x)$ is not defined). We then also set $e(pub, x') = y'$.
Note that the d part of the oracle $\mathcal{O}_1 \diamond_c \mathcal{O}_2$ is deduced from g and e. A useful property of the $\diamond_c$ operator is that for every two possibly partial oracles $\mathcal{O}_1$ and $\mathcal{O}_2$, the oracle $\mathcal{O}_1 \diamond_c \mathcal{O}_2$ has no collisions (i.e. it is a valid partial oracle), and for every α such that $\mathcal{O}_2(\alpha)$ is defined, $(\mathcal{O}_1 \diamond_c \mathcal{O}_2)(\alpha) = \mathcal{O}_2(\alpha)$.

3.2 The Oracle Break

As mentioned in the introduction, the random trapdoor permutations that the oracles (g, e, d) represent are, in fact, correlation secure. This is so because each permutation is random and independent from the other permutations. Thus, we add a weakening oracle Break that allows our adversary to break correlation security of any trapdoor function that makes use of (g, e, d), yet preserves the one-wayness of (g, e, d). For a better exposition we present the oracle and the proof for the case of Trapdoor *Permutations.* However, both easily extend to handle injective trapdoor functions. The details are given in Section 6.

The functionality of Break is defined as follows:

Input. Break takes the following inputs:

1. A triple of oracle circuits $(G^{\mathcal{O}}, E^{\mathcal{O}}, D^{\mathcal{O}})$ that may contain (g, e, d) oracle gates. The functions computed by G, E, D must constitute a valid family of trapdoor permutations.
2. Two strings PUB_1, PUB_2. We think of these strings as public keys that were produced by G.
3. Two strings y_1, y_2. We think of these strings as the outputs of $E_{PUB_1}(x)$ and $E_{PUB_2}(x)$ respectively on some input x.
4. Two partial oracles $\mathcal{O}'_1$ and $\mathcal{O}'_2$, and two strings PRI_1 and PRI_2.

Computation. The following computation is performed by the oracle:

1. Verify that for every $pub \in PK_g(\mathcal{O}'_1) \cap PK_g(\mathcal{O}'_2)$, there exists a pri such that $g(pri) = pub$ and $[g(pri) = pub] \in \mathcal{O}'_1 \cup \mathcal{O}'_2$. Note that the requirement here is that $g(pri) = pub$ under the real oracle $\mathcal{O}$. Therefore, for every pub as above, we require that $\mathcal{O}'_1$ or $\mathcal{O}'_2$ provide us with a real trapdoor for it.
2. Verify that (G, E, D) is a valid family of trapdoor permutations, and for $i \in \{1, 2\}$, for every complete trapdoor permutation oracle $\mathcal{O}''$, if $\mathcal{O}'_i \subseteq \mathcal{O}''$, then for every $x \in \{0,1\}^m$, it holds that $D^{\mathcal{O}''}_{PRI_i}(E^{\mathcal{O}''}_{PUB_i}(x))$ returns x.
3. Let $\mathcal{O}''_i = \mathcal{O} \diamond_c \mathcal{O}'_i$.
4. Run $D^{\mathcal{O}''_1}_{PRI_1}(y_1)$ to obtain an output x. If $x = \perp$, return $\perp$.
5. For $i \in \{1, 2\}$, run $E^{\mathcal{O}''_i}_{PUB_i}(x)$, and return $\perp$ if one of the following events occurs: (i) the output of $E^{\mathcal{O}''_i}_{PUB_i}(x)$ is not equal to y_i, or (ii) $E^{\mathcal{O}''_i}_{PUB_i}(x)$ asks a query $\alpha = [g(pri)]$ such that $\mathcal{O}(\alpha) \neq \mathcal{O}''_i(\alpha)$. Finally, return x.

Complexity. Each query to Break is counted as $3q + |\mathcal{O}'_1| + |\mathcal{O}'_2|$ oracle queries. This is to prevent an adversary from making a very large Break query that gives away too much information about $\mathcal{O}$. The breakdown of the above cost is as follows: $3q$ comes from steps 4 and 5 where Break evaluates D once and E twice. The count $|\mathcal{O}'_1| + |\mathcal{O}'_2|$ is due to steps 1, and 3. In step 1 Break has to compare elements of the form $[g(pri) = pub] \in \mathcal{O}'_1 \cup \mathcal{O}'_2$ to $\mathcal{O}$. In step 3 Break has to know at most $|\mathcal{O}'_1| + |\mathcal{O}'_2|$ entries in $\mathcal{O}$ in order to perform the $\diamond_c$ operation.

This concludes the description of our oracles. In the following two sections we prove the two main lemmas that are required for our theorem: that the oracle Break preserves the one-wayness of $\mathcal{O} = (g, e, d)$, and that there exists an adversary which uses Break that breaks the correlation security of every family of injective trapdoor functions.

4 Breaking Security under Correlated Inputs

In this section we describe an adversary that breaks the correlation-security of any trapdoor permutation, while making only a polynomial number of queries to the oracles $(g, e, d, \mathsf{Break})$.

Let (G, E, D) be a collection of injective trapdoor functions with length parameter m, and maximal number of queries q. For simplicity, and due to lack of space we describe an adversary that breaks only constructions (G, E, D) that do not query Break, but only use (g, e, d). The extension of the adversary and the oracle Break to handle injective trapdoor functions (as opposed to permutations) that make use of Break is quite easy, and is described in Section 6.

4.1 Overview

We start with an informal description of our adversary. The adversary is initially given two independently generated public keys PUB_1 and PUB_2. Recall that in order to make use of the oracle Break the adversary has to come up with two partial oracles $\mathcal{O}'_1$ and $\mathcal{O}'_2$ such that PUB_1 and PUB_2 are valid outputs of $G^{\mathcal{O}'_1}$ and $G^{\mathcal{O}'_2}$ respectively. Since our adversary is computationally unbounded, that is, we count only the number of oracle queries that she makes, it is easy for her to find two such partial oracles. However, there are two issues that arise: (i) in order to pass check 1 of Break the adversary has to know the *correct* trapdoor for each *pub* that is appears in the generation of both PUB_1 and PUB_2; and (ii) if the actual oracle (g, e, d) is not used, it is quite possible that under these partial oracles, y_1 and y_2 will not invert to x. Both issues are dealt with simultaneously by performing a sampling procedure that discovers all the queries that are frequently asked by G, and by $E_{PUB_1}(x)$ and $E_{PUB_1}(x)$ where x is chosen randomly. The adversary then chooses $\mathcal{O}'_1$ and $\mathcal{O}'_2$ offline, without making any further oracle queries, but in a manner that is consistent with the information about (g, e, d) that was learned during the sampling procedure.

To see why the above procedure solves problem (i) recall that PUB_1 and PUB_2 are generated independently. Therefore, with high probability any public keys *pub* that are needed to generate *both* PUB_1 and PUB_2 are also needed to generate many other PUB's that G may output. This means that, with high probability, the adversary will generate at least one such PUB, and discover the correct trapdoor for *pub* in the process. Problem (ii) is solved due to the following simple fact: $\mathcal{O}'_1$ and $\mathcal{O}'_2$ are defined on a polynomial number of points. For each such point, if it is frequently accessed by one of $E_{PUB_1}(x)$ or $E_{PUB_2}(x)$ then the adversary would have discovered the correct value for it during the sampling. If the point is infrequently accessed, then with high probability it was not accessed when y_1 and y_2 were computed. For a similar reason, the adversary's query to Break passes the second check of step 5.

If the adversary manages to make a query to Break that is not answered with $\perp$ then with overwhelming probability the answer to that query is x, which is the inverse of y_1 and y_2. We are now ready to describe the adversary completely and analyze its performance.

4.2 The Adversary

For convenience, and without loss of generality, we assume $q \geq 2$. For any $\varepsilon > 0$ we provide a PPT adversary A, and a constant c such that $\delta_{CS}(A) \geq 1-\varepsilon$ and A makes at most q^c oracle queries. More precisely, given $\varepsilon > 0$ choose two integers c_1, c_2 such that (i) $\left(1 - \frac{1}{q^{c_1}}\right)^q \geq 1 - \frac{1}{q^{c_1-1}}$, and (ii) $\varepsilon \geq \left(\frac{q^{c_1+1}}{e^{q^{c_1}}} + \frac{1}{q^{c_1-1}} + \frac{4}{eq^{c_2}}\right)$. Our adversary proceeds in several steps:

1. The adversary is given PUB_1, PUB_2, y_1, y_2. It starts by initializing tables L, L_1, L_2, which will be used to store points of $\mathcal{O}$ that the adversary discovers. More precisely, these tables are partial oracles that are updated by the adversary in the following steps, and always satisfy $L, L_1, L_2 \subseteq \mathcal{O}$.
2. For $1 \leq i \leq q^{2c_1}$ the adversary chooses $PRI_i \in_{\mathrm{R}} \{0,1\}^n$, and simulates $G(PRI_i)$. For every query α asked by G during the simulation, the adversary adds the entry $(\alpha, \mathcal{O}(\alpha))$ to L.
3. For $1 \leq i \leq q^{c_2}$ the adversary chooses $x_i \in_{\mathrm{R}} \{0,1\}^m$, and simulates $E^{\mathcal{O}}_{PUB_1}(x_i)$ and $E^{\mathcal{O}}_{PUB_2}(x_i)$, recording all queries and answers in L_1 and L_2 respectively.
4. The adversary now selects partial oracles $\mathcal{O}'_i$, and strings PRI_1, PRI_2 for $i \in \{1,2\}$ such that:
   (a) $|\mathcal{O}'_i| \leq |L \cup L_1 \cup L_2| + q$.
   (b) $L_1 \cup L_2 \cup L \subseteq \mathcal{O}'_i$, and $G^{\mathcal{O}'_i}(PRI_i) = PUB_i$.
   (c) For every $pub \in PK_g(\mathcal{O}'_1) \cap PK_g(\mathcal{O}'_2)$ there exists an pri such that $[g(pri) = pub] \in L \cup L_1 \cup L_2$.

   If no such partial oracles exist then the adversary gives up and terminates.
5. The adversary queries $\mathsf{Break}(G, E, D, PUB_1, PUB_2, y_1, y_2, \mathcal{O}'_1, \mathcal{O}'_2, PRI_1, PRI_2)$. If Break returns $\perp$ then the adversary fails (this can be modeled by the adversary returning a random string $x \in \{0,1\}^m$). Otherwise, Break returns x, which the adversary returns as well.

4.3 Analysis

This concludes the description of our adversary. We now turn to proving that our adversary makes a successful query to Break which returns the correct inverse x. In order to prove this we need to show that the following two statements are true with high probability:

1. The adversary's query passes all the checks of Break.
2. Under $\mathcal{O}''_i = \mathcal{O} \diamond_c \mathcal{O}'_i$ it holds that $E^{\mathcal{O}''_1}_{PUB_1}(x) = y_1$ and $E^{\mathcal{O}''_1}_{PUB_2}(x) = y_2$.

We start with the first statement. We will use the following random variables in the statement of the lemma. Consider a run of our adversary in the correlation security experiment. Let $\mathcal{O}, PRI'_1, PRI'_2$ be the TDP oracle, and the private keys respectively, that are selected by the challenger. Let Q_{PUB_i} and $Q_{x,i}$ to be the sets of queries asked during the computations $G^{\mathcal{O}}(PRI'_i)$, and $E^{\mathcal{O}}(PUB_i, x)$ respectively. We define $T_{PUB_i} = \mathcal{O}(Q_{PUB_i})$ and $T_{x,i} = \mathcal{O}(Q_{x,i})$. For the first statement above we are interested in the following event:

Event E**.** For every *pub* for which there exist pri_1, pri_2 such that $[g(pri_1) = pub] \in T_{PUB_1}$ and $[g(pri_2) = pub] \in T_{PUB_2}$ there exists an *pri* such that $[g(pri) = pub] \in L$.

Essentially, the event E states that our adversary has discovered a trapdoor for every public key that was generated in the computation of both $G^{\mathcal{O}}(PRI'_1)$ and $G^{\mathcal{O}}(PRI'_2)$. The following claim shows that this happens with high probability. Intuitively, this is so because PRI'_1 and PRI'_2 are chosen independently at random, and our adversary samples many such computations in step 2. Thus, if a public key is likely to be generated by $G^{\mathcal{O}}(PRI)$ for a random PRI, then our adversary has already found it. If it is unlikely to be generated then it is unlikely to appear in the computation for two independent PRI's.

Claim. At step 5 of the adversary, the event E occurs with probability $\geq 1 - \frac{q^{c_1+1}}{e^{q^{c_1}}} + \frac{1}{q^{c_1-1}}$.

Proof (Sketch). In step 2 the adversary simulates $G(PRI)$ many times, thus learning all the *pub*s that are frequently generated by G on random PRI. In order for a public key *pub* to be likely to appear in the computation of two independent executions of G, it must frequently generated by G. Therefore, with high probability, all the *pub*s that were generated in *both* the computation of $G(PRI_1)$ and $G(PRI_2)$ have already been observed by the adversary during the sampling of step 2. The complete proof is given in the full version of this paper [19].

We now show that it is sufficient for event E to occur in order for our adversary to successfully construct the partial oracles $\mathcal{O}'_1$ and $\mathcal{O}'_2$, and make a Break query that passes checks 1 and 2.

Claim. If event E occurs then the adversary successfully constructs the oracles $\mathcal{O}'_1$ and $\mathcal{O}'_2$ in step 4. Furthermore, the adversary's query to Break successfully passes checks 1 and 2.

Proof. If event E occurs then L contains trapdoors for all *pub* that appear in both the computation of $G^{\mathcal{O}}(PRI_1)$ and $G^{\mathcal{O}}(PRI_2)$. Thus, one possibility for the values of $\mathcal{O}'_1, \mathcal{O}'_2, PRI_1, PRI_2$ in this case is simply the correct PRI_1, PRI_2 and the subset of $\mathcal{O}$ that is used in the generation of PUB_1 and PUB_2 respectively.

Now consider the adversary's query to Break. Check 1 passes because of the conditions imposed on the choice of the partial oracles $\mathcal{O}'_1$ and $\mathcal{O}'_2$. To see why check 2 passes, notice that under $\mathcal{O}'_i$ the made up PRI_i is a correct private key for the public key PUB_i, and so by the correctness of the trapdoor permutation (G, E, D), for every oracle $\mathcal{O}''$ such that $\mathcal{O}'_i \subseteq \mathcal{O}''$, $D^{\mathcal{O}''}(PRI_i, y)$ inverts y correctly.

Our next step is to prove the second property of our adversary: that for for $i \in \{1, 2\}$, with high probability, for every query α that is asked by $E^{\mathcal{O}}(PUB_i, x)$ the answers under the oracle $\mathcal{O}$, and the modified oracle $\mathcal{O}''_i = \mathcal{O} \diamond_c \mathcal{O}'_i$ are identical. The proof of this statement is very similar to Claim 6.9 in [5]. We repeat it here for completeness.

Lemma 2. *Let $\mathcal{O}'_i$, for $i \in \{1,2\}$, be the partial oracles chosen by the adversary, and let $\mathcal{O}''_i = \mathcal{O} \diamond_c \mathcal{O}'_i$. Then, with probability at least $1 - \frac{2}{eq^{c_2}}$, for every query α asked by $E^{\mathcal{O}}_{PUB_i}(x)$, $\mathcal{O}''_i(\alpha) = \mathcal{O}(\alpha)$.*

Proof. From the fact that $\mathcal{O}'_i$ is defined on at most q points that are not in $L \cup L_1 \cup L_2$, and the definition of the $\diamond_c$ operator we know that $\mathcal{O} \diamond_c \mathcal{O}'_i$ differs from $\mathcal{O}$ on at most $2q$ points.

More precisely, for a query α of the form $\alpha = [g(sk)]$ where $\mathcal{O}(\alpha) \neq \mathcal{O}''_i(\alpha) = pk$ it must be that $[g(sk) = pk] \in \mathcal{O}'_i \setminus L \cup L_1 \cup L_2$. Thus, queries of this form contribute at most one point on which $\mathcal{O}''_i$ and $\mathcal{O}$ differ.

If α is of the form $e(pk, x)$, and $[e(pk, x) = y] \in \mathcal{O}$ and $[e(pk, x) = w] \in \mathcal{O}''_i$, where $w \neq y$, then one of the following holds: either $[e(pk, x) = w] \in \mathcal{O}'_i \setminus L \cup L_1 \cup L_2$ or there exists x' such that $e(pk, x') = w$, and $e'_i(pk, x') = y$. Thus, queries of this form can contribute at most two points on which $\mathcal{O}''_i$ and $\mathcal{O}$ differ.

Consider a query α such that $\mathcal{O}(\alpha) \neq \mathcal{O}''_i(\alpha)$. Then, $[\alpha, \mathcal{O}(\alpha)] \notin L_1 \cup L_2$. This means that α did not appear in any of the simulations in step 3. Since the simulations in step 3 are done with independently chosen x_i, we can apply Lemma 1, and so the probability of α appearing during the computations $E^{\mathcal{O}''_i}_{PUB_i}(x) = y_i$ for $i \in \{1,2\}$ is at most $\frac{1}{eq^{c_2}}$.

Applying the union bound over all $\leq 2q$ points on which $\mathcal{O}$ and $\mathcal{O}''_i$ differ, we obtain that the probability that a query α is asked during $E^{\mathcal{O}}_{PUB_i}(,x)$ such that $\mathcal{O}(\alpha) \neq \mathcal{O}''_i(\alpha)$ is at most $\frac{2q}{eq^{c_2}}$.

We are now ready to prove the main theorem of this section: that our adversary successfully breaks the correlation security of any trapdoor permutation with probability which is arbitrarily close to one.

Theorem 1. *Given PUB_1, PUB_2, y_1, y_2 in the correlated one-wayness experiment, our adversary wins with probability $\geq 1 - \left(\frac{q^{c_1+1}}{e^{q^{c_1}}} + \frac{1}{q^{c_1-1}} + \frac{4}{eq^{c_2}}\right)$. Furthermore, it does so by making at most $5q + 3q^{c_1+1} + 6q^{c_2+1}$ oracle queries.*

Proof. The theorem follows by simple calculation from the above claims, and Lemma 2. The complete proof appears in [19].

5 One-Way Trapdoor Permutations Exist Relative to Our Oracle

In this section we show that the trapdoor permutation (G, E, D) where $G^{\mathcal{O}}(pri) = g(pri)$, $E^{\mathcal{O}}_{pub}(x) = e(pub, x)$ and $D^{\mathcal{O}}_{pri}(y) = d(pri, y)$ is a secure one-way trapdoor permutation, even when the adversary is given access to the oracle Break. Let A be an adversary that tries to break the one-wayness of (G, E, D). We show that $\delta \leq \frac{3q}{2^\lambda - q} + \frac{3q}{2^\lambda}$ where $\delta = \delta_{OW}(A)$ is the advantage of A in the one-wayness experiment. In fact, our proof carries through even when the input x is not uniform, but is chosen from a high entropy distribution.

The adversary's input is a pair (pub^*, y^*), and she is given access to oracles $(g, e, d, \mathsf{Break})$. Our proof proceeds in two steps: first, we show that if we modify the oracle Break slightly then the adversary can simulate the modified oracle Break′ on her own with high probability. Since no adversary can break the one-wayness of a random trapdoor permutation, we obtain a bound on the advantage of an adversary that has access to Break′ instead of Break. The second step is to show that, in fact, the oracles Break and Break′ always produce the same answer. Combining the two steps together we get a bound on the advantage of A.

The Modified Oracle Break′. The oracle Break′ is parameterized by a public key pub^*, and is defined as follows: Break′ is the same as Break except step 4, which is modified in Break′ as follows:

4. Let $i \in \{1, 2\}$ such that $pub^* \notin PK_g(\mathcal{O}'_i)$. If $pub^* \in PK_g(\mathcal{O}'_1) \cap PK_g(\mathcal{O}'_2)$ or $pub^* \notin PK_g(\mathcal{O}'_1) \cup PK_g(\mathcal{O}'_2)$ then set $i = 1$. Then, run $D_{PRI_i}^{\mathcal{O}''_i}(y_i)$ to obtain an output x. If $x = \perp$ return $\perp$.

In other words, instead of always inverting y_1, Break′ inverts y_i where pub^* is not generated during the generation of PUB_i. Intuitively, this is a useful property because A is trying to break a single public key pub^*. Relying on the fact that the permutation $e(pub^*, \cdot)$ is random and independent from the rest of the oracle, A can simulate Break′ by generating the rest of the oracle by herself. She runs into trouble only when asked to invert $e(pub^*, \cdot)$. However, this is avoided by the check that is performed in step 5, and by the above modification. The check of step 5 prevents E from making a query that requires a trapdoor for pub^*. Note that although this may seem like a severe restriction, we have shown in Section 4 that it does not prevent us from breaking correlation security. The change in Break′ allows A to invert the one y_i which does not require knowledge of a trapdoor for pub^*.

5.1 Simulating Break′

The simulator itself is very technical, and is given in full detail in the full version [19]. We describe here the main ideas that are used in the construction. As previously mentioned, our adversary can generate all of $\mathcal{O}$ by herself, except for the permutation $e(pub^*, \cdot)$. Thus, if she wanted to simulate Break′ she would run into the following two problems: firstly, she is unable to compute the oracles $\mathcal{O}''_i = \mathcal{O} \diamond_c \mathcal{O}'_i$, and secondly she is unable to answer queries of the form $[d(pri, y)]$ where $[g(pri) = pub^*] \in \mathcal{O}''_i$.

The first problem is caused by the fact that the $\diamond_c$ operator resolves collisions, which requires the knowledge of entries of $\mathcal{O}$ of the form $[e(pub^*, x) = y]$ where $[e(pub^*, x') = y] \in \mathcal{O}'_i$. Our adversary may not possess this knowledge which prevents her from resolving all collisions. Instead, she resolves only the collisions that are *known* to her from the previous queries to the actual oracle $\mathcal{O}$, and the inputs of the Break′ query. Since the rest of $e(pub^*, \cdot)$ is random, she is unlikely to stumble unto any new collisions during the simulation of the Break′ query.

The second problem is caused by the fact that $\mathcal{O}'_i$ are adversarially chosen, and as such may contain an entry $[g(pri) = pub^*]$ which is incorrect according to

$\mathcal{O}$. This allows $D_{PRI_i}^{\mathcal{O}_i''}(y_i)$ and $E_{PUB_i}^{\mathcal{O}_i''}(x)$ to query $d(pri, y)$, which the adversary is unable to answer. This is dealt with as follows: to prevent D from making such queries we introduced the modification in Break′. The only case in which $\mathcal{O}_1''$ and $\mathcal{O}_2''$ may both contain trapdoors for *pub* is when the adversary knows at least one such trapdoor which is correct according to $\mathcal{O}$. Consequently, since it is hard to find a correct trapdoor for *pub**, there is an i such that $\mathcal{O}_i'$ does not define any fake trapdoors for it. It is now safe to simulate $D_{PRI_i}^{\mathcal{O}_i''}(y_i)$ because D is unlikely to find a true trapdoor of *pub**. To prevent E from making such queries, Break (and Break′) perform a check in step 5 that forces the oracle to return $\perp$ if E makes a query of the form $g(pri)$ where $\mathcal{O}_i'$ and $\mathcal{O}$ disagree on the answer.

5.2 Equivalence of Break and Break′

We have shown that Break′ provides very little help to an adversary trying to break the one-wayness of (g, e, d). We now show that Break and Break′ always answer queries identically, which proves that Break does not break the one-wayness of (g, e, d).

Claim. The adversary A has advantage δ when given access to Break′ instead of Break.

Proof (Sketch). The complete proof appears in [19]. Informally, let $x_i = D_{PRI_i}^{\mathcal{O}_i''}(y_i)$. The only difference between Break and Break′ is that in step 4 Break always computes x_1 and Break′ sometimes computes x_2. There are two cases: if $x_1 \neq x_2$ then due to the fact that E_{PUB_i} is a permutation, step 5 will return $\perp$. If $x_1 = x_2$, then both oracles perform the same computation in step 5.

5.3 Main Theorem

We are now ready to prove the main theorem of this section. We prove a strong variant that implies the security of (g, e, d) even when the input x is not uniform, but is chosen from a high entropy distribution. As previously mentioned, this implies that even a strong type of trapdoor functions (deterministic public key encryption) is insufficient to obtain correlation security.

Theorem 2. *Let G, E, D be the trapdoor permutation that forwards its input directly to the oracles g, e, d, and let D be a distribution over $\{0,1\}^\lambda$ such that $H_\infty(D) = k$. Then, for every adversary A that makes at most q oracle queries to $(g, e, d, \mathsf{Break})$, $\delta = \delta_{OW}(A, D) \leq \frac{2q}{2^k - q} + \frac{q}{2^\lambda - q} + \frac{3q}{2^\lambda}$.*

Proof (Sketch). The proof follows by a simple calculation from the equivalence of Break and Break′, and the ability of the adversary to simulate Break′. The complete argument appears in [19].

6 Extensions

In this section we present several strengthenings of our basic theorem, and address the simplifications that we made for our proof.

Injective Trapdoor Functions. The oracle and proof require few changes to handle trapdoor functions that are 1-1 but not necessarily onto. The modifications are as follows:

- Step 2 in the description of Break is modified to verify that (G, E, D) is a valid injective trapdoor function instead of checking that it is a trapdoor permutation. The rest of the oracle stays as before.
- The main issue that arises from changing from permutations to injective functions is the concern that the adversary may design a family of injective trapdoor functions that give away too much information when the inversion algorithm is applied to a string that is not in the range of the function. However, in that case check 5 of both Break and Break′ (which is described in the proof) will always fail since both permutations are evaluated in the forward direction on the inverse obtained in step 4.

Weaker Correlation. Our result easily generalizes to obtain the following stronger theorem: for every $n, k \in \mathbb{N}$, and every distribution $\mathcal{C}$ on elements of $(\{0,1\}^n)^k$ such that, with high probability each of the k coordinates can be found given all the remaining $k-1$ coordinates of the sample, there is no black-box construction of a trapdoor permutation that is correlation secure under $\mathcal{C}$ from a one-way trapdoor permutation. On a very high level, our Break oracle can be generalized to break such constructions due to the following simple fact. In the simulation of Break′ by an adversary that has access only to (g, e, d) (and not to Break′, the simulator is unable to invert only one of the strings y_1, y_2. The simulator can easily be extended to invert all the strings $y_1, \ldots, y_k$ except one. More details are given in [19].

Families of Trapdoor Permutations That Use Break. Our adversary in Section 4 breaks only constructions of trapdoor permutations that only make use of the (g, e, d) part of the oracle, and never query Break. Similarly, our simulator in Section 5 only simulates Break queries that do not make recursive Break queries. We chose to describe the proof in this manner to simplify the presentation. Both Theorem 1 and Theorem 2 extend to the case where G, E, D are allowed to make Break queries.

One modification is to the cost of a Break query. When Break may make recursive queries to itself, a single Break query by the adversary counts as the sum of the costs of all the Break queries in the resulting recursion tree. A second modification is to the adversary that uses Break. The modified adversary keeps track of Break queries and answers that appear during the simulations of G and E in steps 2 and 3. Then, in step 4, she chooses the partial oracles $\mathcal{O}'_1$ and $\mathcal{O}'_2$ to be consistent with the previously observed queries and answers to Break.

The main property of Break that allows us to handle such constructions is that in every call to Break, only one of the values y_i may require a trapdoor for some pub^* which happens to be the public key that our simulator is trying to break. Hence, to extend the simulator of Section 5 to handle constructions that make use of Break, the simulator is modified to recursively simulate Break by running our simulator for Break' for each recursive call.

Adding a PSPACE Oracle. In our proofs the only measure of complexity for algorithms is the number of (g, e, d) queries that they make. This can be interpreted intuitively as ruling out a certain type of reductions between the two primitives in question. However, we are interested in showing that there is an oracle relative to which there exists a secure trapdoor permutation, and yet there exists a polytime adversary that breaks the correlation security of every construction. This is achieved by adding a **PSPACE** oracle. Then, step 4 of our adversary can be implemented in a single step by making a query to the **PSPACE** oracle. The rest of the computation that is performed by the adversary, and by the simulator is done in polynomial time. To complete the proof it is necessary to observe that a random trapdoor permutation remains secure, even when the adversary has access to a **PSPACE** oracle. For more details about the technique of adding a **PSPACE** oracle we direct the reader to [16] and [18].

References

1. Bellare, M., Fischlin, M., O'Neill, A., Ristenpart, T.: Deterministic encryption: Definitional equivalences and constructions without random oracles. In: Wagner, D. (ed.) CRYPTO 2008. LNCS, vol. 5157, pp. 360–378. Springer, Heidelberg (2008)
2. Bellare, M., Boldyreva, A., O'Neill, A.: Deterministic and efficiently searchable encryption. In: Menezes, A. (ed.) CRYPTO 2007. LNCS, vol. 4622, pp. 535–552. Springer, Heidelberg (2007)
3. Boneh, D., Canetti, R., Halevi, S., Katz, J.: Chosen-ciphertext security from identity-based encryption. SIAM J. Comput. 36(5), 1301–1328 (2006)
4. Boneh, D., Franklin, M.K.: Identity-based encryption from the weil pairing. In: Kilian, J. (ed.) CRYPTO 2001. LNCS, vol. 2139, pp. 213–229. Springer, Heidelberg (2001)
5. Boneh, D., Papakonstantinou, P., Rackoff, C., Vahlis, Y., Waters, B.: On the impossibility of basing identity based encryption on trapdoor permutations. In: FOCS 2008: Proceedings of the 2008 49th Annual IEEE Symposium on Foundations of Computer Science, Washington, DC, USA, pp. 283–292. IEEE Computer Society, Los Alamitos (2008)
6. Canetti, R., Goldreich, O., Goldwasser, S., Micali, S.: Resettable zero-knowledge (extended abstract). In: STOC 2000: Proceedings of the thirty-second annual ACM symposium on Theory of computing, pp. 235–244. ACM, New York (2000)
7. Diffie, W., Hellman, M.: New directions in cryptography. IEEE Transactions on Information Theory 22(6), 644–654 (1976)
8. Dolev, D., Dwork, C., Naor, M.: Non-malleable cryptography. In: STOC 1991: Proceedings of the twenty-third annual ACM symposium on Theory of computing, pp. 542–552. ACM, New York (1991)
9. Dolev, D., Dwork, C., Naor, M.: Nonmalleable cryptography. SIAM Review 45(4), 727–784 (2003)

10. Gennaro, R., Gertner, Y., Katz, J., Trevisan, L.: Bounds on the efficiency of generic cryptographic constructions. SIAM J. Comput. 35(1), 217–246 (2005)
11. Gertner, Y., Kannan, S., Malkin, T., Reingold, O., Viswanathan, M.: The relationship between public key encryption and oblivious transfer. In: FOCS 2000: Proceedings of the 41st Annual Symposium on Foundations of Computer Science, Washington, DC, USA, p. 325. IEEE Computer Society, Los Alamitos (2000)
12. Gertner, Y., Malkin, T., Reingold, O.: On the impossibility of basing trapdoor functions on trapdoor predicates. In: FOCS 2001: Proceedings of the 42nd IEEE symposium on Foundations of Computer Science, Washington, DC, USA, p. 126. IEEE Computer Society, Los Alamitos (2001)
13. Gertner, Y., Malkin, T., Myers, S.: Towards a separation of semantic and CCA security for public key encryption. In: Vadhan, S.P. (ed.) TCC 2007. LNCS, vol. 4392, pp. 434–455. Springer, Heidelberg (2007)
14. Gertner, Y., Malkin, T., Myers, S.: Towards a separation of semantic and CCA security for public key encryption. In: Vadhan, S.P. (ed.) TCC 2007. LNCS, vol. 4392, pp. 434–455. Springer, Heidelberg (2007)
15. Impagliazzo, R., Rudich, S.: Limits on the provable consequences of one-way permutations. In: STOC 1989: Proceedings of the twenty-first annual ACM symposium on Theory of computing, pp. 44–61. ACM, New York (1989)
16. Reingold, O., Trevisan, L., Vadhan, S.P.: Notions of reducibility between cryptographic primitives. In: Naor, M. (ed.) TCC 2004. LNCS, vol. 2951, pp. 1–20. Springer, Heidelberg (2004)
17. Rosen, A., Segev, G.: Chosen-ciphertext security via correlated products. In: Reingold, O. (ed.) TCC 2009. LNCS, vol. 5444, pp. 419–436. Springer, Heidelberg (2009)
18. Simon, D.R.: Finding collisions on a one-way street: Can secure hash functions be based on general assumptions? In: Nyberg, K. (ed.) EUROCRYPT 1998. LNCS, vol. 1403, pp. 334–345. Springer, Heidelberg (1998)
19. Vahlis, Y.: Two is a crowd? a black-box separation of one-wayness and security under correlated inputs (2009), `http://www.cs.toronto.edu/~evahlis`

Efficient, Robust and Constant-Round Distributed RSA Key Generation

Ivan Damgård and Gert Læssøe Mikkelsen

Department of Computer Science, Aarhus University

Abstract. We present the first protocol for distributed RSA key generation which is constant round, secure against malicious adversaries and has a negligibly small bound on the error probability, even using only one iteration of the underlying primality test on each candidate number.

1 Introduction

The idea of distributed key generation is to generate a key in secret shared form among a number players such that it is never available in a single location. Together with a protocol for distributed signatures, for instance, one gets a distributed signature scheme that has no single point of attack throughout its lifetime.

Specifically for distributed RSA key generation, the main problem is to generate a modulus such that the prime factors are shared among the players. Two approaches have been suggested in the literature: Boneh and Franklin (BF) [4] suggest to generate a random candidate modulus $N = ab$ where a, b are random and shared among the players. One then runs a so-called biprimality test involving an exponentiation modulo N which is easy to do in a distributed fashion, and will accept an N with more than two prime factors with probability at most $\frac{1}{2}$. An alternative method was suggested by Algesheimer et al.(ACS) [1], where one generates candidate primes separately in shared form and tests each one for primality, by doing a Miller-Rabin primality test securely, i.e., by doing the required exponentiation while base, exponent and modulus are all secret-shared.

We now compare the methods and discuss whether there is room for improvement. Boneh and Franklin's test is very efficient because the modulus N is public. On the other hand, one has to wait until both factors a and b happen to be prime which requires more candidates than the standard method. The error probability is unfortunately very hard to bound: using only the worst-case result of $\frac{1}{2}$ leads to a very poor result that would seem to require many iterations of the biprimality test to bring the error down. For the Miller-Rabin test, it was shown by Damgård et al.[11] that the average case behavior is much better than the worst case, most composites pass the test with probability much smaller than the worst case, and hence for large numbers, only one iteration of the test is necessary for negligible error probability. One might hope for a similar result

D. Micciancio (Ed.): TCC 2010, LNCS 5978, pp. 183–200, 2010.

for the biprimality test, but this turns out to be unclear. The method from [11] relies heavily on the fact that for any prime factor p in a number N to be tested, p does not divide $N-1$. To argue in a similar way for the biprimality test, one would need that if $N = ab$ and p divides a, p does not divide $(a-1)(b-1)$, but this is clearly not true in general.

Algesheimer et al. test each candidate prime individually and so need fewer candidates, and one can use [11] to estimate the error. On the other hand, all exponentiations must be done with a secret modulus which makes them much slower. According to [1], for a small number of players (the interesting case in practice) the computational complexity of BF is somewhat larger than ACS while the communication is smaller. The big difference, however, is that BF is constant-round for checking a candidate while ACS require $\theta(n)$ rounds for checking an n-bit prime. While Algesheimer et al. claim that this is not important, we disagree, and believe the issue is very significant both from a theoretic and a practical point of view (see discussion at the end of the introduction).

Our conclusion is that in many, if not most cases BF is the more attractive approach. It is therefore of interest to construct a protocol with a good bound on the error probability which is as efficient as BF. In this paper we do this by combining the two methods from [1,4] to get, in a sense, the best of both worlds. We are going to compute a public candidate modulus $N = ab$ like Boneh and Franklin, but we are going to test a and b for primality separately, as follows: if we choose $a = b = 3 \bmod 4$, then doing the Miller-Rabin test say on a reduces to testing if $r^{(a-1)/2} \bmod a = \pm 1$ for a random base r. Now, because N is public, we can very efficiently choose a random $g \in Z_n^*$ and compute $y = g^{a-1} \bmod N$ in secret shared form using essentially Boneh and Franklin's protocol. Now we just need to reduce y modulo the secret a and test against $1, -1$. This can be done efficiently and in constant-round using a subprotocol from ACS. Note that there is no need for an exponentiation mod a, we just need a reduction, which is something ACS must do for every secure multiplication.

In this way, we get a protocol that is essentially as fast as Boneh and Franklin's, but where we can directly use [11] to estimate the error. Note that when testing a candidate, we can run the (simpler) biprimality test first without affecting the error probability since it never rejects a good modulus. This way, even if we cannot prove how well it does in the average case, we still get maximal mileage from it.

A second contribution of the paper is an efficient way to get a protocol secure against active (malicious) adversaries. Both BF and ACS were described for passive adversaries. Frankel et al. [14] suggested a way to get active security for the Boneh-Franklin protocol, but estimated themselves that the cost of this would be prohibitive in practice.

We suggest an alternative method where our secure computation is based on replicated integer secret sharing suggested by Damgård and Thorbek [13]. Here, the secret is shared additively over the integers, but each player gets several shares. Because of this replication, the scheme allows for secure multiplication in much the same way as Shamir's scheme. The observation is that because the scheme is also additive over the integers, we can use Algesheimer et al.'s protocol for modular reduction which exactly requires such a sharing scheme. This is in contrast to their original protocol where one uses Shamir's scheme for multiplication and so one must convert back and forth between the schemes throughout the protocol using interactive procedures. Finally, we make the scheme verifiable by keeping players committed to their shares.

The price we pay for the simplifying the protocol is that computational complexity of the scheme does not scale well with the number of players, but we believe this is not a serious issue: in contrast to earlier proposals the protocol is genuinely practical for less than, say, 10 players, and in threshold cryptography, one usually thinks of the number of players as a small constant. For instance, in the framework for distributed RSA signatures suggested by the authors[12], it is natural to run a 3-party protocol where a PC, a server and a mobile device held by the user execute the protocol.

Our protocol is secure against an active and static adversary corrupting any minority of the players, and the cost of going from passive security to active security is a constant factor, both regarding computations, network traffic and the number of rounds. In practice the constant is fairly low, covering committing including local exponentiations, which are already done in the passive protocol, and broadcast of commitments.

We close the introduction by discussing the claim by Algesheimer et al. that the difference in communication and round complexities between BF and ACS do not matter because one can test many candidates in parallel. We disagree with this: It is true that on a network with large round trip time, one can make the average cost of a protocol go down if many instances are to be done in parallel. Each player sends the next message in an instance as soon as he is ready to do so, and if we have enough instances, each player has enough local computation to keep him busy until the other players respond. Ideally, this means that the amortized time per instance can be almost as if there were no network delays. This is the basis of the ACS claim that their large round complexity is not a problem, since of course we can test many candidate primes in parallel. However, on the other hand, the *real time* elapsed from we start until we are done can of course never be smaller than the time it takes to do a single instance stand-alone.

The ACS protocol has some constant number of rounds for every bit in a candidate prime number so for 1000-2000 bit RSA, it will have something like 5000- 10.000 rounds pr. test (as opposed to our protocol with less than 100 rounds pr. test.) If we further assume a malicious adversary and that we are running on a network as the Internet that is basically asynchronous, rounds will tend to take a long time: a corrupt player may not send anything, so to distinguish this from a delay of an honest player's message, one has to wait long enough in each round so that the chance of an honest player's message failing to arrive is negligible. Otherwise, we may exclude an honest player as being corrupt, and then the protocol is no longer secure. If, for instance, we need to set a time-out of 1 second to be sure to avoid mistakes, 5000-10.000 rounds will take between 1 and 3 hours to execute. The conclusion is that when the number of rounds is very large, as for ACS, the parallellization paradigm only make sense on a fast network with very strong guarantees on delivery time.

2 Security Model

The protocols described in this paper are all three player protocols, where we assume that at most one of the players are corrupted. The protocols can be generalized to n players, however, the underlying secret sharing scheme does not scale well in the case of many players. Corruption of players is either considered to be passive, where a corrupt player still follows the protocol, or active where a corrupted player can misbehave arbitrarily.

Our security model only ensure that misbehavior can be detected, it might not be the case that the honest players can tell which player misbehaved. Furthermore, the protocols does not guarantee termination, in case of dishonest behavior. This simplifies the protocols and security proofs. Both detecting which player misbehaved and guaranteed termination can, however, easily be ensured by applying digital signatures such that all messages are signed.

We assume point to point secure communication channels, meaning authenticated and only the length of messages are leaked to the adversary. We also use a broadcast channel, however, since the channel does not have to ensure synchronous broadcast and we allow abort, this can easily be implemented on top of the point to point channels. The player that broadcast a message sends the message to the two other players, they send what they have received to each other to check if they agree. If they do not agree on what they have received they tell the other players and stop the protocol.

Universal Composability and Common Reference String. We use the Universal Composability (UC) framework [7,8] to specify the security of our protocols. The active secure versions of our protocols assumes the *(chosen) common reference string model* (CRS), where all players have access to a common string, which can contain key material used to implement, in our case, commitments. The CRS model is used to improve the power of the simulator. Concretely in our case the reference string contains among others an RSA key N used to implement commitments. By giving the simulator additional information on some elements in $\mathbb{Z}_N$, the commitment scheme is not binding for the simulator, which is needed for our proofs. The CRS model might be circumvented by letting each player choosing its own N that will be used by the other players, when committing values, however, this is conceptually more complicated, and less efficient. Therefore it has been left out. On the other hand, the CRS model might be justifiable in the case of a PKI based on a CA. The CA' public key might be used as N, and as long as the CA does not actively cheats e.g., corrupts one of the players in the protocol, during key generation the protocol remains secure.

Definition 1. *Let $\mathcal{A}_{Pass}$ be the class of passive static adversaries corrupting at most one of the three players; and let $\mathcal{A}_{Act}$ be the class of active static adversaries corrupting at most one of the three players.*

3 Probabilistic Primality Test

In this section we present a probabilistic primality test based on the Miller-Rabin test [19]. The advantage of the test described here is, as we will see later, that it can be very efficiently implemented as a distributed protocol. By requiring that the candidate a being tested fulfills $a \equiv 3 \pmod 4$ the Miller-Rabin test on a reduces to testing if $v^{(a-1)/2} \pmod a \equiv \pm 1$ for a random base v.

ProbPrime. Takes input a and N s.t. $a|N$. We assume $a \in [2^{n-1}, 2^n]$ and $N \in [2^{2n-2}, 2^{2n}]$.

1. $v \in_R \mathbb{Z}_N$.
2. $\gamma_a \leftarrow v^{(a-1)/2} \bmod N$
3. If $\pm 1 \equiv \gamma_a \pmod a$, then output *Probably prime*, else output *Composite*.

The correctness and the error probability are stated in the following theorems. Since the protocol is based on the Miller-Rabin test these theorems are likewise based on error estimates of this test.

Theorem 1. ***ProbPrime*** *is a* Monte Carlo *algorithm with random input v. A correctly formed prime $a \equiv 3$ (mod 4) is always accepted, and for* worst case *input it accepts a composite with probability $< \frac{1}{4}$.*

Proof. **ProbPrime** is essentially the Miller-Rabin test restricted to numbers $a \equiv 3 \pmod 4$ and therefore always accepts correctly formed primes, and has the same worst case error estimate: $\frac{1}{4}$ ([19]).

Theorem 2. *Let* ***ProbPrime*** *be utilized to generate probable primes by inputting randomly chosen n bit integers $a_i \equiv 3 \pmod 4$, running the test t independent times on each a_i and outputting the first number passing all test. Let $P_{n,t}$ denote the probability that a composite number is output. Assuming the Extended Riemann Hypothesis and $a_i > 2.3 \times 10^{10}$ then $P_{n,1} < n^2 4^{3-\sqrt{n}}$ and $P_{n,t} < n^{3/2} 2^t t^{-1/2} 4^{3-\sqrt{tk}}$ for $2 \leq t \leq n/9$.*

Proof. Damgård et al. [11] estimates $P_{n,1} < n^2 4^{2-\sqrt{n}}$ and $P_{n,2 \leq t \leq n/9} < n^{3/2} 2^t t^{-1/2} 4^{2-\sqrt{tk}}$ for the Miller-Rabin test for input chosen uniformly random in the set $I_{odd}(n)$ of n bit odd positive integers. We restrict the set we choose candidates from to $I_{3 \bmod 4}(n)$ the set of n bit positive integers $a \equiv 3 \pmod 4$.

Let $S_{odd}(n)$ and $S_{3 \bmod 4}(n)$ denote the density of the false positives, composite numbers accepted with high probability in $I_{odd}(n)$ and $I_{3 \bmod 4}(n)$ respectively. Since $I_{3 \bmod 4}(n)$ is half the size of $I_{odd}(n)$, $S_{3 \bmod 4}(n)$ is at most the double of $S_{odd}(n)$, which therefore at most doubles the average error probability.

We then consider the density of primes in $I_{3 \pmod 4)}(n)$ compared with $I_{odd}(n)$. Heuristically the density of primes in $I_{odd}(n)$ and $I_{3 \bmod 4}(n)$ are asymptotically the same. However, by assuming the Extended Riemann Hypothesis and that $a_i > 2.3 \times 10^{10}$ the concrete bound [1]: $|\pi(x, 4, 3) - \frac{x}{2 \log x}| < \frac{x}{\log 2x} = \frac{x}{2 \log x} \frac{2}{\log x}$

[1] This bound follows from [2][Theorem 8.8.18].

can be found. This means that the difference between the density of primes in $I_{odd}(n)$ and $I_{3 \pmod 4}(n)$ is at most: $\frac{2}{\ln x}$ which for $2^n = x > 2.3 \times 10^{10}$ means $\frac{2}{\ln(x)} < 2$. This gives us another doubling of the average error probability and our average error probability is therefore 4 times higher than the one from [11] on the original Miller-Rabin test.

Two examples of concrete bounds of the average error probability on $n = 1024$ are $P_{1024,1} < 2^{-38}$ and $P_{1024,4} < 2^{-105}$. This means that running the test only one or a few times on each number is sufficient in practise.

4 Replicated Integer Secret Sharing

Secret sharing [20] is a known primitive used in many cryptographic protocols. This section describes an additive secret sharing scheme over the integers that enables *multi party computation* (MPC) over integers in a given interval. Additive secret sharing over the integers makes it possible to share an integer in some public known interval $[-T, T]$, by choosing the shares $s_0, \ldots, s_n \in_{\mathrm{R}} [-2^\kappa T, 2^\kappa T]$, where κ is the security parameter, such that $s = \sum s_i$. The shares are chosen in the lager interval to make a sharing of s statistically close to a sharing of zero, and therefore an adversary only gains negligible information of s even if all except one share is known to the adversary. Addition of two additive secret shared integers is done by locally adding the shares, however multiplication is not strait forward.

Replicated Integer Secret Sharing (RISS) is a revised variant of additive integer secret sharing, where multiplication and other calculations are made possible by replicating the shares, s.t. each player holds multiple shares, in case of three players they each holds two shares. The product of two secrets s and t can be rewritten as $st = (s_1, s_2, s_3) \times (t_1, t_2, t_3) = s_1t_1 + s_1t_2 + \cdots s_3t_2 + s_3t_3$, and therefore replicating the shares enables multiplication, because each product on the right hand side are known to at least one player.

When a dealer wants to share a secret s it is done as in the nonreplicated case: The dealer generates three uniform random numbers s_1, s_2 and $s_3 \in [-2^\kappa T, 2^\kappa T]$ s.t. $s = s_1 + s_2 + s_3$ then the dealer distributes these shares such that player 1 gets s_2 and s_3, player 2 gets s_1 and s_3 and player 3 gets s_1 and s_2.

In the rest of this section we will see how to implement multi party computation based on RISS, and *Verifiable Replicated Integer Secret Sharing* (VRISS) an active secure version of RISS. The specification of MPC using (V)RISS is defined as the ideal functionality $\mathcal{F}_{\mathrm{RISS}}$ in figure 1. The intuition of $\mathcal{F}_{\mathrm{RISS}}$ is a black box where the players can input values, associated with an index, then the players can do some computations on the values and the box can output results to one or more players. The simulator ideal-world adversary is allowed to delay output from $\mathcal{F}_{\mathrm{RISS}}$ maybe for infinitely long time, however, not to change values inside the functionality nor input or output from honest players.

4.1 Passive Secure Protocol Realizing $\mathcal{F}_{\mathrm{RISS}}$

We will here describe some protocols based on RISS, which together implements the functionality $\mathcal{F}_{\mathrm{RISS}}$. In this section we prove the security of the passive

Ideal functionality $\mathcal{F}_{\mathbf{RISS}}$

When started $\mathcal{F}_{\mathrm{RISS}}$ initializes an empty list $\mathcal{L}$, let $\mathcal{L}(i)$ denote either the value or the memory that can store a value at index i. All output from $\mathcal{F}_{\mathrm{RISS}}$ can be delayed (maybe infinitely) by the adversary.
Input: Upon receiving ($\texttt{Input}, pid, i, x$) from player pid and ($\texttt{Input}, pid, i$) from all other players store x at $\mathcal{L}(i)$.
Output: Upon receiving ($\texttt{Output}, pid, i$) from all players; send ($\texttt{Value}, \mathcal{L}(i)$) to player pid.
Publish: Upon receiving ($\texttt{Publish}, i$) from all players; send ($\texttt{Value}, \mathcal{L}(i)$) to the adversary and afterward to all players. (Maybe output to some or all players is delayed by the adversary)
Addition and Multiplication: Upon receiving ($\texttt{ADD}, i, j, k$) or ($\texttt{MUL}, i, j, k$) from all players store at $\mathcal{L}(i)$ the sum or the product of $\mathcal{L}(j)$ and $\mathcal{L}(k)$.
Constant Addition and Multiplication: Upon receiving ($\texttt{C-ADD}, i, j, x$) or ($\texttt{C-MUL}, i, j, x$) from all players store at $\mathcal{L}(i)$ the sum or product of $\mathcal{L}(j)$ and x.
Detected Misbehavior: Upon receiving ($\texttt{Misbehavior}$) from the adversary, at any point in the protocol. Then output ($\texttt{Misbehavior}$) to all players and halt (Note: This part is only necessary for active secure protocols)

Fig. 1. Ideal functionality defining the security of RISS and VRISS (See section 4.2)

secure protocols, therefore we only consider the security when the players follow the protocols as described. In section 4.2 active secure protocols are described.

It is easy to see that if the shares of $s \in [-T, T]$ has been chosen uniformly s.t: $s_1, s_2 \in_{\mathrm{R}} [-2^\kappa T, 2^\kappa T]$ and $s_3 = s - s_1 - s_2$ and s.t $s_3 \in [-2^\kappa T, 2^\kappa T]$ then two shares of s are indistinguishable from two shares of a sharing of zero, because the distributions are statistically close, with security parameter κ.

Lemma 1. *Generating and distributing shares in RISS UC-realizes* Input *in $\mathcal{F}_{RISS}$ with respect to all $\mathcal{A}_{Pass}$ adversaries.*

Addition and Constant Multiplication. To add shared numbers, each player locally adds the shares. Multiplication by a public known constant is done in the same way by locally multiplying the constant with the shares.

Lemma 2. *Since* addition *and* constant multiplication *in RISS only involves local computations it UC-realizes* Addition, Constant Addition *and* Constant Multiplication *in $\mathcal{F}_{RISS}$ with respect to all $\mathcal{A}_{Pass}$ adversaries.*

Jointly Generating (Pseudo) Random Sharing of Zero. The multiplication protocol has to generate a random nonreplicated integer secret sharing of zero, such that no player know the complete sharing. By using the technique *pseudo random secret sharing* (PRSS) [9] in a novel way, this can be implemented as a noninteractive protocol. If the players pairwise share a secret key for a *pseudo random function* (PRF) in the same way they would share a RISS share, they can use this PRF and the keys to generate three numbers $r_1, r_2, r_3 \in_{\mathrm{R}} [-2^n, 2^n]$

this is a replicated integer secret sharing of $r = r_1 + r_2 + r_3$. 0 can be written as $0 = r - r = (r_1 + r_2 + r_3) - (r_1 + r_2 + r_3) = (r_1 - r_2) + (r_2 - r_3) + (r_3 - r_1)$, and each of the three summands can be calculated by one of the players. The shares of zero have size $n + 2$. If one of the players is corrupt there are $n + 1$ bit uncertainty for the adversary of the two shares the adversary does not know; due to the fact that there are $n + 1$ different equally possible values for the share r_x unknown to the adversary. We will later use PRSS to generate random secret shared values, and publicly known random values. Given point to point secure channels between the players the shared keys can easily be set up beforehand.

Multiplication. To multiply two RISS shared numbers $\langle a \rangle^R$ and $\langle b \rangle^R$, such that $\langle c \rangle^R = \langle a \rangle^R \times \langle b \rangle^R$, each player i locally multiplies the local shares a_{i-1} and a_{i+1} of a and b_{i-1} and $b_{i+1}$of b. Now each player holds some shares of $\langle ab \rangle^R$, however, not all of these shares are replicated, to solve this and to bring the number of shares at each player down to two again, each player sum shares of ab and replicates the shares again. To rerandomize the shares a nonreplicated integer sharing of zero is added to the result before replication.

MUL$(\langle a \rangle^R, \langle b \rangle^R)$ Player i holds a_{i-1}, a_{i+1}, b_{i-1} and b_{i-1}. S.t. $-2^n < ab < 2^n$.

- Calculate $\langle ab \rangle_i^I \leftarrow (a_{i-1} \times b_{i-1}) + (a_{i-1} \times b_{i+1}) + (a_{i+1} \times b_{i-1})$
  (Note that $\langle ab \rangle^I$ is a nonreplicated integer secret sharing of $a \times b$)
- Jointly generate a $\kappa + n$ bit integer secret sharing of zero $\langle 0 \rangle^I$.
- $\langle c \rangle_{i-1}^R \leftarrow \langle ab \rangle_i^I + \langle 0 \rangle_i^I$
- Send $\langle c \rangle_{i-1}^R$ to player P_{i+1}, and wait for $\langle c \rangle_{i+1}^R$ from player P_{i-1}.

Lemma 3. Multiplication *in RISS UC-realizes* Multiplication *in $\mathcal{F}_{RISS}$ with respect to all $\mathcal{A}_{Pass}$ adversaries.*

Proof. We simulate multiplication by using sharings of zero instead of the real values. This is statistically close to the real values because the interval of the shares are κ bit greater than the values. Afterward the simulator can simulate any result by adjusting the share not known to the adversary. This will result in a share in the correct interval, except with negligible probability.

Theorem 3. *MPC with RISS UC-realizes $\mathcal{F}_{RISS}$ with respect to all $\mathcal{A}_{Pass}$ adversaries.*

Proof. lemma 1 - 3

4.2 Verifiable Replicated Integer Secret Sharing

The previous section described how RISS can be used to do multiparty computation securely against a passive adversary. To extend the security to active security, and ensure that secret values are not leaked, and that the adversary cannot influence the output of a protocol except by changing the input of a corrupted machine, we need to force the players to follow the protocol. We note that the protocol we describe in this section does not guarantee termination, and we cannot always determine who has been misbehaving if dishonest behavior is detected. However, to resolve conflicts of which player cheated, digital signatures might be utilized.

To obtain an active secure version of RISS, we need to ensure that, when a player shares a value the player is committed to this value, and that a receiver of a share is committed to the received value. We also need to enforce that each player proves to the others that calculations has been done correctly. To achieve these goals we need to assume the *chosen common reference string* (CRS) model for our commitments.

Commitment Scheme. Fujisaki and Okamoto [15](see Damgård and Fujisaki [10] for a revised version) describes an integer commitment scheme, which is additively homomorph and can be simulated in UC in the CRS model. The common reference string used consists of an RSA modulus N_{CRS} and two elements $g, h \in \mathbb{Z}_{N_{CRS}}$, where the discrete log between g and h is unknown to the players, however, not to the simulator, which allows simulation. When a player wants to commit to a value s a uniform random value $r \in_{\mathrm{R}} \mathbb{Z}_{N_{CRS}}$ is chosen and the commitment of s is: $\mathrm{com}(s, r) \mapsto g^s h^r \mod N_{CRS}$, this scheme is additive homomorphic because: $\mathrm{com}(s+t, r_s+r_t) = (g^s h^{r_s}) \times (g^t h^{r_t}) \mod N_{CRS}$ To open a commitment s and r are revealed. These commitments are statistically hiding and computationally binding, assuming the strong RSA assumption. If the discrete log between g and h is known the commitments are no longer binding.

We also need an other primitive from the commitment scheme, which is the ability to prove that two commitments c_1 and c_2 are commitments of the same value. In the case where the same base (g and h) is used this is an easy task. To show that $c_1 = com(s, r_1)$ and $c_2 = com(s, r_2)$ the prover shows that $c_1 \times c_2^{-1} = com(s - s, r_1 - r_2)$ can be opened to zero. In the case where different bases are used the problem is more difficult, however, in our case with three players and at most one corrupted player, there exists an easy solution. The prover just need to prove to the two others that he is committing correctly if none of the two verifiers are corrupted, because if the prover is corrupted then both verifiers are honest and can thus trust each other, if one of the verifiers is corrupt, then the prover is honest and by assumption committed to the correct value. The actual protocol proving $s = \hat{s}$ for $(g^s h^r)$ and $(\hat{g}^{\hat{s}} \hat{h}^{\hat{r}})$ is the following:

1. Generate:
   s_1, s_2 s.t. $s_1 + s_2 = s = \hat{s}$; r_1, r_2 s.t. $r_1 + r_2 = r$ and $\hat{r}_1, \hat{r}_2$ s.t. $\hat{r}_1 + \hat{r}_2 = \hat{r}$
2. Publish:
   $c_1 = g^{s_1} h^{r_1} \mod N$, $c_2 = g^{s_2} h^{r_2} \mod N$, $\hat{c}_1 = \hat{g}^{s_1} \hat{h}^{\hat{r}_1} \mod N$ and $\hat{c}_2 = \hat{g}^{s_2} \hat{h}^{\hat{r}_1} \mod N$
3. Open c_1 and $\hat{c}_1$ to verifier 1 and c_2 and $\hat{c}_2$ verifier 2. Both accept if c_x and $\hat{c}_x$ opens to the same value.

Generating Shares. When player i wants to share a secret s it is done as in the passive case, with the exception that before the shares are distributed the player broadcasts a commitment of each share, this is also a commitment to s due to the additive homomorphic property of com(). When the shares are distributed as in RISS, player i opens the commitment to each share to the receivers of the share.

Addition, Constant multiplication. As in the passive case *addition* and *constant multiplication* can be computed without communication between the players. This can be done because when a player adds shares of two secrets a and b locally, the other players can calculate $\text{com}(a_i + b_i)$ due to the additive homomorphic property of com(). Analogously with constant multiplication because: $\text{com}(ca_i, cr_i) = g^{ca_i}h^{cr_i} = (g^{a_i}h^{r_i})^c$

Joint Generation of Shares. Utilizing *pseudo random secret sharing* enables active secure generation of a random secret shared number s, by only one broadcast message pr. player. This is done by generating shares s_1, s_2 and s_3 as in the passive case, and in addition generate the randomness r_1, r_2 and r_3 for the commitments by the PRF and the shared keys. Now each player calculates the commitments $c_x = \text{com}(s_x, r_x)$ to the two shares and broadcasts the result. All three players can check if the two commitments to the same share are equal, if not, one of the players misbehaved. Because of the additive homomorphism of com() the joint sharing of zero can also be done in one round with only one broadcast message pr. player.

VRISS Multiplication. Enabling multiplication in VRISS requires that one player can prove to the others that he have multiplied two committed values correctly. This can be done if the prover is committed to a and b with $c_a = \text{com}(a, r_a)$ and $c_b = \text{com}(b, r_y)$ and proves that $c_{ab} = \text{com}(ab, r_{ab})$. First let:

$$c_{ab} \leftarrow (c_a)^b h^r \mod N_{crs} \equiv g^{ab}h^{r_a b + r} \mod N_{crs}$$

This is indeed a commitment to ab with the base g and h. To prove that it is correct the prover proves that c_{ab} base c_a and h is a commitment to the same value as c_b base g and h using the algorithm described earlier. Now the passive protocol is executed with each player committing and proving to the others that the commitments are well formed and that the steps of the protocol has been followed.

Theorem 4. *Assuming the strong RSA assumption and the existence of PRF, then MPC with VRISS UC-realizes $\mathcal{F}_{\text{RISS}}$ with respect to all $\mathcal{A}_{Act}$ adversaries.*

Proof. Theorem 3 proves that we can simulate the protocol, if all players follow the protocol. Adding the commitment scheme and the checks of the commitments, forces a corrupt player to follow the protocol, or the other players will detect the misbehavior. On the other hand, simulation is still possible because we assume the discrete log between g and h is know to the simulator, and it can therefore circumvent the binding property of the commitment scheme.

4.3 Distributed Primality Testing

The ideal functionality $\mathcal{F}_{\text{RISS}}$ and the protocols of RISS and VRISS describes a general secret sharing scheme. However, in addition to this our protocol for RSA key generation needs a protocol implementing a distributed version of the primality test described in section 3. This extended RISS is described as an ideal functionality $\mathcal{F}_{\text{EXT-RISS}}$ (see Fig. 2), which is an extension of $\mathcal{F}_{\text{RISS}}$. The protocols implementing $\mathcal{F}_{\text{EXT-RISS}}$ has additional requirements on the integers used as input and on how these integers are shared. We call these *special form integers.*

Definition 2 (Special Form Integer). *An integer a is a* special form integer *if it has been generated as such and fulfills: $2^{n-1} < a < 2^n$ and $a \equiv 3 \pmod 4$.*

A resharing of a shared integer does not preserve *special form* of an integer, therefore only integers generated as *special form* can be on *special form.* This is because in the realization of $\mathcal{F}_{\text{EXT-RISS}}$ requires that the sharing of the integer is on the following form:

Definition 3 (Special Form Integer Sharing). *A* special form integer sharing *is a sharing of a* special form integer *fulfilling: $a_1 \equiv 3 \pmod 4$ and $a_2 \equiv a_3 \equiv 0 \pmod 4$.*

Ideal functionality $\mathcal{F}_{\text{EXT-RISS}}$

$\mathcal{F}_{\text{EXT-RISS}}$ is identical to $\mathcal{F}_{\text{RISS}}$ except it is extended with the following:
Randomly Generate Special Form Integer Upon receiving (GenSFI, *pid*, i) from all players generate a uniform random *special form integer* a. Store a at $\mathcal{L}(i)$, with a flag specifying that $\mathcal{L}(i)$ holds a *special form integer.*
Trial Division Upon receiving (Div?, *pid*, B, i) from all players, if $\exists \ell < B$ s.t. $\ell | \mathcal{L}(i)$ then output (Fail) to all players, otherwise output (Success) to all players.
Probabilistic Prime Test Upon receiving (ProbPrime?, *pid*, i, j, N) from all players and if $\mathcal{L}(i)$ is a *special form integer*, and $N = \mathcal{L}(i) \times \mathcal{L}(j)$, then let $a \leftarrow \mathcal{L}(i)$, choose $v \in_{\text{R}} \mathbb{Z}_N$ and calculate $\gamma = v^{\frac{a-1}{2}} \bmod a$. If $\gamma = \pm 1$ output (ProbPrime, v) to all players, else send $\{\mathcal{L}(i), \mathcal{L}(j), v\}$ to the adversary and (Composite, v) to all players.

Fig. 2. Ideal functionality defining the security of Extended RISS

Randomly Generate Special Form Integer. Player 1 and 2 each pick a random integer $a^{(i)} \in_{\text{R}} [2^{n-2}, 2^{n-1}]$, s.t. $a^{(1)} \equiv 3 \pmod 4$ and $a^{(2)} \equiv 0 \pmod 4$. This ensures that $a = a^{(1)} + a^{(2)} \in [2^{n-1}, 2^n]$ and $a \equiv 3 \pmod 4$. Both players share them s.t $a_1^{(1)} \equiv 3 \pmod 4$ and for all other shares: $a_j^{(i)} \equiv 0 \pmod 4$. The shares are distributed and added, which ensures that the shares fulfills the congruence requirement for a *special form integer sharing.* The special requirement of the congruency of the shares only leaks the value of a mod 4. The security follows from the security of input and addition of RISS shares.

Lemma 4. Randomly Generate Special Form Integer *UC-realizes this part of $\mathcal{F}_{EXT\text{-}RISS}$ with respect to all $\mathcal{A}_{Pass}$ adversaries.*

Trial Division. To do trial division up to a bound B on a shared number a, the players test if a is divisible by a small prime ℓ by randomly choosing an $n+\kappa$-bit secret shared number r using PRSS as described in section 4.1. $\langle ra \rangle^R$ is calculated by the multiplication protocol and all shares of $\langle ra \rangle^R$ are locally reduced modulo ℓ and afterward broadcast to open $\alpha = (ra \mod \ell) + \beta\ell$, where $0 \le \beta < 3$. If $\alpha \not\equiv 0 \pmod{\ell}$ then $\ell \not| a$, however, if $\alpha \equiv 0 \pmod{\ell}$ then either $\ell | a$ or $\ell | r$. To prevent the protocol from rejecting a when $\ell | r$ the protocol is executed a number of times with new random values r. For optimization reasons local reductions modulo ℓ should be done before and during the calculation of $\langle ra \rangle^R$, and the trial divisions should be executed in parallel.

Lemma 5. Trial Division *UC-realizes this part of* $\mathcal{F}_{EXT\text{-}RISS}$ *with respect to all* $\mathcal{A}_{Pass}$ *adversaries.*

Proof. This can be simulated by simulating the result of $\langle ra \rangle^R$. The leaked value $\alpha = (ra \mod \ell) + \beta\ell$ does not leak information because it can be perfectly simulated, by choosing r and its shares appropriately.

Probabilistic Primality Test. Here we present a distributed version of the primality test described in section 3.

ProbPrime The players holds special form integer shares of the value being tested $\langle a \rangle^R$. The value $\langle b \rangle^R$ is secret shared among the players, and $N = ab$ is publicly known. We assume $a, b \in [2^{n-1}, 2^n]$ and $N \in [2^{2n-2}, 2^{2n}]$.

1. Distributed generate a public known value $v \in_R \mathbb{Z}_N$.
2. The players locally calculates γ_{a_i} s.t.:
   $\gamma_{a_1} = v^{(a_1-1)/2} \mod N$ and
   $\gamma_{a_2} = v^{(a_2)/2} \mod N$
   $\gamma_{a_3} = v^{(a_3)/2} \mod N$
3. The players share the values γ_{a_i} and calculate $\langle \gamma_a \rangle^R = \prod \gamma_{a_i} \equiv v^{(a-1)/2} \pmod{N}$
4. Distributed check if $\pm 1 \equiv \langle \gamma_a \rangle^R \pmod{\langle a \rangle^R}$, if so output *Probably prime*, otherwise output *Composite*.

Generating v can efficiently be done if all players uses the same key for the PRF. To check if $\pm 1 \equiv \gamma_a \pmod{a}$ we use a technique based on the ACS protocols [1]. However, due to a different setting, where we assume $N = ab$ is publicly known, we can improve the protocols from $O(log(n))$ rounds to $O(1)$ rounds, n being the bit length of a. First note that $(\gamma_a \bmod a) = \gamma_a - \lfloor \frac{\gamma_a}{a} \rfloor a$. If we assume the following $2^{n-1} < a, b < 2^n$, $2^{n-2} < N = ab < 2^{2n}$ and $\gamma_a < 2^{2n+2}$ we can approximate $\frac{\gamma_a}{a}$ in the following way [2]:

$$\widetilde{N} = \left\lceil \frac{2^{5n+2}}{N} \right\rfloor \Rightarrow \left| \frac{1}{N} - \frac{\widetilde{N}}{2^{5n+2}} \right| < 2^{-3n+2} \quad \text{and} \quad \widetilde{N} < 2^{3n+4} \tag{1}$$

$$\widetilde{a} = b \times \widetilde{N} \Rightarrow \left| \frac{1}{a} - \frac{\widetilde{a}}{2^{5n+2}} \right| < 2^{-2n+2} \quad \text{and} \quad \widetilde{a} < 2^{4n+4} \tag{2}$$

$$\Rightarrow \left| \frac{\gamma_a}{a} - \frac{\gamma_a \widetilde{a}}{2^{5n+2}} \right| < 1 \quad \text{and} \quad \gamma_a \widetilde{a} < 2^{6n+6} \tag{3}$$

$$\Rightarrow \gamma_a - \lfloor \gamma_a \widetilde{a} 2^{-5n+2} \rfloor a = (\gamma_a \bmod a) + \delta a, \quad -1 \le \delta \le 1 \tag{4}$$

[2] $\lceil x \rfloor$ meaning rounding x to nearest integer ($\lceil x \rfloor = \lfloor x + \frac{1}{2} \rfloor$).

In the above equations $\widetilde{N}$ is calculated locally by each player, $\langle\widetilde{a}\rangle^R$ is calculated by the distributed constant multiplication protocol. The value $\langle\gamma_a\widetilde{a}\rangle^R$ is calculated by the multiplication protocol. Calculating $\lfloor\gamma_a\widetilde{a}2^{-5n+2}\rfloor$ is done by each player locally dividing the shares of $\gamma_a\widetilde{a}$ by 2^{5n+2} and rounding the result downwards: $\lfloor c\rfloor \approx c' = \lfloor c_1\rfloor + \lfloor c_2\rfloor + \lfloor c_3\rfloor, |c'-c| \leq 3$. Therefore we can calculate $y = \gamma_a - \lfloor\gamma_a\widetilde{a}2^{-5n+2}\rfloor = (\gamma_a \bmod a) + \delta a$ s.t. $-4 \leq \delta \leq 4$ and $y < 2^{2n+3}$.

The last step in the protocol to test if $\pm 1 \equiv \gamma_a \pmod a$ is to calculate:

$$z = \left(\prod_{i=-4}^{4}((y+ia)+1 \bmod Q)\right)\left(\prod_{i=-4}^{4}((y+ia)-1 \bmod Q)\right) \bmod Q \quad (5)$$

The number Q is a publicly known prime s.t. $Q > 2^{2n+3} > (y+ia)+1, |i| \leq 4$. The multiplications in (5) are done modulo Q to limit the size of the numbers we are calculating on to $2n+3$ bit numbers. Multiplications modulo Q can be done by doing local modulo reduction on the shares before and after the multiplication protocol. Now the players opens z and if $z = 0$ they output *success*, otherwise they output *failure*. The last step of the protocol is correct because if $\pm 1 \equiv \gamma_a \bmod a$ then $((y+ia)\pm 1)$ is zero when $i = \delta$, and since the numbers we calculate on are less than Q, and thereby relatively prime to Q, then $z \neq 0$ is always the case if $\pm 1 \not\equiv \gamma_a \mod a$.

Lemma 6. *Assuming the existence of PRF's, then on well formed input (special form integer sharing)* **ProbPrime** *UC-realizes this part of $\mathcal{F}_{\text{EXT-RISS}}$ with respect to all $\mathcal{A}_{Pass}$ adversaries.*

Proof. We assume the existence of PRF's, therefore the value v in the ideal and real world cannot be distinguished efficiently. If $\gamma \neq \pm 1$ in $\mathcal{F}_{\text{EXT-RISS}}$ the simulator gets knowledge of all private input and can therefore easily simulate the protocol. If $\gamma = \pm 1$ the protocol can be simulated as follows. Step two only contains local calculations and does therefore not leak any information. Step three can be simulated, see lemma 3. The last step can in the same way be simulated to output $z = 0$, because of lemma 3 (it is easy to see that the lemma still holds modulo Q).

4.4 Active Security Distributed Primality Testing

For active security we need to ensure that the players follow the protocol. This means player 1 and 2 have to prove that there random input during generation of a special form integer sharing is well formed. The correct congruence modulo 4 can easily be tested, because each share will be send to two players, and therefore at least one honest. To prove that $a^{(1)}$ or $a^{(2)}$ is in the correct range we use a technique, from [6], we note that the solution described here is less efficient than [6], however, it is conceptually simpler. Proving that a number $a \in [2^{n-1}, 2^n]$ can be done by proving that $a - 2^{n-1} \geq 0$ and that $2^n - a \geq 0$. Proving that $x \geq 0$ is done by writing x as a sum of squares. Any positive number can be written as the sum of four squares which efficiently can be calculated [17].

A protocol for player i to prove $a \in [2^{n-1}, 2^n]$ is: Player i calculates $\alpha_1, \ldots, \alpha_4$ and $\beta_1, \ldots \beta_4$ s.t. $\sum \alpha_i^2 = a - 2^{n-1}$ and $\sum \beta_i^2 = 2^n - a$. Player i shares the numbers using VRISS and the three players calculates $\tilde{\alpha} = (a - 2^{n-1}) - \prod \alpha_i^2$ and $\tilde{\beta} = (2^n - a) - \prod \beta_i^2$. The values $\tilde{\alpha}$ and $\tilde{\beta}$ is opened and if they are opened to zero then $a \in [2^{n-1}, 2^n]$ is true, otherwise player i is deviating from the protocol.

The protocols also includes local computations on the shares, these do not leak information, and is therefore passively secure. They can be made active secure with one broadcast message pr. player: The players use the PRF to generate three random and replicated values. Now the players uses these random values to commit to the result of the local computation such that each share of the result is committed with the same randomness by the two players calculating the same share. The commitments are broadcast, and if the two commitments of the same value are not equal, one of the players misbehaves.

Theorem 5. *Assuming the strong RSA assumption and existence of PRF, then the above protocols UC-realizes $\mathcal{F}_{EXT\text{-}RISS}$ with respect to all $\mathcal{A}_{Act}$ adversaries.*

Proof. This follows from theorem 4, lemma 4 - 6, and the above description.

5 Distributed RSA Moduli Generation

The security of our RSA moduli generating protocol is given by the ideal functionality $\mathcal{F}_{\text{KeyGen}}$ (Fig. 3). The intuition is that if the players follow the protocol then the factorization of N is secret, however, if misbehavior is detected by all players then N should not be used, and it is secure to reveal p and q.

Ideal functionality $\mathcal{F}_{\text{KeyGen}}$

Key Generation: Upon receiving (`KeyGen`, sid, n) from all players; generate two n-bit primes p and q, s.t. $p \equiv q \equiv 3 \mod 4$ and let $N = pq$.
Send N to the adversary. When the adversary replies with (`Deliver`) then send N to all players and halt.
Detected Misbehavior: Upon receiving (`Misbehavior`) from the adversary, at any point in the protocol. Then send p and q to the adversary, output (`Misbehavior`) to all players and halt

Fig. 3. Ideal functionality for generating an RSA modulus

The protocol Π_{KeyGen} implementing $\mathcal{F}_{\text{KeyGen}}$ is described in Fig. 4. The protocol is based on the BF protocol [4], with an other probabilistic primality test. We start by describing a passive secure protocol, and afterward we extend it to active security. Π_{KeyGen} assume an MPC scheme realizing $\mathcal{F}_{\text{EXT-RISS}}$.

Picking candidates: By the protocol for randomly generating special form integers, the players jointly generates two prime candidates a and b s.t $2^{n-1} < a, b < 2^n$ and $a \equiv b \equiv 3 \pmod 4$)

RSA Moduli Generation Protocol: $\Pi_{\textbf{KeyGen}}$

1. **Pick candidates:** Secretly pick random numbers a and b s.t. $a \equiv b \equiv 3 \mod 4$.
2. **Trial division:** Distributed trial divide a and b up to a bound B.
   Repeat step 1 and 2 until two candidates a and b passes the trial division.
3. **Compute N:** The players calculate and publish $N = ab$.
4. **Primality test:** Run primality test to check a. If a is accepted, b is tested.
   If either a or b was rejected, the protocol is restarted, otherwise output N.
5. **Proof honest behavior** The players prove that they in the earlier steps of the protocol followed the protocol honestly.

Fig. 4. Protocol for distributed generation of RSA moduli

Trial Division: Trial division up to a bound B is performed on a and b. Instead of trial division distributed sieving, which is more efficient, can be utilized, see section 6.

Computing N: To compute N the parties use the multiplication protocol and make the result N public. When N is public the players might do more local trial division before continuing.

Primality Test: To test whether a and b, both having survived trial division, are indeed primes, or at least with overwhelming probability are primes, **ProbPrime** is used to test a and b one or a few times. If a or b is rejected the protocol is restarted, otherwise in the passive case N is output as the RSA modulus. In the active secure case the players need to prove honest behavior before N is output.

Active Security. Extending the protocol to active security, can be done using VRISS instead of RISS. However, a more efficient solution exists. When the players choose the input they commit and broadcasts the commitments. The rest of the protocol is run using the RISS protocol for distributed calculations. If either a or b at some point is rejected as primes the players opens the commitments of a and b publicly and each player can locally test if a or b should have been rejected or if a player is misbehaving. When a modulus N is accepted the players calculates and broadcasts all the proofs of well formed input and of having executed the protocol correctly. If a player cannot broadcast correct proofs, the other players reports that misbehavior is detected.

Lemma 7. *The probability of generating a modulus N which is* not *the product of two primes is the same as in the generic RSA key generation using the Miller-Rabin test to generate Blum integers.*

Proof. In the last round of the protocol, where both a and b passes the test the value a has been chosen completely independent of b and vice versa. Because we choose a and b at the same time there are rounds before the last one where b is rejected and we have to sample a new a. However, since we sample independent in each round these rounds can just be seen as a delay of randomly chosen time inserted in the protocol.

Theorem 6. *Assuming the Extended Riemann Hypothesis, existence of Pseudo Random Functions and the Strong RSA assumption, then Π_{KeyGen} UC-realizes $\mathcal{F}_{KeyGen}$ with respect to all $\mathcal{A}_{Act}$ adversaries.*

Proof. From lemma 7 it follows that the output of the protocol and the ideal functionality is indistinguishable. We also need simulate the transcript of the protocol given N from $\mathcal{F}_{\text{KeyGen}}$. First we assume that the adversary follows the protocol as described. In that case simulating the number of rounds of Π_{KeyGen} where candidates are rejected either by trial division or by the primality test can be done as: The simulator run the real protocols on input a and b not being two primes, such that they are rejected, with the same distribution as in the real world.

The last round, where N is accepted and output from Π_{KeyGen}, can be simulated in the following way: The simulator can simulate trial division and the primality test without knowing the input, this means it can simulate acceptance of the two protocols, without knowing the factorization of N.

If the corrupted player does not follow the protocol there are the following two cases: In one of the rounds where a and b should be rejected, but are not, the adversary cannot present proofs of following the protocol. Therefore in the real world the honest players will detect misbehavior and in the ideal world the simulator will report misbehavior. In the last round where N is supposed to be output, but is rejected, then in the real world the honest players will detect misbehavior when the adversary cannot present shares of a and b making the test fail. In the ideal world the simulator reports misbehavior and is given the factorization of N and can therefor show shares of $a = p$, $b = q$ to the adversary such that the test should have passed.

6 Optimizations

Parallelization. If the bottleneck of the protocol is network latency, then testing many candidates in parallel will decrease effect of the latency.

Distributed Sieving. Instead of first pick candidates to be primes, and thereafter perform trial division. It is possible to do distributed sieving for candidates relatively prime to all small primes less that some bound B. This technique is due to Malkin et al. [18], and in their implementation distributed sieving gave a 10 fold speedup when generating 1024 bit keys. Distributed sieving is done by letting $M = \prod_{\ell \in PRIMES}^{B}(\ell)$ and let the players pick random values $a_i \in \mathbb{Z}_M^*$ and letting the candidate $a = (\prod a_i) + rM$ for a random value r in an appropriate interval. This makes a relatively prime to M and thereby relatively prime to all small primes less than B. After converting a into additive shares the players must initiate a protocol that ensures the additive shares has the right properties (congruence modulo four), this applies [4] and to our protocol, however, not to [18] due to their simpler (only heuristically secure) primality test.

Using Multi Prime RSA Modulus. As mentioned in [4] it is possible to avoid the quadratic slowdown of testing two candidates at the same time instead of testing two candidates independently as done in [1] and when generating RSA keys locally by [19]. The trick is to use a modulus which is a product of three primes, known as multi prime RSA. $N = p_1 p_2 (a_1 + a_2)$ where p_1 is a prime known to player 1 and p_2 is a prime known to player 2 and $a = a_1 + a_2$ is a candidate for a third prime. Unlike [3] that need a special tri-primality test like [5] our protocol can easily be extended to test multi prime moduli due to primality test. It should be noted that the latest PKCS #1 version (v2.1) [16] includes the use of multi prime RSA, although the motivation there is improved speedup when utilizing the Chinese remainder theorem technique.

7 Conclusion and Acknowledgment

We have presented a novel approach to do distributed generation of RSA moduli, with an active secure constant round primality test with a good bound on the average error probability. By using parallelization the complete generation of RSA moduli can made constant round, even when guarantying active security. An second contribution is a novel way to do multi party computations with replicated integer secret sharing. An open question remains, whether a better average case analysis of the Boneh and Franklins biprimality test is possible.

We thank Arjen Lenstra for some useful pointers.

References

1. Algesheimer, J., Camenisch, J., Shoup, V.: Efficient computation modulo a shared secret with application to the generation of shared safe-prime products. In: Yung, M. (ed.) CRYPTO 2002. LNCS, vol. 2442, pp. 417–432. Springer, Heidelberg (2002)
2. Bach, E., Shallit, J.: Algorithmic Number Theory: Efficient algorithms. MIT Press, Cambridge (1996)
3. Boneh, D., Franklin, M.K.: Efficient generation of shared RSA keys (extended abstract). In: Kaliski Jr., B.S. (ed.) CRYPTO 1997. LNCS, vol. 1294, pp. 425–439. Springer, Heidelberg (1997)
4. Boneh, D., Franklin, M.K.: Efficient generation of shared RSA keys. J. ACM 48(4), 702–722 (2001)
5. Boneh, D., Horwitz, J.: Generating a product of three primes with an unknown factorization. In: Buhler, J.P. (ed.) ANTS 1998. LNCS, vol. 1423, pp. 237–251. Springer, Heidelberg (1998)
6. Boudot, F.: Efficient proofs that a committed number lies in an interval. In: Preneel, B. (ed.) EUROCRYPT 2000. LNCS, vol. 1807, pp. 431–444. Springer, Heidelberg (2000)
7. Canetti, R.: Universally composable security: A new paradigm for cryptographic protocols. Cryptology ePrint Archive, Report 2000/067 (2000) (2005 version)
8. Canetti, R.: Universally composable security: A new paradigm for cryptographic protocols. In: FOCS, pp. 136–145 (2001)
9. Cramer, R., Damgård, I., Ishai, Y.: Share conversion, pseudorandom secret-sharing and applications to secure computation. In: Kilian, J. (ed.) TCC 2005. LNCS, vol. 3378, pp. 342–362. Springer, Heidelberg (2005)

10. Damgård, I., Fujisaki, E.: A statistically-hiding integer commitment scheme based on groups with hidden order. In: Zheng, Y. (ed.) ASIACRYPT 2002. LNCS, vol. 2501, pp. 125–142. Springer, Heidelberg (2002)
11. Damgård, I., Landrock, P., Pomerance, C.: Average case error estimates for the strong probable prime test. Mathematics of Computation 61(203), 177–194 (1993)
12. Damgård, I., Mikkelsen, G.L.: On the theory and practice of personal digital signatures. In: Jarecki, S., Tsudik, G. (eds.) PKC 2009. LNCS, vol. 5443, pp. 277–296. Springer, Heidelberg (2009)
13. Damgård, I., Thorbek, R.: Linear integer secret sharing and distributed exponentiation. In: Yung, M., Dodis, Y., Kiayias, A., Malkin, T.G. (eds.) PKC 2006. LNCS, vol. 3958, pp. 75–90. Springer, Heidelberg (2006)
14. Frankel, Y., MacKenzie, P.D., Yung, M.: Robust efficient distributed RSA-key generation. In: STOC, pp. 663–672 (1998)
15. Fujisaki, E., Okamoto, T.: Statistical zero knowledge protocols to prove modular polynomial relations. In: Kaliski Jr., B.S. (ed.) CRYPTO 1997. LNCS, vol. 1294, pp. 16–30. Springer, Heidelberg (1997)
16. RSA Laboratories. PKCS #1 v2.1: RSA cryptography standard. Technical report (2002)
17. Lipmaa, H.: On diophantine complexity and statistical zero-knowledge arguments. In: Laih, C.-S. (ed.) ASIACRYPT 2003. LNCS, vol. 2894, pp. 398–415. Springer, Heidelberg (2003)
18. Malkin, M., Wu, T.D., Boneh, D.: Experimenting with shared generation of RSA keys. In: NDSS. The Internet Society (1999)
19. Rabin, M.O.: Probabilistic algorithm for testing primality. Journal of Number Theory 12(1), 128–138 (1980)
20. Shamir, A.: How to share a secret. Commun. ACM 22(11), 612–613 (1979)

Threshold Decryption and Zero-Knowledge Proofs for Lattice-Based Cryptosystems

Rikke Bendlin and Ivan Damgård

Department of Computer Science, Aarhus University
{rikkeb,ivan}@cs.au.dk

Abstract. We present a variant of Regev's cryptosystem first presented in [Reg05], but with a new choice of parameters. By a recent classical reduction by Peikert we prove the scheme semantically secure based on the worst-case lattice problem GapSVP. From this we construct a threshold cryptosystem which has a very efficient and non-interactive decryption protocol. We prove the threshold cryptosystem secure against passive adversaries corrupting all but one of the players, and againts active adversaries corrupting less than one third of the players. We also describe how one can build a distributed key generation protocol. In the final part of the paper we show how one can, in zero-knowledge - prove knowledge of the plaintext contained in a given ciphertext from Regev's original cryptosystem or our variant. The proof is of size only a constant times the size of the public key.

1 Introduction

Cryptography based on lattice problems is one of the most important examples of techniques holding promise for public-key cryptography that is secure even under quantum attacks and are also interesting in that they can be based on worst-case complexity assumptions. Recently, these techniques have become much more efficient after it has been realized that one can base the actual cryptosystem on the learning with error problem (LWE), and then argue that the (variant of the) LWE problem used is as hard as some lattice related problem, typically computing the shortest vector in a lattice. In the LWE problem, the adversary must compute a secret vector s with entries in some field or ring, given only the inner products of s with some public vectors where, however, some noise has been added to the products. As mentioned, basing a cryptosystem on LWE can lead to quite efficent cryptosystems, see, e.g., [Reg05],[PVW08],[MR08],[Pei09].

As lattice-based cryptography moves closer to practice, it becomes an important research question to investigate whether these cryptosystems can provide the same "extra" functionality we have come to expect from well-known public-key cryptosystems based on factoring or discrete logarithms. For instance, can we have threshold versions of these systems? In other words, we want to share the private key among a set of servers and efficently decrypt a ciphertext while revealing nothing but the plaintext to the adversary. And furthermore, can one

D. Micciancio (Ed.): TCC 2010, LNCS 5978, pp. 201–218, 2010.

prove, in zero-knowledge and efficiently, knowledge of the plaintext contained in a given ciphertext?

In this paper we construct such a threshold cryptosystem, based on a variant of Regev's system [Reg05]. We show our scheme semantically secure based on a worst-case lattice problem using a recent reduction of Peikert [Pei09]. To the best of our knowledge, it is the first lattice-based threshold cryptosystem. We need to use a larger modulus than Regev, thus making ciphertexts larger, on the other hand we get a very efficient and non-interactive decryption protocol: each player needs only to do local computation and announce a single element from the underlying ring. The basic version of the protocol is secure against a passive adversary corrupting all but one of the players. For a small number of players, we show an equally efficent version secure against a malicious adversary corrupting less than a third of the players. We also describe a distributed protocol for generating keys.

Various improvements of Regev's original cryptosystem have been made since its first appearence, e.g. in [PVW08] and [MR08]. Our threshold cryptosystem can be generalized in the same way, but we stick to Regev's original approach here for simplicity.

In the final part of the paper we present a zero-knowledge protocol for proving knowledge of the plaintext contained in a given ciphertext, for Regev's original cryptosystem as well as our variant. The proof is much more efficient than what generic methods would give us: it has size only a constant times the size of the public key, and the computation required is comparable to what is required to generate keys. The protocol is based on the construction from [IKOS07] of zero-knowledge from multiparty computation protocols. Whereas this paradigm has perhaps been perceived primarily as a theoretical tool, we show here that it can also be potentially relevant in practice.

2 Preliminaries

When writing $x \in_R S$ we mean that x is chosen uniformly at random from the set S. Equivalently $x \in_\chi S$ means choosing x from the set S according to the distribution χ. For some distribution χ writing $x \sim \chi$ means that x is distributed according to χ.

Given a probability distribution χ on $\mathbb{Z}_q$, let n be some integer and $\mathbf{s} \in \mathbb{Z}_q^n$. We define $A_{\mathbf{s},\chi}$ as the distribution on $\mathbb{Z}_q^n \times \mathbb{Z}_q$ obtained by choosing $\mathbf{a} \in_R \mathbb{Z}_q^n$, $e \in_\chi \mathbb{Z}_q$ and outputting $(\mathbf{a}, \langle \mathbf{a}, \mathbf{s} \rangle + e)$. We define the decisional learning with errors (LWE) problem as being able to distinguish between a sample from $A_{\mathbf{s},\chi}$ and the uniform distribution on $\mathbb{Z}_q^n \times \mathbb{Z}_q$ with non-negligible probability. We define the search LWE problem as given a sample from $A_{\mathbf{s},\chi}$ finding $\mathbf{s}$ with non-negligible probability.

By $\overline{\Psi}_\alpha$ we denote a discrete Gaussian distribution on $\mathbb{Z}_q$ with mean 0 and standard deviation $\frac{q\alpha}{\sqrt{2\pi}}$. Likewise Ψ_α is a continuous Gaussian distribution on $\mathbb{T} = \mathbb{R}/\mathbb{Z}$ with mean 0 and standard deviation $\frac{\alpha}{\sqrt{2\pi}}$. By χ^{*k} we denote the distribution given by summing k independent samples from χ.

3 Cryptosystem

We first present the underlying cryptosystem which was proposed first in [Reg05], but with a new choice of parameters better suited for the distributed decryption protocol given later.

Let n be the security parameter of the cryptosystem. Then the main parameter is an integer q which is chosen as $q = 2^{O(n)}$. More specifically q will not be a prime but a B-smooth number where B is of polynomial size. That is $q = \prod p_i$ is a product of prime numbers $p_1, \ldots, p_k$, where $p_i < B$ and also $p_i > u$, the number of players in the distributed decryption protocol. The latter requirement on the primes is necessary in order to do secret sharing over the the ring $\mathbb{Z}_q$, more on this later. We also need an integer m which will be chosen to be $O(n^3)$. Finally, we need a distribution χ on $\mathbb{Z}_q$ which will be taken to be the discrete Gaussian distribution $\overline{\Psi}_\alpha$, where $\alpha = q^\beta$ for $\beta = 1/4$.

The cryptosystem is now defined as follows:

- **Secret key:** Choose $\mathbf{s} \in_R \mathbb{Z}_q^n$. The secret key is then $\mathbf{s}$.
- **Public key:** Choose m vectors $\mathbf{a}_1, \ldots, \mathbf{a}_m \in_R \mathbb{Z}_q^n$, m elements $e_1, \ldots, e_m \in_\chi \mathbb{Z}_q$. The public key is then given by $(\mathbf{a}_i, b_i = \langle \mathbf{a}_i, \mathbf{s}\rangle + e_i)_{i=1}^m$.
- **Encryption:** Choose a vector $\mathbf{r} = (r_1, \ldots, r_m) \in_R \{0,1\}^m$. Given a bit γ, the encryption of γ is given by $(\sum_{i=1}^m r_i \mathbf{a}_i, \gamma \cdot \lfloor \frac{q}{2} \rfloor + \sum_{i=1}^m r_i b_i)$.
- **Decryption:** Given a ciphertext $(\mathbf{a}, b)$, calculate $b - \langle \mathbf{a}, \mathbf{s}\rangle$ and determine whether it is closer to 0, the encrypted bit being 0, or closer to $\frac{q}{2}$, the encrypted bit being 1.

The correctness of the decryption protocol is given by the following theorem.

Theorem 1 (Correctness). *If for any $k \in \{0, 1, \ldots, m\}$ it holds that*

$$\Pr_{e \sim \chi^{*k}} (|e| \geq \sqrt[3]{q}) \leq 2^{-O(n)}$$

then the decryption protocol will give correct output except with negligible probability.

A similar theorem is proved in [Reg05] for Regev's original choice of parameters. The intuition is clear, if the noise added is not too big, we will be able to decrypt to the right bit. The correctness with the new parameters follows from the following claim.

Claim (Correctness). For the choice of parameters made, for any $k \in \{0, 1, \ldots, m\}$, a constant $c \in (0, 4)$ and $e \sim \chi^{*k}$ it holds that

$$\Pr_{e \sim \chi^{*k}} (|e| \geq \sqrt[c]{q}) \leq 2^{-O(n)}$$

Proof. We will prove this using the Chebyshev inequality, but first we will reduce the problem from $\overline{\Psi}_\alpha$ to Ψ_α. Given $e \sim \overline{\Psi}_\alpha^{*k}$ we have that $e = \sum_{i=1}^k \lfloor q x_i \rceil \pmod q$, where $x_i \sim \Psi_\alpha$. The value of e is at most $k < m < \sqrt[c]{q}/2$ away from $e' = \sum_{i=1}^k q x_i \pmod q$, so it is sufficient to prove that $|e'| < \sqrt[c]{q}/2$ except with negligible probability. Since e' comes from a distribution with standard deviation $\sqrt{k} \cdot \sqrt[4]{q}$ and mean 0 we get the following result from Chebyshev's inequality,

$$\Pr(|e'| \geq \sqrt[3]{q}/2) \leq \Pr(|e'| \geq t \cdot \sqrt{k}\sqrt[4]{q}) \leq \frac{1}{t^2}$$

where $m = n^3$ and $t = \frac{\sqrt[c]{q}}{2\sqrt{m}\sqrt[4]{q}} \geq \frac{\sqrt[c]{q}}{\sqrt[d]{q}\sqrt[4]{q}} = q^{1/c} \cdot q^{-1/d} \cdot q^{-1/4}$, for some constant d. Now considering $1/t^2$ we see that this will be negligible if $d > -\frac{4c}{c-4}$. But we can always choose such a d since $c < 4$. □

Note that the inequalities used above are not very tight, especially the Chebyshev inequality. Therefore in practice one would expect to be able to choose much better parameters, for instance a bigger standard deviation on the distribution used. This would in turn give us security reductions to the hardness of somewhat bigger lattice problem instances. Furthermore the claim is actually stronger than what is needed for the original decryption protocol to be correct, but we will need this stronger result in the proofs of the distributed decryption protocols described below.

The security of the cryptosystem is given by the following theorem.

Theorem 2 (Security). *The cryptosystem is semantically secure under the assumption that* GapSVP *is hard in the worst case.*

Proof. We sketch the ideas of the proof. The proof of security given in [Reg05] is based on the property that distinguishing between encryptions of 0 and 1 is at least as hard as distinguishing public keys from randomly chosen elements in $\mathbb{Z}_q^n \times \mathbb{Z}_q$. The latter problem being the decision LWE problem. The proof of the reduction does not depend heavily on the values of the parameters, and is therefore still valid with the new choice of parameters.

The decision LWE is then further reduced to search LWE. This reduction in [Reg05] heavily relies on the fact that q is chosen to be polynomial in that it does exhaustive search over all elements in $\mathbb{Z}_q$. But in fact the same idea can be used when q is exponential in size, but B-smooth with B polynomial. The idea being to do the reduction modulo each of the primes p_i in q, and recombine the solutions to a full solution modulo q using the Chinese Remainder Theorem. This was already observed in [Pei09].

The last step is to reduce search LWE to standard lattice problems. Since q is chosen to be exponentially large we can use the reduction to GapSVP made in [Pei09]. □

This is another advantage of choosing an exponentially large q: With the original choice of a polynomial q the reductions to lattice problems are either a quantum reduction as in [Reg05] or a reduction to a special variant of GapSVP, the hardness of which is not completely understood.

4 Distributed Decryption (Passive Adversaries)

In this section we present a distributed decryption protocol for the above cryptosystem involving u players which is secure against a static, passive adversary corrupting up to $t = u - 1$ players. That is, we assume the adversary is able to see all messages and internal data of a corrupted player, but the player still follows the protocol. The adversary must choose which players to corrupt at the start of the protocol.

We assume that communication is synchronous and that the client has access to a broadcast channel to all players. Private channels between players are not necessary since there is no interaction between players in the protocol. We assume the adversary sees all communication between the client and the players.

We use Shamir secret sharing over $\mathbb{Z}_q$ as described in [Sha79] to make secret sharings of various values in the protocol. Normally Shamir secret sharing is done over a field, but since q is not a prime $\mathbb{Z}_q$ is only a ring. This turns out not to be a problem with the choice made of the prime factors in q. The only thing that is needed is that one can do Lagrange interpolation over the points $1, \ldots, u$ which in turn boils down to being able to invert elements in this range. We chose $q = \prod p_i$, where $p_i > u$, so obviously inversion of the points needed is possible.

We furthermore make use of the concept of pseudorandom secret sharing (PRSS) described in [CDI05]. PRSS will enable the players to non-interactively share a common random value from some interval. The idea is as follows. For each subset A of size t of the players we associate a key $K_A \in_R \mathbb{Z}_q$. Such a key is given to player P_j exactly if $P_j \notin A$. Assume we are given a pseudorandom function ϕ that given a key and a ciphertext as input, will output values in the interval $[-\sqrt{q}, \sqrt{q}]$. A player can now compute $\phi_{K_A}(c)$ for all K_A he has been given, and afterwards take an appropriate linear combination of the results. This will result in all players having a Shamir share of the common random value $x = \sum_A \phi_{K_A}(c)$. Since $|A| = t$ there are $\binom{u}{t}$ possibilities for A, so x will be in the interval $\left[-\binom{u}{t}\sqrt{q}, \binom{u}{t}\sqrt{q}\right]$. We note that $\binom{u}{t} = u$ for our choice of t (but we will consider other choices later).

The protocol and proofs will be given in the setting of the Universal Composability (UC) framework proposed by Canetti. For details of this see [Can01].

4.1 Key Generation and Distribution

We assume for now that generation and distribution of keys and key-shares to players are handled by the functionality F_{KeyGen}.

Functionality F_{KeyGen}

1. When receiving "start" from all honest players, choose the secret key $\mathbf{s} = (s_1, ..., s_n)$ and construct the public key $(\mathbf{a}_i, b_i)_{i=1}^m$ as described in section 3. Furthermore for each subset A of size t of the players, choose key $K_A \in_R \mathbb{Z}_q$.
2. Receive from the adversary a set of shares $s_{i,j}, i = 1, \ldots, n$ for each corrupted player P_j. Then construct using Lagrange interpolation a complete set of shares $s_{i,j}, i = 1, \ldots, n, j = 1, \ldots, u$ consistent with the shares received from the adversary, and such that $s_{i,1}, \ldots, s_{i,u}$ form a set of shares of s_i. We write $[\mathbf{s}]$ as short for the set of all shares. Send privately to each player P_j his shares from $[\mathbf{s}]$ and all keys K_A where $P_j \notin A$.
3. Finally send the public key to all players and the adversary.

It may seem strange that this functionality allows the adversary to decide which shares he wants to get of the secret key – why not let the functionality do the sharing on its own? However, we need to define the functionality this way to make sure it can be implemented. The problem is that a simulator trying to simulate a given protocol will have to make sure that the view of the protocol it generates for the adversary is consistent with what the functionality says to the honest players. This is not possible if the functionality decides on all shares on its own. One could say that what we model here is that we don't care which shares the adversary gets, as long as the secret is safe.

4.2 Decryption Protocol

We now describe the decryption protocol. To make things more easily describable we introduce a client, who is the party receiving the ciphertext in the first place, and who wants to decrypt with help from the players.

Protocol *Decrypt*

1. Each player sends "start" to F_{KeyGen} and stores the public key, the share of the secret key and the keys K_A received.
2. When receiving a ciphertext $c = (\mathbf{a}, b)$, the client broadcasts c to all players.
3. The players compute $[e'] = [b - \langle \mathbf{a}, \mathbf{s} \rangle] = [e + \lfloor \frac{q}{2} \rfloor \cdot \gamma]$. Since $(\mathbf{a}, b)$ is public this is a linear operation on $\mathbf{s}$ and only requires the players to locally do the same linear operation on their shares. Then $\phi_{K_A}(c)$ is computed for all the keys K_A the player received and the player takes an appropriate linear combination of the result to obtain a sharing $[x] = [\sum_A \phi_{K_A}(c)]$. Finally the players compute $[x + e']$, and send all these shares to the client.
4. Having received all the shares of $[x + e']$ the client reconstructs $x + e'$, checks whether it is closer to 0 or to $q/2$, and outputs 0 or 1 accordingly.

4.3 Security

To prove security we wish to be able to implement the following functionality.

Functionality $F_{KeyGen-and-Decrypt}$

1. Upon receiving "start" from all honest players, choose the secret key and construct the public key to be used. Send the public key to all players, the client and the adversary.
2. Hereafter on receiving "decrypt $(\mathbf{a}, b)$" from the client, send "decrypt $(\mathbf{a}, b)$" to all players and the adversary.
3. In the next round, send "result γ" to the client and the adversary, where γ is the bit corresponding to the given ciphertext.

Theorem 3 (Security). *When given access to the functionality F_{KeyGen} and assuming that ϕ is a pseudo-random function, the protocol Decrypt securely implements $F_{KeyGen-and-Decrypt}$. The adversary is assumed to be passive and static, corrupting up to $t = u - 1$ of the players.*

Proof. We abbreviate $F_{KeyGen-and-Decrypt}$ by F_{KG-D} in the following. To prove security we must construct a simulator to work on top of the ideal functionality F_{KG-D}, such that an adversary playing with either the simulator and ideal functionality or the real world decryption protocol cannot tell in which case he is. We denote by *Adv* the adversary communicating with the real decryption protocol and must show that we can simulate everything *Adv* sees. The simulation proceeds as follows.

1. Let B denote the set of players corrupted by *Adv*. When receiving "start" to F_{KeyGen} send "start" to F_{KG-D}. Upon receiving the public key, compute a sharing of $\mathbf{0}$, the zero-vector in $\mathbb{Z}_q^n$, to simulate sharing the secret key. Also choose the necessary keys K_A. Then send to the adversary the public key, the shares of the secret key corresponding to B, and the keys K_A that should be send to players in B.
2. When receiving "decrypt $(\mathbf{a}, b)$" from F_{KG-D}, the ciphertext is sent to *Adv* for each player in B. When "result γ" is received in the next round, we have to simulate the shares of $x + e'$ that honest players would send. To play the role of x, we form a value y as the sum of those $\phi_{K_A}(c)$ where the adversary knows K_A, and one uniformly random value from $[-\sqrt{q}, \sqrt{q}]$ for each K_A that the adversary does not know. The idea is to let $y + \lfloor \frac{q}{2} \rfloor \cdot \gamma$ play the role of the value $x + e + \lfloor \frac{q}{2} \rfloor \cdot \gamma$ that would be revealed in the real protocol. Note that from the shares and keys given to the adversary, we can compute the shares corrupted players would send to the client. Using Lagrange interpolation, we can compute a polynomial f of degree at most t that is consistent with these shares and has $f(0) = y + \lfloor \frac{q}{2} \rfloor \cdot \gamma$. We use this polynomial to compute shares for the honest players and give these to the adversary.

The final thing is to prove that no environment is able to distinguish between the real decryption protocol and the simulation presented above. This basically comes down to proving that the decryption protocol is able to recover the bit encrypted and that the distributions of the shares sent to the adversary in both cases are computationally indistinguishable.

The shares of the secret key in step 1 are distributed in the same way in both cases beacuse of the security of the underlying secret sharing scheme used. The keys K_A are also obviously distributed identically in the two cases.

Next, note that in both simulation and real protocol, the shares revealed in the decryption step follow deterministically from the information sent in step 1 and the values $y + \lfloor \frac{q}{2} \rfloor \cdot \gamma$, $x + e + \lfloor \frac{q}{2} \rfloor \cdot \gamma$ used in simulation, respectively real protocol. It is therefore enough to show that these values are computationally indistinguishable in the view of the adversary. For this, note that in the real protocol the adversary is not given all keys K_A, and so, by pseudorandomness of ϕ and construction of y, $y + e + \lfloor \frac{q}{2} \rfloor \cdot \gamma$ is computationally indistinguishable from the $x + e + \lfloor \frac{q}{2} \rfloor \cdot \gamma$ in the view of the adversary. Second since y is a sum including at least one value that is uniform in an interval of size $2\sqrt{q}$, which is exponentially larger than the interval $[-\sqrt[3]{q}, \sqrt[3]{q}]$ in which e is distributed, we find that $y + \lfloor \frac{q}{2} \rfloor \cdot \gamma$ is statistically indistinguishable from $y + e + \lfloor \frac{q}{2} \rfloor \cdot \gamma$.

Finally in both the simulated and the real run the client will output the correctly decrypted value. This is obvious in the simulated case and in the real world it follows from Lemma 1 below. □

Lemma 1 (Correctness). *Let* $\binom{u}{t} < \frac{1}{4}\sqrt{q} - 1$. *Assume that the reconstructed value in the distributed descryption protocol is given by* $e + x$, *and furthermore that the following is satisfied*

$$\Pr\left[|e| \geq \lfloor\sqrt{q}\rfloor\right] \leq 2^{-O(n)}.$$

Then the error probability when decrypting is negligible.

Proof. Given an encryption of 0 the result which is reconstructed is given by $b - \langle \mathbf{a}, \mathbf{s} \rangle = e + x = \sum_{i=1}^{m} r_i e_i + x$. Since $\binom{u}{t} < \frac{1}{4}\sqrt{q} - 1$ according to our assumption, we have that $|x| < \frac{q}{4} - \sqrt{q}$. Combined with the assumption on $|e|$ we get that $|e + x| < \frac{q}{4}$ with probability at least $1 - 2^{-O(n)}$. In this case the result is closer to 0 than $\frac{q}{2}$ and the decryption is correct. A similar proof can be done for an encryption of 1. □

The distribution of e is exactly given by $\chi^{*\sum r_i}$, when F_{KeyGen} has been used to produce the keys, therefore according to the claim of section 3 we know that $|e| < \lfloor\sqrt[3]{q}\rfloor$ with probability at least $1 - 2^{-O(n)}$. And so the assumptions in the lemma is fulfilled.

We note that the correctness puts an upper bound on the possible number of players, which is also to be expected, since there is a limit to how much random noise can be added before an encryption of 0 turns into an encryption of 1.

5 Distributed Decryption for Stronger Adversaries

The protocol for doing distributed decryption against a passive adversary corrupting up to $t = u - 1$ players, can easily be turned into a protocol secure against a stronger adversary. First, if the adversary is semi-honest, i.e. corrupted players follow the protocol but may stop at any point, exactly the same protocol will be secure, if $t < u/2$. The proof is the same, one just notes that at least $t + 1$ players will always complete the protocol.

If the adversary is active, again almost the same protocol and proof applies, if we assume $t < u/3$. The only significant difference to the protocol is that the client must use standard methods for error correction to reconstruct $x + e'$ at the end of the decryption since some players may lie about their shares. This is possible exactly when $t < u/3$.

It should be noted that both variants of the protocol are only feasible to execute for a small number of players, since the number of keys K_A we must give to each player increases exponentially with u whenever t is a constant fraction of u. However, in most realistic applications of threshold cryptography, one indeed expects the number of players to be small.

6 Distributed Key Generation

In this section we will describe how to do key generation and distribution. In some of the parts involving interaction between the players, we will have to assume private communication channels between players. For a passive adversary, key generation is quite straightforward, so we focus on the more interesting case of an active, static adversary corrupting less than $t = u/3$ players.

We will need the following functionality for generating random (shared) values. It offers a number of commands, and a command is executed if all honest players input the same command.

Functionality F_{Rand}

- On input "Random value to B" for a set of players B, choose s at random in $\mathbb{Z}_q$ and send s to all players in B.
- On input "Random shared value", ask the adversary for a set of shares $S = \{s_j | P_j \text{ is corrupt}\}$. Choose s at random in $\mathbb{Z}_q$ and use Lagrange interpolation to construct a set of shares $[s]$ consistent with S, i.e., each corrupt P_j gets share s_j. Send shares from $[s]$ to all honest players.
- On input "Shared value from D", do the same as for "Random shared value", but get s from player D instead of choosing it at random.
- On input "Constrained value from D", do the same as for "Shared value from D", but check that s received from D is in the interval $[-2\binom{u}{t}\sqrt[3]{q}, 2\binom{u}{t}\sqrt[3]{q}]$. If not, send "fail" to all players. Furthermore, if D is honest, he is assumed to choose s in the interval $[-\sqrt[3.5]{q}, \sqrt[3.5]{q}]$. The seemingly strange choice of intervals is dictated by the implementations that are available, see more details below.

"Shared value from D" can be implemented using any protocol for verifiable secret sharing. A simulator would simply run the protocol with the adversary while following the protocol for the honest players (and using a dummy value for s if D is honest), and then send the shares obtained for corrupt players to the functionality. "Random shared value" can be done by calling "Shared value from P_i" for each P_i, asking P_i to supply a random value s_i, and then locally adding the resulting shares, thus obtaining $[\sum_i s_i]$ which we use as $[s]$. "Random value to B" is implemented by calling "Random shared value" and then have players send their shares to all players in B.

Finally, "Constrained value from D" can be implemented using the technique of non-interactive verifiable secret sharing (NIVSS) described in [CDI05] which builds on top of PRSS described earlier. In the protocol for doing NIVSS, a set of keys $\{K_A^D\}$ is assumed to be set up similar as for PRSS, i.e., K_A^D is known to all players not in A. But furthermore player D holds all keys and the value s to be shared. The keys can be set up by calling "Random value to B". The pseudorandom function involved is chosen such that it outputs random values from the interval $[-\sqrt[3]{q}, \sqrt[3]{q}]$. To generate the shared value, each player locally computes random shares of a value r as in PRSS, and D can compute r since he knows all keys. D then broadcasts $s - r$, and each player adds this value to their share in r, thus obtaining $[s]$. D is disqualified if the value broadcast is not in the interval $[-\binom{u}{t}\sqrt[3]{q}, \binom{u}{t}\sqrt[3]{q}]$. This guarantees that s is in the required interval even if D is corrupt, since r is in the interval $[-\binom{u}{t}\sqrt[3]{q}, \binom{u}{t}\sqrt[3]{q}]$. If D is honest, the distribution of $s-r$ is statistically close to uniform in $[-\binom{u}{t}\sqrt[3]{q}, \binom{u}{t}\sqrt[3]{q}]$ since s is smaller than r by an exponentially large factor.

Given F_{Rand}, key generation is for the most part straightforward. The tricky part, however, is that to generate the noise to be added to the public key, shares of non-uniformly distributed values are to be generated and distributed. For this we will invoke "Constrained value from P_j" for each P_j, since we can rely on honest P_j's using the correct distribution, while corrupt P_j's cannot choose values that are large enough to do any damage, as we shall see.

Protocol *KeyGeneration*
The protocol assumes F_{Rand} is available.

1. To generate and distribute the secret key, invoke "Random shared value" n times to form $[\mathbf{s}]$.
2. To generate and distribute the keys K_A for PRSS, invoke for each set A of t players "Random value to B", where B is the complement of A (we assume for simplicity that a random value from $\mathbb{Z}_q$ is sufficient to form a key K_A).
3. To generate the public key, invoke "Random value to P" nm times, where P is the set of all players, and use the output as entries in the vectors $\mathbf{a}_i$.
4. Each player P_j chooses noise contributions $e_{i,j}, i = 1, \ldots, m$ according to the distribution $\bar{\Psi}_\alpha$ and uses these as input to invocations of "Constrained value from P_j". Note that a correctly chosen $e_{i,j}$ will be in the correct interval $[-\sqrt[3.5]{q}, \sqrt[3.5]{q}]$ except with negligible probability. Thus, we obtain $[e_{i,j}]$ for $i = 1 \ldots u, j = 1, \ldots, m$, and players compute by local operations $[e_i] = [\sum_j e_{i,j}]$.
5. Finally the players can compute by local operations $[b_i] = [\mathbf{a}_i \cdot \mathbf{s} + e_i]$, and reconstruct the b_i's by broadcasting the shares.

6.1 Security

For proving security one could show that the protocol *KeyGeneration* securely implements the functionality F_{KeyGen} defined in section 4.1. This functionality however does not reflect the influence an active adversary will have on the public key when using the protocol above. We therefore define a slightly different functionality $F_{KeyGen'}$ and use this in the security proof instead. In the end of this section we will then show that the differences in the two functionalities does not matter in terms of correctness and security.

The main difference from F_{KeyGen} is that we will have the adversary supply additional inputs before constructing and distributing keys. More specifically the adversary will supply the functionality with noise contributions used in generating the public key.

Functionality $F_{KeyGen'}$

1. When receiving "start" from all honest players, also receive from the adversary, for each corrupted player P_j shares $s_{i,j}, i = 1, \ldots, n$ of the secret key to assign to P_j.
2. Choose the secret key $\mathbf{s}$ and for each subset A of size t of the players choose keys $K_A \in_R \mathbb{Z}_q$.
3. For each entry i in the secret key make a complete set of shares $s_{i,j}, i = 1, \ldots, n, j = 1, \ldots, u$ for each player consistent with the shares already received from the adversary. This is done by Lagrange interpolation. To each player P_j privately send his shares from $[\mathbf{s}]$ and all keys K_A where $P_j \notin A$.
4. For each corrupted player P_j receive noise contributions $e_{i,j}, i = 1, \ldots, m$ for generating the public key.
5. To generate the public key choose the m vectors $\mathbf{a}_1, \ldots, \mathbf{a}_m \in_R \mathbb{Z}_q^n$. For each non-corrupted player P_j choose noise contributions $e_{i,j}$ according to the distribution $\overline{\Psi}_\alpha$, the noise elements e_i are now given by $e_i = \sum_{j=1}^{u} e_{i,j}$. The public key is then given by $(\mathbf{a}_i, b_i = \langle \mathbf{a}_i, \mathbf{s} \rangle + e_i)_{i=1}^m$.
6. Finally send the public key to all players and the adversary.

Theorem 4. *Given access to F_{Rand}, the protocol KeyGeneration securely implements the functionality $F_{KeyGen'}$. The adversary is assumed to be active and static, corrupting less than $t = u/3$ of the players.*

Proof. We must construct a simulator to work on top of $F_{KeyGen'}$, such that an adversary playing with either the simulator and $F_{KeyGen'}$ or the real world key generation protocol cannot tell the difference. By *Adv* we denote the adversary communicating with the real world and must show that we can simulate everything *Adv* sees. The simulation proceeds as follows.

1. When receiving the set of shares S from *Adv* in order to invoke "Random shared value", send "start" to $F_{KeyGen'}$ and the shares received from *Adv*.
2. To simulate the generation of the public key, first choose nm random values and send them to all players to simulate running "Random value to P".
3. When receiving the noise contributions $e_{i,j}$ from *Adv*, also give these to $F_{KeyGen'}$. Now we must simulate sharing all the noise contributions in the real protocol from the invocation of "Constrained value from P_j". Again receive the shares that corrupted players will be given from *Adv*, and for the rest simply choose random shares.
4. Finally when given the public key $(\mathbf{a}_i, b_i = \langle \mathbf{a}_i, \mathbf{s} \rangle + e_i)_{i=1}^m$ from F_{KeyGen} we must simulate the broadcast of shares of the b_i's in the real protocol. First compute the shares the corrupted players are holding, based on the shares provided by the adversary during the simulation. Then broadcast shares of the b_i's consistent with the shares of the corrupted players. These can be computed using Lagrange interpolation.

It should be fairly clear from the above steps, that an adversary will not be able to distinguish communicating with the real protocol and the functionality with simulator. Everything that is send back and forth, the secret key shares, public key shares, K_A keys and intermediate shares are distributed exactly the same and in the same order. □

We must also prove that security is still maintained in the original cryptosystem, and furthermore that correctness and security is maintained in the distributed decryption protocol "*Decrypt*". We abbreviate $F_{KeyGen'-and-Decrypt}$ by $F_{KG'-D}$ in the following. By the functionality $F_{KG'-D}$ we denote the functionality F_{KG-D} using $F_{KeyGen'}$ instead of F_{KeyGen}.

Theorem 5. *Assume we use $F_{KeyGen'}$ to generate a key pair pk, sk and the number of players u satisfies $u\binom{u}{t} < \sqrt[10]{q}/(2m)$. If* GapSVP *is hard in the worst case, encryption under pk is semantically secure against any polynomial time adversary who gets to interact with $F_{KeyGen'}$ during key generation. Moreover, the protocol Decrypt securely implements the slightly modified functionality $F_{KG'-D}$ when given access to $F_{KeyGen'}$, in particular, decryption under sk produces the correct plaintext except with negligible probability.*

Proof. For semantic security note that by previous arguments solving decision LWE is at least as hard as solving GapSVP. First note that a ciphertext is made out of the public key $(\mathbf{a}_i, b_i = \langle \mathbf{a}_i, \mathbf{s}\rangle + e_i)_{i=1}^m$, especially if the b_i's are random, ciphertexts contain no information on the plaintext. What we then show is, that if an adversary is able to distinguish a public key generated by $F_{KeyGen'}$ from a sample from the uniform distribution on $\mathbb{Z}_q^n \times \mathbb{Z}_q$, then we could use such an adversary to solve decision LWE. Now given an instance I of LWE pretend to run the $F_{KeyGen'}$ functionality with the adversary. Get the noise contributions from the adversary and add them to the LWE instance I. Return the instance as the public key to the adversary and output exactly what he outputs. If I contains uniform values so will the "public key" given to the adversary, if I is taken from $A_{\mathbf{s},\chi}$ then our output given to the adversary exactly matches the output of the real $F_{KeyGen'}$.

We will now argue that decryption is still correct except with negligible probability. Let $e + x$ be the reconstructed value after running the decryption protocol, we will then look at e. First note that the noise contributed by honest players is much smaller than that by corrupted players. We will look at the worst case where the public key is made entirely by corrupted players. We have $e = \sum_{i=1}^m r_i e_i = \sum_{i=1}^m \sum_{j=1}^u r_i e_{i,j}$, where each $e_{i,j}$ has potential size $2\binom{u}{t}\sqrt[3]{q}$. This leads to a worst case with $|e| = 2um\binom{u}{t}\sqrt[3]{q}$. According to Lemma 1 decryption will be correct if the probability that $|e| \geq \sqrt{q}$ is negligible. Therefore we get that decryption is correct if the equality $u\binom{u}{t} < \sqrt[6]{q}/(2m)$ is fulfilled.

Finally we argue that we can still simulate the execution of the protocol *Decrypt* now using the slightly modified $F_{KG'-D}$. The proof is essentially the same as the proof of Theorem 3, the only difference is that we should argue that the interval from which e is taken is still exponentially much smaller than the interval $[-\sqrt{q}, \sqrt{q}]$ from which x is taken. Following the argument from above we see that if we further limit the number of players, this can still be satisfied. Assume for instance that we limit e to the interval $[-\sqrt[2.5]{q}, \sqrt[2.5]{q}]$, this gives the requirement that the inequality $u\binom{u}{t} < \sqrt[10]{q}/(2m)$ should be fulfilled. □

7 Zero-Knowledge Proof of Plaintext Knowledge

In this section, we consider Regev's original cryptosystem, where the random choices and plaintext are binary and q is a prime. All arithmetic in this section is modulo q. In the appendix we describe a slightly more complicated scheme that works for our variant

We define the relation R_{Regev} as the set of pairs $\{x, w\}$ such that $x = ((\mathbf{a}_i, b_i)_{i=1}^m, (\mathbf{a}, b))$, and $w = (r_1, \ldots, r_m, \gamma)$ such that $(\mathbf{a}, b) = (\sum_{i=1}^m r_i \mathbf{a}_i, \gamma \cdot \lfloor \frac{q}{2} \rfloor + \sum_{i=1}^m r_i b_i)$. The language L_{Regev} will be the set of x for which there exist w with $(x, w) \in R_{Regev}$. Our goal is to build a zero-knowledge interactive proof for L_{Regev} which is also a proof of knowledge for R_{Regev}. In other words, the prover demonstrates that the ciphertext is well-formed and that he knows the plaintext and random coins that were used to form it.

We will use the technique from [IKOS07] where it was shown how to construct zero-knowledge proofs from multiparty computation protocols. We briefly sketch the idea: Assume we have a multiparty computation protocol π for input client I, players $P_1, \ldots, P_u$ and output client O, where I gets the prover's secret witness as input, shares it among the players, who then carry out a secure computation that verifies whether the witness is valid with respect to the public common input. The players send their results to O who outputs 1 or 0 accordingly. The protocol must be secure against a malicious adversary corrupting the clients and/or up to t of the other players. The prover now emulates π "in his head" and commits to the views of all players. Here, a view consists of the inputs and random coins of the player, and all received messages. The verifier selects a random subset of players among those that π can tolerate as corrupted sets[1]. The prover must open the corresponding commitments and the verifier checks that these views are consistent with each other and with the protocol and accepts or rejects accordingly.

The intuition is that the protocol is zero-knowlegde since π is secure even if the set chosen by the verifier is corrupted, and hence no information on the secret witness is released. The protocol is sound since if the witness is invalid, the prover must introduce some inconsistency to make it seem that π accepts the witness.

Indeed, it is shown in [IKOS07] that if π implements the function that checks the witness with perfect sercurity and if both u and t are $\theta(n)$, then the resulting two-party protocol has soundness error $2^{-\Omega(n)}$. It is honest verifier zero-knowledge, and can be made zero-knowledge in general, e.g., by generating the verifier's choice of subset to corrupt via a suitable coinflip protocol.

We make a couple of observations that are helpful in constructing a protocol π for our purposes: first, while broadcast is usually considered an expensive resource, it is virtually for free in this setting - any information π would broadcast can just be sent to the verifier immediately, as he would see it anyway no matter what subset is chosen. This was already noted in [IKOS07]. Second, π does

[1] The protocol must be secure against a corrupt I, but the verifier is of course not allowed to "open" I.

not have to guarantee termination, in the following sense: suppose all players broadcast some message in some round of π, and then all honest players decide (using the same procedure) whether to abort or continue. Suppose further that if all players have behaved honestly so far, we will never abort, and that further π has perfect correctness and privacy conditioned on the event that we do not abort. In this case, we can simply ask the verifier to reject if the prover sends a set of broadcast messages that would cause an abort. This will not hurt the honest prover, but will force a cheating prover to claim that he lets the virtual players behave such that π terminates.

In view of the above, all we have to do is to build an efficient protocol π that checks $r_1, \ldots, r_m, \gamma$ against $(\mathbf{a}_i, b_i)_{i=1}^m$ and $(\mathbf{a}, b)$. In order to do this, we need to borrow two tools from the design of efficient multiparty protocols, namely Packed Secret-Sharing[FY92] and Hyper-Invertible Matrices[BTH08], which we describe below.

7.1 Packed Secret Sharing

Packed Secret-Sharing is a generalization of standard Shamir sharing where secret values are assigned to more than one interpolation point. In other words, the secret to share is in fact a vector $(x_1, \ldots, x_\ell) \in \mathbb{Z}_q^l$. To do the sharing, we construct a random polynomial f of degree at most d, such that $f(0) = x_1, f(-1) = x_2, \ldots, f(-\ell+1) = x_\ell$. The shares are, as usual, $f(1), \ldots, f(u)$. To make this possible, and to guarantee privacy against t corrupted players, d must be at least $t+\ell-1$. In our case, we will choose $\ell = n+1$, and t to be $\theta(n)$. Furthermore, we will need that there are sufficiently many honest players such that their shares alone can determine a polynomial of degree $2d$, i.e., $u - t \geq 2(t+n+1)$. This shows that we can indeed choose u to be $\theta(n)$, as promised above.

Note that to ensure that we have enough distinct evaluation points, we need that if q is a prime, it must be larger than $\ell + u = n+1+u$ which is $\theta(n)$ or, in our construction of q for the threshold scheme, the smallest prime factor must be larger than $\ell + u$. This is already satisfied by the schemes as they stand.

We will write $[\mathbf{z}]_d$ for a set of shares determining a packed sharing of the block $\mathbf{z}$ using a polynomial of degree d.

Note that if players locally add respectively multiply their shares of blocks $\mathbf{z}, \mathbf{z}'$, this results in shares in the coordinate-wise sum respectively product, i.e., we have $[\mathbf{z}]_d + [\mathbf{z}']_d = [\mathbf{z}+\mathbf{z}']_d$, and $[\mathbf{z}]_d * [\mathbf{z}']_d = [\mathbf{z} * \mathbf{z}']_{2d}$, where $*$ denotes the coordinate-wise product.

7.2 Hyper-invertible Matrices

A hyper-invertible matrix M (with entries in $\mathbb{Z}_q$) has the property that any square submatrix of M is invertible. Such matrices can be constructed from Van der Monde matrices and were used in [BTH08] to check consistency of secret sharings with zero error probability. We briefly explain how this works:

Suppose M is a matrix with u rows and $u-t$ columns. Suppose the players hold $u-t$ sets of shares $[\mathbf{z}_1], \ldots, [\mathbf{z}_{u-t}]$, and we want to check that each set of shares is consistent with a polynomial of degree at most e. The players can locally compute u new sets of shares,

$$[M(\mathbf{z}_1,\ldots,\mathbf{z}_{u-t})_1],\ldots,[M(\mathbf{z}_1,\ldots,\mathbf{z}_{u-t})_u] := [\mathbf{y}_1],\ldots,[\mathbf{y}_u],$$

simply by multiplying M on the vector of $u-t$ shares that they hold (thinking of the shares as a column vector). Assume now that for $i = 1,\ldots,u$, each player sends his share in $[\mathbf{y}_i]$ to P_i. This allows P_i to check that the shares he receives are e-consistent, i.e., on a polynomial of degree at most e. P_i can now broadcast whether his check was OK or not.

We can see that if all players are happy, it means in particular that all honest players are happy, and that they therefore agree with all honest players on the set of $u-t$ e-consistent shares that they checked. I.e., $\{[\mathbf{y}_j]\}_{j\in H}$, where H is the set of honest players, are all e-consistent. Let M_H be the matrix we get from M by only taking the rows corresponding to players in H. This matrix is invertible by assumption on M, so we can obtain $[\mathbf{z}_1],\ldots,[\mathbf{z}_{u-t}]$ as a linear function defined by M_H^{-1} of the the shares in $\{[\mathbf{y}_j]\}_{j\in H}$, and hence the $[\mathbf{z}_i]$'s are all e-consistent as well.

Furthermore, if it is important that the shared information is kept secret, one can arrange the input shares such that only $[\mathbf{z}_1],\ldots,[\mathbf{z}_{u-2t}]$ contains information we want to protect, while $[\mathbf{z}_{u-2t+1}],\ldots,[\mathbf{z}_{u-t}]$ are chosen randomly using polynomials of degree at most e. These t random sets of shares will randomize the t sets of shares seen by corrupt players, again by hyper-invertibility of M. This also means that we do not need, for instance, $[\mathbf{z}_1]$ to be a *random* sharing of $\mathbf{z}_1$ to be able to hide it.

Note also that this method can be used to also check if $\mathbf{z}_1,\ldots,\mathbf{z}_{u-t}$ all satisfy some fixed condition, as long as what the condition asks is that each $\mathbf{z}_i$ satisfies some linear equation. For instance, we might want to check that $\mathbf{z}_i = (0,\ldots,0)$ for all i. This is done by having players verify that all $\mathbf{y}_i$ satisfy the same condition.

Regarding the complexity, it is easy to see that a set of shares of total size T bits can be verified while keeping the shared information perfectly private by sending $O(T)$ bits and creating random shares of size $O(T)$ bits.

7.3 The Multiparty Protocol

Recall that the secret witness to be checked consists of binary values $r_1,\ldots,r_m,\gamma$ where $(\mathbf{a},b) = (\sum_{i=1}^m r_i\mathbf{a}_i, \gamma\cdot\lfloor\frac{q}{2}\rfloor + \sum_{i=1}^m r_i b_i)$, and where the public information is public key $(\mathbf{a}_i,b_i)_{i=1}^m$ and ciphertext $(\mathbf{a},b)$. For any $z\in\mathbb{Z}_q$, we set $\overline{z} = (z,z,...,z)$, a vector of length $n+1$. The protocol works as follows:

Protocol VerifyCiphertext

1. The input client I sends shares $[\overline{r_i}]_d$, $i = 1,\ldots,m$ and $[\overline{\gamma}]_d$ to the players. In addition, it also sends random shares as required for the verifications below using the hyper-invertible matrix M.

2. Verify that $[\overline{r_1}]_d, \ldots, [\overline{r_m}]_d, [\overline{\gamma}]_d$ are d-consistent and that in each block shared, all $n+1$ entries are equal. If any player broadcasts "not OK", the protocol aborts.
3. Compute, using local multiplications, $[\overline{r_i(1-r_i)}]_{2d}$ for $i = 1, \ldots, m$ and $[\overline{\gamma(1-\gamma)}]_{2d}$.
4. Form sharings of the public vectors: $[(\mathbf{a}_i, b_i)]_d, i = 1, \ldots, m, [(0, \ldots, 0, \lfloor \frac{q}{2} \rfloor)]_d$, and $[(\mathbf{a}, b)]_d$ (using some default choice of polynomial of degree at most d). We then emulate the encryption on the shared values: compute, using local computation,

$$[\sum_{i=1}^{m} \overline{r_i} * (\mathbf{a}_i, b_i)]_{2d} + [(0, \ldots, 0, \lfloor \frac{q}{2} \rfloor) * \overline{\gamma}]_{2d} = [(\sum_{i=1}^{m} r_i \mathbf{a}_i, \sum_{i=1}^{m} r_i b_i + \gamma \lfloor \frac{q}{2} \rfloor)]_{2d}$$

From this, we locally subtract shares of the ciphertext $[(\mathbf{a}, b)]_d$, so we get

$$[(\sum_{i=1}^{m} r_i \mathbf{a}_i - \mathbf{a}, \sum_{i=1}^{m} r_i b_i + \gamma \lfloor \frac{q}{2} \rfloor) - b]_{2d} := [(\mathbf{z}, v)]_{2d}$$

5. Verify that $[\overline{r_1(1-r_1)}]_{2d}, \ldots, [\overline{r_m(1-r_m)}]_{2d}, [\overline{\gamma(1-\gamma)}]_{2d}$ and $[(\mathbf{z}, v)]_{2d}$ are indeed $2d$-consistent sharings of all-zero blocks. If any player broadcasts "not OK", the protocol aborts. This ensures that the r_i's and γ are binary, and that encryption results in the claimed ciphertext.

Since the verifications of shares work with zero error probability, it is clear that if the protocol terminates successfully, we are guaranteed that the shared values determine the correct ciphertext. No information on the secret is released, since the only communication is what is required for the verification of sharings, and we already argued above that these release no information on the shared values that we verify.

Regarding complexity, it is clear from inspection of the protocol that it is completely determined by the total size T of the sharings $[\overline{r_i}]_d$, $i = 1, \ldots, m$ and $[\overline{\gamma}]_d$, in particular, the total size of communication is $O(T)$. We have that T is $O(mu \log q)$ which is $O(mn \log q)$. Note that the size of the public key is also $O(mn \log q)$.

It is described in [IKOS07] how to transform this protocol into a zero-knowledge proof using an unconditionally binding commitment scheme. If this scheme allows us to commit to strings with an additive length increase that is independent of the string length, we can preserve the efficiency of the multiparty protocol. An unconditionally hiding commitment scheme is also needed, for the verifier to commit to his challenge. This gives us:

Theorem 6. *Given an unconditionally binding and an unconditionally hiding commitment scheme with constant additive overhead, using protocol VerifyCiphertext in the construction from [IKOS07] produces a two-party zero-knowledge proof for L_{Regev}. The protocol has communication complexity $O(mn \log q)$ bits and error probability $2^{-\Omega(n)}$.*

We can base the commitment schemes needed on lattice problems, thus using assumptions we would need anyway. An efficient unconditionally binding scheme follows from the cryptosystem in[PVW08], while an unconditionally hiding scheme can be based on any collision intractable hash function [DPP98], which in turn can be based on lattice assumptions.

In [IKOS07], it was not shown that their construction is a proof of knowledge for R_{Regev}. However, for the honest verifier zero-knowledge version of the protocol, one can do a rewinding argument to show that it is indeed a proof of knowledge with negligible knowledge error. If we go to the version that is zero-knowledge in general, things are different, since the construction from [IKOS07] has the verifier commit to his challenges, which means rewinding the prover is not possible unless the extractor can equivocate these commitments.

However, in the common reference string model, we can easily make the protocol be a proof of knowledge for R_{Regev}, by having a public key for a commitment scheme placed in the reference string, e.g., a public key for the cryptosystem from [PVW08], and the prover uses these for committing to the views. If the extractor knows the corresponding secret key, it can extract all committed views without rewinding and easily compute the secret.

References

[BTH08] Beerliová-Trubíniová, Z., Hirt, M.: Perfectly-secure MPC with linear communication complexity. In: Canetti, R. (ed.) TCC 2008. LNCS, vol. 4948, pp. 213–230. Springer, Heidelberg (2008)

[Can01] Canetti, R.: Universally composable security: A new paradigm for cryptographic protocols. In: FOCS, pp. 136–145 (2001)

[CDI05] Cramer, R., Damgård, I., Ishai, Y.: Share conversion, pseudorandom secret-sharing and applications to secure computation. In: Kilian, J. (ed.) TCC 2005. LNCS, vol. 3378, pp. 342–362. Springer, Heidelberg (2005)

[DPP98] Damgård, I., Pedersen, T.P., Pfitzmann, B.: Statistical secrecy and multibit commitments. IEEE Transactions on Information Theory 44(3), 1143–1151 (1998)

[FY92] Franklin, M.K., Yung, M.: Communication complexity of secure computation (extended abstract). In: STOC, pp. 699–710 (1992)

[IKOS07] Ishai, Y., Kushilevitz, E., Ostrovsky, R., Sahai, A.: Zero-knowledge from secure multiparty computation. In: STOC, pp. 21–30 (2007)

[MR08] Micciancio, D., Regev, O.: Lattice-based cryptography. In: Bernstein, D.J., Buchmann, J. (eds.) Post-quantum Cryprography. Springer, Heidelberg (2008)

[Pei09] Peikert, C.: Public-key cryptosystems from the worst-case shortest vector problem: extended abstract. In: STOC, pp. 333–342 (2009)

[PVW08] Peikert, C., Vaikuntanathan, V., Waters, B.: A framework for efficient and composable oblivious transfer. In: Wagner, D. (ed.) CRYPTO 2008. LNCS, vol. 5157, pp. 554–571. Springer, Heidelberg (2008)

[Reg05] Regev, O.: On lattices, learning with errors, random linear codes, and cryptography. In: STOC, pp. 84–93 (2005)

[Sha79] Shamir, A.: How to share a secret. Commun. ACM 22(11), 612–613 (1979)

A Zero-Knowledge Proof When q Is Not Prime

The only part of the multiparty protocol underlying our zero-knowledge proof that does not work when q is not a prime is the step where it is verified that the r_i are binary, essentially by verifying that $r_i(1 - r_i) \bmod q = 0$. Of course, this check is not good if q is not prime. We sketch a procedure that can be used instead, but has only statistical security:

The input client I supplies $[\overline{r_i}]_d$ and it is checked as in the original protocol that a block has been shared where all entries are equal. Note that if the sharing was correctly formed, it would be the case that $\mathbf{r}' = 2\overline{r_i} - (1, \dots, 1)$ would be $(1, \dots, 1)$ or $(-1, \dots, -1)$. I also supplies a sharing $[\mathbf{z}]_d = [(z_1, \dots, z_{n+1}]_d$ such that all z_i are randomly chosen to be 1 or -1. Finally, a public random challenge is generated: $\mathbf{v} = (v_1, \dots, v_{n+1})$, where each v_i is 0 or 1. (When transforming this to a 2-party protocol, we let the verifier generate the challenge). We compute (locally)

$$[\mathbf{r}' * \mathbf{z} * \mathbf{v} + \mathbf{z} * (\overline{1} - \mathbf{v})]_{3d}.$$

Finally we add a random degree $3d$-sharing of the all-zero block and open the result. The opened block must contain only 1's and -1's. Put another way, the opening shows us, in each coordinate position, an entry from $\mathbf{z}$ or from $\mathbf{r}' * \mathbf{z}$ and they must all be ± 1.

For privacy, the intuition is that by random choice of $\mathbf{z}$, $\mathbf{r}' * \mathbf{z}$ has no information on $\mathbf{r}'$ and neither does $\mathbf{z}$, so seeing, for each index i, the i'th entry of $\mathbf{r}' * \mathbf{z}$ or $\mathbf{z}$ reveals nothing on $\mathbf{r}'$.

For correctness, if there is just a single position in which both $\mathbf{z}$ and $\mathbf{r}' * \mathbf{z}$ are ± 1, $\mathbf{r}'$ will be ± 1 in that position too, and this implies that the original r_i was 0 or 1. On the other hand, if no such position exists, the honest players will accept $[\overline{r_i}]$ with probability only 2^{-n-1}, by the assumed randomness of $\mathbf{v}$.

Ideal Hierarchical Secret Sharing Schemes*

Oriol Farràs and Carles Padró

Universitat Politècnica de Catalunya, Barcelona, Spain

Abstract. Hierarchical secret sharing is among the most natural generalizations of threshold secret sharing, and it has attracted a lot of attention from the invention of secret sharing until nowadays. Several constructions of ideal hierarchical secret sharing schemes have been proposed, but it was not known what access structures admit such a scheme. We solve this problem by providing a natural definition for the family of the hierarchical access structures and, more importantly, by presenting a complete characterization of the ideal hierarchical access structures, that is, the ones admitting an ideal secret sharing scheme. Our characterization deals with the properties of the hierarchically minimal sets of the access structure, which are the minimal qualified sets whose participants are in the lowest possible levels in the hierarchy. By using our characterization, it can be efficiently checked whether any given hierarchical access structure that is defined by its hierarchically minimal sets is ideal. We use the well known connection between ideal secret sharing and matroids and, in particular, the fact that every ideal access structure is a matroid port. In addition, we use recent results on ideal multipartite access structures and the connection between multipartite matroids and integer polymatroids. We prove that every ideal hierarchical access structure is the port of a representable matroid and, more specifically, we prove that every ideal structure in this family admits ideal *linear* secret sharing schemes over fields of all characteristics. In addition, methods to construct such ideal schemes can be derived from the results in this paper and the aforementioned ones on ideal multipartite secret sharing. Finally, we use our results to find a new proof for the characterization of the ideal weighted threshold access structures that is simpler than the existing one.

Keywords: Secret sharing, Ideal secret sharing schemes, Hierarchical secret sharing, Weighted threshold secret sharing, Multipartite secret sharing, Multipartite matroids, Integer polymatroids.

1 Introduction

A *secret sharing scheme* is a method to distribute *shares* of a *secret value* among a set of *participants*. Only the *qualified* subsets of participants can recover the secret value from their shares, while the *unqualified* subsets do not obtain any information about the secret value. The qualified subsets form the *access structure*

* The authors' work was partially supported by the Spanish Ministry of Education and Science under project TSI2006-02731.

D. Micciancio (Ed.): TCC 2010, LNCS 5978, pp. 219–236, 2010.

of the scheme, which is a monotone increasing family of subsets of participants. Only *unconditionally secure perfect* secret sharing schemes are considered in this paper.

Secret sharing was independently introduced by Shamir [32] and Blakley [4] in 1979. They presented two different methods to construct secret sharing schemes for *threshold access structures*, whose qualified subsets are those with at least some given number of participants. These schemes are *ideal*, that is, the length of every share is the same as the length of the secret, which is the best possible situation [14].

There exist scenarios in which non-threshold secret sharing schemes are required because, for instance, some participants should be more powerful than others. The first attempt to overcome the limitation of threshold access structures was made by Shamir in his seminal work [32] by proposing a simple modification of the threshold scheme. Namely, every participant receives as its share a certain number of shares from a threshold scheme, according to its position in the hierarchy. In this way a scheme for a *weighted threshold access structure* is obtained. That is, every participant has a weight (a positive integer) and a set is qualified if and only if its weight sum is at least a given threshold. This new scheme is not ideal because the shares are in general larger than the secret.

Every access structure admits a secret sharing scheme [3,13], but in general the shares must be larger than the secret [7,9]. Very little is known about the optimal length of the shares in secret sharing schemes for general access structures, and there is a wide gap between the best known general lower and upper bounds.

Because of the difficulty (presumably, impossibility) of finding efficient secret sharing schemes for general access structures, the construction of ideal secret sharing schemes for families of access structures with interesting properties for the applications of secret sharing is worth considering. This line of work was initiated by Simmons [33], who proposed two families of access structures, the *multilevel* and the *compartmented* ones, and conjectured them to admit ideal secret sharing schemes. The multilevel and compartmented access structures are *multipartite*, which means that the participants are divided into several parts (levels or compartments) and all participants in the same part play an equivalent role in the structure. In addition, in a multilevel access structure, the participants are hierarchically ordered, and the participants in higher levels are more powerful than the ones in lower levels. Multipartite and, in particular, hierarchical secret sharing are the most natural generalization of threshold secret sharing.

Brickell [5] proposed a general method, based on linear algebra, to construct ideal secret sharing schemes for access structures that are not necessarily threshold, and he applied it to the construction of particular ideal secret sharing schemes proving the conjecture by Simmons. By using different kinds of polynomial interpolation, Tassa [35], and Tassa and Dyn [36] proposed constructions of ideal secret sharing schemes for several families of multipartite access structures, some of them with hierarchical properties. These constructions are based on the general linear algebra method by Brickell [5], but they provide schemes for the multilevel and compartmented access structures that are simpler and more efficient than the particular ones proposed in [5] for those structures. Other constructions of ideal multipartite secret sharing schemes have been presented in [11,26].

In spite of all those constructions of ideal hierarchical secret sharing schemes, it was not known what access structures admit such a scheme. This natural question, which is solved in this paper, is related to the more general problem of determining what access structures admit an ideal secret sharing scheme, that is, the characterization of the *ideal access structures*. This is a very important and long-standing open problem in secret sharing. Brickell and Davenport [6] proved that every ideal secret sharing scheme defines a matroid. Actually, this matroid is univocally determined by the access structure of the scheme. This implies a necessary condition for an access structure to be ideal. Namely, every ideal access structure is a *matroid port*. A sufficient condition is obtained from the method to construct ideal secret sharing schemes by Brickell [5]: the ports of representable matroids are ideal access structures. The results in [6] have been generalized in [16] by proving that, if all shares in a secret sharing scheme are shorter than 3/2 times the secret value, then its access structure is a matroid port. At this point, the remaining open question about the characterization of ideal access structures is determining the matroids that can be defined from ideal secret sharing schemes. Some important results, ideas and techniques to solve this question have been given by Matúš [20,21].

In addition to the search of general results, several authors studied this open problem for particular families of access structures. Some of them deal with families of multipartite access structures. Beimel, Tassa and Weinreb [1] presented a characterization of the ideal weighted threshold access structures that generalizes the partial results in [22,29]. Another important result about weighted threshold access structures have been obtained recently by Beimel and Weinreb [2]. They prove that all such access structures admit secret sharing schemes in which the size of the shares is quasi-polynomial in the number of users. A complete characterization of the ideal bipartite access structures was given in [29], and related results were given independently in [25,27]. Partial results on the characterization of the ideal tripartite access structures appeared in [8,11], and this question was solved in [10]. In every one of these families, all matroid ports are ports of representable matroids, and hence, all ideal access structures are *vector space access structures*, that is, they admit an ideal linear secret sharing scheme constructed by the method proposed by Brickell [5].

The characterization of the ideal tripartite access structures in [10] was obtained actually from the much more general results about ideal multipartite access structures in that paper. Pointing out the close connection between multipartite matroids and integer polymatroids, specially the characterization of this combinatorial object given by Herzog and Hibi [12], and the use for the first time in secret sharing of these concepts are among the main contributions in [10]. The basic definitions and facts about integer polymatroids and the main results in [10] are recalled in Section 4.

This paper deals with the two lines of work in secret sharing that have been discussed previously: first, the construction of ideal secret sharing schemes for useful classes of access structures, in particular the ones with hierarchical properties, and second, the characterization of ideal access structures. In this paper we solve a question that is interesting for both lines of research. Namely, what hierarchical access structures admit an ideal secret sharing scheme?

First of all, we formalize the concept of *hierarchical access structure* by introducing in Section 3 a natural definition for it. Basically, if a participant in a qualified subset is substituted by a *hierarchically superior* participant, the new subset must be still qualified. An access structure is *hierarchical* if, for any two given participants, one of them is hierarchically superior to the other. According to this definition, the family of the hierarchical access structures contains the multilevel access structures [5,33], the hierarchical threshold access structures studied by Tassa [35] and by Tassa and Dyn [36], and also the weighted threshold access structures that were first considered by Shamir [32] and studied in [1,2,22,29]. Duality and minors of access structures are fundamental concepts in secret sharing, as they are in matroid theory. Several important classes of access structures are closed by duality and minors, as for instance, matroid ports or $\mathbb{K}$-vector space access structures. Similarly to multipartite and weighted threshold access structures, the family of the hierarchical access structures is closed by duality and minors. This is discussed in Section 3.

Our main result is Theorem 16, which provides a complete characterization of the ideal hierarchical access structures. In particular, we prove that all hierarchical matroid ports are ports of representable matroids. By combining this with the results in [16], we obtain the following theorem.

Theorem 1. *Let Γ be a hierarchical access structure. The following properties are equivalent:*

1. *Γ admits a vector space secret sharing scheme over every large enough finite field.*
2. *Γ is ideal.*
3. *Γ admits a secret sharing scheme in which the length of every share is less than $3/2$ times the length of the secret value.*
4. *Γ is a matroid port.*

This generalizes the analogous statement that holds for weighted threshold access structures as a consequence of the results in [1,16]. Actually, as an application of our results, we present in Section 8 a new proof of the characterization of the ideal weighted threshold access structures that simplifies the complicated proof given by Beimel, Tassa and Weinreb [1].

Our starting point is the observation that every hierarchical access structure is determined by its *hierarchically minimal sets*, which are the minimal qualified sets that become unqualified if any participant is replaced by another one in a lower level in the hierarchy. Our results strongly rely on the connection between matroids and ideal secret sharing schemes discovered by Brickell and Davenport [6]. Moreover, since hierarchical access structures are in particular multipartite, the results and techniques in [10] about the characterization of ideal multipartite access structures, which are recalled in Section 4, are extremely useful. In particular, integer polymatroids play a fundamental role. Another important tool is the geometric representation introduced in [10,29] for multipartite access structures, which is adapted in Section 3 to the hierarchical case by introducing the *hierarchically minimal points* (or *h-minimal points* for short) that represent the hierarchically minimal sets. Our characterization of the ideal hierarchical access structures is given in terms of some properties of the h-minimal points that can be efficiently checked. By using our results, given a hierarchical access structure that is described by its h-minimal points, one can efficiently determine whether it is ideal or not. If the access structure is described by its minimal qualified subsets, it is easy to determine the h-minimal points. If the access structure is described in another way, one has to find the h-minimal points, but this can be done efficiently most of the times. This is the case, for instance, of weighted threshold access structures that are determined by the weights and the threshold. Moreover, by combining the results in this paper with the ones on ideal multipartite secret sharing in [10], a method to construct an ideal linear secret sharing scheme for every given ideal hierarchical access structure can be obtained. A more detailed study of this method and the analysis of its efficiency is deferred to future work.

2 Ideal Secret Sharing Schemes and Matroids

We recall in this section some facts about the connection between ideal secret sharing schemes and matroids that is derived from the results by Brickell [5] and by Brickell and Davenport [6]. See [16], for instance, for more information on these topics.

We begin by presenting the method by Brickell [5] to construct ideal secret sharing schemes as described by Massey [18,19] in terms of linear codes. Let C be an $[n+1,k]$-linear code over a finite field $\mathbb{K}$ and let M be a generator matrix of C, that is, a $k \times (n+1)$ matrix over $\mathbb{K}$ whose rows span C. Such a code defines an ideal secret sharing scheme on a set $P = \{p_1, \dots, p_n\}$ of participants. Specifically, every random choice of a codeword $(s_0, s_1, \dots, s_n) \in C$ corresponds to a distribution of shares for the secret value $s_0 \in \mathbb{K}$, in which $s_i \in \mathbb{K}$ is the share of the participant p_i. Such an ideal scheme is called a $\mathbb{K}$-*vector space secret sharing scheme* and its access structures is called a $\mathbb{K}$-*vector space access structure*. It is easy to check that a set $A \subseteq P$ is in the access structure Γ of this scheme if and only if the column of M with index 0 is a linear combination of the columns whose indices correspond to the players in A. Therefore, if $Q = P \cup \{p_0\}$ and $\mathcal{M}$ is the representable matroid with ground set Q and rank function r that is defined by the columns of the matrix M, then $\Gamma = \Gamma_{p_0}(\mathcal{M}) = \{A \subseteq P : r(A \cup \{p_0\}) = r(A)\}$. That is, Γ is the *port of the matroid* $\mathcal{M}$ *at the point* p_0. Consequently, a sufficient condition for an access structure to be ideal is obtained. Namely, the ports of representable matroids are ideal access structures. Actually, they coincide with the vector space access structures.

As a consequence the results by Brickell and Davenport [6], this sufficient condition is not very far from being necessary. Specifically, they proved that every ideal access structure is a matroid port.

With a slightly different definition, matroid ports were introduced in 1964 by Lehman [15] to solve the *Shannon switching game*, much before secret sharing was invented by Shamir [32] and Blakley [4] in 1979. A forbidden minor characterization of matroid ports was given by Seymour [31]. Even though the results in [5,6] deal with matroid ports, this terminology was not used in those and many other subsequent works on secret sharing. The old results on matroid ports in [15,31] were rediscovered for secret sharing by Martí-Farré and Padró [16], who used them to generalize the result by Brickell and Davenport by proving that, if all shares in a secret sharing scheme are shorter than 3/2 times the secret, then its access structure is a matroid port.

3 Hierarchical Access Structures

We present here a natural definition for the family of the *hierarchical access structures*, which embraces all possible situations in which there is a hierarchy on the set of participants. For instance, the weighted threshold access structures and the hierarchical threshold access structures [35] are contained in this new family. Hierarchical access structures are in particular multipartite. Therefore, we can take advantage of the results and techniques in [10] about the characterization of ideal multipartite access structures. Moreover, the geometric representation for multipartite access structures that was introduced in [10,29] will be very useful as well for our purposes. This representation is adapted here to hierarchical access structures by introducing the *hierarchically minimal* points.

Let Γ be an access structure on a set P of participants. We say that the participant $p \in P$ is *hierarchically superior* to the participant $q \in P$, and we write $q \preceq p$, if $A \cup \{p\} \in \Gamma$ for every subset $A \subseteq P \smallsetminus \{p,q\}$ with $A \cup \{q\} \in \Gamma$. An access structure is said to be *hierarchical* if all participants are hierarchically related, that is, for every pair of participants $p, q \in P$, either $q \preceq p$ or $p \preceq q$. If $p \preceq q$ and $q \preceq p$, we say that these two participants are *hierarchically equivalent*. Clearly, this is an equivalence relation, and the hierarchical relation $\preceq$ induces an order on the set of the equivalence classes. Observe that an access structure is hierarchical if and only if this is a total order.

For a set P, a sequence $\Pi = (P_1, \ldots, P_m)$ of subsets of P is called here a *partition* of P if $P = P_1 \cup \cdots \cup P_m$ and $P_i \cap P_j = \emptyset$ whenever $i \neq j$. Observe that some of the parts may be empty. An access structure Γ is said to be *Π-partite* if every pair of participants in the same part P_i are hierarchically equivalent. A different but equivalent definition for this concept is given in [10]. If m is the number of parts in Π, such structures are called *m-partite access structures*. The participants that are not in any minimal qualified subset are called *redundant*. An m-partite access structure is said to be *strictly m-partite* if there are no redundant participants, all parts are nonempty, and participants in different parts are not hierarchically equivalent.

A Π-partite access structure is said to be *Π-hierarchical* if $q \preceq p$ for every pair of participants $p \in P_i$ and $q \in P_j$ with $i < j$. That is, the participants in the first level are hierarchically superior to those in the second level and so on. Obviously, an access structure is hierarchical if and only if it is Π-hierarchical for some partition Π of the set of participants. The term *m-hierarchical access structure* applies to every Π-hierarchical access structure with $|\Pi| = m$.

Some notation is needed to recall the geometric representation of multipartite access structures introduced in [10,29]. This notation will be used as well to present in Section 4 the basic facts about integer polymatroids and, because of that, all through the paper. Consider a finite set J. For every two points $u = (u_i)_{i\in J}$ and $v = (v_i)_{i\in J}$ in $\mathbb{Z}^J$, we write $u \le v$ if $u_i \le v_i$ for every $i \in J$. The point $w = u \vee v$ is defined by $w_i = \max\{u_i, v_i\}$ for every $i \in J$. The *modulus* of a point $u \in \mathbb{Z}^J$ is $|u| = \sum_{i\in J} u_i$. For every subset $X \subseteq J$, we notate $u(X) = (u_i)_{i\in X} \in \mathbb{Z}^X$ and $|u(X)| = \sum_{i\in X} u_i$. We notate $\mathbb{Z}_+$ and $\mathbb{Z}_-$ for the sets of the non-negative and the non-positive integers, respectively.

For each partition $\Pi = (P_1, \ldots, P_m)$ of the set P, we consider a mapping $\Pi\colon \mathcal{P}(P) \to \mathbb{Z}_+^m$ defined by $\Pi(A) = (|A \cap P_1|, \ldots, |A \cap P_m|) \in \mathbb{Z}_+^m$. We write $\mathbf{p} = \Pi(P) = (|P_1|, \ldots, |P_m|)$ and $\mathbf{P} = \Pi(\mathcal{P}(P)) = \{u \in \mathbb{Z}_+^m : u \le \mathbf{p}\}$. For a Π-partite access structure $\Gamma \subseteq \mathcal{P}(P)$, consider $\Pi(\Gamma) = \{\Pi(A) : A \in \Gamma\} \subseteq \mathbf{P}$. Observe that $A \in \Gamma$ if and only if $\Pi(A) \in \Pi(\Gamma)$, so Γ is univocally represented by the set of points $\Pi(\Gamma) \subseteq \mathbf{P}$. By an abuse of notation, we will use Γ to denote both a Π-partite access structure on P and the corresponding set $\Pi(\Gamma)$ of points in $\mathbf{P}$.

Let Γ be a Π-partite access structure on P. If two points $u, v \in \mathbf{P}$ are such that $u \le v$ and $u \in \Gamma$, then $v \in \Gamma$. This is due to the fact that Γ is a monotone increasing family of subsets. Therefore, $\Gamma \subseteq \mathbf{P}$ is determined by the family $\min \Gamma \subseteq \mathbf{P}$ of its minimal points. We are using here an abuse of notation as well, because $\min \Gamma$ denotes also the family of minimal subsets of the access structure Γ.

Let Γ be a Π-hierarchical access structure. If a set $B \subseteq P$ is obtained from a set $A \subseteq P$ by replacing some participants by participants in superior levels and $u = \Pi(A)$ and $v = \Pi(B)$, then $\sum_{i=1}^{j} u_i \le \sum_{i=1}^{j} v_i$ for every $j = 1, \ldots, m$. This motivates the following order relation, which was introduced in [36, Definition 4.2], also in the framework of hierarchical secret sharing. We say that the point $v \in \mathbb{Z}_+^m$ is *hierarchically superior* to the point $u \in \mathbb{Z}_+^m$, and we write $u \preceq v$, if $\sum_{i=1}^{j} u_i \le \sum_{i=1}^{j} v_i$ for every $j = 1, \ldots, m$. The points in $\mathbf{P}$ that are minimal according to this order are called the *hierarchically minimal points* (or *h-minimal points* for short) of Γ, and the set of these points is denoted by $\operatorname{hmin} \Gamma$. The *hierarchically minimal sets of Γ* are the sets $A \subseteq P$ such that $\Pi(A)$ is a hierarchically minimal point. Clearly, if $u, v \in \mathbf{P}$ are such that $u \in \Gamma$ and $u \preceq v$, then $v \in \Gamma$. This implies that every Π-hierarchical access structure is determined by the partition Π and its h-minimal points. Since $u \preceq v$ if $u \le v$, we have that $\operatorname{hmin} \Gamma \subseteq \min \Gamma$, and hence describing a hierarchical access structure by its h-minimal points is more compact than doing so by its minimal points. Observe that a subset of participants is hierarchically minimal if and only if it is a minimal qualified subset such that it is impossible to replace a participant in it with another participant in an inferior level and still remain qualified.

We present next three examples of families of hierarchical access structures. For all of them, we consider the same m-partition $\Pi = (P_1, \ldots, P_m)$ of the set p of participants.

Example 2. A *weighted threshold* access structure Γ is defined from a real *weight vector* $w = (w_1, \ldots, w_m) \in \mathbb{R}^m$ with $w_1 > w_2 > \cdots > w_m > 0$ and a positive real *threshold* $T > 0$. Namely, Γ is the Π-partite access structure defined by $\Gamma = \{u \in \mathbf{P} : u_1 w_1 + \cdots + u_m w_m \geq T\} \subseteq \mathbf{P}$. That is, every participant has a weight and a set is qualified if and only if its weight sum is at least the threshold. Clearly, such an access structure is Π-hierarchical.

Example 3. Brickell [5] showed how to construct ideal schemes for the multilevel structures proposed by Simmons [33]. These access structures are of the form $\Gamma = \{A \subseteq P : |A \cap (\cup_{j=1}^{i} P_j)| \geq t_i$ for some $i = 1, \ldots, m\}$ for some monotone increasing sequence of integers $0 < t_1 < \ldots < t_m$. Clearly, such an access structure is Π-hierarchical and, if the number of participants in each level is large enough, its h-minimal points are $\operatorname{hmin} \Gamma = \{t_1 \mathbf{e}^1, \ldots, t_m \mathbf{e}^m\}$, where $\mathbf{e}^i$ is the i-th vector of the canonical basis of $\mathbb{R}^m$.

Example 4. Another family of hierarchical threshold access structures was proposed by Tassa [35]. Given integers $0 < t_1 < \ldots < t_m$, they are defined by $\Gamma = \{A \subseteq P : |A \cap (\cup_{j=1}^{i} P_j)| \geq t_i$ for every $i = 1, \ldots, m\}$. Such an access structure is Π-hierarchical and, if the number of participants in every level is large enough, its only h-minimal point is $(t_1, t_2 - t_1, \ldots, t_m - t_{m-1})$.

Duality and minors are fundamental concepts in secret sharing, as they are in matroid theory. Several important classes of access structures are closed by duality and minors, as for instance, matroid ports or $\mathbb{K}$-vector space access structures. More information about these operations on access structures and their relevance in secret sharing can be found in [16]. The *dual* of an access structure Γ on a set P is the access structure on the same set defined by $\Gamma^* = \{A \subseteq P : P \smallsetminus A \notin \Gamma\}$. For a subset $B \subseteq P$, we define the access structures $\Gamma \backslash B$ and Γ / B on the set $P \smallsetminus B$ by $\Gamma \backslash B = \{A \subseteq P \smallsetminus B : A \in \Gamma\}$ and $\Gamma / B = \{A \subseteq P \smallsetminus B : A \cup B \in \Gamma\}$. Every access structure that can be obtained from Γ by repeatedly applying the operations $\backslash$ and $/$ is called a *minor* of Γ. The proof of the following proposition is straightforward.

Proposition 5. *The class of the hierarchical access structures is minor-closed and duality-closed. The same applies to the class of the weighted threshold access structures.*

Let P' and P'' be two disjoint sets and let Γ' and Γ'' be access structures on P' and P'', respectively. The *composition of* Γ' *and* Γ'' *over* $p \in P'$ is denoted by $\Gamma'[\Gamma''; p]$ and is defined as the access structure on the set of participants $P = P' \cup P'' \smallsetminus \{p\}$ that is formed by all subsets $A \subseteq P$ such that $A \cap P' \in \Gamma'$ and all subsets $A \subseteq P$ such that $(A \cup \{p\}) \cap P' \in \Gamma'$ and $A \cap P'' \in \Gamma''$. The composition of matroid ports is a matroid port, and the same applies to $\mathbb{K}$-vector space access structures. A proof for these facts can be found in [17]. The access structures that can be expressed as the composition of two access structures on sets with at least two participants are called *decomposable*.

Suppose that Γ' is $(P_1, \ldots, P_r)$-partite and Γ'' is $(P_{r+1}, \ldots, P_{r+s})$-partite, and take $p \in P_r$. Then the composition $\Gamma'[\Gamma''; p]$ is $(P'_1, \ldots, P'_{r+s})$-partite with $P'_r = P_r \setminus \{p\}$ and $P'_i = P_i$ if $i \neq r$. If Γ' and Γ'' are hierarchical, then $\Gamma'[\Gamma''; p]$ is also hierarchical. Observe that the composition is made over a participant in the lowest level of Γ'.

4 Multipartite Matroid Ports and Integer Polymatroids

The aim of this and the following sections is to present our main result, Theorem 16, which is a complete characterization of the ideal hierarchical access structures in terms of the properties of their h-minimal points. First we recall here some facts about integer polymatroids and we show the connection between these combinatorial objects and multipartite matroids and their ports. Since all ideal access structures are matroid ports, we obtain in this way some necessary conditions for a hierarchical access structure to be ideal in Section 5. Finally, in Sections 6 and 7 we show that these necessary conditions are also sufficient.

Multipartite matroid ports are ports of *multipartite matroids*, and those matroids are closely related to *integer polymatroids*. We recall here some definitions and basic facts about integer polymatroids and multipartite matroids, the relation between these two combinatorial objects, and their connections to the characterization of multipartite access structures. We use in the following the notation for integer vectors that was introduced in Section 3. More information about these concepts can be found in [10,12].

Similarly to matroids, integer polymatroids can be defined in many different but equivalent ways. We present next the three of those definitions that are needed to present our results. The first one is in terms of an integer submodular rank function. The second one considers an integer polymatroid as a set of integer vectors with certain properties. Finally, the third one is given in terms of the integer bases, which are the maximal elements in that set of integer vectors. The equivalence between these definitions is a consequence of results on submodular functions that are well known in the areas of combinatorial optimization and discrete convex analysis (see, for instance, the works by Murota [23,24]). A full proof of this equivalence has been presented by Herzog and Hibi [12], who used integer polymatroids in commutative algebra. The formalization of these combinatorial concepts presented in [12] has been very useful for our purposes. Actually, a new term (*discrete polymatroid*) was introduced in [12] to denote the set of integer vectors defining an integer polymatroid. In our opinion, this new term is not needed because these sets should be considered as an alternative way to define integer polymatroids, and not as a new combinatorial object. Actually, they are formed by the integer points in the convex polytope associated to the integer polymatroid. See [37], for instance, for more information about polymatroids and their associated polytopes.

We notate $\mathcal{P}(J)$ for the power set of a set J. An *integer polymatroid* is an ordered pair $\mathcal{Z} = (J, h)$, where J is a finite set, the *ground set*, and h, the *rank function*, is a mapping $h \colon \mathcal{P}(J) \to \mathbb{Z}$ satisfying the following properties

1. $h(\emptyset) = 0$.
2. h is *monotone increasing*: if $X \subseteq Y \subseteq J$, then $h(X) \le h(Y)$.
3. h is *submodular*: if $X, Y \subseteq J$, then $h(X \cup Y) + h(X \cap Y) \le h(X) + h(Y)$.

An integer polymatroid with ground set J can be defined as well as a nonempty finite set $\mathcal{D} \subseteq \mathbb{Z}_+^J$ of integer points satisfying the following properties.

1. If $u \in \mathcal{D}$ and $v \in \mathbb{Z}_+^J$ is such that $v \le u$, then $v \in \mathcal{D}$.
2. For every pair of points $u, v \in \mathcal{D}$ with $|u| < |v|$, there exists $w \in \mathcal{D}$ with $u < w \le u \vee v$.

The equivalence between these two definitions can be proved as follows. On the one hand, one has to check that, given an integer polymatroid $\mathcal{Z} = (J, h)$, such a set of integer points is univocally determined by the rank function by $\mathcal{D} = \mathcal{D}(\mathcal{Z}) = \{u \in \mathbb{Z}_+^J : |u(X)| \le h(X) \text{ for every } X \subseteq J\}$. On the other hand, it can be proved that, given a set $\mathcal{D} \subseteq \mathbb{Z}_+^J$ satisfying the properties above, there is a unique integer polymatroid $\mathcal{Z} = (J, h)$ with $\mathcal{D} = \mathcal{D}(\mathcal{Z})$, and its rank function is defined by $h(X) = \max\{|u(X)| : u \in \mathcal{D}\}$ for every $X \subseteq J$.

An *integer basis* of an integer polymatroid $\mathcal{Z}$ is a maximal element in $\mathcal{D}(\mathcal{Z})$, that is, a point $u \in \mathcal{D}$ such that there does not exist any $v \in \mathcal{D}$ with $u < v$. Since we are not going to consider here any other kind of bases of integer polymatroids, from now on integer bases will be called simply bases. Similarly to matroids, all bases have the same modulus, and integer polymatroids are completely determined by their bases. Moreover, a nonempty set $\mathcal{B} \subseteq \mathbb{Z}_+^J$ is the family of bases of an integer polymatroid with ground set J if and only if it satisfies the following *exchange condition*.

- For every $u \in \mathcal{B}$ and $v \in \mathcal{B}$ with $u_i > v_i$, there exists $j \in J$ such that $u_j < v_j$ and $u - \mathbf{e}^i + \mathbf{e}^j \in \mathcal{B}$, where $\mathbf{e}^i \in \mathbb{Z}^J$ is such that $\mathbf{e}^i_k = 0$ if $i \ne k$ and $\mathbf{e}^i_i = 1$.

Because of that, this can be seen as another definition of integer polymatroid.

For an integer polymatroid $\mathcal{Z} = (J, h)$ and a subset $X \subseteq J$, we consider the integer polymatroid $\mathcal{Z}(X) = (X, h')$ defined by $h'(Y) = h(Y)$ for every $Y \subseteq X$. Since h' is a restriction of h, both will be usually denoted by h. Clearly, $\mathcal{D}(\mathcal{Z}(X)) = \{u(X) : u \in \mathcal{D}(\mathcal{Z})\} \subseteq \mathbb{Z}_+^X$. We consider as well the set of points $\mathcal{B}(\mathcal{Z}, X) \subseteq \mathbb{Z}_+^J$ such that $u \in \mathcal{B}(\mathcal{Z}, X)$ if and only if $u(X)$ is a basis of $\mathcal{Z}(X)$ and $u_i = 0$ for every $i \in J \smallsetminus X$.

For a partition $\Pi = (Q_1, \dots, Q_m)$ of the ground set Q, a matroid $\mathcal{M} = (Q, r)$ is said to be *Π-partite* if every permutation σ on Q such that $\sigma(Q_i) = Q_i$ for $i = 1, \dots, m$ is an automorphism of $\mathcal{M}$. From now on, we notate $J_m = \{1, \dots, m\}$ and $J'_m = \{0, 1, \dots, m\}$ for every positive integer m. Then the function $h \colon \mathcal{P}(J_m) \to \mathbb{Z}$ defined by $h(X) = r(\bigcup_{i \in X} Q_i)$ is the rank function of an integer polymatroid $\mathcal{Z}(\mathcal{M}) = (J_m, h)$. Reciprocally, for every integer polymatroid $\mathcal{Z} = (J_m, h)$ with $h(\{i\}) \le |Q_i|$ for $i \in J_m$, there exists a unique Π-partite matroid $\mathcal{M}$ with $\mathcal{Z}(\mathcal{M}) = \mathcal{Z}$.

Consider a partition $\Pi = (P_1, \ldots, P_m)$ of a set P and the partition $\Pi_0 = (\{p_0\}, P_1, \ldots, P_m)$ of the set $Q = P \cup \{p_0\}$. A connected matroid port $\Gamma = \Gamma_{p_0}(\mathcal{M})$ on P is Π-partite if and only if the matroid $\mathcal{M}$ is Π_0-partite. Therefore, multipartite matroids, and hence integer polymatroids, are fundamental in the characterization of ideal multipartite access structures. These connections are in the core of the results in [10]. In particular, we present next a characterization of multipartite matroid ports in terms of integer polymatroids that was proved in [10] and will be extremely useful for our purposes.

Consider a Π-partite matroid port $\Gamma = \Gamma_{p_0}(\mathcal{M})$ and the associated integer polymatroid $\mathcal{Z}' = \mathcal{Z}(\mathcal{M}) = (J'_m, h)$. The Π-partite matroid port Γ is completely determined by the partition Π and the integer polymatroid $\mathcal{Z}'$ and we write $\Gamma = \Gamma_0(\mathcal{Z}')$. As a consequence of this fact, the following characterization of multipartite matroid ports is proved in [10].

Theorem 6 ([10]). *Let $\Pi = (P_1, \ldots, P_m)$ be a partition of a set P and let Γ be an Π-partite access structure on P. Then Γ is a matroid port if and only if there exists an integer polymatroid $\mathcal{Z}' = (J'_m, h)$ with $h(\{0\}) = 1$ and $h(\{i\}) \leq |P_i|$ such that*

$$\min \Gamma = \min \{u \in \mathcal{B}(\mathcal{Z}, X) : X \subseteq J_m \text{ is such that } h(X) = h(X \cup \{0\})\},$$

where $\mathcal{Z} = \mathcal{Z}'(J_m) = (J_m, h)$.

Since every ideal access structure is a matroid port, Theorem 6 provides a necessary condition for a multipartite access structure to be ideal. Several necessary conditions for a hierarchical access structure to be ideal will be deduced from this result in Section 5.

On the other hand, sufficient conditions can be obtained from the fact that the ports of linearly representable matroids are ideal access structures. We present in Theorem 7 an interesting result from [10] connecting the linear representations of multipartite matroids to the ones of integer polymatroids. This result is used in Section 6 to find sufficient conditions for a hierarchical access structure to be ideal.

Let E be a vector space with finite dimension over a finite field $\mathbb{K}$ and, for every $i \in J$, consider a vector subspace $V_i \subseteq E$. It is not difficult to check that the mapping $h \colon \mathcal{P}(J) \to \mathbb{Z}$ defined by $h(X) = \dim(\sum_{i \in X} V_i)$ is the rank function of an integer polymatroid with ground set J. The integer polymatroids that can be defined in this way are said to be $\mathbb{K}$*-linearly representable.*

Theorem 7 ([10]). *For every large enough field $\mathbb{K}$, an m-partite matroid $\mathcal{M}$ is $\mathbb{K}$-linearly representable if and only if its associated integer polymatroid $\mathcal{Z}(\mathcal{M}) = (J_m, h)$ is $\mathbb{K}$-linearly representable.*

5 Hierarchical Matroid Ports

In this section, we use the connection between integer polymatroids and multipartite matroid ports that is discussed in Section 4 to find necessary conditions for hierarchical access structures to be matroid ports. Of course, these will be as well necessary conditions for hierarchical access structures to be ideal.

We present first a technical lemma that apply to every integer polymatroid. Specifical results on integer polymatroids associated to hierarchical matroid ports will be given afterwards. Due to space constraints, the proofs of most of these results are omitted.

For every $i, j \in \mathbb{Z}$ we notate $[i, j] = \{i, i+1, \ldots, j\}$ if $i < j$, while $[i, i] = \{i\}$ and $[i, j] = \emptyset$ if $i > j$. Let $\mathcal{Z} = (J_m, h)$ be an integer polymatroid. For every $i \in J_m$, consider the point $y^i(\mathcal{Z}) \in \mathbb{Z}_+^m$ defined by $y^i_j(\mathcal{Z}) = h([j, i]) - h([j+1, i])$. Observe that $\sum_{j=s}^{i} y^i_j(\mathcal{Z}) = h([s, i])$ for every $s \in [1, i]$. In addition, by the submodularity of the rank function, $y^i_j(\mathcal{Z}) \geq y^{i+1}_j(\mathcal{Z})$ if $1 \leq j \leq i < m$.

Lemma 8. *For every $i = 1, \ldots, m$, the point $y^i(\mathcal{Z})$ is the hierarchically minimum point of $\mathcal{B}(\mathcal{Z}, [1, i])$, that is, $y \in \mathcal{B}(\mathcal{Z}, [1, i])$ and $y \preceq x$ for every $x \in \mathcal{B}(\mathcal{Z}, [1, i])$.*

For the remaining of this section, we assume that Γ is a Π-hierarchical matroid port, where $\Pi = (P_1, \ldots, P_m)$ is an m-partition of the set of participants P. Recall that we notate $\mathbf{P} = \Pi(\mathcal{P}(P)) \subseteq \mathbb{Z}_+^m$. In addition, we assume that the access structure Γ is *connected*, that is, that every participant is in a minimal qualified subset or, equivalently, for every $i \in J_m$, there is a minimal point $x \in \min \Gamma$ such that $x_i > 0$. Consider the integer polymatroid $\mathcal{Z}' = (J'_m, h)$ such that $\Gamma = \Gamma_0(\mathcal{Z}')$, and the integer polymatroid $\mathcal{Z} = \mathcal{Z}'(J_m) = (J_m, h)$. Since Γ is connected, $h(\{i\}) > 0$ for all $i \in J_m$, and hence $y^i_i(\mathcal{Z}) > 0$. For every $x \in \mathbb{Z}_+^m$, we notate $\mathrm{supp}(x) = \{i \in J_m : x_i \neq 0\} \subseteq J_m$ and $s(x) = \max(\mathrm{supp}(x))$. Observe that $s(x)$ is the index of the most inferior hierarchical level represented in the sets $A \subseteq P$ with $\Pi(A) = x$.

Lemma 9. *If $x \in \mathbf{P}$ is a minimal point of Γ, then $x \in \mathcal{B}(\mathcal{Z}, [1, s(x)])$.*

Lemma 10. *If $x \in \mathbf{P}$ is an h-minimal point of Γ, then $x = y^{s(x)}(\mathcal{Z})$.*

Proof. From Lemma 9, $x \in \mathcal{B}(\mathcal{Z}, [1, s(x)])$ and, since $\mathcal{B}(\mathcal{Z}, [1, s(x)]) \subseteq \Gamma$ by Theorem 6, x is h-minimal in $\mathcal{B}(\mathcal{Z}, [1, s(x)])$. By Lemma 8, this implies that $x = y^{s(x)}(\mathcal{Z})$. □

At this point, we have identified the h-minimal points of the hierarchical matroid port Γ. Namely, they are the h-minimal elements in $\{y^1(\mathcal{Z}), \ldots, y^m(\mathcal{Z})\}$.

Lemma 11. *If $x, y \in \mathbf{P}$ are two different h-minimal points of Γ, then $s(x) \neq s(y)$. Moreover, if $s(x) < s(y)$, then $|x| < |y|$ and $x_j \geq y_j$ for all $j = 1, \ldots, s(x)$.*

Proof. Since $s(y^i(\mathcal{Z})) = i$, it is clear that $s(x) \neq s(y)$ if $x \neq y$. Suppose that $s(x) < s(y)$. Since $|x| = h([1, s(x)])$ and $|y| = h([1, s(y)])$, we have that $|x| \leq |y|$. Moreover, if $|x| = |y|$, then $x \in \mathcal{B}(\mathcal{Z}, [1, s(y)])$, and hence $y \preceq x$, a contradiction. Finally, $x_j = y^{s(x)}_j(\mathcal{Z}) \geq y^{s(y)}_j(\mathcal{Z}) = y_j$ for all $j = 1, \ldots, s(x)$. □

As a consequence of Lemma 11, the h-minimal points in a hierarchical matroid port behave as in the hierarchical threshold access structure proposed by Simmons [33] (Example 3). Namely, if A and B are both hierarchically minimal qualified sets, but the least member of B is strictly inferior to the least member of A, then B must be larger than A. The last necessary condition for a hierarchical access structure to be ideal is given in the following lemma, whose proof is also omitted here.

Lemma 12. *Let $x, y \in \mathbf{P}$ be two different h-minimal points of Γ with $s(x) < s(y)$ such that there is not any h-minimal point z with $s(x) < s(z) < s(y)$. If $x_i > y_i$ for some $i \in [1, s(x) - 1]$, then $|P_j| = x_j$ for all $j \in [i + 1, s(x)]$.*

6 A Family of Ideal Hierarchical Access Structures

The results in Section 5 provide necessary conditions for a Π-hierarchical access structure to be a matroid port, and hence to be ideal, in terms of the properties of its h-minimal points. A sufficient condition is given in this section by constructing a new family of hierarchical vector space secret sharing schemes. Specifically, we present a family of linearly representable integer polymatroids and we prove that the multipartite access structures that are obtained from them are actually hierarchical. In addition, they are vector space access structures by Theorem 7.

Consider a finite field $\mathbb{K}$ and a pair of integer vectors $\mathbf{a} = (a_0, \ldots, a_m) \in \mathbb{Z}_+^{m+1}$ and $\mathbf{b} = (b_0, \ldots, b_m) \in \mathbb{Z}_+^{m+1}$ such that $a_0 = a_1 = b_0 = 1$, and $a_i \leq a_{i+1} \leq b_i \leq b_{i+1}$ for every $i = 0, \ldots m-1$. Take $d = b_m$ and consider a basis $\{e^1, \ldots, e^d\}$ of $\mathbb{K}^d$ and, for every $i = 1, \ldots, m$, consider the subspace $V_i = \langle e^{a_i}, \ldots, e^{b_i} \rangle \subseteq \mathbb{K}^d$. Let $\mathcal{Z}' = \mathcal{Z}'(\mathbf{a}, \mathbf{b}) = (J'_m, h)$ be the integer polymatroid that is linearly represented by the subspaces $V_0, V_1, \ldots, V_m$. Observe that the rank function h of $\mathcal{Z}'$ is such that $h(A) = |\cup_{i \in A} [a_i, b_i]|$ for all $A \subseteq J'_m$. In particular, $h([j, i]) = |[a_j, b_i]| = b_i - a_j + 1$ whenever $0 \leq j \leq i \leq m$, and hence $h(\{0\}) = 1$. Therefore, for every set of players P and for every m-partition $\Pi = (P_1, \ldots, P_m)$ of P such that $|P_i| \geq h(\{i\}) = b_i - a_i + 1$, we can consider the Π-partite matroid port $\Gamma = \Gamma_0(\mathcal{Z}')$ that is determined as in Theorem 6. Since $\mathcal{Z}'$ is $\mathbb{K}$-linearly representable for every finite field $\mathbb{K}$, we have from Theorem 7 that Γ is a $\mathbb{K}$-vector space access structure for every large enough finite field $\mathbb{K}$.

Consider the integer polymatroid $\mathcal{Z} = \mathcal{Z}(\mathbf{a}, \mathbf{b}) = \mathcal{Z}'(J_m) = (J_m, h)$ and, for $i = 1, \ldots, m$, the points $y^i = y^i(\mathcal{Z}) \in \mathbb{Z}_+^m$. Observe that $y_j^i = h([j, i]) - h([j + 1, i]) = a_{j+1} - a_j$ if $j < i$ while $y_i^i = b_i - a_i + 1$. Therefore, $y^i = (a_2 - a_1, \ldots, a_i - a_{i-1}, b_i - a_i + 1, 0, \ldots, 0)$. A proof for the following lemma, which is the key result in this section, will be given in the full version.

Lemma 13. *The access structure Γ is Π-hierarchical.*

By taking into account Lemma 13 and the fact that the h-minimal points of Γ are of the form $y^i(\mathcal{Z}(\mathbf{a}, \mathbf{b}))$, the next proposition can be proved. It provides a sufficient condition for a hierarchical access structure to be ideal.

Proposition 14. *Let $\Pi = (P_1, \ldots, P_m)$ be an m-partition of a set P and let Γ be a Π-hierarchical access structure on P. Let $x^1, \ldots, x^r \in \mathbb{Z}_+^m$ be the h-minimal points of Γ and consider $s_i = s(x^i) = \max(\operatorname{supp}(x^i))$. Suppose that the following properties are satisfied.*

1. *If $i < j$, then $s_i < s_j$ and $x_k^i = x_k^j$ for all $k = 1, \ldots, s_i - 1$.*
2. *If $s_{j-1} < i \leq s_j$, then $|P_i| \geq \sum_{\ell=i}^{s_j} x_\ell^j$.*

Then Γ is ideal and, moreover, it admits a $\mathbb{K}$-vector space secret sharing scheme for every finite field $\mathbb{K}$ with $|\mathbb{K}| > \binom{|P|+1}{|x^r|}$.

The bound on the size of the field is a consequence of the results in [10] (full version) about the representability of multipartite matroids. Observe that, in particular, all hierarchical access structures that have only one h-minimal point are vector space access structures. Because of that, it can be proved by using well known basic decomposition techniques (see [34], for instance) that every hierarchical access structure admits a linear secret sharing scheme in which the length of every share is at most m times the length of the secret, being m the number of h-minimal points.

7 A Characterization of Ideal Hierarchical Access Structures

By using the results in Sections 5 and 6, we present here a complete characterization of ideal hierarchical access structures. Moreover, we prove that every ideal hierarchical access structure is a $\mathbb{K}$-vector space access structure for every large enough finite field $\mathbb{K}$. The next result is a consequence of Proposition 14 and the necessary conditions for a hierarchical access structure to be ideal given in Section 5. It provides a characterization of hierarchical access structures in which the number of participants in every hierarchical level is large enough in relation to the h-minimal points. The proof of this result is omitted here.

Theorem 15. *Let $\Pi = (P_1, \ldots, P_m)$ be an m-partition of a set P and let Γ be a Π-hierarchical access structure on P with $\operatorname{hmin} \Gamma = \{x^1, \ldots, x^r\}$. For $i = 1, \ldots, r$, consider $s_i = s(x^i) = \max(\operatorname{supp}(x^i))$ and suppose that $|P_{s_i}| > x_{s_i}^i$. Then Γ is ideal if and only if*

1. *$s_i \neq s_j$ if $i \neq j$, and*
2. *if $s_i < s_j$, then $x_k^i = x_k^j$ for all $k = 1, \ldots, s_i - 1$.*

Moreover, in this situation Γ is a $\mathbb{K}$-vector space access structure for every finite field $\mathbb{K}$ with $|\mathbb{K}| > \binom{|P|+1}{|x^r|}$.

Finally, we present our complete characterization of ideal hierarchical access structures in terms of the properties of the h-minimal points. Actually, we prove that a hierarchical access structure is ideal if and only if it is a minor of an access structure in the family that is presented in Section 6. Therefore every ideal hierarchical access structure is a $\mathbb{K}$-vector access structure for all large enough finite fields $\mathbb{K}$, and this proves Theorem 1. The proof of this result will be presented in the full version.

Theorem 16. *Let $\Pi = (P_1, \ldots, P_m)$ be an m-partition of a set P and let Γ be a Π-hierarchical access structure on P with* $\min_H \Gamma = \{x^1, \ldots, x^r\}$. *For $i = 1, \ldots, m$, consider $s_i = s(x^i) = \max(\operatorname{supp}(x^i))$ and suppose that the h-minimal points are ordered in such a way that $s_i \leq s_{i+1}$. Then Γ is ideal if and only if*

1. *$s_i < s_{i+1}$ and $|x^i| < |x^{i+1}|$ for all $i = 1, \ldots, r-1$, and*
2. *$x^i_j \geq x^{i+1}_j$ if $1 \leq i \leq r-1$ and $1 \leq j \leq s_i$, and*
3. *if $x^i_j > x^r_j$ for some $1 \leq i < r$ and $1 \leq j < s_i$, then $|P_k| = x^i_k$ for all $k = j+1, \ldots, s_i$.*

We present in the following a few examples of applications of our characterization of the ideal hierarchical access structures.

Example 17. Consider a set P with a 4-partition $\Pi = (P_1, P_2, P_3, P_4)$ with $|P_i| = 4$ for every $i = 1, \ldots, 4$. Let Γ be the weighted threshold access structure defined as in Example 2 by the weight vector $w = (7, 5, 4, 3)$ and the threshold $T = 13$. The h-minimal points of Γ are $x^1 = (2, 0, 0, 0)$, $x^2 = (0, 1, 2, 0)$, and $x^3 = (0, 0, 1, 3)$. Since $x^2_2 > x^3_2$ and $|P_3| > x^2_3$, it follows from Theorem 16 that Γ is not ideal.

Example 18. For a 4-partition $\Pi = (P_1, P_2, P_3, P_4)$ of the set P of participants and positive integers $0 < t_1 < t_2 < t_3 < t_4$, consider the Π-hierarchical access structure Γ that is formed by the sets with at least one participant from P_1 that, in addition, have t_1 participants in P_1, or t_2 participants in $P_1 \cup P_2$, or t_3 participants in $P_1 \cup P_2 \cup P_3$, or t_4 participants in total. If the number of participants in each part is large enough, then Γ is ideal by Theorem 16 because its h-minimal points are $(1, 0, 0, t_4)$, $(1, 0, t_3, 0)$, $(1, t_2, 0, 0)$, and $(t_1, 0, 0, 0)$. In any other case, Γ is a minor of a 4-hierarchical access structure having those h-minimal points, and hence it is ideal as well.

Example 19. From the constructions by Brickell [5] and by Tassa [35], we know that the access structures described in Examples 3 and 4 are ideal. Actually, this fact is proved very easily from our results. The h-minimal points of the access structures in Example 3 are $\operatorname{hmin} \Gamma = \{t_1 \mathbf{e}^1, \ldots, t_m \mathbf{e}^m\}$, which clearly satisfy the conditions in Theorem 16. Since the access structures in Example 4 have only one h-minimal point, they are ideal as well.

Example 20. Tassa [35] proposed an open problem on hierarchical access structures that can be solved by using our results. For a partition $\Pi = (P_1, \ldots, P_m)$ of the set P of participants, a sequence of integers $0 < t_1 < \cdots < t_m$, and an integer $\ell \in J_m$, consider the Π-hierarchical access structure Γ defined as follows: A point $u \in \mathbf{P}$ is in Γ if and only if $|\{i \in J_m : \sum_{j=1}^{i} u_j \geq k_i\}| \geq \ell$. The open problem proposed by Tassa [35] is to determine what access structures of this form are ideal. Observe that the extreme cases $\ell = 1$ and $\ell = m$ correspond to the ideal hierarchical access structures in Examples 3 and 4, respectively. By using the results in this paper it can be proved that, if Γ is connected, then it is ideal if and only if $\ell = 1$ or $\ell = m$. This is proved by finding, for every connected access structure of this form with $1 < \ell < m$, two different h-minimal points $x, y \in \text{hmin}\,\Gamma$ with $s(x) = s(y)$.

8 Ideal Weighted Threshold Access Structures

Beimel, Tassa and Weinreb [1] presented a characterization of the ideal weighted threshold access structures. Their proof is long and complicated. By using our characterization of ideal hierarchical access structures, we obtained a simpler proof for the result in [1]. Due to space constraints, we can only present here a sketch of it. The complete proof will be given in the full version of the paper.

As was noticed in [1], an ideal weighted threshold access structure can be the composition smaller such ideal structures Because of that, we focus on the indecomposable structures in this family.

First, we describe several families of ideal weighted threshold access structures such that, as is stated in Theorem 21, they contain all indecomposable ideal weighted threshold access structures. The (t, n)-threshold access structures form the first of those families. Of course, they are ideal weighted threshold access structures. We consider as well three families of ideal bipartite hierarchical access structures, that is, ideal Π-hierarchical access structures for some partition $\Pi = (P_1, P_2)$ of the set of participants. The family $\mathbf{B}_1$ consists of the access structures with $\text{hmin}\,\Gamma = \{(x_1, x_2)\}$, where $0 < x_1 < |P_1|$ and $0 < x_2 = |P_2| - 1$. The family $\mathbf{B}_2$ is formed by the access structures with $\text{hmin}(\Gamma) = \{(x_1, 0), (0, x_1 + 1)\}$ for some integer $x_1 > 1$. The family $\mathbf{B}_3$ contains the access structures with $\text{hmin}\,\Gamma = \{(y_1 + y_2 - 1, 0), (y_1, y_2)\}$, where $y_1 > 0$, $y_2 > 2$, and $|P_2| \leq y_2 \leq |P_2| + 1$. In addition, we consider three families of ideal tripartite hierarchical access structures. The family $\mathbf{T}_1$ consists of the structures with $\text{hmin}\,\Gamma = \{(x_1, 0, 0), (0, y_2, y_3)\}$, where $0 < y_2 < |P_2|$ and $1 < y_3 = |P_3| - 1$, and $x_1 = y_2 + y_3 - 1$. We consider as well the family $\mathbf{T}_2$ of the structures such that $\text{hmin}\,\Gamma = \{(x_1, 0, 0), (y_1, y_2, y_3)\}$ with $0 < y_2 = |P_2|$ and $1 < y_3 = |P_3| - 1$, and $x_1 = y_1 + y_2 + y_3 - 1$. Finally, the family $\mathbf{T}_3$ contains the access structures with $\text{hmin}\,\Gamma = \{(x_1, x_2, 0), (y_1, y_2, y_3)\}$, where $0 < y_1 < x_1$, and $1 < y_3 = |P_3|$, and $0 < x_2 = y_2 + 1 = |P_2|$, and $x_1 + x_2 = y_1 + y_2 + y_3 - 1$. It can be proved that all the members of these families are weighted threshold access structures. At this point, we can state the characterization of the ideal weighted threshold access structures.

Theorem 21. *A weighted threshold access structure is ideal if and only if*

1. *it is a threshold access structure, or*
2. *it is a bipartite access structure in one of the families* $\mathbf{B}_1$, $\mathbf{B}_2$ *or* $\mathbf{B}_3$, *or*

3. *it is a tripartite access structure in one of the families* $\mathbf{T}_1$*,* $\mathbf{T}_2$ *or* $\mathbf{T}_3$*, or*
4. *it is a composition of smaller ideal weighted threshold access structures.*

We present next a sketch of our proof for this result. To begin with, several technical results on the properties of h-minimal points in indecomposable hierarchical access structures are needed. Then, several properties that must be satisfied by every ideal indecomposable weighted threshold access structure Γ are proved. First, if Γ is strictly bipartite, then it is in one of the families $\mathbf{B}_1$, $\mathbf{B}_2$ or $\mathbf{B}_3$. Second, if Γ is strictly m-partite with $m \geq 3$, then it has exactly two h-minimal points. Third, if Γ is strictly tripartite, then it is in one of the families $\mathbf{T}_1$, $\mathbf{T}_2$ or $\mathbf{T}_3$. Finally, it is proved that such an access structure cannot be strictly m-partite with $m > 3$.

Acknowledgements

The authors thank Ronald Cramer and Enav Weinreb for useful discussions, comments and suggestions. The authors thank as well the anonymous referees for their careful revision of the paper and their valuable comments that greatly improved the presentation of the paper.

References

1. Beimel, A., Tassa, T., Weinreb, E.: Characterizing Ideal Weighted Threshold Secret Sharing. SIAM J. Discrete Math. 22, 360–397 (2008)
2. Beimel, A., Weinreb, E.: Monotone Circuits for Monotone Weighted Threshold Functions. Information Processing Letters 97, 12–18 (2006)
3. Benaloh, J., Leichter, J.: Generalized secret sharing and monotone functions. In: Goldwasser, S. (ed.) CRYPTO 1988. LNCS, vol. 403, pp. 27–35. Springer, Heidelberg (1990)
4. Blakley, G.R.: Safeguarding cryptographic keys. In: AFIPS Conference Proceedings, vol. 48, pp. 313–317 (1979)
5. Brickell, E.F.: Some ideal secret sharing schemes. J. Combin. Math. and Combin. Comput. 9, 105–113 (1989)
6. Brickell, E.F., Davenport, D.M.: On the classification of ideal secret sharing schemes. J. Cryptology 4, 123–134 (1991)
7. Capocelli, R.M., De Santis, A., Gargano, L., Vaccaro, U.: On the size of shares of secret sharing schemes. J. Cryptology 6, 157–168 (1993)
8. Collins, M.J.: A Note on Ideal Tripartite Access Structures. Cryptology ePrint Archive, Report 2002/193, `http://eprint.iacr.org/2002/193`
9. Csirmaz, L.: The size of a share must be large. J. Cryptology 10, 223–231 (1997)
10. Farràs, O., Martí-Farré, J., Padró, C.: Ideal Multipartite Secret Sharing Schemes. In: Naor, M. (ed.) EUROCRYPT 2007. LNCS, vol. 4515, pp. 448–465. Springer, Heidelberg (2007), `http://eprint.iacr.org/2006/292`
11. Herranz, J., Sáez, G.: New Results on Multipartite Access Structures. In: IEE Proceedings of Information Security, vol. 153, pp. 153–162 (2006)
12. Herzog, J., Hibi, T.: Discrete polymatroids. J. Algebraic Combin. 16, 239–268 (2002)

13. Ito, M., Saito, A., Nishizeki, T.: Secret sharing scheme realizing any access structure. In: Proc. IEEE Globecom 1987, pp. 99–102 (1987)
14. Karnin, E.D., Greene, J.W., Hellman, M.E.: On secret sharing systems. IEEE Trans. Inform. Theory 29, 35–41 (1983)
15. Lehman, A.: A solution of the Shannon switching game. J. Soc. Indust. Appl. Math. 12, 687–725 (1964)
16. Martí-Farré, J., Padró, C.: On Secret Sharing Schemes, Matroids and Polymatroids. In: Vadhan, S.P. (ed.) TCC 2007. LNCS, vol. 4392, pp. 273–290. Springer, Heidelberg (2007), the full version of this paper is available at the Cryptology ePrint Archive, http://eprint.iacr.org/2006/077
17. Martí-Farré, J., Padró, C.: Ideal secret sharing schemes whose minimal qualified subsets have at most three participants. Des. Codes Cryptogr. 52, 1–14 (2009)
18. Massey, J.L.: Minimal codewords and secret sharing. In: Proceedings of the 6th Joint Swedish-Russian Workshop on Information Theory, Molle, Sweden, August 1993, pp. 269–279 (1993)
19. Massey, J.L.: Some applications of coding theory in cryptography. In: Codes and Ciphers: Cryptography and Coding IV, pp. 33–47 (1995)
20. Matúš, F.: Matroid representations by partitions. Discrete Math. 203, 169–194 (1999)
21. Matúš, F.: Two Constructions on Limits of Entropy Functions. IEEE Trans. Inform. Theory 53, 320–330 (2007)
22. Morillo, P., Padró, C., Sáez, G., Villar, J.L.: Weighted Threshold Secret Sharing Schemes. Inf. Process. Lett. 70, 211–216 (1999)
23. Murota, K.: Discrete convex analysis. Math. Programming 83, 313–371 (1998)
24. Murota, K.: Discrete convex analysis. SIAM Monographs on Discrete Mathematics and Applications. SIAM, Philadelphia (2003)
25. Ng, S.-L.: A Representation of a Family of Secret Sharing Matroids. Des. Codes Cryptogr. 30, 5–19 (2003)
26. Ng, S.-L.: Ideal secret sharing schemes with multipartite access structures. IEE Proc.-Commun. 153, 165–168 (2006)
27. Ng, S.-L., Walker, M.: On the composition of matroids and ideal secret sharing schemes. Des. Codes Cryptogr. 24, 49–67 (2001)
28. Oxley, J.G.: Matroid theory. Oxford Science Publications. The Clarendon Press, Oxford University Press, New York (1992)
29. Padró, C., Sáez, G.: Secret sharing schemes with bipartite access structure. IEEE Trans. Inform. Theory 46, 2596–2604 (2000)
30. Padró, C., Sáez, G.: Correction to Secret Sharing Schemes With Bipartite Access Structure. IEEE Trans. Inform. Theory 50, 1373–1373 (2004)
31. Seymour, P.D.: A forbidden minor characterization of matroid ports. Quart. J. Math. Oxford Ser. 27, 407–413 (1976)
32. Shamir, A.: How to share a secret. Commun. of the ACM 22, 612–613 (1979)
33. Simmons, G.J.: How to (Really) Share a Secret. In: Goldwasser, S. (ed.) CRYPTO 1988. LNCS, vol. 403, pp. 390–448. Springer, Heidelberg (1990)
34. Stinson, D.R.: An explication of secret sharing schemes. Des. Codes Cryptogr. 2, 357–390 (1992)
35. Tassa, T.: Hierarchical Threshold Secret Sharing. J. Cryptology 20, 237–264 (2007)
36. Tassa, T., Dyn, N.: Multipartite Secret Sharing by Bivariate Interpolation. J. Cryptology 22, 227–258 (2009)
37. Welsh, D.J.A.: Matroid Theory. Academic Press, London (1976)

A Hardcore Lemma for Computational Indistinguishability: Security Amplification for Arbitrarily Weak PRGs with Optimal Stretch

Ueli Maurer and Stefano Tessaro

Department of Computer Science, ETH Zurich, 8092 Zurich, Switzerland
{maurer,tessaros}@inf.ethz.ch

Abstract. It is well known that two random variables X and Y with the same range can be viewed as being equal (in a well-defined sense) with probability $1 - d(X, Y)$, where $d(X, Y)$ is their statistical distance, which in turn is equal to the best distinguishing advantage for X and Y. In other words, if the best distinguishing advantage for X and Y is ϵ, then with probability $1 - \epsilon$ they are completely indistinguishable. This statement, which can be seen as an information-theoretic version of a hardcore lemma, is for example very useful for proving indistinguishability amplification results.

In this paper we prove the computational version of such a hardcore lemma, thereby extending the concept of hardcore sets from the context of computational *hardness* to the context of computational *indistinguishability*. This paradigm promises to have many applications in cryptography and complexity theory. It is proven both in a non-uniform and a uniform setting.

For example, for a weak pseudorandom generator (PRG) for which the (computational) distinguishing advantage is known to be bounded by ϵ (e.g. $\epsilon = \frac{1}{2}$), one can define an event on the seed of the PRG which has probability at least $1 - \epsilon$ and such that, conditioned on the event, the output of the PRG is essentially indistinguishable from a string with almost maximal min-entropy, namely $\log(1/(1-\epsilon))$ less than its length. As an application, we show an optimally efficient construction for converting a weak PRG for any $\epsilon < 1$ into a strong PRG by proving that the intuitive construction of applying an extractor to an appropriate number of independent weak PRG outputs yields a strong PRG. This improves strongly over the best previous construction for security amplification of PRGs which does not work for $\epsilon \geq \frac{1}{2}$ and requires the seed of the constructed strong PRG to be very long.

1 Introduction

1.1 (Weak) Pseudorandomness

Randomness is a central resource in cryptography. In many applications, true randomness must be replaced by *pseudorandomness*, for example when it needs to be reproduced at a second location and one can only afford to transmit a short

D. Micciancio (Ed.): TCC 2010, LNCS 5978, pp. 237–254, 2010.

value to be used as the seed of a so-called *pseudorandom generator* (PRG). An example are cryptographic applications where a key agreement protocol yields only a short key. More generally, PRGs are a central building block in cryptographic protocols and are used in different applications where a random functionality (e.g. a uniform random function) must be realized from a short secret key.

The concept of a PRG was first proposed by Blum and Micali [2], initiating a large body of literature dealing with various aspects of pseudorandomness: More formally, a random variable X is said to be pseudorandom if it is *computationally* indistinguishable from a uniformly distributed random variable U with the same range, i.e., no computationally bounded (i.e., polynomial time) distinguisher can tell X and U apart with better than negligible advantage. In particular, a PRG $G : \{0,1\}^k \to \{0,1\}^\ell$ (for $\ell > k$) extends a uniform random string U_k of length k into a pseudorandom string $G(U_k)$ of length ℓ.

Computational infeasibility is at the core of cryptographic security. In contrast to cryptographic primitives (like a one-way function f) assuring that a certain value (e.g. the input of f) cannot efficiently be computed, the notion of computational indistinguishability is substantially more involved. It is hence not a surprise that all constructions (cf. e.g. [6,5,9]) of a PRG from an arbitrary[1] one-way function f are too inefficient (in terms of the number of calls to f) to be of any practical use.

Therefore, it appears much more difficult to propose a cryptographic function that can be believed to be a PRG than one that can be believed to be a one-way function. As a consequence, a prudent approach in cryptography is to make weaker assumptions about a concrete proposal for a PRG G. One possible way[2] to achieve this is by considering a so-called *ϵ-pseudorandom generator (ϵ-PRG)*, where the best distinguishing advantage of an efficient distinguisher is not necessarily negligible, but instead bounded by some noticeable quantity ϵ, such as a constant (e.g. $\epsilon = 0.75$), or even a function in the security parameter k mildly converging to 1 (e.g. $1 - \frac{1}{p(k)}$ for some polynomial p).[3]

1.2 Security Amplification of PRGs

Security Amplification. In order to deploy some ϵ-PRG within a particular cryptographic application, we need to find an efficient construction transforming it into a fully secure PRG. This is an instance of the general problem of *security amplification*, which was first considered by Yao [17] in the context of one-way

[1] i.e. without any particular assumption on the combinatorial structure of the function.

[2] An alternative approach to modeling a weak PRG is to assume its output to be computationally indistinguishable from a random variable with only moderate min-entropy. However, this approach does not capture certain failure types, such as a function G that with some substantial probability may output a constant value. In contrast, the notion of an ϵ-PRG captures this case. One of the contributions of this paper is to show a tight relation between these two approaches.

[3] An ϵ-PRG $G : \{0,1\}^k \to \{0,1\}^\ell$ is only interesting in the case $\ell > k + \log\left(\frac{1}{1-\epsilon}\right)$, as otherwise an unconditionally secure ϵ-PRG is given by the mapping $x \mapsto x \| 0^{\log\left(\frac{1}{1-\epsilon}\right)}$.

functions, and has subsequently been followed by a prolific line of research considering a wide range of other cryptographic primitives.

PREVIOUS WORK. The *only* known security amplification result for PRGs considers the construction $\mathsf{SUM}^G : \{0,1\}^{mk} \to \ell$ (for any $m > 1$) which outputs

$$\mathsf{SUM}^G(x) := G(x_1) \oplus \cdots \oplus G(x_m).$$

for all inputs $x = x_1 \| \ldots \| x_m \in \{0,1\}^{km}$ (with $x_1, \ldots, x_m \in \{0,1\}^k$). As pointed out in [14], Yao's XOR-lemma [17,4] yields a direct proof of security amplification for the construction SUM, and an improved bound can be obtained using the tools from [14]. (An independent proof with a weaker bound was also given in [3].) Namely, one can show that if G is an ϵ-PRG, then SUM^G is a $(2^{m-1}\epsilon^m + \nu)$-PRG, where ν is a negligible function. Also, the result extends to the case where $\oplus$ is replaced by any quasi-group operation $\star$.

However, this construction has two major disadvantages: First, security amplification is inherently limited to the case $\epsilon < \frac{1}{2}$. For instance, the security of a PRG with a very large stretch and with one constant output bit is not amplified by the SUM construction, even if all other output bits are pseudorandom. Second, the construction is expanding only when $\ell > k \cdot m$. Note that this issue cannot be overcome by first extending the output size of the weak PRG, due to the high security loss in the extension which would yield an ϵ'-PRG with ϵ' close to one.

OUR CONSTRUCTION. In this paper, we provide the first solution which amplifies the security of an ϵ-PRG $G : \{0,1\}^k \to \{0,1\}^\ell$ for any $\epsilon < 1$. Our construction, called *concatenate and extract* (CaE), takes input $x = x_1 \| \ldots \| x_m \| r$, where $x_1, \ldots, x_m \in \{0,1\}^k$ and $r \in \{0,1\}^d$, and outputs

$$\mathsf{CaE}^G(x) := \mathsf{Ext}(G(x_1) \,\|\, \ldots \,\|\, G(x_m), r) \,\|\, r,$$

where $\mathsf{Ext} : \{0,1\}^{m\ell} \times \{0,1\}^d \to \{0,1\}^n$ is a sufficiently good strong randomness extractor. In particular, a good instantiation (for instance using two-universal hash functions or even appropriate deterministic extractors) allows to achieve $n \approx (1-\epsilon)m \cdot \left[\ell - \log\left(\frac{1}{1-\epsilon}\right)\right]$, and we show the resulting output length $n + d$ to be optimal with respect to constructions combining m outputs of an ϵ-PRG.

We provide security proofs both in the non-uniform and in the uniform models, which follow as an application of a new characterization of computational indistinguishability that we present in this paper, and which we outline in the next section.

Finally, we point out that the idea of concatenating strings with weaker pseudorandomness guarantees and then extracting the resulting computational entropy was previously used (most notably in constructions of PRGs from one-way functions [6,9,5]): However, all these previous results only consider individual independent *bits* which are hard to compute (given some other part of the concatenation), whereas our result is the first to deal with the more general case of weakly pseudorandom *strings*.

1.3 A Tight Characterization of Computational Indistinguishability

Let X and Y be random variables with the same range $\mathcal{U}$. Assume that we can show that there exist events $\mathcal{A}$ and $\mathcal{B}$ defined on the choices of X and Y by some conditional probability distributions $\mathsf{P}_{\mathcal{A}|X}$ and $\mathsf{P}_{\mathcal{B}|Y}$ such that $\mathsf{P}[\mathcal{A}] \geq 1-\epsilon$, $\mathsf{P}[\mathcal{B}] \geq 1-\epsilon$, and X and Y are identically distributed when conditioned on $\mathcal{A}$ and $\mathcal{B}$, respectively. Then this implies that the advantage $\Delta^D(X,Y) := |\mathsf{P}[D(X)=1] - \mathsf{P}[D(Y)=1]|$ is upper bounded by ϵ for *every* distinguisher D. However, is the converse also true? Namely, if the best distinguishing advantage is upper bounded by ϵ, do such two events always exist?

An affirmative answer is known to exist if we maximize over *all* distinguishers: In this case, the best advantage is the *statistical distance*

$$d(X,Y) := \frac{1}{2}\sum_{u\in\mathcal{U}} |\mathsf{P}_X(u) - \mathsf{P}_Y(u)|,$$

and it is *always* possible to define two such events $\mathcal{A}$ and $\mathcal{B}$ by the joint probabilities $\mathsf{P}_{\mathcal{A}X}(u) = \mathsf{P}_{\mathcal{A}Y}(u) = \min\{\mathsf{P}_X(u), \mathsf{P}_Y(u)\}$. Because $d(X,Y) = 1 - \sum_{u\in\mathcal{U}} \min\{\mathsf{P}_X(u), \mathsf{P}_Y(u)\}$, it is easy to see that $\mathsf{P}[\mathcal{A}] = \mathsf{P}[\mathcal{B}] = \sum_u \mathsf{P}_{\mathcal{A}X}(u) = 1 - d(X,Y)$. This can be interpreted as saying the the random variables X and Y are equal with probability $1-\epsilon$. A generalization of this property to discrete systems was considered by Maurer, Pietrzak, and Renner [13].

However, the quantity of interest in the cryptographic setting (as for example in the definition of a PRG) is the best distinguishing advantage of a *computationally bounded* (i.e. polynomial-time) distinguisher, which in general is substantially smaller than the statistical distance $d(X,Y)$, and hence the above property is of no help in the context of computational indistinguishability.

The main technical and conceptual contribution of this paper is a computational version of the above characterization, which we prove both in the uniform and the non-uniform settings. Roughly speaking, we show that if the advantage of every computationally bounded distinguisher is bounded by ϵ (and the statistical distance may be considerably higher), there exist events $\mathcal{A}$ and $\mathcal{B}$ occurring each with probability $1-\epsilon$ such that X and Y are *computationally indistinguishable* when conditioned on $\mathcal{A}$ and $\mathcal{B}$. This can be seen as a hardcore lemma for the setting of computational indistinguishability, and hence solves, for the case of random variables, an open question stated by Myers [15].

The security of the aforementioned concatenate-and-extract approach follows then from the simple observation, due to our characterization, that the output of an ϵ-PRG can be shown to have high computational min-entropy with probability $1-\epsilon$, and hence the concatenation of sufficiently many such outputs always contains enough randomness to be extracted.

1.4 Outline of This Paper

The main part of this paper is Section 3, which is devoted to discussing the characterizations of computational indistinguishability in terms of events in both the uniform and the non-uniform computational models. Furthermore, Section 4 is devoted to proving the soundness of the concatenate-and-extract approach for security amplification of PRGs. All tools employed throughout this paper are introduced in Section 2, where in particular we discuss the hardcore lemma in the uniform and non-uniform computational models, which is a central component of our main proofs.

2 Preliminaries

2.1 Notational Preliminaries and Computational Model

NOTATION. Recall that a function is *negligible* if it vanishes faster than the inverse of any polynomial. We use both notations poly and negl as placeholders for some polynomial and negligible function, respectively. In particular, a function $\gamma = \frac{1}{\mathsf{poly}}$ is called *noticeable.*

Throughout this paper, we use calligraphic letters $\mathcal{X}, \mathcal{Y}, \ldots$ to denote sets, upper-case letters $X, Y, \ldots$ to denote random variables, and lower-case letters $x, y, \ldots$ denote the values they take on. Moreover, $\mathsf{P}[\mathcal{A}]$ stands for the probability of the event $\mathcal{A}$, while we use the shorthands $\mathsf{P}_X(x) := \mathsf{P}[X = x]$, $\mathsf{P}_{\mathcal{A}|X}(x) := \mathsf{P}[\mathcal{A}|X = x]$, and $\mathsf{P}_{X|\mathcal{A}}(x) := \mathsf{P}[X = x|\mathcal{A}]$. Also, P_X, $\mathsf{P}_{\mathcal{A}|X}$ and $\mathsf{P}_{X|\mathcal{A}}$ are the corresponding (conditional) probability distributions, and $x \xleftarrow{\$} \mathsf{P}$ is the action of sampling a concrete value x according to the distribution P. (We use $x \xleftarrow{\$} \mathcal{X}$ in the case where P is the uniform distribution on $\mathcal{X}$.) Finally, $\mathsf{E}[X]$ is the expected value of the (real-valued) random variable X. Also, we use $\|$ to denote the concatenation of binary strings.

COMPUTATIONAL MODEL. The notation $A^{\mathcal{O}}(x, x', \ldots)$ denotes the (oracle) algorithm A which runs on inputs $x, x', \ldots$ with access to the oracle $\mathcal{O}$. In the asymptotic setting, a *uniform* algorithm A always obtains the unary representation 1^k of the current security parameter k as its first input and is said to run in time $t : \mathbb{N} \to \mathbb{N}$ (or to have *time complexity* t) if for all $k > 0$ the worst-case number of steps it takes (counting oracle queries as single steps) on first input 1^k, taken over all randomness values, all compatible additional inputs and oracles, is at most $t(k)$. In particular, we say that a family of functions $F = \{f_k\}_{k \in \mathbb{N}}$, where $f_k : \mathcal{X}_k \to \mathcal{Y}_k$ is *efficiently* (or polynomial-time) computable if there exists a uniform algorithm which for every security parameter k computes f_k. Finally, we model as usual *non-uniform* algorithms in terms of (families of) circuits $C : \{0,1\}^m \to \{0,1\}^n$ with bounded size.

For ease of notation, we do not make asymptotics explicit in this paper (in particular, we omit the input 1^k), despite the formal statements being asymptotic in nature.

2.2 Pseudorandom Generators and Randomness Extractors

DISTANCE MEASURES. The *distinguishing advantage* of the distinguisher D in distinguishing random variables X and Y with equal range $\mathcal{U}$ is

$$\Delta^D(X, Y) := |\mathsf{P}[D(X) = 1] - \mathsf{P}[D(Y) = 1]|,$$

whereas the *statistical distance between* X *and* Y is defined as $d(X,Y) := \frac{1}{2}\sum_{u\in\mathcal{U}} |\mathsf{P}_X(u) - \mathsf{P}_Y(u)| = \sum_{u:\mathsf{P}_X(u)\leq \mathsf{P}_Y(u)} \mathsf{P}_Y(u) - \mathsf{P}_X(u)$.

PSEUDORANDOM GENERATORS. An efficiently computable function $G : \{0,1\}^k \to \{0,1\}^\ell$ is a (t,ϵ)-PRG if for all distinguishers D with time complexity t we have $\Delta^D(G(U_k), U_\ell) \leq \epsilon$, where U_k and U_ℓ are uniformly distributed k- and ℓ-bit strings, respectively. (In the non-uniform setting we rather use the notation (s,ϵ)-PRG, maximizing over all circuits with size at most s.) Furthermore, we use the short-hands ϵ-PRG and PRG for (poly,ϵ)- and $(\mathsf{poly},\mathsf{negl})$-PRGs, respectively.

RANDOMNESS EXTRACTORS. A *source* $\mathcal{S}$ is a set of probability distributions, and an ϵ*-extractor for* $\mathcal{S}$ is an efficiently computable function $\mathsf{Ext} : \{0,1\}^m \times \{0,1\}^d \to \{0,1\}^n$ such that for a uniformly distributed d-bit string R, we have $d(\mathsf{Ext}(X,R), U_n) \leq \epsilon$ for all m-bit random variables X with $\mathsf{P}_X \in \mathcal{S}$ and a uniformly distributed n-bit string U_n. Furthermore, the extractor is called *strong* if the stronger condition $d\left((\mathsf{Ext}(X,R),R),(U_n,R)\right) \leq \epsilon$ holds.

Also recall that the *min-entropy of* X is $\mathsf{H}_\infty(X) := -\log\left(\max_{x\in\mathcal{X}} \mathsf{P}_X(x)\right)$. A two-parameter function $h : \{0,1\}^m \times \{0,1\}^d \to \{0,1\}^n$ is called *two-universal* if $\mathsf{P}[h(x,K) = h(x',K)] = 2^{-n}$ for any two distinct m-bit x and x' and a uniform d-bit K. An example with $d = m$ is the function $h(x,k) := (x \odot k)|_n$, where $\odot$ is the multiplication of binary strings interpreted as elements of $GF(2^m)$, and $|_n$ outputs the first n bits of a given string. Two-universal hash functions are good extractors:

Lemma 1 (Leftover Hash Lemma [1,11]). *For any* $\epsilon > 0$*, every two-universal hash function* $h : \{0,1\}^m \times \{0,1\}^d \to \{0,1\}^n$ *is a strong* ϵ*-extractor for the source of* m*-bit random variables with min-entropy at least* $n + 2\log\left(\frac{1}{\epsilon}\right)$.

We also note that extractors with smaller seed exist for the source of random variables with guaranteed min-entropy. We refer the reader to [16] for a survey.

DETERMINISTIC EXTRACTORS. An extractor is *deterministic* if $d = 0$, i.e., no additional randomness is needed. (Note that such extractors are vacuously strong.) A class of sources allowing for deterministic extraction are so-called (m,ℓ,k)*-total-entropy independent sources* [12], consisting of random variables of the form $(X_1,\ldots,X_m)$, where $X_1,\ldots,X_m$ are *independent* ℓ-bit strings, and the total min-entropy of $(X_1,\ldots,X_m)$ is at least k.[4] In particular, the following extractor from [12] will be useful for our purposes. (Unconditional constructions requiring a higher entropy rate δ are also given in [12].)

Theorem 1 ([12]). *Under the assumption that primes with length in* $[\tau, 2\tau]$ *can be found in time* $\mathsf{poly}(\tau)$*, there is a constant* η *such that for all* $m,\ell \in \mathbb{N}$ *and all* $\delta > \zeta > (m\ell)^{-\eta}$*, there exists a polynomial-time computable* ϵ*-extractor* $\mathsf{Ext} : \left(\{0,1\}^\ell\right)^m \to \{0,1\}^n$ *for* $(m,\ell,\delta\cdot m\ell)$*-total-entropy independent sources, where* $n = (\delta-\zeta)m\ell$ *and* $\epsilon = e^{-(m\ell)^{\Omega(1)}}$.

[4] Note that in this case $\mathsf{H}_\infty(X_1,\ldots,X_m) = \sum_{i=1}^m \mathsf{H}_\infty(X_i)$.

2.3 Measures and the Hardcore Lemma

GUESSING ADVANTAGES. Let (X, B) be a pair of correlated random variables with joint probability distribution P_{XB}, where B is binary, and let A be an adversary taking an input in the range $\mathcal{X}$ of X and outputting a bit (i.e., A has the same form as a distinguisher): The *guessing advantage* of A in guessing B given X is

$$\mathrm{Guess}^A(B \mid X) := 2 \cdot \mathsf{P}[A(X) = B] - 1.$$

In particular, $\mathrm{Guess}^A(B \mid X) = 1$ means that A is always correct in guessing B given X, whereas it always errs if $\mathrm{Guess}^A(B \mid X) = -1$.[5]

MEASURES. A *measure* $\mathcal{M}$ on a set $\mathcal{X}$ is a mapping $\mathcal{M} : \mathcal{X} \to [0, 1]$. Intuitively, it captures the notion of a "fuzzy" characteristic function of a subset of $\mathcal{X}$. Consequently, its *size* $|\mathcal{M}|$ is defined as $\sum_{x \in \mathcal{X}} \mathcal{M}(x)$, and its *density* is $\mu(\mathcal{M}) := |\mathcal{M}|/|\mathcal{X}|$. Also one associates with each measure $\mathcal{M}$ the probability distribution $\mathsf{P}_{\mathcal{M}}$ such that $\mathsf{P}_{\mathcal{M}}(x) := \mathcal{M}(x)/|\mathcal{M}|$, and we say that a random variable M is *sampled according to* $\mathcal{M}$ if $M \stackrel{\$}{\leftarrow} \mathsf{P}_{\mathcal{M}}$. The following lemma shows that such random variables have high min-entropy, as long as $\mathcal{M}$ is sufficiently dense.

Lemma 2. *Let $\mathcal{M} : \mathcal{X} \to [0, 1]$ be a measure with density $\mu(\mathcal{M}) \geq \delta$, and let M be sampled according to $\mathcal{M}$. Then, $\mathsf{H}_\infty(M) \geq \log |\mathcal{X}| - \log \left(\frac{1}{\delta}\right)$.*

Proof. We have $\mathsf{P}_M(x) = \frac{\mathcal{M}(x)}{|\mathcal{M}|} \leq \frac{\mathcal{M}(x)}{\delta \cdot |\mathcal{X}|} \leq \frac{1}{\delta} \cdot \frac{1}{|\mathcal{X}|}$ due to $\mathcal{M}(x) \leq 1$, which implies $\mathsf{H}_\infty(M) = -\log \max_{x \in \mathcal{X}} \mathsf{P}_{\mathcal{M}}(x) \geq \log |\mathcal{X}| - \log(1/\delta)$. □

THE HARDCORE LEMMA. For a set $\mathcal{W}$, let $g : \mathcal{W} \to \mathcal{Y}$ be a function, and let $P : \mathcal{W} \to \{0, 1\}$ be a predicate. The so-called *hardcore lemma* shows that, roughly speaking, if $\mathrm{Guess}^A(P(W) \mid g(W)) \leq \delta$ (for W uniform in $\mathcal{W}$) for all efficient A, then for all $\gamma > 0$ there exists a measure $\mathcal{M}$ on $\mathcal{W}$ with $\mu(\mathcal{M}) \geq 1 - \delta$ such that $\mathrm{Guess}^{A'}(P(W') \mid g(W')) \leq \gamma$ for all efficient adversaries A' and for W' sampled according to $\mathcal{M}$. This result was first introduced and proven by Impagliazzo [10]. However, his original proof only ensures $\mu(\mathcal{M}) \geq \frac{1-\delta}{2}$. The following theorem, due to Holenstein [8], gives a *tight* version of the lemma for the non-uniform setting.

Theorem 2 (Non-Uniform Hardcore Lemma). *Let $g : \mathcal{W} \to \mathcal{X}$ and $P : \mathcal{W} \to \{0, 1\}$ be functions, and let $\delta, \gamma \in (0, 1)$ and $s > 0$ be given. If for all circuits C with size s we have*

$$\mathrm{Guess}^C(P(W) \mid g(W)) \leq \delta$$

for $W \stackrel{\$}{\leftarrow} \mathcal{W}$, then there exists a measure $\mathcal{M}$ on $\mathcal{W}$ (called the hardcore measure*) such that $\mu(\mathcal{M}) \geq 1 - \delta$ and such that all circuits C' with size $s' = \frac{s \cdot \gamma^2}{32 \log |\mathcal{W}|}$ satisfy $\mathrm{Guess}^{C'}(P(W') \mid g(W')) \leq \gamma$, where $W' \stackrel{\$}{\leftarrow} \mathsf{P}_{\mathcal{M}}$.*

[5] In particular, flipping the output bit of such an A yields an adversary that is always correct.

A slightly weaker statement holds in the *uniform* setting, where we can only show that for every polynomial-time adversary A' there exists a measure $\mathcal{M}$ for which $\mathsf{Guess}^{A'}(P(W') \mid g(W')) \leq \gamma$ *even* if A' is allowed to query the measure $\mathcal{M}$ as an oracle[6] *before* obtaining $g(W')$. This is captured by the following theorem also due to Holenstein [8].

Theorem 3 (Uniform Hardcore Lemma). *Let $g : \mathcal{W} \to \mathcal{X}$, $P : \mathcal{W} \to \{0,1\}$, $\delta : \mathbb{N} \to [0,1]$, and $\gamma : \mathbb{N} \to [0,1]$ be functions computable in time $\mathsf{poly}(k)$, where δ and γ are both noticeable. Assume that for all polynomial-time adversaries A we have*

$$\mathsf{Guess}^{A}(P(W) \mid g(W)) \leq \delta$$

for $W \xleftarrow{\$} \mathcal{W}$, then for all polynomial-time adversaries $A'^{(\cdot)}$, whose oracle queries are independent of their input[7], there exists a measure $\mathcal{M}$ on $\mathcal{W}$ with $\mu(\mathcal{M}) \geq 1 - \delta$ such that $\mathsf{Guess}^{A'^{\mathcal{M}}}(P(W') \mid g(W')) \leq \gamma$, where $W' \xleftarrow{\$} \mathsf{P}_{\mathcal{M}}$.

The independence requirement on oracle queries is due to the hardcore lemma of [8] considering a model with uniform adversaries A' which are given oracle access to $\mathcal{M}$ (with no input) and subsequently output a *circuit* for guessing $P(W')$ out of $g(W')$ (which in particular does not make queries to $\mathcal{M}$). The simpler statement of Theorem 3 follows by standard techniques.

Note that in contrast to [10,8], and the traditional literature on the hardcore lemma, we swap the roles of δ and $1 - \delta$ in order to align our statements with the (natural) information-theoretic intuition. Also, note that both theorems have equivalent versions in terms of hardcore *sets* (i.e., where $\mathcal{M}(x) \in \{0,1\}$), yet we limit ourselves to considering the measure versions in this paper.

Efficient Sampling from Measures. Sometimes, we need to sample a random element according to a measure $\mathcal{M}$ on $\mathcal{X}$ with $\mu(\mathcal{M}) \geq \delta$ (for a noticeable δ) given only oracle access to this measure. A solution to this is to sample a random element $x \xleftarrow{\$} \mathcal{X}$ and then output x with probability $\mathcal{M}(x)$, and otherwise go to the next iteration (and abort if a maximal number of iterations k is achieved.) It is easy to see that if an output is produced, it has the right distribution, whereas the probability that no output is produced is at most $(1-\delta)^k < e^{-\delta k}$, and can hence be made smaller than any $\alpha > 0$ by choosing $k = \frac{1}{\delta} \ln\left(\frac{1}{\alpha}\right)$.

In the following, we assume that the sampling can be done perfectly, neglecting the inherent small error probability in the analysis.

3 Characterizing Computational Indistinguishability Via Hardcore Theorems

3.1 Non-uniform Case

This section considers a setting with two efficiently computable functions $E : \mathcal{U} \to \mathcal{X}$ and $F : \mathcal{V} \to \mathcal{X}$, and we define the random variables $X := E(U)$ and

[6] That is, the oracle $\mathcal{M}$ answers a query x with $\mathcal{M}(x) \in [0,1]$.

[7] In particular, they only depend on the randomness of the distinguisher and previous oracle queries.

$Y := F(V)$, where U and V are uniformly[8] distributed on $\mathcal{U}$ and $\mathcal{V}$, respectively. Note that this is the usual way to capture that X and Y are *efficiently samplable*, where typically $\mathcal{U}$ and $\mathcal{V}$ both consist of bit strings of some length.

Let us now assume that $\Delta^D(X,Y) \leq \epsilon$ for every efficient distinguisher D. In full analogy with the information-theoretic setting [13], we aim at *extending* the random experiments where $E(U)$ and $F(V)$ are sampled by adjoining, for all $\gamma > 0$, corresponding events $\mathcal{A}$ and $\mathcal{B}$ defined by conditional probability distributions $\mathsf{P}_{\mathcal{A}|U}$ and $\mathsf{P}_{\mathcal{B}|V}$, such that both events occur with probability roughly $1-\epsilon$, and, conditioned on $\mathcal{A}$ and $\mathcal{B}$, respectively, the random variables $E(U)$ and $F(V)$ can be distinguished with advantage at most γ by an efficient distinguisher. Note that for notational convenience (and in order to interpret the result as a hardcore lemma), we describe the conditional probability distributions $\mathsf{P}_{\mathcal{A}|U}$ and $\mathsf{P}_{\mathcal{B}|V}$ in terms of measures $\mathcal{M} : \mathcal{U} \to [0,1]$ and $\mathcal{N} : \mathcal{V} \to [0,1]$. In particular, the values $\mathcal{M}(u)$ and $\mathcal{N}(v)$ take the roles of $\mathsf{P}_{\mathcal{A}|U}(u)$ and $\mathsf{P}_{\mathcal{B}|V}(v)$, and note that $\mu(\mathcal{M}) = \frac{1}{|\mathcal{U}|}\sum_{u\in\mathcal{U}} \mathcal{M}(u) = \sum_{u\in\mathcal{U}} \mathsf{P}_U(u)\cdot \mathsf{P}_{\mathcal{A}|U}(u) = \mathsf{P}[\mathcal{A}]$ and hence $\mathsf{P}_M(u) = \frac{\mathcal{M}(u)}{|\mathcal{M}|} = \frac{\mathsf{P}_{\mathcal{A}|U}(u)\mathsf{P}_U(u)}{\mathsf{P}[\mathcal{A}]} = \mathsf{P}_{U|\mathcal{A}}(u)$.

This is summarized by the following theorem. We refer the reader to Section 3.2 for its proof.

Theorem 4 (Non-Uniform Indistinguishability Hardcore Lemma). *Let $E : \mathcal{U} \to \mathcal{X}$ and $F : \mathcal{V} \to \mathcal{X}$ be functions, and let $\epsilon, \gamma \in (0,1)$ and $s > 0$ be given. If for all distinguishers D with size s we have*

$$\Delta^D(E(U), F(V)) \leq \epsilon$$

for $U \overset{\$}{\leftarrow} \mathcal{U}$ and $V \overset{\$}{\leftarrow} \mathcal{V}$, then there exist measures $\mathcal{M}$ on $\mathcal{U}$ and $\mathcal{N}$ on $\mathcal{V}$ with $\mu(\mathcal{M}) \geq 1-\epsilon$ and $\mu(\mathcal{N}) \geq 1-\epsilon$ such that

$$\Delta^{D'}(E(U'), F(V')) \leq \gamma,$$

for all distinguishers D' with size $s' := \frac{s\cdot\gamma^2}{128(\log|\mathcal{U}|+\log|\mathcal{V}|+1)}$, where $U' \overset{\$}{\leftarrow} \mathsf{P}_{\mathcal{M}}$ and $V' \overset{\$}{\leftarrow} \mathsf{P}_{\mathcal{N}}$.

We stress that the measures $\mathcal{M}$ and $\mathcal{N}$ given by the theorem generally depend on γ and s.

PRGs and Computational Entropy. As an example application of Theorem 4, we instantiate the function E by an (s,ϵ)-PRG $G : \{0,1\}^k \to \{0,1\}^\ell$ (in particular $\mathcal{U} := \{0,1\}^k$ and $\mathcal{X} := \{0,1\}^\ell$), whereas F is the identity function and $\mathcal{V} = \mathcal{X} = \{0,1\}^\ell$. For any $\gamma > 0$, Theorem 4 implies that we can define an event $\mathcal{A}$ on the choice of the seed of the PRG (with $\mathsf{P}_{\mathcal{A}|U}(u) := \mathcal{M}(u)$) occurring with probability $\mathsf{P}[\mathcal{A}] = \mu(\mathcal{M}) \geq 1-\epsilon$ such that conditioned on $\mathcal{A}$, no distinguisher with size s' can achieve advantage higher than γ in distinguishing the ℓ-bit PRG

[8] In fact, our results can naturally be generalized to the case where U and V have arbitrary distributions by considering a slightly more general version of Theorem 2 with arbitrary distributions for W.

output from an ℓ-bit string U'_ℓ sampled according to $\mathcal{N}$, which, by Lemma 2, has min-entropy at least $\ell - \log\left(\frac{1}{1-\epsilon}\right)$.

In other words, the output of every ϵ-PRG $G : \{0,1\}^k \to \{0,1\}^\ell$ exhibits (with probability $1-\epsilon$) high computational min-entropy. Note that the achieved form of computational entropy is somewhat weaker than the traditional notion of HILL min-entropy [6], where the random variable U'_ℓ is the *same* for all polynomially bounded s and all noticeable $\gamma > 0$. Still, it is strong enough to allow for the use of G's output in place of some string which has high min-entropy with probability $1 - \epsilon$.

3.2 Proof of Theorem 4

We start by defining a function $g : \mathcal{U} \times \mathcal{V} \times \{0,1\} \to \mathcal{X}$ and a predicate $P : \mathcal{U} \times \mathcal{V} \times \{0,1\} \to \{0,1\}$ such that

$$g(u,v,b) := \begin{cases} E(u) & \text{if } b = 0, \\ F(v) & \text{if } b = 1. \end{cases}$$

and $P(u,v,b) := b$ for all $u \in \mathcal{U}$ and $v \in \mathcal{V}$. It is well known that for any two random variables $\widetilde{U}$ and $\widetilde{V}$, and a distinguisher D, the distinguishing advantage $\Delta^D(E(\widetilde{U}), F(\widetilde{V}))$ can equivalently be characterized in terms of the probability that D guesses the uniform random bit B in a game where it is given $E(\widetilde{U})$ if $B = 0$ and $F(\widetilde{V})$ otherwise: In particular, we have

$$\Delta^D(E(\widetilde{U}), F(\widetilde{V})) = \left|\text{Guess}^D(B \mid g(\widetilde{U}, \widetilde{V}, B))\right|$$

for a uniform random bit B, where $\widetilde{U}$, $\widetilde{V}$, and B are sampled independently.

We now assume towards a contradiction that for all pairs of measures $\mathcal{M}$ and $\mathcal{N}$, both with density at least $1 - \epsilon$, there exists a distinguisher D' of size at most s' with $\Delta^{D'}(E(U'), F(V')) \geq \gamma$, for $U' \xleftarrow{\$} \mathsf{P}_{\mathcal{M}}$ and $V' \xleftarrow{\$} \mathsf{P}_{\mathcal{N}}$.

We prove that, under this assumption, for all measures $\mathcal{M}$ on $\mathcal{U} \times \mathcal{V} \times \{0,1\}$ with $\mu(\mathcal{M}) \geq 1 - \epsilon$ there exists a circuit C' with size s' such that $\text{Guess}^{C'}(B' \mid g(U', V', B')) \geq \frac{\gamma}{2}$, for $(U', V', B') \xleftarrow{\$} \mathsf{P}_{\mathcal{M}}$. As this contradicts the statement of the non-uniform hardcore lemma (Theorem 2) for $\frac{\gamma}{2}$ (instead of γ), this implies that there must be a circuit C with size s such that

$$\Delta^C(E(U), F(V)) \geq \text{Guess}^C(B \mid g(U, V, B)) > \epsilon.$$

In turn, this contradicts the assumed indistinguishability of $E(U)$ and $F(V)$, concluding the proof.

Reduction to the Hardcore Lemma. In the remainder of this proof, let us assume that we are given a measure $\mathcal{M}$ on $\mathcal{U} \times \mathcal{V} \times \{0,1\}$ with $\mu(\mathcal{M}) \geq 1 - \epsilon$. We first define the measures $\mathcal{M}_0$ and $\mathcal{M}_1$ on $\mathcal{U}$ and $\mathcal{V}$, respectively, such that

$$\mathcal{M}_0(u) := \frac{1}{|\mathcal{V}|} \sum_{v \in \mathcal{V}} \mathcal{M}(u,v,0) \quad \text{and} \quad \mathcal{M}_1(v) := \frac{1}{|\mathcal{U}|} \sum_{u \in \mathcal{U}} \mathcal{M}(u,v,1)$$

Furthermore, let $m_b := \sum_{u,v} \mathcal{M}(u,v,b)$ for $b \in \{0,1\}$, and let $m := |\mathcal{M}| = m_0 + m_1$. Note that in particular $\mu(\mathcal{M}_b) = \frac{m_b}{|\mathcal{U}|\cdot|\mathcal{V}|}$ and $\mu(\mathcal{M}) = \frac{m}{2\cdot|\mathcal{U}|\cdot|\mathcal{V}|}$.

We consider two cases in the following, both leading to a circuit C'.

Case $|\frac{m_0}{m} - \frac{1}{2}| > \frac{\gamma}{4}$. Assume that $\frac{m_0}{m} - \frac{1}{2} > \frac{\gamma}{4}$. (The other case is symmetric.) Then, for the circuit C' *always* outputting the bit 0,

$$\mathrm{Guess}^{C'}(B' \mid g(U', V', B')) = 2 \cdot \mathsf{P}[B' = 0] - 1 = 2 \cdot \tfrac{m_0}{m} - 1 > \frac{\gamma}{2}.$$

Case $|\frac{m_0}{m} - \frac{1}{2}| \le \frac{\gamma}{4}$. We assume that $\frac{1}{2} \ge \frac{m_0}{m} \ge \frac{1}{2}(1 - \frac{\gamma}{2})$ and hence also $\frac{1}{2}(1 + \frac{\gamma}{2}) \ge \frac{m_1}{m} \ge \frac{1}{2}$ (once again the other case is symmetric). This yields in particular that $(1 - \gamma/2)\mu(\mathcal{M}) \le \mu(\mathcal{M}_0) \le \mu(\mathcal{M})$ and $\mu(\mathcal{M}) \le \mu(\mathcal{M}_1) \le (1 + \gamma/2)\mu(\mathcal{M})$.

The goal is to define two measures $\widetilde{\mathcal{M}}_0$ on $\mathcal{U}$ and $\widetilde{\mathcal{M}}_1$ on $\mathcal{V}$, *both* with density at least $1 - \epsilon$, such that a distinguisher D' achieving advantage larger than γ in distinguishing $E(\widetilde{U}')$ and $F(\widetilde{V}')$ for $\widetilde{U}' \xleftarrow{\$} \mathsf{P}_{\widetilde{\mathcal{M}}_0}$ and $\widetilde{V}' \xleftarrow{\$} \mathsf{P}_{\widetilde{\mathcal{M}}_1}$ also achieves advantage higher than $\gamma/2$ in guessing B' given $g(U', V', B')$. Ideally, we would set $\widetilde{\mathcal{M}}_b := \mathcal{M}_b$, but note that $\mu(\mathcal{M}_0) < 1 - \epsilon$ possibly holds. We slightly modify $\mathcal{M}_0$ in order to satisfy this property, i.e., we define for all $u \in \mathcal{U}$ and $v \in \mathcal{V}$

$$\widetilde{\mathcal{M}}_0(u) := \frac{1 - \mu(\mathcal{M})}{1 - \mu(\mathcal{M}_0)} \cdot \mathcal{M}_0(u) + \frac{\mu(\mathcal{M}) - \mu(\mathcal{M}_0)}{1 - \mu(\mathcal{M}_0)} \quad \text{and} \quad \widetilde{\mathcal{M}}_1(v) := \mathcal{M}_1(v).$$

(We tacitly assume $\mu(\mathcal{M}_0) < 1$, otherwise we can simply set $\widetilde{\mathcal{M}}_0 := \mathcal{M}_0$.) It is easy to verify that $\mathcal{M}_0(u) \le \widetilde{\mathcal{M}}_0(u) \le 1$. Moreover,

$$\mu(\widetilde{\mathcal{M}}_0) = \frac{1 - \mu(\mathcal{M})}{1 - \mu(\mathcal{M}_0)} \cdot \mu(\mathcal{M}_0) + \frac{\mu(\mathcal{M}) - \mu(\mathcal{M}_0)}{1 - \mu(\mathcal{M}_0)} = \mu(\mathcal{M}) \ge 1 - \epsilon.$$

This implies, by our assumption, that for $\widetilde{U}'$ and $\widetilde{V}'$ sampled according to $\widetilde{\mathcal{M}}_0$ and $\widetilde{\mathcal{M}}_1$ there exists D' such that

$$\mathsf{P}[D'(F(\widetilde{V}')) = 1] - \mathsf{P}[D'(E(\widetilde{U}')) = 1] > \gamma. \tag{1}$$

We now show that the advantage of D' in guessing B' given $g(U', V', B')$ is larger than $\frac{\gamma}{2}$. To this aim, we introduce the following two probability distributions P_1 and P_2, both with range $(\mathcal{U} \times \{0\}) \cup (\mathcal{V} \times \{1\})$. The former distribution is the distribution of $(g(\widetilde{U}', \widetilde{V}', B), B)$ for $\widetilde{U}' \xleftarrow{\$} \mathsf{P}_{\widetilde{\mathcal{M}}_0}$, $\widetilde{V}' \xleftarrow{\$} \mathsf{P}_{\widetilde{\mathcal{M}}_1}$, and $B \xleftarrow{\$} \{0,1\}$, that is

$$\mathsf{P}_1(u, 0) := \frac{\widetilde{\mathcal{M}}_0(u)}{2|\widetilde{\mathcal{M}}_0|} \quad \text{and} \quad \mathsf{P}_1(v, 1) := \frac{\widetilde{\mathcal{M}}_1(v)}{2|\widetilde{\mathcal{M}}_1|} \text{ for all } u \in \mathcal{U} \text{ and } v \in \mathcal{V}.$$

The latter is the distribution of $(g(U', V', B'), B')$ for $(U', V', B') \xleftarrow{\$} \mathsf{P}_{\mathcal{M}}$, i.e.,

$$\mathsf{P}_2(u, 0) := \frac{|\mathcal{V}| \cdot \mathcal{M}_0(u)}{|\mathcal{M}|} \quad \text{and} \quad \mathsf{P}_2(v, 1) := \frac{|\mathcal{U}| \cdot \mathcal{M}_1(v)}{|\mathcal{M}|} \text{ for all } u \in \mathcal{U} \text{ and } v \in \mathcal{V}.$$

We prove the following two lemmas for $(X_1, B_1) \xleftarrow{\$} \mathsf{P}_1$ and $(X_2, B_2) \xleftarrow{\$} \mathsf{P}_2$.

Lemma 3. $\mathrm{Guess}^{D'}(B' \mid g(U',V',B')) > \gamma - 2 \cdot d((X_1,B_1),(X_2,B_2))$.

Proof. Consider the distinguisher $\overline{D}$ which given a pair $(x,b) \in (\mathcal{U}\times\{0\}) \cup (\mathcal{V}\times\{1\})$ outputs 1 if $b=0$ *and* $D'(E(x))=0$, or if $b=1$ *and* $D'(F(x))=1$. Then, note that by (1)

$$\mathsf{P}[\overline{D}(X_1,B_1)=1] = \tfrac{1}{2}\left(\mathsf{P}[D'(E(\widetilde{U}'))=0] + \mathsf{P}[D'(F(\widetilde{V}'))=1]\right)$$
$$= \tfrac{1}{2}\left(1 + \mathsf{P}[D'(F(\widetilde{V}'))=1] - \mathsf{P}[D'(E(\widetilde{U}'))=1]\right) > \tfrac{1}{2} + \tfrac{\gamma}{2}.$$

Furthermore, $\mathsf{P}(\overline{D}(X_2,B_2)=1] \leq \frac{1}{2} + \frac{\mathrm{Guess}^{D'}(B' \mid g(U',V',B'))}{2}$ by the definition of g. The fact that

$$\mathsf{P}[\overline{D}(X_1,B_1)=1] - \mathsf{P}[\overline{D}(X_2,B_2)=1] \leq \Delta^{\overline{D}}((X_1,B_1),(X_2,B_2))$$
$$\leq d((X_1,B_1),(X_2,B_2))$$

implies the lemma. □

Lemma 4. $d((X_1,B_1),(X_2,B_2)) \leq \frac{\gamma}{4}$.

Proof. For all $v \in \mathcal{V}$ we have $\mathsf{P}_1(v,1) \leq \mathsf{P}_2(v,1)$, since $|\mathcal{M}| \leq 2\cdot|\mathcal{U}|\cdot|\mathcal{M}_1|$. Furthermore, for all $u \in \mathcal{U}$ we have

$$\mathsf{P}_1(u,0) = \frac{\widetilde{\mathcal{M}}_0(u)}{2|\widetilde{\mathcal{M}}_0|} = \frac{\widetilde{\mathcal{M}}_0(u)}{2|\mathcal{U}|\mu(\mathcal{M})} \geq \frac{\mathcal{M}_0(u)}{2|\mathcal{U}|\mu(\mathcal{M})} = \frac{|\mathcal{V}|\cdot\mathcal{M}_0(u)}{|\mathcal{M}|} = \mathsf{P}_2(u,0)$$

using the fact that $|\mathcal{M}| = 2\cdot\mu(\mathcal{M})\cdot|\mathcal{U}|\cdot|\mathcal{V}|$. This yields

$$d((X_1,B_1),(X_2,B_2)) = \sum_{v\in\mathcal{V}} \mathsf{P}_2(v,1) - \mathsf{P}_1(v,1)$$
$$= \frac{1}{2|\mathcal{V}|}\sum_{v\in\mathcal{V}} \mathcal{M}_1(v)\cdot\left(\frac{1}{\mu(\mathcal{M})} - \frac{1}{\mu(\mathcal{M}_1)}\right)$$
$$= \frac{1}{2}\left(\frac{\mu(\mathcal{M}_1)}{\mu(\mathcal{M})} - 1\right) \leq \frac{\gamma}{4},$$

since $\mu(\mathcal{M}_1) \leq (1+\frac{\gamma}{2})\cdot\mu(\mathcal{M})$. □

Therefore, we conclude the proof of Theorem 4 by combining both lemmas, which show that the advantage of $C' := D'$ is larger than $\gamma/2$, as desired.

3.3 The Uniform Case

In this section, we prove a *uniform* version of Theorem 4 in the same spirit as the uniform hardcore lemma (Theorem 3): If $E(U)$ and $F(V)$ can only be distinguished with advantage ϵ by a polynomial-time distinguisher, then for all noticeable $\gamma > 0$ and for all polynomial-time oracle distinguishers $D^{(\cdot,\cdot)}$ (making input-independent oracle queries), there exist two measures $\mathcal{M}$ and $\mathcal{N}$ on $\mathcal{U}$ and $\mathcal{V}$, each with density $1-\epsilon$, such that $D^{\mathcal{M},\mathcal{N}}$ cannot achieve advantage better than γ in telling $E(U')$ and $F(V')$ apart, where $U' \xleftarrow{\$} \mathsf{P}_{\mathcal{M}}$ and $V' \xleftarrow{\$} \mathsf{P}_{\mathcal{N}}$.

Theorem 5 (Uniform Indistinguishability Hardcore Lemma). *Let $E : \mathcal{U} \to \mathcal{X}$ and $F : \mathcal{V} \to \mathcal{X}$, $\epsilon : \mathbb{N} \to [0,1]$, and $\gamma : \mathbb{N} \to [0,1]$ be functions computable in time* $\mathsf{poly}(k)$, *where ϵ and γ are both noticeable. Assume that for all polynomial-time distinguishers D we have*

$$\Delta^D(E(U), F(V)) \leq \epsilon$$

for $U \xleftarrow{\$} \mathcal{U}$ and for $V \xleftarrow{\$} \mathcal{V}$, then for all polynomial time distinguishers $D'^{(\cdot,\cdot)}$ whose oracle queries are independent of their input, there exist measures $\mathcal{M}$ on $\mathcal{U}$ and $\mathcal{N}$ on $\mathcal{V}$ with $\mu(\mathcal{M}) \geq 1-\epsilon$ and $\mu(\mathcal{N}) \geq 1-\epsilon$ such that

$$\Delta^{D'^{\mathcal{M},\mathcal{N}}}(E(U'), F(V')) \leq \gamma,$$

where $U' \xleftarrow{\$} \mathsf{P}_{\mathcal{M}}$ and $V' \xleftarrow{\$} \mathsf{P}_{\mathcal{N}}$.

Due to lack of space, the proof of the theorem (which follows the lines of the non-uniform case, but with extra difficulties) can be found in the full version.

4 Security Amplification of PRGs

4.1 The Concatenate-and-Extract Construction

This section presents, as an application of Theorems 4 and 5, the first construction achieving security amplification of arbitrarily weak PRGs.

CONSTRUCTION. Let $G : \{0,1\}^k \to \{0,1\}^\ell$ and $\mathsf{Ext} : \{0,1\}^{m\ell} \times \{0,1\}^d \to \{0,1\}^n$ be efficiently computable functions. We consider the *Concatenate-and-Extract* (CaE) construction $\mathsf{CaE}^{G,\mathsf{Ext}} : \{0,1\}^{mk+d} \to \{0,1\}^{n+d}$ such that

$$\mathsf{CaE}^{G,\mathsf{Ext}}(x_1 \| \cdots \| x_m \| r) := \mathsf{Ext}(G(x_1) \| \cdots \| G(x_m), r) \| r.$$

for all $x_1, \ldots, x_m \in \{0,1\}^k$ and $r \in \{0,1\}^d$.

PARAMETERS AND MAIN SECURITY STATEMENT. The intuition justifying the security of the CaE construction relies on the simple observation that, provided that G is an ϵ-PRG, each individual and *independent* PRG output in the concatenation $G(X_1)\| \cdots \|G(X_m)$ (for uniform $X_1, \ldots, X_m$) has computational min-entropy at least $\ell - \log\left(\frac{1}{1-\epsilon}\right)$ with probability at least $1-\epsilon$, and thus we can expect the whole concatenation to have computational min-entropy $\approx m \cdot (1-\epsilon) \cdot \left[\ell - \log\left(\frac{1}{1-\epsilon}\right)\right]$ with very high probability, which can be extracted if Ext is an appropriate extractor. Note that the resulting construction is expanding if $n/mk > 1$, and for an optimal extractor this ratio is roughly $(1-\epsilon)\left[\ell - \log\left(\frac{1}{1-\epsilon}\right)\right]/k$ (we ignore the entropy loss of the extractor for simplicity), or, turned around, our construction expands if the underlying ϵ-PRG satisfies $\ell/k > \frac{1}{1-\epsilon} + \log\left(\frac{1}{1-\epsilon}\right)/k$ holds. In particular, this value is independent of m. In Section 4.3, we show that for a large class of natural constructions this is essentially optimal. For example, for $\epsilon = \frac{1}{2}$, the output length ℓ of the given generator G needs to be slightly larger than $2k$ in order to achieve expansion. For comparison, the SUM construction is expanding if $\ell/k > m$, where $m = \omega(k/\log(1/\epsilon))$ in order for the construction to be secure.

Also, the fact that all ℓ-bit blocks are independent allows for using *deterministic* extractors in the CaE construction, such as the one given by Theorem 1, as long as $(1-\epsilon)\left(1 - \frac{1}{\ell}\log\left(\frac{1}{1-\epsilon}\right)\right)$ is bounded from below by $(m\ell)^{-\eta}$.

The following theorem proves the soundness of the above intuition.

Theorem 6 (Strong Security Amplification for PRGs). *Let $\rho, \delta, \epsilon : \mathbb{N} \rightarrow [0,1]$ be functions, and let $G : \{0,1\}^k \rightarrow \{0,1\}^\ell$ (for $k < \ell$) be an ϵ-PRG. Furthermore, let $\mathsf{Ext} : \{0,1\}^{m\ell} \times \{0,1\}^d \rightarrow \{0,1\}^n$ be a strong δ-extractor for $\left(m, \ell, (1-\epsilon-\rho)\cdot m \cdot \left[\ell - \log\left(\frac{1}{1-\epsilon}\right)\right]\right)$-total-entropy independent sources.*

Then the function $\mathsf{CaE}^{G,\mathsf{Ext}} : \{0,1\}^{mk+d} \rightarrow \{0,1\}^{n+d}$ is a $(e^{-\rho^2 m} + \delta + \nu)$-PRG, where ν is a negligible function.

The theorem is proven by means of a *uniform reduction* using Theorem 5, and hence holds both in the uniform *and* in the non-uniform settings. However, the next paragraph gives an ad-hoc proof for the non-uniform case which follows the above simple intuition and which is also tighter than the more involved uniform reduction, which we defer to Section 4.2.

Non-Uniform Proof. In the following, let us fix $s, \gamma > 0$, and assume G is an (s, ϵ)-PRG. Also consider the $m\ell$-bit string $G(X_1)\|\ldots\|G(X_m)$, where $X_1, \ldots, X_m$ are independent uniform k-bit strings.

By Theorem 4, there exist independent events $\mathcal{A}_1, \ldots, \mathcal{A}_m$ such that $\mathcal{A}_i$ can be adjoined to X_i and $\mathsf{P}[\mathcal{A}_i] \geq 1 - \epsilon$, and, conditioned on theses events, no size s' distinguisher can distinguish $G(X_i)$ from some variable U'_i with min-entropy $\mathsf{H}_\infty(U'_i) \geq \ell - \log\left(\frac{1}{1-\epsilon}\right)$ with advantage larger than γ. In particular, by Hoeffding's inequality (Lemma 5), the events $\mathcal{A}_i$ occur for a subset $\mathcal{I} \subseteq \{1, \ldots, m\}$ of indices such that $|\mathcal{I}| \geq (1-\epsilon-\rho)\cdot m$, except with probability $e^{-\rho^2 m}$. In this case, for a uniform random d-bit string R, a standard hybrid argument yields that every distinguisher of size s'' (where s'' is only slightly smaller than s') can achieve advantage at most $m\gamma$ in distinguishing $\mathsf{CeE}^{G,\mathsf{Ext}}(X_1\|\ldots\|X_m\|R) = \mathsf{Ext}(G(X_1)\|\ldots\|G(X_m), R)\|R$ from the string $\mathsf{Ext}(U', R)\|R$, where U' is obtained by replacing each $G(X_i)$ with $i \in \mathcal{I}$ with the corresponding U'_i. In particular, since U' has min-entropy at least $(1-\epsilon-\rho)\cdot m \cdot \left[\ell - \log\left(\frac{1}{1-\epsilon}\right)\right]$, the variable $\mathsf{Ext}(U', R)\|R$ has distance at most δ from a uniform random $(n+d)$-bit string U_{n+d}.

Thus, using the triangle inequality and adding the three advantages, we obtain that $\mathsf{CaE}^{G,\mathsf{Ext}}$ is a $(s'', m\gamma + \delta + e^{-\rho^2 m})$-PRG. The asymptotic bound follows by applying the same argument to any polynomially bounded s and to all noticeable γ.

Distinguisher $D'^{\mathcal{M},\mathcal{N}}(z)$ *// On input* $z \in \{0,1\}^\ell$

$x_1, \ldots, x_m \xleftarrow{\$} \{0,1\}^k$, $r \xleftarrow{\$} \{0,1\}^d$
for all $i = 1, \ldots, m$ **do**
  $\mathcal{G} := \mathcal{G} \cup \{i\}$ with probability $\mathcal{M}(x_i)$
$i^* \xleftarrow{\$} \{1, \ldots, m\}$
for all $i = 1, \ldots m$ **do**
  if $i \in \mathcal{G}$ *and* $i < i^*$ **then** $y_i \xleftarrow{\$} \mathsf{P}_{\mathcal{N}}$ **else** $y_i := G(x_i)$
if $i^* \in \mathcal{G}$ **then**
  return $D'_{\mathcal{M},\mathcal{N}}(\cdot) := D(\mathsf{Ext}(y_1 \| \ldots \| y_{i^*-1} \| z \| y_{i^*+1} \| \cdots \| y_m, r) \| r)$
else
  return $D'_{\mathcal{M},\mathcal{N}}(\cdot) := D(\mathsf{Ext}(y_1 \| \ldots \| y_m, r) \| r)$ *//* $D'_{\mathcal{M},\mathcal{N}}$ *ignores its input*

Fig. 1. The distinguisher $D'^{(\cdot,\cdot)}$ in the proof of Theorem 6

4.2 Proof of Theorem 6

Assume, towards a contradiction, that there exists a polynomial-time distinguisher D and a noticeable function η such that for infinitely many values of the security parameter k (which we omit) we have

$$\Delta^D(\mathsf{CaE}^{G,\mathsf{Ext}}(X_1 \| \cdots \| X_m \| R), U_{n+d}) > e^{-\rho^2 m} + \delta + \eta,$$

where $X_1, \ldots, X_m$ are uniformly distributed k-bit string, R is a uniformly distributed d-bit string, and U_{n+d} is a uniformly distributed $(n+d)$-bit string.

THE DISTINGUISHER $D'^{(\cdot,\cdot)}$. We give a distinguisher $D'^{(\cdot,\cdot)}$ (which is fully specified in Figure 1), which on input $z \in \{0,1\}^\ell$ and given oracle access to any two measures $\mathcal{M} : \mathcal{U} \to [0,1]$ and $\mathcal{N} : \mathcal{V} \to [0,1]$ operates as follows: First, it chooses m k-bit strings $x_1, \ldots, x_m$ independently and uniformly at random, and for each $i \in \{1, \ldots, m\}$ an independent coin is flipped (taking value 1 with probability $\mathcal{M}(x_i)$, and 0 otherwise), and if the coin flip returns 1, the position i is marked as "being in the measure". Let $\mathcal{G}$ be the set of marked positions. Subsequently, an index i^* is chosen uniformly random from $\{1, \ldots, m\}$. Then, a string $y_1 \| \ldots \| y_m \in \{0,1\}^{m\ell}$ (where $y_1, \ldots, y_m \in \{0,1\}^\ell$) is built as follows: Each y_i is set to an independent element sampled according to $\mathsf{P}_{\mathcal{N}}$ if $i \in \mathcal{G}$ and $i < i^*$, and in any other case it is set to $G(x_i)$. Finally, the distinguisher chooses the seed r for the extractor uniformly at random, and outputs the bit $D(\mathsf{Ext}(y_1 \| \ldots \| y_{i^*-1} \| z \| y_{i^*+1} \| \ldots \| y_m, r) \| r)$ if $i^* \in \mathcal{G}$ holds, or it outputs $D(\mathsf{Ext}(y_1 \| \ldots \| y_m, r) \| r)$ else (in particular, the input z is ignored in this latter case).

ANALYSIS. In the following, let $\mathcal{M}$ and $\mathcal{N}$ both have density $1 - \epsilon$, let X' be sampled according to $\mathsf{P}_{\mathcal{M}}$, and let U' be sampled according to $\mathsf{P}_{\mathcal{N}}$. We compute the average advantage $\Delta^{D'}(G(X'), U')$ of the distinguisher $D' = D'^{\mathcal{M},\mathcal{N}}$.

It is convenient to use the shorthands $\mathsf{P}[D'(\cdot) \,|\, g] := \mathsf{P}[D'(X) = 1 \,|\, |\mathcal{G}| = g]$ to denote the conditional probability of D' outputting 1 on input X given that $|\mathcal{G}| = g \in \{0, 1, \ldots, m\}$. Similarly, we denote $\mathsf{P}[D'(X) \,|\, g, i] := \mathsf{P}[D'(X) = 1 \,|\, |\mathcal{G}| = g \wedge i^* = i]$ when additionally conditioned on $i^* = i$. Then,

$$\begin{aligned}\Delta^{D'}(G(X'), U') &= |\mathsf{P}[D'(G(X')) = 1] - \mathsf{P}[D'(U') = 1]| \\ &= \left|\sum_{g=0}^{m} \mathsf{P}_{|\mathcal{G}|}(g)\,(\mathsf{P}[D'(G(X')) \,|\, g] - \mathsf{P}[D'(U') \,|\, g])\right| \\ &= \left|\sum_{g=0}^{m} \mathsf{P}_{|\mathcal{G}|}(g) \cdot \frac{1}{m}\sum_{i^*=1}^{m} (\mathsf{P}[D'(G(X')) \,|\, g, i^*] - \mathsf{P}[D'(U') \,|\, g, i^*])\right|\end{aligned}$$

By construction $\mathsf{P}[D'(G(X')) \,|\, g, i^*] = \mathsf{P}[D'(U') \,|\, g, i^* - 1]$ for $g \in \{1, \ldots, m\}$ and $i^* = \{2, \ldots, m\}$, and we hence obtain

$$\Delta^{D'}(G(X'), U') = \frac{1}{m}\left|\sum_{g=0}^{m} \mathsf{P}_{|\mathcal{G}|}(g) \cdot (\mathsf{P}[D'(G(X')) \,|\, g, 1] - \mathsf{P}[D'(U') \,|\, g, m])\right|$$

On the one hand, we now remark that

$$\sum_{g=0}^{m} \mathsf{P}_{|\mathcal{G}|}(g) \cdot \mathsf{P}[D'(G(X')) \,|\, g, 1] = \mathsf{P}[D(\mathsf{CaE}^{G,\mathsf{Ext}}(X_1\| \ldots \|X_m\|R)) = 1].$$

On the other hand, because $\mu(\mathcal{N}) \geq 1 - \epsilon$, whenever $g \geq (1 - \epsilon - \rho)m$ and $i^* = m$, the distribution of $y_1\| \ldots \|y_m$ belongs to an $\left(m, \ell, (1 - \epsilon - \rho)\left(\ell - \log\left(\frac{1}{1-\epsilon}\right)\right)\right)$-total-entropy independent source, and as Ext is a δ-extractor for this source, we obtain $|\mathsf{P}[D'(U') \,|\, g, m] - \mathsf{P}[D(U_{n+d}) = 1]| \leq \delta$, whereas $\mathsf{P}[|\mathcal{G}| < (1 - \epsilon - \rho)m] < e^{-\rho^2 m}$ by Hoeffding's inequality (Lemma 5) and that fact that $\mu(\mathcal{M}) \geq 1 - \epsilon$. We can finally infer

$$\Delta^{D'}(G(X'), U') \geq \frac{\Delta^{D}(\mathsf{CaE}^{G,\mathsf{Ext}}(X_1\| \ldots \|X_m\|R), U_{n+d}) - \delta - e^{-\rho^2 m}}{m} > \frac{\eta}{m}$$

by our assumption on D.

As the queries of D' do not depend on the inputs, and the above lower bound on its advantage holds for all measures $\mathcal{M}$ and $\mathcal{N}$ with density at least $1 - \epsilon$, the distinguisher D' contradicts Theorem 5 for $\gamma := \frac{\eta}{m}$, which is noticeable, and implies that G is *not* an ϵ-PRG, which is a contradiction.

4.3 Optimality of the Output Length

This final section discusses the optimality of the output length of the concatenate-and-extract construction with respect to the class of constructions which operate by combining a number of independent outputs from weak PRGs, and such that the corresponding security reduction is black-box. In particular, the reduction only exploits the capability of efficiently sampling a given distribution.[9] This is formally summarized by the following definition.

[9] In particular, note that the proof itself uses black-box access to some function sampling the PRG output which is not required to be expanding. All known proofs have this form.

Definition 1. *A* black-box (ℓ, ϵ)-indistinguishability amplifier *consists a pair of polynomial-time algorithms* (C, S) *with the following two properties:*

(i) For some functions m, d, and h, the algorithm C *implements a function family* $(\{0,1\}^\ell)^m \times \{0,1\}^d \to \{0,1\}^h$*, where the second input parameter models explicitly the d-bit randomness used by the algorithm* C*.*

(ii) Let P_X *be an arbitrary distribution on the* ℓ*-bit strings which is sampled by an algorithm* X*, let* $X_1, \ldots, X_m$ *be independent samples of* P_X*, and let* R *and* U_h *be uniformly distributed d- and h-bit strings, respectively. Then, for every distinguisher D such that*

$$\Delta^D(\mathsf{C}(X_1, \ldots, X_m, R), U_h) > \gamma$$

for infinitely many values of the security parameter and a noticeable function γ*, we have* $\Delta^{\mathsf{S}^{D,\mathsf{X}}}(X, U_\ell) > \epsilon$ *for infinitely many values of the security parameter, where* $X \xleftarrow{\$} \mathsf{P}_X$ *and* U_ℓ *is a uniform* ℓ*-bit string.*

The following theorem (proven in the full version) shows that the output length achieved by concatenate-and-extract is essentially optimal.

Theorem 7. *For all* $\ell \in \mathbb{N}$*, for all constants* $0 < \rho < \epsilon < 1$*, there exists no black-box* (ℓ, ϵ)*-indistinguishability amplifier if* $h \geq (1-\epsilon+\rho) \cdot m \cdot \left[\ell - \log\left(\frac{1}{1-\epsilon}\right)\right] + d + 1$.

Acknowledgments. We thank Russell Impagliazzo for helpful discussions. This research was partially supported by the Swiss National Science Foundation (SNF), project no. 200020-113700/1.

References

1. Bennett, C.H., Brassard, G., Robert, J.-M.: Privacy amplification by public discussion. SIAM Journal on Computing 17(2), 210–229 (1988)
2. Blum, M., Micali, S.: How to generate cryptographically strong sequences of pseudo random bits. In: FOCS 1982: Proceedings of the 23rd IEEE Annual Symposium on Foundations of Computer Science, pp. 112–117 (1982)
3. Dodis, Y., Impagliazzo, R., Jaiswal, R., Kabanets, V.: Security amplification for interactive cryptographic primitives. In: Reingold, O. (ed.) TCC 2009. LNCS, vol. 5444, pp. 128–145. Springer, Heidelberg (2009)
4. Goldreich, O., Nisan, N., Wigderson, A.: On Yao's XOR-lemma. In: Electronic Colloquium on Computational Complexity (ECCC), vol. 2(50) (1995)
5. Haitner, I., Harnik, D., Reingold, O.: On the power of the randomized iterate. In: Dwork, C. (ed.) CRYPTO 2006. LNCS, vol. 4117, pp. 22–40. Springer, Heidelberg (2006)
6. Håstad, J., Impagliazzo, R., Levin, L.A., Luby, M.: A pseudorandom generator from any one-way function. SIAM Journal on Computing 28(4), 1364–1396 (1999)
7. Hoeffding, W.: Probability inequalities for sums of bounded random variables. Journal of the American Statistical Association 58(301), 13–30 (1963)

8. Holenstein, T.: Key agreement from weak bit agreement. In: STOC 2005: Proceedings of the 37th Annual ACM Symposium on Theory of Computing, pp. 664–673 (2005)
9. Holenstein, T.: Pseudorandom generators from one-way functions: A simple construction for any hardness. In: Halevi, S., Rabin, T. (eds.) TCC 2006. LNCS, vol. 3876, pp. 443–461. Springer, Heidelberg (2006)
10. Impagliazzo, R.: Hard-core distributions for somewhat hard problems. In: FOCS 1995: Proceedings of the 36th IEEE Annual Symposium on Foundations of Computer Science, pp. 538–545 (1995)
11. Impagliazzo, R., Levin, L.A., Luby, M.: Pseudo-random generation from one-way functions (extended abstracts). In: STOC 1989: Proceedings of the 21st Annual ACM Symposium on Theory of Computing, pp. 12–24 (1989)
12. Kamp, J., Rao, A., Vadhan, S., Zuckerman, D.: Deterministic extractors for small-space sources. In: STOC 2006: Proceedings of the 38th Annual ACM Symposium on Theory of Computing, pp. 691–700 (2006)
13. Maurer, U., Pietrzak, K., Renner, R.: Indistinguishability amplification. In: Menezes, A. (ed.) CRYPTO 2007. LNCS, vol. 4622, pp. 130–149. Springer, Heidelberg (2007)
14. Maurer, U., Tessaro, S.: Computational indistinguishability amplification: Tight product theorems for system composition. In: Halevi, S. (ed.) CRYPTO 2009. LNCS, vol. 5677, pp. 350–368. Springer, Heidelberg (2009)
15. Myers, S.: Efficient amplification of the security of weak pseudo-random function generators. Journal of Cryptology 16, 1–24 (2003)
16. Shaltiel, R.: Recent developments in explicit constructions of extractors. Bulletin of the EATCS 77, 67–95 (2002)
17. Yao, A.C.: Theory and applications of trapdoor functions. In: FOCS 1982: Proceedings of the 23rd IEEE Annual Symposium on Foundations of Computer Science, pp. 80–91 (1982)

A Tail Estimates

The following well-known result from probability theory [7] is repeatedly used throughout this paper.

Lemma 5 (Hoeffding's Inequalities). *Let* $X_1, \ldots, X_m$ *be independent random variables with range* $[0,1]$, *and let* $\overline{X} := \frac{1}{m}\sum_{i=1}^{m} X_i$. *Then, for all* $\rho > 0$ *we have*

$$\mathsf{P}[\overline{X} \geq \mathsf{E}[\overline{X}] + \rho] \leq e^{-m\rho^2} \text{ and } \mathsf{P}[\overline{X} \leq \mathsf{E}[\overline{X}] - \rho] \leq e^{-m\rho^2}.$$

In particular,

$$\mathsf{P}\left[\left|\overline{X} - \mathsf{E}[\overline{X}]\right| \geq \rho\right] \leq 2 \cdot e^{-m\rho^2}.$$

On Related-Secret Pseudorandomness*

David Goldenberg and Moses Liskov

The College of William and Mary
{dcgold,mliskov}@cs.wm.edu

Abstract. Related-key attacks are attacks against constructions which use a secret key (such as a blockcipher) in which an attacker attempts to exploit known or chosen relationships among keys to circumvent security properties. Security against related-key attacks has been a subject of study in numerous recent cryptographic papers. However, most of these results are attacks on specific constructions, while there has been little positive progress on constructing related-key secure primitives.

In this paper, we attempt to address the question of whether related-key secure blockciphers can be built from traditional cryptographic primitives. We develop a theoretical framework of "related-secret secure" cryptographic primitives, a class of primitives which includes related-key secure blockciphers and PRFs. We show that while a single related-secret pseduorandom bit is sufficient and necessary to create related-key secure blockciphers, hard-core bits with typical proofs are *not* related-secret psuedorandom. Since the pseudorandomness of hard-core bits is the essential technique known to make pseudorandomness from assumptions of simple hardness, this presents a very strong barrier to the development of provably related-key secure blockciphers based on standard hardness assumptions.

1 Introduction

Related-key attacks are attacks against constructions using a secret key (such as a blockcipher) in which an attacker attempts to exploit known or chosen relationships among keys to circumvent security properties. Several related-key attacks on primitives have been developed [1,2,3], including attacks on AES [4,5,6,7]. While the realism of an adversary's ability to directly influence a secret key is questionable, the issue of related-key security has implications beyond such a setting. For instance, weakness in a blockcipher's key scheduling algorithm may result in known likely relationships amongst round keys, which could lead to an attack against the cipher [8]. As another example, blockcipher based hash functions are only proven secure in the ideal cipher model [9]; in this strong model, related-key security is implied [8]. Thus, the use of a real blockcipher for hashing that is not related-key secure is theoretically questionable: in many such constructions, the adversary's ability to choose the message to be hashed implies an ability to launch related-key attacks on the underlying cipher. Indeed, a recent

* Supported by NSF grant CNS-0845662.

D. Micciancio (Ed.): TCC 2010, LNCS 5978, pp. 255–272, 2010.

paper by Biryukov et al has made substantial progress on attacking AES-256 in Davies-Meyer mode via a strong related-key attack on AES [7]. Finally, there are settings in which related-key security has been put to good use: several papers make use of schemes with one-time related-key security properties in order to make fuzzy extractors robust against adversarial modification [10,11,12].

Positive results concerning related-key security are few. Bellare and Kohno [8] develop a theoretical framework for defining related-key security, show that some notions of related-key security are inherently impossible, and prove that an ideal cipher is related-key secure for a general class of relations. Lucks [13] shows how to achieve "partial" related-key security (meaning, that only part of the key can be varied), and also gave two proposed constructions of related-key secure pseudorandomness from novel, very strong number theoretic assumptions.

Defining related-secret security. Bellare and Kohno define related-key security as follows. If $\mathsf{F}_K(x)$ is a pseudorandom function (or permutation) then it is related-key secure if an adversary cannot distinguish between (1) an oracle that, on input x and a perturbation δ, returns $\mathsf{F}_{\delta(K)}(x)$ and (2) an oracle that implements a random function (or permutation) on x independently for each distinct δ. In order to study the relationship between such strong primitives and simpler ones, we broaden the concept of related-key attacks to "related-secret" attacks, which extends Bellare and Kohno's notion of related-key security to allow adversarially-specified perturbations on any inputs unknown to the adversary, whether or not these inputs are considered "keys". Simpler related-secret secure primitives are good candidates for intermediate steps between basic primitives and related-key security. We give definitions for many different related-secret secure primitives such as related-secret secure one way functions/permutations, hardcore bits, pseudorandom generators, pseudorandom functions, and pseudorandom permutations.[1]

Our results. We attempt to mirror the development of pseudorandom permutations from one-way functions or permutations in the related-secret setting, without assuming (as Lucks does [13]) that any part of any secret cannot be modified by the adversary. Because of this strong requirement, in most cases, we expect that strong related-secret secure primitives will require simple related-secret secure building blocks.

Most steps in the strengthening of basic primitives into pseudorandom permutations can be translated to the related-secret setting. We show related-secret security for any homomorphic one-way function, for instance, modular exponentiation (under the discrete log assumption). The idea is that homomorphic perturbations can be calculated without knowing the secret. We also show that the critical step is to obtain related-secret pseudorandomness: we show that even a 1-bit related-secret pseudorandom generator is sufficient to build related-secret pseudorandom permutations. Moreover, we show that these "pseudorandom bits" are necessary for related-key blockciphers.

[1] Our definitions of RK-PRF's and RK-PRP's are the same as [8].

In the standard model, a hard-core bit fills this role simply: a hard-core bit is hard to learn, and thus, hard to distinguish from a random bit [14,15]. However, in the related-secret setting things are different, because of the ability of a related-secret attacker to obtain multiple related values. So we distinguish between a related-secret hard-core bit (in which no bits are shown to the adversary, and one must be learned), and a pseudorandom bit (in which all the bits are shown to the adversary, who must distinguish them from random).

Here, we give a negative result: any "strong" hard-core bit (that is, a hard-core bit with a reduction from learning the bit to inverting the associated function) is *not* a related-secret pseudorandom bit. More specifically, we give a transformation from the reduction between the hard-core bit finder and the function inverter, to an attack against the same hard-core bit as a pseudorandom bit in the related-secret model. This transformation assumes that the reduction is a "black-token" reduction, in other words, that it simply manipulates the function output with known tools while being blind of the actual value of the function output. The notion of a black-token algorithm is akin to the concept of *algebraic reductions* and *generic ring algorithms* [16,17,18,19], except that we do not require the same level of algebraic structure.

This leads us to the conclusion that if related-secret pseudorandomness (including related-key blockciphers) are possible, they must be proven either based on other related-secret pseudorandomness assumptions[2], or a dramatically new way of creating pseudorandomness from hardness must be developed.

2 Definitions

If $f : \mathbb{N} \rightarrow \mathbb{R}$ is a function, we say that f is *negligible* if $\forall c$, $\exists n_0$ such that for all $n > n_0$, $f(n) < \frac{1}{n^c}$.

We use A to denote an adversary or algorithm. We denote the set of all probabilistic polynomial time adversaries as PPT. If an adversary A takes an oracle O we denote that as A^{O}.

We denote a value x randomly sampled from a set $\mathcal{X}$ as $x \leftarrow \mathcal{X}$. For a function f we denote $\mathcal{D}_f$ as the domain of f and $\mathcal{R}_f$ as the range. We denote the j'th bit of x as x_j and the i'th through the j'th bit of x as $x_{i,\ldots,j}$. For two inputs x, r we denote the inner product $(\sum_{i=0}^{k} x_i r_i)$ of x and r as $\langle x, r \rangle$. If x and r come from a metric space $\mathcal{M}$ we denote the distance between x and r as $||x - r||$.

Definition 1. *A (k, ρ, p) list decodable code is a triple of algorithms $\mathsf{C}, \mathsf{C}^{-1}, \mathsf{R}$ where $\mathsf{C}: \{0,1\}^k \rightarrow \mathcal{C} \subset \{0,1\}^n$, $\mathsf{C}^{-1} : \mathcal{C} \rightarrow \{0,1\}^k$ and R has the property that if $x \in \{0,1\}^k$ is the message, and y is the corrupted encoding of x, then $Pr[\mathcal{S} \leftarrow \mathsf{R}^{\mathsf{Y}} : \forall x \in \mathcal{S}, ||\mathsf{C}(x) - y|| \leq \rho n] \geq p$ as long as $||\mathsf{C}(x) - y|| \leq \rho n$ where Y is the oracle which on input i returns y_i. and where $|\mathcal{S}|$ is polynomial in k. We say a list decodable code allows for local decoding if R only queries Y a polynomial number of times.*

[2] Such as the related-key pseudorandom constructions of Lucks [13], based on novel assumptions effectively as strong as related-secret pseudorandomness.

Definition 2. *A list decodable code* $\mathsf{C}, \mathsf{C}^{-1}, \mathsf{R}$ *is* linear *when* $\mathcal{C}$ *is a linear subspace of* $\{0,1\}^n$.

2.1 Related-Secret Security

We consider primitives to be *related-secret secure* if they maintain their security even under an adversary which can exploit known or chosen relationships between related secret inputs. We specify the notion of interacting with the primitives using secrets "related to" the secret input $x \in \mathcal{D}$ by allowing the adversary to interact with the primitive under $\delta(x)$, where δ is a perturbation (function) from $\mathcal{D} \rightarrow \mathcal{D}$, which is specified by the adversary. For example, a block-cipher is related-key secure if no distinguishing adversary exists, even when we allow the adversary to make queries in which a particular transformation of the secret key is specified. While we might hope to design related-secret primitives for any possible set of perturbations, it has been shown in [8] that related-secret attacks inherently exist when the set of perturbations is "invalid", defined as follows [8]:

Definition 3 (Valid sets). *We say a set of functions* $\Delta : \mathcal{D} \rightarrow \mathcal{D}$ *is valid if it satisfies the following two properties:*

- *Output unpredictable:* Δ *is output unpredictable if* $\forall S \subset \Delta$, $\forall X \subset \mathcal{D}$, $Pr[x \leftarrow \mathcal{D}; \{\delta(x) : \delta \in S\} \cap X \neq \emptyset]$ *is negligible as long as* $|X|$ *and* $|S|$ *are polynomial in* $\log |\mathcal{D}|$.
- *Collision resistant:* Δ *is collision resistant if* $\forall S \subset \Delta$, $Pr[x \leftarrow \mathcal{D}; |\{\delta(x) : \delta \in S\}| < |S|]$ *is negligible when* $|S|$ *is polynomial in* $\log |\mathcal{D}|$.

Due to the power of the attack in [8], it is impossible to design cryptographic primitives that remain secure under arbitrary perturbations. As such, we will assume that Δ, the set of "allowable" perturbations, is a valid set. We also require that Δ has a minimal level of additional structure: we assume it is closed under function composition and contains the identity perturbation $\delta_{ident} \in \Delta : \delta_{ident}(x) = x$.[3]

We also say that Δ is *complete* in that $\forall x, y \in \mathcal{D}_f$, $\exists \delta : \delta(x) = y$. We note that related-key or related-secret security against an *incomplete* Δ is a far easier problem [13].

Two standard examples of Δ classes that meet these criteria:

- $\Delta_+ = \{\delta_c : x \mapsto x + c\}$
- $\Delta_\oplus = \{\delta_c : x \mapsto x \oplus c\}$

In both cases we identify the function δ_c with the value c. This sort of perturbation class is the most relevant: in many published related-key attack results, perturbations are from $\Delta_\oplus$ (for example, [7]).

We start by stating the definitions of a related key secure pseudorandom permutation and a related key secure pseudorandom function in the above notation. The definitions are taken from [8].

---

[3] For any Δ not closed under composition, or not including the identity, we can expand it to its closure under composition, and add the identity.

Definition 4 (Related key secure pseudorandom permutation (RK-PRP)). *An efficiently computable function* $\mathsf{E} : \{0,1\}^{p(k)} \times \{0,1\}^k \to \{0,1\}^{p(k)}$ *where p is a polynomial is considered a related key secure pseudorandom permutation under* Δ *if* $\mathsf{E}(\cdot, K)$ *is a permutation for all* $K \in \{0,1\}^k$ *and if* $\forall \mathsf{A}$, $\exists \nu$ *negligible:* $\forall k \; |Pr[K \leftarrow \{0,1\}^k : \mathsf{A}^{\mathsf{F}_{\mathsf{E}(\cdot,K),K}} = 1] - Pr[K \leftarrow \mathcal{K} : \mathsf{A}^{\mathsf{F}_{\mathcal{P}(\cdot,K),K}} = 1]| \leq \nu(k)$ *where k is the security parameter,* $\mathcal{P} : \{0,1\}^{p(k)} \times \{0,1\}^k \to \{0,1\}^{p(k)}$ *is a family of random permutations indexed by its second parameter, and where* $\mathsf{F}_{f(\cdot,K),K}$ *is the oracle which on input* $x, \delta \in \Delta$ *returns* $f(x, \delta(K))$.

Definition 5 (Related key secure pseudorandom function (RK-PRF)). *An efficiently computable function* $\mathsf{R} : \{0,1\}^{p(k)} \times \{0,1\}^k \to \{0,1\}^{p'(k)}$, *where* p *and* p' *are polynomials, is considered to be a related-key secure pseudorandom function under* Δ *if* $\forall \mathsf{A}$, $\exists \nu$ *negligible:* $\forall k \; |Pr[K \leftarrow \{0,1\}^k : \mathsf{A}^{\mathsf{F}_{\mathsf{R}(\cdot,K),K}} = 1] - Pr[K \leftarrow \{0,1\}^k : \mathsf{A}^{\mathsf{F}_{\mathcal{F}(\cdot,K),K}} = 1]| \leq \nu(k)$ *where k is the security parameter and where* $\mathcal{F} : \{0,1\}^{p(k)} \times \{0,1\}^k \to \{0,1\}^{p'(k)}$ *is a family of random functions indexed by its second parameter, and where* $\mathsf{F}_{f(\cdot,K),K}$ *is the oracle which on input* $x, \delta \in \Delta$ *returns* $f(x, \delta(K))$.

We can extend the notion of an RK-PRP / PRF to the notion of a related secret pseudorandom generator (RS-PRG). Like an RK-PRP or RK-PRF, an RS-PRG generates pseudorandomness even under an adversary who can affect the "secret" input. However, an RS-PRG only has one input, the secret seed to the generator.

Definition 6 (Related-secret secure pseudorandom generator (RS-PRG)). *Let* $\mathsf{F}_{f,x}$ *be the oracle which on input* $\delta \in \Delta$ *returns* $f(\delta(x))$. *An efficiently computable function* $g(x)$ *which takes n bits to* $l(n)$ *bits is a related-secret secure pseudorandom generator under* Δ *if* $\forall \mathsf{A} \in PPT$, $\exists \nu$ *negligible:* $\forall k$

$$Pr[x \leftarrow \{0,1\}^k; \mathsf{A}^{\mathsf{F}_{g,x}} = 1] - Pr[x \leftarrow \{0,1\}^k; \mathsf{A}^{\mathsf{F}_{\mathcal{O},x}} = 1] \leq \nu(k)$$

where $\mathcal{O}$ *returns a random string from* $\{0,1\}^{l(k)}$ *on input* x.

Note that for plain-model PRGs, it is normally required that $l(n) > n$, as the identity function is a PRG for $l(n) = n$, and because we anticipate using PRGs repeatedly to obtain arbitrary amounts of randomness. Those reasons for requring $l(n) > n$ do not apply to the related-secret setting: (1) $l(n) = n$ does not imply a trivial function here, and (2) we anticipate using ordinary PRGs to stretch randomness, and RS-PRGs to provide related-secret security.

We can extend the notion of related secret security to even simpler primitives such as one way functions.

Definition 7 (Related-secret secure one way function family (RS-OWFF)). *An indexed family of functions* $\{\mathcal{F}_k\}$, *where each function* $f_s \in \mathcal{F}_k$ *goes from* $\mathcal{D}_s \to \mathcal{R}_s$ *is considered a related secret secure one way function family under* Δ_s *if* $\forall \mathsf{A} \in PPT$, $\exists \nu : \forall k$, $Pr[f_s \leftarrow \{\mathcal{F}\}_k; x \leftarrow \mathcal{D}_{f_s}; x' \leftarrow \mathsf{A}^{\mathsf{F}_{f_s,x}}(f_s) : f_s(x) = f_s(x')] \leq \nu(k)$ *where* ν *is negligible and where* $\mathsf{F}_{f_s,x}$ *on input* $\delta \in \Delta_s$ *returns* $f_s(\delta(x))$.

Definition 8 (Related-secret secure one way permutation family (RS-OWPF)). *A related-secret secure one-way permutation family is a related-secret secure one-way function family in which each individual f_s is a permutation.*

2.2 Hard-Core Bits

The technique that has been used to create pseudorandomness from the existence of hard problems has been the idea of the *hard core bit.*

Definition 9 (Hard-core bit). *A function $B(x)$: $\{0,1\}^k \to \{0,1\}$ is considered* hard-core *for a function f if:*

1. *$B(x)$ is polynomial time computable.*
2. *$\forall \mathsf{A} \in PPT$, $\exists \nu$ negligible: $\forall k$, $Pr[x \leftarrow \{0,1\}^k; b \leftarrow \mathsf{A}(f(x)) : B(x) = b] \leq \frac{1}{2} + \nu(k)$.*

A hardcore bit $B(x)$ for a function f can easily be shown to be pseudorandom even given the value $f(x)$. This does not imply however, that the pseudorandomness of the bit is related to the hardness of any particular problem. Consider $f'(x) = f(x_{2,\dots,k})$. It is clear that x_1 is a hard-core bit for f however it is hard-core due to information loss. As such, no matter what the properties of the function f are, $B(x)$ can never be recovered with probability bounded away from $\frac{1}{2}$.[4]

With this in mind, we define the notion of a *strong hard-core bit*, a bit whose security is directly related to the one way security of the function f.

Definition 10 (Strong hard-core bit). *A function $B(x)$: $\{0,1\}^k \to \{0,1\}$ is considered a* strong hard-core bit *for a function f if for any $\mathsf{A} \in PPT$ such that $\exists$ non-negligible ϵ where $\forall k$ $Pr[x \leftarrow \{0,1\}^k; b \leftarrow \mathsf{A}(f(x)) : b = B(x)] \geq \frac{1}{2} + \epsilon(k)$ then $\exists$ $\mathsf{A}' \in PPT$ and ϵ' non-negligible such that $\forall k, Pr[x \leftarrow \{0,1\}^k; x' \leftarrow \mathsf{A}'(f(x)) : f(x) = f(x')] \geq \epsilon'(k)$.*

In addition we note that all known non-trivial hard-core bits are in fact strong hard core bits, as their security is proven via a reduction between the ability to predict $B(x)$ and the one-way security of f.

We can extend these definitions to the ideas of a related-secret secure hard-core bit and a related-secret secure strong hard-core bit:

Definition 11 (Related-secret secure hard core bit (RS-HCB)). *A function $B(x)$ is a related-secret secure hard core bit for a function f secure under Δ, if $\forall A \in PPT$, $\exists \nu$ negligible: $\forall k$, $Pr[x \leftarrow \{0,1\}^k; b \leftarrow \mathsf{A}^{\mathsf{F}_{f,x}} : b = B(x)] \leq \frac{1}{2} + \nu(k)$ where $\mathsf{F}_{f,x}$ returns $f(\delta(x))$ on input δ.*

Definition 12 (Related-secret secure strong hard core bit (RS-SHCB)). *A function $B(x)$ is considered a related-secret secure strong hard core bit for a function f secure under Δ, if $\forall \mathsf{A} \in PPT$ such that if $\epsilon(k)$ non-negligible: where $\forall k$ $Pr[x \leftarrow \{0,1\}^k; b \leftarrow \mathsf{A}^{\mathsf{F}_{f,x}} : b = B(x)] \geq \frac{1}{2} + \epsilon(k)$ $\exists \mathsf{A}'$ and $\epsilon'(k)$ non-negligible : $Pr[x \leftarrow \{0,1\}^k; x' \leftarrow \mathsf{A}'^{\mathsf{F}_{f,x}}(f(x)) : f(x) = f(x')] \geq \epsilon'(k)$ for non-negligible ϵ' where $\mathsf{F}_{f,x}$ returns $f(\delta(x))$ on input δ.*

[4] Note, however, that a *permutation* with a hard-core bit is necessarily one-way.

We finally introduce the idea of a related-secret secure pseudorandom bit or RS-PRB. As noted before, normal hardcore bits are inherently pseudorandom. As we will see in the next section however, when the adversary is allowed to see $f(\delta(x))$ and $B(\delta(x))$ for adversarially chosen δ, $B(\delta(x))$ is no longer necessarily indistinguishable from random. With this in mind, we define the notion of a related-secret secure pseudorandom bit as a bit which is pseudorandom even when the adversary gets to see adaptively perturbed bits $B(\delta(x))$.

Definition 13 (Related-secret secure pseudorandom bit (RS-PRB)). *A function $B(x)$ is considered a related-secret secure pseudorandom bit for a function f, under Δ, if $\forall \mathsf{A} \in PPT$, $\exists \nu$ negligible: $\forall k$*

$$Pr[x \leftarrow \{0,1\}^k; \mathsf{A}^{\mathsf{F}_{f||B,x}} = 1] - Pr[x \leftarrow \{0,1\}^k; \mathsf{A}^{\mathsf{F}_{f||\mathcal{R},x}} = 1] \leq \nu(k)$$

where $\mathsf{F}_{g,x}$ is an oracle that on input $\delta \in \Delta$, returns $g(\delta(x))$, and where $f||g$ denotes the function $f||g : x \mapsto f(x)||g(x)$, and where $\mathcal{R}(x)$ is a random function from $\{0,1\}^k$ to $\{0,1\}$.

2.3 Black Token Algorithms

In this paper we introduce the idea of a *black token* algorithm. Informally, an algorithm is black token if it operates, or could equivalently operate, oblivious to the value of its input, but rather uses a set of allowed operations to manipulate that value.

Explicitly, in the black token model of computation, an algorithm A_{BT} works with two types of values: public values, which are known fully, and private values, which A_{BT} must work with while ignorant of the actual value. For every private value x, A_{BT} sees only a pseudonym for x, id_x.

When A_{BT} receives a private input, it is given only id_x rather than the actual value x; similarly, when A_{BT} makes a private output, the output is taken to be the value for which A_{BT} specified the pseudonym. That is, if A_{BT} outputs id_y, this is interpreted as outputting y. If A_{BT} makes a private output of a pseudonym that has not been determined externally to A_{BT}, we interpret this as outputting a special error symbol $\perp$. The input and output wires of A (both its initial inputs and final outputs, and its means of communicating with its oracles) are each classified as either public or private and always treated in this manner. As such, A_{BT} cannot send a pseudonym down a public channel, or vice versa. Note that A is not inherently given a way to get pseudonyms for values it knows (or chooses) completely. All pseudonyms A receives, it receives from some outside source (as input, or as the output from some oracle).

If there is a class of allowable operations that are polynomially time computable given the actual values of the pseudonyms, we allow A to perform these functions by providing A_{BT} with a "private operation oracle" P, which can be used to perform such operations without revealing the actual values of the inputs to A. P returns outputs which may be public or private.

If such a machine A_{BT} exists in a black token model, we can create a black token algorithm in the standard model by creating a machine T to act as a tokenizer. T has oracle access to A_{BT}, as well as any oracle A_{BT} might possess. As such, $(T^{A_{BT}})^O(x)$ is the machine that runs A_{BT}^O where T translates things to and from pseudonyms as appropriate as it gives input to A, passes messages back and forth between A and O, and receives final output from A.

For example, $B(x)$, defined to be the parity of x, is a strong hard-core bit for RSA. The reduction from a hard-core bit finder to inverting RSA is blac-token: the x^e can be treated as a pseudonym, and the algorithm requires only the ability to calculate $(-x)^e$ and $(x \cdot 2^{-1})^e$, both of which can be done via private operations with private outputs. For details, see Theorem 11 in the appendix.

Effectively, the assumption that an algorithm is black-token is akin to assuming that the algorithm "knows" how to derive the ultimate output from the input using the allowed private operations. This is a stronger assumption than "knowledge of exponent assumptions," which do not restrict the allowable private operations, and it is a stronger than the claim that the algorithm is *algebraic* or a *generic ring algorithm* [16,17,18,19], which do restrict allowable operations but do not explicitly require this sort of knowledge. As such, assuming that any *adversary* is a black-token algorithm is an uncomfortably strong assumption. However, unlike prior results that use assumptions of this type, we never assume this of an adversary.

3 The Importance of RS-PRBs

In this section we attempt to mirror the construction of a pseudorandom permutation from simpler primitives in the related-secret security model. Since we address related-secret security for *complete* perturbation sets Δ, we cannot expect to build any related-secret secure constructions without an underlying component with related-secret security. As such, we show that homomorphic one-way functions or permutations are related-secret one-way under a Δ compatible with the homomorphism.

Theorem 1. *Let f be one-way and homomorphic, so that there exist efficiently computable binary operations $\odot$ and $\star$, such that for all x, y, $f(x \odot y) = f(x) \star f(y)$. Then f is related-secret one-way under $\Delta_{\odot}$.*

Proof. We use the homomorphic property of f to simulate related-secret queries $f(\delta_c(x)) = f(c \odot x)$ by making queries $f(c)$, $f(x)$ and outputting $f(c) \star f(x)$.

For instance, if f is defined as $f(x) = g^x \bmod p$ for prime p and where g is a generator for $\mathbb{Z}_p^*$, then f is homomorphic where $\odot$ is addition and $\star$ is multiplication. So if f is a one-way permutation, f is also a RS-OWP.

In the standard model, the next step would be to find hard-core bits of hard functions. We give a general construction of related-secret strong hard-core bits. Our construction will use similiar techniques to the ones found in [20].

Theorem 2. *Let $\mathsf{C}, \mathsf{C}^{-1}, \mathsf{R}$ be a (k, ρ, p) linear list decodable code over $\mathbb{F}_2^n$. Define $B(x, r)$ as $\mathsf{C}(x)_r$, the rth bit of $\mathsf{C}(x)$. Let f be an RS-OWF secure under $\Delta_{\oplus}$. Define $f'(x, r)$ as $f(x), r$. $B(x, r)$ is an RS-SHCB for f'.*

Proof. To prove that $B(x,r)$ is an RS-SHCB for f' we will create a black-box reduction $\mathsf{M}^{\mathsf{F}_{f',(x,r)},\mathsf{A}}$ which will invert $f(x)$ with non-negligible probability as long as $\mathsf{A}^{\mathsf{F}_{f',(x,r)}}$ returns $B(x,r)$ with probability non-negligibly better than 1/2. We will build $\mathsf{M}^{\mathsf{F}_{f',(x,r)}}$ to use R to reconstruct x. As such, M needs to simulate oracle access to Y for a corrupted codeword y such that $||C(x) - y|| \leq \rho n$ where $|C(x)| = n$. M has access to an oracle A which returns $B(x,r) = \mathsf{C}(x)_r$ with probability $\frac{1}{2} + \epsilon$ for non-negligible ϵ. A difficulty in this proof is that $1 - \rho$, the probability that R requires $B(x,r) = \mathsf{C}(x)_r$ to be correct, will often be much larger than $\frac{1}{2} + \epsilon$, the probability that $\mathsf{A}^{\mathsf{F}_{f',(x,r)}}$ returns $B(x,r)$ correctly. As such, we cannot directly use the answers returned by A.

To amplify the success probability of A, M will use the fact that Δ is closed under composition. As such, M given $\mathsf{F}_{f',(x,r)}$ can simulate $\mathsf{F}_{f',(\delta_c(x),r)}$ for random δ_c as $\mathsf{F}_{f',(\delta_c(x),r)}(\delta, \delta_{c'}) = \mathsf{F}_{f',(x,r)}(\delta\delta_c, \delta_{c'})$ When $\mathsf{A}^{\mathsf{F}_{f',(\delta_c(x),r)}}$ returns its guess at g at $B(\delta_c(x), r)$, M can compute $B(x,r) = g \oplus \mathsf{C}(c)_r$ which is correct as long as $g = B(\delta_c(x), r)$ as the code is linear. For random $\delta_c \in \Delta_\oplus$ $\delta_c(x)$ is random. This allows $\mathsf{M}^{\mathsf{F}_x,\mathsf{A}}$ to find many independent votes for $B(x,r)$, where each individual vote is correct with non-negligible probability. As such, M can find $\mathsf{C}(x)_r = B(x,r)$ with high enough probability to simulate Y and thus run the reconstruction program R. When R returns $\mathcal{S}$, M computes $f(x_i)$ $\forall x_i \in \mathcal{S}$ until he finds x_* :, $f(x_*) = f(x)$.

This can be seen as a generalization of the Goldreich-Levin bit to the related-secret case, and expanded to capture the use of other list-decodable codes. As most known list decodable codes are linear, this suggests that list decodable codes in general imply a RS-HCB for any p secure under $\Delta_\oplus$.[5] In Appendix B we prove a partial converse to this theorem, showing that certain well behaved strong hard core bits can be used to create list decodable codes.

In the standard model, the next step towards a pseudorandom permutation would be to show that hard-core bits are pseudorandom. Unfortunately, this does not hold in general in the related-secret case. In particular, we have shown the Goldreich-Levin hardcore bit $\langle x, r\rangle$ to be an RS-SHCB, but it is trivially seen to not be a RS-PRB under $\oplus$ for $f'(x,r) = f(x)||r$.

Theorem 3. *The Goldreich Levin hardcore bit $\langle x, r\rangle$ for the function $f'(x,r) = f(x), r$ is not an RS-PRB for f' under $\Delta_\oplus$.*

Proof. Just query the oracle under $(\delta_{ident}, \delta_c)$, $(\delta_{ident}, \delta_{c'})$, and $(\delta_{ident}, \delta_{c \oplus c'})$ receiving either $b_1 = \langle x, r \oplus c\rangle$, $b_2 = \langle x, r \oplus c'\rangle$ and $b_3 = \langle x, r \oplus c \oplus c'\rangle$ or random bits b_1, b_2, b_3. Output 1 if $b_1 \oplus b_2 = b_3$. If the bits are the inner products (outputs of $B(x,r)$) then this equation will hold with probability 1. If the bits are random, the equation will hold with probability $\frac{1}{2}$.

Thus we see a potential separation between the difficulty in predicting the bit $B(x)$ and the bit being pseudorandom in the related secret attack setting, a separation that does not exist with regards to normal hardcore bits.

[5] Also note that if we have a linear list decodable code that takes words from $\mathbb{F}_q^k$ to $\mathbb{F}_2^n$, this gives us an RS-SHCB for any RS-OWF secure under vector addition, where the vectors are in $\mathbb{F}_q^k$.

We go on to show that related-secret pseudorandom bits can be used to construct related-key secure blockciphers. First, we show how to construct an RS-PRG from an RS-PRB.

Theorem 4. *Let f be a function from $\{0,1\}^k$ to $\{0,1\}^{p(k)}$, such that there is a B that is a RS-PRB for f under Δ. Then there exists a g that is an RS-PRG from k^2 to k^2 bits.*

Proof. Since f has an RS-PRB, f must be an RS-OWF: if an adversary were able to invert f with probability ϵ, we could attack the PRB by inverting f, and, if successful, checking the outputs of B. Since f is an RS-OWF, f must in particular be a OWF.

Let $g'(x)$ be a k-bit to k^2 bit PRG; we know g' exists because we have shown that OWFs exist. Define $g : \{0,1\}^{k^2} \to \{0,1\}^{k^2}$ as follows. On input x, parse x as k k-bit blocks $x_1, \ldots, x_k$, compute $y = B(x_1)|| \ldots ||B(x_k)$, and output $g'(y)$.

Then g is an RS-PRG for Δ^k where $(\delta_1, \ldots, \delta_k)(x_1|| \ldots ||x_k) = (\delta_1(x_1)|| \ldots || \delta_k(x_k))$.

This proof comes easily from the idea that $y_i = B(\delta_i(x_i))$ is indistinguishable from random for all δ_i selected by A, due to the fact that $B()$ is an RS-PRB and x_i is random. As each x_i is random and independent of any other x_j, and each δ_i is independent from the other δ_j, we can consider each bit $B(\delta_i(x_i))$ to be indistinguishable from random, even given the other bits $B(\delta(x_j))$. The *normal* PRG expands the pseudorandom string to the correct length, finishing the proof.

The proof of the following corollary is obvious from the proof of the previous theorem.

Corollary 1. *Let $g()$ be a RS-PRG that takes n bits to $p(n)$ bits. Then $f_x(\delta) = g(\delta(x))$ is a PRF from $\Delta \to \{0,1\}^{p(n)}$.*

This proof illustrates an important trick, namely, that if related-key pseudorandomness is applied directly to a random secret, we can achieve security using traditional techniques afterwards.

We now give two proofs, together showing that the existence of an RS-PRG implies the existence of an RK-PRP and RK-PRF. The proofs illustrate an easy way to gain related-key security from a related-secret secure pseudorandom generator. We may use a related-secret pseudorandom generator to eliminate any advantage an adversary may gain from a related-secret attack on a standard construction. The adversary's choice of δ has no effect as it gets translated to a key that looks random and independent of other keys.

Theorem 5. *If $RS-PRGs$ exist under Δ, $RK-PRFs$ exist under Δ.*

Proof. Say g is an RS-PRG that takes k bits to $l(k)$ bits. If g is an RS-PRG, g must be a one-way function. If not, an adversary could distinguish the output of g from random by inverting g and checking if the result is correct and consistent with queries to g. From [21] and [22], we know that if one-way functions exist, so do (standard) pseudorandom functions. Let f be a pseudorandom function taking a $p(k)$ bit input and an $l(k)$-bit seed to a $p'(k)$-bit output. Let $f'(x, K) = f(x, g(K))$. Then f' is a related-key secure pseudorandom function under Δ.

We prove this by a simple hybrid argument. If A can distinguish between $\mathsf{F}_{f',K}$ and $\mathsf{F}_{\mathcal{F},K}$ for random K, then either A can distinguish between $\mathsf{F}_{f'(\cdot,K),K}$ and $\mathsf{F}_{f''(\cdot,K),K}$ or between $\mathsf{F}_{f''(\cdot,K),K}$ and $\mathsf{F}_{\mathcal{F}(\cdot,K),K}$, where $f''(x, K) = f(x, \mathcal{R}(K))$ where $\mathcal{R}$ is a random function from k bits to $l(k)$ bits. If the former, then A breaks the related-key security of g. If the latter, then A distinguishes between a random function and f with many independent random seeds.

If $|Pr[K \leftarrow \{0,1\}^k : \mathsf{A}^{\mathsf{F}_{f'(\cdot,K),K}} = 1] - Pr[K \leftarrow \{0,1\}^k : \mathsf{A}^{\mathsf{F}_{f''(\cdot,K),K}} = 1]|$ is non-negligible, then we can use A to break g. Given an oracle O, we run A in its attack with oracle $\mathsf{F}_{\mathsf{f}'_{\mathsf{O}},\mathsf{K}}$, where $f'_{\mathsf{O}}(x, K) = f(x, \mathsf{O}(K))$.

If $|Pr[K \leftarrow \{0,1\}^k : \mathsf{A}^{\mathsf{F}_{f''(\cdot,K),K}} = 1] - Pr[\mathsf{A}^{\mathsf{F}_{\mathcal{F}(\cdot,K),K}} = 1]| \geq \epsilon(k)$ non-negligible, then we can use A to break f. Simply put, $\mathsf{F}_{\mathcal{F}(\cdot,K),K}$ differs from $\mathsf{F}_{f''(\cdot,K),K}$ only in that f'' runs f on a random seed for each distinct $\delta(K)$, while $\mathcal{F}$ is a random function for each distinct $\delta(K)$. By a simple hybrid reduction, we obtain a probability of at least $\epsilon(k)/T(\mathsf{A})$, where $T(\mathsf{A})$ is the running time of A, which is polynomial.

We state the next theorem, showing that RS-PRG's imply RK-PRP's, as a corollary. We omit the proof, but it is essentially identical to the proof of Theorem 5.

Corollary 2. *Let $\mathsf{E}(x, K)$ be a pseudorandom permutation family taking a $p(k)$-bit input and an $l(k)$-bit seed to a $p(k)$-bit output. Let g be a RS-PRG taking a k-bit input to an $l(k)$-bit output. Then $\mathsf{E}(x, g(K))$ is a RK-PRP.*

Remark 1. A related-secret PRG effectively allows us to "harden" any secret-key-based construction to be secure against related-key attacks, by using $g(K)$ in place of K.

This should apply to any construction X for which security implies security when an attacker may query X with many independent random secret keys. For such constructions (such as PRFs, as in the proof of 5), related-key security follows because no adversary can distinguish between the modified construction and querying the original consturction on many independent random secret keys, one for each perturbation.

We have shown that RS-PRBs are sufficient to construct PK-PRPs. We end this section by showing that RS-PRBs are "necessary" for the existence of related key secure pseudorandom functions and permutations. We show that RK-PRP's imply RS-PRG's, and RS-PRG's imply a one way function f and an RS-PRB $B()$ for f.

Theorem 6. *Let $\mathsf{E}(x, K)$ be an RK-PRP under Δ. Let $g(x')$ be the function which parses x' as (x, K) and outputs $\mathsf{E}(x, K), \mathsf{E}(x+1, K), \mathsf{E}(x+2, K)$. Then for any valid set of perturbations Δ' on $\{0,1\}^{p(k)}$, $g(x)$ is an RS-PRG under $\Delta' \times \Delta$.*

Proof. The proof follows from the fact that $g(x)$ is simulatable given access to oracle access to $\mathsf{E}(x, K)$. As such, if an adversary A can distinguish between

g and random values, we can build an adversary A' which picks a random x, queries its E oracle on $(\delta_1(x), \delta_2)$, $(\delta_1(x+1), \delta_2)$, $(\delta_1(x+2), \delta_2)$ when A queries on $\delta_{12} = (\delta_1 || \delta_2)$. If A's oracle is the oracle to the pseudorandom permutation then A simulates g, otherwise A' simply returns a random value for each distinct $\delta_{12}(x, K)$. As such, A' can use A to distinguish between the two cases.

We now demonstrate that RS-PRG's give us a one way function $f(x)$ and an RS-PRB $B(x)$ for that one way function $B()$.

Theorem 7. *Let $g(x)$ be a k bit to $l(k) > k$ bit RS-PRG. Let $f(x)$ be the first $l(k)-1$ bits of $g(x)$ and let $B(x)$ be the last bit of $g(x)$. Then $f(x)$ is an RS-OWF, and $B(x)$ is an RS-PRB for f.*

Proof. If f is not an RS-OWF, then it is easy to show that g is not an RS-PRG. If A attacks the one way security of f, A' takes the δ queries made by A queries it's oracle, chops off the last bit of the result and returns it to A. If A returns a value x, A' can check the outputs of its oracle to see if they are equal to $g(\delta(x))$.

The fact that $B(x)$ is an RS-PRB for $f()$ comes from the fact that $g(x) = f(x)||B(x)$ is an RS-PRG.

4 Difficulties in Constructing an RS-PRB

In the previous section we demonstrated that related-secret pseudorandom bits were both necessary and sufficient for the existence of provably secure related-key pseudorandom permutations. While we discussed related-secret one-wayness and constructed related-secret strong hard-core bits we were unable to give any construction of an RS-PRB. In fact, the homomorphic properties of known hardcore bits mean that they cannot be pseudorandom.

In this section we give a surprising result concerning possible constructions of RS-PRB's, in that we show that black token, black box strong hard-core bits cannot be related-secret pseudorandom bits.

Definition 14. *If M is a black-token algorithm, its private operation oracle P is said to be perturbation-private if all operations performed by P with private outputs are of the form $(id_{f(x)}, \delta) \mapsto id_{f(\delta(x))}$ for δ in a class Δ, for some f.*

Theorem 8. *Let $B(x)$ be a strong hard-core bit for a one way function f that has a black token, black-box reduction M where the private operation oracle P is efficiently computable, and perturbation-private for a class Δ that is closed under composition. Then $B(x)$ is not an RS-PRB under any $\Delta' \supset \Delta$.*

For example, the reduction that proves that parity is a hard-core bit for RSA is a black-token, perturbation-private reduction, where the class $\Delta = \{\delta_r : x \mapsto x \cdot r\}$.

Proof. If $B(x)$ is a black token, black box SHCB for a one-way function f, then there exists a black token algorithm M such that $\mathsf{M}^{\mathsf{P},\mathsf{A}}(id_{f(x)})$ finds x with non-negligible probability as long as $\mathsf{A}(f(x))$ outputs $B(x)$ with probability $1/2 + \epsilon$ where ϵ is non-negligible.

We construct A' to attack B as an RS-PRB under Δ'. We can view A' as having access to two oracles, O and $\mathsf{F}_{f,x}$, where O either returns $B(x)$ or a random bit. First A' queries $\mathsf{F}_{f,x}(\delta_{ident})$ to obtain $f(x)$, creates a token $id_{f(x)}$ and then uses as $id_{f(x)}$ as input to M. A' then runs M, acting as the tokenizer T. A' uses O to answer M's queries to A, and uses $\mathsf{F}_{f,x}$ to answer M's queries to P, when necessary. When M returns x', A' checks if $f(x') = f(x)$; if so, it outputs 1, otherwise it outputs 0.

Since M is black token it can only obtain new pseudonyms from P. Since Δ is closed under composition we can always associate every pseudonym id_y M has with a specific $\delta \in \Delta$ such that $y = f(\delta(x))$. Since A' keeps track of this association, when M asks for $\mathsf{A}(id_{f(\delta(x))})$, A' can query $\mathsf{O}(\delta)$ and return the result as the answer. A' answers queries to P of the form $id_{f(\delta(x))}, \delta'$ by querying $\mathsf{F}_{f,x}(\delta' \circ \delta)$, and returning a pseudonym of the result. Other queries to P may be possible, but always produce "public" results. For such queries, the answers are computable given the actual values of all inputs. Thus, A' need only translate pseudonyms to real values and then perform the computation.

When $\mathsf{O} = \mathsf{B}$, A' provides a faithful simulation of A and P. As such, the probability that A' will output 1 is non-negligible, since the probability that M outputs the correct x is non-negligible (since B is always right, it has advantage $\epsilon = 1/2$). On the other hand, if $\mathsf{O} = \mathcal{R}_x$, the bits given to M are random. If $\mathsf{M}^{\mathsf{P},\mathsf{A}}(f(x))$ could output the correct value x with non-negligible probability where A returns only random bits, then $\mathsf{M}^{\mathsf{P},\mathsf{S}}(id_{f(x)})$ can output x with non-negligible probability where S just outputs random bits. As such M can be used by itself to break the one-wayness of $f(x)$. Thus, with all but negligible probability, if $\mathsf{O} = \mathcal{R}_x$, A' will receive x' which is not a preimage of $f(x)$, and will thus return 0. As such, A' is a successful distinguisher.

Note that we make only the most general restriction on the types of acceptable reduction in Theorem 8, the main limitation being that we require that M only ask A for the hardcore bit of a $\delta(x)$ where we know the δ. All known examples of strong hard-core bits have reductions that can be seen as black-token and perturbation-private. Note that the conditions of Theorem 8 only require such a reduction for A that is correct with probability 1: a very useful observation, since the probability-1 reductions are often far simpler. As an example, consider the Goldreich-Levin hard-core bit, and its reduction assuming A is always correct:

Theorem 9. *Let $f(x)$ be a one way function. Let $f'(x,r) = f(x), r$. Let $B(x,r) = \langle x, r\rangle$. $B(x,r)$ is a black token SHCB for f' with $\{I\} \times \Delta_{\oplus}$*

Proof. The machine M is simple to construct. M begins by receiving $id_{f(x),r}$. M makes use of P only to compute $P(id_{f(x),r}) = r$ and to compute $P(id_{f(x),\delta(r)}, \delta') = id_{f(x),\delta'(\delta(r))}$. It aims to query A on $id_{f(x),r_i}$ for different r_i until it can obtain x via standard linear algebra. In order to do this, it learns r, and calculates $id_{f(x),r_i}$ by querying P on $id_{f(x),r}, \delta_{r_i \oplus r}$.

This reduction can easily extend to an A whose success probability is less than 1. The proof of Goldreich and Levin involves querying $f(x), r_i$ for many random pairs of r_i with specified differences. This allows for us to determine $\langle x, r_i \rangle$ with high probability. Since the $f(x)$ value is left untouched, the same argument applies; the general reduction is also black token.

See Appendix A for discussion of some other hard-core bit reductions.

We have shown that black token strong hard-core bits cannot be related-secret pseudorandom. We further show that all related-secret pseudroandom bits are hard-core bits:

Theorem 10. *If B is an RS-PRB for a function f under Δ then B is a hard-core bit for f.*

Proof. Suppose there exists an A such that $\mathsf{A}(f(x))$ returns $B(x)$ with probability $1/2 + \epsilon$. Then we can attack B as an RS-PRB by obtaining $f(x)$ and $\mathsf{O}(x)$ and checking whether $\mathsf{O}(x) = \mathsf{A}(f(x))$; if so, we return 1, otherwise, we return 0. Then the difference in probabilities is ϵ, which for non-negligible ϵ is enough to break $B(x)$ as an RS-PRB.

5 Discussion

We know of only two ways to construct pseudorandom primitives: directly from hard-core bits in the standard model, or from other standard-model pseudorandom primitives. Our impossibility result shows that related-secret pseudorandomness based off of hard-core bits is unlikely. Nor is it plausible to construct a related-secret pseudorandom bit directly from standard-model pseudorandomness: all such primitives have a secret seed or key, and thus, in any construction, the adversary (because of the completeness we require of Δ) must be able to make queries that require modifications of those secrets. In other words, either the construction will fail, or the pseudorandomness we are using must already be related-secret secure.

Since any RS-PRB is inherently a hard-core bit, Theorem 8 leaves open two potential ways in which an RS-PRB might yet be possible. The RS-PRB might be a hard-core bit, but not a strong one, or it might be a strong hard-core bit but not one with a reduction that is black-token and perturbation-private. Normally, one proves that a bit is hard-core by providing a reduction to the hardness of inverting the associated function: in other words, normally, hard-core bits are always strong hard-core bits. This is natural, since the associated function must be one-way in any case, and thus any proof not requiring extra assumptions would reduce to its one-wayness.

If B is a strong hard-core bit, then its reduction must not meet the conditions of Theorem 8. As we discussed, all known examples of strong hard-core bits have black-token reductions of the type necessary for our impossibility proof. Note also, that it is difficult to see what useful information M can receive by running A on input that is not a valid $f(\delta(x))$ value for some δ, or when δ is unknown to M.

At a higher level, our restrictions on the type of reductions used in the proof are reasonable ones. Since no assumptions can be made about observable properties of $f(x)$, these values are mostly ignored in any proof involving generic one-way functions or permutations. As such, any "private operations" in those proofs are kept to a strict minimum because they must be efficiently computable given only $f(x)$. Also, in the case of bits that are *generic* - that is, secure for a variety of functions - it is hard to imagine a proof of their security that is not black-token and black-box. Finally note that the proof applies if there *exists* a reduction of the specified type. As such, even if there is a very unusual reduction that does not meet the conditions of Theorem 8, if other more usual reductions exist, the theorem still applies.

The major open problem in related-secret security is whether or not related-key secure blockciphers exist. We have shown that related-secret pseudorandom bits are necessary and sufficient for higher forms of related-secret pseudorandomness. However, related-secret pseudorandom bits cannot be constructed using traditional techniques. This leaves a significant open problem: are related-secret pseudorandom bits possible under only basic assumptions? Or alternatively, can fundamentally new techniques be found to create related-secret pseudorandom bits?

References

1. Kelsey, J., Schneier, B., Wagner, D.: Related-key cryptanalysis of 3-way, biham-des, cast, des-x, newdes, rc2, and tea. In: Han, Y., Quing, S. (eds.) ICICS 1997. LNCS, vol. 1334, pp. 233–246. Springer, Heidelberg (1997)
2. Poorvi, L., Vora, D.J.M.: Related-key linear cryptanalysis. In: 2006 IEEE International Symposium on Information Theory, pp. 1609–1613 (2006)
3. Mir, D.J., Vora, P.L.: Related-key statistical cryptanalysis. Cryptology ePrint Archive, Report 2007/227 (2007), `http://eprint.iacr.org/`
4. Gorski, M., Lucks, S.: New related-key boomerang attacks on AES. In: Chowdhury, D.R., Rijmen, V., Das, A. (eds.) INDOCRYPT 2008. LNCS, vol. 5365, pp. 266–278. Springer, Heidelberg (2008)
5. Biham, E.: New types of cryptanalytic attacks using related keys. Journal of Cryptology 7(4), 229–246 (Fall 1994), `citeseer.nj.nec.com/biham94new.html`
6. Zhang, W., Zhang, L., Wu, W., Feng, D.: Related-key differential-linear attacks on reduced AES-192. In: Srinathan, K., Rangan, C.P., Yung, M. (eds.) INDOCRYPT 2007. LNCS, vol. 4859, pp. 73–85. Springer, Heidelberg (2007)
7. Biryukov, A., Khovratovich, D., Nikolic, I.: Distinguisher and related-key attack on the full AES-256. In: Halevi, S. (ed.) CRYPTO 2009. LNCS, vol. 5677, pp. 231–249. Springer, Heidelberg (2009)
8. Bellare, M., Kohno, T.: A Theoretical Treatment of Related-Key Attacks: PKA-PRPs, RKA-PRFs, and Applications. In: Biham, E. (ed.) EUROCRYPT 2003. LNCS, vol. 2656, pp. 491–506. Springer, Heidelberg (2003)
9. Black, J., Cochran, M., Shrimpton, T.: On The Impossibility of Highly-Efficient Blockcipher-Based Hash Functions. In: Cramer, R. (ed.) EUROCRYPT 2005. LNCS, vol. 3494, pp. 526–541. Springer, Heidelberg (2005)
10. Dodis, Y., Katz, J., Reyzin, L., Smith, A.: Robust fuzzy extractors and authenticated key agreement from close secrets. In: Dwork, C. (ed.) CRYPTO 2006. LNCS, vol. 4117, pp. 232–250. Springer, Heidelberg (2006)

11. Cramer, R., Dodis, Y., Fehr, S., Padro, C., Wichs, D.: Detection of algebraic manipulation with applications to robust secret sharing and fuzzy extractors. In: Smart, N.P. (ed.) EUROCRYPT 2008. LNCS, vol. 4965, pp. 471–488. Springer, Heidelberg (2008)
12. Kanukurthi, B., Reyzin, L.: An improved robust fuzzy extractor. In: Ostrovsky, R., De Prisco, R., Visconti, I. (eds.) SCN 2008. LNCS, vol. 5229, pp. 156–171. Springer, Heidelberg (2008)
13. Lucks, S.: Ciphers secure against related-key attacks. In: Roy, B., Meier, W. (eds.) FSE 2004. LNCS, vol. 3017, pp. 359–370. Springer, Heidelberg (2004)
14. Akavia, A., Goldwasser, S., Vaikuntanathan, V.: Simultaneous hardcore bits and cryptography against memory attacks. In: Reingold, O. (ed.) TCC 2009. LNCS, vol. 5444, pp. 474–495. Springer, Heidelberg (2009)
15. Goldreich, O., Levin, L.: A hard-core predicate for all one-way functions. In: Proceedings of the Twenty First Annual ACM Symposium on Theory of Computing, Seattle, Washington, May 15-17, pp. 25–32 (1989)
16. Paillier, P., Vergnaud, D.: Discrete-log-based signatures may not be equivalent to discrete log. In: Roy, B. (ed.) ASIACRYPT 2005. LNCS, vol. 3788, pp. 1–20. Springer, Heidelberg (2005)
17. Boneh, D., Venkatesan, R.: Breaking RSA may not be equivalent to factoring. In: Nyberg, K. (ed.) EUROCRYPT 1998. LNCS, vol. 1403, pp. 59–71. Springer, Heidelberg (1998)
18. Aggarwal, D., Maurer, U.: Breaking rsa generically is equivalent to factoring. Cryptology ePrint Archive, Report 2008/260 (2008), `http://eprint.iacr.org/`
19. Maurer, U.: Abstract models of computation in cryptography. In: Smart, N.P. (ed.) Cryptography and Coding 2005. LNCS, vol. 3796, pp. 1–12. Springer, Heidelberg (2005)
20. Akavia, A., Goldwasser, S., Safra, S.: Proving hard-core predicates using list decoding. In: FOCS 2003: Proceedings of the 44th Annual IEEE Symposium on Foundations of Computer Science, Washington, DC, USA, p. 146. IEEE Computer Society, Los Alamitos (2003)
21. Goldreich, O., Goldwasser, S., Micali, S.: How to construct random functions. Journal of the ACM 33(4), 792–807 (1986)
22. Håstad, J., Impagliazzo, R., Levin, L., Luby, M.: Construction of pseudorandom generator from any one-way function. SIAM Journal on Computing 28(4), 1364–1396 (1999)
23. Goldwasser, S., Micali, S., Tong, P.: Why and how to establish a private code on a public network. In: SFCS 1982: Proceedings of the 23rd Annual Symposium on Foundations of Computer Science, Washington, DC, USA, pp. 134–144. IEEE Computer Society, Los Alamitos (1982)
24. Blum, M., Micali, S.: How to generate cryptographically strong sequences of pseudo-random bits. SIAM Journal on Computing 13(4), 850–863 (1984)

A Black Token Reductions for Known Hardcore Bits

In this section we examine two other well-known hardcore bits, the hardcore bits for RSA, discrete log, (the generic hardcore bit of Goldreich and Levin was examined in the body of the paper). For each one we demonstrate that these hardcore bits are black token hardcore bits that fit the requirements of Theorem 8. Many of these proofs will simply be restatements of previous reductions, or slightly modified versions to emphasize the fact that they are black token. For each proof, we only formally show that the necessary reduction exists and is black token for an adversary which always returns the correct hardcore bit perfectly, however we also informally state how the proof is changed to adapt an imperfect adversary and how this is also black token.

We next address the hardcore bit for the RSA function. The reduction is taken from [23].

Theorem 11. *Let* $\mathsf{RSA}_{N,e}$ *be the RSA function, that is* $\mathsf{RSA}_{N,e}(x) = x^e \bmod N$. *Define* $B(x)$ *as the parity of* x. *Then* $B(x)$ *is a black token SHCB for* $\mathsf{RSA}_{N,e}$

Proof. P allows two transformations, $\delta_{\frac{1}{2}}(x) = x(2^{-1}) \bmod N$ and $\delta_{-1}(x) = -x \bmod N$. These can be viewed as two specific transformation among a more general class of multiplicative transformations $\delta_r(x) = xr \bmod N$. In general, to calculate $(rx)^e$ given x^e, we need only multiply x^e by r^e.

M asks A for the parity (LSB) of id_{x^e}. If 0, then M runs P on $(id_{x^e}, \delta_{\frac{1}{2}})$ to obtain $id_{(x/2)^e}$. If 1, then M runs P on $(id_{x^e}, \delta_{\frac{1}{2}} \circ \delta_{-1})$ to obtain $id_{(-x/2)^e}$.

Since $-x$ has the opposite parity of x (since N is odd), we always divide an *even* residue by 2, thus effectively shifting the unknown bits down by one. We collect one bit of x at a time, keeping track of the number of times we have applied δ_{-1}, as these reverse our results.

For A with probability of success less than 1, the reduction is far more complicated, but still can be viewed as a sequence of applications of multiplicative transformations of x by manipulating x^e.

We next address the hardcore bit for the discrete log function. The reduction is taken from [24].

Theorem 12. *Let* $\mathsf{f}_{g,p}$ *be the modular exponentiation function, where* g *is a generator of the group* $\mathbb{Z}_p^*$. *Let* $B_p(x)$ *be the function that outputs 1 if* $x \leq \frac{p-1}{2}$, *0 otherwise.* $B_p(x)$ *is a black-token SHCB for* $\mathsf{f}_{g,p}$.

Proof. P computes several transformations, $\delta_{-1}(x) = x - 1$, $\delta_{\frac{1}{2}}(x)$ which returns either $\frac{x}{2}$ or $\frac{x}{2} + \frac{p-1}{2}$, $\delta_{+\frac{p-1}{2}}$ which returns $x + \frac{p-1}{2}$, and $p(x)$, a predicate which returns the least significant bit of x. These can all be seen as computable using multiplicative and/or additive transformations on x, which are efficiently computable. $g^{\delta_{+r}(x)} = g^{x+r} = g^x g^r$, and $g^{\delta_{*r}(x)} = g^{xr} = (g^x)^r$. Take Δ to be the resulting class, closed under composition.

M proceeds as follows. It first obtains id_y for $y = g^x \bmod p$. It then queries $p(id_y)$ and obtains a bit. If 1, M queries P on (id_y, δ_{-1}) obtaining a pseudonym for $y' = g^{x-1}$. If 0, $\mathcal{M}$ considers $y' = y$ and $id_y = id_{y'}$. M then makes a query to $\mathsf{P}(id_{y'}, \delta_{\frac{1}{2}})$ and $\mathsf{P}(\cdot, \delta_{+\frac{p-1}{2}})$, obtaining pseudonyms for the two square roots, g^s and $g^{s+\frac{p-1}{2}}$ where $g^{2s} = y'$, in unknown order. M then sends both pseudonyms to A which enables him to find the pseudonym for $g^{\frac{x'}{2}}$ where $x' = x$ or $x - 1$.

This allows M to obtain the least significant bit of x, then shift the bits of x one to the right. By repeating this process we may obtain all the bits of x.

We note that dealing with imperfect A is done by simply computing multiplication mod p by known quadratic residues. As such, the full proof still remains black token.

A.1 Other Hardcore Bits

There are too many examples of hardcore bits to analyze all known proofs. Hardcore bits for specific functions tend to work via homomorphic properties of $f(x)$; the RSA and discrete log examples show how these can be viewed as black token. Generalized hardcore bits are extensions of Goldreich-Levin, and, as such, ignore the value of $f(x)$ completely. Note that virtually any algorithm can be viewed as a black-token algorithm of this sort, for the appropriate class Δ. One question may be whether Δ is "valid," but actually the answer is irrelevant: all perturbatinos in the Δ that arise are in fact secret perturbation we can calculate efficiently.

B Black Token RS-SHCB's to Codes

In this section we prove a corollary to Theorem 2, demonstrating that certain well behaved strong hard core bits can be viewed as error correcting codes.

Theorem 13. *Let $B(x)$ be a black token RS-SHCB for a function f secure under Δ where Δ is of finite size and where* M *only makes queries to* P *of the form* $\mathsf{P}(id_{f(\delta(x))}, \delta')$. *Establish an ordering on Δ, $\delta_1, \delta_2, \cdots \delta_{|\Delta|}$. Define $\mathsf{C}(x)_i$ as $B(\delta_i(x))$. Then we can create a C^{-1} and a R such that $\mathsf{C}, \mathsf{C}^{-1}, \mathsf{R}$ is a (k, ρ, p) list decodable code where $1 - \rho = \frac{1}{2} + \epsilon$, where p is the success probability of* M *and where l is the number of queries* M *makes to* A.

Proof. The machine R will use the machine M so it needs to be able to simulate P, $\mathsf{F}_{f,x}$ and A. The simulations of $\mathsf{F}_{f,x}$ and P will be relatively easy as their outputs in the black token model are random pseudonyms $id_{f(\delta(x))}$. The simulation of A is accomplished by using the oracle Y.

R^{Y} first creates a random value $id_{f(x)}$ as the "token" for $f(x)$ (which it does not know) and passes $id_{f(x)}$ to M. When M makes a query $\mathsf{F}_{f,x}(\delta)$ or $\mathsf{P}(id_{f(\delta(x))}, \delta')$, R returns a randomly generated token which it associates with $id_{f(\delta(x))}$ / $id_{f(\delta'\delta(x))}$. When M makes a query $\mathsf{A}(id_{f(\delta_i(x))})$, R^{Y} simulates A by querying Y for $B(\delta_i(x)) = \mathsf{C}(x)_i$, which is correct with probability $\frac{1}{2} + \epsilon$. Since M operates in the black token model, and receives only $id_{f(x)}$ as input, and only queries P on $id_{f(\delta(x))}, \delta'$, R can simulate A perfectly and as such $\mathsf{M}^{\mathsf{A}, \mathsf{F}_{f,x}}$ will output x with probability p.

A Domain Extender for the Ideal Cipher

Jean-Sébastien Coron[2], Yevgeniy Dodis[1], Avradip Mandal[2], and Yannick Seurin[3]

[1] New York University
[2] University of Luxembourg
[3] Orange Labs

Abstract. We describe the first domain extender for ideal ciphers, *i.e.* we show a construction that is indifferentiable from a $2n$-bit ideal cipher, given a n-bit ideal cipher. Our construction is based on a 3-round Feistel, and is more efficient than first building a n-bit random oracle from a n-bit ideal cipher (as in [9]) and then a $2n$-bit ideal cipher from a n-bit random oracle (as in [10], using a 6-round Feistel). We also show that 2 rounds are not enough for indifferentiability by exhibiting a simple attack. We also consider our construction in the standard model: we show that 2 rounds are enough to get a $2n$-bit tweakable block-cipher from a n-bit tweakable block-cipher and we show that with 3 rounds we can get beyond the birthday security bound.

Keywords: Ideal cipher model, indifferentiability, tweakable block-cipher.

1 Introduction

A block cipher is a primitive that encrypts a n-bit string using a k-bit key. The standard security notion for block-ciphers is to be indistinguishable from a random permutation, for a polynomially bounded adversary, when the key is generated at random in $\{0,1\}^k$. A block-cipher is said to be a strong pseudo-random permutation (or chosen-ciphertext secure) when computational indistinguishability holds even when the adversary has access to the inverse permutation.

When dealing with block-ciphers, it is sometimes useful to work in an idealized model of computation, in which a concrete block-cipher is replaced by a publicly accessible random block-cipher (or ideal cipher); this is a block cipher with a k-bit key and a n-bit input/output, that is chosen uniformly at random among all block ciphers of this form; this is equivalent to having a family of 2^k independent random permutations. All parties including the adversary can make both encryption and decryption queries to the ideal block cipher, for any given key; this is called the Ideal Cipher Model (ICM). Many schemes have been proven secure in the ICM [5,11,13,15,19,20,27]; however, it is possible to construct artificial schemes that are secure in the ICM but insecure for any concrete block cipher (see [4]). Still, a proof in the ideal cipher model seems useful because it

D. Micciancio (Ed.): TCC 2010, LNCS 5978, pp. 273–289, 2010.

shows that a scheme is secure against generic attacks, that do not exploit specific weaknesses of the underlying block cipher.

It was shown in [9,10] that the Ideal Cipher Model and the Random Oracle Model are equivalent; the random oracle model is similar to the ICM in that a concrete hash function is replaced by a publicly accessible random function (the random oracle). The authors of [9] proved that a random oracle (taking arbitrary long inputs) can be replaced by a block cipher-based construction, and the resulting scheme will remain secure in the ideal cipher model. Conversely, it was shown in [10] that an ideal cipher can be replaced by a 6-round Feistel construction, and the resulting scheme will remain secure in the random oracle model. Both directions were obtained using an extension of the classical notion of indistinguishability, called *indifferentiability*, introduced by Maurer *et al.* in [24].

Since a block cipher can only encrypt a string of fixed length, one must consider the encryption of longer strings. A *mode of operation* of a block-cipher is a method used to extend the domain of applicability from fixed length strings to variable length strings. Many modes of operations have been defined that provide both privacy and authenticity (such as OCB [28]). A mode of operation can also be a permutation; in this case, one obtains an extended block cipher that must satisfy the same property as the underlying block-cipher, *i.e.* it must be a (strong) pseudo-random permutation. Many constructions of domain extender for block-ciphers have been defined that satisfy this security notion, for example PEP [6], XCB [14], HCTR [30], HCH [7] and TET [18].

However, it is easy to see that none of those constructions provide the indifferentiability property that enables to get a $2n$-bit ideal cipher from a n-bit ideal cipher. This is because these constructions were proposed with privacy concerns in mind (mainly for disk encryption purposes) and proven secure only in the classical pseudo-random permutation model. Therefore, these constructions cannot be used when security must hold under the random permutation model (or ideal cipher model). Consider for example the public-key encryption scheme described by Phan and Pointcheval in [27]. The scheme requires a public random permutation with the same size as the RSA modulus, say 1024 bits. In order to replace a 1024-bit random permutation by a construction based on a smaller primitive (for example a 128-bit block cipher), indifferentiability with respect to a 1024-bit random permutation is required. Given a 128-bit block-cipher, none of the previous constructions can provide such property; therefore if one of these constructions is plugged into the Phan and Pointcheval scheme, nothing can be said about the security of the resulting scheme.

In this paper we construct the first domain extender for the ideal cipher; that is we provide a construction of an ideal cipher with $2n$-bit input from an ideal cipher with n-bit input. Given an ideal cipher with n-bit input/output, one could in principle use the construction in [9] to get a random oracle with n-bit output, and then use the 6-round Feistel in [10] to obtain an ideal cipher with $2n$-bit input/output, but that would be too inefficient. Moreover the security bound in [10] is rather loose, which implies that the construction only works for large values of n.[1] In this paper we describe a more efficient construction, based on a 3-round Feistel only, and with a better security bound; we view this as the main result of the paper. More precisely, we show that the 3-round construction in Figure 1 (left) is enough to get a $2n$-bit random permutation from a n-bit ideal cipher, and that its variant in Figure 1 (right) provides a $2n$-bit ideal cipher. We also show that 2 rounds are not enough by providing a simple attack. Interestingly, in the so called honest-but-curious model of indifferentiability [12], we show that 2 rounds are sufficient.

Our construction is similar to that of Luby-Rackoff [23]. However we stress that the "indifferentiable construction" security notion is very different from the classical indistinguishability notion. The well known Luby-Rackoff result that 4 rounds are enough to obtain a strong pseudo-random permutation from pseudo-random functions [23], is proven under the classical indistinguishability notion. Under this notion, the adversary has only access to the input/output of the Luby-Rackoff construction, and tries to distinguish it from a random permutation; in particular it does not have access to the input/output of the inner pseudo-random functions. On the contrary, in our setting, the distinguisher can make oracle calls to the inner block-ciphers E_i's (see Fig. 1); the indifferentiability notion enables to accommodate these additional oracle calls in a coherent definition.

The indifferentiability security notion still requires a (small) ideal component. We stress that it is unknown how to instantiate such ideal component (be it a random oracle or an ideal cipher, as opposed to a PRF or a PRP) and that the security guarantee does not hold anymore once that component is instantiated. Moreover the recent related-key attacks on AES [2,3] show that AES-192 and AES-256 do not behave as ideal ciphers; as of 2009 it is unclear if we have a candidate block-cipher with key-size larger than block-size that behaves like an ideal cipher.

Finally, we also analyze our construction in the standard model. In this case, we use a *tweakable* block-cipher as the underlying primitive. Tweakable block-ciphers were introduced by Liskov, Rivest and Wagner in [22] and provide an additional input - the tweak - that enables to get a *family* of independent block-ciphers; efficient constructions of tweakable block-ciphers were described in [22], given ordinary block-ciphers. In this paper we show that our construction with 2 rounds enables to get a $2n$-bit tweakable block-cipher from a n-bit tweakable block-cipher. Moreover we show that with 3 rounds we achieve a security guarantee beyond the birthday paradox.

1.1 Related Work

At FSE 2009, Minematsu [25] provided two constructions of a $2n$-bit block-cipher from an n-bit tweakable block-cipher :

[1] The security bound in [10] for the 6-round Feistel random oracle based construction is $q^{16}/2^n$, where q is the number of distinguisher's queries. This implies that for $q = 2^{64}$, one must take at least $n = 1024$, which corresponds to a 2048-bit permutation.

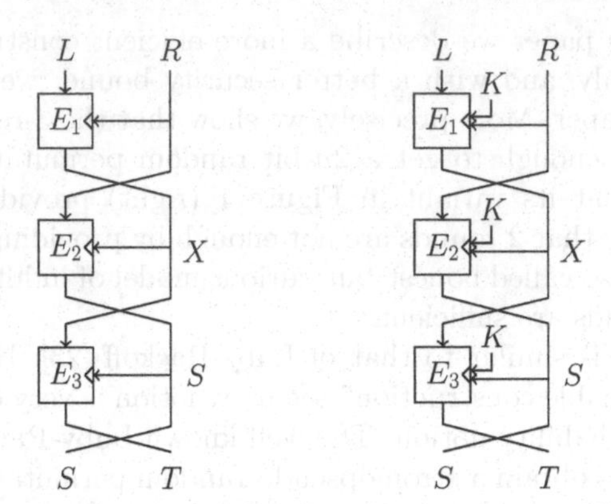

Fig. 1. Construction of a $2n$-bit permutation given a n-bit ideal cipher with n-bit key (left). Construction of a $2n$-bit ideal cipher with k-bit key, given a n-bit ideal cipher with $(n + k)$-bit key (right).

1. A 3-round Feistel construction with universal hashing in the 1st round and tweakable block ciphers in the 2nd and the 3rd rounds. This construction is a secure pseudo-random permutation beyond the birthday bound.
2. A 4-round Feistel with universal hashing in the 1st and the 4th rounds and tweakable block ciphers in the 2nd and the 3rd rounds. This construction is a secure strong pseudo-random permutation beyond the birthday bound.

On the other hand, our construction in this paper is a 3-round Feistel, with tweakable block ciphers in every round, and it gives a secure (tweakable) strong pseudo-random permutation beyond the birthday bound. Therefore, the construction in [25] is more efficient as only 2 calls are required to the underlying tweakable block-cipher, instead of 3 calls in our construction (this is assuming very fast universal hashing, *e.g.* [21]). However, we stress that the constructions in [25] are secure only in the symmetric-key setting; it is easy to see that none of the two constructions from [25] can achieve the indifferentiability property (the attack is similar to the attack against 2-round Feistel described in Section 3).

2 Definitions

We first recall the notion of indifferentiability of random systems, introduced by Maurer *et al.* in [24]. This is an extension of the classical notion of indistinguishability, where one or more oracles are publicly available, such as random oracles or ideal ciphers.

As in [24], we define an *ideal primitive* as an algorithmic entity which receives inputs from one of the parties and delivers its output immediately to the querying party. In this paper, we consider ideal primitives such as random oracle, random permutation and ideal cipher. A *random oracle* [1] is an ideal primitive which provides a random output for each new query; identical input queries are given the same answer. A *random permutation* is an ideal primitive that provides oracle access to a random permutation $P : \{0,1\}^n \to \{0,1\}^n$ and to P^{-1}. An *ideal cipher* is a generalization of a random permutation that models a random block cipher $E : \{0,1\}^k \times \{0,1\}^n \to \{0,1\}^n$. Each key $k \in \{0,1\}^k$ defines an independent random permutation $E_k = E(k, \cdot)$ on $\{0,1\}^n$. The ideal primitive also provides oracle access to E and E^{-1}; that is, on query $(0, k, m)$, the primitive answers $c = E_k(m)$, and on query $(1, k, c)$, the primitive answers m such that $c = E_k(m)$. We stress that in the ideal cipher model, the adversary has oracle access to a publicly available ideal cipher and must send both the key and the plaintext in order to obtain the ciphertext; this is different from the standard model in which the key is privately generated by the system.

The notion of indifferentiability [24] enables to show that an ideal primitive $\mathcal{P}$ (for example, a random permutation) can be replaced by a construction C that is based on some other ideal primitive E; for example, C can be the Feistel construction illustrated in Fig. 1 (left).

Definition 1 ([24]). *A Turing machine C with oracle access to an ideal primitive E is said to be $(t_D, t_S, q, \varepsilon)$-indifferentiable from an ideal primitive $\mathcal{P}$ if there exists a simulator S with oracle access to $\mathcal{P}$ and running in time at most t_S, such that for any distinguisher D running in time at most t_D and making at most q queries, it holds that:*

$$\left|\Pr\left[D^{C^E, E} = 1\right] - \Pr\left[D^{\mathcal{P}, S^{\mathcal{P}}} = 1\right]\right| < \varepsilon$$

C^E is simply said to be indifferentiable from $\mathcal{P}$ if ε is a negligible function of the security parameter n, for polynomially bounded q, t_D and t_S.

The previous definition is illustrated in Figure 2, where C is our 3-round construction of Figure 1 (left), E is an ideal cipher, $\mathcal{P}$ is a random permutation and $\mathcal{S}$ is the simulator. In this paper, for a 3-round construction, we denote these ideal ciphers by E_1, E_2, E_3 (see Fig. 1). Equivalently, one can consider a single ideal cipher E and encode in the first 2 key bits which round ideal cipher E_1, E_2, or E_3 is actually called. The distinguisher has either access to the system

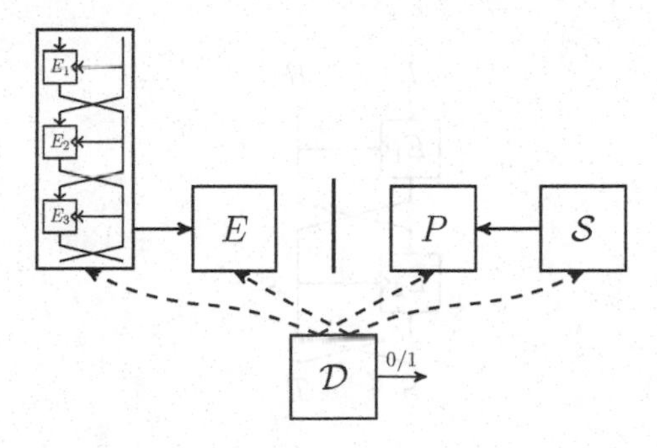

Fig. 2. The indifferentiability notion

formed by the construction C and the ideal cipher E, or to the system formed by the random permutation P and a simulator $\mathcal{S}$. In the first system (left), the construction C computes its output by making calls to the ideal cipher E (equivalently the 3 ideal ciphers E_1, E_2 and E_3); the distinguisher can also make calls to E directly. In the second system (right), the distinguisher can either query the random permutation P, or the simulator that can make queries to P. If the distinguisher first makes a call to the construction C, and then makes the corresponding calls to ideal cipher E, he will get the same answer. This must remain true when the distinguisher interacts with permutation P and simulator $\mathcal{S}$. The role of simulator $\mathcal{S}$ is then to simulate the ideal ciphers E_i's so that 1) the output of $\mathcal{S}$ should be indistinguishable from that of ideal ciphers E_i's and 2) the output of $\mathcal{S}$ should look "consistent" with what the distinguisher can obtain independently from P. We note that in this model the simulator does not see the distinguisher's queries to P; however, it can call P directly when needed for the simulation.

It is shown in [24] that the indifferentiability notion is the "right" notion for substituting one ideal primitive with a construction based on another ideal primitive. That is, if C^E is indifferentiable from an ideal primitive $\mathcal{P}$, then C^E can replace $\mathcal{P}$ in any cryptosystem, and the resulting cryptosystem is at least as secure in the E model as in the $\mathcal{P}$ model; see [24] or [9] for a proof.

3 An Attack against 2 Rounds

In this section we show that 2 rounds are not enough when the inner ideal ciphers are publicly accessible, that is we exhibit a property for 2 rounds that does not exist for a random permutation.

Formally, the 2 round construction is defined as follows (see Fig. 3). Let $E_1 : \{0,1\}^n \times \{0,1\}^n \to \{0,1\}^n$ be a block cipher, where $c = E_1(K, m)$ is the n-bit ciphertext corresponding to n-bit key K and n-bit input message m; let E_2 be defined similarly. We define the permutation $\Psi_2 : \{0,1\}^{2n} \to \{0,1\}^{2n}$ as:

$$\Psi_2(L, R) := (E_1(R, L), E_2(E_1(R, L), R))$$

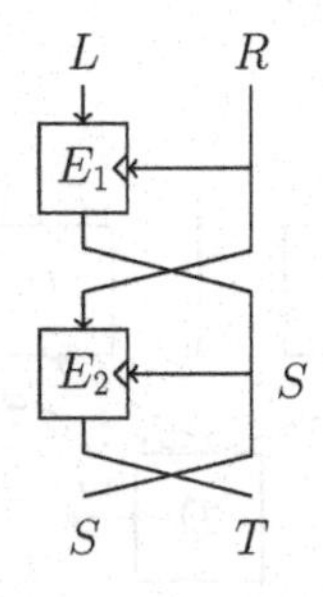

Fig. 3. The 2-round Feistel construction $\Psi_2(L, R)$

It is easy to see that this defines an invertible permutation over $\{0,1\}^{2n}$. Namely, given a ciphertext (S,T) the value R is recovered by "decrypting" T with block-cipher E_2 and key S, and the value L is recovered by "decrypting" S with block-cipher E_1 and key R.

The attack against permutation Ψ_2 is straightforward; it is based on the fact that the attacker can arbitrarily choose both R and S. More precisely, the attacker selects $R = 0^n$ and $S = 0^n$ and queries $L = E_1^{-1}(R,S)$ and $T = E_2(S,R)$. This gives $\Psi_2(L,R) = (S,T)$ as required. However, it is easy to see that with a random permutation P and a polynomially bounded number of queries, it is impossible to find L, R, S, T such that $P(L\|R) = S\|T$ with both $R = 0^n$ and $S = 0^n$, except with negligible probability. Therefore, the 2-round construction cannot replace a random permutation.

Theorem 1. *The 2-round Feistel construction Ψ_2 is not indifferentiable from a random permutation.*

In the full version of the paper [8] we also analyse existing constructions of domain extender for block ciphers and show that they are not indifferentiable from an ideal cipher; more precisely, we show that the CMC [16] and EME [17] constructions are not indifferentiable from an ideal cipher. We stress that our observations do not imply anything concerning their security in the standard pseudo-random permutation model.

4 Indifferentiability of 3-Round Feistel Construction

We now prove our first main result: the 3-round Feistel construction is indifferentiable from a random permutation. To get an ideal cipher, it suffices to prepend a key K to the 3 ideal ciphers E_1, E_2 and E_3; one then gets a family of independent random permutation, parametrised by K, *i.e.* an ideal cipher (see Fig. 1 for an illustration).

Formally, the 3 round permutation $\Psi_3 : \{0,1\}^{2n} \to \{0,1\}^{2n}$ is defined as follows, given block ciphers E_1, E_2 and E_3 with n-bit key (first variable) and n-bit input/output (second variable):

$$\begin{aligned} X &= E_1(R,L) \\ S &= E_2(X,R) \\ T &= E_3(S,X) \\ \Psi_3(L,R) &:= (S,T) \end{aligned}$$

The 3 round block cipher $\Psi_3' : \{0,1\}^k \times \{0,1\}^{2n} \to \{0,1\}^{2n}$ is defined as follows, given block ciphers E_1, E_2 and E_3 with $(k+n)$-bit key and n-bit input/output:

$$\begin{aligned} X &= E_1(K\|R,L) \\ S &= E_2(K\|X,R) \\ T &= E_3(K\|S,X) \\ \Psi_3'(K,(L,R)) &:= (S,T) \end{aligned}$$

Theorem 2. *The 3-round Feistel construction Ψ_3 is $(t_D, t_S, q, \varepsilon)$-indifferentiable from a random permutation, with $t_S = \mathcal{O}(qn)$ and $\varepsilon = 5q^2/2^n$. The 3-round block-cipher construction Ψ_3' is $(t_D, t_S, q, \varepsilon)$-indifferentiable from an ideal cipher, with $t_S = \mathcal{O}(qn)$ and $\varepsilon = 5q^2/2^n$.*

Proof. We only consider the 3-round permutation Ψ_3; the extension to block-cipher Ψ_3' is straightforward. We must construct a simulator $\mathcal{S}$ such that the two systems formed by (Ψ_3, E) and $(P, \mathcal{S})$ are indistinguishable (see Fig. 2).

Our simulator maintains an history of already answered queries for E_1, E_2 and E_3. Formally, when the simulator answers X for a $E_1(R, L)$ query, it stores $(1, R, L, X)$ in history; the simulator proceeds similarly for E_2 and E_3 queries. We write that the simulator "simulates" $E_1(R, L) \leftarrow X$ when it first generates a random $X \in \{0,1\}^n \setminus \mathcal{B}$, where $\mathcal{B}$ is the set of already defined values for $E_1(R, \cdot)$, and then stores $(1, R, L, X)$ in history, meaning that $E_1(R, L) = X$; we use similar notations for E_2 and E_3. The distinguisher's queries are answered as follows by the simulator:

$E_1(R, L)$ query:
1. Simulate $E_1(R, L) \leftarrow X$
2. $(S, T) \leftarrow \mathsf{Adapt}(L, R, X)$
3. Return X

$E_1^{-1}(R, X)$ query
1. Simulate $E_1^{-1}(R, X) \leftarrow L$
2. $(S, T) \leftarrow \mathsf{Adapt}(L, R, X)$
3. Return L

$E_2(X, R)$ query:
1. Simulate $E_1^{-1}(R, X) \leftarrow L$
2. $(S, T) \leftarrow \mathsf{Adapt}(L, R, X)$
3. Return S

$\mathsf{Adapt}(L, R, X)$:
1. $S \| T \leftarrow P(L \| R)$
2. Store $E_2(X, R) = S$ in history
3. Store $E_3(S, X) = T$ in history.
4. Return (S, T).

The procedure for answering the other queries is essentially symmetric; we provide it for completeness:

$E_3^{-1}(S, T)$ query:
1. Simulate $E_3^{-1}(S, T) \leftarrow X$
2. $(L, R) \leftarrow \mathsf{Adapt}^{-1}(S, T, X)$
3. Return X

$E_3(S, X)$ query
1. Simulate $E_3(S, X) \leftarrow T$
2. $(L, R) \leftarrow \mathsf{Adapt}^{-1}(S, T, X)$
3. Return T

$E_2^{-1}(X, S)$ query:
1. Simulate $E_3(S, X) \leftarrow T$
2. $(L, R) \leftarrow \mathsf{Adapt}^{-1}(S, T, X)$
3. Return R

$\mathsf{Adapt}^{-1}(S, T, X)$:
1. $L \| R \leftarrow P^{-1}(S \| T)$
2. Store $E_2(X, R) = S$ in history.
3. Store $E_1(R, L) = X$ in history.
4. Return (L, R)

Finally, the simulator aborts if for some E_i and some key K, it has not defined a permutation for $E_i(K, \cdot)$; that is the simulator aborts if it has defined $E_i(K, X) = E_i(K, Y)$ for some $X \neq Y$ or it has defined $E_i^{-1}(K, X) = E_i^{-1}(K, Y)$ for some $X \neq Y$. This completes the description of the simulator.

As a consistency check, it is easy to see that if the distinguisher makes a single query for $P(L \| R)$ and then queries the simulator for $X \leftarrow E_1(R, L)$, $S \leftarrow$

$E_2(X,R)$ and $T \leftarrow E_3(S,X)$, then the distinguisher obtains $S\|T = P(L\|R)$ as required.

We now proceed to prove that the systems (Ψ_3, E) and $(P, \mathcal{S})$ are indistinguishable. We consider a distinguisher $\mathcal{D}$ making at most q queries to the system (Ψ_3, E) or $(P, \mathcal{S})$ and outputting a bit γ. We define a sequence $\texttt{Game}_0$, $\texttt{Game}_1, \ldots$ of modified distinguisher games. In the first game the distinguisher interacts with the system (Ψ_3, E). We incrementally modify the system so that in the last game the distinguisher interacts with the system $(P, \mathcal{S})$, where $\mathcal{S}$ is the previously defined simulator. We denote by S_i the event that in game i the distinguisher outputs $\gamma = 1$.

• $\texttt{Game}_0$: the distinguisher interacts with Ψ_3 and the ideal ciphers E_i.

• $\texttt{Game}_1$: we modify the way E_i queries are answered, without actually changing the value of the answer. We also maintain an history of already answered queries for E_1, E_2 and E_3. We proceed as follows:

$E_1(R,L)$ query:	$E_1^{-1}(R,X)$ query
1. Let $X \leftarrow E_1(R,L)$	1. Let $L \leftarrow E_1^{-1}(R,X)$
2. $(S,T) \leftarrow \mathsf{Adapt}'(L,R,X)$	2. $(S,T) \leftarrow \mathsf{Adapt}'(L,R,X)$
3. Return X	3. Return L
$E_2(X,R)$ query:	$\mathsf{Adapt}'(L,R,X)$:
1. Let $L \leftarrow E_1^{-1}(R,X)$	1. $S\|T \leftarrow \Psi_3(L\|R)$
2. $(S,T) \leftarrow \mathsf{Adapt}'(L,R,X)$	2. Store $E_2(X,R) = S$ in history.
3. Return S	3. Store $E_3(S,X) = T$ in history.
	4. Return (S,T)

The queries to $E_2^{-1}(X,S)$, $E_3(S,X)$ and $E_3^{-1}(S,T)$ are answered symmetrically.

For example, when given a query to $E_1(R,L)$, we first query ideal cipher E_1 for $X \leftarrow E_1(R,L)$; then instead of X being returned immediately as in $\texttt{Game}_0$, we let $S\|T = \Psi_3(L\|R)$, which gives $S = E_2(X,R)$ and $E_3(S,X) = T$; we then store $(2,X,R,S)$ and $(3,S,X,T)$ in history. Therefore, the value that get stored in history is exactly the same as the value from ideal ciphers E_2 and E_3; the only difference is that this value was obtained indirectly by querying Ψ_3 instead of directly by querying E_2 and E_3. It is easy to see that this holds for any query made by the distinguisher, who receives exactly the same answers in $\texttt{Game}_0$ and $\texttt{Game}_1$; this implies:

$$\Pr[S_1] = \Pr[S_0]$$

As illustrated in Fig. 4, we have actually constructed a simple simulator $\mathcal{S}'$ that makes queries to a subsystem $\mathcal{T}$ that comprises the construction Ψ_3 and the ideal ciphers E_1, E_2 and E_3. The difference between $\mathcal{S}'$ in $\texttt{Game}_1$ and the main simulator $\mathcal{S}$ defined previously is that 1) $\mathcal{S}'$ calls ideal cipher $E_1(R,L)$ instead of simulating it and 2) $\mathcal{S}'$ makes calls to $\Psi_3(L\|R)$ instead of $P(L\|R)$.

• $\texttt{Game}_2$: we modify the way the permutation queries are answered. Instead of using Ψ_3 as in system $\mathcal{T}$, we use the random permutation P in the new system $\mathcal{T}'$ (see Fig. 4).

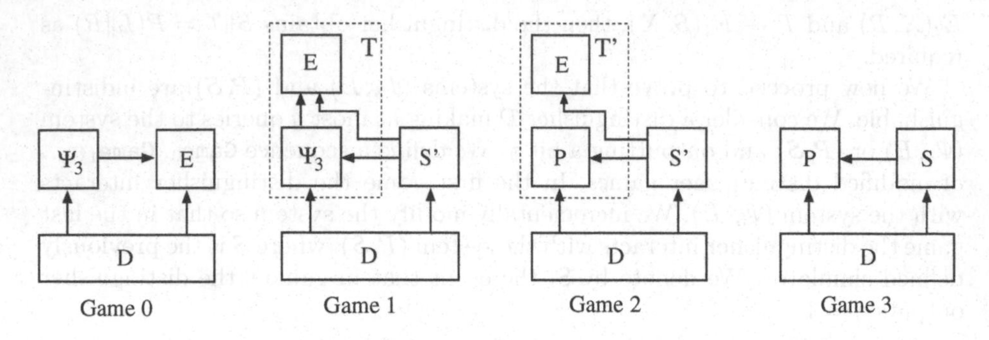

Fig. 4. Sequence of games for proving indifferentiability

We must show that the distinguisher's view has statistically close distribution in $\texttt{Game}_1$ and $\texttt{Game}_2$. For this, we consider the subsystem $\mathcal{T}$ with the 3-round Feistel Ψ_3 and the ideal ciphers E_i's in $\texttt{Game}_1$, and the subsystem $\mathcal{T}'$ with the random permutation P and ideal ciphers E_i's in $\texttt{Game}_2$. We show that the output of systems $\mathcal{T}$ and $\mathcal{T}'$ is statistically close; this in turn shows that the distinguisher's view has statistically close distribution in $\texttt{Game}_1$ and $\texttt{Game}_2$. Note that the indistinguishability of $\mathcal{T}$ and $\mathcal{T}'$ only holds for the particular set of queries made by the distinguisher and the simulator; it could not hold for any possible set of queries.

In the following, we assume that the distinguisher eventually makes a sequence of E_i queries corresponding to all previous Ψ_3 queries that he has made. More precisely, if the distinguisher has made a $\Psi_3(L, R)$ query, then eventually the distinguisher makes the sequence of queries $X \leftarrow E_1(R, L)$, $S \leftarrow E_2(X, R)$ and $T \leftarrow E_3(S, X)$ to the simulator; the same holds for $\Psi_3^{-1}(S, T)$ queries. This is without loss of generality, because from any distinguisher $\mathcal{D}$ we can build a distinguisher $\mathcal{D}'$ with the same output that satisfies this property.

The outputs to E_i queries provided by subsystem $\mathcal{T}$ in $\texttt{Game}_1$ and by subsystem $\mathcal{T}'$ in $\texttt{Game}_2$ are the same, since in both cases these queries are answered by ideal ciphers E_i. Therefore, we must show that the output to P/P^{-1} queries provided by $\mathcal{T}$ and $\mathcal{T}'$ have statistically close distribution, when the outputs to E_i queries provided by $\mathcal{T}$ or $\mathcal{T}'$ are fixed.

We consider a forward permutation query $L\|R$ made by either the distinguisher or the simulator $\mathcal{S}'$. If this $L\|R$ query is made by the distinguisher, since we have assumed that the distinguisher eventually makes the E_i queries corresponding to all his permutation queries, this $L\|R$ query will also be made by the simulator $\mathcal{S}'$, by definition of $\mathcal{S}'$. Therefore we can consider $L\|R$ queries made by the simulator $\mathcal{S}'$ only.

We first consider the answer to $S\|T = \Psi_3(L\|R)$ in $\texttt{Game}_1$. In this case the answer $S\|T$ is computed as follows:

$$X = E_1(R, L)$$
$$S = E_2(X, R)$$
$$T = E_3(S, X)$$

By definition of the simulator $\mathcal{S}'$, when the simulator $\mathcal{S}'$ makes a query for $\Psi_3(L\|R)$, it must have made an ideal cipher query to $E_1(R, L)$ before, or an ideal cipher query to $E_1^{-1}(R, X)$ before, with $L = E_1^{-1}(R, X)$.

If the simulator $\mathcal{S}'$ has made an ideal cipher query for $E_1(R, L)$ to subsystem $\mathcal{T}$, then from the definition of the simulator a call to $\mathsf{Adapt}'(L, R, X)$ has occurred, where $X = E_1(R, L)$; in this Adapt' call the values $E_2(X, R)$ and $E_3(S, T)$ are defined by the simulator; therefore the simulator does not make these queries to sub-system $\mathcal{T}$. This implies that the values of $E_2(X, R)$ and $E_3(S, X)$ are not included in the subsystem $\mathcal{T}$ output; therefore these values are not fixed in the probability distribution that we consider; only the value $X = E_1(R, L)$ is fixed.

Moreover, for fixed X, R the distribution of $S = E_2(X, R)$ is uniform in $\{0,1\}^n \backslash \mathcal{B}$, where $\mathcal{B}$ is the set of already defines values for $E_2(X, \cdot)$. Since there are at most q queries, the statistical distance between the distribution of $E_2(X, R)$ and the uniform distribution in $\{0,1\}^n$ is at most $2q/2^n$; the same holds for the distribution of $T = E_3(S, X)$. Therefore, we obtain that for a fixed X, the distribution of (S, T) is statistically close to the uniform distribution in $\{0,1\}^{2n}$, with statistical distance at most $4q/2^n$.

If the simulator has made an ideal cipher query for $E_1^{-1}(R, X)$, then the same analysis applies and we obtain that for a fixed $L = E_1^{-1}(R, X)$ the distribution of (S, T) is statistically close to the uniform distribution in $\{0,1\}^{2n}$, with statistical distance at most $4q/2^n$. Therefore we obtain that in $\mathtt{Game}_1$ the statistical distance of $S\|T = \Psi_3(L\|R)$ with the uniform distribution is always at most $4q/2^n$.

In $\mathtt{Game}_2$, the output to permutation query $L\|R$ is $S\|T = P(L\|R)$; since there are at most q queries to P/P^{-1}, the statistical distance between $P(L\|R)$ and the uniform distribution in $\{0,1\}^{2n}$ is at most $2q/2^{2n}$.

Therefore the statistical distance between $\Psi_3(L, R)$ in $\mathtt{Game}_1$ and $P(L\|R)$ in $\mathtt{Game}_2$ is at most $4q/2^n + 2q/2^{2n} \leq 5q/2^n$. The same argument applies to inverse permutation queries. This holds for a single permutation query; since there are at most q such queries, we obtain that the statistical distance between outputs of systems $\mathcal{T}$ and $\mathcal{T}'$ to permutation queries and E_i queries, is at most $5q^2/2^n$; this implies:

$$|\Pr[S_2] - \Pr[S_1]| \leq \frac{5q^2}{2^n}$$

• $\mathtt{Game}_3$: eventually the distinguisher interacts with system $(P, \mathcal{S})$. The only difference between the simulator $\mathcal{S}'$ in $\mathtt{Game}_2$ and the simulator $\mathcal{S}$ in $\mathtt{Game}_3$ is that instead of querying ideal ciphers E_i in $\mathtt{Game}_2$, these ideal ciphers are simply simulated in $\mathtt{Game}_3$, while the answer to permutation queries are exactly the same. Therefore, the distinguisher's view has the same distribution in $\mathtt{Game}_2$ and $\mathtt{Game}_3$, which gives:

$$\Pr[S_2] = \Pr[S_3]$$

and finally:

$$|\Pr[S_3] - \Pr[S_0]| \leq \frac{5q^2}{2^n}$$

which terminates the proof of Theorem 2. □

We note that the security bound in $q^2/2^n$ for our 3-round ideal cipher based construction is much better than the security bound in $q^{16}/2^n$ obtained for the 6-round Feistel construction in [10] (based on random oracles).

4.1 Practical Considerations

EXTENDING THE KEY. So far, we showed how to construct an ideal cipher Ψ_3 with $2n$-bit message and k-bit key from three ideal ciphers E_1, E_2, E_3 on n-bit message and $(n+k)$-bit key. As already mentioned, we can actually implement E_1, E_2, E_3 from a single n-bit ideal cipher E whose key length is $n+k+2$.

However, if only a block-cipher with n-bit key and n-bit message is available (for example AES-128), we need a procedure to extend the key size. To handle such cases, we notice that it suffices to first hash the key using a random oracle, and the resulting block cipher remains indifferentiable from an ideal cipher.

Lemma 1. *Assume $E : \{0,1\}^k \times \{0,1\}^n \to \{0,1\}^n$ is an ideal cipher and $H : \{0,1\}^t \to \{0,1\}^k$ is a random oracle. Define $E' : \{0,1\}^t \times \{0,1\}^n \to \{0,1\}^n$ by $E'(K', X) = E(H(K'), X)$, $E'^{-1}(K', Y) = E^{-1}(H(K'), Y)$. Then E' is $(t_D, t_S, q, \varepsilon)$-indifferentiable from an ideal cipher, where $t_S = \mathcal{O}(q(n+t))$ and $\varepsilon = \mathcal{O}(q^2/2^k)$.*

Proof. See the full version of the paper [8].

Using this observation, given a single ideal cipher E on n-bit messages and k-bit key and a random oracle H with output size k bits, we can first build an ideal cipher E' with n-bit message and $(n+k'+2)$-bit key, and then from Theorem 2 we can obtain an ideal cipher Ψ_3 on $2n$-bit messages and k'-bit key. It remains to remove the assumption of having random oracle H; this can easily be accomplished by sacrificing 1 key bit from E, and then using one of the two resulting (independent) ideal ciphers to efficiently implement H using any of the methods from [9].

GOING BEYOND DOUBLE? Another natural question is to extend the domain of the ideal cipher beyond doubling it. One way to accomplish this task is to apply our 3-round construction recursively, each time doubling the domain. However, in this case it is not hard to see that, to extend the domain by a factor of t, the original block cipher E will have to be used $\mathcal{O}(t^{\log_2 3})$ times.[2] This makes the resulting constructions somewhat impractical for large t.

In contrast, assume that we use the 2-step construction: first build a length-preserving random oracle H on $nt/2$ bits (using [9]), and then use the 6-round Feistel construction [10] to get a nt-bit permutation. To construct a random oracle from $nt/2$-bit to $nt/2$-bit, only $\mathcal{O}(t)$ calls to the n-bit ideal cipher are

[2] In essence, this is because we call E three times for each doubling. Actually, this is not counting the calls to the independent variable length random oracle H to hash down the key, as above. However, because the constructions of such an H in [9] are so efficient, it is not hard to see that, even when implementing H using E itself, the dominant term remains $\mathcal{O}(t^{\log_2 3})$ (although the constant is slightly worse).

required (first hash from $nt/2$-bit to n-bit using [9], then expand back to $nt/2$-bits using counter mode). Therefore the 2-step construction requires only $\mathcal{O}(t)$ calls to E, instead of $\mathcal{O}(t^{\log_2 3})$ when iterating our construction. This implies that for large t, the 2-step construction is more efficient.

To give a practical example, let us consider the applications of [15,27], where one needs to apply a random permutation to the domain of an RSA modulus. We take the length of modulus N to be 1024 bits and the underlying block-cipher E to be $n = 128$ with 128-bit key (as in AES-128). One can see that to obtain a 1024-bit permutation from E, only 48 calls to E are required for the 2-step construction, instead of 243 when iterating our construction. However for 1024-bit, the exact security of the 2-step construction is dominated by the term $\mathcal{O}(q^{16}/2^{512})$ from [10], which requires $q \ll 2^{32}$, whereas the exact security of the recursive construction is $\mathcal{O}(q^2/2^{128})$, which requires $q \ll 2^{64}$. Therefore, for a 1024-bit permutation our recursive construction still provides a better security bound; however, for any size larger than 2048 bits, the two constructions have the same $q \ll 2^{64}$ bound [3].

To summarize, our construction is more efficient than the 2-step construction when doubling only once ($t = 2$). However for a large expansion factor t the 2-step construction is more efficient than the recursive method.

4.2 Indifferentiability for 2 Rounds in the Honest-But-Curious Model

In the full version of the paper we also consider the *honest-but-curious* model of indifferentiability introduced by Dodis and Puniya [12], which is a variant of the general indifferentiability model. We show that in the honest-but-curious model, 2 rounds as depicted in Fig 3 are actually sufficient to get indifferentiability.

5 Domain Extension of Tweakable Block Cipher

In this section, we also analyse our construction in the standard model, and we use a *tweakable* block-cipher as the underlying primitive. The main result of this section is that a 3-round Feistel enables to get a security guarantee beyond the birthday paradox.

Tweakable block-ciphers were introduced by Liskov, Rivest and Wagner in [22] and provide an additional input - the tweak - that enables to get a *family* of independent block-ciphers. Efficient constructions of tweakable block-ciphers were described in [22], given ordinary block-ciphers.

Definition 2. *A tweakable block-cipher is an efficiently computable function* $\tilde{E} : \{0,1\}^k \times \{0,1\}^\omega \times \{0,1\}^n \rightarrow \{0,1\}^n$ *that takes as input a key* $K \in \{0,1\}^k$*, a tweak* $W \in \{0,1\}^\omega$ *and a message* $m \in \{0,1\}^n$ *and returns a ciphertext* $c \in \{0,1\}^n$*. For every* $K \in \{0,1\}^k$ *and* $W \in \{0,1\}^\omega$*, the function* $\tilde{E}(K, W, \cdot)$ *is a permutation over* $\{0,1\}^n$*.*

[3] The length-preserving random oracle used in the 6-round Feistel has the birthday bound of $q^2/2^{128}$.

The security notion for a tweakable block-cipher is a straightforward extension of the corresponding notion for block-ciphers. A classical block-cipher E is a strong pseudo-random permutation if no adversary can distinguish $E(K,\cdot)$ from a random permutation, where $\mathcal{A}$ can make calls to both E and E^{-1}, and $K \leftarrow \{0,1\}^k$. For tweakable block-ciphers, the adversary can additionally choose the tweak, and $E(K,\cdot,\cdot)$ should be indistinguishable from a *family* of random permutations, parametrised by $W \in \{0,1\}^{\omega}$:

Definition 3. *A tweakable block-cipher is said to be (t,q,ε)-secure if for any adversary $\mathcal{A}$ running in time at most t and making at most q queries, the adversary's advantage in distinguishing $\tilde{E}(K,\cdot,\cdot)$ with $K \leftarrow \{0,1\}^k$ from a family of independent random permutation $\tilde{\Pi}(\cdot,\cdot)$ is at most ε, where $\mathcal{A}$ can make calls to both $\tilde{E}$ and $\tilde{E}^{-1}$.*

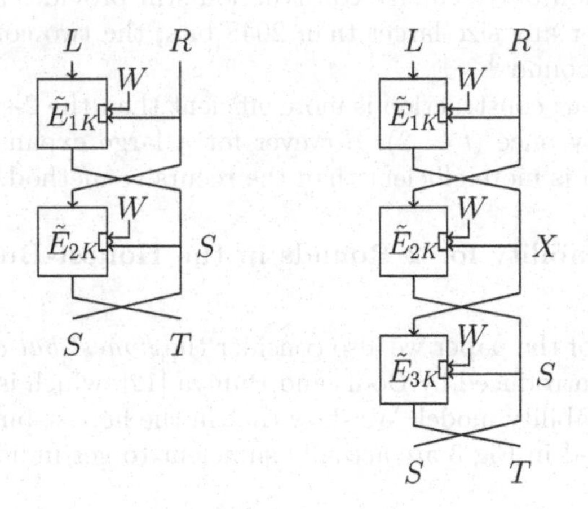

Fig. 5. The tweakable block ciphers $\tilde{\Psi}_2$ (left) and $\tilde{\Psi}_3$ (right), with key K and tweak W

We first show that 2 rounds are enough to get a $2n$-bit tweakable block-cipher from a n-bit tweakable block-cipher (see Fig. 5, left). Formally, our 2-round domain extender for tweakable block-cipher works as follows. Let E_1 and E_2 be two tweakable block-ciphers with the same signature:

$$\tilde{E}_i : \{0,1\}^k \times \{0,1\}^{\omega} \times \{0,1\}^n \to \{0,1\}^n$$

The tweakable block cipher $\tilde{\Psi}_2 : \{0,1\}^k \times \{0,1\}^{\omega-n} \times \{0,1\}^{2n} \to \{0,1\}^{2n}$ is then defined as follows; the difference with Fig. 3 is that the R and S inputs go to the tweak (concatenated with the main tweak W) instead of the key.

$$S = E_1(K, W\|R, L)$$
$$T = E_2(K, W\|S, R)$$
$$\tilde{\Psi}_2(K, W, (L,R)) = (S,T)$$

Theorem 3. *The tweakable block-cipher $\tilde{\Psi}_2$ is a (t', q, ε')-secure tweakable block-cipher, if $\tilde{E}_1$ and $\tilde{E}_2$ are both (t, q, ε)-secure tweakable block-ciphers, where $\varepsilon' = 2 \cdot \varepsilon + q^2/2^n + q^2/2^{2n}$ and $t' = t - \mathcal{O}(qn)$.*

Proof. See the full version of the paper [8].

Now we consider the 3 round tweakable block cipher $\tilde{\Psi}_3$, defined in a similar manner as $\tilde{\Psi}_2$ (see Fig. 5 for an illustration). The 3-round construction enables to go beyond the birthday security bound. Namely instead of having a bound in $q^2/2^n$ as in the 2-round construction, the bound for the 3-round construction is now $q^2/2^{2n}$, which shows that the construction remains secure until $q < 2^n$ instead of $q < 2^{n/2}$.

Theorem 4. *The tweakable block-cipher $\tilde{\Psi}_3$ is a (t', q, ε')-secure tweakable block-cipher, if $\tilde{E}_1$, $\tilde{E}_2$ and $\tilde{E}_3$ are all (t, q, ε)-secure tweakable block-ciphers, where $\varepsilon' = 3 \cdot \varepsilon + q^2/2^{2n}$ and $t' = t - \mathcal{O}(qn)$.*

Proof. See the full version of the paper [8].

One drawback of our construction is that it shrinks the tweak size from ω bits to $\omega - n$ bits. We show a simple construction that extends the tweak size, using a keyed universal hash function; this construction can be of independent interest.

Definition 4. *A family $\mathcal{H}$ of functions with signature $\{0,1\}^{\omega'} \to \{0,1\}^{\omega}$ is said to be ε-almost universal if $\Pr_h[h(x) = h(y)] \leq \varepsilon$ for all $x \neq y$, where the probability is taken over h chosen uniformly at random from $\mathcal{H}$.*

Let $\tilde{E}$ be a tweakable block-cipher with tweak in $\{0,1\}^{\omega}$. Given a family $\mathcal{H}$ of hash functions h with signature $\{0,1\}^{\omega'} \to \{0,1\}^{\omega}$ and $\omega' > \omega$, our tweakable block-cipher $\tilde{E}$ with extended tweak length ω' is defined as:

$$\tilde{E}'((K, h), W', m) = \tilde{E}(K, h(W'), m)$$

Theorem 5. *The tweakable block cipher $\tilde{E}'$ is a (q, t', ε')-secure tweakable block cipher if $\tilde{E}$ is a (q, t, ε_1)-secure tweakable block cipher and the hash function family $\mathcal{H}$ is ε_2-almost universal, with $\varepsilon' = \varepsilon_1 + q^2 \cdot \varepsilon_2$ and $t' = t - \mathcal{O}(q)$.*

Proof. See the full version of the paper [8].

We note that many efficient constructions of universal hash function families are known, with $\varepsilon_2 \simeq 2^{-\omega}$. Therefore the new tweakable block-cipher can have the same level of security as the original one, up to the birthday bound for the tweak, *i.e.* for $q \leq 2^{\omega/2}$.

6 Conclusion

We have described the first domain extender for ideal ciphers, *i.e.* we have showed a construction that is indifferentiable from a $2n$-bit ideal cipher, given a n-bit ideal cipher. Our construction is based on a 3-round Feistel, and is more efficient and more secure than first building a n-bit random oracle from a n-bit ideal cipher (as in [9]) and then a $2n$-bit ideal cipher from a n-bit random oracle (as in [10]). We have also shown that in the standard model, our construction with 2 rounds enables to get a $2n$-bit tweakable block-cipher from a n-bit tweakable block-cipher and that with 3 rounds we get a security guarantee beyond the birthday paradox.

References

1. Bellare, M., Rogaway, P.: Random oracles are practical: A paradigm for designing efficient protocols. In: Proceedings of the 1st ACM Conference on Computer and Communications Security, pp. 62–73 (1993)
2. Biryukov, A., Khovratovich, D., Nikolic, I.: Distinguisher and Related-Key Attack on the Full AES-256. In: Halevi, S. (ed.) CRYPTO 2009. LNCS, vol. 5677, pp. 231–249. Springer, Heidelberg (2009)
3. Biryukov, A., Khovratovich, D.: Related-key Cryptanalysis of the Full AES-192 and AES-256. In: Matsui, M. (ed.) ASIACRYPT 2009. LNCS, vol. 5912, pp. 1–18. Springer, Heidelberg (2009)
4. Black, J.: The Ideal-Cipher Model, Revisited: An Uninstantiable Blockcipher-Based Hash Function. In: Robshaw, M.J.B. (ed.) FSE 2006. LNCS, vol. 4047, pp. 328–340. Springer, Heidelberg (2006)
5. Black, J., Rogaway, P., Shrimpton, T.: Black-Box Analysis of the Block Cipher-Based Hash-Function Constructions from PGV. In: Yung, M. (ed.) CRYPTO 2002. LNCS, vol. 2442, p. 320. Springer, Heidelberg (2002)
6. Chakraborty, D., Sarkar, P.: A new mode of encryption providing a tweakable strong pseudo-random permutation. In: Robshaw, M.J.B. (ed.) FSE 2006. LNCS, vol. 4047, pp. 293–309. Springer, Heidelberg (2006)
7. Chakraborty, D., Sarkar, P.: HCH: A new tweakable enciphering scheme using the hash-encrypt-hash approach. In: Barua, R., Lange, T. (eds.) INDOCRYPT 2006. LNCS, vol. 4329, pp. 287–302. Springer, Heidelberg (2006)
8. Coron, J.S., Dodis, Y., Mandal, A., Seurin, Y.: A Domain Extender for the Ideal Cipher. Full version of this paper. Cryptology ePrint Archive, Report 2009/356, `http://eprint.iacr.org/`
9. Coron, J.S., Dodis, Y., Malinaud, C., Puniya, P.: Merkle-Damgård Revisited: How to Construct a Hash Function. In: Shoup, V. (ed.) CRYPTO 2005. LNCS, vol. 3621, pp. 430–448. Springer, Heidelberg (2005)
10. Coron, J.S., Patarin, J., Seurin, Y.: The Random Oracle Model and the Ideal Cipher Model are Equivalent. In: Wagner, D. (ed.) CRYPTO 2008. LNCS, vol. 5157, pp. 1–20. Springer, Heidelberg (2008), Full version available at Cryptology ePrint Archive, Report 2008/246, `http://eprint.iacr.org/`
11. Desai, A.: The security of all-or-nothing encryption: Protecting against exhaustive key search. In: Bellare, M. (ed.) CRYPTO 2000. LNCS, vol. 1880, p. 359. Springer, Heidelberg (2000)
12. Dodis, Y., Puniya, P.: On the Relation Between the Ideal Cipher and the Random Oracle Models. In: Halevi, S., Rabin, T. (eds.) TCC 2006. LNCS, vol. 3876, pp. 184–206. Springer, Heidelberg (2006)

13. Even, S., Mansour, Y.: A construction of a cipher from a single pseudorandom permutation. In: Matsumoto, T., Imai, H., Rivest, R.L. (eds.) ASIACRYPT 1991. LNCS, vol. 739, pp. 210–224. Springer, Heidelberg (1993)
14. Fluhrer, S.R., McGrew, D.A.: The extended codebook (XCB) mode of operation. Technical Report 2004/078, IACR eprint archive (2004)
15. Granboulan, L.: Short signature in the random oracle model. In: Zheng, Y. (ed.) ASIACRYPT 2002. LNCS, vol. 2501, pp. 364–378. Springer, Heidelberg (2002)
16. Halevi, S., Rogaway, P.: A tweakable enciphering mode. In: Boneh, D. (ed.) CRYPTO 2003. LNCS, vol. 2729, pp. 482–499. Springer, Heidelberg (2003)
17. Halevi, S., Rogaway, P.: A parallelizable enciphering mode. In: Okamoto, T. (ed.) CT-RSA 2004. LNCS, vol. 2964, pp. 292–304. Springer, Heidelberg (2004)
18. Halevi, S.: Invertible Universal hashing and the TET Encryption Mode. In: Menezes, A. (ed.) CRYPTO 2007. LNCS, vol. 4622, pp. 412–429. Springer, Heidelberg (2007)
19. Jonsson, J.: An OAEP variant with a tight security proof, `http://eprint.iacr.org/2002/034/`
20. Kilian, J., Rogaway, P.: How to protect DES against exhaustive key search (An analysis of DESX). Journal of Cryptology 14(1), 17–35 (2001)
21. Krovetz, T.: Message Authentication on 64-Bit Architectures. In: Biham, E., Youssef, A.M. (eds.) SAC 2006. LNCS, vol. 4356, pp. 327–341. Springer, Heidelberg (2007)
22. Liskov, M., Rivest, R., Wagner, D.: Tweakable Block Ciphers. In: Yung, M. (ed.) CRYPTO 2002. LNCS, vol. 2442, p. 31. Springer, Heidelberg (2002)
23. Luby, M., Rackoff, C.: How to construct pseudorandom permutations from pseudorandom functions. SIAM Journal of Computing 17(2), 373–386 (1988)
24. Maurer, U., Renner, R., Holenstein, C.: Indifferentiability, Impossibility Results on Reductions, and Applications to the Random Oracle Methodology. In: Naor, M. (ed.) TCC 2004. LNCS, vol. 2951, pp. 21–39. Springer, Heidelberg (2004)
25. Minematsu, K.: Beyond-Birthday-Bound Security Based on Tweakable Block Cipher. In: Dunkelman, O. (ed.) FSE 2009. LNCS, vol. 5665, pp. 308–326. Springer, Heidelberg (2009)
26. Naor, M., Reingold, O.: On the construction of pseudorandom permutations: Luby-Rackoff revisited. J. of Cryptology (1999); Preliminary Version: STOC 1997
27. Phan, D.H., Pointcheval, D.: Chosen-Ciphertext Security without Redundancy. In: Laih, C.-S. (ed.) ASIACRYPT 2003. LNCS, vol. 2894, pp. 1–18. Springer, Heidelberg (2003)
28. Rogaway, P., Bellare, M., Black, J.: OCB: A block-cipher mode of operation for efficient authenticated encryption. In: ACM Conference on Computer and Communication Security, pp. 196–205 (2001)
29. Shoup, V.: Sequences of games: a tool for taming complexity in security proofs, `http://eprint.iacr.org/2004/332/`
30. Wang, P., Feng, D., Wu, W.: HCTR: A variable-input-length enciphering mode. In: Feng, D., Lin, D., Yung, M. (eds.) CISC 2005. LNCS, vol. 3822, pp. 175–188. Springer, Heidelberg (2005)

Delayed-Key Message Authentication for Streams

Marc Fischlin and Anja Lehmann

Darmstadt University of Technology, Germany
www.minicrypt.de

Abstract. We consider message authentication codes for streams where the key becomes known only at the end of the stream. This usually happens in key-exchange protocols like SSL and TLS where the exchange phase concludes by sending a MAC for the previous transcript and the newly derived key. SSL and TLS provide tailor-made solutions for this problem (modifying HMAC to insert the key only at the end, as in SSL, or using upstream hashing as in TLS). Here we take a formal approach to this problem of delayed-key MACs and provide solutions which are "as secure as schemes where the key would be available right away" but still allow to compute the MACs online even if the key becomes known only later.

1 Introduction

With the final step in key exchange protocols the parties usually authenticate the previous communication. This is typically achieved by exchanging message authentication codes $\mathsf{Mac}(\mathsf{K}, \mathsf{transcript})$ computed over the transcript of the communication. Examples include the final message in the handshake protocol of SSL and TLS [17], as well as many other key exchange protocols [4,13,14,8].

The intriguing observation here is that the key for the MAC computations becomes only known after the transcript is provided. We call this *delayed-key* authentication. For such schemes, even MACs which potentially allow to authenticate streams may need to store the entire transcript before the MAC can be derived. One well-known example is HMAC where the (inner) key is prepended to the message before hashing, $H(\mathsf{K}_{\text{out}}, H(\mathsf{K}_{\text{in}}, m))$. In this case the key must be available *before* processing the message in order to take advantage of the iterated hash function structure.

For computational efficiency and, especially, for storage reasons it is often desirable to compute the MAC iteratively, though. This has been acknowledged by popular protocols like SSL, which uses a variant of HMAC where the key is *appended* to the message instead, and TLS which first hashes the transcript iteratively and then runs the MAC on the hash value only. Similarly, for the key exchange protocols for machine readable travel documents (MRTD) by the German government [5] the final MAC computation omits large parts of the

D. Micciancio (Ed.): TCC 2010, LNCS 5978, pp. 290–307, 2010.

transcript and only inputs the messages of the final rounds. This allows the resource-bounded passport to free memory immediately. The protocol is under standardization for ISO/IEC JTC1/SC17.

The SSL and TLS solution to the problem both rely on the collision resistance of the underlying hash function for HMAC.[1] For TLS collision resistance suffices to show security (assuming HMAC is secure), but introduces another requirement on the hash function. Recall that HMAC (resp. its theoretical counterpart NMAC) can be shown to be secure if the compression function is pseudorandom [1] or non-malleable [7]. For SSL it is still unclear how the security of the modified HMAC relates to the security of the original HMAC. As for the MRTD protocol for German passports, in most key exchange protocols it is recommended to include the whole transcript (yet, we are not aware of any concrete attack if only parts of the transcript enter the computation).

An additional constraint originates from the implementation of the MAC algorithm. Key-exchange protocols are often used as building blocks in more complex cryptographic protocols which, in turn, also use the same MAC algorithm for subsequent authentication (e.g., the record protocol in TLS/SSL). To be applicable to resource-bounded devices a delayed-key MAC should therefore draw on the same implementation as the regular MAC. This is particularly true if the implementation has been designed to resist side-channel attacks. Hence, instead of designing delayed-key MACs from scratch, a "lightweight" transformation given an arbitrary MAC algorithm is preferable.

Our Results. We initiate a study of solutions for the delayed-key MAC problem. There are two reasonable scenarios, originating from the key-exchange application: The most relevant case in practice is the *one-sided* case where one party is resource-bounded while the other party is more powerful, e.g., a TLS/SSL secured connection between a mobile device and a server, or an authentication procedure between a smart card and a card reader. Then, ideally, the constraint device should benefit from solutions with low storage, whereas we can still assume that the server is able to store the entire transcript. If both parties have storage limitations, e.g., two mobile devices communicating with each other, then we are interested in *two-sided* solutions. Since the one-sided case allows for the weaker devices in terms of resource constraints, the necessity of storage-optimized protocols in this scenario is usually higher than in the two-sided case.

Thus, we focus on the one-sided case for which we present efficient solutions which are all based on the same seemingly obvious principle: to compute a MAC the sending party first picks an ephemeral key L and computes the MAC for this key and the data stream. Then, in addition to the MAC under this key, the party also transmits an "encryption" (or a "pointer") P allowing the other party to recover the ephemeral key L from P and the meanwhile available long-term

[1] The weaker requirement of preimage resistance does not suffice, because the transcript that gets authenticated, is partially determined by both the sender and the receiver of the MAC.

key K.[2] Note that since verification is usually done by re-computing a MAC the idea also applies to the verification of the other's party MAC, i.e., one of the parties in a key-exchange protocol can both compute its own MAC and verify the other party's MAC with low storage requirements.

From an efficiency and implementation viewpoint the instantiations of this principle should interfere as little as possible with the underlying protocol such that we get a universal solution. Note that this general approach already allows to obtain a delayed-key solution starting from a regular MAC, such that both variants can be used conveniently even on severely constraint devices. In terms of security we require the solution to be as secure as the original scheme. The latter condition at foremost demands that the instantiation inherits the unforgeability property of the original MAC. But since the long-term key K is subsequently used in protocols (like encryption with the derived keys from the master secret in SSL and TLS), unforgeability alone is not sufficient.

We also demand that the modified scheme only leaks "as much about the key K as the original scheme would" and call this notion *leakage-invariance*. The idea behind this notion is that, in the original key-exchange protocol, the MAC for K leaks some information about the key itself, and that the subsequent usage of the key (derivation, direct encryption etc.) should be still be secure. Following the idea of semantically secure encryption [10] we require that a solution for the delayed-key problem allows to compute at most the information about K that one could derive from a $\mathsf{Mac}(\mathsf{K}, \cdot)$ oracle (used in the original protocol).

We discuss four solutions which are secure according to our notion (and which come with different efficiency/security trade-offs). Roughly, these are:

Encrypt-then-MAC: We assume that the underlying (deterministic) MAC is a pseudorandom function (which is a widely used assumption about HMAC) and then compute the MAC $\sigma \leftarrow \mathsf{Mac}(\mathsf{L}, m, \ell)$ for the ephemeral key and then encrypt L under K and MAC this data, $\mathsf{P} = (c, t) = (\mathsf{Mac}(\mathsf{K}, 0||\ell) \oplus \mathsf{L}, \mathsf{Mac}(\mathsf{K}, 1||\ell||c))$ for a label ℓ which can either be the server or client constant as in SSL or a random session identifier. The receiver can then recover L from the encryption and verify the MAC σ.

Pseudorandom Permutation: We again assume that the MAC is a pseudorandom function and use a four-round Feistel structure to build a pseudorandom permutation $\pi(\mathsf{K}, \cdot)$ out of it. Then $\sigma \leftarrow \mathsf{Mac}(\mathsf{L}, m)$ and $\mathsf{P} = \pi^{-1}(\mathsf{K}, \mathsf{L})$ such that the receiver can re-obtain $\mathsf{L} = \pi(\mathsf{K}, \mathsf{P})$ and verify the MAC σ. The communication overhead here is smaller than in the previous case but the construction requires more MAC computations.

Encrypt-only: For the pseudorandom MAC we simply let $\mathsf{P} = (\ell, \mathsf{Mac}(\mathsf{K}, \ell) \oplus \mathsf{L})$ for random label ℓ. In this case the security condition is that an adversary

[2] This approach is more general than it may seem at first glance: One can think of the MAC computation for key L as a (probabilistic) processing of the message and the final computation of the pointer (from K, L and the value from the first stage) as an "enveloping" transformation involving the key. It comprises for example the SSL/TLS solutions (with empty L). We finally remark that sending L in clear usually violates the secure deployment of such MACs in key agreement protocols.

attacking this modified scheme can only make a limited number of verification requests (which corresponds to the common case that in two-party key-exchange protocols for each exchanged key K the server and the client compute and verify only one MAC each). Also, we can only show that the adversary is unable to recover the entire key K from the modified scheme (in contrast to any information about the key, as in the previous cases). This is sufficient to provide security if the key is afterwards hashed (assuming that the hash functions is a good randomness extractor or even behaves like a random oracle).

XOR: In the most simple case we let $\mathsf{P} = \mathsf{K} \oplus \mathsf{L}$ be the one-time pad encryption of L under K. Assuming that MAC remains pseudorandom under related-key attacks [3] this is again an unforgeable, leakage-invariant MAC (if the adversary task is to recover the whole key K). The leakage-invariance also relies on the assumption that the adversary can only make a limited number of verification queries, and gets to see at most one MAC. The latter is justified in schemes where only one of the party sends a MAC or where one party immediately aborts without sending its MAC if the received MAC is invalid.

As mentioned before all proposed solutions above support the one-sided case where one of the parties can store the message easily. In contrast, the TLS/SSL solutions also work in the two-sided case of two resource-constraint parties, but both rely on the collision-resistance of the underlying hash function whereas our solutions can in principle be implemented based on one-way functions. We therefore address the question whether or not collision-resistance is necessary for the two-sided case or not, and show that one-way functions suffice. However, as our solution make use of digital signatures it is mainly a proof of concept and it remains an interesting open problem to find more efficient constructions for this case.

Related Results. To the best of our knowledge the delayed-key problem has not undergone a comprehensive formal treatment so far. The solution in TLS can be shown to be secure according to our model, but relies on collision-resistance. As attacks have shown, however, this appears to be a stronger assumption than pseudorandomness, especially in light of the deployed hash functions MD5 and SHA-1 in TLS (see also the discussion in [1]). We note that relaxing the requirement of collision-resistance is also a goal in other areas like hash-and-sign schemes [12].

Closest to our setting here comes the scenario of broadcast authentication of streams via the TESLA protocol [16]. There, the two parties share a one-way chain of keys and authenticate each packet in time t with the t-th key of the chain. Hence, TESLA also deals with authentication of streams and supports limited buffering, but in contrast to our setting TESLA covers immediate authentication of packets, requiring synchronization between the parties, and assumes shared keys right away (whereas our key is delayed).

Analogously to TESLA, all other works on stream authentication refer to immediate verification of each packet, e.g. [11].

In a recent work, Garay et al. [9] also address the problem of MAC precomputations. However, they consider MACs in the context of hardware security and show how to perform most of a MAC computation offline, before the message is available.

2 Preliminaries

In this section we introduce the basic notions for message authentication codes. In the key exchange application the two parties at the end usually compute the MAC for the same message m but include their identity in the message. For instance, SSL includes the server and client constant in the computation of the finished message. Alternatively, the label can also be a random value chosen by the party computing the MAC. In any case we assume that the label is known at the outset of the MAC computation. We thus introduce labels in the model such that each message m is escorted by a label $\ell \in \{0,1\}^n$ and the authentication code covers both parts. We note that, for regular MACs, this is rather a syntactic modification and becomes important only for the case of delayed-key MACs.

Definition 1. *A message authentication code scheme* $\mathsf{MAC} = (\mathsf{KGen}, \mathsf{Mac}, \mathsf{Vf})$ *(with labels) is a triple of efficient algorithms where*

Key Generation. $\mathsf{KGen}(1^n)$ *gets as input the security parameter* 1^n *and returns a key* k*.*

Authentication. *The authentication algorithm* $\sigma \leftarrow \mathsf{Mac}(k, m, \ell)$ *takes as input the key* k*, a message* m *from a space* $\mathcal{M}_n$ *and a label* $\ell \in \{0,1\}^n$ *and returns a tag* σ *in a range* $\mathcal{R}_n$*.*

Verification. $\mathsf{Vf}(k, m, \ell, \sigma)$ *returns a bit.*

It is assumed that the scheme is complete, *i.e., for all* $k \leftarrow \mathsf{KGen}(1^n)$*, any* $(m, \ell) \in \mathcal{M}_n$*, and any* $\sigma \leftarrow \mathsf{Mac}(k, m, \ell)$ *we have* $\mathsf{Vf}(k, m, \ell, \sigma) = 1$*.*

A MAC is called deterministic if algorithm Mac is deterministic. Unforgeability of MACs demands that it is infeasible to produce a valid tag for a new message:

Definition 2. *A message authentication code* $\mathsf{MAC} = (\mathsf{KGen}, \mathsf{Mac}, \mathsf{Vf})$ *(with labels) is called* unforgeable under chosen message attacks *if for any efficient algorithm* $\mathcal{A}$ *the probability that the experiment* $\mathsf{Forge}_{\mathcal{A}}^{\mathsf{MAC}}$ *evaluates to* 1 *is negligible (as a function of* n*), where*

Experiment $\mathsf{Forge}_{\mathcal{A}}^{\mathsf{MAC}}(n)$
$k \leftarrow \mathsf{KGen}(1^n)$
$(m^*, \ell^*, \sigma^*) \leftarrow \mathcal{A}^{\mathsf{MAC}(k,\cdot,\cdot), \mathsf{Vf}(k,\cdot,\cdot,\cdot)}(1^n)$
Return 1 *iff*
$\mathsf{Vf}(k, m^*, \ell^*, \sigma^*) = 1$ *and* $\mathcal{A}$ *has never queried* $\mathsf{Mac}(k, \cdot, \cdot)$ *about* (m^*, ℓ^*)*.*

Note that for deterministic MACs where, in addition, the verification algorithm recomputes the tag and compares it to the given tag, the verification oracle $\mathsf{Vf}(k, \cdot, \cdot, \cdot)$ can be omitted [2] while decreasing the adversary's success probability by at most the number of verification queries. This particularly holds for HMAC.

For some of our security proofs it is necessary to assume that the MAC is a pseudorandom function. We note again that HMAC (or, to be precise, NMAC) has this property as long as the underlying compression function is pseudorandom [1].

Definition 3. *A message authentication code* MAC *is a* pseudorandom function *if for any efficient distinguisher $\mathcal{D}$ the advantage*

$$\left|\text{Prob}\left[\mathcal{D}^{\mathsf{Mac}(k,\cdot)}(1^n) = 1\right] - \text{Prob}\left[\mathcal{D}^{f(\cdot)}(1^n) = 1\right]\right|$$

is negligible, where the probability in the first case is over $\mathcal{D}$'s coin tosses and the choice of $k \leftarrow \mathsf{KGen}(1^n)$, and in the second case over $\mathcal{D}$'s coin tosses and the choice of the random function $f : \mathcal{M}_n \to \mathcal{R}_n$.

3 Defining Delayed-Key MACs for Streams

As explained in the introduction in the setting of MACs for streams where the key K is only available at the end of the communication, we augment the MAC by a function Point which maps the ephemeral key L (used to derive the MAC for the stream) via K to a pointer P, and such that the verifier can recover the ephemeral key from this pointer and K by the "inverse" Point^{-1}. We let Point also depend on the MAC σ computed with the ephemeral key to capture general solutions as in TLS and since this information is available when computing the pointer (see also the remark after the definition). If Point does not depend on σ we usually omit it from the algorithm's input.

Definition 4. *A delayed-key message authentication code scheme* $\mathsf{DKMAC} = (\mathsf{KGen}, (\mathsf{Mac}, \mathsf{Point}), \mathsf{Vf})$ *(with labels) is a tuple of efficient algorithms where*

Key Generation. $\mathsf{KGen}(1^n)$ *gets as input the security parameter* 1^n *and returns a secret key* K.

Authentication. *Algorithm* Mac *on input an ephemeral key* L*, a message* m *and a label* ℓ *returns a tag* σ*, and algorithm* Point *for input two keys* K *and* L *and the label* ℓ *returns a pointer* P*. An augmented tag for key* K *and* (m, ℓ) *then consists of the pair* $(\sigma, \mathsf{P}) \leftarrow (\mathsf{Mac}(\mathsf{L}, m, \ell), \mathsf{Point}(\mathsf{K}, \mathsf{L}, \ell, \sigma))$ *for random* $\mathsf{L} \xleftarrow{\$} \mathsf{KGen}(1^n)$.

Verification. $\mathsf{Vf}(\mathsf{K}, \mathsf{P}, m, \ell, \sigma)$ *returns a bit.*

It is assumed that the scheme is complete, i.e., for any $\mathsf{K} \leftarrow \mathsf{KGen}(1^n)$*, any* $(m, \ell) \in \mathcal{M}_n \times \{0,1\}^n$*, any augmented tag* $(\sigma, \mathsf{P}) \leftarrow (\mathsf{Mac}(\mathsf{L}, m, \ell), \mathsf{Point}(\mathsf{K}, \mathsf{L}, \ell))$ *for* $\mathsf{L} \leftarrow \mathsf{KGen}(1^n)$ *we have* $\mathsf{Vf}(\mathsf{K}, \mathsf{P}, m, \ell, \sigma) = 1$.

Both the SSL as well as the TLS solution can be mapped trivially to the definition above. Namely, in both cases the ephemeral key L is the empty string and the "MAC" σ is merely the hash value of the message. The pointer P is then the result of the actual MAC computations for K (i.e., HMAC with appended key in SSL and HMAC for the hash value in TLS).

We remark that in key exchange protocols usually both parties send a MAC of the transcript, possibly adding some distinct public identifiers. Our notion of delayed-key MACs can be easily used to model the one-sided case with a bounded client and a powerful server such that the client can *compute* its own MAC and *verify* the server's MAC with limited storage only (assuming that the underlying MAC implements verification by recomputing the MAC and comparing the outcome to the given tag): Namely, the client uses an ephemeral key L to compute its own MAC, and another ephemeral key L' to start computing the server's MAC for verification. At the end, the client transmits the pointers P and P' for the two MACs and the server derives L, L' through K and verifies the client MAC and computes and sends its own MAC. The client then only needs to verify that this received MAC matches the previously computed value.

3.1 Security of Delayed-Key MACs

We adapt the security requirement of unforgeable MACs to our scenario of delayed-key MACs, i.e., we grant the adversary access to an oracle $\mathcal{O}_{\mathsf{MAC}}(\mathsf{K}, \cdot)$ that is initialized with a secret key K and mimics the authentication process, returning augmented tags. Thus, for every query the oracle first chooses a fresh ephemeral key L_i and then returns the augmented tag $(\sigma_i, \mathsf{P}_i) \leftarrow (\mathsf{Mac}(\mathsf{L}_i, m_i, \ell_i),$ $\mathsf{Point}(\mathsf{K}, \mathsf{L}_i, \ell_i, \sigma_i,))$. After learning several tags the adversary eventually halts and outputs a tuple $(\mathsf{P}^*, m^*, \ell^*, \sigma^*)$. The adversary is successful if the output verifies as true under key K and the oracle has never been invoked on (m^*, ℓ^*).

Definition 5. *A delayed-key message authentication code* $\mathsf{DKMAC} = (\mathsf{KGen},$ $(\mathsf{Mac}, \mathsf{Point}), \mathsf{Vf})$ *(with labels) is called* unforgeable under chosen message attacks *if for any efficient algorithm $\mathcal{A}$ the probability that the experiment* $\mathsf{Forge}_{\mathcal{A}}^{\mathsf{DKMAC}}$ *evaluates to* 1 *is negligible (as a function of n), where*

Experiment $\mathsf{Forge}_{\mathcal{A}}^{\mathsf{DKMAC}}(n)$
 $\mathsf{K} \leftarrow \mathsf{KGen}(1^n)$
 $(\mathsf{P}^*, m^*, \ell^*, \sigma^*) \leftarrow \mathcal{A}^{\mathcal{O}_{\mathsf{MAC}}(\mathsf{K},\cdot)}(1^n)$
  where $\mathcal{O}_{\mathsf{MAC}}(\mathsf{K}, \cdot)$ *for every query* (m_i, ℓ_i) *samples a fresh* $\mathsf{L}_i \leftarrow \mathsf{KGen}$
  and returns $(\sigma_i, \mathsf{P}_i) \leftarrow (\mathsf{Mac}(\mathsf{L}_i, m_i, \ell_i, \sigma_i), \mathsf{Point}(\mathsf{K}, \mathsf{L}_i, \ell_i))$
 Return 1 *iff*
  $\mathsf{Vf}(\mathsf{K}, \mathsf{P}^*, m^*, \ell^*, \sigma^*) = 1$
  and $\mathcal{A}$ has never queried $\mathcal{O}_{\mathsf{MAC}}(\mathsf{K}, \cdot)$ *about* (m^*, ℓ^*).

When a MAC is used in a stand-alone fashion the security guarantee of unforgeability usually suffices. However, when applied as a building block in protocols like TLS or SSL the MAC is computed for a key which is subsequently used to derive further keys or to encrypt data. Besides the regular unforgeability requirement it is thus also necessary to ensure that any delayed-key MAC is "as secure as applying the original MAC". That is, the delayed-key MAC should leak at most the information about the key K as the deployment of the original MAC does.

We therefore introduce the notion of *leakage-invariance*, basically saying that MACs may leak information about the key, but this information does not depend on the specific key value. In our setting this means that the leakage of the ephemeral keys and of the long-term key for each MAC computation are identical (yet, since we augment the tag by the pointer we still need to ensure that this extra information does not violate security). More formally, we compare the success probability of an adversary $\mathcal{A}$ predicting some information $f(\mathsf{K})$ about key K after learning several tuples $(\mathsf{P}_i, m_i, \ell_i, \sigma_i)$ with the success probability of an adversary $\mathcal{B}$ given only access to the plain underlying authentication algorithm $\mathsf{Mac}(\mathsf{K}, \cdot, \cdot)$. For a leakage-invariant delayed-key MAC these probabilities should be close.

Definition 6. *A delayed-key* $\mathsf{DKMAC} = (\mathsf{KGen}, (\mathsf{Mac}, \mathsf{Point}), \mathsf{Vf})$ *(with labels) is called* leakage-invariant *if for any probabilistic polynomial-time algorithm* $\mathcal{A}$ *there exists a probabilistic polynomial-time algorithm* $\mathcal{B}$ *such that for any (probabilistic) function* f *the difference*

$$\text{Prob}\left[\boldsymbol{Exp}^{leak\text{-}inv}_{\mathcal{A},\mathsf{DKMAC}}(n) = 1\right] - \text{Prob}\left[\boldsymbol{Exp}^{leak\text{-}inv}_{\mathcal{B},\mathsf{DKMAC}}(n) = 1\right]$$

is negligible, where:

Experiment $\boldsymbol{Exp}^{leak\text{-}inv}_{\mathcal{A},\mathsf{DKMAC}}(n)$	***Experiment*** $\boldsymbol{Exp}^{leak\text{-}inv}_{\mathcal{B},\mathsf{DKMAC}}(n)$
$\mathsf{K} \leftarrow \mathsf{KGen}(1^n)$	$\mathsf{K} \leftarrow \mathsf{KGen}(1^n)$
$a \leftarrow \mathcal{A}^{\mathcal{O}_{\mathsf{MAC}}(\mathsf{K},\cdot),\mathsf{Vf}(\mathsf{K},\cdots)}(1^n)$	$a \leftarrow \mathcal{B}^{\mathsf{Mac}(\mathsf{K},\cdot,\cdot),\mathsf{Vf}(\mathsf{K},\cdot,\cdot)}(1^n)$
where $\mathcal{O}_{\mathsf{MAC}}(\mathsf{K}, m_i)$ *samples a key*	
$\mathsf{L}_i \leftarrow \mathsf{KGen}(1^n)$ *and returns* (σ_i, P_i)	
$\leftarrow (\mathsf{Mac}(\mathsf{L}_i, m_i, \ell_i), \mathsf{Point}(\mathsf{K}, \mathsf{L}_i, \ell_i, \sigma_i))$	
output 1 if and only if	*output 1 if and only if*
$a = f(\mathsf{K})$	$a = f(\mathsf{K})$

If the function f *is from a set* $\mathcal{F}$ *of functions and* $\mathcal{A}$ *makes at most* q_{Mac} *queries to oracle* $\mathcal{O}_{\mathsf{MAC}}$ *and at most* q_{Vf} *queries to oracle* Vf*, then we say that the MAC is* $(q_{\mathsf{Mac}}, q_{\mathsf{Vf}}, \mathcal{F})$-leakage-invariant. *The scheme is called* leakage-invariant for distinct labels *if* $\mathcal{A}$ *only submits queries with distinct labels to oracle* $\mathcal{O}_{\mathsf{MAC}}(\mathsf{K}, \cdot, \cdot)$. *It is called* leakage-invariant for random labels *if the labels are chosen at random by oracle* $\mathcal{O}_{\mathsf{MAC}}$ *(instead of being picked by the adversary).*

We can even strengthen our definition by bounding the adversary $\mathcal{B}$ to the number of $\mathcal{A}$'s queries, i.e., if $\mathcal{A}$ can derive some information $f(\mathsf{K})$ in $q = (q_{\mathsf{Mac}}, q_{\mathsf{Vf}})$ queries, then $\mathcal{B}$ should be able to deduce $f(\mathsf{K})$ in at most q queries as well. We call such schemes *strongly leakage-invariant.* We do not impose such a restriction per se, since there can be leakage-invariant solutions where $\mathcal{B}$ can safely make more queries (e.g., if MACs are pseudorandom, except that they always leak the first three bits of the key).

Above we do not put any restriction on the function f, i.e., it could even be not efficiently computable. For our more efficient solution we weaken the notion above and demand that the adversary computes the identity function $f(\mathsf{K}) = \mathsf{K}$, i.e., predicts the entire key. Formally, we then let $\mathcal{F} = \{\text{ID}\}$. If, as done in most key exchange protocols, the key is subsequently piped through a hash function modeled as a random oracle, then the adversary needs to query the random oracle about the entire key (and thus needs to predict it). Else the adversary is completely oblivious about the random hash value and the derived key. In other words, in this scenario considering the identity function suffices.

We remark that we refrain from using Canetti's universal composition (UC) model [6] although we are interested in how the key is subsequently used. The second experiment with adversary $\mathcal{B}$ of our notion of leakage-invariance already resembles the notion of an ideal functionality and the ideal-world scenario, and the actual attack on the concrete scheme mimics the real-world setting. However, the UC model introduces additional complications like session IDs and seems to provide more than what is often needed in the applications we have in mind (i.e., one typically asks for more than that the adversary cannot recover the entire key, even though this may be sufficient).

We finally note that the "TLS solution" to first compute $H(m)$ and then $\mathsf{Mac}(\mathsf{K}, H(m), \ell)$ is clearly strongly leakage-invariant if H is collision-resistant (essentially because the ephemeral key L is empty, $\sigma_i = H(m_i)$ is publicly known and the pointer P is the MAC for σ_i). In addition, it is also unforgeable, providing a secure solution under the stronger assumption.

Leakage-Invariance vs. Unforgeability. In general, the notions of unforgeability and leakage-invariance are somewhat incomparable, as we show by separating examples in the full version of the paper. However, in the case that the leakage invariance is limited to the function $f = \mathrm{ID}$ which is the prediction of the entire key, an adversary against leakage-invariance trivially gives an adversary against the unforgeability, as well.

4 One-Sided Delayed-Key MACs: The Unbounded Case

In this section we present our first construction of a delayed-key MAC, that uses a pseudorandom MAC as building block. We show that this approach is unforgeable and leakage-invariant if the underlying MAC is a pseudorandom function. This is independent of any bound on the number of MAC or verification queries and of any assumption about the function f. We present our second construction for the unbounded case in the full version of the paper.

4.1 Pseudorandom Permutation

The idea of our construction $\mathsf{DKMAC}_{\mathrm{PRP}}$ is to authenticate a message m for a random key L and to derive the pointer $\mathsf{P} = \mathsf{Point}(\mathsf{K}, \mathsf{L})$ by applying the inverse of a four-round Feistel permutation $\pi^{-1}(\mathsf{K}, \cdot)$ on the ephemeral key L. For the Feistel permutation we use $\mathsf{Mac}(\mathsf{K}, \langle i \rangle_2 || \cdot)$ as round function, where $\langle i \rangle_2$ denotes the fixed-length binary representation of the round number $i = 0, 1, 2, 3$ with two bits. To verify a given tuple $(\mathsf{K}, \mathsf{P}, \sigma, m)$ one first recovers L by evaluating the permutation on P and then verifies if (L, σ, m) validates as true. The pseudorandomness of the MAC ensures that the pointer leaks no information about the secret key, nor the ephemeral key.

The construction $\mathsf{DKMAC}_{\mathrm{PRP}}$ is optimal in terms of output length (assuming that keys are uniform bit strings and that at least $|\mathsf{L}|$ additional bits must be communicated for L). Yet, it slightly increases the computational costs, as the

Mac algorithm is now also invoked four times to derive the pointer information (but only on short strings). The construction also shows that neither randomized encryption nor labels are necessary.

For (keyed) pseudorandom round functions f_1, f_2, f_3, f_4 and input $x_0||y_0$ (of equal length parts x_0, y_0), let $x_{i+1}||y_{i+1} = y_i||(x_i \oplus f_i(y_i))$ for $i = 0, 1, 2, 3$. This defines a permutation π (with the round functions and keys given implicitly) mapping input $x_0||y_0$ to output $x_4||y_4$. For our solution here we assume for simplicity that keys L are of even length, such that they can be written as $\mathsf{L} = x_0||y_0$. Instead of using independent round functions we use quasi-independent round functions $f_i = \mathsf{Mac}(\mathsf{K}, \langle i \rangle_2 ||\cdot)$ by prepending the round number i in binary (represented with the fixed length of two bits).

Construction 1. *Let* $\mathsf{MAC} = (\mathsf{KGen}, \mathsf{Mac}, \mathsf{Vf})$ *be a (deterministic) message authentication code. Define* $\mathsf{DKMAC}_{PRP} = (\mathsf{KGen}_{PRP}, (\mathsf{Mac}, \mathsf{Point})_{PRP}, \mathsf{Vf}_{PRP})$ *as follows:*

Key Generation $\mathsf{KGen}_{\mathbf{PRP}}$**.** *The key generation algorithm gets a security parameter* 1^n *and outputs a key* $\mathsf{K} \leftarrow \mathsf{KGen}(1^n)$*.*

Authentication $(\mathsf{Mac}, \mathsf{Point})_{\mathbf{PRP}}$**.** *The authentication procedure takes as input a secret key* K*, a message* m *and first samples a fresh ephemeral key* $\mathsf{L} \leftarrow \mathsf{KGen}(1^n)$ *by running the key generation of the underlying MAC scheme. For key* L *and input message* m *it computes the tag* $\sigma \leftarrow \mathsf{Mac}(\mathsf{L}, m)$ *and the pointer* $\mathsf{P} \leftarrow \mathsf{Point}(\mathsf{K}, \mathsf{L})$*, where* Point *computes* $\mathsf{P} \leftarrow \pi^{-1}(\mathsf{K}, \mathsf{L})$ *for a four-round Feistel permutation* π *that uses* $\mathsf{Mac}(\mathsf{K}, \langle i \rangle_2 ||\cdot)$ *as the round functions for* $i = 0, 1, 2, 3$ *and* L *as input. The output of* $(\mathsf{Mac}, \mathsf{Point})_{PRP}$ *is the pair* (σ, P)*.*

Verification $\mathsf{Vf}_{\mathbf{PRP}}$**.** *Upon input a secret key* K*, a pointer* P*, a message* m *and a tag* σ*, it first derives the ephemeral key* $\mathsf{L} = \mathsf{Point}^{-1}(\mathsf{K}, \mathsf{P}) = \pi(\mathsf{K}, \mathsf{P})$ *and outputs* $\mathsf{Vf}(\mathsf{L}, m, \sigma)$*.*

Correctness of this MAC follows easily form the correctness of the underlying MAC.

Lemma 1. *If* $\mathsf{MAC} = (\mathsf{KGen}, \mathsf{Mac}, \mathsf{Vf})$ *is a pseudorandom message authentication code then the delayed-key message authentication scheme* $\mathsf{DKMAC}_{PRP} = (\mathsf{KGen}_{PRP}, (\mathsf{Mac}, \mathsf{Point})_{PRP}, \mathsf{Vf}_{PRP})$ *in Construction 1 is unforgeable against chosen message attacks.*

As for concrete security, the advantage of any adversary $\mathcal{A}_{\mathsf{DKMAC}}$ making q_{MAC} queries of bit length at most l is bounded by q_{MAC} times the advantage of an adversary $\mathcal{A}_{\mathsf{MAC}}$ against the pseudorandomness of MAC that makes $4q_{\mathsf{MAC}}$ queries of length at most $\max(n+2, l)$. Again, the running times of both algorithms are comparable.

Proof. Assume towards contradiction that an adversary $\mathcal{A}$ making q queries $m_1, \ldots, m_q$ to the $\mathcal{O}_{\mathsf{MAC}}(\mathsf{K}, \cdot)$ oracle outputs with non negligible probability a tuple $(\mathsf{P}^*, m^*, \sigma^*)$, s.t. $\mathsf{Vf}^*(\mathsf{K}, \mathsf{P}^*, m^*, \sigma^*)$ but m^* was never submitted to the oracle. Then we can distinguish between two cases:

– $\mathsf{P}^* \neq \mathsf{P}_1, \ldots \mathsf{P}_q$, i.e., the adversary has created a valid forgery for a fresh pointer and thus for a fresh ephemeral key $\mathsf{L}^* \neq \mathsf{L}_1, \ldots \mathsf{L}_q$, since the pointer algorithm is a permutation. Denote the event by E_1.
– $\mathsf{P}^* = \mathsf{P}_i$ for some $i \in \{1, \ldots, q\}$, i.e., the pointer P^* has already appeared in one of the oracle replies. Thus, the adversary $\mathcal{A}$ has successfully forged a MAC for a key L^* after seeing at least one tag $\sigma_i \leftarrow \mathsf{Mac}(\mathsf{L}^*, m_i)$. We denote this event by E_2.

As one of the two cases has to occur if $\mathcal{A}$ is successful —which we denote as the event WIN— we have that $\text{Prob}[\text{WIN}] \leq \text{Prob}[E_1] + \text{Prob}[E_2]$ (note that events E_1, E_2 both require a success). We show in the full paper that in both cases we can construct an adversary that breaks the underlying MAC scheme. □

Lemma 2. *The delayed-key MAC scheme* DKMAC_{PRP} *in Construction 1 is leakage-invariant.*

Proof. To prove leakage-invariance we have to show that for every adversary $\mathcal{A}$ with oracle access to $\mathcal{O}_{\mathsf{MAC}}(\mathsf{K}, \cdot)$ and $\mathsf{Vf}(\mathsf{K}, \cdots)$ that predicts with noticeable probability some information $f(\mathsf{K})$ about the key K, we can derive an adversary $\mathcal{B}$ that only has access to $\mathsf{Mac}(\mathsf{K}, \cdot)$ and $\mathsf{Vf}(\mathsf{K}, \cdot)$ but predicts $f(\mathsf{K})$ with the same advantage as $\mathcal{A}$.

Assume that $\mathcal{A}$ is able to derive some non-trivial information about K after sending q queries to its $\mathcal{O}_{\mathsf{MAC}}$ and Vf oracles, which implements the authentication process of our delayed-key MAC. Then we can construct an adversary $\mathcal{B}$ that successfully determines $f(\mathsf{K})$ when sending $4q$ queries to its $\mathsf{Mac}(\mathsf{K}, \cdot)$ and $\mathsf{Vf}(\mathsf{K}, \cdots)$ oracles. To this end, $\mathcal{B}$ mimics the $\mathcal{O}_{\mathsf{MAC}}$ oracle by computing the tag $\sigma_i \leftarrow \mathsf{Mac}(\mathsf{L}_i, m_i)$ for any query m_i and some self-chosen key L_i and calculating P_i with the help of its own oracle (and analogously for verification requests). Thus, for each of $\mathcal{A}$'s queries, $\mathcal{B}$ has to invoke $\mathsf{Mac}(\mathsf{K}, \cdot)$ four times to simulate $\mathcal{O}_{\mathsf{MAC}}$ or Vf. If $\mathcal{A}$ outputs some information a, $\mathcal{B}$ forwards it as its own output. Since the simulation is perfect from $\mathcal{A}$'s point of view the success probabilities of $\mathcal{B}$ and $\mathcal{A}$ are identical. □

The construction $\mathsf{DKMAC}_{\mathsf{PRP}}$ is already optimal concerning the communication overhead (assuming, that at least $|\mathsf{L}|$ additional bits have to be communicated) but increases the computational costs by four additional evaluations of the underlying MAC. Our second construction of an unbounded delayed-key MAC, which we discuss in detail in the full version, requires less Mac computations (two instead of four) but comes with larger output lengths.

5 One-Sided Delayed-Key MACs: The Bounded Case

In this section we show that, by reducing the security requirements for unforgeability and leakage-invariance, we can construct key-delayed MACs that require less Mac invocations than our previous constructions or are even optimal in both, computational costs and output length. In other words, we can trade in security for efficiency. First, we bound the adversaries against unforgeability and leakage-invariance to make at most $O(\log(n))$ many verification queries, which allows to obtain a construction that requires only two MAC computations and is almost optimal in terms of output length. We present the construction in the full version of the paper, where we also show that the scheme is even strongly leakage-invariant (meaning that $\mathcal{B}$ does not make more queries than $\mathcal{A}$), as long as we only demand that $\mathcal{A}$ is unable to predict the entire key.

By further restricting the adversary against the leakage-invariance to make only a single authentication query, we obtain our most efficient solution that requires no additional Mac computations and has optimal output length . Note that the underlying MAC is then assumed to be secure against related-key attacks.

As already mentioned in the introduction, limiting the number of verification queries corresponds to the common approach that in key-exchange protocols, both server and client verify only a single MAC each. Leakage-invariance for only $\mathcal{F} = \{\text{ID}\}$ is sufficient, if the key gets afterwards hashed by a hash function that behaves like a random oracle.

5.1 XOR-Construction

In our most simple and efficient construction, we use the shared key K to directly mask the ephemeral key. That is, by computing the one-time-pad encryption of L under K, i.e., $\mathsf{P} = \mathsf{K} \oplus \mathsf{L}$. Thus, for any authentication query, $\mathsf{DKMAC}_{\oplus}$ makes only a single Mac computation.

Definition 7. *Let* $\mathsf{MAC} = (\mathsf{KGen}, \mathsf{Mac}, \mathsf{Vf})$ *be a message authentication code. Define the delayed-key* $\mathsf{DKMAC}_{\oplus} = (\mathsf{KGen}_{\oplus}, (\mathsf{Mac}, \mathsf{Point})_{\oplus}, \mathsf{Vf}_{\oplus})$ *as follows*

Key Generation $\mathsf{KGen}_{\oplus}$. *The key generation algorithm gets a security parameter* 1^n *and outputs a key* $\mathsf{K} \leftarrow \mathsf{KGen}(1^n)$.

Authentication $(\mathsf{Mac}, \mathsf{Point})_{\oplus}$. *The authentication procedure takes as input a shared secret key* K, *a message* m *and outputs* $\sigma \leftarrow \mathsf{Mac}(\mathsf{L}, m)$ *and pointer* $\mathsf{P} = \mathsf{K} \oplus \mathsf{L}$ *for a randomly chosen* $\mathsf{L} \leftarrow \mathsf{KGen}(1^n)$.

Verification $\mathsf{Vf}_{\oplus}$. *Upon input a secret key* K, *a pointer* P, *a message* m *and a tag* σ *it outputs* $\mathsf{Vf}(\mathsf{P} \oplus \mathsf{K}, m, \sigma)$.

Correctness of $\mathsf{DKMAC}_{\oplus}$ *follows from the correctness of the underlying MAC.*

In order to prove the unforgeability of our $\mathsf{DKMAC}_{\oplus}$ construction, we require a stronger assumption on the underlying MAC, namely that it is a related-key secure pseudorandom function. The first formal security model for related key attacks was introduced by Bellare and Kohno in [3]. Inter alia, they have shown that PRFs that are provably secure against those attacks can be achieved when the set of relations is restricted to some non-trivial class of key transformation functions, denoted by Φ. The notion for Φ-related-key security then extends the notion of standard PRF's and grants the adversary access to a related-key oracle that is either $\mathsf{Mac}_{\text{RK}(\cdot,k)}(\cdot)$ or $f_{\text{RK}(\cdot,k)}(\cdot)$. In both cases a key k is chosen at random and in the random world, also a function f gets chosen randomly. Each query of the adversary then consists of a key transformation function $\phi : \mathcal{K} \rightarrow \mathcal{K}$

and an input value m. The query is answered by $\mathsf{Mac}(\phi(k), m)$ and $f(\phi(k), m)$ respectively.

Definition 8. *Let Φ be a set of key transformation functions, and $\mathcal{D}$ an adversary with access to related-key oracles that is allowed to send queries $(\phi, m) \leftarrow \Phi \times \mathcal{M}$. A pseudorandom* Mac *is called* secure against related-key attacks *if for any efficient algorithm $\mathcal{D}$ the advantage*

$$\left|\text{Prob}\left[\mathcal{D}^{\mathsf{Mac}_{RK(\cdot,k)}(\cdot)}(1^n) = 1\right] - \text{Prob}\left[\mathcal{D}^{f_{RK(\cdot,k)}(\cdot)}(1^n) = 1\right]\right|$$

is negligible, where the probability in the first case is over $\mathcal{D}$'s coin tosses and the choice of $k \leftarrow \mathsf{KGen}(1^n)$, and in the second case over $\mathcal{D}$'s coin tosses, the choice of the random function $f : \mathcal{K}_n \times \mathcal{M}_n \rightarrow \mathcal{R}_n$ and random $k \leftarrow \mathcal{K}_n$.

Note that related-key secure pseudorandom MACs are unforgeable with respect to related-key attacks, too.

For our construction we need related-key security only for one class of transformations, that is the function that adds a given value $\Delta \in \{0,1\}^n$ to the hidden key K. Sticking to the notation of [3] we denote this function by $\text{XOR}_\Delta : \mathcal{K} \rightarrow \mathcal{K}$ and the resulting class of functions by $\Phi_n^{\oplus} = \{\text{XOR}_\Delta : \Delta \in \{0,1\}^n\}$. Constructions for $\Phi_n^{\oplus}$-related-key secure pseudorandom functions were proposed in [15].

Lemma 3. *If* $\mathsf{MAC} = (\mathsf{KGen}, \mathsf{Mac}, \mathsf{Vf})$ *is a pseudorandom message authentication code secure against related-key attacks for the relation $\Phi_n^{\oplus}$, then the delayed-key MAC scheme* $\mathsf{DKMAC}_\oplus = (\mathsf{KGen}_\oplus, (\mathsf{Mac}, \mathsf{Point})_\oplus, \mathsf{Vf}_\oplus)$ *in Construction 7 is unforgeable against chosen message attacks, if the adversary makes at most $O(\log(n))$ verification queries.*

A closer look at the concrete security reveals that the advantage of any adversary $\mathcal{A}_{\mathsf{DKMAC}}$ making $q_{\mathsf{MAC}}, q_{\mathsf{Vf}}$ queries each of length at most l, is bounded by $2^{q_{\mathsf{Vf}}}$ times the advantage of an adversary $\mathcal{A}_{\mathsf{MAC}}$ against the related-key pseudorandomness of MAC that makes q_{MAC} queries of length at most l.

Proof. Assume towards contradiction that an adversary $\mathcal{A}$ after learning several tags $(\sigma_1, \mathsf{P}_1), \ldots, (\sigma, \mathsf{P}_q)$ from its oracle $\mathcal{O}_{\mathsf{MAC}}(\mathsf{K}, \cdot)$ is able to compute a forgery $(\mathsf{P}^*, m^*, \sigma^*)$ with $m^* \neq m_1 \ldots m_q$. Then we can construct an adversary $\mathcal{A}_{\mathsf{MAC}}$ breaking the related-key unforgeability of the underlying MAC.

Our adversary $\mathcal{A}_{\mathsf{MAC}}$ has black-box access to a related-key oracle $\mathsf{Mac}_{\mathsf{RK}(\cdot,\mathsf{L})}(\cdot)$ and uses $\mathcal{A}$ to produce a forgery $(\Delta^*, m^*, \sigma^*)$ for some key $\mathsf{L} \oplus \Delta^*$. For the sake of readability it is assumed, that the real key transformation XOR is already included in the oracle and the adversary has only to provide some value $\Delta \in \{0,1\}^n$.

When $\mathcal{A}$ sends the first authentication query m_1, $\mathcal{A}_{\mathsf{MAC}}$ invokes its own oracle on $(0^n, m_1)$ receiving $\sigma_1 = \mathsf{Mac}(\mathsf{L}, m_1)$ which he passes together with a randomly chosen P back to $\mathcal{A}$. The value P can also be seen as $\mathsf{L} \oplus \mathsf{K}$ for some unknown K. Due to the pseudorandomness of Mac, the tag σ_1 does not leak any information about the applied key L. Thus, from $\mathcal{A}$'s point of view the value P is indistinguishable from a real one-time-pad encryption of some secret key K. For any further authentication query m_i of $\mathcal{A}$, our adversary chooses a random Δ_i and sends (Δ_i, m_i) to its own oracle. The adversary $\mathcal{A}_{\mathsf{MAC}}$ then responds with the answer σ_i and a pointer $\mathsf{P}_i = \mathsf{P} \oplus \Delta_i$.

When $\mathcal{A}$ wants to query its verification oracle, our adversary $\mathcal{A}_{\mathsf{MAC}}$ has to guess the answer bit, otherwise it might send the message of the potential forgery to his tagging oracle, thereby nullifying the message for its own output. Thus, whenever $\mathcal{A}$ makes a verification query, $\mathcal{A}_{\mathsf{MAC}}$ halts $\mathcal{A}$ and then runs two instantiations for the answer bit $b = 0$, resp. $b = 1$. Hence, for efficiency reasons we allow $\mathcal{A}$ to make at most $O(\log(n))$ queries to the verification oracle.

If, at the end, each of the at most n instantiations of $\mathcal{A}$ holds with a forgery $(\mathsf{P}_j^*, m_j^*, \sigma_j^*)$, our adversary $\mathcal{A}_{\mathsf{MAC}}$ guesses an index $j \in \{1, \ldots, n\}$. It then computes $\Delta^* = \mathsf{P}_j^* \oplus \mathsf{P}$ and outputs $(\Delta^*, m_j^*, \sigma_j^*)$ as its own forgery. Overall, $\mathcal{A}_{\mathsf{MAC}}$ succeeds with probability $1/\mathrm{poly}(n)$ times the success probability of $\mathcal{A}$, which contradicts the assumption that MAC is related-key unforgeable. □

Lemma 4. *The delayed-key MAC scheme* $\mathsf{DKMAC}_\oplus$ *in Construction 7 is* $(1, O(\log(n)), \{ID\})$*-leakage invariant.*

Proof. If there exists an adversary $\mathcal{A}$ that outputs with non-negligible probability the complete secret key K after it received a tag $(\sigma, \mathsf{P}) \leftarrow \langle \mathsf{Mac}(\mathsf{L}, m), \mathsf{K} \oplus \mathsf{L} \rangle$ for some random L and chosen m, we can derive an adversary $\mathcal{B}$ that is able to extract K only from $\sigma \leftarrow \mathsf{Mac}(\mathsf{K}, m)$ for some chosen m as well.

The idea is that by determining K, also the key L can be obtained unambiguously. Thus, when we construct the adversary $\mathcal{B}$ that uses $\mathcal{A}$, its target key K actually plays the role of L in the game of $\mathcal{A}$. Thus, when $\mathcal{B}$ receives the authentication query m from $\mathcal{A}$ it triggers its oracle $\mathsf{Mac}(\mathsf{K}, \cdot)$ on m and passes the answer σ together with a randomly chosen pointer P back to $\mathcal{A}$. The pointer value then corresponds to the one-time-pad encryption of K with some random, secret key L.

For any verification query $(\mathsf{P}_i, m_i, \sigma_i)$ of $\mathcal{A}$, the adversary $\mathcal{B}$ first checks whether $\mathsf{P}_i = \mathsf{P}$. If so, it forwards the query to its $\mathsf{Vf}(\mathsf{K}, \cdot)$ oracle, otherwise it has to "guess" the answer bit. To this end, $\mathcal{B}$ runs two instantiations of $\mathcal{A}$, for each $b = 0, 1$. Since we allow $\mathcal{A}$ to make only at most $O(\log(n))$ verification queries, $\mathcal{B}$ starts at most n instantiations.

Finally, each instantiation of $\mathcal{A}$ stops, outputting its guess a_j that corresponds to some L_j in $\mathcal{B}$'s game. To determine the right key, adversary $\mathcal{B}$ computes for each $j = 1, 2, \ldots, n$ the potential counterpart $\mathsf{K}_j = \mathsf{P} \oplus \mathsf{L}_j$ and outputs K_j where $\sigma = \mathsf{Mac}(\mathsf{K}_j, m)$.

Due to the limitation of a single authentication query, our adversary $\mathcal{B}$ is able to simulate the oracle $\mathcal{O}_{\mathsf{MAC}}$ of $\mathcal{A}$ perfectly, such that $\mathcal{B}$ succeeds with the same probability as $\mathcal{A}$. □

6 Two-Sided Delayed-Key MACs: A Feasibility Result

In this section we discuss that two-sided delayed-key MACs are realizable without relying on collision-resistance. The idea —explained in the setting of key exchange— is to use a signature scheme to authenticate each transmitted message immediately (such that both parties basically only have to store keys for the MAC), and to finally MAC the public key of the signature scheme.

Note that the existence of one-way functions is shown to be necessary and sufficient for the existence of secure signature schemes in [18]. As we, in addition, only require unforgeability from the underlying MAC, the security of our construction formally relies only on one-way functions. Yet, applying a signature scheme for each message is very expensive, of course. Hence, this construction should be seen as a feasibility result only. We leave it as an interesting open problem to find an efficient construction for this scenario.

Note that in order to turn the idea above into a formal solution we need to change the notion of unforgeability and leakage-invariant slightly. Namely, we assume that the adversary $\mathcal{A}$ in both cases now can pass another parameter `keep` or `pointer` (besides m_i, ℓ_i) to oracle $\mathcal{O}_{\mathsf{MAC}}$. For parameter `keep` the oracle returns tags σ_i for the previously selected ephemeral key L and only if queried for `pointer` it returns the pointer P and generates a new ephemeral key. An adversary $\mathcal{A}$ against the unforgeability is then deemed successful if it outputs a tuple $(\mathsf{P}^*, \bar{m}^*, \bar{\ell}^*, \sigma^*)$ with $\mathsf{Vf}(\mathsf{K}, \mathsf{P}^*, \bar{m}^*, \bar{\ell}^*, \sigma^*) = 1$ and $\mathcal{A}$ has never issued $(\bar{m}^*, \bar{\ell}^*) = ((m_1{}^*, \ell_1{}^*), \ldots, (m_n{}^*, \ell_n{}^*))$ between two `pointer` queries to $\mathcal{O}_{\mathsf{MAC}}(\mathsf{K}, \cdot)$.

The $\mathsf{DKMAC_{two}}$ *Construction.* Recall the notion of signature schemes: a signature scheme consists of three efficient algorithms $(\mathsf{SKGen}, \mathsf{SSign}, \mathsf{SVf})$ where SKGen on input 1^n returns a key pair (sk, pk); algorithm SSign on input sk and a

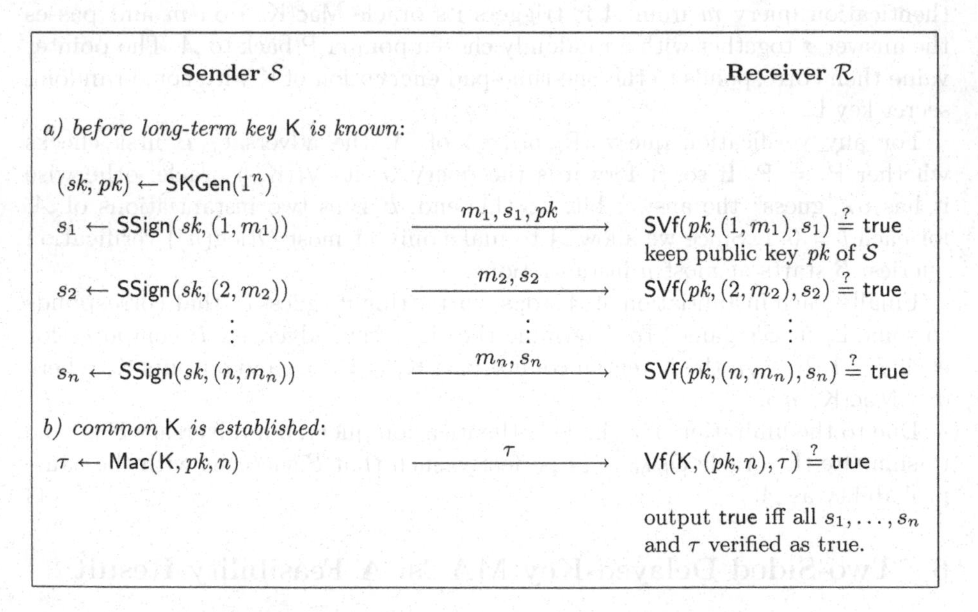

Fig. 1. $\mathsf{DKMAC_{two}}$: Two-sided Delayed-Key MAC

message $m \in \{0,1\}^*$ returns a signature s; and algorithm SVf for input pk, m, s returns a decision bit. We assume completeness in the sense that any signature generated via SSign is also accepted by SVf. *Unforgeability* of signature schemes is defined analogously to unforgeability of MACs, but now the adversary gets as input the public key pk instead of the security parameter 1^n and has access to a signing oracle $\mathsf{SSign}(sk, \cdot)$.

Our construction $\mathsf{DKMAC_{two}}$ (incorporated into a key exchange protocol) is given in Figure 1. Note that the sender only needs to store the key pair (sk, pk) and the receiver merely stores pk and a bit indicating any error in the verifications so far. Formally, we can let $\mathsf{Mac}(\mathsf{L}, m, \ell)$ be the algorithm which for $\mathsf{L} = (sk, pk) \leftarrow \mathsf{SKGen}(1^n)$ outputs $\sigma = (pk, \mathsf{SSign}(sk, m, \ell))$. The point algorithm $\mathsf{Point}(\mathsf{K}, \mathsf{L}, \ell)$ returns a MAC value P of pk under key K for an unforgeable MAC. Then an adversary against the key exchange protocol can be easily cast in our extended unforgeability and leakage-invariance model. This adversary calls $\mathcal{O}_{\mathsf{MAC}}$ several times with (i, m_i, ℓ_i) for parameter `keep` and subsequently eventually calls the oracle about parameter `pointer` to retrieve the MAC of the public key under K.

Unforgeability and Leakage-Invariance of $\mathsf{DKMAC_{two}}$. The $\mathsf{DKMAC_{two}}$ construction is unforgeable if the underlying signatures scheme is unforgeable against chosen-message attacks and the underlying MAC is unforgeable as well. The unforgeability of the MAC and the fact that collisions among independently generated keys are unlikely implies that the adversary can only use a previously chosen public key by $\mathcal{O}_{\mathsf{MAC}}$ (or else forges a MAC under K for a new key pk^*). But then the adversary must forge a signature for a tuple (i^*, m^*, ℓ^*) which has not been signed before under this public key. By the unforgeability of the signature scheme this cannot happen with more than negligible probability.

Obviously, the scheme $\mathsf{DKMAC_{two}}$ is strongly leakage-invariant, as it uses the secret long-term key K only for a single computation of the underlying MAC.

Online Verification with Immediate Abort. In the context of online verification it might be desirable that the verifier can abort the authentication process as soon as he receives the first invalid tag. To this end, we augment the usual verification algorithm Vf of DKMAC's such that it allows online processing: $\mathsf{Vf}'(\mathsf{K}, \mathsf{P}, m, \ell, \sigma, \mathsf{st})$ now also expects some state information st which can either be `keep` or `pointer`. On input `keep` the algorithm Vf' returns $\mathsf{Vf}(m, \ell, \sigma)$ and for `pointer` it outputs $\mathsf{Vf}(\mathsf{K}, \mathsf{P}, m, \ell, \sigma)$. Thus, as long as the long-term key K is unknown, the verifier runs $\mathsf{Vf}'(\perp, \perp, m_i, \ell_i, \sigma_i, \mathtt{keep})$ and aborts when it receives 0, indicating an invalid tag. Obviously, our construction $\mathsf{DKMAC_{two}}$ allows for online verification with immediate abort as the verifier can check, while being in `keep`-mode, if $\mathsf{SVf}(pk, (i, m_i), s_i) = \mathtt{true}$ and abort the authentication as soon as the first verification fails.

Acknowledgments

We thank Yevgeniy Dodis, Stefan Lucks and the anonymous reviewers for valuable comments. Both authors are supported by the Emmy Noether Program Fi 940/2-1 of the German Research Foundation (DFG).

References

1. Bellare, M.: New Proofs for NMAC and HMAC: Security without Collision- Resistance. In: Dwork, C. (ed.) CRYPTO 2006. LNCS, vol. 4117, pp. 602–619. Springer, Heidelberg (2006)
2. Bellare, M., Goldreich, O.: A. Mityagin The Power of Verification Queries in Message Authentication and Authenticated Encryption. Number 2004/309 in Cryptology eprint archive (2004), eprint.iacr.org
3. Bellare, M., Kohno, T.: A Theoretical Treatment of Related-Key Attacks: RKA-PRPs, RKA-PRFs, and Applications. In: Biham, E. (ed.) EUROCRYPT 2003. LNCS, vol. 2656, pp. 491–506. Springer, Heidelberg (2003)
4. Bellare, M., Pointcheval, D., Rogaway, P.: Authenticated Key Exchange Secure against Dictionary Attacks. In: Preneel, B. (ed.) EUROCRYPT 2000. LNCS, vol. 1807, pp. 139–155. Springer, Heidelberg (2000)
5. Advanced Security Mechanism for Machine Readable Travel Documents Extended Access Control (EAC). Technical Report (BSI-TR-03110) Version 2.0 Release Candidate, Bundesamt fuer Sicherheit in der Informationstechnik, BSI (2008)
6. Canetti, R.: Universally Composable Security: A new Paradigm for Cryptographic Protocols. In: Proceedings of the Annual Symposium on Foundations of Computer Science (FOCS) 2001. IEEE Computer Society Press, Los Alamitos (2001), for an updated version see: eprint.iacr.org
7. Fischlin, M.: Security of NMAC and HMAC Based on Non-malleability. In: Malkin, T.G. (ed.) CT-RSA 2008. LNCS, vol. 4964, pp. 138–154. Springer, Heidelberg (2008)
8. Gennaro, R.: Faster and Shorter Password-Authenticated Key Exchange. In: Canetti, R. (ed.) TCC 2008. LNCS, vol. 4948, pp. 589–606. Springer, Heidelberg (2008)
9. Garay, J.A., Kolesnikov, V., McLellan, R.: MAC Precomputation with Applications to Secure Memory. In: Samarati, P., et al. (eds.) ISC 2009. LNCS, vol. 5735, pp. 427–442. Springer, Heidelberg (2009)
10. Goldwasser, S., Micali, S.: Probabilistic Encryption. Journal of Computer and System Science 28(2), 270–299 (1984)
11. Gennaro, R., Rohatgi, P.: How to Sign Digital Streams. In: Kaliski Jr., B.S. (ed.) CRYPTO 1997. LNCS, vol. 1294, pp. 180–197. Springer, Heidelberg (1997)
12. Halevi, S., Krawczyk, H.: Strengthening Digital Signatures Via Randomized Hashing. In: Dwork, C. (ed.) CRYPTO 2006. LNCS, vol. 4117, pp. 41–59. Springer, Heidelberg (2006)
13. Jablon, D.: Strong password-only authenticated key exchange. ACM Computer Communications Review 26(5), 5–26 (1996)
14. Katz, J., Ostrovsky, R., Yung, M.: Efficient Password- Authenticated Key Exchange Using Human-Memorable Passwords. In: Pfitzmann, B. (ed.) EUROCRYPT 2001. LNCS, vol. 2045, p. 475. Springer, Heidelberg (2001)

15. Lucks, S.: Ciphers Secure against Related-Key Attacks. In: Roy, B., Meier, W. (eds.) FSE 2004. LNCS, vol. 3017, pp. 359–370. Springer, Heidelberg (2004)
16. Perrig, A., Canetti, R., Song, D., Tygar, J.D.: The TESLA Broadcast Authentication Protocol. In: CryptoBytes, vol. 5, pp. 2–13. RSA Security (2002)
17. Rescorla, E.: SSL and TLS: designing and building secure systems. Addison-Wesley, Reading (2001)
18. Rompel, J.: One-Way Functions are Necessary and Sufficient for Secure Signatures. In: Proceedings of the Annual Symposium on the Theory of Computing (STOC) 1990, pp. 387–394. ACM Press, New York (1990)

Founding Cryptography on Tamper-Proof Hardware Tokens

Vipul Goyal[1,*], Yuval Ishai[2,**], Amit Sahai[3,***], Ramarathnam Venkatesan[4], and Akshay Wadia[3]

[1] UCLA and MSR India
vipul.goyal@gmail.com
[2] Technion and UCLA
yuvali@cs.technion.ac.il
[3] UCLA
{sahai,awadia}@cs.ucla.edu
[4] Microsoft Research, India and Redmond
venkie@microsoft.com

Abstract. A number of works have investigated using tamper-proof hardware tokens as tools to achieve a variety of cryptographic tasks. In particular, Goldreich and Ostrovsky considered the problem of software protection via oblivious RAM. Goldwasser, Kalai, and Rothblum introduced the concept of *one-time programs*: in a one-time program, an honest sender sends a set of *simple* hardware tokens to a (potentially malicious) receiver. The hardware tokens allow the receiver to execute a secret program specified by the sender's tokens exactly once (or, more generally, up to a fixed t times). A recent line of work initiated by Katz examined the problem of achieving UC-secure computation using hardware tokens.

Motivated by the goal of unifying and strengthening these previous notions, we consider the general question of basing secure computation on hardware tokens. We show that the following tasks, which cannot be realized in the "plain" model, become feasible if the parties are allowed to generate and exchange tamper-proof hardware tokens.

- UNCONDITIONAL AND NON-INTERACTIVE SECURE COMPUTATION. We show that by exchanging simple *stateful* hardware tokens, any functionality can be realized with *unconditional* security against malicious parties. In the case of two-party functionalities $f(x, y)$ which take their inputs from a sender and a receiver and deliver their output to the receiver, our protocol is non-interactive and only requires a unidirectional communication of simple stateful tokens from the

* Research supported in part by a Microsoft Research Graduate Fellowship and the grants of Amit Sahai mentioned below.

** Supported in part by ISF grant 1310/06, BSF grants 2008411, and NSF grants 0830803, 0716835, 0627781.

*** Research supported in part from NSF grants 0916574, 0830803, 0627781, and 0716389, BSF grant 2008411, an equipment grant from Intel, and an Okawa Foundation Research Grant.

D. Micciancio (Ed.): TCC 2010, LNCS 5978, pp. 308–326, 2010.

sender to the receiver. This strengthens previous feasibility results for one-time programs both by providing *unconditional* security and by offering general protection against *malicious senders*. As is typically the case for unconditionally secure protocols, our protocol is in fact *UC-secure*. This improves over previous works on UC-secure computation based on hardware tokens, which provided computational security under cryptographic assumptions.
- INTERACTIVE SECURE COMPUTATION FROM STATELESS TOKENS BASED ON ONE-WAY FUNCTIONS. We show that *stateless* hardware tokens are sufficient to base general secure (in fact, UC-secure) computation on the existence of *one-way functions*.
- OBFUSCATION FROM STATELESS TOKENS. We consider the problem of realizing non-interactive secure computation from stateless tokens for functionalities which allow the receiver to provide an arbitrary number of inputs (these are the only functionalities one can hope to realize non-interactively with *stateless* tokens). By building on recent techniques for resettably secure computation, we obtain a general positive result under standard cryptographic assumptions. This gives the first general feasibility result for program obfuscation using *stateless* tokens, while strengthening the standard notion of obfuscation by providing security against a malicious sender.

1 Introduction

A number of works (e.g. [1,2,3,4,5,6,7,8,9,10,11,12,13]) have investigated using tamper-proof hardware tokens[1] as tools to achieve a variety of cryptographic goals. There has been a surge of research activity on this front of late. In particular, the recent work of Katz [9] examined the problem of achieving UC-secure [14] two party computation using tamper-proof hardware tokens. A number of follow-up papers [10,11,12] have further investigated this problem. In another separate (but related) work, Goldwasser et al. [13] introduced the concept of *one-time programs*: in a one-time program, a (semi-honest) sender sends a set of very simple hardware tokens to a (potentially malicious) receiver. The hardware tokens allow the receiver to execute a program specified by the sender's tokens exactly once (or, more generally, up to a fixed t times). This question is related to the more general goal of software protection using hardware tokens, which was first addressed by Goldreich and Ostrovsky [1] using the framework of oblivious RAM.

The present work is motivated by the observation that several of these previous goals and concepts can be presented in a unified way as instances of one general goal: realizing *secure computation* using tamper-proof hardware tokens. The lines of work mentioned above differ in the types of functionalities being

[1] Informally, a tamper-proof hardware token provides the holder of the token with black-box access to the functionality of the token. We will often omit the words "tamper-proof" when referring to hardware tokens, but all of the hardware tokens referred to in this paper are assumed to be tamper-proof.

considered (e.g., non-reactive vs. reactive), the type of interaction between the parties (interactive vs. non-interactive protocols), the type of hardware tokens (stateful vs. stateless, simple vs. complex), and the precise security model (standalone vs. UC, semi-honest vs. malicious parties). This unified point of view also gives rise to strictly stronger notions than those previously considered, which in turn give rise to new feasibility questions in this area.

The introduction of tamper-proof hardware tokens to the model of secure computation, as formalized in [9], invalidates many of the fundamental impossibility results in cryptography. Taking a step back to look at this general model from a foundational perspective, we find that a number of natural feasibility questions regarding secure computation with hardware tokens remain open. In this work we address several of these questions, focusing on goals that are impossible to realize in the plain model without tamper-proof hardware tokens:

- **Is it possible to achieve unconditional security for secure computation with hardware tokens?** We note that this problem is open even for stand-alone security, let alone UC security, and impossible in the plain model [15]. While in the semi-honest model this question is easy to settle by relying on unconditional protocols based on oblivious transfer (OT) [16,17,18,19], this question is more challenging when both parties as well as the tokens they generate can be malicious. (See Sections 1.2 and 3.1 for relevant discussion.) In the case of *stateless* tokens, which may be much easier to implement, security against unbounded adversaries cannot be generally achieved, since an unbounded adversary can "learn" the entire description of the token. A natural question in this case is **whether stateless tokens can be used to realize (UC) secure computation based on the assumption that one-way functions exist.**

  Previous positive results for secure two-party computation with hardware tokens relied either on specific number theoretic assumptions [9] or the existence of oblivious transfer protocols in the plain model [10,11], or alternatively offered weaker notions of security [20].

  A related question is: **is it possible to obtain unconditionally secure one-time programs for all polynomial-time computable functions?** The previous work of [13] required the existence one-way functions in order to construct one-time programs.
- **Is it possible to realize non-interactive secure two-party computation with simple hardware tokens?** Again, this problem is open[2] even for stand-alone security, and impossible in the plain model. Constructions of oblivious RAM [1] and one-time programs [13] provide partial solutions to

[2] All the previous questions were open even without any restriction on the size of the tokens. In the current and the following questions we restrict the tokens to be *simple* in the sense that the size of each token can only depend on the security parameter. This rules out a trivial solution of building a token which realizes a party in a secure two-party computation protocol.

this problem; however, in these models the sender is semi-honest.[3] Thus, in the context of one-time programs we ask: **is it possible to achieve one-time programs tolerating a malicious sender?** We note that [13] make partial progress towards this question by constructing one-time zero knowledge proofs, where the prover can be malicious. However, in the setting of hardware tokens, the GMW paradigm [21] of using zero knowledge proofs to compile semi-honest protocols into protocols tolerating malicious behavior does not apply, since one would potentially need to prove statements about hardware tokens (as opposed to ordinary NP statements).

- **Which notions of program obfuscation can be realized using simple hardware tokens?** Again, this problem can be captured in an elegant way within the framework of secure two-party computation, except that here we need to consider *reactive* functionalities which may take a single input from the "sender" and a sequence of (possibly adaptively chosen) inputs from the "receiver". Obfuscation can be viewed as a non-interactive secure realization of such functionalities. While this general goal is in some sense realized by the construction of oblivious RAM [1] (which employs stateful tokens), several natural questions remain: **Is it possible to achieve obfuscation using only stateless tokens? Is it possible to offer a general protection against a malicious sender using stateless or even stateful tokens?** To illustrate the motivation for the latter question, consider the goal of obfuscating a poker-playing program. The receiver of the obfuscated program would like to be assured that the sender did not violate the rules of the game (and in particular cannot bias the choice of the cards).
- **What are the simplest kinds of tamper-proof hardware tokens needed to realize the above goals?** For example, Goldwasser et al. [13] introduce a very simple kind of stateful token that they call an OTM (one-time memory) token.[4] An OTM token stores two strings s_0 and s_1, takes a single bit b as input, and then outputs s_b and stops working (or self-destructs). Note that an OTM token essentially implements the one-out-of-two string OT functionality; a subtle distinction between OTM and traditional OT is discussed in Section 3.1. An even simpler type of token is a *bit-OTM* token, where the strings s_0 and s_1 are restricted to be single bits. **Is it possible to realize unconditional, non-interactive, or UC-secure two-party computation using only bit-OTM tokens?** We note that previous works on secure two-party computation with hardware tokens [9,10,11,20] all make use of more complicated hardware tokens.

---

[3] In these models, the sender is allowed to arbitrarily specify the functionality of the oblivious RAM or the one-time program, and the receiver knows nothing about this functionality except an upper bound on its circuit size or running time. (Thus, the issue of dishonest senders does not arise in these models.) In the present work, by a one-time program tolerating a malicious sender, we mean that the receiver knows some partial specification of the functionality – modeled in the usual paradigm of secure two-party computation.

[4] The use of OTM tokens in [13] is motivated in part by the goal of achieving *leakage resilience,* a feature that our constructions using such tokens inherit as well.

1.1 Our Results

We show that the following tasks, which cannot be realized in the "plain" model, become feasible if the parties are allowed to generate and exchange simple tamper-proof hardware tokens.

- **Unconditional non-interactive secure computation.** We show that by exchanging *stateful* hardware tokens, any functionality can be realized with *unconditional* security against malicious parties. In the case of two-party functionalities $f(x, y)$ which take their inputs from a sender and a receiver and deliver their output to the receiver, our protocol is non-interactive and only requires a unidirectional communication of tokens from the sender to the receiver (in case an output has to be given to both parties, adding a reply from the receiver to the sender is sufficient). This result strengthens previous feasibility results for one-time programs by providing *unconditional* security, by offering general protection against *malicious senders*, and by using only bit-OTM tokens.

  As is typically the case for unconditionally secure protocols, our protocol is in fact *UC-secure*. This improves over previous works on UC-secure computation based on hardware tokens, which provided computational security under cryptographic assumptions.

  See Sections 3.1 and 3.2 for details of this result and a high level overview of techniques.
- **Interactive secure computation from stateless tokens based on one-way functions.** We show that *stateless* hardware tokens are sufficient to base general secure (in fact, UC-secure) computation on the existence of *one-way functions*. One cannot hope for security against unbounded adversaries with stateless tokens since an unbounded adversary could query the token multiple times to "learn" the functionality it contains. See Section 4 for details.
- **Obfuscation from stateless tokens.** We consider the problem of realizing non-interactive secure computation from stateless tokens for reactive functionalities which take a single input from the sender and an arbitrary sequence of inputs from the receiver (these are the only functionalities one can hope to realize non-interactively with *stateless* tokens). By building on recent techniques for resettably secure computation [22], we obtain a general positive result under standard cryptographic assumptions. This gives the first general feasibility result for program obfuscation using *stateless* tokens, while strengthening the standard notion of obfuscation by providing security against a malicious sender. We also propose constructions of non-interactive secure computation for general reactive functionalities with *stateful* tokens. See the full version for details.

In all of the above results, the size of each hardware token is either constant or polynomial in the security parameter, and its code is independent of the inputs of the parties. Thus, the tokens could theoretically be "mass-produced" before being used in any particular protocol with any particular inputs.

We stress that in contrast to some previous results along this line (most notably, [1,13,20]), our focus is almost entirely on *feasibility* questions, while only briefly discussing more refined efficiency considerations. However, in most cases our stronger feasibility results can be realized while also meeting the main efficiency goals pursued in previous works.

The first two results above are obtained by utilizing previous protocols for secure computation based on OT [18,19], and thus a main ingredient in our constructions is showing how to securely implement OT using hardware tokens. Note that in the case of non-interactive secure computation, additional tools are needed since the protocols of [18,19] are (necessarily) interactive.

1.2 Related Work

The use of tamper-proof hardware tokens for cryptographic purposes was first explored by Goldreich and Ostrovsky [1] in the context of software protection (one-time programs [13] is a relaxation of this goal, generally called program obfuscation [23]), and by Chaum, Pederson, Brands, and Cramer [2,3,4] in the context of e-cash. Ishai, Sahai, and Wagner [5] and Ishai, Prabhakaran, Sahai and Wagner [24] consider the question of how to construct tamper-proof hardware tokens when the hardware itself does not guarantee complete protection against tampering. Gennaro, Lysyanskaya, Malkin, Micali, and Rabin [6] consider a similar question, when the underlying hardware guarantees that part of the hardware is tamper-proof but readable, while the other part of the hardware is unreadable but susceptible to tampering. Moran and Naor [8] considered a relaxation of tamper-proof hardware called "tamper-evident seals," and given number of constructions of graphic tasks based on this relaxed notion. Hofheinz, Müller-Quade, and Unruh [25] consider a model similar to [9] in the context of UC-secure protocols where tamper-proof hardware tokens (signature cards) are issued by a trusted central authority.

The model that we primarily build on here is due to Katz [9], who considers a setting in which users can create and exchange tamper-proof hardware tokens where malicious users have full control over the functionality realized by each token they create. The main result of [9] is a general protocol for UC-secure two-party computation using stateful tokens, under the DDH assumption. Chandran, Goyal, Sahai [10] implement UC-secure two-party computation using stateless tokens, under the assumption that oblivious transfer protocols exist in the plain model. Aside from just considering stateless tokens, [10] also introduce a variant of the model of [9] that allows for the adversary to pass along tokens, and in general allows the adversary not to know the code of the tokens he produces. We do not consider this model here. Moran and Segev [11] also implement UC-secure two-party computation under the same assumption as [10], but using stateful tokens, and only requiring tokens to be passed in one direction. Damgård, Nielsen, and Wichs [12] show how to relax the "isolation" requirement of tamper-proof hardware tokens, and consider a model in which tokens can communicate a fixed number of bits back to its creator. Hazay and Lindell [20] propose constructions of practical protocols for various problems of interest using trusted stateful tokens. Very recently and independently of our work, practical oblivious transfer protocols using stateless tokens and relying only on one-way functions were suggested by Kolesnikov [26]. In contrast to the corresponding feasibility result from our work, these protocols either provide a weaker security guarantee or assume that tokens are well-formed, but on the other hand they offer better practical efficiency.

Goldwasser, Kalai, and Rothblum [13] introduced the notion of one-time programs, and showed how to realize it under the assumption that one-way functions exist, as we have already discussed. They also construct one-time zero-knowledge proofs under the same assumption. Their results focus mainly on achieving efficiency in terms of the number of tokens needed, and a non-adaptive use of the tokens by the receiver.

Finally, in a seemingly unrelated work which is motivated by quantum physics, Buhrman, Christandl, Unger, Wehner and Winter [27] consider the application of *non-local boxes* to cryptography. Using non-local boxes, Buhrman et al. show an unconditional construction for oblivious transfer in the interactive setting. A non-local box implements a *trusted* functionality taking input and giving output to both the parties (as opposed to OTM tokens which could be prepared maliciously). However, the key problem faced by Buhrman et al. is similar to a problem we face as well: delayed invocation of the non-local box by a malicious party. Indeed, one can give a simple interactive protocol (omitted here) for building a trusted non-local-box using OTM tokens. This provides an alternative to the interactive variant of our construction of unconditional secure computation from hardware tokens described in Section 3.1.

2 Preliminaries

In this section we briefly discuss some of the underlying definitions and concepts. The reader is referred to the full version for the details.

We use the UC-framework of Canetti [28] to capture the general notion of secure computation of (possibly reactive) functionalities. Our main focus is on the two-party case. We will usually refer to one party as a "sender" and to another as a "receiver". A *non-reactive* functionality may receive an input from each party and deliver output to each party (or only to the receiver). A *reactive* functionality may have several rounds of inputs and outputs, possibly maintaining state information between rounds.

Our model for tamper-proof hardware is similar to that of Katz [9]. As we consider both stateful and stateless tokens, we define different ideal functionalities for the two. By $\mathcal{F}_{wrap}^{single}$ we denote an ideal functionality that allows a sender to generate a "one-time token" which can be invoked by its designated receiver. A one-time token is a stateful token which takes an input from the receiver and returns a function which is specified in advance by the sender. (Note that if the sender is malicious, this function can be arbitrary.) After being invoked by the receiver, such a token "self-destructs". Thus, the only state these tokens keep is a flag which indicates whether the token has been run or not. Simple tokens of this type were used in [13].

We also define an ideal functionality $\mathcal{F}_{wrap}^{stateless}$ for *stateless* tokens. Here the token computes some (deterministic) function specified by the sender, and the receiver can query the token an unbounded number of times. Note that this makes stateless tokens useless if the receiver has enough resources to "learn" the token's description (either because the token is too small or the receiver is too powerful).[5]

By a *non-interactive* protocol we refer to a protocol in which the communication only involves a single batch of tokens, possibly along with an additional message, communicated from a sender to a receiver.

3 Unconditional Non-interactive Secure Computation Using Stateful Tokens

In this section we establish the feasibility of unconditionally non-interactive secure computation based on *stateful* hardware tokens. As is typically the case for unconditionally secure protocols, our protocols are in fact *UC secure.*

This section is organized as follows. In Subsection 3.1 we present an interactive protocol for arbitrary functionalities, which requires the parties to engage in multiple rounds of interaction. This gives an unconditional version of previous protocols for UC-secure computation based on hardware tokens [9,10,11], which all relied on computational assumptions.[6] This subsection also introduces some useful building blocks that are used for the non-interactive solution in the next subsection.

In Subsection 3.2 we consider the case of secure evaluation of two-party functionalities which deliver output to only one of the parties (the "receiver"). We strengthen the previous result in two ways. First, we show that in this case interaction can be completely eliminated: it suffices for the sender to non-interactively send tokens to the receiver, without any additional communication. Second, we show that even very simple, constant-size stateful tokens are sufficient for this purpose. This strengthens previous feasibility results for one-time programs [13] by providing *unconditional* security (in fact, UC-security), by offering general protection against *malicious senders*, and by using constant-size tokens.

3.1 The Interactive Setting

Unconditionally secure two-party computation is impossible to realize for most nontrivial functionalities, even with semi-honest parties [29,30]. However, if the parties are given oracle access to a simple ideal functionality such as Oblivious

[5] While the formal definition of this functionality forces a malicious sender to also use only *stateless* tokens, this requirement can be relaxed without affecting the security of our protocols. See Section 4 for details.

[6] The work of [11] realizes an unconditionally UC-secure *commitment* from stateful tokens. This does not directly yield protocols for secure computation without additional computational assumptions.

Transfer (OT) [16,17], then it becomes possible not only to obtain unconditionally secure computation with semi-honest parties [31,32,33], but also unconditional *UC-security* against *malicious* parties [18,19]. This serves as a natural starting point for our construction.

In the OT-hybrid model, the two parties are given access to the following ideal OT functionality: the input of P_1 (the "sender") consists of a pair of k-bit strings (s_0, s_1), the input of P_2 (the "receiver") is a choice bit c, and the receiver's output is the chosen string s_c. The natural way to implement a single OT call using stateful hardware tokens is by having the sender send to the receiver a token which, on input c, outputs s_c and erases s_{1-c} from its internal state. The use of such hardware tokens was first suggested in the context of one-time programs [13]. Following the terminology of [13], we refer to such tokens as OTM (one-time-memory) tokens.

An appealing feature of OTM tokens is their simplicity, which can also lead to better resistance against side-channel attacks (see [13] for discussion). This simplicity feature served as the main motivation for using OTM tokens as a basis for one-time programs. Another appealing feature, which is particularly important in our context, is that the OTM functionality does not leave room for bad sender strategies: whatever badly formed token a malicious sender may send is equivalent from the point of view of an honest receiver to having the sender send a well-formed OTM token picked from some probability distribution. (This is not the case for tokens implementing more complex functionalities, such as 2-out-of-3 OT or the extended OTM functionality discussed below, for which badly formed tokens may not correspond to any distribution over well-formed tokens.)

Given the above, it is tempting to hope that our goal can be achieved by simply taking any unconditionally secure protocol in the OT-hybrid model, and using OTM tokens to implement OT calls. However, as observed in [13], there is a subtle but important distinction between the OT-hybrid model and the OTM-hybrid model: while in the former model the sender knows the point in the protocol in which the receiver has already made its choice and received its output, in the latter model invoking the token is entirely at the discretion of the receiver. This may give rise to attacks in which the receiver adaptively invokes the OTM tokens "out of order," and such attacks may have a devastating effect on the security of protocols even in the case of unconditional security. A more detailed discussion of such attacks and simple solution ideas (that do not work) is included in the full version.

Extending the OTM functionality. To solve the above problem, we will realize an extended OTM functionality which takes from the sender a pair of strings (s_0, s_1) along with an auxiliary string r, takes from the receiver a choice bit c, and delivers to the receiver both s_c and r. We denote this functionality by ExtOTM. What makes the ExtOTM functionality nontrivial to realize using hardware tokens is the need to protect the receiver from a malicious sender who may try to make the received r depend on the choice bit c while *at the same* *time* protecting the sender from a malicious receiver who may try to postpone its choice c until after it learns r.

Using the ExtOTM functionality, it is easy to realize a UC-style version of the OT functionality which not only delivers the chosen string to the receiver (as in the OTM functionality) but also delivers an acknowledgement to the sender. This flavor of the OT functionality, which we denote by $\mathcal{F}^{\mathrm{OT}}$, can be realized by having the sender invoke ExtOTM with (s_0, s_1) and a randomly chosen r, and having the receiver send r to the sender. In contrast to OTM, the $\mathcal{F}^{\mathrm{OT}}$functionality allows the sender to force any subset of the OT calls to be completed before proceeding with the protocol. This suffices for instantiating the OT calls in the unconditionally secure protocols from [18,19]. We refer the reader to the full version of this paper for a UC-style definition of the OTM, ExtOTM, and $\mathcal{F}^{\mathrm{OT}}$functionalities.

Realizing ExtOTM using general[7] stateful tokens. As discussed above, we cannot directly use a stateful token for realizing the ExtOTM functionality, because this allows the sender to correlate the delivered r with the choice bit c. On the other hand, we cannot allow the sender to directly reveal r to the receiver, because this will allow the receiver to postpone its choice until after it learns r. In the following we sketch our protocol for realizing ExtOTM using stateful tokens. This protocol is non-interactive (i.e., it only involves tokens sent from the sender to the receiver) and will also be used as a building block towards the stronger results in the next subsection. We refer the reader to the full version of this paper for a formal description of the protocol and its proof of security. Below we include a detailed overview.

As mentioned above, at a high level, the challenge we face is to prevent unwanted correlations in an information-theoretic way for both malicious senders and malicious receivers. This is a more complex situation than a typical similar situation where only one side needs to be protected against (c.f. [34,35]). To accomplish this goal, we make use of secret-sharing techniques combined with additional token-based "verification" techniques to enforce honest behavior.

Our ExtOTM protocol Π_{ExtOTM} starts by having the sender break its auxiliary string r into $2k$ additive shares r^i, and pick $2k$ pairs of random strings (q_0^i, q_1^i). (Each of the strings q_b^i and r^i is k-bit long, where k is a statistical security parameter.) It then generates $2k$ OTM tokens, where the i-th token contains the pair $(q_0^i \circ r^i, q_1^i \circ r^i)$ (where '$\circ$' is the concatenation operator). Note that a malicious sender may generate badly formed OTM tokens which correlate r^i with the i-th choice of the receiver; we will later implement a token-based verification strategy that convinces an honest receiver that the sender did not cheat (too much) in this step.

Now the receiver breaks its choice bit c into $2k$ additive shares c^i, and invokes the $2k$ OTM tokens with these choice bits. Let $(\hat{q}^i, \hat{r}^i)$ be the pair of k-bit strings obtained by the receiver from the i-th token. Note that if the sender is honest, the

[7] Here, we make use of general tokens. Later in this section, we will show how to achieve the ExtOTM functionality (and in fact every poly-time functionality) using only very simple tokens – just bit OTM tokens.

receiver can already learn r. We would like to allow the receiver to learn its chosen string s_c while convincing it that the sender did not correlate all of the auxiliary strings $\hat{r}^i$ with the corresponding choice bits c_i. (The latter guarantee is required to assure an honest receiver that $\hat{r} = \sum \hat{r}^i$ is independent of c as required.)

This is done as follows. The sender prepares an additional single-use hardware token which takes from the receiver its $2k$ received strings $\hat{q}^i$, checks that for each $\hat{q}^i$ there is a valid selection $\hat{c}_i$ such that $\hat{q}^i = q^i_{\hat{c}_i}$ (otherwise the token returns $\perp$), and finally outputs the chosen string $s_{\hat{c}^1 \oplus \ldots \oplus \hat{c}^{2k}}$. (All tokens in the protocol can be sent to the receiver at one shot.) Note that the additive sharing of r in the first $2k$ tokens protects an honest sender from a malicious receiver who tries to learn $s_{\hat{c}}$ where $\hat{c}$ is significantly correlated with r, as it guarantees that the receiver effectively commits to c before obtaining any information about r. The receiver is protected against a malicious sender because even a badly formed token corresponds to some (possibly randomized) ideal-model strategy of choosing (s_0, s_1).

Finally, we need to provide to the receiver the above-mentioned guarantee that a malicious sender cannot correlate the receiver's auxiliary output $\hat{r} = \sum \hat{r}^i$ with the choice bit c. To explain this part, it is convenient to assume that both the sender and the badly formed tokens are deterministic. (The general case is handled by a standard averaging argument.) In such a case, we call each of the first $2k$ tokens well-formed if the honest receiver obtains the same r^i regardless of its choice c^i, and we call it badly formed otherwise. By the additive sharing of c, the only way for a malicious sender to correlate the receiver's auxiliary output with c is to make *all* of the first $2k$ tokens badly formed. To prevent this from happening, we require the sender to send a final token which proves that it knows all of the $2k$ auxiliary strings $\hat{r}^i$ obtained by the receiver. This suffices to convince the receiver that not all of the first $2k$ tokens are badly formed. Note, however, that we cannot ask the sender to send these $2k$ strings r^i in the clear, since this would (again) allow a malicious receiver to postpone its choice c until after it learns r.

Instead, the sender generates and sends a token which first verifies that the receiver knows r (by comparing the receiver's input to the k-bit string r) and only then outputs all $2k$ shares r^i. The verification step prevents correlation attacks by a malicious receiver. The final issue to worry about is that the string r received by the token (which may be correlated with the receiver's choices c_i) does not reveal to the sender enough information to pass the test even if all of its first $2k$ tokens are badly formed. This follows by a simple information-theoretic argument: in order to pass the test, the token must correctly guess *all* $2k$ bits c_i, but this cannot be done (except with $2^{-\Omega(k)}$ probability) even when given arbitrary k bits of information about the c_i.

The above protocol shows the following (see full version for proof):

Claim. Protocol Π_{ExtOTM} realizes ExtOTM with statistical UC-security in the $\mathcal{F}^{single}_{wrap}$-hybrid model.

We are now ready to prove the main feasibility result of this subsection.

Theorem 1 (Interactive unconditionally secure computation using stateful tokens). *Let f be a (possibly reactive) polynomial-time computable functionality. Then there exists an efficient, statistically UC-secure interactive protocol which realizes f in the $\mathcal{F}_{wrap}^{single}$-hybrid model.*

Proof. We compose three reductions. The protocols of [18,19] realize unconditionally secure two-party (and multi-party) computation of general functionalities using $\mathcal{F}^{\mathrm{OT}}$. A trivial reduction described above reduces $\mathcal{F}^{\mathrm{OT}}$ to ExtOTM. Finally, the above Claim reduces ExtOTM to $\mathcal{F}_{wrap}^{single}$.

3.2 The Non-interactive Setting

In this subsection we restrict the attention to the case of securely evaluating two-party functionalities $f(x, y)$ which take an input x from the sender and an input y from the receiver, and deliver $f(x, y)$ to the receiver. We refer to such functionalities as being *sender-oblivious.* Note that here we consider only *non-reactive* sender-oblivious functionalities, which interact with the sender and the receiver in a single round. The reactive case will be discussed in the full version.

Unlike the case of general functionalities, here one can hope to obtain *non-interactive* protocols in which the sender unidirectionally send tokens (possibly along with additional messages[8]) to the receiver.

For sender-oblivious functionalities, the main result of this subsection strengthens the results of Section 3.1 in two ways. First, it shows that a non-interactive protocol can indeed realize such functionalities using stateful tokens. Second, it pushes the simplicity of the tokens to an extreme, relying only on OTM tokens which contain pairs of *bits.*

Below we provide only a high-level description of the construction and the underlying ideas. We refer the reader to the full version for the full description of the protocols and their analysis.

One-time programs. Our starting point is the concept of a *one-time program* (OTP) [13]. A one-time program can be viewed in our framework as a non-interactive protocol for $f(x, y)$ which uses only OTM tokens, and whose security only needs to hold for the case of a *semi-honest sender* (and a malicious receiver).[9] The main result of [13] establishes the feasibility of computationally-secure OTPs for any polynomial-time computable f, based on the existence of one-way functions. The construction is based on Yao's garbled circuit technique [37]. Our initial observation is that if f is restricted to the complexity class NC^1, one can replace Yao's construction by an efficient perfectly secure variant (cf. [38]). This yields perfectly secure OTPs for NC^1. Alternatively, we

[8] Since our main focus is on establishing feasibility results, the distinction between the "hardware" part and the "software" part is not important for our purposes.

[9] The original notion of OTP from [13] is syntactically different in that it views f as a function of the receiver's input, where a description of f is given to the sender. This can be captured in our framework by letting $f(x, y)$ be a universal functionality.

also present a general construction of a OTP from any "decomposable randomized encoding" of f. This can be used to derive perfectly secure OTPs for larger classes of functions (including NL) based on randomized encoding techniques from [39,38]. See the full version for further details.

A next natural step is to construct unconditionally secure OTPs for any polynomial-time computable function f. In the full version of this paper, we describe a direct and self-contained construction which uses the perfect OTPs for NC^1 described above to build a statistically secure construction for any f. However, this result will be subsumed by our main result, which can be proved (in a less self-contained way) without relying on the latter construction.

Handling malicious senders. As in Section 3.1, the main ingredient in our solution is an interactive secure protocol Π for f. The high level idea of our construction is obtain a non-interactive protocol for f which emulates Π by having the sender generate and send a one-time token which computes the sender's next message function for each round of Π (a similar idea was used in [13] to construct one time proofs). Using the above procedure, we transform Π into a non-interactive protocol Π' which uses very complex one-time tokens (for implementing the next message functions of Π). The next idea is that we can break each such complex token into simple OTM tokens by using a one-time program realization of each complex token. More details are provided in the full version.

From the plain model to the OT-hybrid model. So far we assumed the protocol Π to be secure in the plain model. This rules out unconditional security as well as UC-security, which are our main goals in this section. A natural approach for obtaining unconditional UC-security is to extend the above compiler to protocols in the OT-hybrid model. This introduces a subtle difficulty which was already encountered in Section 3.1: the sender cannot directly implement the OT calls by using OTM tokens. To solve this problem, we build on the (non-interactive) ExtOTM protocol from Section 3.1. See full version for details.

From string-OTM to bit-OTMs. As a final optimization, in the full version we show how to use an unconditionally UC-secure non-interactive implementation of a string-OTM token using bit-OTM tokens.

This yields the following main result of this section:

Theorem 2 (Non-interactive unconditionally secure computation using bit-OTM tokens). *Let $f(x, y)$ be a non-reactive, sender-oblivious, polynomial-time computable two-party functionality. Then there exists an efficient, statistically UC-secure non-interactive protocol which realizes f in the $\mathcal{F}_{wrap}^{single}$-hybrid model in which the sender only sends bit-OTM tokens to the receiver.*

4 Two-Party Computation with Stateless Tokens

In this section, we again address the question of achieving interactive two-party computation protocols, but asking the following questions: (1) Can we rely on *stateless* tokens while only assuming that one-way functions exist? (2) Can the above be achieved without requiring that the complexity or number of the tokens grows with the complexity of the function being computed, as was the case in the previous section? We show how to positively answer both questions: We use stateless tokens, whose complexity is polynomial in the security parameter, to implement the OT functionality. Since (as discussed earlier) secure protocols for any two-party task exist given OT, this suffices to achieve the claimed result.

Before turning to our protocols, we make a few observations about stateless tokens to set the stage. First, we observe that with stateless tokens, it is always possible to have protocols where tokens are exchanged *only at the start of the protocol*. This is simply because each party can create a "universal" token that takes as input a pair (c, x), where c is a (symmetric authenticated/CCA-secure) encryption[10] of a machine M, and outputs $M(x)$. Then, later in the protocol, instead of sending a new token T, a party only has to send the encryption of the code of the token, and the other party can make use of that encrypted code and the universal token to emulate having the token T. The proof of security and correctness of this construction is straightforward.

Dealing with dishonestly created *stateful* tokens. The above discussion, however, assumes that dishonest players also only create stateless tokens. If that is not the case, then re-using a dishonestly created token may cause problems with security. If we allow dishonest players to create stateful tokens, then a simple solution is to repeat the above construction and send separate universal tokens for each future *use* of any token by the other player, where honest players are instructed to only use each token once. Since this forces all tokens to be used in a stateless manner, this simple fix is easily shown to be correct and secure; however, it may lead to a large number of tokens being exchanged. To deal with this, as was discussed in the previous section, we observe that by Beaver's OT extension result [36] (which requires only one-way functions), it suffices to implement $O(k)$ OTs, where k is the security parameter, in order to implement any polynomial number of OTs. Thus, it suffices to exchange only a polynomial number of tokens even in the setting where dishonest players may create stateful tokens.

Convention for intuitive protocol descriptions. In light of the previous discussions, in our protocol descriptions, in order to be as intuitive as possible, we describe tokens as being created at various points during the protocol. However, as noted above, our protocols can be immediately transformed into ones where a bounded number of tokens (or in the model where statelessness is guaranteed, only one token each) are exchanged in an initial setup phase.

4.1 Protocol Intuition

We now discuss the intuition behind our protocol for realizing OT using stateless tokens; due to the complexity of the protocol, we do not present the intuition

[10] An "encrypt-then-MAC" scheme would suffice here.

for the entire protocol all at once, but rather build up intuition for the different components of the protocol and why they are needed, one component at a time. For this intuition, we will assume that the sender holds two *random* strings s_0 and s_1, and the receiver holds a choice bit b. Note that OT of random strings is equivalent to OT for chosen strings [41].

The Basic Idea. Note that, since stateless tokens can be re-used by malicious players, if we naively tried to create a token that output s_b on input the receiver's choice bit b, the receiver could re-use it to discover both s_0 and s_1. A simple idea to prevent this reuse would be the following protocol, which is our starting point:

1. Receiver sends a commitment $c = \mathsf{com}(b; r)$ to its choice bit b.
2. Sender sends a token, that on input (b, r), checks if this is a valid decommitment of c, and if so, outputs s_b.
3. Receiver feeds (b, r) to the token it received, and obtains $w = s_b$

Handling a Malicious Receiver. Similar to the problem discussed in the previous section, there is a problem that the receiver may choose not to use the token sent by the sender until the end of the protocol (or even later!). In our context, this can be dealt with easily. We can have the sender commit to a random string π at the start of the protocol, and require that the sender's token must, in addition to outputting s_b, also output a valid decommitment to π. We then add a last step where the receiver must report π to the sender. Only upon receipt of the correct π value does the sender consider the protocol complete.

Proving Knowledge. While this protocol seems intuitive, we note that it is actually *insecure* for a fairly subtle reason. A dishonest sender could send a token that on input (b, r), simply outputs (b, r) (as a string). This means that at the end of the protocol, the dishonest sender can output a specific commitment c, such that the receiver's output is a decommitment of c showing that it was a commitment to the receiver's choice bit b. It is easy to see that this is impossible in the ideal world, where the sender can only call an ideal OT functionality.

To address the issue above, we need a way to prevent the sender from creating a token that can adaptively decide what string it will output. Thinking about it in a different way, we want the sender to "prove knowledge" of two strings before he sends his token. We can accomplish this by adding the following preamble to the protocol above:

1. Receiver chooses a pseudo-random function (PRF) $f_\gamma : \{0,1\}^{5k} \to \{0,1\}^k$, and then sends a token that on input $x \in \{0,1\}^{5k}$, outputs $f_\gamma(x)$.
2. Sender picks two strings $x_0, x_1 \in \{0,1\}^{5k}$ at random, and feeds them (one-at-a-time) to the token it received, and obtains y_0 and y_1. The sender sends (y_0, y_1) to the receiver.
3. Sender and receiver execute the original protocol above with x_0 and x_1 in place of s_0 and s_1. The receiver checks to see if the string w that it obtains from the sender's token satisfies $f_\gamma(w) = y_b$, and aborts if not.

The crucial feature of the protocol above is that a dishonest sender is effectively committed to two values x_0 and x_1 after the second step (and in fact the simulator can use the PRF token to extract these values), such that later on it must output x_b on input b, or abort.

Note that a dishonest receiver may learn k bits of useful information about x_0 and x_1 each from its token, but this can be easily eliminated later using the Leftover Hash Lemma (or any strong extractor).

Preventing correlated aborts. A final significant subtle obstacle remains, however. A dishonest sender can still send a token that causes an abort to be correlated with the receiver's input, *e.g.* it could choose whether or not to abort based on the inputs chosen by the receiver (see full version for a discussion of why this is a problem).

To prevent a dishonest sender from correlating the probability of abort with the receiver's choice, the input b of the receiver is additively shared into bits $b_1, \ldots, b_k$ such that $b_1 + b_2 + \cdots + b_k = b$. The sender, on the other hand, chooses strings $z_1, \ldots, z_k$ and r uniformly at random from $\{0,1\}^{5k}$. Then the sender and receiver invoke k parallel copies of the above protocol (which we call the *Quasi-OT* protocol), where for the ith execution, the sender's inputs are $(z_i, z_i + r)$, and the receiver's input is b_i. Note that at the end of the protocol, the receiver either holds $\sum z_i$ if $b = 0$, or $r + \sum z_i$ if $b = 1$.

Intuitively speaking, this reduction (variants of which were previously used by, e.g. [34,35]) forces the dishonest sender to make one of two bad choices: If each token that it sends aborts too often, then with overwhelming probability at least one token will abort and therefore the entire protocol will abort. On the other hand, if few of the sender's tokens abort, then the simulator will be able to perfectly simulate the probability of abort, since the bits b_i are $(k-1)$-wise independent (and therefore all but one of the Quasi-OT protocols can be perfectly simulated from the receiver's perspective). We make the receiver commit to its bits b_i using a statistically hiding commitment scheme (which can be constructed from one-way functions [42]) to make this probabilistic argument go through.

This completes the intuition behind our protocol. The result of this section is summarized by the following theorem, whose proof appears in full version.

Theorem 3 (Interactive UC-secure computation using stateless tokens). *Let f be a (possibly reactive) polynomial-time computable functionality. Then, assuming one-way functions exist, there exists a computationally UC-secure interactive protocol which realizes f in the $\mathcal{F}_{wrap}^{stateless}$-hybrid model. Furthermore, the protocol only makes a black-box use of the one-way function.*

Oblivious Reactive Functionalities in the Non-Interactive Setting. In the full version, we generalize our study of non-interactive secure computation to the case of reactive functionalities. Roughly speaking, reactive functionalities are the ones for which in the ideal world, the parties might invoke the ideal trusted party multiple times and this trusted party might possibly keep state between different invocations. For the interactive setting (i.e. when the parties are allowed multiple rounds of interaction in the $\mathcal{F}_{wrap}$-hybrid models) there are standard techniques using which, given protocol for non-reactive functionality, protocol for securely realizing reactive functionality can be constructed. However, these techniques fail in the non-interactive setting. In the full version, we study what class of reactive functionalities can be securely realized in the non-interactive setting for the case of stateless as well as stateful hardware token.

Acknowledgements. We thank Jürg Wullschleger for pointing out the relevance of [27] and for other helpful comments. We thank Guy Rothblum for useful discussions.

References

1. Goldreich, O., Ostrovsky, R.: Software protection and simulation on oblivious rams. J. ACM 43(3), 431–473 (1996)
2. Chaum, D., Pedersen, T.P.: Wallet databases with observers. In: Brickell, E.F. (ed.) CRYPTO 1992. LNCS, vol. 740, pp. 89–105. Springer, Heidelberg (1993)
3. Brands, S.: Untraceable off-line cash in wallets with observers (extended abstract). In: Stinson, D.R. (ed.) CRYPTO 1993. LNCS, vol. 773, pp. 302–318. Springer, Heidelberg (1994)
4. Cramer, R., Pedersen, T.P.: Improved privacy in wallets with observers (extended abstract). In: Helleseth, T. (ed.) EUROCRYPT 1993. LNCS, vol. 765, pp. 329–343. Springer, Heidelberg (1994)
5. Ishai, Y., Sahai, A., Wagner, D.: Private circuits: Securing hardware against probing attacks. In: Boneh, D. (ed.) CRYPTO 2003. LNCS, vol. 2729, pp. 463–481. Springer, Heidelberg (2003)
6. Gennaro, R., Lysyanskaya, A., Malkin, T., Micali, S., Rabin, T.: Algorithmic tamper-proof (ATP) security: Theoretical foundations for security against hardware tampering. In: Naor, M. (ed.) TCC 2004. LNCS, vol. 2951, pp. 258–277. Springer, Heidelberg (2004)
7. Hofheinz, D., Müller-quade, J., Unruh, D.: Universally composable zero-knowledge arguments and commitments from signature cards. In: Proc. of the 5th Central European Conference on Cryptology MoraviaCrypt 2005, Mathematical Publications (2005)
8. Moran, T., Naor, M.: Basing cryptographic protocols on tamper-evident seals. In: Caires, L., Italiano, G.F., Monteiro, L., Palamidessi, C., Yung, M. (eds.) ICALP 2005. LNCS, vol. 3580, pp. 285–297. Springer, Heidelberg (2005)
9. Katz, J.: Universally composable multi-party computation using tamper-proof hardware. In: Naor, M. (ed.) EUROCRYPT 2007. LNCS, vol. 4515, pp. 115–128. Springer, Heidelberg (2007)
10. Chandran, N., Goyal, V., Sahai, A.: New constructions for UC secure computation using tamper-proof hardware. In: Smart, N.P. (ed.) EUROCRYPT 2008. LNCS, vol. 4965, pp. 545–562. Springer, Heidelberg (2008)
11. Moran, T., Segev, G.: David and Goliath commitments: UC computation for asymmetric parties using tamper-proof hardware. In: Smart, N.P. (ed.) EUROCRYPT 2008. LNCS, vol. 4965, pp. 527–544. Springer, Heidelberg (2008)

12. Damgård, I., Nielsen, J.B., Wichs, D.: Isolated proofs of knowledge and isolated zero knowledge. In: Smart, N.P. (ed.) EUROCRYPT 2008. LNCS, vol. 4965, pp. 509–526. Springer, Heidelberg (2008)
13. Goldwasser, S., Kalai, Y.T., Rothblum, G.: One-time programs. In: Wagner, D. (ed.) CRYPTO 2008. LNCS, vol. 5157, pp. 39–56. Springer, Heidelberg (2008)
14. Canetti, R.: Universally composable security: A new paradigm for cryptographic protocols. In: FOCS, pp. 136–145 (2001)
15. Chor, B., Kushilevitz, E.: A zero-one law for boolean privacy. SIAM J. Discrete Math. 4(1), 36–47 (1991)
16. Rabin, M.O.: How to exchange secrets with oblivious transfer (1981)
17. Even, S., Goldreich, O., Lempel, A.: A randomized protocol for signing contracts. Commun. ACM 28(6), 637–647 (1985)
18. Kilian, J.: Founding cryptography on oblivious transfer. In: STOC, pp. 20–31 (1988)
19. Ishai, Y., Prabhakaran, M., Sahai, A.: Founding cryptography on oblivious transfer - efficiently. In: Wagner, D. (ed.) CRYPTO 2008. LNCS, vol. 5157, pp. 572–591. Springer, Heidelberg (2008)
20. Hazay, C., Lindell, Y.: Constructions of truly practical secure protocols using standardsmartcards. In: Ning, P., Syverson, P.F., Jha, S. (eds.) ACM Conference on Computer and Communications Security, pp. 491–500. ACM, New York (2008)
21. Goldreich, O., Micali, S., Wigderson, A.: How to play any mental game or a completeness theorem for protocols with honest majority. In: STOC, pp. 218–229 (1987)
22. Goyal, V., Sahai, A.: Resettably secure computation. In: Joux, A. (ed.) EUROCRYPT 2009. LNCS, vol. 5479, pp. 54–71. Springer, Heidelberg (2009)
23. Barak, B., Goldreich, O., Impagliazzo, R., Rudich, S., Sahai, A., Vadhan, S.P., Yang, K.: On the (im)possibility of obfuscating programs. In: Kilian, J. (ed.) CRYPTO 2001. LNCS, vol. 2139, pp. 1–18. Springer, Heidelberg (2001)
24. Ishai, Y., Prabhakaran, M., Sahai, A., Wagner, D.: Private circuits ii: Keeping secrets in tamperable circuits. In: Vaudenay, S. (ed.) EUROCRYPT 2006. LNCS, vol. 4004, pp. 308–327. Springer, Heidelberg (2006)
25. Hofheinz, D., Müller-Quade, J., Unruh, D.: Universally composable zero-knowledge arguments and commitments from signature cards. In: 5th Central European Conference on Cryptology (2005), `http://homepages.cwi.nl/~hofheinz/card.pdf`
26. Kolesnikov, V.: Truly efficient string oblivious transfer using resettable tamper-proof tokens. In: Micciancio, D. (ed.) TCC 2010. LNCS, vol. 5978. Springer, Heidelberg (2010)
27. Buhrman, H., Christandl, M., Unger, F., Wehner, S., Winter, A.: Implications of superstrong nonlocality for cryptography. Proceedings of the Royal Society A 462(2071), 1919–1932
28. Canetti, R.: Universally composable security: A new paradigm for cryptographic protocols. In: FOCS, pp. 136–145 (2001)
29. Ben-Or, M., Goldwasser, S., Wigderson, A.: Completeness theorems for non-cryptographic fault-tolerant distributed computation (extended abstract). In: STOC, pp. 1–10 (1988)
30. Kushilevitz, E.: Privacy and communication complexity. SIAM J. Discrete Math. 5(2), 273–284 (1992)
31. Goldreich, O., Vainish, R.: How to solve any protocol problem - an efficiency improvement. In: Pomerance, C. (ed.) CRYPTO 1987. LNCS, vol. 293, pp. 73–86. Springer, Heidelberg (1988)

32. Galil, Z., Haber, S., Yung, M.: Cryptographic computation: Secure faut-tolerant protocols and the public-key model. In: Pomerance, C. (ed.) CRYPTO 1987. LNCS, vol. 293, pp. 135–155. Springer, Heidelberg (1988)
33. Goldreich, O.: Foundations of Cryptography: Basic Applications. Cambridge University Press, Cambridge (2004)
34. Kilian, J.: Uses of Randomness in Algorithms and Protocols. MIT Press, Cambridge (1990)
35. Lindell, Y., Pinkas, B.: An efficient protocol for secure two-party computation in the presence of malicious adversaries. In: Naor, M. (ed.) EUROCRYPT 2007. LNCS, vol. 4515, pp. 52–78. Springer, Heidelberg (2007)
36. Beaver, D.: Correlated pseudorandomness and the complexity of private computations. In: STOC, pp. 479–488 (1996)
37. Yao, A.: How to generate and share secrets. In: FOCS, pp. 162–167 (1986)
38. Ishai, Y., Kushilevitz, E.: Perfect constant-round secure computation via perfect randomizing polynomials. In: Widmayer, P., Triguero, F., Morales, R., Hennessy, M., Eidenbenz, S., Conejo, R. (eds.) ICALP 2002. LNCS, vol. 2380, pp. 244–256. Springer, Heidelberg (2002)
39. Feige, U., Kilian, J., Naor, M.: A minimal model for secure computation (extended abstract). In: STOC, pp. 554–563 (1994)
40. Brassard, G., Crépeau, C., Santha, M.: Oblivious transfers and intersecting codes. IEEE Transactions on Information Theory 42(6), 1769–1780 (1996)
41. Beaver, D., Goldwasser, S.: Multiparty computation with faulty majority (extended announcement). In: FOCS, pp. 468–473. IEEE, Los Alamitos (1989)
42. Haitner, I., Reingold, O.: Statistically-hiding commitment from any one-way function. In: STOC, pp. 1–10 (2007)
43. Anderson, W.E.: On the secure obfuscation of deterministic nite automata. Cryptology ePrint Archive, Report 2008/184 (2008)
44. Canetti, R., Goldreich, O., Goldwasser, S., Micali, S.: Resettable zero-knowledge (extended abstract). In: STOC, pp. 235–244 (2000)

Truly Efficient String Oblivious Transfer Using Resettable Tamper-Proof Tokens

Vladimir Kolesnikov

Alcatel-Lucent Bell Laboratories, Murray Hill, NJ 07974, USA
kolesnikov@research.bell-labs.com

Abstract. SFE requires expensive public key operations for each input bit of the function. This cost can be avoided by using tamper-proof hardware. However, all known efficient techniques require the hardware to have long-term secure storage and to be resistant to reset or duplication attacks. This is due to the intrinsic use of counters or erasures. Known techniques that use resettable tokens rely on expensive primitives, such as generic concurrent ZK, and are out of reach of practice.

We propose a *truly efficient* String Oblivious Transfer (OT) technique relying on *resettable* (actually, *stateless*) tamper-proof token. Our protocols require between 6 and 27 *symmetric key* operations, depending on the model. Our OT is secure against covert sender and malicious receiver, and is sequentially composable.

If the token is semi-honest (e.g. if it is provided by a trusted entity, but adversarially initialized), then our protocol is secure against malicious adversaries in concurrent execution setting.

Only one party is required to provide the token, which makes it appropriate for typical asymmetric client-server scenarios (banking, TV, etc.)

1 Introduction

We propose efficiency improvements of two-party Secure Function Evaluation (SFE). We take advantage of tamper-proof hardware issued by one of the participants of the computation. We restrict ourselves to *stateless* (thus resettable) tokens to avoid the cost of adding long-term secure storage and to protect against a class of physical attacks on the hardware.

Two-party general (SFE) allows two parties to evaluate any function on their respective inputs x and y, while maintaining privacy of both x and y. SFE is (justifiably) a subject of an immense amount of research, e.g. [28,29,23]. Efficient SFE algorithms enable a variety of electronic transactions, previously impossible due to mutual mistrust of participants. Examples include auctions [27,8,12], contract signing [11], set intersection [18], etc. As computation and communication resources have increased, SFE of many useful functions has become practical for common use.

Still, SFE of most of today's functions of interest is either completely out of reach of practicality, or carries costs sufficient to deter would-be adopters, who

D. Micciancio (Ed.): TCC 2010, LNCS 5978, pp. 327–342, 2010.

instead choose stronger trust models, entice users to give up their privacy with incentives, or use similar crypto-workarounds. We believe that truly practical efficiency is required for SFE to see more use in real-life applications.

To achieve this, in addition to improving protocols in the standard model, it is useful to "give ourselves" some help in the form of less demanding (yet acceptable) security properties, such as the recently proposed covert adversaries model [3]. When it fits the setting, we could also rely on additional assumptions that the world has been long using, such as simple tamper-proof (or tamper-resistant) tokens. We note that token-supported protocols received a lot of attention recently, e.g., [21, 7, 18, 14, 9, 10, 15]. To allow cheaper tokens and higher confidence in the security of the system, it is desired to minimize the assumptions on the hardware, while still reaping the performance benefit.

A weaker hardware model that recently received a lot of attention, e.g. [7, 17, 15], is that of *resettable* token. Here, the adversary, e.g., by interrupting power supply or applying highly targeted laser or electro-magnetic radiation, is able to manipulate the token and reset its internal variables (e.g., a counter) to the initial state[1]. This realistic capability trivially breaks most of currently known protocols taking advantage of secure tokens. Similar effect is achieved if an attacker is able to obtain a clone of the card, e.g. by insider attacks during manufacturing process, etc. From another perspective, it may be convenient to allow legitimate clone cards to allow the user operate several instances of itself independently. For example, a person may sign up for a telephone, wireless, and TV services from the same provider. It may be convenient to simply provide him with several identical tokens, which he can use interchangeably in all of his devices. It is easy to see that such deployment clearly requires resettable tokens.

We consider even weaker *stateless* tokens. This gives an important advantage of avoiding the manufacturing cost of long-term secure storage.

Our setting, goals and approach. We consider two-party SFE aided by stateless tamper-proof tokens. In fact, our protocols work in the very important client-server setting, where only one party has the capability to issue tokens. In practice, this occurs in TV, phone, cellular and internet service provision, banking, etc. We aim to enable efficient computation of a variety of functions on moderate-size inputs. On the client side, for example, inputs could be viewing preferences, browsing history, etc. Server's inputs may include content or other digital rights to be transferred to the client. We stress that our solutions are general and can be used to compute any function.

We note the inherently asymmetric trust model in the client-server setting. Servers are usually established businesses who are likely unwilling to cheat, especially if there is a chance of being caught. The risk of loss of business and public embarrassment is a strong deterrent. Therefore, it is natural and sufficient to model servers as covert adversaries. Also, servers are capable of issuing tamper-proof tokens, and, in many cases, already routinely do. Clients, on the other

---

[1] Not all is necessarily lost with this adversarial capability. For example, it is *much* harder to reprogram the token, simply by resetting certain bits, to, e.g., output its keys. Thus the resettable token model appears reasonable.

hand, have much less credibility, and may be more willing to attempt interfering with the protocol and the provided token. Therefore, it is appropriate to view them as malicious adversaries, and to aim to reduce the trust assumptions on the token.

Oblivious Transfer (OT) is often a bottleneck in SFE, due to the high cost of the required public key operations. Our main contribution is, we believe, the first truly efficient protocol to use a resettable token to replace public key operations in OT (and SFE) with symmetric primitives.

The hardware assumption: costs and security comparison with number-theoretic OT. At first, it may seem that the cost of deployment of tokens is greater than that of using standard OT based on public key primitives. We argue that this is often not the case. As noted above, tamper-resistant tokens are often already deployed in the form of cell phones' SIM cards, TV cable smart cards, etc. In these important scenarios, tokens are "free". Otherwise, they cost from $2 in retail (e.g., [1]). At the same time, using standard OT may necessitate much higher costs in terms of increased CPU requirements, much slower processing, and decreased battery life (in mobile devices). Further, evaluating garbled circuits (our main application) involves transferring them (megabytes or gigabytes of data, especially in the malicious model) to the client. In large-scale deployment of SFE (e.g., by banks and service providers), these communication costs cannot be afforded; fortunately, they can be avoided by generating circuits from a seed by the server-issued token [20]. This solution requires reliance on tamper-proof/tamper-resistant tokens, forces their use, fits well with our setting, and further justifies it.

One may also question the hardware assumption and rightfully presume that a sufficiently strong attacker can always break into the token. We argue that in many applications we envision, the cost of the break far exceeds the gain (e.g., free cable TV for one user). Therefore, the barrier raised by even weak tamper resistance is high enough for typical applications, and constitutes a reasonable assumption.

1.1 Our Contributions and Outline of the Work

We consider two-party SFE, where one party (Sender S) is able to issue a token T to the receiver R. T is assumed to be tamper-proof, but resettable. Our main contribution is a new efficient string OT protocol in the covert adversaries model [3], which takes advantage of T. To our knowledge, ours is the first truly efficient protocol that gets rid of public key operations in this setting. Our protocol requires a total of 6 symmetric key operations if the token is semi-honest and S and R are malicious, and 27 such operations in the covert adversaries model (with deterrence $\epsilon = 1/2$). This immediately leads to corresponding improvements in 2-party SFE.

We start with the overview of related work in Sect. 2 and preliminaries in Sect. 3. For clarity of presentation, we first present the semi-honest variant of the OT scheme in Sect. 4. We then show how to achieve security against covert sender and fully malicious receiver in Sect. 4.1. In Sect. 4.2, we discuss aspects of simultaneous execution and sequential composition, and show security in this case.

Under an additional reasonable assumption that the tokens are trusted to run the specified code (e.g. when standard tokens are provided by a trusted manufacturer), our protocol is composable concurrently and secure against malicious S and R (Sect. 4.3).

1.2 Main Idea of Our Approach

Our main tool is Strong Pseudorandom Permutation Generators (SPRPG). We capitalize on the observation that it is difficult to find a collision in a SPRPG F under independent secret keys k_0, k_1. At the same time, a value in the range of F does not reveal which key was used to arrive at this value (via the evaluation of F). That is, R can know a preimage of the (random to S) value under (only one) key of R's choice. This is naturally used to construct efficient OT protocols. We use T to securely store the keys and provide R the evaluation interface to F (but not F^{-1}).

We stress that the use of SPRPG (vs PRPG) does not introduce additional assumptions, and has no performance impact (Sect. 3.2).

We give additional intuition for semi-honest and covert setting protocols in Sect. 4 and 4.1, before presenting corresponding protocols.

2 Related Work

There is a massive body of work on SFE, and, in particular, two-party SFE, of which most relevant to this work is Yao's garbled circuit [28, 29, 23] and the techniques of its implementation in the malicious setting. The complete solutions are excellently presented in [24, 3, 16]. We work in the recently introduced security model of covert adversaries [3], which we find to help significantly in design of efficient protocols. SFE solutions of [3, 16] are presented in this model. Our work is an improvement of an important building block used by above (and many other) solutions.

There has been a recent surge of interest in SFE supported by tamper-proof tokens [21, 7, 18, 14, 10, 15], and, specifically, resettable tokens [7, 15]. Our work is different from the above, as follows.

Firstly, of the above, only [7, 15] consider resettable tokens. Work of Katz [21] was the first to formalize tamper-proof token model, and show sufficiency for Universally Composable (UC) security. This is mainly a feasibility and definitional result. Chandran et al. [7] extended the results of Katz, considered resettable tokens, and constructed UC-secure protocols. Goyal and Sahai [17] constructed protocols secure against resettable adversaries (and not just tokens). Very recently, Goyal et al. [15] considered the general question of basing cryptography on tamper-proof tokens, under minimal computational assumptions. As one of their results, they showed that stateless tokens and one-way functions are sufficient for UC-secure computation. Damgård et al. [10] consider the setting where the tamper-proof token may leak a limited amount of information back to its issuer. They show how to achieve UC-secure computation in this setting. We remark that the protocols of [21,7,10,17,15] achieve stronger UC security. Our protocols either have weaker security guarantee, or rely on a semi-honest token. In exchange, we use only a few (6 to 27) symmetric key operations and thus offer truly practical performance.

Hazay and Lindell [18] give a very practical protocol which takes advantage of secure tokens (standard smart cards). As compared to our work, they solve a specific problem, using a much stronger assumption (non-resettable semi-honest smart card) than we do.

Goldwasser et al. [14] consider one-time (or, generally, k-time) programs and proofs, and rely on tamper-proof tokens to achieve it. Their token security model is different from ours – it is non-resettable, but it is assumed that data that is ever accessed by the token's program may be leaked to the adversary. This assumption makes impossible the use of pseudorandomness in their construction, and all the random values (i.e. wire keys used by garbled circuit) ever used by the program need to be preloaded explicitly. While opening a significant new application of secure hardware, these tokens can only be used for a predetermined number of executions. The most important difference, however, is that we achieve security with resettable token.

3 Preliminaries

3.1 Notation and Security Model

We will use Pseudorandom Permutation Generators (PRPG) and Strong PRPG (SPRPG). While we don't need SPRPG properties in all uses of PRPG, for simplicity we refer to all of them as SPRPG, and denote by F. We denote the domain and range of F by D, and the key space of F by K. For simplicity, we set $K = D = \{0,1\}^n$, where n is the security parameter. We denote OT Sender by S, OT Receiver by R and the token by T. In our protocols, R will be testing the correctness of the actions of S and T. Objects associated with testing would often have subscript $_t$, for example, D_t is the subdomain of D, reserved for test queries. By "OT" we mean "string OT".

We prove security against covert adversaries. Aumann and Lindell [3] propose several definitions of security in this setting, with various guarantees. Our protocols are secure in their strongest model (strong explicit cheat formulation). Informally, this guarantees that covert adversary does not learn anything about honest party's input if he is caught.

We stress that this notion of security requires *full simulation* in the ideal world (where cheating attempts are modeled by a designated query). Further, this notion is modularly sequentially composable [3]. In particular, this means that our OT can be composed, e.g., with a garbled circuit protocol, and result in a corresponding SFE protocol in the covert model.

For simplicity of presentation, we often omit the introduction of the deterrence parameter ϵ (probability of being caught when cheating) in our calculation. Our constructions are given w.r.t. $\epsilon = 1/2$, but are readily generalized to ϵ polynomially close to 1.

We assume that tokens cannot be internally observed or tampered with; we only allow R to reset the token at will. We leave it as an interesting direction to investigate efficient security in presence of both resets and tampering (perhaps using the techniques of Gennaro et al. [13]).

3.2 Strong Pseudorandom Permutation Generators (SPRPG)

We assume reader's familiarity with PRPG. SPRPG (sometimes also referred to as *Super PRPG*) is a natural extension of the notion of PRPG which considers efficiently invertible permutations. Informally, in terms of security, the difference between the two notions is that while PRPG allows the adversary to query the "forward evaluation" oracle, SPRPG allows to query both "forward evaluation" and "inversion" oracles. Luby and Rackoff [25] showed how to construct SPRPG from PRFG. Therefore, our use of SPRPG does not require an additional assumption. Further, invertibility has been a design requirement of almost all block ciphers, notably, of AES. It appears that the trend will continue in the future as well. Therefore, in practical terms, our reliance on SPRPG does not incur any performance penalty. We envision instantiation of SPRPG in our constructions with AES.

We note that most of cryptographic protocols literature relies on PRFG/PRPG, and largely ignores SPRPG. The novelty of our approach is in part in using tools just outside of standard "crypto toolbox". Use of invertible PRPG seems to be promising in tamper-proof token designs.

3.3 Oblivious Transfer

Garbled Circuit (GC), excellently presented in [23], is the standard and the most efficient method of two-party SFE of boolean circuits. An important (and often the most computationally expensive) step of GC and other SFE techniques is the oblivious transfer (OT) of a secret (one of the two held by the sender), depending on the choice of the receiver.

The 1-out-of-2 OT is a two-party protocol. The *sender* S has two secrets s_0, s_1, and the *receiver* R has an selection bit $i \in \{0, 1\}$. Upon completion, R learns s_i, but nothing about s_{1-i}, and S learns nothing about i. OT is a widely studied primitive in the standard model [5, 2], with improved implementations in the Random Oracle model [26, 6].

OT Using Tamper-Proof Hardware. While existing OT constructions are simple, they are not very efficient due to use of several public key operations for each OT. If parties possess tamper-proof hardware tokens, such as smart cards, the use of public key operations can be avoided as follows (shown in the semi-honest model).

Suppose, S creates and gives R the following token T. Equipped with a non-resettable counter (initially set to 0), and seeded with the secret key k chosen by S, T has the following interface. Let F be a PRPG. On input i, where $i \in \{0,1\}$, T sets $counter = counter + 2$ and outputs $F_k(counter + i - 2)$. Now, to execute the j-th instance of OT, the sender sends to the receiver $(F_k(counter_j) \oplus s_0^j, F_k(counter_j + 1) \oplus s_1^j)$, where $counter_j$ is the value of the counter used for evaluation of j-th OT. Receiver obtains $F_k(counter_j + i)$ by calling T with argument i, and obtains s_i^j. Because of the properties of PRPG and since the token guarantees that F_k will not be evaluated on both $counter_j$ and $counter_j + 1$, the receiver will be able to obtain only one of s_0, s_1. Further, S does not learn the receiver's input i, since nothing is sent to S.

This efficient protocol can be naturally modified to withstand covert adversaries, but it (and other natural protocols) fails trivially if T is reset.

4 OT with Resettable Tamper-Proof Cards

We build our presentation incrementally. We find it instructive to first present the protocol for a hybrid semi-honest model, where the only allowed malicious behavior is for R to arbitrarily query the token T, and to reset T to the state T was in when it was received by R. (This additional power is necessary to exclude trivial solutions secure in the semi-honest model, such as relying on the semi-honest parties to not reset the token.) We show how to efficiently handle malicious behavior in Sect. 4.1.

In our construction, we use Strong PRPG (SPRPG), i.e. a PRPG that allows efficient inversion. As discussed in Sect. 3.2, this does not constitute an additional assumption and causes no performance penalty (we envision using AES as SPRPG).

The main idea of our construction is to limit the functionality of the token to evaluate the SPRPG F *in the forward direction only*, to keep no state, and to have S load two random SPRPG keys on the token. Then, the simple but crucial observation is that it is infeasible for the token receiver to find two preimages (under the two keys) of any element in the range D of F. (In fact at least one of the preimages will look random to him.) At the same time, he can trivially generate a random element on the range with the preimage under either of the keys. The token creator S, who knows both keys, can invert F and compute both preimages. If he uses them to encrypt the respective input secrets, the receiver can recover the secret of his choice. We discuss the intuition for protection against malicious actions before presenting our main protocol in Sect. 4.1.

Construction 1. *(OT, stateless token, augmented semi-honest model)*

1. Initialization. *The token T is created by sender S, seeded with keys k_0, k_1, randomly chosen by S, and given to receiver R. T answers any number of queries of the form $Q(i,x) = F_{k_i}(x)$.*
2. OT Protocol execution. *S has inputs $s_0, s_1 \in D$. R has input $i \in \{0,1\}$.*
   (a) *R chooses $x \in_R D$, and queries the token $v = Q(i,x) = F_{k_i}(x)$. R sends v to S.*

(b) *S computes encryption keys $ek_0 = F_{k_0}^{-1}(v)$, and $ek_1 = F_{k_1}^{-1}(v)$. He then sends $(e_0, e_1) = (F_{ek_0}(s_0), F_{ek_1}(s_1))$ to R.*
(c) *R computes and outputs $s_i = F_x^{-1}(e_i)$.*

Observation 1. *We need to be careful with the choice of the encryption scheme used to encrypt Sender's secrets s_0, s_1, since R has access to the forward evaluation oracle $F_{k_i}(x)$. For example, the random pad would not work here, since this would allow R to check the unknown secret (also transferred, but masked) for equality to constants of his choice. Jumping ahead, we note that to efficiently handle malicious attacks (e.g. R reusing x for different executions) and multiple executions of the protocol, we need a stronger primitive, such as semantically secure encryption (which, e.g., can be implemented simply by allocating some of the PRPG domain for randomness used for encryption).*

Theorem 1. *Assume F is a SPRPG. Then the protocol of Construction 1 evaluates the String OT functionality securely in the semi-honest model, where R is additionally allowed to arbitrarily query and reset the token T to its original state (i.e. as received by R).*

Proof (Sketch). Correctness is straightforward. To prove security, we first show the simulator Sim_S which simulates the view of the semi-honest S. It is easy to see that all S sees is a uniformly distributed element in D, which is trivial to simulate.

We now show the simulator Sim_R of the view of semi-honest R, who additionally has oracle access to T. Given input i and output s, Sim_S outputs $(i, x', (e_0', e_1'))$, where $x' \in_R D$, $e_i' = F_{x'}(s)$, and $e_{1-i}' \in_R D$. It is easy to see that the output of Sim_S is computationally indistinguishable from the real execution. Indeed, the only "fake" part is e_{1-i}'. In the real execution, $e_{1-i} = F_{F_{k_{1-i}}^{-1}}(s_{1-i})$. A hybrid argument shows that existence of a distinguisher implies a distinguisher for F. □

4.1 Protocols for Malicious (Covert) Setting

The most practical solution to move to malicious model would be to use tokens manufactured by a third party, which are trusted to run the specified functionality (loading the keys, and answering queries as above), and be non-reprogrammable. In this semi-honest token case, Construction 1 works, with the small modification of using semantically secure encryption to transfer secrets in Step 2b (see also Observation 1). In fact, the resulting protocol is concurrently secure (see Sect. 4.3).

However, it is highly desirable to avoid this trust assumption. With the exception of Sect. 4.3, we consider tokens running arbitrary code.

Intuition. The main avenue of attack for S and T is to try to combine their views. This is done by T incorporating a side-channel in its answers to R (recall, Construction 1 provides no channel from S to T). We note that R never knows when this attack occurs, so he must continuously check on T. We handle this by a strong variant of a cut-and-choose technique, which we introduce in this work. The main avenue of attack for R is to adversarily choose his queries to T. We handle this by using semantic encryption to encrypt secrets being passed back (see Observation 1).

We start with Construction 1, and extend it as follows. First, we embed an exponentially large number of keys in the token, by pseudorandomly generating them. This allows R to test keys of his choice at any time in the lifetime of T, a what we call *continuous cut-and-choose*. To achieve this, the keys k_0, k_1, as used in Construction 1, will be replaced with derived keys $F_{k_0}(y), F_{k_1}(y)$, where y is chosen by R. The token query function Q would now take an extra argument and return $Q(y,x) = (F_{F_{k_0}(y)}(x), F_{F_{k_1}(y)}(x))$. To test T's response, R will ask S to reveal $F_{k_0}(y), F_{k_1}(y)$. Of course this y cannot be used for "live" OT transfer. To avoid S storing large history, R and S agree, *after the token has been received by* R, on the test domain $D_t \subset D$. Now, S will only reveal keys for test queries $y_t \in D_t$, and will only use $y \in D \setminus D_t$ for "live" OT. Of course, D_t must be unpredictable to T. This is easily achieved by R defining it pseudorandomly, e.g. by R choosing k_D and setting $D_t = \{F_{k_D}(x) | x \in D, x \text{ is even}\}$.

Further, we unconditionally hide the input i of R from T by having R choose a random bit b and flipping i iff $b = 1$. S is notified of b to allow for the corresponding flip of his inputs s_0, s_1.

The above changes are sufficient. After presenting the protocol, we give additional intuition of why security holds, and then state and prove the corresponding theorem. Our construction is for deterrence factor $\epsilon = 1/2$, since the probability of catching the attempted cheat is 1/2. This probability can be naturally modified for any ϵ polynomially close to 1, simply by performing more test queries. Namely, k test queries provide for $\epsilon = 1 - 1/k$.

Let $E = (Gen, Enc, Dec)$ be a CPA-secure encryption scheme, such that every element of the ciphertext domain is uniquely decrypted. Such schemes can be easily constructed from a SPRPG.

Construction 2. *(OT using resettable tokens, covert adversaries model, deterrence* $\epsilon = 1/2$*)*

1. Initialization. *Let Enc be an encryption scheme as described above. The token T is created by sender S, seeded with keys k_0, k_1, randomly chosen by S, and given to receiver R. T answers any number of queries of the form $Q(y,x) = (F_{F_{k_0}(y)}(x), F_{F_{k_1}(y)}(x))$. R chooses a random string $k_D \in_R \{0,1\}^n$ and sends to S. Parties set $D_t = \{F_{k_D}(x) | x \in D, x$ is even$\}$.*
2. OT Protocol execution. *S has inputs $s_0, s_1 \in D.R$ has input $i \in \{0,1\}$.*
   (a) *R chooses $y_t \in_R D_t$ and sends to S.*
   (b) *S checks that $y_t \in D_t$ and if so, responds with $k_t^0 = F_{k_0}(y_t)$ and $k_t^1 = F_{k_1}(y_t)$. Otherwise, S outputs* corrupted$_R$ *and halts.*
   (c) *R chooses $b \in_R \{0,1\}, x, x_t \in_R D, y \in_R D \setminus D_t$. Then R queries T, in random order*
   $$(v_0, v_1) = Q(y,x) = (F_{F_{k_0}(y)}(x), F_{F_{k_1}(y)}(x))$$
   $$(c_0, c_1) = Q(y_t, x_t) = (F_{F_{k_0}(y_t)}(x_t), F_{F_{k_1}(y_t)}(x_t))$$
   R checks whether the check values c_0, c_1 match the evaluation of F based

on keys received from S. If so, R sends $b, y, v_{i \oplus b}$ to S. Otherwise, R outputs corrupted$_S$ *and halts.*

(d) S checks that $y \notin D_t$. If not, S outputs corrupted$_R$ *and halts. If so, S computes encryption keys $ek_0 = F^{-1}_{F_{k_0}(y)}(v_{i \oplus b})$, and $ek_1 = F^{-1}_{F_{k_1}(y)}(v_{i \oplus b})$, and sends $(e_0, e_1) = (Enc_{ek_0}(s_{0 \oplus b}), Enc_{ek_1}(s_{1 \oplus b}))$ to R.*

(e) R computes and outputs $s_i = Dec_x(e_{i+b})$.

Note, only R's (selective) check is related to catching the possible covert cheating by S and T. The protocol is secure against malicious R.

Intuition for security. Observe that T only sees two Q queries with easily simulatable random-looking arguments. While S receives one message from T, T does not know which of the two queries it sees is the test one, and which is safe to attack. If T guesses the test query, then it can pass information with the other query, and S learns R's input and can set R's output. However, if T guesses incorrectly and attempts to pass information in the test query, he is caught w.h.p., and without revealing R's input. (This is because S "fixes" the test SPRPs by revealing their keys, and T must answer the query according to the fixed keys not to be caught.) Since the two queries are indistinguishable for T, the deterrence factor is $\epsilon = 1/2$. We stress that T/S cannot base their decision to cheat on R's input, since they commit to this decision (and are checked by R) before R sends any input-dependent messages.

If T behaves semi-honestly, S learns no information from seeing $v_{i \oplus b}$, since this value could have been generated with either of the k_0, k_1 since F is a permutation. Other than the above attack, S and T are limited to oblivious input substitution.

Security against malicious R's attempts to manipulate his queries follows from the use of semantically secure scheme in Step 2d. Now even if S sends back multiple encryptions under the same key (e.g., when OT is composed), R will not be able to relate them without knowing the key. As before, T keeps no state, and thus resetting T does not help.

Theorem 2. *Assume that F is a SPRPG. Then Construction 2 evaluates the String OT function securely against malicious receiver R and covert sender S with deterrence factor $\epsilon = 1/2$. The security against S is in the strongest covert setting (strong explicit cheat formulation).*

Proof (Sketch). We treat each corruption case separately. We give detailed description of the simulator of the protocol of Construction 2 against covert adversaries.

The OT Sender S is corrupted. We construct an ideal-model simulator Sim_S of the view of S that works with a trusted party computing the OT functionality. Let $\mathcal{A}_S$ and $\mathcal{A}_T$ be (arbitrary polytime) adversaries controlling S and T respectively. (We consider $\mathcal{A}_S$ and $\mathcal{A}_T$ in full generality. In particular, in contrast with, e.g., [18], we do not assume that $\mathcal{A}_S$ initializes $\mathcal{A}_T$, and thus do not allow Sim_S to intercept the corresponding initialization message sent by $\mathcal{A}_S$. Further, $\mathcal{A}_S$ may not even know the code of $\mathcal{A}_T$.) In its execution, Sim_S interacts with $\mathcal{A}_S$ and $\mathcal{A}_T$ in a black-box manner. We model the physical separation of S and T

by the fact that $\mathcal{A}_S$ and $\mathcal{A}_T$ cannot exchange messages with each other once protocol execution begins.

1. Sim_S will determine whether $\mathcal{A}_T$ wants to cheat. For this, Sim_S first obtains both opening keys from $\mathcal{A}_S$, by rewinding $\mathcal{A}_S$ and setting up different test domains. More specifically:
   (a) Sim_S chooses two random strings k'_D and k''_D, which define two testing domains $D'_t, D''_t \subset D$. Sim_S chooses $b' \in_R \{0,1\}, x', y', x'_t$, $y'_t \in_R D$, such that $y' \in D''_t \setminus D'_t$ and $y'_t \in D'_t \setminus D''_t$.
   (b) Sim_S gives k''_D, and then y' to $\mathcal{A}_S$, and receives two keys from $\mathcal{A}_S$.
   (c) Sim_S rewinds $\mathcal{A}_S$, gives him k'_D, then y'_t, and receives another two keys from $\mathcal{A}_S$.
2. Sim_S calls $\mathcal{A}_T$ with queries $(v'_0, v'_1) = Q(y', x'), (c'_0, c'_1) = Q(y'_t, x'_t)$, in random order. (Recall, Sim_S has obtained from $\mathcal{A}_S$ keys to verify both of $\mathcal{A}_T$'s responses.)
   (a) If neither of $\mathcal{A}_T$'s two responses are valid (where by validity we mean a response that would not cause R to output *corrupted*$_S$), Sim_S sends *corrupted*$_S$ to the trusted party, simulates the honest R aborting due to detected cheating, and outputs whatever $\mathcal{A}_S$ outputs. Since so far Sim_S only sent y'_t to $\mathcal{A}_S$, it is easy to see that the simulation is statistical.
   (b) If $\mathcal{A}_T$ sends exactly one valid response, Sim_S sends *cheat*$_S$ to the trusted party.
      i. If the trusted party replies with *corrupted*$_S$, then Sim_S rewinds $\mathcal{A}_S$, and hands it the query for which $\mathcal{A}_T$'s invalid response could legally be the test query. Sim_S then simulates honest R aborting due to detecting cheating, and outputs whatever $\mathcal{A}_S$ outputs.
      ii. If the trusted party replies with *undetected*$_S$ and the honest R's input i, then Sim_S rewinds $\mathcal{A}_S$, and hands it the query for which $\mathcal{A}_T$'s valid response could legally be the test query. In the remainder of execution, Sim_S plays honest R with input i. At the conclusion, Sim_S outputs whatever $\mathcal{A}_S$ outputs.
   (c) If both of $\mathcal{A}_S$' responses are valid, then we know that $\mathcal{A}_T$ is not passing any information to S. Then Sim_S proceeds with the simulation by playing honest R on a randomly chosen input $i \in_R \{0,1\}$. That is, Sim_S sends $b', y', v'_{i \oplus b'}$ to $\mathcal{A}_S$ (recall, $y' \notin D'_t$, so this message looks proper to $\mathcal{A}_S$). Once Sim_S receives the final message from $\mathcal{A}_S$, using the extracted keys, Sim_S correctly recovers both of $\mathcal{A}_S$'s inputs, and sends them to the trusted party. Sim_S outputs whatever $\mathcal{A}_S$ outputs. This simulation is also statistical.

This completes the description of Sim_S. It is not hard to see that Sim_S simulates the ideal view of covert S with the (ideal model) deterrent factor $\epsilon = 1/2$. In particular, ϵ is equal to $1/2$, since the probability of the honest R catching cheating in the real execution is $1/2$. The simulation is computational since, especially in a batched execution, an unbounded T can compute extra information on the test domain, and have better than 1/2 guess of which one is the test query.

The OT Receiver R is corrupted. We construct an ideal-model simulator Sim_R that works with a trusted party computing the OT functionality. Let $\mathcal{A}_R$ be the adversary controlling R. In its execution, Sim_R plays honest S and T in their interaction with $\mathcal{A}_R$ in a black-box manner. Sim_R does not need to rewind $\mathcal{A}_R$, since $\mathcal{A}_R$'s input can be extracted from the messages sent to S and T (in real execution, S and T don't communicate, so the privacy of R's input is preserved).

1. Sim_R runs $\mathcal{A}_R$ and acts as honest S and T until the computation of e_0, e_1 in Step 2d of Construction 2. (Sim_R honestly answers as many queries to T as $\mathcal{A}_R$ requests.) At this point, Sim_R needs to provide the right answer to $\mathcal{A}_R$ in its message of Step 2d.
2. If v' output by $\mathcal{A}_R$ in simulation of Step 2c was computed as v'_i by Sim_R while answering a query $Q(y', x')$ to T, then Sim_R can compute the input i used by $\mathcal{A}_R$. Sim_R sets $i = i' \oplus b'$ (where b' was sent by $\mathcal{A}_R$ in Step 2c), calls the trusted party with input i, and receives output s. Sim_R randomly chooses encryption key r, computes random encryptions $e_{i \oplus b'} = Enc_{x'}(s)$, $e_{i \oplus b' \oplus 1} = Enc_r(0)$ and sends e_0, e_1 to $\mathcal{A}_R$. Sim_R outputs whatever $\mathcal{A}_R$ outputs.
3. If v' was never computed by Sim_R in his simulation of T, $\mathcal{A}_R$ w.h.p. cannot know its preimage. Sim_R chooses two random encryptions of 0 under random keys, sends to $\mathcal{A}_R$, and outputs whatever $\mathcal{A}_R$ outputs. □

4.2 Discussion: Protocol Composition and Practical Considerations

In this section, we discuss the issues that arise in OT protocol composition and its use in SFE.

Reuse of the token. We have argued that in many settings (especially in established client-server commercial relationships, such as TV and banking) it is realistic for a player (e.g., a service provider) to provide a token to the other party. At the same time, the same token must be sent only once, and then reused for multiple invocation of the protocol, for it to be cost-effective. In case of SFE, "multiple invocation" usually means sequential execution of simultaneous executions of OT. This is easy to achieve with our protocol, as discussed below.

Simultaneous OT. As in [3], we define simultaneous OT functionality naturally as

$$((s_0^1, s_1^1), .., (s_0^m, s_1^m), (i_1, .., i_m)) \mapsto (\perp, (s_{i_1}, .., s_{i_m})) \quad (1)$$

It is easy to see that natural self-composition of Construction 2 works. Further, efficiency may be improved, as compared to independent parallel execution, by choosing y and y_t once for the entire simultaneous OT execution (This saves about a half of computation by both S and T since $F_{k_i}(y)$ and $F_{k_i}(y_t)$ can be precomputed and reused.) The resulting protocol is a secure implementation of simultaneous OT secure against covert adversaries. The proof is almost identical to the proof of Theorem 2.

Sequential composition. Even though the token is resettable, it may still maintain state across executions. That means that if the token successfully cheats, it may be able to help S compute R's input in a *previous* OT execution. Furthermore, since T in this case is able to transfer a whole element of D to S, several bit inputs of R may be compromised. In Section 4.2, we argue that the compromise is limited. However, to simplify the claims and arguments, we "give up everything" in case of successful cheat, and allow the adversary to learn the entire previous input of R. The following discussion of sequential composition and the reuse of the token are with respect to this ideal model.

Even with the above simplification, it is not immediately clear how to prove sequential composition, since the simulator needs to rewind all the way to the initialization phase to obtain the second opening key. Therefore, an easy way to achieve sequential composition is to amend the protocol to run the test domain selection for each (simultaneous) OT execution. Clearly, this would create a problem if the keys k_0, k_1 are reused and R chooses a different test domain. To prevent this, we instead derive k_0, k_1 for each execution. This can be done, e.g. by using $k_0^j = F_{k_0}(j), k_1^j = F_{k_1}(j)$ in the j-th batch in place of k_0, k_1. Of course, S would need to keep track of the counter, and T would then take an additional input j. It is easy to see that sequential composition holds in this case, since the simulator now only needs to rewind to the beginning of the current execution. It is also clear that T holds no state, and as such is still resettable.

Practical implications and considerations. Clearly, the ability of the token to leak a few bits of information (at the risk of being caught) on a previous execution is undesirable. This means, for example, that S and R cannot have both "high-cheat-consequence" and "low-cheat-consequence" transactions, since T could conceivably help leak high-consequence inputs in low-consequence transactions.

Still, we believe that our protocols are applicable in the majority of practical situations where the covert model is applicable, namely, where the value of the privacy of the inputs is not worth the risk associated with being caught. This is the case especially often in the settings we consider, i.e. where it is cost-effective for one party to provide a token to the other. Usually, this is a Client-Server setting, with a long-term association of parties. Examples include Client's relationships with Banks, Internet, TV, phone and cellular service providers. In these settings, there is need in protecting client's input such as browsing history, TV channel preferences, list of movies watched, and in general, the profile of user's habits and preferences. While Service Providers are interested in obtaining this information, the potential loss of business far outweighs the benefit.

Amount of the leak. Even though, for the ease of analysis, we generously "gave up" R's entire input in case of a successful cheat, this is clearly not the case in real execution. The amount of data transferred from T to S in this case is roughly the security parameter, say 128 bits. Recall, T never learns R's input, so it can only help S by transferring (compressed) parts of the history of queries

T saw. Observe that T must compress random values provided by R, since this is what the query history consists of from the point of view of S. Therefore, S must spend at least "a few" bits to reasonably let S match the information and recover a bit of R's input with some degree of confidence. Therefore, with all on the attacker's side, S can potentially recover ≈ 20 input bits of R per each cheating attempt.

Leak reduction. Our protocol allows querying T out-of-synch with protocol executions. Further, T can be queried by R arbitrarily, with only a small fraction of these queries being used for the actual protocol execution. Thus, essentially for free, R can overwhelm T with queries (which are not seen by S and may not be even tested by R), so that T is likely to send information on unused queries, and thus effectively would be able to transfer little beyond a single bit.

Precomputation and OT extension. It is easy to see that much of the work of R can be precomputed. Further, as shown by Beaver [4], almost all of OT computation can be shifted to a setup phase. Additionally, the efficient technique of Ishai at al. [19] can be used to implement an arbitrary number of OT's, given a small (security parameter) number of OT's[2].

Deterrence factor. If we allow R the option to request additional test keys, this effectively increases ϵ, even if the option not used, since R can test-query T and delay the verification request.

4.3 Concurrent Composition with the Semi-honest Token

We observe that a realistic change in the setup assumptions adds security in concurrent executions. This stronger security property holds if both parties trust that the token executes the prescribed code. This could be the case, e.g. if the token is supplied by a third party, and S is limited to loading the keys k_0, k_1 onto the token. This is a reasonable setting in practice, and this assumption was used, e.g., by Hazay and Lindell [18], to efficiently achieve concurrent composition.

The simpler Construction 1, modified as discussed in Observation 1, is secure against malicious S and R. Recall, the modification simply requires using a semantically secure scheme, instead of direct application of SPRPG, to encrypt messages e_0, e_1 in Step 2b of Construction 1.

Theorem 3. *Assume that F is a SPRPG. Then modified Construction 1, as described above, evaluates the String OT function securely against malicious receiver R and sender S. The security against S is statistical.*

We explicate this construction and prove Theorem 3 in the full version.

Further, security holds for concurrent composition. Intuitively, this is because the simulator will not need to rewind. Indeed, the simulator of the covert case rewinds twice – to extract the keys loaded into token T, and in relation with

[2] [19] requires the use of correlation-robust functions (a weak form of random oracle). In practice, even the RO assumption seems to be much more solid than that of tamper-resistance. Thus, its use will not have effect on security.

testing T. Now, however, the token is modeled as a separate entity from S, and S must explicitly output the keys loaded into T. Then the need for the first rewinding disappears, since Sim_S simply receives the keys from S as the first step of the simulation. Since T is semi-honest, we do not test it, and thus there is no need for second rewinding. Then, as shown in [22], this is sufficient to achieve concurrent general composition (equivalently, universal composability). In [22], this is only shown for protocols that have additional property of *initial synchronization*. Informally, a protocol is said to have initial synchronization if all parties announce that they are ready to start before they actually start. As pointed out in [18], this property always holds for two-party protocols.

Acknowledgements. I would like to thank the anonymous referees of this paper for their valuable comments.

References

1. http://www.scdeveloper.com/cards.htm (retrieved on August 28, 2009)
2. Aiello, W., Ishai, Y., Reingold, O.: Priced oblivious transfer: How to sell digital goods. In: Pfitzmann, B. (ed.) EUROCRYPT 2001. LNCS, vol. 2045, pp. 119–135. Springer, Heidelberg (2001)
3. Aumann, Y., Lindell, Y.: Security against covert adversaries: Efficient protocols for realistic adversaries. In: Vadhan, S.P. (ed.) TCC 2007. LNCS, vol. 4392, pp. 137–156. Springer, Heidelberg (2007)
4. Beaver, D.: Precomputing oblivious transfer. In: Coppersmith, D. (ed.) CRYPTO 1995. LNCS, vol. 963, pp. 97–109. Springer, Heidelberg (1995)
5. Bellare, M., Micali, S.: Non-interactive oblivious transfer and applications. In: Brassard, G. (ed.) CRYPTO 1989. LNCS, vol. 435, pp. 547–557. Springer, Heidelberg (1990)
6. Bellare, M., Rogaway, P.: Random oracles are practical: A paradigm for designing efficient protocols. In: ACM CCS, pp. 62–73 (1993)
7. Chandran, N., Goyal, V., Sahai, A.: New constructions for uc secure computation using tamper-proof hardware. In: Smart, N.P. (ed.) EUROCRYPT 2008. LNCS, vol. 4965, pp. 545–562. Springer, Heidelberg (2008)
8. Di Crescenzo, G.: Private selective payment protocols. In: Frankel, Y. (ed.) FC 2000. LNCS, vol. 1962, pp. 72–89. Springer, Heidelberg (2001)
9. Damgård, I., Nielsen, J.B., Wichs, D.: Isolated proofs of knowledge and isolated zero knowledge. In: Smart, N.P. (ed.) EUROCRYPT 2008. LNCS, vol. 4965, pp. 509–526. Springer, Heidelberg (2008)
10. Damgård, I., Nielsen, J.B., Wichs, D.: Universally composable multiparty computation with partially isolated parties. In: Reingold, O. (ed.) TCC 2009. LNCS, vol. 5444, pp. 315–331. Springer, Heidelberg (2009)
11. Even, S., Goldreich, O., Lempel, A.: A randomized protocol for signing contracts. Commun. ACM 28(6), 637–647 (1985)
12. Fischlin, M.: A cost-effective pay-per-multiplication comparison method for millionaires. In: Naccache, D. (ed.) CT RSA 2001. LNCS, vol. 2020, pp. 457–471. Springer, Heidelberg (2001)
13. Gennaro, R., Lysyanskaya, A., Malkin, T., Micali, S., Rabin, T.: Algorithmic tamper-proof (ATP) security: Theoretical foundations for security against hardware tampering. In: Naor, M. (ed.) TCC 2004. LNCS, vol. 2951, pp. 258–277. Springer, Heidelberg (2004)

14. Goldwasser, S., Kalai, Y.T., Rothblum, G.N.: One-time programs. In: Wagner, D. (ed.) CRYPTO 2008. LNCS, vol. 5157, pp. 39–56. Springer, Heidelberg (2008)
15. Goyal, V., Ishai, Y., Sahai, A., Venkatesan, R., Wadia, A.: Founding cryptography on tamper-proof hardware tokens. In: Micciancio, D. (ed.) TCC 2010. LNCS, vol. 5978, pp. 308–326. Springer, Heidelberg (2010)
16. Goyal, V., Mohassel, P., Smith, A.: Efficient two party and multi party computation against covert adversaries. In: Smart, N.P. (ed.) EUROCRYPT 2008. LNCS, vol. 4965, pp. 289–306. Springer, Heidelberg (2008)
17. Goyal, V., Sahai, A.: Resettably secure computation. In: Joux, A. (ed.) EUROCRYPT 2009. LNCS, vol. 5479, pp. 54–71. Springer, Heidelberg (2009)
18. Hazay, C., Lindell, Y.: Constructions of truly practical secure protocols using standard smartcards. In: ACM Conference on Computer and Communications Security, pp. 491–500 (2008)
19. Ishai, Y., Kilian, J., Nissim, K., Petrank, E.: Extending oblivious transfers efficiently. In: Boneh, D. (ed.) CRYPTO 2003. LNCS, vol. 2729, pp. 145–161. Springer, Heidelberg (2003)
20. Järvinen, K., Kolesnikov, V., Sadeghi, A.-R., Schneider, T.: Embedded SFE: Offloading server and network using hardware tokens. In: FC 2010 (2010)
21. Katz, J.: Universally composable multi-party computation using tamper-proof hardware. In: Naor, M. (ed.) EUROCRYPT 2007. LNCS, vol. 4515, pp. 115–128. Springer, Heidelberg (2007)
22. Kushilevitz, E., Lindell, Y., Rabin, T.: Information-theoretically secure protocols and security under composition. In: Proc. 38th ACM Symp. on Theory of Computing, pp. 109–118 (2006)
23. Lindell, Y., Pinkas, B.: A proof of Yao's protocol for secure two-party computation. Cryptology ePrint Archive, Report 2004/175 (2004), `http://eprint.iacr.org/`
24. Lindell, Y., Pinkas, B.: An efficient protocol for secure two-party computation in the presence of malicious adversaries. In: Naor, M. (ed.) EUROCRYPT 2007. LNCS, vol. 4515, pp. 52–78. Springer, Heidelberg (2007)
25. Luby, M., Rackoff, C.: How to construct pseudorandom permutations from pseudorandom functions. SIAM J. Comput. 17(2), 373–386 (1988)
26. Naor, M., Pinkas, B.: Efficient oblivious transfer protocols. In: SODA 2001: Proceedings of the twelfth annual ACM-SIAM symposium on Discrete algorithms, Philadelphia, PA, USA. Society for Industrial and Applied Mathematics (2001)
27. Naor, M., Pinkas, B., Sumner, R.: Privacy preserving auctions and mechanism design. In: 1st ACM Conf. on Electronic Commerce, pp. 129–139 (1999)
28. Yao, A.C.: Protocols for secure computations. In: Proc. 23rd IEEE Symp. on Foundations of Comp. Science, Chicago, pp. 160–164. IEEE, Los Alamitos (1982)
29. Yao, A.C.: How to generate and exchange secrets. In: Proc. 27th IEEE Symp. on Foundations of Comp. Science, Toronto, pp. 162–167. IEEE, Los Alamitos (1986)

Leakage-Resilient Signatures

Sebastian Faust[1,⋆], Eike Kiltz[2,⋆⋆], Krzysztof Pietrzak[2], and Guy N. Rothblum[3]

[1] K.U. Leuven ESAT-COSIC/IBBT, Belgium
[2] CWI, Amsterdam, The Netherlands
[3] MIT, Boston, USA

Abstract. The strongest standard security notion for digital signature schemes is unforgeability under chosen message attacks. In practice, however, this notion can be insufficient due to "side-channel attacks" which exploit leakage of information about the secret internal state. In this work we put forward the notion of "leakage-resilient signatures," which strengthens the standard security notion by giving the adversary the additional power to learn a bounded amount of *arbitrary information* about the secret state that was accessed during *every signature generation*. This notion naturally implies security against all side-channel attacks as long as the amount of information leaked on each invocation is bounded and "only computation leaks information."

The main result of this paper is a construction which gives a (tree-based, stateful) leakage-resilient signature scheme based on any 3-time signature scheme. The amount of information that our scheme can safely leak *per signature generation* is 1/3 of the information the underlying 3-time signature scheme can leak *in total*. Signature schemes that remain secure even if a bounded total amount of information is leaked were recently constructed, hence instantiating our construction with these schemes gives the first constructions of provably secure leakage-resilient signature schemes.

The above construction assumes that the signing algorithm can sample truly random bits, and thus an implementation would need some special hardware (randomness gates). Simply generating this randomness using a leakage-resilient stream-cipher will in general not work. Our second contribution is a sound general principle to replace uniform random bits in any leakage-resilient construction with pseudorandom ones: run two leakage-resilient stream-ciphers (with independent keys) in parallel and then apply a two-source extractor to their outputs.

1 Introduction

Traditionally, provable security treats cryptographic algorithms as black-boxes. An adversary may have access to inputs and outputs, but the computation within

⋆ Supported in part by Microsoft Research through its PhD Scholarship Programme and FWO grant G.0225.07. This work has partly be done while visiting CWI.

⋆⋆ Supported by the research program Sentinels (http://www.sentinels.nl). Sentinels is being financed by Technology Foundation STW, the Netherlands Organization for Scientific Research (NWO), and the Dutch Ministry of Economic Affairs.

D. Micciancio (Ed.): TCC 2010, LNCS 5978, pp. 343–360, 2010.

the box stays secret. In particular, the standard security notion of digital signatures is existential unforgeability under chosen message attacks [16] (UF-CMA), where one requires that an adversary cannot forge a valid signature even when given access to a signing oracle.

Unfortunately, this traditional security model often does not match reality where an adversary can attack the algorithm's *implementation* with more powerful attacks. An important example in this context are side-channel attacks, which provide an adversary with a partial view on the inner secret state (e.g., a secret signing key) of an algorithm's execution due to physical leakage during computation. In the last two decades a vast number of ingenious side-channel attacks have been invented and used to break implementations of schemes which were provably secure in the traditional model. Examples of side-channels include information derived from running-time [20], electromagnetic radiation [30,14], power consumption [21], and many more (see, e.g., [31,27]).

1.1 Leakage-Resilient Cryptography

Classical research in side-channel attacks sometimes resembles a cat-and-mouse game. New side-channel attacks are discovered, and then heuristic countermeasures are proposed to prevent the specific new attack. This yields countermeasures that are tailored specifically for the class of attacks they intend to defeat. Not very surprisingly, these countermeasures are often later found to be vulnerable to new attacks. This state of affairs is fundamentally different from the design principles of "modern cryptography," where one usually requires that the system is secure against all adversaries from some well defined resource bounded class[1] and for a broad and well-defined attack scenario. (E.g., existential unforgeability for signature schemes or IND-CCA2 security for encryption.)

As this situation is clearly not very satisfying, in an influential paper Micali and Reyzin [24] suggest a framework for adapting the methodology of modern cryptography to the scenario of side-channel attacks.

A FORMAL SECURITY DEFINITION. Inspired by the framework of Micali-Reyzin and Maurer's bounded storage model (and the subsequent bounded-retrieval model), in [10] the notion of *leakage-resilience* was proposed.[2] A cryptographic primitive (or protocol) is said to be leakage-resilient, if it is secure in the traditional (black-box) sense but now the adversary may additionally obtain *arbitrary side-channel information* (also called *leakage*) during the execution of the

[1] In complexity based cryptography one usually bounds the running time. Other bounds that often are used include the size of the memory an adversary can use or the number of queries the adversary can make to some oracle.

[2] The primary contribution of [10] was not proposing a new model, their model combined ideas that were explicit and implicit in prior work. Rather, the primary contribution was actually constructing a primitive (a stream-cipher) and proving it secure in this model.

security experiment. The side-channel information given to the adversary only has to satisfy the following two restrictions

LR1 (bounded leakage): the *amount* of leakage *in each invocation* is bounded (but overall can be arbitrary large).

LR2 (only computation leaks information): the internal state that is not accessed during an invocation ("passive state") does not leak.

At a technical level this is modeled by considering adversaries that, when attacking the primitive, additionally to the regular input specify a *leakage function* f with bounded range $\{0,1\}^\lambda$ and then (besides the regular output) also obtain $\Lambda = f(s^+, r)$, where s^+ denotes the part of the internal secret state that has been accessed during this invocation ("active state") and r are the internal coin tosses that were made by the cryptosystem during the invocation.

MOTIVATION OF THE LEAKAGE RESTRICTIONS. It is clear that one has to restrict the class of leakage functions, as if we would allow the identity function $f(s) = s$ (where s is the cryptographic algorithm's internal state) , no security whatsoever can be achieved.

In this work we focus on leakage functions that are restricted in terms of their output length. This is a natural resource bound, and allows to model a rich class of side-channel attacks (e.g. timing or hamming-weight attacks, which exploit only a polylogarithmic amount of information on each invocation. This is much smaller than the constant-fraction leakage for which we can still prove security in this work.) We remark that we could use a more relaxed restriction than a bound on the leakage function,[3] but we will stick to *bounded leakage* (LR1) which is more intuitive and simpler to work with.

Bounded leakage alone might not be a *sufficiently* strong restriction, and we use a further restriction on the leakage function, which still seems to allow a rich and very natural family of side-channel attacks. Following [10], we use LR2 ("only computation leaks information"), originally put forward as one of the axioms of "physically observable cryptography" by Micali and Reyzin [24]. The original axiom requires that if a primitive with secret internal state s is invoked, then on this particular invocation, only the part $s^+ \subseteq s$ of the memory leaks that was accessed during this invocation.

Let us stress that the restrictions LR1 and LR2 on the leakage are sufficient, but not necessary to imply security of a leakage-resilient primitive. For example, in a so called "cold-boot attack" [17], the adversary learns (at some point in time) a random subset of the bits of the *entire* secret state, even when no computation is going on. Such an attack clearly does not satisfy the "only computation leaks information" restriction, but nonetheless, leakage-resilience does imply security against cold-boot attacks for any primitive which satisfies some

[3] In particular, we can consider the class $\mathcal{F}$ of leakage functions such that the degradation of the HILL-pseudoentropy of the internal state S due to leakage of $f(S)$ (where $f \in \mathcal{F}$) is sufficiently bounded.

natural additional properties on how the memory is accessed. In particular, these properties are satisfied by our construction and [10,29].[4]

1.2 Leakage-Resilient Signatures

Previous work has shown how to build stream-ciphers that are provably resistant to continual leakage in the standard model [10,29]. In this paper we construct a leakage-resilient public-key primitive in the plain model, a signature-scheme.

Digital signatures are a central cryptographic primitive and are widely implemented on computational devices that are especially vulnerable to side-channel attacks (such as smart cards). Starting with the seminal work by Kocher [20], there have been a great number of theoretical and practical side-channel attacks on signature schemes (e.g., [20,21,32,13]).

SECURITY DEFINITION. The standard notion for secure signatures schemes is unforgeability under adaptive chosen-message attacks [16]. Here one requires that an adversary cannot forge a signature of any message m, even when given access to a signing oracle. We strengthen this notion by giving the adversary access to a more powerful oracle, which not only outputs signatures for chosen messages, but as an additional input takes a leakage function $f : \{0,1\}^* \to \{0,1\}^\lambda$ and outputs $f(s^+, r)$ where s^+ is the state that has been accessed during computation of the signature and (if the scheme is probabilistic) r is the randomness that was sampled. Note that if we want the signature scheme to sign a large number of messages (i.e., more than the state length), then this security definition inherently requires the signature scheme to update its internal state. We call signature schemes which are secure in the above sense UF-CMLA (unforgeable under chosen message/leakage attacks) or simply leakage resilient. We also define a notion called UF-CMTLA (unforgeability under chosen message *total* leakage attacks), which is defined similarly to UF-CMLA but is significantly weaker, as here the total amount of leakage (and not the leakage per invocation) is bounded.

OVERVIEW OF OUR CONSTRUCTION. Our construction of leakage resilient signature schemes is done in two steps. First, we give a number of instantiations of 3-time UF-CMTLA signature schemes offering different trade-offs. Then, we present a generic tree-based transformation from any UF-CMTLA secure 3-time signature scheme (i.e., a signature scheme that can securely sign up to 3 messages) to a UF-CMLA signature scheme.

FROM UF-CMTLA TO UF-CMLA SECURITY. Following the construction of Naor and Yung [26] and the ideas of Lamport [22] and Merkle [23], we propose a simple tree-based leakage-resilient signature scheme SIG^* that is constructed from any leakage resilient 3-time signature scheme SIG. The scheme we propose

[4] One requirement is that ultimately the entire initial secret state is touched. (The only setting we are aware of where the entire state will *not* be touched, are the the tokens used in the construction of one-time programs [15].) The second requirement is that the memory access is "oblivious", in the sense that which parts of the memory are accessed does not depend on the secret state.

strongly resembles the construction of a forward-secure signature scheme [3] from [4], but let us stress that leakage-resilience and forward-security are orthogonal concepts. In particular, our construction is *not* forward-secure, but could be made so in a straight forward way, at the cost of having a more complicated description.

For any a-priori fixed $d \in \mathbb{N}$, our construction can sign up to $2^{d+1}-2$ messages and one can think of the (stateful) signing algorithm as traversing the $2^{d+1}-1$ nodes of a binary tree of depth d in a depth-first manner. Suppose the signing algorithm of SIG^* wants to sign the i-th message m and its state points to the i-th node $\tilde{w}$ in a depth-first traversal of the tree. It first computes a fresh public/secret-key pair $(pk_{\tilde{w}}, sk_{\tilde{w}})$ of SIG for this node. Next, the signature (σ, Γ) for m is computed, where σ is a signature on m according to the 3-time signature scheme SIG using the secret key $sk_{\tilde{w}}$ of the current node $\tilde{w}$, and Γ contains a signature path from the root of the tree to the node $\tilde{w}$: for each node w on the path it contains a signature on pk_w using the secret key $sk_{\mathsf{par}(w)}$, where $\mathsf{par}(w)$ denotes the parent of w in the tree. The public-key of SIG^* is the public-key associated to the root node and verification of a signature of SIG^* is done by verifying all the 3-time signatures on the path from $\tilde{w}$ to the root.

The crucial observation that will allow us to prove leakage-resilience of our construction, is that for each node w in the tree, the secret key sk_w associated to this node is only accessed a constant number of times (at most three times). The security we prove roughly states that if SIG is a UF-CMTLA secure 3-time signature scheme which is secure even after leaking a total of λ_{total} bits, then SIG^* is a UF-CMLA secure signature scheme that can tolerate $\lambda = \lambda_{\mathsf{total}}/3$ bits of leakage per signature query. The loss in security is a factor of q.

INSTANTIATIONS UF-CMTLA SECURE 3-TIME SIGNATURE SCHEMES. It is not hard to see that every signature scheme looses at most an exponential factor $2^{\lambda_{\mathsf{total}}}$ in security (compared to the standard UF-CMA security) when λ_{total} bits about the secret key are leaked (as the UF-CMA adversary can simply guess the leakage, and a random guess will be correct with probability $2^{-\lambda_{\mathsf{total}}}$). Recently, much better constructions have been proposed. Alwen, Dodis, and Wichs [2] show that the Okamoto-Schnorr signature-scheme [28,33] remains secure even if almost $n/2$ bits (where n is the length of the secret key) of information about the secret-key are leaked. Instantiating our construction with Okamoto-Schnorr signatures thus gives a leakage resilient signature scheme which can leak a constant fraction (almost 1/6) of the accessed state on each invocation. Due to the Fiat-Shamir heuristic used in the Okamoto-Schnorr signature scheme, this scheme can only be proven secure in the random-oracle model. Recently, Katz and Vaikuntanathan [19] showed how to construct signature schemes in the standard model (and under standard assumptions) which can tolerate leakage of as much as $\lambda_{\mathsf{total}} = n - n^{\epsilon}$ bits ($\epsilon > 0$). With this construction we get a leakage resilient signature scheme in the standard model. Unfortunately it is not practical due to the use of general NIZK proofs.

In the same paper [19], Katz et al. also construct an efficient *one-time* signature scheme that tolerates leakage of $\lambda_{\mathsf{total}} = (1/4 - \epsilon)n$ bits (for any $\epsilon > 0$).

This scheme is easily generalized to a (stateful) 3-time signature schemes where one can leak $\lambda_{\text{total}} = (1/12 - \epsilon)n$ bits.[5] This construction fits well into our general transformation, yielding a UF-CMLA secure scheme where one can leak $\lambda_{\text{total}} = (1/36 - \epsilon)n$ bits (here n is the size of the accessed state on each invocation). As the construction only assumes universal one-way hash functions (UOWHF), we get that it is secure in the standard model under the minimal [26] assumption that one-way functions exist.

1.3 Replacing Randomness in Leakage-Resilient Primitives

In the construction of SIG^* we silently assumed that the device could sample uniformly random bits to be used in the key-generation and signing steps of the underlying scheme SIG. This , however, would require special hardware for generating random bits (such as noise generating gates). In the non-leakage setting one can avoid the necessity for such special hardware by using pseudorandomness (generated by a stream-cipher) instead of truly random bits.

Unfortunately, in the leakage-setting the simple analogous idea of replacing the random bits with the output of a *leakage-resilient* stream-cipher (as defined in [10]) does not work (at least we do not know how to prove it). The reason is that an output block of a leakage-resilient stream-cipher is only guaranteed to have high HILL-pseudoentropy *when given the leakage* that was generated while computing this block.

A sound approach to replace uniform random bits in *any* leakage-resilient construction while provably preserving leakage-resilience is as follows: run two leakage-resilient stream ciphers with independent keys in parallel and feed their output to a two-source extractor. For lack of space, this can be found in the full version [11]. Intuitively, the reason is that now the outputs X, X' of the two stream ciphers are indistinguishable from having high min-entropy (given the leakage), and thus applying a two source extractor ext gives a (indistinguishable from) uniform $Y = \mathsf{ext}(X, X')$ which then can be used in the signature scheme.

While we do not know how to prove in general the security of the simpler approach of using a single leakage-resilient stream cipher to generate the random bits, in some special cases this simpler approach does go through. For example:

- If the scheme (for which we want to replace the uniform random bits) already can be proven leakage-resilient assuming only that the random bits have high min-entropy (as opposed to being uniform), this is e.g. the case for the (generalized) Okamoto signature scheme from [2].
- The output of the particular leakage-resilient stream-cipher from [10] can always be used directly. Informally, the reason is that here (unlike e.g. in [29]) the final output already was generated by applying an extractor.

[5] They propose a general transformation to t-time schemes using cover free sets which can leak $\lambda_{\text{total}} = \Omega(n/t^2)$ bits (which for $t = 3$ is $\Omega(n)$). We note however, that (while this leakage bound is worse than ours) their scheme enjoys the advantage of being stateless, whereas ours is stateful.

1.4 Related Work

A body of prior work has considered countermeasures against different classes of side-channel attacks. Most works consider security against some particular attack, like "template attacks" [34]. Below we mention some work on "provable security" in the context of side-channel attacks, where only the class of leakage functions is restricted, but not the adversaries ability to exploit the leakage.

Ishai *et al.* [18] show how to securely implement *any* (efficiently computable) function even when the attacker can probe a bounded number of wires in the implementation. This result has been recently extendend [12] to allow leakage functions that get as input the values carried by all the wires in the circuit, as long as the output of the leakage functions is short, and the leakage functions is from some low complexity class like AC_0.

Micali and Reyzin [24] proposed the influential theoretical framework of "physically observable cryptography" to model side-channel attacks. In particular, they explicitly state and motivate the "only computation leaks information" axiom used in leakage-resilient cryptography [10,29].

Several recent works [1,25,19,7] propose (stateless) constructions which are secure against so called "memory attacks". This means that they remain secure even after a bounded total amount of information has leaked (this is sufficient against attacks like "cold-boot" attacks [17], but not for most other side-channel attacks which leak on each invocation). Unlike in leakage-resilient cryptography, here the leakage functions need not obey the only "computation leaks information" restriction. Akavia et. al [1] and Naor and Segev [25] construct symmetric/public-key encryption schemes that are secure in this model. Katz et al [19] and Alwen et al [2] construct digital signatures in this setting (see the discussion above). The "bounded retrieval model" (BRM) [5,8,9,2] is an extension of "memory attacks" where the key is made artificially huge and thus the tolerated leakage can also be made arbitrary large (but still a priory bounded by the key size). The difficulty in this model (as compared to memory attacks), is that in the BRM model the efficiency of a scheme must only depend on some security parameter, but not on the size of the (potentially huge) secret key. Dodis et al. [7,6] consider the case where the range of $f(\cdot)$ is not necessarily bounded, but instead one only requires that it is (exponentially) hard to recover sk from $f(sk)$.

2 Preliminaries

NOTION. If x is a string, then $|x|$ denotes its length, while if S is a set then $|S|$ denotes its size. If $k \in \mathbb{N}$ then 1^k denotes the string of k ones. For $n \in \mathbb{N}$, we write $[n]$ as shorthand for $\{1, \ldots, n\}$. If S is a set then $s \stackrel{\$}{\leftarrow} S$ denotes the operation of picking an element s of S uniformly at random. With PPT we denote probabilistic polynomial time.

ALGORITHMS. We write $y \leftarrow \mathcal{A}(x)$ to indicate that $\mathcal{A}$ is an algorithm which runs on input x and outputs y. If $\mathcal{A}$ is probabilistic, $y \stackrel{\$}{\leftarrow} \mathcal{A}(x)$ denotes running the algorithms using fresh randomness.

To model *stateful algorithms* we will in particular consider algorithms with a special input/output syntax. We split the input into three disjoint syntactic parts: a query x, the state s, and (in case the algorithm is probabilistic) randomness r. Similarly, the output is split into the output y and the new state s'. We write $(y, s') \leftarrow \mathcal{B}(x, s, r)$ to make this explicit. Here one can think of the query x as being chosen (or at least known) to the adversary. The state s and s' is the secret internal state of the primitive before and after execution of the algorithm on input x, respectively.

If we consider the execution $(y, s') \leftarrow \mathcal{B}(x, s, r)$ of an algorithm, we can split the state in two parts $s = s^+ \cup s^-$. The *active state*, s^+, denotes the part that is accessed by $\mathcal{B}$ in order to compute y and update its state.[6] The *passive state*, $s^- = s \setminus s^+$, is the part of the state that is not accessed (i.e., read and/or overwritten) during the current execution. We use the notation

$$(y, s') \overset{s^+}{\hookleftarrow} \mathcal{B}(x, s, r) .$$

to make explicit that s^+ is the active state of the execution of $\mathcal{B}$ with inputs x, s, r. This is illustrated in Figure 1. Note that the passive state s^- is completely contained in s', i.e., state information that is never accessed is contained entirely in the next state s'.

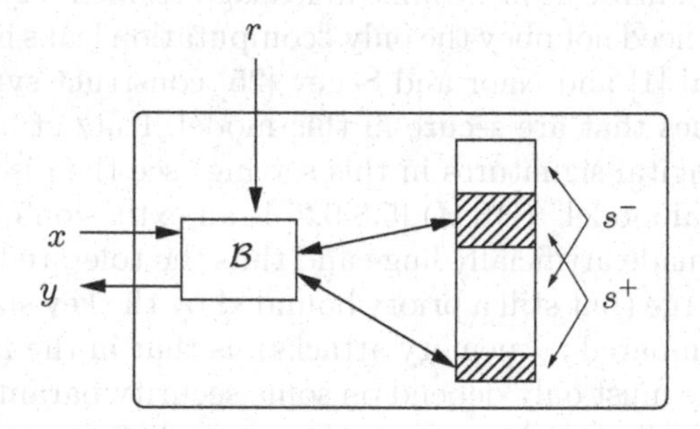

Fig. 1. Illustration of the execution of a stateful algorithm $(y, s') \overset{s^+}{\hookleftarrow} \mathcal{B}(x, s, r)$. The secret state s splits into the active state s^+ (that is accessed during the execution of $\mathcal{B}$) and the passive state s^-.

3 Leakage Resilient Signatures

3.1 Standard Signatures

A (stateful) digital signature scheme $\mathsf{SIG} = (\mathsf{Kg}, \mathsf{Sign}, \mathsf{Vfy})$ consists of three PPT algorithms. The key generation algorithm Kg generates a secret signing key sk and a public verification key pk. The signing algorithm Sign get as input the

[6] For this to be well defined, we really need that $\mathcal{B}$ is given as an algorithm, e.g. in pseudocode, and not just as a function.

signing key sk and a message m and returns a signature and a new state sk' which replaces the old signing key. The deterministic verification algorithm Vfy inputs the verification key and returns 1 (accept) or 0 (reject). We demand the usual correctness properties.

We recall the definition for unforgeability against chosen-message attacks (UF-CMA) for stateful signatures. To an adversary $\mathcal{F}$ and a signature scheme $\mathsf{SIG} = (\mathsf{Kg}, \mathsf{Sign}, \mathsf{Vfy})$ we assign the following experiment.

Experiment $\mathbf{Exp}_{\mathsf{SIG}}^{\text{uf-cma}}(\mathcal{F}, k)$
  $(pk, sk_0) \xleftarrow{\$} \mathsf{Kg}(1^k)$; $i \leftarrow 1$
  $(m^*, \sigma^*) \xleftarrow{\$} \mathcal{F}^{\mathcal{O}_{sk_{i-1}}}(pk)$
  If $\mathsf{Vfy}(pk, m^*, \sigma^*) = 1$ and $m^* \notin \{m_1, \ldots m_i\}$
    then return 1 else return 0.

Oracle $\mathcal{O}_{sk_{i-1}}(m_i)$
  $(\sigma_i, sk_i) \xleftarrow{\$} \mathsf{Sign}(sk_{i-1}, m_i)$
  Return σ_i and
  set $i \leftarrow i + 1$

We remark that for the special case where the signature scheme is stateless (i.e., $sk_{i+1} = sk_i$), we can consider a simpler experiment where the signing oracle $\mathcal{O}_{sk_i}(\cdot)$ is replaced by $\mathsf{Sign}(sk, \cdot)$. With $\mathbf{Adv}_{\mathsf{SIG}}^{\text{uf-cma}}(\mathcal{F}, k)$ we denote the probability that the above experiment returns 1. Forger $\mathcal{F}$ (t, q, ϵ)-breaks the UF-CMA security of SIG if $\mathbf{Adv}_{\mathsf{SIG}}^{\text{uf-cma}}(\mathcal{F}, k) \geq \epsilon$, its running time is bounded by $t = t(k)$, and it makes at most $q = q(k)$ signing queries. We call SIG *UF-CMA secure* (or simply *secure*) if no forger can (t, q, ϵ)-break the UF-CMA security of SIG for polynomial t and q and non-negligible ϵ.

3.2 Leakage Resilient Signatures

We now define the notion of unforgeability against chosen-message/leakage attacks (UF-CMLA) for stateful signatures. This extends the UF-CMA security notion as now the adversary can learn λ bits of leakage with every signature query. With the ith signature query, the adversary can adaptively choose any leakage function f_i (described by a circuit[7]) with range $\{0,1\}^\lambda$ and then learns the output Λ_i of f_i which as input gets everything the signing algorithm gets, that is the active state S_{i-1}^+ and the random coins r_i. To an adversary $\mathcal{F}$ and a signature scheme $\mathsf{SIG} = (\mathsf{Kg}, \mathsf{Sign}, \mathsf{Vfy})$ we assign the following experiment.

Experiment $\mathbf{Exp}_{\mathsf{SIG}}^{\text{uf-cmla}}(\mathcal{F}, k, \lambda)$
  $(PK, SK_0) \xleftarrow{\$} \mathsf{Kg}(1^k)$; $i \leftarrow 1$
  $(m^*, \sigma^*) \xleftarrow{\$} \mathcal{F}^{\mathcal{O}_{SK_{i-1}}}(PK)$
  If $\mathsf{Vfy}(PK, m^*, \sigma^*) = 1$ and
    $m^* \notin \{m_1, \ldots m_i\}$
    then return 1 else return 0.

Oracle $\mathcal{O}_{SK_{i-1}}(m_i, f_i)$
  Sample fresh randomness r_i
  $(\sigma_i, SK_i) \overset{SK_{i-1}^+}{\hookleftarrow} \mathsf{Sign}(SK_{i-1}, m_i, r_i)$
  $\Lambda_i \leftarrow f_i(SK_{i-1}^+, r_i)$
  if $|\Lambda_i| \neq \lambda$ then $\Lambda_i \leftarrow 0^\lambda$
  Return (σ_i, Λ_i) and set $i \leftarrow i + 1$

[7] We could also model the f_i's as Turing machines, but then we would have to require that the output length is independent of the input, as otherwise information could be encoded in the output length itself.

With $\mathbf{Adv}_{\mathsf{SIG}}^{\text{uf-cmla}}(\mathcal{F}, k, \lambda)$ we denote the probability that the above experiment returns 1. Forger $\mathcal{F}$ $(t, q, \epsilon, \lambda)$-breaks the UF-CMLA security of SIG if its running time is bounded by $t = t(k)$, it makes at most $q = q(k)$ signing queries and $\mathbf{Adv}_{\mathsf{SIG}}^{\text{uf-cmla}}(\mathcal{F}, k, \lambda) \geq \epsilon(k)$. We call SIG *UF-CMLA secure with* λ *leakage* (or simply λ*-leakage resilient*) if no forger can $(t, q, \epsilon, \lambda)$-break the UF-CMLA security of SIG for polynomial t and q and non-negligible ϵ.

3.3 Signatures with Bounded Total Leakage

In the previous section we defined signatures that remain secure even if λ bits leak on each invocation. We will construct such signatures using as building block signature schemes that can only sign a constant number (we will need 3) of messages, and are unforgeable assuming that a *total* of λ_{total} bits are leaked (including from the randomness r_0 that was used at key-generation). Following [19], we augment the standard UF-CMA experiment with an oracle $\mathcal{O}_{\text{leak}}$ which the adversary can use to learn up to λ_{total} arbitrary bits about the randomness used in the entire key generation and signing process. This oracle will use a random variable state that contains all the random coins used by the signature scheme so far and a counter λ_{cnt} to keep track how much has already been leaked. Note that we do not explicitly give the leakage functions access to the key sk_i, as those can be efficiently computed given $r_0 \in \mathtt{state}$.

Experiment $\mathbf{Exp}_{\mathsf{SIG}}^{\text{uf-cmtla}}(\mathcal{F}, k, \lambda_{\text{total}})$
$(pk, sk_0) \overset{r_0}{\leftarrow} \mathsf{Kg}(1^k)$; $i \leftarrow 1$; $\lambda_{\text{cnt}} \leftarrow 0$; $\mathsf{state} \leftarrow r_0$
$(m^*, \sigma^*) \overset{\$}{\leftarrow} \mathcal{F}^{\mathcal{O}_{sk_{i-1}}, \mathcal{O}_{\text{leak}}}(pk)$
If $\mathsf{Vfy}(pk, m^*, \sigma^*) = 1$ and $m^* \notin \{m_1, \ldots m_i\}$
then return 1 else return 0.

Oracle $\mathcal{O}_{sk_{i-1}}(m_i)$	**Oracle** $\mathcal{O}_{\text{leak}}(f)$
Sample fresh randomness r_i	$\Lambda \leftarrow f(\mathtt{state})$
$\mathsf{state} \leftarrow \mathsf{state} \cup r_i$	If $\lambda_{\text{cnt}} + \lvert\Lambda\rvert > \lambda_{\text{total}}$ Return $\perp$
$(\sigma_i, sk_i) \leftarrow \mathsf{Sign}(sk_{i-1}, m_i, r_i)$	$\lambda_{\text{cnt}} \leftarrow \lambda_{\text{cnt}} + \lvert\Lambda\rvert$
Return σ_i and set $i \leftarrow i + 1$	Return Λ

With $\mathbf{Adv}_{\mathsf{SIG}}^{\text{uf-cmtla}}(\mathcal{F}, k, \lambda_{\text{total}})$ we denote the probability that the above experiment returns 1. Forger $\mathcal{F}$ $(t, d, \epsilon, \lambda_{\text{total}})$-breaks the UF-CMTLA security of SIG if its running time is bounded by $t = t(k)$, it makes at most $d = d(k)$ signing queries and $\mathbf{Adv}_{\mathsf{SIG}}^{\text{uf-cmtla}}(\mathcal{F}, k, \lambda_{\text{total}}) \geq \epsilon(k)$. We call SIG *UF-CMTLA secure with* λ_{total} *leakage* if no forger can $(t, d, \epsilon, \lambda_{\text{total}})$-break the UF-CMTLA security of SIG for polynomial t and non-negligible ϵ.

4 Construction of Leakage Resilient Signature Schemes

We first discuss three constructions of UF-CMTLA secure 3-time signature schemes. We then prove our main result which shows how to get a leakage-resilient signature scheme from any UF-CMTLA 3-time signatures scheme using a tree based construction.

4.1 Signatures with Bounded Leakage Resilience

GENERIC CONSTRUCTION WITH EXPONENTIAL LOSS. We first present a simple lemma showing that *every* d-time UF-CMA secure signature scheme is also a d-time UF-CMTLA secure signature scheme, where the security loss is exponential in λ_{total}.

Lemma 1. *For any security parameter* k, $t = t(k)$, $\epsilon = \epsilon(k)$, $d = d(k)$, *and* λ_{total}, *if* SIG *is* (t, d, ϵ) *UF-CMA secure, then* SIG *is* $(t', d, 2^{\lambda_{\mathsf{total}}}\epsilon, \lambda_{\mathsf{total}})$ *UF-CMTLA secure where* $t' \approx t$.

Proof. For contradiction, assume there exists an adversary $\mathcal{F}_{\lambda_{\mathsf{total}}}$ who breaks the $(t', d, 2^{\lambda_{\mathsf{total}}}\epsilon, \lambda_{\mathsf{total}})$ UF-CMTLA security. We will show how to construct an adversary $\mathcal{F}$ which on input a public-key pk breaks the (t, d, ϵ) UF-CMA security of SIG in a chosen message attack. $\mathcal{F}^{\mathcal{O}_{sk_{i-1}}}(pk)$ simply runs $\mathcal{F}_{\lambda_{\mathsf{total}}}^{\mathcal{O}_{sk_{i-1}}, \mathcal{O}_{\mathsf{leak}}}(pk)$, where it randomly guesses the output of the leakage oracle $\mathcal{O}_{\mathsf{leak}}$. As $\mathcal{O}_{\mathsf{leak}}$ outputs at most λ_{total} bits, $\mathcal{F}$ will guess all the leakage correctly with probability $2^{-\lambda_{\mathsf{total}}}$. Conditioned on $\mathcal{F}$ guessing correctly, $\mathcal{F}_{\lambda_{\mathsf{total}}}$ will output a forgery with probability at least ϵ, thus $\mathcal{F}$ will output a forgery with probability at least $\epsilon \cdot 2^{-\lambda_{\mathsf{total}}}$.

AN EFFICIENT SCHEME IN THE RANDOM ORACLE MODEL. The security loss in the above reduction is exponential in λ_{total}. Recently, Alwen, Dodis and Wichs [2] proposed a signature scheme which can leak a substantial bounded amount λ_{total} of information without suffering an exponential decrease in security. More precisely, [2,19] show that in the random oracle model (a variant of) the Okamoto-Schnorr signature scheme [28,33] is still secure even if a constant fraction λ_{total} of the total secret key is leaked to the adversary. For concreteness we now recall the variant $\mathsf{SIG}_\ell^{\mathsf{OS}} = (\mathsf{Kg}_\ell^{\mathsf{OS}}, \mathsf{Sign}_\ell^{\mathsf{OS}}, \mathsf{Vfy}_\ell^{\mathsf{OS}})$ of the Okamoto-Schnorr signature scheme.

Let $\mathsf{G}(1^k)$ be a group sampling algorithm which outputs a tuple $(p, \mathbb{G})$, where p is a prime of size $\log p = 2k$ and $\mathbb{G}$ is a group of order p in which the discrete logarithm problem is hard.[8] Let $H : \{0,1\}^* \to \mathbb{Z}_p$ be a hash function that will be modeled as a random oracle. The scheme is given in Figure 2.

In [2,19] the authors show that $\mathsf{SIG}_\ell^{\mathsf{OS}}$ is secure under the hardness of the ℓ-representation problem (c.f. [2,19] and the references therein for its description and its equivalence to the DL problem). More precisely, they prove the following lemma.

Lemma 2. *For any* $\delta > 0$ *and* $\ell \in \mathbb{N}$, *security parameter* k, $t = t(k)$, $\epsilon = \epsilon(k)$, $d = d(k)$, $\lambda_{\mathsf{total}} = (1/2 - 1/2\ell - \delta)n$ *where* $n = 2k\ell$ *is the length of the secret key, if the* ℓ*-representation problem is* (t, ϵ)*-hard then the signature scheme* $\mathsf{SIG}_\ell^{\mathsf{OS}}$

[8] For technical reasons we assume that elements of $\mathbb{G}$ can be sampled "obliviously", this means, there exists an efficient algorithm $\mathsf{samp}_{\mathbb{G}}$ that outputs random elements of $\mathbb{G}$ with the property that, given $g \in \mathbb{G}$, one can sample uniformly from the set of coins ω for which $g := \mathsf{samp}_{\mathbb{G}}(\omega)$. See [19] for more details.

Algorithm $\mathsf{Kg}_\ell^{\mathsf{OS}}(1^k)$ $(\mathbb{G}, p) \xleftarrow{\$} \mathsf{G}(1^k)$ $(g_1, \ldots, g_\ell) \xleftarrow{\$} \mathbb{G}^\ell;\ (x_1, \ldots, x_\ell) \xleftarrow{\$} \mathbb{Z}_p^\ell$ $h \leftarrow \prod_i g_i^{x_i}$ return $(pk, sk) = ((\mathbb{G}, p, g_1, \ldots, g_\ell, h), (x_1, \ldots, x_\ell))$	**Algorithm** $\mathsf{Sign}_\ell^{\mathsf{OS}}(sk, m)$ $(r_1, \ldots, r_\ell) \xleftarrow{\$} \mathbb{Z}_q^\ell$ $A \leftarrow \prod_i g_i^{r_i}$ $c \leftarrow H(A, m)$ return $\sigma = (A, cx_1 + r_1, \ldots, cx_\ell + r_\ell)$
Algorithm $\mathsf{Vfy}_\ell^{\mathsf{OS}}(pk, \sigma, m)$ Parse σ as $(A, \alpha_1, \ldots, \alpha_\ell)$ $c \leftarrow H(A, m)$ Iff $\prod g_i^{\alpha_i} \stackrel{?}{=} Ah^c$ return 1; else return 0	

Fig. 2. $\mathsf{SIG}_\ell^{\mathsf{OS}} = (\mathsf{Kg}_\ell^{\mathsf{OS}}, \mathsf{Sign}_\ell^{\mathsf{OS}}, \mathsf{Vfy}_\ell^{\mathsf{OS}})$

from Figure 2 is $(t', d, \epsilon', \lambda_{\mathsf{total}})$ UF-CMTLA secure in the random oracle model, where $t' \approx t$ and $\epsilon' = (q_H \cdot (2 \cdot \epsilon + 1/p + q_H/p^{2\delta\ell}))^{1/2}$, where q_H is the number of random oracle queries made by the adversary.

A SCHEME IN THE STANDARD MODEL. From a universal one-way hash function (UOWHF) H, [19] constructs an efficient *one-time* signature scheme that tolerates leakage of a $(1 - \delta)/4$ fraction of the secret key. Using sequential composition this scheme is easily generalized to a stateful d-time signature schemes SIG_δ^K which can leak up to a $(1 - \delta)/4d$ fraction of the secret-key.

Lemma 3. *For any $\delta > 0$, security parameter k, $t = t(k)$, $\epsilon = \epsilon(k)$, $d = d(k)$, if H is a (t, ϵ)-secure UOWHF, then SIG_δ^K is $(t', d, \epsilon', \lambda_{\mathsf{total}})$ UF-CMTLA secure, where $\epsilon' = d\epsilon$, $t' \approx t$ and $\lambda_{\mathsf{total}} = n \cdot \frac{1-\delta}{4d}$ where $n = O(dk^2/\delta)$ is the length of the secret key.*

4.2 Construction of Leakage Resilient Signature Schemes

In this section we show how to construct a UF-CMLA secure signature scheme $\mathsf{SIG}^* = (\mathsf{Kg}^*, \mathsf{Sign}^*, \mathsf{Vfy}^*)$ from any UF-CMTLA 3-time signature scheme $\mathsf{SIG} = (\mathsf{Kg}, \mathsf{Sign}, \mathsf{Vfy})$.

We first introduce some notation related to binary trees that will be useful for the description of our signature scheme. For $d \in \mathbb{N}$, we denote with $\{0,1\}^{\leq d} = \bigcup_{i=0}^{d} \{0,1\}^i \cup \varepsilon$ the set of size $2^{d+1} - 1$ containing all binary bitstrings of length at most d including the empty string ε. We will think of $\{0,1\}^{\leq d}$ as the labels of a binary tree of depth d. The left and right child of an internal node $w \in \{0,1\}^{\leq d-1}$ are $w0$ and $w1$, respectively. For a node $w \in \{0,1\}^{\leq d} \setminus 1^d$, we denote with $\mathsf{DF}(w)$ the node visited after w in a depth-first traversal.

$$\mathsf{DF}(w) := \begin{cases} w0 & \text{if } \ |w| < d \qquad\qquad\qquad (w \text{ is an internal node}) \\ \hat{w}1 & \text{if } \ |w| = d, \text{ where } w = \hat{w}01^t \quad (w \text{ is the root}) \end{cases}$$

We define the mapping $\varphi : \{0,1\}^{\leq d} \to [2^{d-1} - 1]$ where $\varphi(w) = i$ if w is the i-th node to be visited in a depth first traversal, i.e. $\varphi(\varepsilon) = 1, \varphi(0) = 2, \varphi(00) = 3, \ldots$.

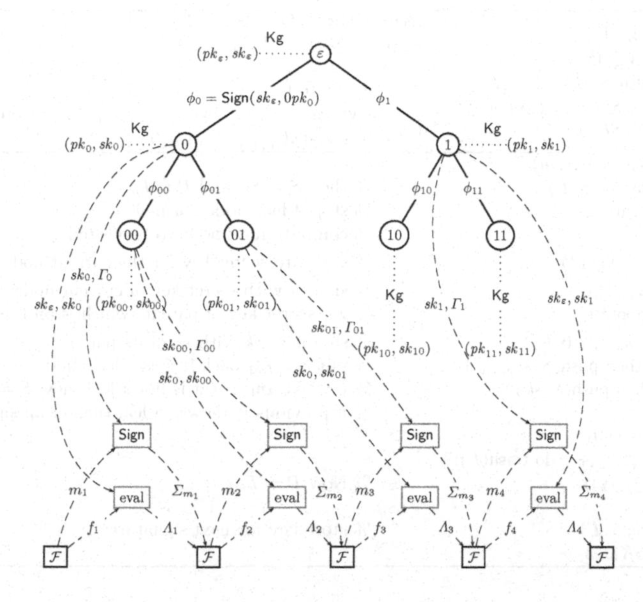

Fig. 3. Illustration of the execution of SIG^* in the UF-CMLA experiment. This figure shows the first 4 rounds of interaction between the adversary $\mathcal{F}$ and Sign. The dotted edges associate a public/secret key to each node. The dashed arrows represent $\mathcal{F}$'s oracle queries. $\mathcal{F}$ queries for a message m_i and a leakage function f_i, and obtains the signature Σ_{m_i}. Additionally, it obtains the leakage function f_i evaluated on the active state S_i^+, which, for instance for $i = 1$, includes the keys sk_ε, sk_0.

We now give the construction of our leakage resilient signature scheme. To simplify the exposition, we will assume that SIG is a stateless signature scheme, but this is not required. We fix some $d \in \mathbb{N}$ such that $q = 2^{d+1} - 2$ is an upper bound on the number of messages that SIG can sign. The signing algorithm Sign^* traverses a tree (depth first), "visiting" the node w and associating to it a key-pair (pk_w, sk_w) generated from the underlying signature scheme SIG. We will use the following notational conventions for a node $w = w_1 w_2 \ldots w_t$.

- $\Gamma_w = [(pk_{w_1}, \phi_{w_1}), (pk_{w_1 w_2}, \phi_{w_1 w_2}), \ldots, (pk_w, \phi_w)]$ is a *signature path* from w to the root, where $\phi_{w'}$ always denotes the signature of $pk_{w'}$ with its parent secret key $sk_{\mathsf{par}(w')}$.
- $S_w = \{sk_{w_1 w_2 \ldots w_i} : w_{i+1} = 0\}$ denotes a subset of the secret keys on the path from the root ε to w. S_w contains $sk_{w'}$, if the path goes to the left child $w'0$ at some node w' on the path. (The reason is, that in this case the right child $w'1$ will be visited after w in a depth first search, and we will then need $sk_{w'}$ to sign the public key $pk_{w'1}$ of that child.)

The secret key of SIG^* will always be of the form (w, S_w, Γ_w), and we will use stacks S and Γ to keep track of the state. We denote an empty stack with $\emptyset$. For a stack A, $\mathsf{push}(A, a)$ denotes putting element a on the stack A, $a \leftarrow \mathsf{pop}(A)$

Algorithm $\mathsf{Kg}^*(1^k)$
$(pk, sk) \xleftarrow{\$} \mathsf{Kg}(1^k)$
$S \leftarrow \emptyset$; $\mathsf{push}(S, sk)$; $\Gamma \leftarrow \emptyset$
$SK_0 \leftarrow (w_\varepsilon, S, \Gamma)$; $PK \leftarrow pk$
return (PK, SK_0)

Algorithm $\mathsf{Vfy}^*(PK, m, \Sigma)$
parse Σ as $(\sigma, \Gamma_{w_1 w_2 \ldots w_t})$
$pk_\varepsilon \leftarrow PK$
for i = 1 to t do
if $\mathsf{Vfy}(pk_{w_1 \ldots w_{i-1}}, 0pk_{w_1 \ldots w_i}, \phi_{w_1 \ldots w_i}) = 0$ return 0
return $\mathsf{Vfy}(pk_{w_1 w_2 \ldots w_t}, 1m, \sigma)$

Algorithm $\mathsf{Sign}^*(SK_i, m)$
parse SK_i as (w, S, Γ) % then $S = S_w$ and $\Gamma = \Gamma_w$
if $w = 1^d$ return $\perp$ % stop if last node reached
$\hat{w} \leftarrow \mathsf{DF}(w)$ % compute next node to be visited
$(sk_{\hat{w}}, pk_{\hat{w}}) \xleftarrow{\$} \mathsf{Kg}(1^n)$ % generate secret key for the current node
$\sigma \xleftarrow{\$} \mathsf{Sign}(sk_{\hat{w}}, 1m)$ % sign m with secret key of current node
$sk_{\mathsf{par}(\hat{w})} \leftarrow \mathsf{pop}(S)$ % get secret key of parent (which is on top of S)
$\phi_{\hat{w}} \xleftarrow{\$} \mathsf{Sign}(sk_{\mathsf{par}(\hat{w})}, 0pk_{\hat{w}})$ % sign new pk with sk of its parent
if $\hat{w}_{|\hat{w}|} = 0$ then $\mathsf{push}(S, sk_{\mathsf{par}(\hat{w})})$ % put $sk_{\mathsf{par}(\hat{w})}$ back if $\hat{w}$ is a left child
if $|\hat{w}| < d$ then $\mathsf{push}(S, sk_{\hat{w}})$ % put $sk_{\hat{w}}$ on S if it is not a leaf, now $S = S_{\hat{w}}$
if $|w| = d$ % if previous node was a leaf then clean signature chain
parse w as $w'01^j$
for $i = 1, \ldots, j+1$ do $\mathsf{trash}(\Gamma)$;
$\mathsf{push}(\Gamma, (pk_{\hat{w}}, \phi_{\hat{w}}))$ % Now $\Gamma = \Gamma_{\hat{w}}$
$\Sigma \leftarrow (\sigma, \Gamma)$
$SK_{i+1} \leftarrow (\hat{w}, S, \Gamma)$ % store key for next signature
return (Σ, SK_{i+1})

Fig. 4. The leakage resilient signature scheme SIG^*

denotes removing the topmost element from A and assigning it to a, and $\mathsf{trash}(A)$ denotes removing the topmost element from A (without assigning it). To avoid confusion we will always use upper case letters (PK, SK) for keys of SIG^* and lower case letters (pk, sk) for keys used by the underlying signature scheme SIG. To ease exposition, we use the secret key of the node 0, and not the root to sign the first message. The scheme SIG^* is defined in Figure 4.

Theorem 1. *For any security parameter k, $t = t(k)$, $\epsilon = \epsilon(k), q = q(k), \lambda = \lambda(k)$, if SIG is $(t, 3, \epsilon, \lambda_{\text{total}})$ UF-CMTLA secure, then SIG^* is $(t', q-1, q\epsilon, \lambda)$ UF-CMLA secure where $t' \approx t$ and $\lambda = \lambda_{\text{total}}/3$.*

Proof. We will show how to construct an adversary $\mathcal{F}$ which breaks the UF-CMTLA security of SIG (with $\lambda_{\text{total}} = 3 \cdot \lambda$ bits of total leakage) using as a subroutine the adversary $\mathcal{F}_\lambda$ who breaks the UF-CMLA security of SIG^* (with λ bits of leakage in each of the q observations) with advantage at least

$$\mathbf{Adv}^{\text{uf-cmtla}}_{\mathsf{SIG}}(\mathcal{F}, k, \lambda_{\text{total}}) \geq \frac{1}{q} \cdot \mathbf{Adv}^{\text{uf-cmla}}_{\mathsf{SIG}^*}(\mathcal{F}_\lambda, k, \lambda) \,. \tag{1}$$

The adversary $\mathcal{F}(pk)$ (attacking the UF-CMTLA security of SIG) simulates $\mathcal{F}_\lambda(PK)$ attacking the UF-CMLA security of SIG^*, embedding its challenge public-key pk into one of the nodes of SIG^*. That is, $\mathcal{F}(pk)$ simulates the following experiment (as defined in Section 3.2, cf. also Figure 3 for a graphical illustration.)

Experiment $\mathbf{Exp}^{\text{uf-cmla}}_{\mathsf{SIG}^*}(\mathcal{F}_\lambda, k, \lambda)$	**Oracle** $\mathcal{O}_{SK_{i-1}}(m_i, f_i)$
$(PK, SK_0) \stackrel{\$}{\leftarrow} \mathsf{Kg}^*(1^k)$; $i \leftarrow 1$	Sample fresh randomness r_i
$(m, \Sigma) \stackrel{\$}{\leftarrow} \mathcal{F}_\lambda^{\mathcal{O}_{SK_{i-1}}(\cdot,\cdot)}(PK)$	$(\Sigma_i, SK_i) \stackrel{SK^+_{i-1}}{\hookleftarrow} \mathsf{Sign}^*(SK_{i-1}, m_i, r_i)$
If $\mathsf{Vfy}^*(PK, m, \Sigma) = 1$	$\Lambda_i \leftarrow f_i(SK^+_{i-1}, r_i)$
and $m \notin \{m_1, \ldots m_i\}$	if $\|\Lambda_i\| \neq \lambda$ then $\Lambda_i \leftarrow 0^\lambda$
then return 1 else return 0.	Return (Σ_i, Λ_i) and set $i \leftarrow i + 1$

Simulation of PK**.** On input pk, $\mathcal{F}$ samples a node $\tilde{w}$ at random from the first q nodes (i.e., $\tilde{i} \stackrel{\$}{\leftarrow} \{1, \ldots, q\}$ and $\tilde{w} \leftarrow \varphi^{-1}(\tilde{i})$). The key $(pk_{\tilde{w}}, sk_{\tilde{w}})$ used by Sign will be the challenge key (pk, sk). Note that $sk = sk_{\tilde{w}}$ is unknown to $\mathcal{F}$. Next, $\mathcal{F}$ generates the other keys $(pk_w, sk_w), w \in \{0,1\}^{\leq d} \setminus \tilde{w}$ by calling $\mathsf{Kg}(1^k)$ using fresh randomness for each call. (Of course, these keys will only be computed when needed during the simulation of the signing oracle.) $\mathcal{F}$ defines $PK = pk_\varepsilon$ and runs $\mathcal{F}_\lambda$ on PK.

Simulation of the signing oracle. Let (m_i, f_i) be the i-th query to oracle $\mathcal{O}_{SK_{i-1}}(m_i, f_i)$ and let SK^+_{i-1} be the active state information in an execution of the real signing algorithm (i.e., $(\Sigma_i, SK_i) \stackrel{SK^+_{i-1}}{\hookleftarrow} \mathsf{Sign}^*(SK_{i-1}, m_i, r_i)$). Depending if $sk_{\tilde{w}} \in SK^+_{i-1}$ or not, adversary $\mathcal{F}$ distinguishes the two cases.

Case 1: $sk_{\tilde{w}} \notin SK^+_{i-1}$ ($\mathsf{Sign}(SK_{i-1}, m_i, r_i)$ does not access $sk_{\tilde{w}}$.) In this case the adversary $\mathcal{F}$ computes $\sigma_i \stackrel{\$}{\leftarrow} \mathsf{Sign}(SK_{i-1}, m_i, r_i)$ and $\Lambda_i = f_i(SK^+_{i-1}, r_i)$ itself and outputs (σ_i, Λ_i).

Case 2: $sk_{\tilde{w}} \in SK^+_{i-1}$ ($\mathsf{Sign}(SK_{i-1}, m_i, r_i)$ does access $sk_{\tilde{w}} \in SK^+_{i-1}$.) In this case $\mathcal{F}$ can compute (σ_i, Λ_i) without knowing $sk_{\tilde{w}} = sk$ as it has access to the signing oracle $\mathcal{O}_{sk_{\tilde{w}}}$ and the leakage oracle $\mathcal{O}_{\mathsf{leak}}$ as defined in the CMTLA attack game. As $sk_{\tilde{w}} \in SK^+_{i-1}$ for at most three different i, and on for each i the range of f_i is λ bits, the total leakage will be $\lambda_{\mathsf{total}} = 3 \cdot \lambda$ bits, which is what we assume $\mathcal{F}$ can get from $\mathcal{O}_{\mathsf{leak}}$.

The simulation of the UF-CMLA experiment by $\mathcal{F}$ is perfect (i.e. has the right distribution). As $\mathcal{F}$ perfectly simulates the UF-CMLA experiment, by assumption, $\mathcal{F}_\lambda$ does output a forgery with probability $\mathbf{Adv}^{\text{uf-cmla}}_{\mathsf{SIG}}(\mathcal{F}_\lambda, k, \lambda)$. We now show that from $\mathcal{F}$'s forgery one can extract a forgery for at least one of the keys (pk_w, sk_w) of the underlying signature scheme SIG.

Claim. If $\mathcal{F}_\lambda$ outputs a forgery (σ, Σ) in the UF-CMLA experiment, then one can extract a forgery for SIG with respect to at least one of the public-keys $(pk_w, sk_w), w \in \{\varphi_d^{-1}(1), \ldots, \varphi_d^{-1}(q)\}$.

Proof. Let $W = \{\varphi_d^{-1}(0), \ldots, \varphi_d^{-1}(q)\}$ be the set of nodes that have been visited during the query phase of the UF-CMLA experiment. Further, let $U := \{\Gamma_w\}_{w \in W}$ be the set of all signature chains that have been generated during the experiment. We distinguish two cases.

Case 1: $\Gamma \in U$. Then $\Gamma = \Gamma_w$ for one $w \in W$. If $\Sigma = (\sigma, \Gamma)$ is a valid forgery, then $\sigma \in \mathsf{Sign}(sk_w, 1m)$, where $m \neq m_{\varphi_d^{-1}(w)}$. Thus, $(1m, \sigma)$ is a valid forgery of SIG for public key pk_w.

Case 2: $\Gamma \notin U$. Then there must exist a node $w \in W$ such that $\phi \in \Gamma$ with $\phi \in \mathsf{Sign}(sk_w, 0pk_{w^*})$, where $pk_{w^*} \neq pk_{w0}$ and $pk_{w^*} \neq pk_{w1}$.[9] It follows that ϕ is a valid signature for key pk_w and message $0pk_{w^*}$ that has not been queried before.

The claim follows. △

With this claim and the fact that the simulation is perfect, it follows that we can extract a forgery for SIG with respect to the challenge public-key pk with probability $\mathbf{Adv}^{\text{uf-cmla}}_{\mathsf{SIG}^*}(\mathcal{F}_\lambda, k, \lambda)/q$ (namely when the w from the claim is $\tilde{w}$). This proves (1) and completes the proof. □

4.3 Efficiency and Trade-offs

We analyze the performance of our basic leakage resilient signature scheme and provide some efficiency trade-offs. For $d \in \mathbb{N}$ let $D = 2^{d+1} - 2$ be the upper bound on the number of messages that can be signed.

For simplicity, we assume that for SIG key generation, signing and verification all take approximately the same time, and further that public keys, secret keys and signatures are all of the same length. Let us now analyze the efficiency of SIG*. Public key size and key generation are as in the underlying scheme.

In the signing process, Sign* has to run at most two instances of Sign (i.e., to sign the message and to certify the next public key) and one run of the underlying key generation algorithm Kg. This adds up to an overhead of 3 compared to SIG. In our scheme, a signature consists of the signature of the actual message together with a signature chain from the current signing node to the root. Thus, the size of a signature increases in the worst case (if we sign with a leaf node) by a factor of $\approx 2d$. For the verification of a signature, in Vfy* we have to first verify the signature chain, and only if all those verifications pass, we check the signature on the actual message. This results in an overhead of d compared to the the underlying verification algorithm Vfy. Finally, in contrast to SIG our scheme requires storage of $\approx d$ secret keys, $\approx d$ public keys and $\approx d$ signatures, whereas in a standard signature scheme one only has to store a single secret key. Note however that only the storage for the secret keys needs to be kept secret.

In the special case, when we instantiate SIG* with $\mathsf{SIG}^{\mathsf{OS}}$ and set $\delta = 1/2$ (thus, $\ell = 3$), then SIG* is quite efficient[10]: signing requires only 9 exponentiations and 2 evaluations of a hash function. Verification is slightly less efficient and needs in the worst case $4d$ exponentiations and d evaluations of the underlying hash function. Finally, in the worst case a signature contains $18d$ group elements. Notice that our construction instantiated with $\mathsf{SIG}^{\mathsf{OS}}$ allows us to leak a 1/36th fraction of the secret key in each observation. It is easy to increase this to a 1/24th fraction by only using the leafs of SIG* to sign actual messages.

[9] Wlog assume that $w0$ and $w1$ are both in W.

[10] Only counting exponentiations and hash function evaluations.

References

1. Akavia, A., Goldwasser, S., Vaikuntanathan, V.: Simultaneous hardcore bits and cryptography against memory attacks. In: Reingold, O. (ed.) TCC 2009. LNCS, vol. 5444, pp. 474–495. Springer, Heidelberg (2009)
2. Alwen, J., Dodis, Y., Wichs, D.: Leakage-resilient public-key cryptography in the bounded-retrieval model. In: Halevi, S. (ed.) CRYPTO 2009. LNCS, vol. 5677, pp. 36–54. Springer, Heidelberg (2009)
3. Anderson, R.: Two remarks on public-key cryptology. Manuscript. Relevant material presented by the author in an invited lecture at the 4th ACM Conference on Computer and Communications Security, CCS 1997, Zurich, Switzerland, April 1–4, September 2000 (1997)
4. Bellare, M., Miner, S.K.: A forward-secure digital signature scheme. In: Wiener, M. (ed.) CRYPTO 1999. LNCS, vol. 1666, pp. 431–448. Springer, Heidelberg (1999)
5. Crescenzo, G.D., Lipton, R.J., Walfish, S.: Perfectly secure password protocols in the bounded retrieval model. In: Halevi, S., Rabin, T. (eds.) TCC 2006. LNCS, vol. 3876, pp. 225–244. Springer, Heidelberg (2006)
6. Dodis, Y., Goldwasser, S., Kalai, Y., Peikert, C., Vaikuntanathan, V.: Public-key encryption schemes with auxiliary inputs. In: Micciancio, D. (ed.) TCC 2010. LNCS, vol. 5978. Springer, Heidelberg (2010)
7. Dodis, Y., Kalai, Y.T., Lovett, S.: On cryptography with auxiliary input. In: 41st ACM STOC. ACM Press, New York (2009)
8. Dziembowski, S.: On forward-secure storage (extended abstract). In: Dwork, C. (ed.) CRYPTO 2006. LNCS, vol. 4117, pp. 251–270. Springer, Heidelberg (2006)
9. Dziembowski, S., Pietrzak, K.: Intrusion-resilient secret sharing. In: 48th FOCS, pp. 227–237. IEEE Computer Society Press, Los Alamitos (2007)
10. Dziembowski, S., Pietrzak, K.: Leakage-resilient cryptography. In: FOCS 2008. ACM Press, New York (2008)
11. Faust, S., Kiltz, E., Pietrzak, K., Rothblum, G.: Leakage-resilient signatures. Cryptology ePrint Archive, Report 2009/282 (2009), `http://eprint.iacr.org/`
12. Faust, S., Reyzin, L., Tromer, E.: Protecting circuits from computationally-bounded leakage. Cryptology ePrint Archive, Report 2009/379 (2009), `http://eprint.iacr.org/`
13. Fouque, P.-A., Martinet, G., Poupard, G.: Attacking unbalanced RSA-CRT using SPA. In: Walter, C.D., Koç, Ç.K., Paar, C. (eds.) CHES 2003. LNCS, vol. 2779, pp. 254–268. Springer, Heidelberg (2003)
14. Gandolfi, K., Mourtel, C., Olivier, F.: Electromagnetic analysis: Concrete results. In: Koç, Ç.K., Naccache, D., Paar, C. (eds.) CHES 2001. LNCS, vol. 2162, pp. 251–261. Springer, Heidelberg (2001)
15. Goldwasser, S., Kalai, Y.T., Rothblum, G.N.: One-time programs. In: Wagner, D. (ed.) CRYPTO 2008. LNCS, vol. 5157, pp. 39–56. Springer, Heidelberg (2008)
16. Goldwasser, S., Micali, S., Rivest, R.L.: A digital signature scheme secure against adaptive chosen-message attacks. SIAM Journal on Computing 17(2), 281–308 (1988)
17. Halderman, J.A., Schoen, S.D., Heninger, N., Clarkson, W., Paul, W., Calandrino, J.A., Feldman, A.J., Appelbaum, J., Felten, E.W.: Lest we remember: Cold boot attacks on encryption keys. In: USENIX Security Symposium, pp. 45–60 (2008)
18. Ishai, Y., Sahai, A., Wagner, D.: Private circuits: Securing hardware against probing attacks. In: Boneh, D. (ed.) CRYPTO 2003. LNCS, vol. 2729, pp. 463–481. Springer, Heidelberg (2003)

19. Katz, J., Vaikuntanathan, V.: Signature schemes with bounded leakage resilience. In: Matsui, M. (ed.) ASIACRYPT 2009. LNCS, vol. 5912, pp. 703–720. Springer, Heidelberg (2009)
20. Kocher, P.C.: Timing attacks on implementations of Diffie-Hellman, RSA, DSS, and other systems. In: Koblitz, N. (ed.) CRYPTO 1996. LNCS, vol. 1109, pp. 104–113. Springer, Heidelberg (1996)
21. Kocher, P.C., Jaffe, J., Jun, B.: Differential power analysis. In: Wiener, M. (ed.) CRYPTO 1999. LNCS, vol. 1666, pp. 388–397. Springer, Heidelberg (1999)
22. Lamport, L.: Constructing digital signatures from a one-way function. Technical Report SRI-CSL-98, SRI International Computer Science Laboratory (October 1979)
23. Merkle, R.C.: A certified digital signature. In: Brassard, G. (ed.) CRYPTO 1989. LNCS, vol. 435, pp. 218–238. Springer, Heidelberg (1990)
24. Micali, S., Reyzin, L.: Physically observable cryptography (extended abstract). In: Naor, M. (ed.) TCC 2004. LNCS, vol. 2951, pp. 278–296. Springer, Heidelberg (2004)
25. Naor, M., Segev, G.: Public-key cryptosystems resilient to key leakage. In: Halevi, S. (ed.) CRYPTO 2009. LNCS, vol. 5677, pp. 18–35. Springer, Heidelberg (2009)
26. Naor, M., Yung, M.: Universal one-way hash functions and their cryptographic applications. In: 21st ACM STOC, pp. 33–43. ACM Press, New York (1989)
27. European Network of Excellence (ECRYPT). The side channel cryptanalysis lounge, `http://www.crypto.ruhr-uni-bochum.de/en_sclounge.html` (retrieved on 29.03.2008)
28. Okamoto, T.: Provably secure and practical identification schemes and corresponding signature schemes. In: Brickell, E.F. (ed.) CRYPTO 1992. LNCS, vol. 740, pp. 31–53. Springer, Heidelberg (1993)
29. Pietrzak, K.: A leakage-resilient mode of operation. In: Joux, A. (ed.) EUROCRYPT 2009. LNCS, vol. 5479, pp. 462–482. Springer, Heidelberg (2009)
30. Quisquater, J.-J., Samyde, D.: Electromagnetic analysis (EMA): Measures and counter-measures for smart cards. In: Attali, I., Jensen, T. (eds.) E-smart 2001. LNCS, vol. 2140, pp. 200–210. Springer, Heidelberg (2001)
31. Quisquater, J.-J., Koene, F.: Side channel attacks: State of the art. [27](October 2002)
32. Schindler, W.: A timing attack against RSA with the chinese remainder theorem. In: Paar, C., Koç, Ç.K. (eds.) CHES 2000. LNCS, vol. 1965, pp. 109–124. Springer, Heidelberg (2000)
33. Schnorr, C.-P.: Efficient identification and signatures for smart cards. In: Brassard, G. (ed.) CRYPTO 1989. LNCS, vol. 435, pp. 239–252. Springer, Heidelberg (1990)
34. Standaert, F.-X., Malkin, T., Yung, M.: A unified framework for the analysis of side-channel key recovery attacks. In: Joux, A. (ed.) EUROCRYPT 2009. LNCS, vol. 5479, pp. 443–461. Springer, Heidelberg (2009)

Public-Key Encryption Schemes with Auxiliary Inputs

Yevgeniy Dodis[1,*], Shafi Goldwasser[2,**], Yael Tauman Kalai[3], Chris Peikert[4,***], and Vinod Vaikuntanathan[5]

[1] New York University
[2] MIT and Weizmann Institute
[3] Microsoft Research
[4] Georgia Institute of Technology
[5] IBM Research

Abstract. We construct *public-key* cryptosystems that remain secure even when the adversary is given any *computationally uninvertible function* of the secret key as auxiliary input (even one that may reveal the secret key information-theoretically). Our schemes are based on the decisional Diffie-Hellman (DDH) and the Learning with Errors (LWE) problems.

As an independent technical contribution, we extend the Goldreich-Levin theorem to provide a hard-core (pseudorandom) value over *large* fields.

1 Introduction

Modern cryptographic algorithms are designed under the assumption that keys are perfectly secret and independently chosen for the algorithm at hand. Still, in practice, information about secret keys does get compromised for a variety of reasons, including side-channel attacks on the physical implementation of the cryptographic algorithm, or the use of the same secret key or the same source of randomness for keys across several applications.

In recent years, starting with the works of [5,15,18], a new goal has been set within the theory of cryptography community to build a general theory of physical security against large classes of side channel attacks. A large body of work has accumulated by now in which different classes of side channel attacks have been defined and different cryptographic primitives have been designed to provably withstand these attacks (See [5,15,18,9,1,2,20,8,24,23,14,9,10] and the references therein).

Placing the current paper within this body of work, we focus on side channel attacks which result from "memory leakages" [1,8,2,20,16]. In this class of attacks, the attacker chooses an arbitrary, efficiently computable function h (possibly as a function of the public parameters of the system), and receives the result of h applied on the secret key SK. Clearly, to have some secrecy left, we must restrict the attacker to choose

* Supported in part by NSF grants CNS-0831299 and CNS-0716690.
** Supported in part by NSF grants CCF-0514167, CCF-0635297, NSF-0729011, the Israel Science Foundation 700/08 and the Chais Family Fellows Program.
*** Supported in part by NSF grant CNS-0716786.

D. Micciancio (Ed.): TCC 2010, LNCS 5978, pp. 361–381, 2010.

a function h that "does not fully reveal the secret". The challenge is to model this necessary constraint in a clean and general manner, which both captures real attacks and makes the definition achievable. As of now, several models have appeared trying to answer this question.

Akavia, Goldwasser and Vaikuntanathan [1] considered a model in which the leakage function h is an arbitrary polynomial-time computable function *with bounded output length*. Letting k denote the length (or more generally, the min-entropy) of the secret key SK, the restriction is that h outputs $\ell(k) < k$ bits. In particular, this ensures that the leakage does not fully reveal the secret key. Akavia et al. [1] show that the public-key encryption scheme of Regev [25] is secure against $\ell(k)$-length bounded leakage functions as long as $\ell(k) < (1 - \epsilon)k$ for some constant $\epsilon > 0$, under the intractability of the learning with error problem (LWE).

Subsequent work of Naor and Segev [20] relaxed the restriction on h so that the leakage observed by the adversary may be longer than the secret key, but *the min-entropy of the secret drops by at most $\ell(k)$ bits* upon observing $h(SK)$; they call this the *noisy leakage* requirement. The work of [20] also showed how to construct a public-key encryption scheme which resists noisy leakage as long as $\ell(k) < k - k^{\epsilon}$ for some constant $\epsilon > 0$, under the decisional Diffie Hellman (DDH) assumption. They also showed a variety of other public-key encryption schemes tolerating different amounts of leakage, each under a different intractability assumption: Paillier's assumption, the quadratic residuosity assumption, and more generally, the existence of any universal hash-proof system [7]. We refer the reader to [20] for a detailed discussion of these results. (Finally, we note that the proof of [1] based on the LWE assumption generalizes to the case of noisy leakage.).

The bottom line is that both [1] and [20] (and the results that use the models therein) interpret the necessary restriction on the leakage function h by insisting that it is (information-theoretically) *impossible* to recover SK given the leakage $h(SK)$.

1.1 The Auxiliary Input Model

The natural question that comes out of the modeling in [1,20] is whether this restriction is essential. For example, is it possible to achieve security if the leakage function h is a one-way permutation? Such a function information-theoretically reveals the entire secret key SK, but still it is computationally infeasible to recover SK from $h(SK)$.

The focus of this work is the model of *auxiliary input* leakage functions, introduced by Dodis, Kalai, and Lovett [8], generalizing [1,20]. They consider the case of *symmetric encryption* and an adversary who can learn an arbitrary polynomial time computable function h of the secret key, provided that the secret key SK is *hard to compute* given $h(SK)$ (but not necessarily impossible to determine, as implied by the definitions of [1,20]). Formally, the restriction imposed by [8] on the leakage function h is that any polynomial time algorithm attempting to invert $h(SK)$ will succeed with probability at most $2^{-\lambda(k)}$ for some function $\lambda(\cdot)$ (i.e., a smaller $\lambda(k)$ allows for a larger class of functions, and thus a stronger security result). The ultimate goal is to capture all polynomially *uninvertible* functions: namely, all functions for which the probability of inversion by a polynomial time algorithm is bounded by some negligible function in k.[1]

The work of [8] constructed, based on a non-standard variant of the learning parity with noise (LPN) assumption, a symmetric-key encryption scheme that remains secure w.r.t. any auxiliary input $h(SK)$, as long as no polynomial time algorithm can invert h with probability more than $2^{-\epsilon k}$ for some $\epsilon > 0$.

In the same work [8], it was observed that in addition to generalizing the previous leakage models, the auxiliary input model offers additional advantages, the main one being *composition*. Consider a setting where a user prefers to use the same secret key for multiple tasks, as is the case when using biometric keys [4,8]. Suppose we construct an encryption scheme that is secure w.r.t. any auxiliary input which is an uninvertible function of the secret key. Then, one can safely use his secret and public key pair to run arbitrary protocols, as long as these protocols together do not (computationally) reveal the entire secret key.

1.2 Our Results

In this paper, we focus on designing *public key* encryption algorithms which are secure in the presence of auxiliary input functions. We adapt the definition of security w.r.t. auxiliary input from [8] to the case of public-key encryption algorithms. To address the issue of whether the function h is chosen by the adversary after seeing the corresponding public key PK (so called adaptive security in [1]), we allow the adversary to receive $h(SK, PK)$. In other words, we allow the leakage function to depend on PK.

We prove auxiliary input security for two public-key encryption schemes based on different assumptions.

1. We show that the encryption scheme of Boneh, Halevi, Hamburg and Ostrovsky [3] (henceforth called the BHHO encryption scheme), suitably modified, is CPA-secure in the presence of auxiliary input functions h that can be inverted with probability *at most* 2^{-k^ϵ} for any $\epsilon > 0$. The underlying hard problem for our auxiliary-input CPA-secure scheme is again the same as that of the original BHHO scheme, i.e, the decisional Diffie-Hellman assumption. Previously, [20] showed that the BHHO scheme is secure w.r.t. bounded length leakage of size at most $k - k^\epsilon$ (and more generally, noisy leakages).
2. We show that the "dual" of Regev's encryption scheme [25], first proposed by Gentry, Peikert and Vaikuntanathan [12], when suitably modified is CPA-secure in the presence of auxiliary input functions h that can be inverted with probability *at most* 2^{-k^ϵ} for any $\epsilon > 0$. The underlying hard problem for our auxiliary-input CPA-secure scheme is the same as that for (standard) CPA-security, i.e, the "learning with errors" (LWE) problem. This result, in particular, implies that the scheme is

[1] It is instructive to contrast *uninvertible* functions with the standard notion of *one way* functions. In the former, we require the adversary who is given $h(SK)$ to come up with the actual pre-image SK itself, whereas in the latter, the adversary need only output an SK' such that $h(SK') = h(SK)$. Thus, a function h that outputs nothing is an uninvertible function, but not one-way! The right notion to consider in the context of leakage and auxiliary input is that of *uninvertible* functions.

secure w.r.t. bounded length leakage of size at most $k - k^{\epsilon}$ (and more generally, noisy leakages). This improves on the previous bound of [1] for the [25] system.

We note that we can prove security of both the dual Regev encryption scheme and the BHHO encryption scheme w.r.t. a richer class of auxiliary inputs, i.e., those that are hard to invert with probability $2^{-\text{polylog}(k)}$. However, then the assumptions we rely on are that LWE/DDH are secure against an adversary that runs in subexponential time.

Of course, the holy grail in this line of work would be a public-key encryption scheme secure against *polynomially hard auxiliary inputs*, that is functions that no polynomial-time adversary can invert with probability better than negligible (in k). Is this in fact achievable? We give two answers to this question.

A Negative Answer. We show that the holy grail is unattainable for public-key encryption schemes. In particular, for every *public-key encryption scheme*, we show an auxiliary input function h such that it is hard to compute SK given $h(PK, SK)$, and yet the scheme is completely insecure in the presence of the leakage given by h. The crux of the proof is to use the fact that the adversary that tries to break the encryption scheme gets *the public key* PK *in addition to the leakage* $h(PK, SK)$. Thus, it suffices to come up with a leakage function h such that:

– Given $h(PK, SK)$, it is hard to compute SK (i.e., h is uninvertible), and yet
– Given $h(PK, SK)$ and PK, it is easy to compute SK.

We defer the details of the construction of such a function to the full version.

A Weaker Definition and a Positive Answer. To complement the negative result, we construct a public key encryption schemes (under DDH and LWE) that are secure if the leakage function h is polynomially hard to invert *even given the public key*. This is clearly a weaker definition, but still it is advantageous in the context of composition. See Section 3.1 for definitional details, and the full version for the actual scheme.

We end this section by remarking that the complexity of all of the the encryption schemes above depends on the bound on the inversion probability of h which we desire to achieve. For example, in the case of the BHHO scheme, the size of the secret key (and the complexity of encrypting/decrypting) is $k^{1/\epsilon}$, where k is the length of the secret key (or more generally, the min-entropy) and security is w.r.t. auxiliary inputs that are hard to invert with probability $2^{-k^{\epsilon}}$. (In fact, this is the case for most of the known schemes, in particular [1,2,20,16].)

1.3 Overview of Techniques

We sketch the main ideas behind the auxiliary input security of the GPV encryption scheme (slightly modified). The scheme is based on the hardness of the learning with error (decisional LWE) problem, which states that for a security parameter n and any polynomially large m, given a uniformly random matrix $\mathbf{A} \in \mathbb{Z}_q^{n \times m}$, the vector $\mathbf{A}^T\mathbf{s} + \mathbf{x}$ is pseudorandom where $\mathbf{s} \leftarrow \mathbb{Z}_q^n$ is uniformly random, and each component of $\mathbf{x}$ is chosen from a "narrow error distribution".

Let us first recall how the GPV scheme works. The secret key in the scheme is a vector $\mathbf{e} \in \{0,1\}^m$, and the public key is a matrix $\mathbf{A} \in \mathbb{Z}_q^{n \times m}$ together with $\mathbf{u} = \mathbf{Ae} \in \mathbb{Z}_q^n$. Here n is the security parameter of the system, q is a prime (typically polynomial in n, but in our case slightly superpolynomial), and m is a sufficiently large polynomial in n and $\log q$. (The min-entropy of the secret key k, in this case, is m.) The basic encryption scheme proceeds bit by bit. To encrypt a bit b, we first choose $\mathbf{s} \leftarrow \mathbb{Z}_q^n$ uniformly at random, $\mathbf{x} \in \mathbb{Z}_q^m$ from a "narrow" error distribution, and $x' \in \mathbb{Z}_q$ from a "much wider" error distribution. The ciphertext is

$$(\mathbf{A}^T\mathbf{s} + \mathbf{x}, \mathbf{u}^T\mathbf{s} + x' + b\lfloor q/2 \rfloor) \in \mathbb{Z}_q^m \times \mathbb{Z}_q.$$

Given the secret key $\mathbf{e}$, the decryption algorithm computes $\mathbf{e}^T(\mathbf{A}^T\mathbf{s} + \mathbf{x}) \approx (\mathbf{Ae})^T\mathbf{s} = \mathbf{u}^T\mathbf{s}$ (where the approximation holds because $\mathbf{e}^T\mathbf{x}$ and x' are all small compared to q) and uses this to recover b from the second component.

The first idea we use to show auxiliary input security for this scheme is that the intractability assumption (i.e, the hardness of LWE mentioned above) refers "almost entirely" to the "preamble" of the ciphertext $\mathbf{A}^T\mathbf{s} + \mathbf{x}$, and not to the secret key at all. This suggests considering an alternative encryption algorithm (used by the simulator) which generates a ciphertext using knowledge of the secret key $\mathbf{e}$ rather than the secret $\mathbf{s}$. The advantage of this in the context of leakage is that knowing the secret key enables the simulator to compute and reveal an arbitrary (polynomial-time computable) leakage function h of the secret key.[2] More specifically, we consider an alternate encryption algorithm that, given a preamble $\mathbf{y} = \mathbf{A}^T\mathbf{s} + \mathbf{x}$, encrypts the bit b using the secret key $\mathbf{e}$ as:

$$(\mathbf{y}, \mathbf{e}^T\mathbf{y} + x' + b\lfloor q/2 \rfloor).$$

The distribution thus produced is statistically "as good as" that of the original encryption algorithm; in particular, $\mathbf{e}^T\mathbf{y} + x' = \mathbf{u}^T\mathbf{s} + (\mathbf{e}^T\mathbf{x} + x')$, where $\mathbf{e}^T\mathbf{x} + x'$ is distributed (up to negligible statistical distance) like a sample from the "wide" error distribution when $\mathbf{e}^T\mathbf{x}$ is negligible compared to x'. (This is why we need to use a slightly super-polynomial q, and to choose $\mathbf{e}$ and $\mathbf{x}$ to be negligible relative to the magnitude of x').

Next, by the LWE assumption, we can replace $\mathbf{y}$ with a uniformly random vector over $\mathbb{Z}_q^m$. We would then like to show that the term $\mathbf{e}^T\mathbf{y} = \langle \mathbf{e}, \mathbf{y} \rangle$ in the second component of the ciphertext is pseudorandom given the rest of the adversary's view, namely $(\mathbf{A}, \mathbf{Ae}, h(\mathbf{A}, \mathbf{e}), \mathbf{y})$ where $\mathbf{y} \in \mathbb{Z}_q^m$ is uniformly random. Assuming that the function $h'(\mathbf{A}, \mathbf{e}) = (\mathbf{A}, \mathbf{Ae}, h(\mathbf{A}, \mathbf{e}))$ is uninvertible, this suggests using a *Goldreich-Levin type theorem over the large field* $GF(q)$. Providing such a theorem is the first technical contribution of this work.

The original Goldreich-Levin theorem over the binary field $GF(2)$ says that for an uninvertible function $h : GF(2)^m \to \{0,1\}^*$, the inner product $\langle \mathbf{e}, \mathbf{y} \rangle \in GF(2)$ is pseudorandom given $h(\mathbf{e})$ and uniformly random $\mathbf{y} \in GF(2)^m$. The later work of [13] extends this result to deal with inner products over $GF(q)$ for a general prime q. In particular, it shows that any PPT algorithm that distinguishes between $\langle \mathbf{y}, \mathbf{e} \rangle$ and

[2] This kind of technique for proving security of public-key encryption was already used in [12], in the context of leakage in [20,16], and to our knowledge, traces at least as far back as the Cramer-Shoup CCA-secure encryption scheme [6].

uniform, given $h(\mathbf{e})$ and $\mathbf{y}$, gives rise to a poly(q)-*time* algorithm that inverts $h(\mathbf{e})$ with probability $1/\text{poly}(q)$ (for a more detailed comparison of our result with [13], see Remark 2). When q is super-polynomial in the main security parameter n, the running time of the inverter is superpolynomial, which we wish to avoid. We consider a special class of functions (which is exactly what is needed in our applications) where each coordinate of $\mathbf{e}$ comes from a much smaller subdomain $H \subseteq GF(q)$. For this class of functions, we show how to make the running time of the inverter *polynomial in* n (and independent of q). We state the result informally below.

Informal Theorem 1. *Let q be prime, and let H be a* poly(m)*-sized subset of $GF(q)$. Let $h : H^m \to \{0,1\}^*$ be any (possibly randomized) function. If there is a PPT algorithm $\mathcal{D}$ that distinguishes between $\langle \mathbf{e}, \mathbf{y} \rangle$ and the uniform distribution over $GF(q)$ given $h(\mathbf{e})$ and $\mathbf{y} \leftarrow GF(q)^m$, then there is a PPT algorithm $\mathcal{A}$ that inverts h with probability* $1/(q^2 \cdot \text{poly}(m))$.

Applying this variant of the Goldreich-Levin theorem over $GF(q)$, we get security against auxiliary input functions h that are hard to invert given $(\mathbf{A}, \mathbf{Ae}, h(\mathbf{A}, \mathbf{e}))$; we call this weak auxiliary-input security in the rest of the paper. Obtaining strong auxiliary input security, i.e., security against functions h that are hard to invert given only $(\mathbf{A}, h(\mathbf{A}, \mathbf{e}))$, is very easy in our case: since the public key $\mathbf{Ae}$ has bit-length $n \log q = m^\epsilon \ll |SK|$, the reduction can simply guess the value of $PK = \mathbf{Ae}$ and lose only a factor of 2^{-m^ϵ} in the inversion probability.

The proof of security for the BHHO encryption scheme follows precisely the same line of argument, but with two main differences: (1) the proof is somewhat simpler because one does not have to deal with any pesky error terms (and the resulting statistical deviation between encrypting with the public key versus the secret key), and (2) we use the Goldreich-Levin theorem over an exponentially large field $GF(q)$, rather than a superpolynomial one.

2 Preliminaries

Throughout this paper, we denote the security parameter by n. We write negl(n) to denote an arbitrary negligible function, i.e., one that vanishes faster than the inverse of any polynomial.

The Decisional Diffie Hellman Assumption. Let $\mathcal{G}$ be a probabilistic polynomial-time "group generator" that, given the security parameter n in unary, outputs the description of a group G that has prime order $q = q(n)$. The decisional Diffie Hellman (DDH) assumption for $\mathcal{G}$ says that the following two ensembles are computationally indistinguishable:

$$\{(g_1, g_2, g_1^r, g_2^r) : g_i \leftarrow G, r \leftarrow \mathbb{Z}_q\} \approx_c \{(g_1, g_2, g_1^{r_1}, g_2^{r_2}) : g_i \leftarrow G, r_i \leftarrow \mathbb{Z}_q\}$$

We will use a lemma of Naor and Reingold [19] which states that a natural generalization of the DDH assumption which considers $m > 2$ generators is actually equivalent to DDH. The proof follows from the self-reducibility of DDH.

Lemma 1 ([19]). *Under the DDH assumption on $\mathcal{G}$, for any positive integer m,*

$$\left\{(g_1,\ldots,g_m,g_1^r,\ldots,g_m^r) : g_i \leftarrow G, r \leftarrow \mathbb{Z}_q\right\} \approx_c$$

$$\left\{(g_1,\ldots,g_m,g_1^{r_1},\ldots,g_m^{r_m}) : g_i \leftarrow G, r_i \leftarrow \mathbb{Z}_q\right\}$$

The Learning with errors (LWE) Assumption. The LWE problem was introduced by Regev [25] as a generalization of the "learning noisy parities" problem. For positive integers n and $q \geq 2$, a vector $\mathbf{s} \in \mathbb{Z}_q^n$, and a probability distribution χ on $\mathbb{Z}_q$, let $A_{\mathbf{s},\chi}$ be the distribution obtained by choosing a vector $\mathbf{a} \in \mathbb{Z}_q^n$ uniformly at random and a noise term $x \leftarrow \chi$, and outputting $(\mathbf{a}, \langle \mathbf{a}, \mathbf{s}\rangle + x) \in \mathbb{Z}_q^n \times \mathbb{Z}_q$.

Definition 1. *For an integer $q = q(n)$ and an error distribution $\chi = \chi(n)$ over $\mathbb{Z}_q$, the (worst-case) learning with errors problem $\mathsf{LWE}_{n,m,q,\chi}$ in n dimensions is defined as follows. Given m independent samples from $A_{\mathbf{s},\chi}$ (where $\mathbf{s} \in \mathbb{Z}_q^n$ is arbitrary), output $\mathbf{s}$ with noticeable probability.*

The (average-case) decisional variant of the LWE problem, denoted $\mathsf{DLWE}_{n,m,q,\chi}$, is to distinguish (with non-negligible advantage) m samples chosen according to $A_{\mathbf{s},\chi}$ for uniformly random $\mathbf{s} \in \mathbb{Z}_q^n$, *from m samples chosen according to the uniform distribution over $\mathbb{Z}_q^n \times \mathbb{Z}_q$.*

For cryptographic applications we are primarily interested in the average-case decision problem DLWE. Fortunately, Regev [25] showed that for a prime modulus q, the (worst-case) LWE and (average-case) DLWE problems are equivalent, up to a $q \cdot \text{poly}(n)$ factor in the number of samples m. We say that $\mathsf{LWE}_{n,m,q,\chi}$ (respectively, $\mathsf{DLWE}_{n,m,q,\chi}$) is hard if no PPT algorithm can solve it for infinitely many n.

At times, we use a compact matrix notation to describe the LWE problem $\mathsf{LWE}_{n,m,q,\chi}$: given $(\mathbf{A}, \mathbf{A}^T\mathbf{s} + \mathbf{x})$ where $\mathbf{A} \leftarrow \mathbb{Z}_q^{n\times m}$ is uniformly random, $\mathbf{s} \leftarrow \mathbb{Z}_q^n$ is the LWE secret, and $\mathbf{x} \leftarrow \chi^m$, find $\mathbf{s}$. We also use a similar notation for the decision version DLWE.

Gaussian error distributions. We are primarily interested in the LWE and DLWE problems where the error distribution χ over $\mathbb{Z}_q$ is derived from a Gaussian. For any $r > 0$, the density function of a one-dimensional Gaussian probability distribution over $\mathbb{R}$ is given by $D_r(x) = 1/r \cdot \exp(-\pi(x/r)^2)$. For $\beta > 0$, define $\overline{\Psi}_\beta$ to be the distribution on $\mathbb{Z}_q$ obtained by drawing $y \leftarrow D_\beta$ and outputting $\lfloor q \cdot y \rceil \pmod q$. We write $\mathsf{LWE}_{n,m,q,\beta}$ as an abbreviation for $\mathsf{LWE}_{n,m,q,\overline{\Psi}_\beta}$. Here we state two basic facts about Gaussians (tailored to the error distribution $\overline{\Psi}_\beta$); see, e.g. [11].

Lemma 2. *Let $\beta > 0$ and $q \in \mathbb{Z}$. Let the vector $\mathbf{x} \in \mathbb{Z}^n$ be arbitrary, and $\mathbf{y} \leftarrow \overline{\Psi}_\beta^n$. With overwhelming probability over the choice of $\mathbf{y}$, $|\mathbf{x}^T\mathbf{y}| \leq ||\mathbf{x}|| \cdot \beta q \cdot \omega(\sqrt{\log n})$.*

Lemma 3. *Let $\beta > 0$, $q \in \mathbb{Z}$ and $y \in \mathbb{Z}$. The statistical distance between the distributions $\overline{\Psi}_\beta$ and $\overline{\Psi}_\beta + y$ is at most $|y|/(\beta q)$.*

Evidence for the hardness of $\mathsf{LWE}_{n,m,q,\beta}$ follows from results of Regev [25], who gave a *quantum* reduction from approximating certain problems on n-dimensional lattices in the worst case to within $\tilde{O}(n/\beta)$ factors to solving $\mathsf{LWE}_{n,m,q,\beta}$ for any desired $m = \text{poly}(n)$, when $\beta \cdot q \geq 2\sqrt{n}$. Recently, Peikert [21] also gave a related *classical* reduction for similar parameters.

3 Security against Auxiliary Inputs

We start by defining security of public-key encryption schemes w.r.t. auxiliary input.

Definition 2. *A public-key encryption scheme* $\Pi = (\mathsf{Gen}, \mathsf{Enc}, \mathsf{Dec})$ *with message space* $\mathcal{M} = \{\mathcal{M}_n\}_{n\in\mathbb{N}}$ *is* `CPA secure w.r.t. auxiliary inputs from` $\mathcal{H}$ *if for any* PPT *adversary* $\mathcal{A} = (\mathcal{A}_1, \mathcal{A}_2)$*, any function* $h \in \mathcal{H}$*, any polynomial* p*, and any sufficiently large* $n \in \mathbb{N}$*,*

$$\mathsf{Adv}_{\mathcal{A},\Pi,h} \stackrel{def}{=} \left| \Pr[\text{CPA}_0(\Pi, \mathcal{A}, n, h) = 1] - \Pr[\text{CPA}_1(\Pi, \mathcal{A}, n, h) = 1] \right| < \frac{1}{p(n)},$$

where $\text{CPA}_b(\Pi, \mathcal{A}, n, h)$ *is the output of the following experiment:*

$$(SK, PK) \leftarrow \mathsf{Gen}(1^n)$$
$$(M_0, M_1, \mathsf{state}) \leftarrow \mathcal{A}_1(1^n, PK, h(SK, PK)) \text{ s.t. } |M_0| = |M_1|$$
$$c_b \leftarrow \mathsf{Enc}(PK, M_b)$$
$$\textit{Output } \mathcal{A}_2(c_b, \mathsf{state})$$

3.1 Classes of Auxiliary Input Functions

Of course, we need to decide which function families $\mathcal{H}$ we are going to consider. We define two such families. For future convenience, we will parametrize these families by the min-entropy k of the secret key, as opposed to the security parameter n. (Note, in our schemes the secret key will be random, so k is simply the length of the secret key.)

The first family $\mathcal{H}_{\text{bdd}}$ is the length-bounded family studied by the prior work [1,20],[3] while the second family $\mathcal{H}_{\text{ow}}$ is the auxiliary-input family we introduce and study in this work, where we only assume that the secret key is "hard to compute" given the leakage.

- Let $\mathcal{H}_{\text{bdd}}(\ell(k))$ be the class of all polynomial-time computable functions $h : \{0,1\}^{|SK|+|PK|} \to \{0,1\}^{\ell(k)}$, where $\ell(k) \leq k$ is the number of bits the attacker is allowed to learn. If a public-key encryption scheme Π is CPA secure w.r.t. this family of functions, it is called $\ell(k)$-`LB-CPA (length-bounded CPA) secure`.
- Let $\mathcal{H}_{\text{ow}}(f(k))$ be the class of all polynomial-time computable functions $h : \{0,1\}^{|SK|+|PK|} \to \{0,1\}^*$, such that given $h(SK, PK)$ (for a randomly

[3] For simplicity, we do not define a more general family corresponding to the noisy leakage model of [20]. However, all the discussion, including Lemma 4, easily holds for noisy-leakage instead of length-bounded leakage.

generated (SK, PK)), no PPT algorithm can find SK with probability greater than $f(k)$, where $f(k) \geq 2^{-k}$ is the hardness parameter. If a public-key encryption scheme Π is CPA secure w.r.t. this family of functions, it is called $f(k)$-`AI-CPA (auxiliary input CPA) secure`. Our goal is to make $f(k)$ as large (i.e., as close to $\mathrm{negl}(k)$) as possible.

We also consider a weaker notion of auxiliary input security, called $f(k)$-`wAI-CPA (weak auxiliary input CPA) security`, where the class of functions under consideration is uninvertible *even given the public key of the scheme*. This notion is used as a stepping stone to achieving the stronger $f(k)$-`AI-CPA` notion.

- Let $\mathcal{H}_{\texttt{pk-ow}}(f(k))$ be the class of all polynomial-time computable functions $h : \{0,1\}^{|SK|+|PK|} \rightarrow \{0,1\}^*$, such that given $(PK, h(SK, PK))$ (for a randomly generated (SK, PK)), no PPT algorithm can find SK with probability greater than $f(k)$, where $f(k) \geq 2^{-k}$ is the hardness parameter. If a public-key encryption scheme Π is CPA secure w.r.t. this family of functions, it is called $f(k)$-`wAI-CPA (weak auxiliary input CPA) secure`.

The following lemma shows various relations between these notions of security. The proof follows directly from the definitions, and is omitted.

Lemma 4. *Assume Π is a public-key encryption scheme whose public key is of length $t(k)$.*

1. *If Π is $f(k)$-`AI-CPA secure`, then Π is $f(k)$-`wAI-CPA secure`.*
2. *If Π is $f(k)$-`wAI-CPA secure`, then Π is $(2^{-t(k)}f(k))$-`AI-CPA secure`.*
3. *If Π is $f(k)$-`AI-CPA secure`, then Π is $(k - \log(1/f(k)))$-`LB-CPA secure`.*

We now examine our new notions or strong and weak auxiliary input security ($f(k)$-`AI-CPA` and $f(k)$-`wAI-CPA`, respectively).

Strong Notion. We start with $f(k)$-`AI-CPA` security, which is the main notion we advocate. It states that as long as the leakage $y = h(SK, PK)$ did not reveal SK (with probability more than $f(k)$), the encryption remains secure. First, as shown in Part 3. of Lemma 4, it immediately implies that it is safe to leak $(k - \log(1/f(k)))$ arbitrary bits about the secret key. Thus, if $\log(1/f(k)) \ll k$, it means that we can leak almost the entire (min-)entropy of the secret key! This motivates our convention of using k as the min-entropy of the secret key, making our notion intuitive to understand in the leakage setting of [1,20]. Second, it implies very strong composition properties. As long as other usages of SK make it $f(k)$-hard to compute, these usages will not break the security of our encryption scheme.

Weak Notion. We next move to the more subtle notion of $f(k)$-`wAI-CPA` security. Here, we further restrict the leakage functions h to ones where SK remains hard to compute *given both the leakage $y = h(PK, SK)$ and the public key PK*. While this might sound natural, it has the following unexpected "anti-monotonicity" property. By making PK contain *more information* about the secret key, we could sometimes make the scheme *more secure* w.r.t. to this notion (i.e., the function $f(k)$ becomes larger), which seems unnatural. At the extreme, setting $PK = SK$ would make the scheme wAI-CPA "secure", since we now ruled out all "legal" auxiliary functions, making the notion vacuously true. (In the full version, we also give more convincing examples.)

Although this shows that the wAI-CPA security should be taken with care, we show it is still very useful. First, Lemma 4 shows that it is useful *when the scheme has a short public key*. In particular, this will be the case for all the schemes that we construct, where we will first show good wAI-CPA security, and then deduce almost the same AI-CPA security. Second, even if the scheme does not have a very short public key, wAI-CPA security might be useful in composing different schemes *sharing the same public-key infrastructure*. For example, assume we have a signature and an encryption scheme having the same pair (PK, SK). And assume that the signature scheme is shown to be $f(k)$-secure against key recovery attacks. Since the auxiliary information obtained by using the signature scheme certainly includes the public key, we can conclude that our $f(k)$-wAI-CPA secure encryption scheme is still secure, despite being used together with the signature scheme. In other words, while strong auxiliary input security would allow us to safely compose with any $f(k)$-secure signature scheme, even using a different PK, weak auxiliary input security is still enough when the PKI is shared, which is one of the motivating settings for auxiliary input security. Finally, we are able to construct $f(k)$-wAI-CPA secure schemes with the *optimal* value of $f(k)$, namely $f(k) = \text{negl}(k)$. We defer the details to the full version.

Public Parameters. For simplicity, when defining auxiliary input security, we did not consider the case when the encryption schemes might depend on system-wide parameters params. However, the notions of strong and weak auxiliary input security naturally generalize to this setting, as follows. First, to allow realistic attacks, the leakage function h can also depend on the parameters. Second, for both AI-CPA and wAI-CPA notions, SK should be hard to recover given params and $h(SK, PK, \texttt{params})$ (resp. $(PK, h(SK, PK, \texttt{params}))$).[4] Correspondingly, when applying Lemma 4, the length of the parameters is not counted towards the length $t(k)$ of the public key.

4 Goldreich-Levin Theorem for Large Fields

In this section, we prove a Goldreich-Levin theorem over any field $GF(q)$ for a prime q. In particular, we show:

Theorem 1. *Let q be prime, and let H be an arbitrary subset of $GF(q)$. Let $f : H^n \to \{0,1\}^*$ be any (possibly randomized) function. If there is a distinguisher $\mathcal{D}$ that runs in time t such that*

$$\Big| \Pr[\mathbf{s} \leftarrow H^n, y \leftarrow f(\mathbf{s}), \mathbf{r} \leftarrow GF(q)^n : \mathcal{D}(y, \mathbf{r}, \langle \mathbf{r}, \mathbf{s} \rangle) = 1] - \Pr[\mathbf{s} \leftarrow H^n, y \leftarrow f(\mathbf{s}), \mathbf{r} \leftarrow GF(q)^n, u \leftarrow GF(q) : \mathcal{D}(y, \mathbf{r}, u) = 1] \Big| = \epsilon$$

[4] Notice, unlike the case of wAI-CPA security, the inclusion of params as part of the leakage does not result in the "anti-monotonicity" problem discussed earlier, since params are independent of the secret key.

then there is an inverter $\mathcal{A}$ that runs in time $t' = t \cdot \mathsf{poly}(n, |H|, 1/\epsilon)$ such that

$$\Pr[\mathbf{s} \leftarrow H^n, y \leftarrow f(\mathbf{s}) : \mathcal{A}(y) = \mathbf{s}] \geq \frac{\epsilon^3}{512 \cdot n \cdot q^2} \tag{1}$$

Remark 1. Assume that the distinguisher $\mathcal{D}$ is a PPT algorithm and the distinguishing advantage ϵ is non-negligible in n. When q is polynomial in n, the running time of $\mathcal{A}$ is polynomial, and the success probability is inverse-polynomial in n, irrespective of H. When q is super-polynomial in n, but $|H|$ *is polynomial in* n, the running time of $\mathcal{A}$ *remains polynomial in* n, but the success-probability is dominated by the $1/q^2$ factor.

Remark 2. We briefly compare our new variant of the GL Lemma for general q with the similar-looking extension of Goldreich, Rubinfeld and Sudan [13]. The extensions are incomparable, in the following sense. In [13], the authors assume an *ϵ-predictor* for the inner product $\langle \mathbf{r}, \mathbf{s} \rangle$, which is a stronger assumption than the existence of an *ϵ-distinguisher*, especially as q grows. On the other, in [13] both the running time and the inversion probability of the inverter they construct depends only on n/ϵ and is *independent of* q (and, hence, $|H|$, if one considers restricting the domain as we do). Unfortunately, if one generically converts a distinguisher into a predictor, this conversion makes the prediction advantage equal to ϵ/q, which means that applying [13] would make both the inversion probability and the *running time of the inverter* depend on q. In contrast, we directly work with the distinguisher, and manage to only make the inversion probability dependent on q, while the *running time dependent only on* $|H|$.

Overview of the Proof. As in the standard Goldreich-Levin proof, we concentrate on the vectors $\mathbf{s}$ on which the distinguisher has $\Omega(\epsilon)$-advantage for random $\mathbf{r}$. Let c be a parameter that depends on $|H|, n, \epsilon$ (this parameter will be specified precisely in the formal proof below). As in the standard proof, our inverter $\mathcal{A}$ will guess c inner products $\langle \mathbf{s}, \mathbf{z}_i \rangle$ for random vectors $\mathbf{z}_1 \ldots \mathbf{z}_c$, losing $1/q^c$ factor in its success probability in the process. Then, assuming all our c guessed inner products are correct, for each $i = 1 \ldots n$ and $a \in H$, we use the assumed distinguisher $\mathcal{D}$ to design an efficient procedure to test, with high probability, if $s_i = a$. The details of this test, which is the crux of the argument, is given in the formal proof below, but the above structure explains why our running time only depends on $|H|$ and not q.

Proof. We will actually prove a tighter version of the bound stated in Equation (1), and design an inverter $\mathcal{A}$ with success probability $\epsilon/4q^c$, where $c \geq 2$ is the smallest integer such that $q^c > 128|H|n/\epsilon^2$. The general bound in Equation (1) follows since

$$4q^c \leq 4\max(q^2, q \cdot (128|H|n/\epsilon^2)) = 4q \cdot \max(q, 128|H|n/\epsilon^2) \leq 512q^2n/\epsilon^2.$$

Thus,

- If $q > 128|H|n/\epsilon^2$ then $c = 2$ and the above probability is at least $\epsilon/4q^2$.
- If $q \leq 128|H|n/\epsilon^2$ then $q^c \leq q \cdot (128|H|n/\epsilon^2)$, and thus the above probability is at least $\epsilon^3/512nq|H| \geq \epsilon^3/512nq^2$.

Without loss of generality, we will drop the absolute value from the condition on $\mathcal{D}$, by flipping the decision of $\mathcal{D}$, if needed. Also, for a fixed value $\mathbf{s} \in H^\ell$ and fixed randomness of f (in case f is randomized), let $y = f(\mathbf{s})$ and let

$$\alpha_{\mathbf{s},y} = \Pr[\mathbf{r} \leftarrow GF(q)^n : \mathcal{D}(y, \mathbf{r}, \langle \mathbf{r}, \mathbf{s} \rangle) = 1]$$
$$\beta_{\mathbf{s},y} = \Pr[\mathbf{r} \leftarrow GF(q)^n, u \leftarrow GF(q) : \mathcal{D}(y, \mathbf{r}, u) = 1]$$

Thus, we know that $\mathbb{E}_{\mathbf{s}}[\alpha_{\mathbf{s},y} - \beta_{\mathbf{s},y}] \geq \epsilon$ (note, this expectation also includes possible randomness of f, but we ignore it to keep the notation uncluttered). Let us call the pair $(\mathbf{s}, y)$ *good* if $\alpha_{\mathbf{s},y} - \beta_{\mathbf{s},y} \geq \epsilon/2$. Since $\alpha_{\mathbf{s},y} - \beta_{\mathbf{s},y} \leq 1$, a simple averaging argument implies that

$$\Pr[\mathbf{s} \leftarrow H^n, y \leftarrow f(\mathbf{s}) : (\mathbf{s}, y) \text{ is good}] \geq \epsilon/2 \tag{2}$$

Below, we will design an algorithm $\mathcal{A}(y)$ which will succeed to recover $\mathbf{s}$ from y with probability $1/2q^c$, whenever the pair $(\mathbf{s}, y)$ is good. Coupled with Equation (2), this will establish that $\mathcal{A}$'s overall probability of success (for random $\mathbf{s}$ and y) is at least $\epsilon/4q^c$, as required. Thus, in the discussion below, we will assume that $(\mathbf{s}, y)$ is fixed and good.

Before describing $\mathcal{A}(y)$, we will also assume that $\mathcal{A}(y)$ can compute, with overwheliming probability, a number $\gamma_{\mathbf{s},y}$ such that $(\alpha_{\mathbf{s},y} - \epsilon/8 \geq \gamma_{\mathbf{s},y} \geq \beta_{\mathbf{s},y} + \epsilon/8)$. Indeed, by sampling $O(n/\epsilon^2)$ random and independent vectors $\mathbf{r}$ and u, $\mathcal{A}(y)$ can compute an estimate e for $\beta_{\mathbf{s},y}$, such that $\Pr(|e - \beta_{\mathbf{s},y}| > \epsilon/8) \leq 2^{-n}$ (by the Chernoff's bound), after which one can set $\gamma_{\mathbf{s},y} = e + \epsilon/4$. So we will assume that $\mathcal{A}(y)$ can compute such an estimate $\gamma_{\mathbf{s},y}$.

Let $m \stackrel{\text{def}}{=} 128|H|n/\epsilon^2$. By assumption, $c \geq 2$ is such that $q^c > m$. Let us fix an arbitrary subset $S \subseteq GF(q)^c \backslash \{0^c\}$ of cardinality m, such that every two elements in S are linearly independent. This can be achieved for example by choosing $S \subseteq GF(q)^c \backslash \{0^c\}$ to be an arbitrary set of cardinality m such that the first coordinate of each element in S equals 1. The algorithm $\mathcal{A}(y)$ works as follows.

---

1. Compute the value $\gamma_{\mathbf{s},y}$ such that $(\alpha_{\mathbf{s},y} - \epsilon/8 \geq \gamma_{\mathbf{s},y} \geq \beta_{\mathbf{s},y} + \epsilon/8)$, as described above.
2. Choose c random vectors $\mathbf{z}_1, \ldots, \mathbf{z}_c \leftarrow GF(q)^n$, and c random elements $g_1, \ldots, g_c \leftarrow GF(q)$.
   [Remark: Informally, the g_i are A's "guesses" for the values of $\langle \mathbf{z}_i, \mathbf{s} \rangle$.]
3. For every tuple $\bar{\rho} = (\rho_1, \ldots, \rho_c) \in S$, compute

$$\mathbf{r}_{\bar{\rho}} := \sum_{j=1}^{c} \rho_j \mathbf{z}_j \quad \text{and} \quad h_{\bar{\rho}} := \sum_{j=1}^{c} \rho_j g_j \tag{3}$$

   [Remark: If the guesses g_i are all correct, then for every $\bar{\rho}$, we have $h_{\bar{\rho}} = \langle \mathbf{r}_{\bar{\rho}}, \mathbf{s} \rangle$. Also notice that the vectors $\mathbf{r}_{\bar{\rho}}$ are pairwise independent since $c \geq 2$ and $\bar{\rho} \neq 0^c$.]
4. For each $i \in [n]$, do the following:
   - For each $a \in H$, guess that $s_i = a$, and run the following procedure to check if the guess is correct:

- For each $\bar{\rho} \in S$, choose a random $\tau_{\bar{\rho}}^{(i,a)} \in GF(q)$ and run

$$\mathcal{D}(y, \mathbf{r}_{\bar{\rho}} + \tau_{\bar{\rho}}^{(i,a)} \cdot \mathbf{e}_i, h_{\bar{\rho}} + \tau_{\bar{\rho}}^{(i,a)} \cdot a)$$

where $\mathbf{e}_i$ is the ith unit vector. Let $p^{(i,a)}$ be the fraction of $\mathcal{D}$'s answers which are 1.
- If $p^{(i,a)} \geq \gamma_{\mathbf{s},y}$, set $s_i := a$ and move to the next $i + 1$. Otherwise, move to the next guess $a \in H$.

5. Output $\mathbf{s} = s_1 s_2 \ldots s_n$ (or fail if some s_i was not found).

The procedure invokes the distinguisher $\mathcal{D}$ at most $O(nm|H|)$ times (not counting the estimation step for $\gamma_{\mathbf{s},y}$ which is smaller), and thus the running time is $O(t \cdot nm|H|) = t \cdot \mathsf{poly}(n, |H|, 1/\epsilon)$, where t is the running time of $\mathcal{D}$. Let us now analyze the probability that the procedure succeeds.

First, define the event E to be the event that for all $\bar{\rho} \in S$, we have $h_{\bar{\rho}} = \langle \mathbf{r}_{\bar{\rho}}, \mathbf{s} \rangle$.

$$\Pr[E] = \Pr[\forall \bar{\rho} \in S, h_{\bar{\rho}} = \langle \mathbf{r}_{\bar{\rho}}, \mathbf{s} \rangle] \geq \Pr[\forall i \in [1, \ldots, c], g_i = \langle \mathbf{z}_i, \mathbf{s} \rangle] = \frac{1}{q^c}$$

where the last equality follows from the fact that A's random guess of g_i are all correct with probability $1/q^c$. For the rest of the proof, we condition on the event E (and, of course, on the goodness of $(\mathbf{s}, y)$), and show that $\mathcal{A}$'s success is at least $1/2$ in this case, completing the proof.

We next prove two claims. First, in Claim 4 we show that if A's guess a for s_i is correct, then each individual input to $\mathcal{D}$ is distributed like $(y, \mathbf{r}, \langle \mathbf{r}, \mathbf{s} \rangle)$, for a random $\mathbf{r}$. Thus, the probability that $\mathcal{D}$ answers 1 on these inputs is exactly $\alpha_{\mathbf{s},y}$. Moreover, the inputs to $\mathcal{D}$ are *pairwise independent*. This follows from their definition in Equation 3, since every two elements in S are linearly independent and we excluded $\bar{\rho} = 0^c$. Thus, by Chebyshev's inequality, the probability that the average $p^{i,a}$ of m pairwise independent estimations of $\alpha_{\mathbf{s},y}$ is smaller than $\gamma_{\mathbf{s},y}$, which is more than $\epsilon/8$ smaller than the true average $\alpha_{\mathbf{s},y}$, is at most $1/(m(\epsilon/8)^2) = 1/2|H|n$, where we recall that $m = 128|H|n/\epsilon^2$.

Secondly, in Claim 4 we show that for every incorrect guess a for s_i, each individual input to $\mathcal{D}$ is distributed like $(y, \mathbf{r}, u)$ for random $\mathbf{r}$ and u. Thus, the probability that $\mathcal{D}$ answers 1 on these inputs is exactly $\beta_{\mathbf{s},y}$. And, as before, different values $\mathbf{r}, u$ are pairwise independent.[5] By an argument similar to the above, in this case the probability that the average $p^{i,a}$ of m pairwise independent estimations of $\beta_{\mathbf{s},y}$ is larger than $\gamma_{\mathbf{s},y}$, which is more than $\epsilon/8$ larger than the true average $\beta_{\mathbf{s},y}$, is at most $1/(m(\epsilon/8)^2) = 1/2|H|n$.

This suffices to prove our result, since, by the union bound over all $i \in [1, \ldots, n]$ and $a \in |H|$, the chance that $\mathcal{A}$ will incorrectly test any pair (i, a) (either as false positive or false negative) is at most $|H|n \cdot 1/(2|H|n) = 1/2$. Thus, it suffices to prove the two claims.

[5] The argument is the same for $\mathbf{r}$ and for the values u, Claim 4 shows that they are in fact completely independent.

Claim. If A's guess is correct, i.e, $s_i = a$, the inputs to $\mathcal{D}$ are distributed like $(y, \mathbf{r}, \langle \mathbf{r}, \mathbf{s} \rangle)$.

Proof: Each input to $\mathcal{D}$ is of the form $(y, \mathbf{r}_{\bar{\rho}} + \tau_{\bar{\rho}}^{(i,a)} \cdot \mathbf{e}_i, h_{\bar{\rho}} + \tau_{\bar{\rho}}^{(i,a)} \cdot a)$. Since we already conditioned on E, we know that $h_{\bar{\rho}} = \langle \mathbf{r}_{\bar{\rho}}, \mathbf{s} \rangle$. Also, we assumed that $s_i = a$. Thus,

$$h_{\bar{\rho}} + \tau_{\bar{\rho}}^{(i,a)} \cdot a = \langle \mathbf{r}_{\bar{\rho}}, \mathbf{s} \rangle + \tau_{\bar{\rho}}^{(i,a)} \cdot s_i = \langle \mathbf{r}_{\bar{\rho}}, \mathbf{s} \rangle + \langle \tau_{\bar{\rho}}^{(i,a)} \mathbf{e}_i, \mathbf{s} \rangle = \langle \mathbf{r}_{\bar{\rho}} + \tau_{\bar{\rho}}^{(i,a)} \cdot \mathbf{e}_i, \mathbf{s} \rangle$$

Since $\mathbf{r}_{\bar{\rho}} + \tau_{\bar{\rho}}^{(i,a)} \cdot \mathbf{e}_i$ is uniformly random by itself (see Equation 3 and remember $\bar{\rho} \neq 0^c$), the input of $\mathcal{D}$ is indeed of the form $(y, \mathbf{r}, \langle \mathbf{r}, \mathbf{s} \rangle)$, where $\mathbf{r} := \mathbf{r}_{\bar{\rho}} + \tau_{\bar{\rho}}^{(i,a)} \cdot \mathbf{e}_i$ is uniformly random. □

Claim. If A's guess is incorrect, i.e, $s_i \neq a$, the inputs to $\mathcal{D}$ are distributed like $(y, \mathbf{r}, u)$ for a uniformly random $u \in GF(q)$.

Proof: The proof proceeds similar to Claim 4. As before, each input to $\mathcal{D}$ is of the form $(y, \mathbf{r}_{\bar{\rho}} + \tau_{\bar{\rho}}^{(i,a)} \cdot \mathbf{e}_i, h_{\bar{\rho}} + \tau_{\bar{\rho}}^{(i,a)} \cdot a)$. Now, however, $s_i \neq a$, so suppose $a - s_i = t_i \neq 0$. Then

$$h_{\bar{\rho}} + \tau_{\bar{\rho}}^{(i,a)} \cdot a = \langle \mathbf{r}_{\bar{\rho}}, \mathbf{s} \rangle + \tau_{\bar{\rho}}^{(i,a)} \cdot (s_i + t_i) = \langle \mathbf{r}_{\bar{\rho}} + \rho_{\bar{\rho}}^{(i,a)} \cdot \mathbf{e}_i, \mathbf{s} \rangle + \tau_{\bar{\rho}}^{(i,a)} \cdot t_i$$

Let $\mathbf{r} := \mathbf{r}_{\bar{\rho}} + \tau_{\bar{\rho}}^{(i,a)} \cdot \mathbf{e}_i$ and $u := h_{\bar{\rho}} + \tau_{\bar{\rho}}^{(i,a)} \cdot a$, so that the input to $\mathcal{D}$ is $(y, \mathbf{r}, u)$. By the equation above, we have $u = \langle \mathbf{r}, \mathbf{s} \rangle + \rho_{\bar{\rho}}^{(i,a)} \cdot t_i$. Also, since $\mathbf{r}_{\bar{\rho}}$ is uniformly random, it perfectly hides $\tau_{\bar{\rho}}^{(i,a)}$ in the definition of $\mathbf{r}$. Thus, $\mathbf{r}$ is independent from $\tau_{\bar{\rho}}^{(i,a)}$. Finally, since we assumed that $t_i \neq 0$ and the value $\tau_{\bar{\rho}}^{(i,a)}$ was random in $GF(q)$, this means that $u = \langle \mathbf{r}, \mathbf{s} \rangle + \tau_{\bar{\rho}}^{(i,a)} \cdot t_i$ is random and independent of $\mathbf{r}$, as claimed. □

This concludes the proof of Theorem 1.

5 Auxiliary Input Secure Encryption Schemes

5.1 Construction Based on the DDH Assumption

We show that the BHHO encryption scheme [3] is secure against subexponentially hard-to-invert auxiliary input.

The BHHO Cryptosystem. Let n be the security parameter. Let $\mathcal{G}$ be a probabilistic polynomial-time "group generator" that, given the security parameter n in unary, outputs the description of a group G that has prime order $q = q(n)$.

– $\mathsf{KeyGen}(1^n, \epsilon)$: Let $m := (4 \log q)^{1/\epsilon}$, and let $G \leftarrow \mathcal{G}(1^n)$. Sample m random generators $g_1, \ldots, g_m \leftarrow G$. Let $\mathbf{g} = (g_1, \ldots, g_m)$. Choose a uniformly random m-bit string $\mathbf{s} = (s_1, \ldots, s_m) \in \{0,1\}^m$, and define

$$y := \prod_{i=1}^{m} g_i^{s_i} \in G.$$

Let the secret key $SK = \mathrm{s}$, and let the public key $PK = (\mathrm{g}, y)$ (plus the description of G). Note, g can be viewed as public parameters, meaning only y can be viewed as the "user-specific" public key.

- $\mathsf{Enc}(PK, M)$: Let the message $M \in G$. Choose a uniformly random $r \in \mathbb{Z}_q$. Compute $f_i := g_i^r$ for each i, and output the ciphertext

$$C := (f_1, \dots, f_m, y^r \cdot M) \in G^{m+1}.$$

- $\mathsf{Dec}(SK, C)$: Parse the ciphertext C as $(f_1, \dots, f_m, c)$, and the secret key $SK = (s_1, \dots, s_m)$. Output

$$M' := c \cdot \left(\prod_{i=1}^{m} f_i^{s_i} \right)^{-1} \in G.$$

To see the correctness of the encryption scheme, observe that if $f_i = g_i^r$ for all i, then the decryption algorithm outputs

$$M' = c \cdot \left(\prod_{i=1}^{m} f_i^{s_i} \right)^{-1} = c \cdot \left(\prod_{i=1}^{m} g_i^{s_i} \right)^{-r} = c \cdot y^{-r} = M$$

We now show that the BHHO scheme is secure against subexponentially hard auxiliary inputs, under the DDH assumption.

Theorem 2. *Assuming that the Decisional Diffie-Hellman problem is hard for $\mathcal{G}$, the encryption scheme described above is* $(2^{-m^{\epsilon}})$-`AI-CPA` *secure (when* g *is viewed as a public parameter).*

Remark. We can actually handle a richer class of auxiliary functions, namely, any $h(\mathrm{g}, \mathrm{s})$ that (given g) is to hard invert with probability $1/2^k$, where k can be as small as $\mathrm{polylog}(m)$. However, then the assumption we rely on is that DDH is hard for adversaries that run in subexponential time. For the sake of simplicity, we only state the theorem for $k = m^{\epsilon}$ in which case we can rely on the standard DDH hardness assumption.

Proof (of Theorem 2). By Lemma 4 (Part 2) and because the length of "user-specific" public key y is $\log q$ bits, to prove the theorem it suffices to show that our encryption scheme is $(q2^{-m^{\epsilon}})$-`wAI-CPA` secure. Fix any auxiliary-input function h, so that s is still $(q \cdot 2^{-m^{\epsilon}})$-hard given $(\mathrm{g}, y, h(\mathrm{g}, \mathrm{s}))$, and a PPT adversary $\mathcal{A}$ with advantage $\delta = \delta(n) = \mathsf{Adv}_{\mathcal{A},h}(n)$.

We consider a sequence of experiments, letting $\mathsf{Adv}_{\mathcal{A},h}^{(i)}(n)$ denote the advantage of the adversary in experiment i.

Experiment 0*:* This is the experiment in Definition 2. The adversary $\mathcal{A}$ gets as input $PK = (\mathrm{g}, y)$ and the auxiliary input $h(\mathrm{g}, \mathrm{s})$. $\mathcal{A}$ chooses two messages M_0 and M_1, and receives $C = \mathsf{Enc}(PK, M_b)$ where $b \in \{0, 1\}$ is uniformly random. $\mathcal{A}$ succeeds in the experiment if he succeeds in guessing b. By assumption, $\mathsf{Adv}_{\mathcal{A},h}^{(0)}(n) = \mathsf{Adv}_{\mathcal{A},h}(n) = \delta$.

Experiment 1*:* In this experiment, the challenge ciphertext C is generated by "encrypting with the secret key," rather than with the usual $\mathsf{Enc}(PK, M_b)$ algorithm. In particular, define the algorithm $\mathsf{Enc}'(\mathbf{g}, \mathbf{s}, M_b)$ as follows.

1. Choose $r \leftarrow \mathbb{Z}_q$, and compute the first m components of the ciphertext $(f_1, \ldots, f_m) = (g_1^r, \ldots, g_m^r)$.
2. Compute the last component of the ciphertext as $c = \prod_{i=1}^m f_i^{s_i} \cdot M_b$.

Claim. The distribution produced by Enc' is identical to the distribution of a real ciphertext; in particular, $\mathsf{Adv}^{(0)}_{\mathcal{A},h}(n) = \mathsf{Adv}^{(1)}_{\mathcal{A},h}(n)$.

Proof. Fix $\mathbf{g}$ and $\mathbf{s}$ (and hence y). Then for uniformly random $r \in \mathbb{Z}_q$, both Enc and Enc' compute the same $f_1, \ldots, f_m$, and their final ciphertext components also coincide:

$$C = \prod_{i=1}^{m} f_i^{s_i} \cdot M = \prod_{i=1}^{m} g_i^{rs_i} \cdot M = \Big(\prod_{i=1}^{m} g_i^{s_i}\Big)^r \cdot M = y^r \cdot M.$$

Experiment 2*:* In this experiment, the vector $(f_1, \ldots, f_m)$ in the ciphertext is taken from the *uniform* distribution over G^m, i.e., each $f_i = g^{r_i}$ for uniformly random and independent $r_i \in \mathbb{Z}_q$ (where g is some fixed generator of G), and $C = \prod_i f_i^{s_i} \cdot M_b$ as before. Under the DDH assumption and by Lemma 1, it immediately follows that the advantage of the adversary changes by at most a negligible amount.

Claim. If the DDH problem is hard for $\mathcal{G}$, then for every PPT algorithm $\mathcal{A}$ and for every function $h \in \mathcal{H}$, $\left|\mathsf{Adv}^{(1)}_{\mathcal{A},h}(n) - \mathsf{Adv}^{(2)}_{\mathcal{A},h}(n)\right| \le \mathrm{negl}(n)$.

Experiment 3*:* In this experiment, the final component of the ciphertext is replaced by a uniformly random element $u \leftarrow G$. Namely, the ciphertext is generated as $(g^{r_1}, \ldots, g^{r_m}, g^u)$, where $r_i \in \mathbb{Z}_q$ and $u \in \mathbb{Z}_q$ are all uniformly random and independent.

Claim. For every PPT algorithm $\mathcal{A}$ and for every $h \in \mathcal{H}$, $\left|\mathsf{Adv}^{(2)}_{\mathcal{A},h}(n) - \mathsf{Adv}^{(3)}_{\mathcal{A},h}(n)\right| \le \mathrm{negl}(n)$.

Proof. We reduce the task of inverting h (with suitable probability) to the task of gaining some non-negligible $\delta = \delta(n)$ distinguishing advantage between experiments 2 and 3.

We wish to construct an efficient algorithm that, given $PK = (\mathbf{g}, y)$ and $h(\mathbf{g}, \mathbf{s})$, outputs $\mathbf{s} \in H^m = \{0,1\}^m$ with probability at least

$$q \cdot \frac{\delta^3}{512n \cdot q^3} > q \cdot \frac{1}{512n \cdot 2^{3m^\epsilon/4} \cdot \mathrm{poly}(n)} > q \cdot 2^{-m^\epsilon},$$

for large enough n. By Theorem 1, it suffices to reduce δ-distinguishing

$$(PK, h(\mathbf{g}, \mathbf{s}), \mathbf{r} \in \mathbb{Z}_q^m, \langle \mathbf{r}, \mathbf{s}\rangle) \text{ from } (PK, h(\mathbf{g}, \mathbf{s}), \mathbf{r}, u \in \mathbb{Z}_q)$$

to δ-distinguishing between experiments 2 and 3.

The reduction $\mathcal{B}$ that accomplishes this simulates the view of the adversary $\mathcal{A}$ as follows. On input $(PK = (\mathbf{g}, y), h(\mathbf{g}, \mathbf{s}), \mathbf{r}, z \in \mathbb{Z}_q)$, give PK to $\mathcal{A}$ and get back two messages M_0, M_1; choose a bit $b \in \{0,1\}$ at random and give $\mathcal{A}$ the ciphertext $(g^{r_1}, \ldots, g^{r_m}, g^z \cdot M_b)$. Let b' be the output of $\mathcal{A}$; if $b = b'$ then $\mathcal{B}$ outputs 1, and otherwise $\mathcal{B}$ outputs 0.

By construction, it may be checked that when $\mathcal{B}$'s input component $z = \langle \mathbf{r}, \mathbf{s} \rangle \in \mathbb{Z}_q$, $\mathcal{B}$ simulates experiment 2 to $\mathcal{A}$ perfectly. Likewise, when z is uniformly random and independent of the other components, $\mathcal{B}$ simulates experiment 3 perfectly. It follows that $\mathcal{B}$'s advantage equals $\mathcal{A}$'s.

Now the ciphertext in experiment 3 is independent of the bit b that selects which message is encrypted. Thus, the adversary has no advantage in this experiment, i.e, $\mathsf{Adv}^{(3)}_{\mathcal{A},h}(n) = 0$. Putting together the claims, we get that $\mathsf{Adv}_{\mathcal{A},h}(n) \leq \mathrm{negl}(n)$.

5.2 Constructions Based on LWE

First, we present (a modification of) the GPV encryption scheme [12]. We then show that the system is secure against sub-exponentially hard auxiliary input functions, assuming the hardness of the learning with error (LWE) problem.

The GPV Cryptosystem. Let n denote the security parameter, and let $0 < \epsilon \leq 1$. Let $f(n) = 2^{\omega(\log n)}$ be some superpolynomial function. Let the prime $q \in (f(n), 2 \cdot f(n)]$, the integer $m = ((n+3)\log q)^{1/\epsilon}$ and the error-distributions $\overline{\Psi}_\beta$ and $\overline{\Psi}_\gamma$ where $\beta = 2\sqrt{n}/q$ and $\gamma = 1/(8 \cdot \omega(\sqrt{\log n}))$ be parameters of the system.

$\mathsf{Gen}(1^n)$: Choose a uniformly random matrix $\mathbf{A} \leftarrow \mathbb{Z}_q^{n \times m}$ and a random vector $\mathbf{e} \leftarrow \{0,1\}^m$. Compute $\mathbf{u} = \mathbf{A}\mathbf{e}$. The public key $PK := (\mathbf{A}, \mathbf{u})$ and the secret key $SK := \mathbf{e}$. We notice that the matrix $\mathbf{A}$ could be viewed as a public parameter, making $\mathbf{u}$ the only "user-specific" part of PK for the purposes of Lemma 4.

$\mathsf{Enc}(PK, b)$, where b is a bit, works as follows. Choose a random vector $\mathbf{s} \leftarrow \mathbb{Z}_q^n$, a vector $\mathbf{x} \leftarrow \overline{\Psi}_\beta^m$ and $x' \leftarrow \overline{\Psi}_\gamma$. Output the ciphertext

$$\left(\mathbf{A}^T\mathbf{s} + \mathbf{x}, \mathbf{u}^T\mathbf{s} + x' + b\left\lfloor \frac{q}{2} \right\rfloor\right)$$

$\mathsf{Dec}(SK, c)$: Parse the ciphertext as $(\mathbf{y}, c) \in \mathbb{Z}_q^m \times \mathbb{Z}_q$ and compute $b' = (c - \mathbf{e}^T\mathbf{y})/q$. Output 1 if $b' \in [\frac{1}{4}, \frac{3}{4}] \bmod 1$, and 0 otherwise.

Remark 3. There are two main differences between the cryptosystem in [12] and the variant described here. First, we choose the error parameter β to be superpolynomially small in n (and the modulus q to be superpolynomially large), whereas in [12], both are polynomially related to n. Second, the secret-key distribution in our case is the uniform distribution over $\{0,1\}^m$, whereas in [12], it is the discrete Gaussian distribution $D_{\mathbb{Z}^m,r}$ for some $r > 0$. The first modification is essential to our proof of auxiliary-input security. The second modification is not inherent; using a discrete Gaussian distribution also results in an identity-based encryption scheme (as in [12]) secure against auxiliary inputs. We defer the details to the full version of this paper.

Remark 4. Although the encryption scheme described here is a bit-encryption scheme, it can be modified to encrypt $O(\log q)$ bits with one invocation of the encryption algorithm, using the ideas of [17,22]. However, we note that another optimization proposed in [22] that achieves *constant* ciphertext expansion does not seem to lend itself to security against auxiliary inputs. Roughly speaking, the reason is that the optimization enlarges the secret key by repeating the secret key of the basic encryption scheme polynomially many times; this seems to adversely affect auxiliary input security.

We now show that the (modified) GPV scheme is secure against subexponentially hard auxiliary inputs, under the decisional LWE assumption.

Theorem 3. *Let the superpolynomial function $f(n)$ and the parameters m, q, β and γ be as in the encryption scheme described above. Assuming that the $\mathsf{DLWE}_{n,m,q,\beta}$ problem is hard, the encryption scheme above is (2^{-m^ϵ})-*AI-CPA *secure (when $\mathbf{A}$ is viewed as a public parameter).*

Remark. We can actually prove security even for auxiliary functions $h(\mathbf{A}, \mathbf{e})$ that (given $\mathbf{A}$) are hard to invert with probability 2^{-k}, where k can be as small as $\text{polylog}(m)$. However, then the assumption we rely on is that LWE is hard for adversaries that run in subexponential time. For the sake of simplicity, we only state the theorem for $k = m^\epsilon$ in which case we can rely on the standard LWE hardness assumption.

Proof (of Theorem 3). By Lemma 4 (Part 2) and because the length of "user-specific" public key $\mathbf{u}$ is $n \log q$ bits, to show Theorem 3 it suffices to show that our encryption scheme is $(q^n 2^{-m^\epsilon})$-wAI-CPA secure. Fix any auxiliary-input function h, so that $\mathbf{e}$ is still $(q^n \cdot 2^{-m^\epsilon})$-hard given $(\mathbf{A}, \mathbf{u}, h(\mathbf{A}, \mathbf{e}))$, and a PPT adversary $\mathcal{A}$ with advantage $\delta = \delta(n) = \mathsf{Adv}_{\mathcal{A},h}(n)$.

We consider a sequence of experiments, and let $\mathsf{Adv}^{(i)}_{\mathcal{A},h}(n)$ denote the advantage of the adversary in experiment i.

Experiment 0*:* This is the experiment in Definition 2. The adversary $\mathcal{A}$ gets as input $PK = (\mathbf{A}, \mathbf{u})$ and the auxiliary input $h(\mathbf{A}, \mathbf{e})$. $\mathcal{A}$ receives $\mathsf{Enc}(PK, b)$ where $b \in \{0, 1\}$ is uniformly random. $\mathcal{A}$ succeeds in the experiment if he succeeds in guessing b. By assumption, $\mathsf{Adv}^{(0)}_{\mathcal{A},h}(n) = \mathsf{Adv}_{\mathcal{A},h}(n) = \delta$.

Experiment 1*:* In this experiment, the challenge ciphertext is generated by "encrypting with the secret key," rather than with the usual $\mathsf{Enc}(PK, b)$ algorithm. In particular, define the algorithm $\mathsf{Enc}'(\mathbf{A}, \mathbf{e}, b)$ as follows.

1. Choose $\mathbf{s} \leftarrow \mathbb{Z}_q^n$ at random and $\mathbf{x} \leftarrow \overline{\Psi}_\beta^m$, and compute the first component of the ciphertext $\mathbf{y} = \mathbf{A}^T\mathbf{s} + \mathbf{x}$.
2. Choose $x' \leftarrow \overline{\Psi}_\gamma$ and compute the second component of the ciphertext as

$$c = \mathbf{e}^T\mathbf{y} + x' + b\lfloor q/2 \rfloor$$

Claim. The distribution produced by Enc' is statistically close to the distribution of a real ciphertext; in particular, there is a negligible function negl such that

$$\left|\mathsf{Adv}^{(0)}_{\mathcal{A},h}(n) - \mathsf{Adv}^{(1)}_{\mathcal{A},h}(n)\right| \leq \mathsf{negl}(n).$$

Proof. Fix $\mathbf{A}$ and $\mathbf{e}$ (and hence $\mathbf{u}$). Then for uniformly random $\mathbf{s} \in \mathbb{Z}_q^n$ and $\mathbf{x} \leftarrow \overline{\Psi}_\beta^m$, both Enc and Enc′ compute the same $\mathbf{y}$. Given $\mathbf{y}$, the second component of the ciphertext produced by Enc′ is

$$c = \mathbf{e}^T\mathbf{y} + x' + b\lfloor q/2 \rfloor = \mathbf{e}^T A\mathbf{s} + (\mathbf{e}^T\mathbf{x} + x') + b\lfloor q/2 \rfloor = \mathbf{u}^T\mathbf{s} + (\mathbf{e}^T\mathbf{x} + x') + b\lfloor q/2 \rfloor$$

It suffices to show that the distribution of $\mathbf{e}^T\mathbf{x} + x'$ is statistically indistinguishable from $\overline{\Psi}_\gamma$. This follows from Lemma 3 and the fact that

$$\mathbf{e}^T\mathbf{x}/(\gamma q) \leq ||\mathbf{e}|| \cdot ||\mathbf{x}||/(\gamma q) \leq \sqrt{m} \cdot \beta q \cdot \omega(\sqrt{\log n})/(\gamma q) = 2 \cdot \sqrt{mn} \cdot \omega(\log n)/q$$

is a negligible function of n.

Experiment 2*:* In this experiment, the vector $\mathbf{y}$ in the ciphertext is taken from the uniform distribution over $\mathbb{Z}_q^m$. Assuming the $\mathsf{DLWE}_{n,m,q,\beta}$ problem is hard, it immediately follows that the advantage of the adversary changes by at most a negligible amount.

Claim. If the $\mathsf{DLWE}_{n,m,q,\beta}$ problem is hard, then for every PPT algorithm $\mathcal{A}$ and for every function $h \in \mathcal{H}$, there is a negligible function negl such that $\left|\mathsf{Adv}_{\mathcal{A},h}^{(1)}(n) - \mathsf{Adv}_{\mathcal{A},h}^{(2)}(n)\right| \leq \mathsf{negl}(n)$.

Experiment 3*:* In this experiment, the second component of the ciphertext is replaced by a uniformly random element $r \leftarrow \mathbb{Z}_q$. Namely, the ciphertext is generated as $(\mathbf{y}, r)$, where $\mathbf{y} \leftarrow \mathbb{Z}_q^m$ is uniformly random, and $r \leftarrow \mathbb{Z}_q$ is uniformly random.

Claim. For every PPT algorithm $\mathcal{A}$ and for every function $h \in \mathcal{H}$, there is a negligible function negl such that $\left|\mathsf{Adv}_{\mathcal{A},h}^{(2)}(n) - \mathsf{Adv}_{\mathcal{A},h}^{(3)}(n)\right| \leq \mathsf{negl}(n)$.

Proof. We reduce the task of inverting h to the task of gaining a non-negligible distinguishing advantage between experiments 2 and 3. Suppose for the sake of contradiction that there exists a PPT algorithm $\mathcal{A}$, a function $h \in \mathcal{H}$, and a polynomial p such that for infinitely many n's, $\left|\mathsf{Adv}_{\mathcal{A},h}^{(2)}(n) - \mathsf{Adv}_{\mathcal{A},h}^{(3)}(n)\right| \geq 1/p(n)$. We show that this implies that there exists a PPT algorithm $\mathcal{B}$ so that for infinitely many n's,

$$\left|\Pr[\mathcal{B}(\mathbf{A}, \mathbf{u}, h(\mathbf{A}, \mathbf{e}), \mathbf{y}, \mathbf{e}^T\mathbf{y}) = 1] - \Pr[\mathcal{B}(\mathbf{A}, \mathbf{u}, h(\mathbf{A}, \mathbf{e}), \mathbf{y}, r) = 1]\right| \geq 1/p(n) \quad (4)$$

The adversary $\mathcal{B}$ will simulate $\mathcal{A}$, as follows. On input $(\mathbf{A}, \mathbf{u}, h(\mathbf{A}, \mathbf{e}), \mathbf{y}, c)$, algorithm $\mathcal{B}$ will choose a random bit $b \in \{0, 1\}$ and will start emulating $\mathcal{A}(PK, h(\mathbf{A}, \mathbf{e}))$, where $PK = (\mathbf{A}, \mathbf{u})$. The algorithm $\mathcal{B}$ will then sample $x' \leftarrow \overline{\Psi}_\gamma$ and a uniformly random bit $b \leftarrow \{0, 1\}$ and feed $\mathcal{A}$ the ciphertext $(\mathbf{y}, r + x' + b\lfloor q/2 \rfloor \pmod q)$. Let b' be the output of $\mathcal{A}$. If $b = b'$ then $\mathcal{B}$ outputs 1, and otherwise $\mathcal{B}$ outputs 0.

By definition

$$\Pr[\mathcal{B}(\mathbf{A}, \mathbf{u}, h(\mathbf{A}, \mathbf{e}), \mathbf{y}, \mathbf{e}^T\mathbf{y}) = 1] = \mathsf{Adv}_{\mathcal{A},h}^{(2)}(n),$$

and

$$\Pr[\mathcal{B}(\mathbf{A}, \mathbf{u}, h(\mathbf{A}, \mathbf{e}), \mathbf{y}, r) = 1] = \mathsf{Adv}_{\mathcal{A},h}^{(3)}(n).$$

This, together with the assumption that $|\mathsf{Adv}^{(2)}_{\mathcal{A},h}(n) - \mathsf{Adv}^{(3)}_{\mathcal{A},h}(n)| \geq 1/p(n)$, implies that Equation (4) holds. Now, we use Goldreich-Levin theorem over the (large) field $\mathbb{Z}_q$ and $H = \{0,1\} \subseteq \mathbb{Z}_q$ (Theorem 1). By Theorem 1, there is an algorithm that, given $PK = (\mathbf{A}, \mathbf{u})$, inverts $h(\mathbf{A}, \mathbf{e})$ with probability greater than

$$\frac{\delta^3}{512 \cdot n \cdot q^2} = q^n \cdot \frac{\delta^3 \cdot q}{512 \cdot n \cdot q^{n+3}} > q \cdot 2^{-m^\epsilon}$$

since $q^{n+3} = 2^{m^\epsilon}$ and $512 \cdot n/\delta^3 \cdot q < 1$ for large enough n. This provides the desired contradiction.

The ciphertext in experiment 3 contains no information about the message. Thus, the adversary has no advantage in this experiment, i.e, $\mathsf{Adv}^{(3)}_{\mathcal{A},h}(n) = 0$. Putting together the claims, we get that $\mathsf{Adv}_{\mathcal{A},h}(n) \leq \mathsf{negl}(n)$. This concludes the proof.

References

1. Akavia, A., Goldwasser, S., Vaikuntanathan, V.: Simultaneous hardcore bits and cryptography against memory attacks. In: Reingold, O. (ed.) TCC 2009. LNCS, vol. 5444, pp. 474–495. Springer, Heidelberg (2009)
2. Alwen, J., Dodis, Y., Wichs, D.: Leakage-resilient public-key cryptography in the bounded-retrieval model. In: Halevi, S. (ed.) CRYPTO 2009. LNCS, vol. 5677, pp. 36–54. Springer, Heidelberg (2009)
3. Boneh, D., Halevi, S., Hamburg, M., Ostrovsky, R.: Circular-secure encryption from decision diffie-hellman. In: Wagner, D. (ed.) CRYPTO 2008. LNCS, vol. 5157, pp. 108–125. Springer, Heidelberg (2008)
4. Boyen, X.: Reusable cryptographic fuzzy extractors. In: ACM Conference on Computer and Communications Security, pp. 82–91 (2004)
5. Canetti, R., Dodis, Y., Halevi, S., Kushilevitz, E., Sahai, A.: Exposureresilient functions and all-or-nothing transforms. In: Preneel, B. (ed.) EUROCRYPT 2000. LNCS, vol. 1807, pp. 453–469. Springer, Heidelberg (2000)
6. Cramer, R., Shoup, V.: A practical public key cryptosystem provably secure against adaptive chosen ciphertext attack. In: Krawczyk, H. (ed.) CRYPTO 1998. LNCS, vol. 1462, pp. 13–25. Springer, Heidelberg (1998)
7. Cramer, R., Shoup, V.: Universal hash proofs and a paradigm for adaptive chosen ciphertext secure public-key encryption. In: Knudsen, L.R. (ed.) EUROCRYPT 2002. LNCS, vol. 2332, pp. 45–64. Springer, Heidelberg (2002)
8. Dodis, Y., Kalai, Y.T., Lovett, S.: On cryptography with auxiliary input. In: STOC, pp. 621–630 (2009)
9. Dziembowski, S., Pietrzak, K.: Leakage-resilient cryptography. In: FOCS, pp. 293–302 (2008)
10. Faust, S., Kiltz, E., Pietrzak, K., Rothblum, G.: Leakage-resilient signatures (2009), `http://eprint.iacr.org/2009/282`
11. Feller, W.: An Introduction to Probability Theory and Its Applications, vol. 1. Wiley, Chichester (1968)
12. Gentry, C., Peikert, C., Vaikuntanathan, V.: Trapdoors for hard lattices and new cryptographic constructions. In: STOC, pp. 197–206 (2008)

13. Goldreich, O., Rubinfeld, R., Sudan, M.: Learning polynomials with queries: The highly noisy case. SIAM J. Discrete Math. 13(4), 535–570 (2000)
14. Goldwasser, S., Kalai, Y.T., Rothblum, G.N.: One-time programs. In: Wagner, D. (ed.) CRYPTO 2008. LNCS, vol. 5157, pp. 39–56. Springer, Heidelberg (2008)
15. Ishai, Y., Sahai, A., Wagner, D.: Private circuits: Securing hardware against probing attacks. In: Boneh, D. (ed.) CRYPTO 2003. LNCS, vol. 2729, pp. 463–481. Springer, Heidelberg (2003)
16. Katz, J., Vaikuntanathan, V.: Signature schemes with bounded leakage. In: Matsui, M. (ed.) ASIACRYPT 2009. LNCS, vol. 5912, pp. 703–720. Springer, Heidelberg (2009)
17. Kawachi, A., Tanaka, K., Xagawa, K.: Multi-bit cryptosystems based on lattice problems. In: Okamoto, T., Wang, X. (eds.) PKC 2007. LNCS, vol. 4450, pp. 315–329. Springer, Heidelberg (2007)
18. Micali, S., Reyzin, L.: Physically observable cryptography (extended abstract). In: Naor, M. (ed.) TCC 2004. LNCS, vol. 2951, pp. 278–296. Springer, Heidelberg (2004)
19. Naor, M., Reingold, O.: Number-theoretic constructions of efficient pseudo-random functions. J. ACM 51(2), 231–262 (2004)
20. Naor, M., Segev, G.: Public-key cryptosystems resilient to key leakage. In: Halevi, S. (ed.) CRYPTO 2009. LNCS, vol. 5677, pp. 18–35. Springer, Heidelberg (2009)
21. Peikert, C.: Public-key cryptosystems from the worst-case shortest vector problem: extended abstract. In: STOC, pp. 333–342 (2009)
22. Peikert, C., Vaikuntanathan, V., Waters, B.: A framework for efficient and composable oblivious transfer. In: Wagner, D. (ed.) CRYPTO 2008. LNCS, vol. 5157, pp. 554–571. Springer, Heidelberg (2008)
23. Petit, C., Standaert, F.-X., Pereira, O., Malkin, T., Yung, M.: A block cipher based pseudo random number generator secure against side-channel key recovery. In: ASIACCS, pp. 56–65 (2008)
24. Pietrzak, K.: A leakage-resilient mode of operation. In: Joux, A. (ed.) EUROCRYPT 2009. LNCS, vol. 5479, pp. 462–482. Springer, Heidelberg (2009)
25. Regev, O.: On lattices, learning with errors, random linear codes, and cryptography. In: STOC, pp. 84–93 (2005)

Public-Key Cryptographic Primitives Provably as Secure as Subset Sum

Vadim Lyubashevsky[1,*], Adriana Palacio[2], and Gil Segev[3,**]

[1] Tel-Aviv University, Tel-Aviv, Israel
vlyubash@cs.ucsd.edu
[2] Bowdoin College, Brunswick, ME USA
apalacio@bowdoin.edu
[3] Weizmann Institute of Science, Rehovot, Israel
gil.segev@weizmann.ac.il

Abstract. We propose a semantically-secure public-key encryption scheme whose security is polynomial-time equivalent to the hardness of solving random instances of the subset sum problem. The subset sum assumption required for the security of our scheme is weaker than that of existing subset-sum based encryption schemes, namely the lattice-based schemes of Ajtai and Dwork (STOC'97), Regev (STOC'03, STOC'05), and Peikert (STOC'09). Additionally, our proof of security is simple and direct. We also present a natural variant of our scheme that is secure against key-leakage attacks, and an oblivious transfer protocol that is secure against semi-honest adversaries.

1 Introduction

Since the early days of modern cryptography, the presumed intractability of the subset sum problem has been considered an interesting alternative to hardness assumptions based on factoring and the discrete logarithm problem. The appeal of the subset sum problem stems from the fact that it is simple to describe, and computing the subset sum function requires only a few addition operations. Another attractive feature is that the subset sum problem seems to be rather different in nature from number-theoretic problems. In fact, while there are polynomial-time quantum algorithms that break virtually all number-theoretic cryptographic assumptions [Sho97], there are currently no known quantum algorithms that perform better than classical ones on the subset sum problem.

The *subset sum problem*, $SS(n, M)$, is parameterized by two integers n and M. An instance of $SS(n, M)$ is created by picking a uniformly random vector $\mathbf{a} \in \mathbb{Z}_M^n$, a uniformly random vector $\mathbf{s} \in \{0,1\}^n$, and outputting $\mathbf{a}$ together with $T = \mathbf{a} \cdot \mathbf{s} \bmod M$. The problem is to find $\mathbf{s}$, given $\mathbf{a}$ and T. The hardness of

* Research supported by the Israel Science Foundation and a European Research Council (ERC) Starting Grant.

** Research supported by the Adams Fellowship Program of the Israel Academy of Sciences and Humanities.

D. Micciancio (Ed.): TCC 2010, LNCS 5978, pp. 382–400, 2010.

breaking $SS(n, M)$ depends on the ratio between n and $\log M$, which is usually referred to as the *density* of the subset sum instance. When $n/\log M$ is less than $1/n$ or larger than $n/\log^2 n$, the problem can be solved in polynomial time [LO85, Fri86, FP05, Lyu05, Sha08]. However, when the density is constant or even as small as $O(1/\log n)$, there are currently no algorithms that require less than $2^{\Omega(n)}$ time. It is also known that the subset sum problem can only get harder as its density gets closer to one [IN96].

Starting with the Merkle-Hellman cryptosystem [MH78], there have been many proposals for constructions of public-key encryption schemes that were somewhat based on subset sum. Unfortunately, all of these proposals have subsequently been broken (see [Odl90] for a survey). While efforts to build subset-sum based public-key encryption schemes were met with little success, Impagliazzo and Naor were able to construct provably-secure primitives such as universal one-way hash functions, pseudorandom generators and bit-commitment schemes, based on the subset sum problem, that remain secure until this day [IN96]. The main difference between the public-key constructions and the "minicrypt" constructions in [IN96] is that the latter could be proved secure based on random instances of the standard subset sum problem, whereas the former modified the subset sum instances in order to allow decryption. Unfortunately, these modifications always seemed to introduce fatal weaknesses.

A provably-secure cryptosystem based on subset sum was finally constructed by Ajtai and Dwork [AD97], who showed that their scheme is as hard to break as solving worst-case instances of a lattice problem called the "unique shortest vector problem." The reduction of subset sum to breaking their scheme is then obtained via the classic reduction from random subset sum to the unique shortest vector problem [LO85, Fri86]. While the Ajtai-Dwork and the subsequent lattice-based cryptosystems [Reg03, Reg05, Pei09] are as hard to break as the average-case subset sum problem, these schemes are based on subset sum in a somewhat indirect way, and this causes their connection to the subset sum problem to not be as tight as possible.

In this work, we present a cryptosystem whose security is *equivalent* to the hardness of the $SS(n, q^n)$ problem, where q is a positive integer of magnitude $\tilde{O}(n)$. Compared to the lattice-based cryptosystems, the subset sum assumption required for the security of our scheme is weaker, and the proof of security is much simpler. We direct the reader to Section 1.2 for a more in-depth comparison between our scheme and the lattice-based ones.

In addition to our semantically-secure public-key encryption scheme, we present a semi-honest oblivious transfer protocol based on the same hardness assumption. We also show that a natural variant of our encryption scheme is resilient to key-leakage attacks (as formalized by Akavia et al. [AGV09]), but under slightly stronger assumptions than our basic cryptosystem.

1.1 Our Contributions and Techniques

Semantically-secure public-key encryption. Our main contribution is a semantically secure public-key encryption scheme whose security is based directly on the hardness of the subset sum problem. The construction of our scheme is similar in spirit to the cryptosystem of Alekhnovich based on the Learning Parity with Noise (LPN) problem [Ale03], and that of Regev based on the Learning With Errors (LWE) problem [Reg05]. Both of the aforementioned schemes are built from the assumption that for a randomly chosen matrix $\mathbf{A} \in \mathbb{Z}_q^{m \times n}$, a random vector $\mathbf{s} \in \mathbb{Z}_q^n$, and some "small" noise vector $\mathbf{c} \in \mathbb{Z}_q^m$, the distribution $(\mathbf{A}, \mathbf{As} + \mathbf{c})$ is computationally indistinguishable from the uniform distribution over $\mathbb{Z}_q^{m \times (n+1)}$. To construct our scheme, we show that the subset sum problem can be made to look very similar to the LWE problem. Then the main ideas (with a few technical differences) used in constructing cryptosystems based on LWE [Reg05, GPV08, Pei09] can be transferred over to subset sum.

Consider instances of the subset sum problem $SS(n, q^m)$ where q is some small integer. If $\mathbf{a}$ is a vector in $\mathbb{Z}_{q^m}^n$ and $\mathbf{s}$ is a vector in $\{0,1\}^n$, then $\mathbf{a} \cdot \mathbf{s} \bmod q^m$, written in base q, is equal to $\mathbf{As} + \mathbf{c} \bmod q$, where $\mathbf{A} \in \mathbb{Z}_q^{m \times n}$ is a matrix whose i-th column corresponds to $\mathbf{a}_i$ written in base q, and $\mathbf{c}$ is a vector in $\mathbb{Z}_q^m$ that corresponds to the carries when performing "grade-school" addition. For example, let $q = 10$, $m = n = 3$, $\mathbf{a} = (738, 916, 375)$, and $\mathbf{s} = (0, 1, 1)$. Then

$$\mathbf{a} \cdot \mathbf{s} \bmod 10^3 = 916 + 375 \bmod 10^3 = 291,$$

which can be written as addition in base q as follows:

$$\begin{bmatrix} 7 & 9 & 3 \\ 3 & 1 & 7 \\ 8 & 6 & 5 \end{bmatrix} \begin{bmatrix} 0 \\ 1 \\ 1 \end{bmatrix} + \begin{bmatrix} 0 \\ 1 \\ 0 \end{bmatrix} = \begin{bmatrix} 2 \\ 9 \\ 1 \end{bmatrix}$$

where all operations are performed over $\mathbb{Z}_q$.

The key observation is that the magnitude of the entries in the carries vector $[0\ 1\ 0]^T$ is at most $n - 1$, and so if $q \gg n$, then $\mathbf{As} + \mathbf{c} \bmod q \approx \mathbf{As} \bmod q$. In fact, the elements of the vector $\mathbf{c}$ are distributed normally around $n/2$ with standard deviation $\sqrt{n}$. In the instantiation of our scheme described in Section 3, we generate the elements in $\mathbf{A}$ from the range $[-\frac{q-1}{2}, \frac{q-1}{2}]$ and so the entries in the carries vector are normally distributed around 0 with standard deviation $\sqrt{n}$. Readers familiar with the cryptosystems based on LWE [Reg05, GPV08, Pei09] should recognize the resemblance of the carry vector $\mathbf{c}$ to the noise vector in the LWE-based schemes. The main difference is that in the latter the noise vector is chosen independently at random, whereas in our scheme, the carries vector $\mathbf{c}$ occurs "naturally" and is completely determined by the matrix $\mathbf{A}$ and the vector $\mathbf{s}$. The fact that the "noise" vector is not random is of no consequence to us, since it was shown by Impagliazzo and Naor that distinguishing $(\mathbf{a}, \mathbf{a} \cdot \mathbf{s} \bmod q^m)$ from uniform is as hard as recovering $\mathbf{s}$ [IN96]. Thus the distribution $(\mathbf{A}, \mathbf{As} + \mathbf{c} \bmod q)$, which is just the base q representation of the previous distribution, is also computationally indistinguishable from uniform, based on the hardness of subset sum. The following theorem summarizes our main result:

Theorem 1.1. *For any integer $q > 10n \log^2 n$, there exists a semantically secure cryptosystem encrypting k bits whose security is polynomial-time equivalent to the hardness of solving the $SS(n, q^{n+k})$ problem.*

Leakage-resilient public-key encryption. We show that a natural variant of our encryption scheme is resilient to any non-adaptive leakage of $L(1-o(1))$ bits of its secret key, where L is the length of the secret key (see Appendix A.2 for the formal definition of non-adaptive key-leakage attacks). In this paper we deal with the non-adaptive setting of key leakage, and note that this notion of leakage is still very meaningful as it captures many realistic attacks in which the leakage does not depend on the parameters of the encryption scheme. For example, it captures the cold boot attacks of Halderman et al. [HSH+08], in which the leakage depends only on the properties of the hardware devices that are used for storing the secret key. We note that although Naor and Segev [NS09] presented a generic and rather simple construction that protects any public-key encryption scheme from non-adaptive leakage attacks, we show that for our specific scheme an even simpler modification suffices.

Oblivious transfer. We use our original encryption scheme to construct an oblivious transfer (OT) protocol that provides security for the receiver against a cheating sender and security for the sender against an honest-but-curious receiver. Our protocol is an instance of a natural construction used by Gertner et al. [GKM+00], based on ideas of Even et al. [EGL82, EGL85], to show that public-key encryption with a certain property implies two-message semi-honest OT. The property is roughly that public keys can be sampled "separately of private keys," while preserving the semantic security of the encryption. Pseudorandomness of subset sum implies that our encryption scheme satisfies this property.

1.2 Comparisons with Lattice-Based Schemes

To the best of our knowledge, the only other cryptosystems based on subset sum are those that are based on the worst-case hardness of the approximate unique shortest vector problem (uSVP_γ) [AD97, Reg03, Reg05, Pei09]. The cryptosystems of Regev [Reg03] and Peikert [Pei09] are both based on the hardness of $\mathrm{uSVP}_{n^{1.5}}$ (the latter is based on uSVP via a reduction in [LM09]). What this means is that an algorithm that breaks these cryptosystems can be used to find the shortest vector in any lattice whose shortest vector is a factor of $n^{1.5}$ shorter than the next shortest vector that is not a multiple of it.

A reduction from the random subset sum problem to uSVP_γ was given in [LO85, Fri86]. The exact parameter γ depends on the density of the subset sum instance. The smaller the density, the larger the γ can be, and the easier the uSVP_γ problem becomes. The reduction from an instance of $SS(n, M)$ to uSVP_γ is as follows:

Given an instance of $SS(n, M)$ consisting of a vector $\mathbf{a} \in \mathbb{Z}_M^n$ and an element $T \in \mathbb{Z}_M$, we define the lattice $\mathcal{L}$ as

$$\mathcal{L} = \{\mathbf{x} \in \mathbb{Z}^{n+1} : [\mathbf{a}|| - T] \cdot \mathbf{x} \bmod M = 0\}.$$

Notice that the vector $\mathbf{x} = [\mathbf{s}||1]$ is in $\mathcal{L}$ for the $\mathbf{s}$ for which $\mathbf{a} \cdot \mathbf{s} \bmod M = T$, so the ℓ_2 norm of the shortest vector is approximately $\sqrt{n}$. The next shortest non-parallel vector is the vector that meets the Minkowski bound of $\sqrt{n+1} \cdot det(L)^{\frac{1}{n+1}} \approx \sqrt{n}M^{1/n}$, which is a factor $M^{1/n}$ larger than the shortest vector. Therefore solving uSVP$_{n^{1.5}}$ allows us to solve instances of $SS(n, M)$ where $M \approx n^{1.5n}$.

The cryptosystem that we construct in this paper is based on the hardness of $SS(n, M)$ where $M \approx n^n$. In order to have a lattice scheme based on the same subset sum assumption, it would need to be based on uSVP$_n$. The construction of such a scheme is currently not known and would be considered a breakthrough.

We want to point out that we are not claiming that just because our scheme is based on a weaker instance of subset sum, it is somehow more secure than the lattice-based schemes. All we are claiming is that the connection of our scheme to the subset sum problem is better. In terms of security, the lattice-based schemes based on LWE [Reg05, Pei09] and our scheme are actually very similar because the LWE and subset sum problems can both be viewed as average-case instances of the "bounded distance decoding" problem, with essentially the same parameters but different distributions. Unfortunately, we do not know of any tight reduction between the two distributions, so there is no clear theoretical connection between LWE and subset sum.

In practice, though, there may be some advantages of our scheme over the lattice-based ones. The secret key in our scheme is an n-bit vector $\mathbf{s} \in \{0,1\}^n$, whereas the secret keys in lattice-based schemes are on the order of $n \log n$ bits. Also, the public key in our scheme is a matrix $\mathbf{A} \in \mathbb{Z}_q^{n \times n}$, whereas lattice-based schemes use an $n \times n \log n$ matrix. The reason for the savings of a factor of $\log n$ in the size of both the secret and public keys in our scheme has to do with the fact that the distribution $(\mathbf{A}, \mathbf{A}\mathbf{s} + \mathbf{c})$ is indistinguishable from random, where $\mathbf{s} \in \{0,1\}^n$, based on the subset sum assumption. But in order to get a proof of security based on lattices, the vector $\mathbf{s}$ has to be chosen uniformly from $\mathbb{Z}_q^n$ (see [ACPS09] for a slight improvement), and is thus $\log n$ times longer. One can thus view our proof of security based on subset sum as justification that having $\mathbf{s}$ come from a smaller set and having the "noise" be a deterministic function of $\mathbf{A}$ and $\mathbf{s}$, is still secure.

1.3 Open Problems

Our construction of the subset sum cryptosystem involves transforming the subset sum problem into something that very much resembles the LWE problem. It would be interesting to see whether the same type of idea could be used to transform other problems into LWE-type problems upon which semantically-secure cryptosystems can be built.

Another open problem concerns weakening the computational assumption underlying the multi-bit version of our scheme. While our one-bit cryptosystem is based on the hardness of solving instances of $SS(n, q^n)$ for some $q = \tilde{O}(n)$, when simultaneously encrypting k bits using the same randomness our cryptosystem becomes equivalent to the easier $SS(n, q^{n+k})$ problem (clearly, it is possible to encrypt k bits bit-by-bit, but this is less efficient). This is somewhat peculiar since one can simultaneously encrypt polynomially-many bits using the LWE cryptosystem without making the underlying assumption stronger [PVW08], while simultaneously encrypting $\Omega(n^2)$ bits in our scheme is completely insecure (since the $SS(n, q^{n^2})$ problem can be solved in polynomial time [LO85, Fri86]). We believe that this weakness in the subset sum construction is due to the fact that the noise in the LWE schemes is generated independently, whereas in our scheme, the "noise" is just the carry bits. It is an interesting open problem to see whether one can modify our scheme so that its security does not depend on the number of bits being simultaneously encrypted using the same randomness.

Another interesting open problem concerns security against leakage attacks. First, we were not able to prove the security of our scheme against *adaptive* key-leakage attacks, in which the leakage can be chosen as a function of the public key as well. Although our scheme is somewhat similar to that of Akavia et al. [AGV09], it seems that their approach for proving security against adaptive attacks does not immediately apply to our setting. Second, our leakage-resilient scheme relies on a slightly stronger assumption than our basic scheme, and it will be interesting to minimize the required computational assumption.

Finally, we leave it as an open problem to construct a CCA-secure scheme in the standard model based directly on subset sum. While there are CCA-secure encryption schemes based on lattice problems (and thus on subset sum as well) [PW08, Pei09], it would be interesting to build one directly based on subset sum that will hopefully require weaker assumptions than the lattice based ones.

2 Preliminaries

2.1 The Subset Sum Problem

The subset sum problem with parameters n and q^m, where n and m are integers and q is a positive integer such that $2^n < q^m$, is defined as follows: Given n numbers $a_1, \ldots, a_n \in \mathbb{Z}_{q^m}$ and a target $T \in \mathbb{Z}_{q^m}$, find a subset $S \subseteq \{1, \ldots, n\}$ such that $\sum_{i \in S} a_i = T \bmod q^m$. This can be viewed as the problem of inverting the function $f_{\mathbf{a}} : \{0,1\}^n \to \mathbb{Z}_{q^m}$ defined as

$$f_{\mathbf{a}}(s_1, \ldots, s_n) = \sum_{i=1}^{n} s_i a_i \bmod q^m ,$$

where $\mathbf{a} = (a_1, \ldots, a_n) \in \mathbb{Z}_{q^m}^n$ is its index (i.e., this is a collection of functions, where a function is sampled by choosing its index $\mathbf{a}$ uniformly at random).

We denote by $SS(n, q^m)$ the subset sum problem with parameters n and q^m. Using the above notion, the hardness of the subset sum problem is the assumption that $\{f_{\mathbf{a}}\}_{\mathbf{a} \in \mathbb{Z}_{q^m}^n}$ is a collection of one-way functions. We now state two properties of the subset sum problem that were proved by Impagliazzo and Naor [IN96] and are used in analyzing the security of our constructions. The first property is that subset sum instances with larger moduli are not harder than subset sum instances with smaller moduli. The second property is that if the subset sum is a one-way function, then it is also a pseudorandom generator. In the following two statements, we fix n, m and q as above.

Lemma 2.1 ([IN96]). *For any integers i and j such that $i < j$, if $q^{m+i} > 2^n$, then the hardness of $SS(n, q^{m+j})$ implies the hardness of $SS(n, q^{m+i})$.*

Lemma 2.2 ([IN96]). *The hardness of $SS(n, q^m)$ implies that the distributions $(\mathbf{a}, f_{\mathbf{a}}(\mathbf{s}))$ and $(\mathbf{a}, t)$ are computationally indistinguishable, where $\mathbf{a} \in \mathbb{Z}_{q^m}^n$, $\mathbf{s} \in \{0,1\}^n$, and $t \in \mathbb{Z}_{q^m}$ are chosen independently and uniformly at random.*[1]

2.2 Notation

We represent vectors by bold-case letters and all vectors will be assumed to be column vectors. Unless stated otherwise, all scalar and vector operations are performed modulo q. For simplicity, we will assume that q is odd, but our results follow for all q with minimal changes. We represent elements in $\mathbb{Z}_q$ by integers in the range $[-(q-1)/2, (q-1)/2]$. For an element $e \in \mathbb{Z}_q$, its length, denoted by $|e|$ is the absolute value of its representative in the range $[-(q-1)/2, (q-1)/2]$. For a vector $\mathbf{e} = (e_1, \ldots, e_m) \in \mathbb{Z}_q^m$, we define $\|\mathbf{e}\|_\infty = \max_{1 \le i \le m} |e_i|$.

We now present some notation that is convenient for describing the subset sum function. For a matrix $\mathbf{A} \in \mathbb{Z}_q^{m \times n}$ and a vector $\mathbf{s} \in \{0,1\}^n$, we define $\mathbf{A} \odot \mathbf{s}$ as the vector $\mathbf{t}^T = (t_0, \ldots, t_{m-1})$ such that $|t_i| \le (q-1)/2$ for every $1 \le i \le m$, and

$$\sum_{i=0}^{m-1} t_i q^i \equiv \left(\sum_{j=0}^{n-1} s_j \sum_{i=0}^{m-1} A_{i,j} q^i \right) \bmod q^m .$$

In other words, we interpret the n columns of $\mathbf{A}$ as elements in $\mathbb{Z}_{q^m}$ represented in base q, and sum all the elements in the columns j where $s_j = 1$. The result is an element in $\mathbb{Z}_{q^m}$, which we write in base q using coefficients between $-(q-1)/2$ and $(q-1)/2$. We then write the coefficients of the base q representation as an m-dimensional vector $\mathbf{t}$. It will sometimes be more convenient to consider the subset sum of the numbers represented by the rows of $\mathbf{A}$, and to this effect we naturally define $\mathbf{r}^T \odot \mathbf{A} = \left(\mathbf{A}^T \odot \mathbf{r}\right)^T$.

3 The Encryption Scheme

In this section we present our main contribution: a public-key encryption scheme that is based directly on the hardness of the subset sum problem. Given a security parameter n, we set $q(n)$ to be some number greater than $10n \log^2 n$, let $k \in \mathbb{N}$ be the number of bits we want to encrypt, and define the following encryption scheme:

- **Key generation:** On input 1^n sample $\mathbf{A}' \in \mathbb{Z}_q^{n \times n}$ and $\mathbf{s}_1, \ldots, \mathbf{s}_k \in \{0,1\}^n$ independently and uniformly at random. For every $1 \le i \le k$ let $\mathbf{t}_i = \mathbf{A}' \odot \mathbf{s}_i$, and let $\mathbf{A} = [\mathbf{A}' \| \mathbf{t}_1 \| \cdots \| \mathbf{t}_k]$. Output $\mathsf{pk} = \mathbf{A}$ and $\mathsf{sk} = (\mathbf{s}_1, \ldots, \mathbf{s}_k)$.

[1] Impagliazzo and Naor [IN96] only prove their result for q's that are prime or a power of 2, but their results extend to all q.

- **Encryption:** On input a message $\mathbf{z} \in \{0,1\}^k$, sample $\mathbf{r} \in \{0,1\}^n$ uniformly at random, and output the ciphertext $\mathbf{u}^T = \mathbf{r}^T \odot \mathbf{A} + (\frac{q-1}{2})[0^n||\mathbf{z}^T]$.
- **Decryption:** On input a ciphertext $\mathbf{u}^T = [\mathbf{v}^T||w_1||\cdots||w_k]$ where $\mathbf{v} \in \mathbb{Z}_q^n$ and $w_1, \ldots, w_k \in \mathbb{Z}_q$, for every $1 \le i \le k$ compute $y_i = \mathbf{v}^T\mathbf{s}_i - w_i$. If $|y_i| < q/4$ then set $z_i = 0$ and otherwise set $z_i = 1$. Output $\mathbf{z}^T = (z_1, \ldots, z_k)$.

The intuition for the semantic security of the scheme is fairly simple. Because the vectors $\mathbf{t}_i$ are subset sums of the numbers represented by the columns of $\mathbf{A}'$, the public key $\mathbf{A}$ is computationally indistinguishable from random. Therefore, to an observer, the vector $\mathbf{r}^T \odot \mathbf{A}$, which is a subset sum of numbers represented by the rows of $\mathbf{A}$, is again computationally indistinguishable from uniform. The formal proof is in Section 3.1.

The intuition for decryption is based on the fact that $\mathbf{A}' \odot \mathbf{s}_i \approx \mathbf{A}'\mathbf{s}_i$ and $\mathbf{r}^T \odot \mathbf{A} \approx \mathbf{r}^T\mathbf{A}$. For simplicity, assume that $\mathbf{A}' \odot \mathbf{s}_i = \mathbf{A}'\mathbf{s}_i$ and $\mathbf{r}^T \odot \mathbf{A} = \mathbf{r}^T\mathbf{A}$. Then it is not hard to see that

$$|\mathbf{v}^T\mathbf{s}_i - w_i| = \left|(\mathbf{r}^T\mathbf{A}')\mathbf{s}_i - \left(\mathbf{r}^T(\mathbf{A}'\mathbf{s}_i) + \frac{q-1}{2}z_i\right)\right| = \frac{q-1}{2}z_i\,,$$

and we recover z_i. Because the subset sum function does not quite correspond to a vector/matrix multiplication, decryption will recover $\frac{q-1}{2}z_i +$ error. What we will need to show is that this error term is small enough so that we can still tell whether z_i was 0 or 1. The proof is in Section 3.2.

3.1 Proof of Security

Our scheme enjoys a rather simple and direct proof of security. The proof consists of two applications of the pseudorandomness of the subset sum function, which by Lemma 2.2 is implied by the hardness of the subset sum problem. Informally, the first application allows us to replace the values $\mathbf{A}' \odot \mathbf{s}_1, \ldots, \mathbf{A}' \odot \mathbf{s}_k$ in the public key with k vectors that are sampled independently and uniformly at random. Then, the second application allows us to replace the value $\mathbf{r}^T \odot \mathbf{A}$ in the challenge ciphertext with an independently and uniformly chosen vector. In this case, the challenge ciphertext is statistically independent of the encrypted message and the security of the scheme follows. More formally, the following theorem establishes the security of the scheme:

Theorem 3.1. *Assuming the hardness of the $SS(n, q^{n+k})$ problem, where n is the security parameter and k is the plaintext length, the above public-key encryption scheme is semantically secure.*

Proof. We show that for any two messages $\mathbf{m}_0, \mathbf{m}_1 \in \{0,1\}^k$, the ensembles $(\mathsf{pk}, \mathcal{E}_{\mathsf{pk}}(\mathbf{m}_0))$ and $(\mathsf{pk}, \mathcal{E}_{\mathsf{pk}}(\mathbf{m}_1))$ are computationally indistinguishable. In fact, we prove an even stronger statement, namely that $(\mathbf{A}, \mathbf{r}^T \odot \mathbf{A})$ is computationally indistinguishable from $(\mathbf{M}, \mathbf{v})$, where $\mathbf{M} \in \mathbb{Z}_q^{n\times(n+k)}$ and $\mathbf{v} \in \mathbb{Z}_q^{n+k}$ are sampled independently and uniformly at random. This, in turn, implies that for every $b \in \{0,1\}$, the distribution $(\mathsf{pk}, \mathcal{E}_{\mathsf{pk}}(\mathbf{m}_b))$ is computationally indistinguishable from a distribution that perfectly hides the message $\mathbf{m}_b$. Therefore, any probabilistic polynomial-time adversary attacking the scheme will have a negligible cpa-advantage.

The hardness of the $SS(n, q^n)$ problem, Lemmas 2.1 and 2.2, and a standard hybrid argument imply that the distributions $(\mathbf{A}', \mathbf{A}' \odot \mathbf{s}_1, \ldots, \mathbf{A}' \odot \mathbf{s}_k)$ and $(\mathbf{A}', \mathbf{b}_1, \ldots, \mathbf{b}_k)$, where $\mathbf{b}_1, \ldots, \mathbf{b}_k \in \mathbb{Z}_q^n$ are sampled independently and uniformly at random, are computationally indistinguishable. Letting $\mathbf{M} = [\mathbf{A}'||\mathbf{b}_1|| \cdots ||\mathbf{b}_k]$, it then follows that the distributions $(\mathbf{A}, \mathbf{r}^T \odot \mathbf{A})$ and $(\mathbf{M}, \mathbf{r}^T \odot \mathbf{M})$, are computationally indistinguishable. Now, the hardness of the $SS(n, q^{n+k})$ problem and Lemma 2.2 imply that the latter distribution is computationally indistinguishable from $(\mathbf{M}, \mathbf{v})$, where $\mathbf{v} \in \mathbb{Z}_q^{n+k}$ is sampled uniformly at random, independently of $\mathbf{M}$. This concludes the proof of the theorem. □

3.2 Proof of Correctness

We will use the following bound due to Hoeffding [Hoe63] throughout our proof.

Lemma 3.2 (Hoeffding Bound). *Let $X_1, \ldots, X_n$ be independent random variables in the range $[a, b]$ and let $X = X_1 + \ldots + X_n$. Then*

$$Pr[|X - E[X]| \geq t] \leq 2e^{-\left(\frac{2t^2}{n(a-b)^2}\right)}.$$

The next lemma shows that the carries during the subset sum operation $\mathbf{r}^T \odot \mathbf{A}$ are distributed with mean 0 and their absolute value is bounded (with high probability) by $\sqrt{n} \log n$. In addition, the carries are almost independent of each other. The slight dependency comes from the fact that a carry element can cause the following carry to increase by 1.

Lemma 3.3. *For any $n, m \in \mathbb{N}$ and $\mathbf{r} \in \{0,1\}^n$,*

$$Pr_{\mathbf{A} \xleftarrow{\$} \mathbb{Z}_q^{n \times m}} \left[\|\mathbf{r}^T \odot \mathbf{A} - \mathbf{r}^T \mathbf{A}\|_\infty < \sqrt{n} \log n \right] = 1 - n^{-\omega(1)} .$$

Furthermore, the vector $\mathbf{r}^T \odot \mathbf{A} - \mathbf{r}^T \mathbf{A}$ can be written as a sum of two vectors $\mathbf{x}, \mathbf{y} \in \mathbb{Z}_q^m$ where all the coordinates of $\mathbf{x}$ are independently distributed with mean 0, while all the coordinates of $\mathbf{y}$ have absolute value at most 1 (but could be dependent among themselves).

Proof. Computing $\mathbf{r}^T \odot \mathbf{A}$ can be done via the following algorithm, where $\mathbf{a}_i$ is the i-th column of $\mathbf{A}$:

$c_0 = 0$
for $i = 0$ to $m - 1$
  $b_i = (c_i + \mathbf{r}^T \mathbf{a}_i) \bmod q$
  $c_{i+1} = \left\lceil \frac{c_i + \mathbf{r}^T \mathbf{a}_i}{q} \right\rfloor$
output $\mathbf{b}^T = (b_0, \ldots, b_{m-1})$

Notice that this algorithm is just performing addition in base q, where all the coefficients are between $-(q-1)/2$ and $(q-1)/2$. The difference $\mathbf{r}^T \odot \mathbf{A} - \mathbf{r}^T\mathbf{A}$ is simply the "carries" c_i. Note that the only dependency among the c_i's is that c_{i+1} slightly depends on c_i. We can rewrite the above algorithm by writing each c_i as $x_i + y_i$ such that all the x_i's are independent among themselves, whereas the y_i's could be dependent but are very small.

$x_0 = 0$; $y_0 = 0$
for $i = 0$ to $m-1$
  $b_i = (x_i + y_i + \mathbf{r}^T\mathbf{a}_i) \bmod q$
  $x_{i+1} = \left\lceil \frac{\mathbf{r}^T\mathbf{a}_i}{q} \right\rfloor$
  $y_{i+1} = \left\lceil \frac{x_i+y_i+\mathbf{r}^T\mathbf{a}_i}{q} \right\rfloor - \left\lceil \frac{\mathbf{r}^T\mathbf{a}_i}{q} \right\rfloor$
output $\mathbf{b}^T = (b_0, \ldots, b_{m-1})$

Observe that in the second algorithm, the x_i's are completely independent among themselves. We now bound the absolute value of the x_i's. Each vector $\mathbf{a}_i$ consists of numbers uniformly distributed between $-(q-1)/2$ and $(q-1)/2$. Applying the Hoeffding bound (Lemma 3.2), we obtain that

$$Pr[|\mathbf{r}^T\mathbf{a}_i| \geq q\sqrt{n}\log n] \leq 2e^{-2\log^2 n} .$$

Therefore, with probability $1 - n^{-\omega(1)}$, $|x_i| \leq \sqrt{n}\log n$ for all $0 \leq i \leq m-1$. Also notice that by symmetry, $E[x_i] = 0$. By induction, we will now show that $|y_i| \leq 1$. This is true for y_0, and assume it is true for y_i. Then,

$$|y_{i+1}| = \left| \left\lceil \frac{x_i + y_i + \mathbf{r}^T\mathbf{a}_i}{q} \right\rfloor - \left\lceil \frac{\mathbf{r}^T\mathbf{a}_i}{q} \right\rfloor \right| \leq \left| \left\lceil \frac{x_i + y_i}{q} \right\rfloor + 1 \right| \leq 1 ,$$

where the last inequality follows because $|x_i| \leq \sqrt{n}\log n < q/2 - 1$ and $|y_i| \leq 1$, and so $\left\lceil \frac{x_i+y_i}{q} \right\rfloor = 0$. □

Lemma 3.4. *For any* $\mathbf{r}, \mathbf{s} \in \{0,1\}^n$,

$$Pr_{\mathbf{A} \xleftarrow{\$} \mathbb{Z}_q^{n\times n}} [\|(\mathbf{r}^T \odot \mathbf{A})\mathbf{s} - \mathbf{r}^T\mathbf{A}\mathbf{s}\|_\infty < n\log^2 n] = 1 - n^{-\omega(1)} .$$

Proof. Using Lemma 3.3, we can rewrite $\mathbf{r}^T \odot \mathbf{A}$ as $\mathbf{r}^T\mathbf{A} + \mathbf{x}^T + \mathbf{y}^T$ where each element of $\mathbf{x}$ is independently distributed around 0 with magnitude at most $\sqrt{n}\log n$, and each element of $\mathbf{y}$ has magnitude at most 1. Multiplying by $\mathbf{s}$, we obtain $(\mathbf{r}^T \odot \mathbf{A})\mathbf{s} - \mathbf{r}^T\mathbf{A}\mathbf{s} = \mathbf{x}^T\mathbf{s} + \mathbf{y}^T\mathbf{s}$.

Because $\|\mathbf{y}\|_\infty \leq 1$, we have $|\mathbf{y}^T\mathbf{s}| \leq n$. By the Hoeffding bound (Lemma 3.2), we obtain that

$$Pr[|\mathbf{x}^T\mathbf{s}| \geq n\log^2 n] \leq 2e^{-\frac{\log^2 n}{2}},$$

and the lemma is proved. □

Theorem 3.5. *Decryption succeeds with probability* $1 - n^{-\omega(1)}$.

Proof. The encryption of a message $\mathbf{z}$ is the vector $\mathbf{u}^T = \mathbf{r}^T \odot \mathbf{A} + (\frac{q-1}{2})(0^n||\mathbf{z}^T)$. To decrypt bit i, we write $\mathbf{u}^T = [\mathbf{v}^T||w_1||\ldots||w_k]$ and compute $\mathbf{v}^T\mathbf{s}_i - w_i$. Observe that $\mathbf{v}^T$ is equal to $\mathbf{r}^T \odot \mathbf{A}' + (0^{n-1}||\nu)$ and $w_i = \mathbf{r}^T\mathbf{t}_i + \frac{q-1}{2}z_i + \eta$, where ν, η are carries whose magnitudes are less than n (actually, we can show that with high probability $\nu, \eta < \sqrt{n}\log n$, but the looser bound suffices here). Therefore, if s_n is the last element of $\mathbf{s}_i$, then

$$\begin{aligned}\mathbf{v}^T\mathbf{s}_i - w_i &= (\mathbf{r}^T \odot \mathbf{A}' + (0^{n-1}||\nu))\mathbf{s}_i - \left(\mathbf{r}^T\mathbf{t}_i + \frac{q-1}{2}z_i + \eta\right) \\ &= (\mathbf{r}^T \odot \mathbf{A}')\mathbf{s}_i + \nu s_n - \left(\mathbf{r}^T(\mathbf{A}' \odot \mathbf{s}_i) + \frac{q-1}{2}z_i + \eta\right).\end{aligned}$$

We will now show that $\frac{q-1}{2}z_i$ is the dominant term in the second equation. Thus, if $z_i = 0$, the result will be close to 0, and if $z_i = 1$, the result will be close to $-(q-1)/2$. We will show this by bounding the magnitude of the other terms.

$$\begin{aligned}&|(\mathbf{r}^T \odot \mathbf{A}')\mathbf{s}_i + \nu s_n - \mathbf{r}^T(\mathbf{A}' \odot \mathbf{s}_i) - \eta| \\ &\qquad \le \left|(\mathbf{r}^T \odot \mathbf{A}')\mathbf{s}_i - \mathbf{r}^T\mathbf{A}'\mathbf{s}_i - \mathbf{r}^T(\mathbf{A}' \odot \mathbf{s}_i) + \mathbf{r}^T\mathbf{A}'\mathbf{s}_i\right| + |\nu s_n| + |\eta| \\ &\qquad \le \left|(\mathbf{r}^T \odot \mathbf{A}')\mathbf{s}_i - \mathbf{r}^T\mathbf{A}'\mathbf{s}_i\right| + \left|\mathbf{r}^T(\mathbf{A}' \odot \mathbf{s}_i) - \mathbf{r}^T\mathbf{A}'\mathbf{s}_i\right| + 2n \\ &\qquad \le n\log^2 n + n\log^2 n + 2n\,,\end{aligned}$$

where the last inequality follows from applying Lemma 3.4 twice to bound $\left|(\mathbf{r}^T \odot \mathbf{A}')\mathbf{s}_i - \mathbf{r}^T\mathbf{A}'\mathbf{s}_i\right|$ and $\left|\mathbf{r}^T(\mathbf{A}' \odot \mathbf{s}_i) - \mathbf{r}^T\mathbf{A}'\mathbf{s}_i\right|$. So if $z_i = 0$, we will have

$$|\mathbf{v}^T\mathbf{s}_i - w_i| \le 2n\log^2 n + 2n < q/4$$

with probability $1 - n^{-\omega(1)}$, and we will decrypt to 0. If $z_i = 1$, we will decrypt to 1 since

$$|\mathbf{v}^T\mathbf{s}_i - w_i| \ge (q-1)/2 - 2n\log^2 n - 2n > q/4\,. \qquad \square$$

4 Security against Key-Leakage Attacks

In this section we prove that a natural variant of the scheme described in Section 3 is resilient to any non-adaptive leakage of $L(1 - o(1))$ bits, where L is the length of the secret key (see Appendix A.2 for the formal definition of non-adaptive key-leakage attacks). Given a security parameter n and a leakage parameter $\lambda = \lambda(n)$, set $q = O\left(\left(n + \frac{\lambda}{\log n}\right) n\log^2 n\right)$, $T = \sqrt{q}$, and $m \ge (\lceil n\log q\rceil + \lambda + \omega\log n) / \log T$. Consider the following encryption scheme:

- **Key generation:** On input 1^n sample $\mathbf{A}' \in \mathbb{Z}_q^{n\times m}$ and $\mathbf{s} \in \{-(T-1)/2, \ldots, (T-1)/2\}^m$ uniformly and independently at random, and let $\mathbf{A} = [\mathbf{A}'||\mathbf{A}'\mathbf{s}]$. Output $\mathsf{pk} = \mathbf{A}$ and $\mathsf{sk} = \mathbf{s}$.

– **Encryption:** On input a bit b, sample $\mathbf{r} \in \{0,1\}^n$ uniformly at random, and output the ciphertext $\mathbf{u}^T = \mathbf{r}^T \odot \mathbf{A} + (\frac{q-1}{2})[0^m||b]$.
– **Decryption:** On input a ciphertext $\mathbf{u}^T = [\mathbf{v}^T||w]$ where $\mathbf{v} \in \mathbb{Z}_q^m$ and $w \in \mathbb{Z}_q$, compute $y = \mathbf{v}^T\mathbf{s} - w$. If $|y| < q/4$ then output 0. Otherwise, output 1.

The main idea underlying the scheme is that the min-entropy of the secret key is $m \log T \geq \lceil n \log q \rceil + \lambda + \omega \log n$, and thus even given any leakage of λ bits it still has average min-entropy at least $\lceil n \log q \rceil + \omega \log n$. Since the leakage is independent of the public key, we can apply the leftover hash lemma and argue that $\mathbf{A} = [\mathbf{A}'||\mathbf{A}'\mathbf{s}]$ is statistically close to uniform, even given the leakage.

We note that in this scheme, unlike in the scheme presented in Section 3, we use matrix-vector multiplication instead of the subset sum operation in forming the public key. The proof of correctness in this case is similar to that presented in Section 3. Specifically, a generalization of Lemma 3.4 shows that for every $\mathbf{r} \in \{0,1\}^n$ and $\mathbf{s} \in \{-(T-1)/2, \ldots, (T-1)/2\}^m$, with overwhelming probability over the choice of $\mathbf{A} \xleftarrow{\$} \mathbb{Z}_q^{n\times m}$ it holds that $\|(\mathbf{r}^T \odot \mathbf{A})\mathbf{s} - \mathbf{r}^T\mathbf{A}\mathbf{s}\|_\infty < \sqrt{Tmn}\log^2 n + Tm$. As in the proof of Theorem 3.5, this implies that $\mathbf{v}^T\mathbf{s} - w = \gamma + \frac{q-1}{2}z$, where $|\gamma| \leq \sqrt{Tmn}\log^2 n + (T+2)m$. Therefore, we need to set q to be an integer such that $q/4 > \sqrt{Tmn}\log^2 n + (T+2)m$. By setting roughly $q = \left(n + \frac{\lambda}{\log n}\right) n \log^2 n$ (ignoring a small leading constant) and $T = \sqrt{q}$, we can base the security of the scheme on the hardness of the $SS(n, q^m)$ problem, where

$$q^m = q^{\frac{n\log q + \lambda + \omega \log n}{\log T}} = q^{2\left(n + \frac{\lambda + \omega \log n}{\log q}\right)} = \left(\left(n + \frac{\lambda}{\log n}\right) n \log^2 n\right)^{2n} \cdot 4^{\lambda + \omega \log n} .$$

The following theorem establishes the security of the scheme:

Theorem 4.1. *Assuming the hardness of the $SS(n, q^{m+1})$ problem for $q = q(n)$ and $m = m(n)$ as above, the scheme is semantically secure against non-adaptive $\lambda(n)$-key-leakage attacks, where n is the security parameter.*

Proof. We show that for any efficiently computable leakage function f mapping secret keys into $\{0,1\}^\lambda$, the ensembles $(\mathsf{pk}, \mathcal{E}_{\mathsf{pk}}(0), f(\mathsf{sk}))$ and $(\mathsf{pk}, \mathcal{E}_{\mathsf{pk}}(1), f(\mathsf{sk}))$ are computationally indistinguishable. In fact, we prove a stronger statement, namely that $(\mathbf{A}, \mathbf{r}^T \odot \mathbf{A}, f(\mathbf{s}))$ is computationally indistinguishable from $(\mathbf{M}, \mathbf{v}, f(\mathbf{s}))$, where $\mathbf{M} \in \mathbb{Z}_q^{n\times(m+1)}$, $\mathbf{v} \in \mathbb{Z}_q^{m+1}$ are sampled independently, uniformly at random.

Lemma A.1 guarantees that the average min-entropy of $\mathbf{s}$ given $f(\mathbf{s})$ is at least $m \log T - \lambda \geq n \log q + \omega \log n$. The leftover hash lemma (when adapted to the notion of average min-entropy – see Lemma A.3) then implies that the statistical distance between the distributions $(\mathbf{A}', \mathbf{A}'\mathbf{s}, f(\mathbf{s}))$ and $(\mathbf{A}', \mathbf{t}, f(\mathbf{s}))$, where $\mathbf{t} \in \mathbb{Z}_q^n$ is sampled uniformly at random, is negligible in n. Letting $\mathbf{M} = [\mathbf{A}'||\mathbf{t}]$ and noting that applying a deterministic function cannot increase the statistical distance between distributions, it follows that the statistical distance between $(\mathbf{A}, \mathbf{r}^T \odot \mathbf{A}, f(\mathbf{s}))$ and $(\mathbf{M}, \mathbf{r}^T \odot \mathbf{M}, f(\mathbf{s}))$, where $\mathbf{M} \in \mathbb{Z}_q^{n\times(m+1)}$ is sampled uniformly at random, is negligible. Now, the hardness of the $SS(n, q^{m+1})$ problem implies that the latter distribution is computationally indistinguishable from $(\mathbf{M}, \mathbf{v}, f(\mathbf{s}))$, where $\mathbf{v} \in \mathbb{Z}_q^{m+1}$ is sampled uniformly at random, independently of $\mathbf{M}$. This concludes the proof of the theorem. □

5 Oblivious Transfer Protocol

In this section we present an oblivious transfer (OT) protocol based on subset sum that provides security for the receiver against a cheating sender, and security for the sender against an honest-but-curious receiver. (See Appendix A.3 for the formal definition of OT.) Our protocol is an instance of a construction proposed by Gertner et al. [GKM+00], based on protocols by Even et al. [EGL82, EGL85], to show that a special property of public-key encryption is sufficient for the construction of two-message semi-honest OT. Informally, the property is that it is possible to efficiently sample a string pk with a distribution indistinguishable from that of a properly generated public key, while preserving the semantic security of the encryption $\mathcal{E}_{\mathsf{pk}}$. Our cryptosystem satisfies this property, by pseudorandomness of subset sum. For the sake of self-containment, however, we provide direct proofs of our OT protocol's correctness and security.

5.1 OT Based on Subset Sum

Our oblivious transfer protocol is a simple application of our encryption scheme. We denote by $\mathcal{G}$, $\mathcal{E}$ and $\mathcal{D}$, respectively, the key-generation, encryption and decryption algorithms of the public-key encryption scheme described in Section 3. The receiver with inputs $1^n, b$ first sends a properly generated public key pk_b and a uniformly random fake public key $\mathsf{pk}_{1-b} \in \mathbb{Z}_q^{n\times(n+k)}$. The sender with inputs $1^n, \mathbf{z}_0, \mathbf{z}_1$ uses each key pk_i to encrypt its input $\mathbf{z}_i$ and replies with the ciphertexts $\mathbf{u}_0^T, \mathbf{u}_1^T$. The receiver can then retrieve $\mathbf{z}_b$ by decrypting $\mathbf{u}_b^T$, using the secret key corresponding to pk_b. Details follow.

Let $n, k \in \mathbb{N}$, $b \in \{0,1\}$, and $\mathbf{z}_0, \mathbf{z}_1 \in \{0,1\}^k$

<u>Receiver</u> $R(1^n, b)$: $(\mathsf{pk}_b, \mathsf{sk}_b) \stackrel{\$}{\leftarrow} \mathcal{G}(1^n)$; $\mathsf{pk}_{1-b} \stackrel{\$}{\leftarrow} \mathbb{Z}_q^{n\times(n+k)}$; Send $\mathsf{pk}_0, \mathsf{pk}_1$

<u>Sender</u> $S(1^n, \mathbf{z}_0, \mathbf{z}_1)$: $\mathbf{u}_0^T \leftarrow \mathcal{E}_{\mathsf{pk}_0}(\mathbf{z}_0)$; $\mathbf{u}_1^T \leftarrow \mathcal{E}_{\mathsf{pk}_1}(\mathbf{z}_1)$; Send $\mathbf{u}_0^T, \mathbf{u}_1^T$

<u>Receiver</u> R: $\mathbf{z}_b \leftarrow \mathcal{D}_{\mathsf{sk}_b}(\mathbf{u}_b^T)$; Return $\mathbf{z}_b$

5.2 Proofs of Correctness and Security

We now show that correctness follows from correctness of the cryptosystem.

Theorem 5.1. *If the sender and receiver both follow the protocol, then the former outputs nothing and the latter outputs* $\mathbf{z}_b$ *with probability* $1 - n^{-\omega(1)}$.

Proof. Since pk_b is a properly generated public key corresponding to secret key sk_b, $\mathbf{u}_b^T$ is a valid encryption of message $\mathbf{z}_b$ under pk_b, and the receiver computes the decryption of $\mathbf{u}_b^T$ using sk_b, the proof follows from Theorem 3.5. □

Security for the receiver is proved based on the pseudorandomness of subset sum. A properly generated public key is indistinguishable from a uniformly random element in $\mathbb{Z}_q^{n\times(n+k)}$. Therefore, for any input bit, the receiver's message consists of two elements from computationally indistinguishable distributions. It follows that the distribution of the receiver's message when the input is 0 is computationally indistinguishable from the distribution when the input is 1. The precise statement of this result is the following.

Theorem 5.2. *Assuming the hardness of the $SS(n, q^n)$ problem, where n is the security parameter, the above OT protocol is secure for the receiver.*

Proof. Let $R(1^n, b)$ denote the message sent by the honest receiver with inputs $1^n, b$. We show that the ensembles $R(1^n, 0)$ and $R(1^n, 1)$ are computationally indistinguishable.

As in the proof of Theorem 3.1, the hardness of the $SS(n, q^n)$ problem implies that the distributions pk_0 and pk_1 are computationally indistinguishable. This implies that ensembles $R(1^n, 0)$ and $R(1^n, 1)$ are indistinguishable as well. □

The protocol is not secure against malicious receivers. Indeed, a malicious receiver can properly generate two key pairs $\mathsf{pk}_0, \mathsf{sk}_0$ and $\mathsf{pk}_1, \mathsf{sk}_1$, and then use the secret keys to decrypt both ciphertexts $\mathbf{u}_0^T, \mathbf{u}_1^T$. The protocol is, however, secure for the sender against honest-but-curious receivers, as we now show.

Theorem 5.3. *Assuming the hardness of the $SS(n, q^{n+k})$ problem, where n is the security parameter and k is the length of the sender's input messages, the above OT protocol is secure for the sender against an honest-but-curious receiver.*

Proof. Let $R(1^n, b)$ denote the message sent by the honest receiver with inputs $1^n, b$, and $S(1^n, \mathbf{z}_0, \mathbf{z}_1, R(1^n, b))$ denote the reply of the honest sender with inputs $1^n, \mathbf{z}_0, \mathbf{z}_1$. We show that the ensembles $(S(1^n, \mathbf{z}_0, \mathbf{z}_1, R(1^n, 0)), R(1^n, 0))$ and $(S(1^n, \mathbf{z}_0, 0^k, R(1^n, 0)), R(1^n, 0))$ are computationally indistinguishable, and the ensembles $(S(1^n, \mathbf{z}_0, \mathbf{z}_1, R(1^n, 1)), R(1^n, 1))$ and $(S(1^n, 0^k, \mathbf{z}_1, R(1^n, 1)), R(1^n, 1))$ are computationally indistinguishable.

In the proof of Theorem 3.1, we showed that for any $\mathbf{m}_0, \mathbf{m}_1 \in \{0, 1\}^k$ the ensembles $(\mathsf{pk}, \mathcal{E}_{\mathsf{pk}}(\mathbf{m}_0))$ and $(\mathsf{pk}, \mathcal{E}_{\mathsf{pk}}(\mathbf{m}_1))$ are computationally indistinguishable. This is true when pk is a properly generated public key and also when pk is a random element in $\mathbb{Z}_q^{n\times(n+k)}$. Therefore, the ensembles $(\mathsf{pk}_{1-b}, \mathcal{E}_{\mathsf{pk}_{1-b}}(\mathbf{z}_{1-b}))$ and $(\mathsf{pk}_{1-b}, \mathcal{E}_{\mathsf{pk}_{1-b}}(0^k))$ are computationally indistinguishable. Hence for $b \in \{0, 1\}$ the ensembles $(S(1^n, \mathbf{z}_b, \mathbf{z}_{1-b}, R(1^n, b)), R(1^n, b))$ and $(S(1^n, \mathbf{z}_b, 0^k, R(1^n, b)), R(1^n, b))$ are computationally indistinguishable. This completes the proof. □

Acknowledgements

We would like to thank Phong Nguyen for pointing out a mistake in our connection between uSVP and subset sum in Section 1.2, and Petros Mol for pointing out that the result of Impagliazzo and Naor stated in Lemma 2.2 generalizes for all q. The first author would also like to thank Richard Lindner, Markus Rückert, and Michael Schneider for useful conversations.

References

[ACPS09] Applebaum, B., Cash, D., Peikert, C., Sahai, A.: Fast cryptographic primitives and circular-secure encryption based on hard learning problems. In: Halevi, S. (ed.) CRYPTO 2009. LNCS, vol. 5677, pp. 595–618. Springer, Heidelberg (2009)

[AD97] Ajtai, M., Dwork, C.: A public-key cryptosystem with worst-case/average-case equivalence. In: STOC (1997); An improved version is described in ECCC 2007

[AGV09] Akavia, A., Goldwasser, S., Vaikuntanathan, V.: Simultaneous hardcore bits and cryptography against memory attacks. In: Reingold, O. (ed.) TCC 2009. LNCS, vol. 5444, pp. 474–495. Springer, Heidelberg (2009)

[Ale03] Alekhnovich, M.: More on average case vs approximation complexity. In: FOCS (2003)

[Cré87] Crépeau, C.: Equivalence between two flavours of oblivious transfers. In: Pomerance, C. (ed.) CRYPTO 1987. LNCS, vol. 293, pp. 350–354. Springer, Heidelberg (1988)

[DORS08] Dodis, Y., Ostrovsky, R., Reyzin, L., Smith, A.: Fuzzy extractors: How to generate strong keys from biometrics and other noisy data. SIAM J. Computing 38(1) (2008)

[EGL82] Even, S., Goldreich, O., Lempel, A.: A randomized protocol for signing contracts. In: CRYPTO (1982)

[EGL85] Even, S., Goldreich, O., Lempel, A.: A randomized protocol for signing contracts. Communications of the ACM 28(6) (1985)

[FP05] Flaxman, A., Przydatek, B.: Solving medium-density subset sum problems in expected polynomial time. In: Diekert, V., Durand, B. (eds.) STACS 2005. LNCS, vol. 3404, pp. 305–314. Springer, Heidelberg (2005)

[Fri86] Frieze, A.: On the Lagarias-Odlyzko algorithm for the subset sum problem. SIAM Journal on Computing 15 (1986)

[GKM+00] Gertner, Y., Kannan, S., Malkin, T., Reingold, O., Viswanathan, M.: The relationship between public key encryption and oblivious transfer. In: FOCS (2000)

[GMW87] Goldreich, O., Micali, S., Wigderson, A.: How to play a mental game - a completeness theorem for protocols with honest majority. In: STOC (1987)

[Gol04] Goldreich, O.: Foundations of Cryptography - Volume 2 (Basic Applications). Cambridge University Press, Cambridge (2004)

[GPV08] Gentry, C., Peikert, C., Vaikuntanathan, V.: Trapdoors for hard lattices, and new cryptographic constructions. In: STOC (2008)

[Hai08] Haitner, I.: Semi-honest to malicious oblivious transfer – The black-box way. In: Canetti, R. (ed.) TCC 2008. LNCS, vol. 4948, pp. 412–426. Springer, Heidelberg (2008)

[Hoe63] Hoeffding, W.: Probability inequalities for sums of bounded random variables. Journal of the American Statistical Association 58(301) (1963)

[HSH+08] Halderman, J.A., Schoen, S.D., Heninger, N., Clarkson, W., Paul, W., Calandrino, J.A., Feldman, A.J., Appelbaum, J., Felten, E.W.: Lest we remember: Cold boot attacks on encryption keys. In: USENIX Security (2008)

[IN96] Impagliazzo, R., Naor, M.: Efficient cryptographic schemes provably as secure as subset sum. Journal of Cryptology 9(4) (1996)

[Kil88] Kilian, J.: Founding cryptography on oblivious transfer. In: STOC (1988)

[LM09] Lyubashevsky, V., Micciancio, D.: On bounded distance decoding, unique shortest vectors, and the minimum distance problem. In: Halevi, S. (ed.) CRYPTO 2009. LNCS, vol. 5677, pp. 577–594. Springer, Heidelberg (2009)

[LO85] Lagarias, J.C., Odlyzko, A.M.: Solving low density subset sum problems. Journal of the ACM 32 (1985)

[LPS09] Lyubashevsky, V., Palacio, A., Segev, G.: Public-key cryptographic primitives provably as secure as subset sum. ePrint (2009)

[Lyu05] Lyubashevsky, V.: The parity problem in the presence of noise, decoding random linear codes, and the subset sum problem. In: Chekuri, C., Jansen, K., Rolim, J.D.P., Trevisan, L. (eds.) APPROX 2005 and RANDOM 2005. LNCS, vol. 3624, pp. 378–389. Springer, Heidelberg (2005)

[MH78] Merkle, R.C., Hellman, M.E.: Hiding information and signatures in trapdoor knapsacks. IEEE Trans. on Inf. Theory IT-24 (1978)

[NS09] Naor, M., Segev, G.: Public-key cryptosystems resilient to key leakage. In: Halevi, S. (ed.) CRYPTO 2009. LNCS, vol. 5677, pp. 18–35. Springer, Heidelberg (2009)

[Odl90] Odlyzko, A.: The rise and fall of knapsack cryptosystems. In: Symposia of Applied Mathematics (1990)

[Pei09] Peikert, C.: Public-key cryptosystems from the worst-case shortest vector problem. In: STOC (2009)

[PVW08] Peikert, C., Vaikuntanathan, V., Waters, B.: A framework for efficient and composable oblivious transfer. In: Wagner, D. (ed.) CRYPTO 2008. LNCS, vol. 5157, pp. 554–571. Springer, Heidelberg (2008)

[PW08] Peikert, C., Waters, B.: Lossy Trapdoor Functions and Their Applications. In: STOC (2008)

[Rab81] Rabin, M.O.: How to exchange secret keys by oblivious transfer. In: Technical Report TR-81. Harvard Aiken Computation Laboratory (1981)

[Reg03] Regev, O.: New lattice based cryptographic constructions. In: STOC (2003)

[Reg05] Regev, O.: On lattices, learning with errors, random linear codes, and cryptography. In: STOC (2005)

[Sha08] Shallue, A.: An improved multi-set algorithm for the dense subset sum problem. In: van der Poorten, A.J., Stein, A. (eds.) ANTS-VIII 2008. LNCS, vol. 5011, pp. 416–429. Springer, Heidelberg (2008)

[Sho97] Shor, P.: Polynomial-time algorithms for prime factorization and discrete logarithms on a quantum computer. SIAM J. Comput. 26(5) (1997)

[Yao86] Yao, A.C.: How to generate and exchange secrets. In: FOCS (1986)

A Cryptographic Definitions

Do to space constraints, the well-known definition of semantically-secure public-key encryption is presented in the full version of our paper [LPS09].

A.1 Randomness Extraction

We say that two variables are *ϵ-close* if their statistical distance is at most ϵ. The *min-entropy* of a random variable X is $\mathrm{H}_\infty(X) = -\log(\max_x \Pr[X = x])$.

Dodis et al. [DORS08] formalized the notion of *average min-entropy* that captures the remaining unpredictability of a random variable X conditioned on the value of a random variable Y, formally defined as follows:

$$\widetilde{\mathrm{H}}_{\infty}(X|Y) = -\log\left(E_{y\leftarrow Y}\left[2^{-\mathrm{H}_{\infty}(X|Y=y)}\right]\right).$$

The average min-entropy corresponds exactly to the optimal probability of guessing X, given knowledge of Y. The following bound on average min-entropy was proved in [DORS08]:

Lemma A.1 ([DORS08]). *If* Y *has* 2^r *possible values and* Z *is any random variable, then*

$$\widetilde{\mathrm{H}}_{\infty}(X|(Y,Z)) \geq \mathrm{H}_{\infty}(X|Z) - r.$$

A main tool in our constructions in this paper is a strong randomness extractor. The following definition naturally generalizes the standard definition of a strong extractor to the setting of average min-entropy:

Definition A.2 ([DORS08]). *A function* $\mathsf{Ext}: \{0,1\}^n \times \{0,1\}^d \rightarrow \{0,1\}^m$ *is an* average-case (k,ϵ)-strong extractor *if for all random variables* X *and* I *such that* $X \in \{0,1\}^n$ *and* $\widetilde{\mathrm{H}}_{\infty}(X|I) \geq k$ *it holds that*

$$\mathrm{SD}\left((\mathsf{Ext}(X,S),S,I),(U_m,S,I)\right) \leq \epsilon,$$

where S *is uniform over* $\{0,1\}^d$.

Dodis et al. proved the following lemma stating that any strong extractor is in fact also an average-case strong extractor:

Lemma A.3 ([DORS08]). *For any* $\delta > 0$, *if* Ext *is a (worst-case)* $(m - \log(1/\delta),\epsilon)$*-strong extractor, then* Ext *is also an average-case* $(m,\epsilon+\delta)$*-strong extractor.*

A.2 Key-Leakage Attacks

We follow the framework introduced by Akavia et al. [AGV09] and recall their notion of a key-leakage attack. Informally, an encryption scheme is secure against key-leakage attacks if it is semantically secure even when the adversary obtains sensitive leakage information. This is modeled by allowing the adversary to submit any function f and receive $f(\mathsf{sk})$, where sk is the secret key, as long as the output length of f is bounded by a predetermined parameter λ.

Akavia et al. defined two notions of key-leakage attacks: adaptive attacks and non-adaptive attacks. In an adaptive key-leakage attack, the adversary is allowed to choose the leakage function after seeing the public key, and in a non-adaptive key-leakage attack the adversary has to choose the leakage function in advance. In this paper we deal with the non-adaptive setting, and note that this notion of leakage is still very meaningful as it captures many realistic attacks in which the leakage does not depend on the parameters of the encryption scheme. For example, it captures the cold boot attacks of Halderman et al. [HSH+08], in which the leakage depends only on the properties of the hardware devices that are used for storing the secret key.

Formally, for a public-key encryption scheme $(\mathcal{G}, \mathcal{E}, \mathcal{D})$ we denote by sk_n and pk_n the sets of secret keys and public keys that are produced by $\mathcal{G}(1^n)$. That is, $\mathcal{G}(1^n) : \{0,1\}^* \to \mathsf{sk}_n \times \mathsf{pk}_n$ for every $n \in \mathbb{N}$. The following defines the notion of a non-adaptive key-leakage attack:

Definition A.4 (non-adaptive key-leakage attacks). *A public-key encryption scheme* $(\mathcal{G}, \mathcal{E}, \mathcal{D})$ *is* semantically secure against non-adaptive $\lambda(n)$-key-leakage attacks *if for any collection* $\mathcal{F} = \left\{ f_n : \mathsf{sk}_n \to \{0,1\}^{\lambda(n)} \right\}_{n \in \mathbb{N}}$ *of efficiently computable functions and any two messages* m_0 *and* m_1*, the distributions* $(\mathsf{pk}, \mathcal{E}_{\mathsf{pk}}(m_0), f_n(\mathsf{sk}))$ *and* $(\mathsf{pk}, \mathcal{E}_{\mathsf{pk}}(m_1), f_n(\mathsf{sk}))$ *are computationally indistinguishable, where* $(\mathsf{sk}, \mathsf{pk}) \xleftarrow{\$} \mathcal{G}(1^n)$.

A.3 Oblivious Transfer

Oblivious transfer is a cryptographic primitive, introduced by Rabin [Rab81], which has been shown to be sufficiently strong to enable any multiparty computation [Yao86, GMW87, Kil88]. There are several equivalent formulations of OT in the literature. We use the version of Even, et al. [EGL85] known as *1-out-of-2 oblivious transfer*, and refer to it as simply OT. Crépeau [Cré87] showed that this variant is equivalent to the original definition of oblivious transfer.

A 1-out-of-2 oblivious transfer is a two-party protocol in which a sender has two secret strings $\mathbf{z}_0, \mathbf{z}_1$ and a receiver has a secret bit b. At the end of the interaction, the receiver learns $\mathbf{z}_b$ but has no information about $\mathbf{z}_{1-b}$, and the sender learns nothing about b. General OT guarantees security even in the face of cheating parties who deviate from the prescribed protocol. *Honest* OT, on the other hand, guarantees security only against honest-but-curious parties. These are parties that follow the protocol, but keep a record of all intermediate results and may perform any computation to extract additional information from this record, once the protocol ends. Any honest OT protocol can be transformed into a general OT protocol, using either black-box techniques [Hai08], or using zero-knowledge proofs to force parties to behave in an honest-but-curious manner [Gol04]. The formal definition of OT follows.

Definition A.5. *Oblivious Transfer (OT) is a two-party protocol involving a sender* S *with inputs* 1^n *and* $\mathbf{z}_0, \mathbf{z}_1 \in \{0,1\}^k$*, where* k *is a constant, and a receiver* R *with inputs* 1^n *and* $b \in \{0,1\}$*.* S *and* R *are polynomial-time randomized algorithms such that if both follow the protocol, then the former outputs nothing and the latter outputs* $\mathbf{z}_b$ *(with overwhelming probability). We consider the following security requirements:*

Security for the receiver. *Let* $R(1^n, b)$ *denote the message sent by the honest receiver with inputs* $1^n, b$*. Then the ensembles* $\{R(1^n, 0)\}_{n \in \mathbb{N}}$ *and* $\{R(1^n, 1)\}_{n \in \mathbb{N}}$ *are computationally indistinguishable.*

Security for the sender. *Let $S(1^n, \mathbf{z}_0, \mathbf{z}_1, m)$ denote the message sent by the honest sender with inputs $1^n, \mathbf{z}_0, \mathbf{z}_1$ when the (possibly cheating, polynomial time) receiver's message is m. Then for every $\mathbf{z}_0, \mathbf{z}_1 \in \{0,1\}^k$ and every polynomial-length message $m \in \{0,1\}^*$, either the ensembles $\{S(1^n, \mathbf{z}_0, \mathbf{z}_1, m), m\}_{n \in \mathbb{N}}$ and $\{S(1^n, \mathbf{z}_0, 0^k, m), m\}_{n \in \mathbb{N}}$* **or** *the ensembles $\{S(1^n, \mathbf{z}_0, \mathbf{z}_1, m), m\}_{n \in \mathbb{N}}$ and $\{S(1^n, 0^k, \mathbf{z}_1, m), m\}_{n \in \mathbb{N}}$ are computationally indistinguishable.*

Security against honest-but-curious (a.k.a. "semi-honest") receivers relaxes the second condition above to consider only a receiver that faithfully follows the protocol, but keeps a record of all intermediate results and may perform any computation, after the protocol is completed, to extract additional information from this record.

Rationality in the Full-Information Model

Ronen Gradwohl*

Kellogg School of Management, Northwestern University, Evanston, IL 60208
r-gradwohl@kellogg.northwestern.edu

Abstract. We study rationality in protocol design for the full-information model, a model characterized by computationally unbounded adversaries, no private communication, and no simultaneity within rounds. Assuming that players derive some utility from the outcomes of an interaction, we wish to design protocols that are faithful: following the protocol should be an optimal strategy for every player, for various definitions of "optimal" and under various assumptions about the behavior of others and the presence, size, and incentives of coalitions. We first focus on leader election for players who only care about whether or not they are elected. We seek protocols that are both faithful and resilient, and for some notions of faithfulness we provide protocols, whereas for others we prove impossibility results. We then proceed to random sampling, in which the aim is for the players to jointly sample from a set of m items with a distribution that is a function of players' preferences over them. We construct protocols for $m \geq 3$ that are faithful and resilient when players are single-minded. We also show that there are no such protocols for 2 items or for complex preferences.

1 Introduction

The full-information model of Ben-Or and Linial [8] is one of the classically-studied settings for protocol design. In this model there are no computational limits on the adversary, there is no private communication, and there is no guarantee of simultaneity within rounds of a protocol. Three famous problems are collective coin-flipping, leader election, and random sampling. In the first, players jointly flip a coin; in the second, they jointly select a random player; and in the third, they jointly select a random element from some universe of m items. In general, the goal is to design protocols that are resilient: the outcome should be random even in the presence of an adversary who corrupts and coordinates the behavior of a fraction of the players.

In this paper we explore the role of preferences in the design of such protocols. While preferences are not explicitly considered in the well-studied formulations of the problems, they are implicitly present. For example, leader election has a fairness criterion, which requires each player to be elected with roughly equal

* Much of this research was done while the author was a graduate student at the Weizmann Institute of Science, supported by US-Israel Binational Science Foundation Grant 2002246 and ISF Grant 334/08.

D. Micciancio (Ed.): TCC 2010, LNCS 5978, pp. 401–418, 2010.

probability (presumably because everybody wants to be the leader). A leader election protocol is resilient if an adversary can not force the elected leader to be a member of his coalition (or at least will fail to do so with constant probability). Again, the adversary *wants* a coalition-member to be elected. For collective coin-flipping and random sampling, resilience is also measured as a bound on the probability that an adversary succeeds at something. It is implicitly assumed that the adversary wants to do this, and that the honest (non-adversarial) players do not wish him to achieve his goal.

The study of preferences in the design of protocols is primarily the domain of mechanism design. In mechanism design a planner wishes to implement some function of players' private information. His goal is to design a mechanism and provide incentives for the players so that their optimal strategy is to truthfully reveal their private information, and more generally to adhere to the mechanism. The optimality of players' strategies is measured via some solution concept: following the mechanism should be in some equilibrium, most commonly Nash, ex post Nash, or dominant strategy. In this paper we take a similar approach – we define new solution concepts appropriate for the full-information model, and seek protocols that are *faithful*: following them is optimal for players with respect to these solution concepts (in addition to the usual resilience guarantees).

For any problem of protocol design, making the structure of preferences explicit has two potential benefits, both of which we achieve in this paper. First, it can result in better protocols – protocols are arguably of little use if players have no incentives to follow them. If one can obtain faithful protocols without harming the original guarantees of the protocol, then one has only gained. Second, it may be possible to sidestep some impossibility results of the original problem, since often these impossibility results are based on arbitrary play by the adversary. If players do not play arbitrarily but rather obey some preference structure, then many of these results no longer hold.

The model. In the full-information model all communication is by broadcast. In each round, some of the players send a message, which may depend on messages sent in previous rounds. The main difficulty is that adversarial players are allowed to "rush" – to wait until all messages have been sent within a round, and only then to send their own messages.

This paper. We are largely motivated by recent work in rational cryptography, in which the aim is to design cryptographic protocols that participants *want* to follow. Two of the main difficulties encountered when attempting a game theoretic analysis of cryptographic protocols are computational limits and potentially adversarial timing. In this paper we focus solely on the latter issue by considering a model in which (adversarial) players may be computationally unbounded, and the guaranteed security (i.e. resilience) is information-theoretic. We highlight the various challenges and subtleties caused by a combination of rational and adversarial players, particularly in the presence of adversarial timing. We also draw a possibility-impossibility border for various problems and requirements in this setting. Finally, we believe that this paper is an illuminating stepping-stone towards a game theoretic analysis of more general cryptographic protocols.

1.1 Our Results and Organization

Definitions (Section 2). The initial difficulty encountered when considering preferences in the full-information model is to precisely formulate a notion of equilibrium. The first notion to consider is Nash equilibrium (NE), in which each player's strategy must be expected utility maximizing assuming others also follow their Nash strategies. If the protocol is such that only one player sends a message in each round, then this suffices. One such protocol is Baton Passing [32], a protocol that is resilient and in fact satisfies our weaker solution concept[1]. However, state-of-the-art protocols are often round-efficient, and allow multiple players to broadcast within a round. Because of the lack of synchrony within rounds, however, NE does not suffice. In the Lightest Bin protocol [14], for example, a player may increase his chance of winning from $1/n$ (where n is the number of players) to a constant by deviating. To deal with asynchrony, we will require that for *any ordering* of the players within each round, the protocol is in a NE. In Section 2 we formalize this and other notions of what it means to be faithful and faithful in the presence of adversarial players.

Impossibility with complex preferences (Section 3). In Section 3 we encounter our first impossibility result. Theorem 1 states that no random selection protocol can satisfy even our weakest solution concept if players have a full preference order over the outcomes of the protocol. One implication of this impossibility result is that collective coin-flipping is impossible with players who have some preference about the outcome. For leader election and random sampling, this result forces us to concentrate on more restricted preferences for players. For the former, we assume that players care only about whether or not they are elected, and are indifferent otherwise. For the latter, we assume players are single-minded: each prefers one of the items, and is indifferent about the others.

Faithfulness with resilience (Section 4.2). The standard aim of selection protocols in the full-information model is resilience: if an adversary corrupts and coordinates the actions of a fraction of the players, he still fails to force his desired outcome with non-negligible probability. In Section 4.2 we construct optimal protocols that both satisfy a notion of equilibrium and are resilient. Players wish to faithfully adhere to the protocol if the others also do, and there is a resilience guarantee in the presence of an adversary.

Faithfulness in the face of an adversary (Section 4.3). In Section 4.3 we consider the problem of constructing leader election protocols that are in equilibrium even when not all others follow the protocol. We show that it is impossible to construct such protocols in the presence of a malicious adversary, even if the adversary has his own objective of maximizing the probability that a coalition-member wins. However, if the adversary maximizes this probability, but also only deviates from the protocol if he strictly gains from doing so (i.e. if deviating is costly), then we do design a resilient protocol.

[1] More specifically, it is in a full-information ex post NE – see Definition 5. It is not, however, in a full-information dominant strategy equilibrium (Definition 3).

Resilience to rational coalitions (Section 4.4). A different form of resilience against adversarial play is when there is no controlling adversary, but instead players may form "rational coalitions" to benefit all members. In Section 4.4 we give an impossibility result for one notion of a "rational coalition", but for a weaker notion provide a protocol that is resilient against all such coalitions of size at most $n-2$.[2]

Random sampling (Section 5). Our final set of results concerns random sampling. Each player has some preferences over a universe of m items, and the goal is to design a protocol in which an item is sampled with a probability distribution that is a function of those preferences. We design protocols that are simultaneously in a full-information ex post Nash equilibrium (in which truthful revelation of one's preferences is optimal) and resilient against adversarial coalitions.

1.2 Related Work

This paper draws from three different literatures – protocol design in cryptography and distributed computing, and algorithmic mechanism design. The extensive literature on collective coin-flipping, random sampling, and leader election in the full-information model includes [32,16,27,3,11,12,17,31,14,13,33,6]. The paper most closely related to ours is that of Antonakopoulos [6], who also considers 1-round protocols in which individual players have no incentive to deviate. However, his protocols all attain either faithfulness or resilience, but never both. Similarly, Ben-Or and Linial [8] have a 2-round protocol that is faithful but not resilient to larger coalitions. The paper most closely related to ours from the mechanism design literature is that of Altman and Tennenholtz [5], who construct 1-round protocols that are faithful (but also not resilient). Their goal is to attain arbitrary distributions over the players. Also related is the literature on ranking games [10], in which players have preferences about their rankings in some game.

While we believe that we are the first to study notions of rationality tailored specifically for the full-information model, such notions have been studied in other settings for distributed computing. For example, Monderer and Tennenholtz [25] consider an implementation problem in a distributed network. Shneidman and Parkes [34] introduce the idea that protocols should be faithful. Additionally, the field of Distributed Algorithmic Mechanism Design (DAMD) focuses on implementing mechanisms for various problems in a distributed setting. In a general "mission statement" for DAMD, Feigenbaum and Shenker [15] argue that it would be desirable to incorporate various fault models into the DAMD framework. Also, Halpern [21,22] has expressed the need to incorporate faulty or malicious behavior into distributed settings with rational players. Some papers that address this issue are Aiyer et al. [2], Abraham et al. [1], and Gradwohl [18].

Finally, as mentioned in the introduction, this work is closely related to the growing literature on rational cryptography (see, for example, Katz [24] and

[2] Compare this with the fact that there are *no* protocols that are resilient against an adversary of size $n/2$ [32].

the references therein). Many works in this literature study rational behavior in a cryptographic setting, for which the full-information model is a special case. However, due to computational issues, the definitions in the general setting are messier (and often also weaker). We note that the way we model rushing is closely related to an idea of Ong et al. [29], who adopt the methods of Kalai [23] to a protocol design setting. The idea of considering rational coalitions was also explored in this context by Ong et al. [28].

Our notions of stability of coalitions are related to similar notions in the game theory literature, such as the strong Nash equilibrium of Aumann [7] and the coalition-proof equilibrium notions of Bernheim et al. [9], Moreno and Wooders [26], and Abraham et al. [1].

2 Protocols and Games

For any vector $X = (X_1, \ldots, X_n)$ and $S \subset [n]$, we denote by $X_S = \{X_i\}_{i \in S}$ and by $X_{-S} = \{X_i\}_{i \notin S}$.

2.1 Resilient Protocols

We are interested in protocols involving many players and the incentives of players in following these protocols. Thus, we will assume that players have preferences over possible outcomes, as well as other private information. As in the game theory literature, all this information is collectively called a player's type. Player i's type is denoted by t_i, and the vector $t = (t_1, \ldots, t_n)$ is called the type profile. The space of possible types of player i is T_i.

Definition 1 (selection protocol). *An n-player* selection protocol *$\mathcal{P}$ is specified by a function f, a natural number q, and, for each of the n players, a set of q randomized functions $\{S_i^1, \ldots, S_i^q\}_{i \in [n]}$. The protocol proceeds as follows:*

- *At round j, the i'th player broadcasts a random message M_i^j obtained by applying the randomized function S_i^j to all previous messages sent, namely $\{M_k^l : k \in [n], l < j\}$, as well as player i's type t_i. The randomness of the function comes from the player's independent coins.*
- *After q rounds, the players output $f(\{M_k^l : k \in [n], l \in [q]\})$ which is an element of $[m]$ in an m-item* random sampling protocol *and an element of $[n]$ in a* leader election protocol. *If all players follow the protocol then the output is a uniformly random element (unless stated otherwise).*

In any round j, a player i's *legal* messages are those in $\bigcup_{t_i \in T_i} supp(S_i^j(t_i, \{M_k^l : k \in [n], l < j\}))$[3]. We assume that if a player noticeably deviates from the protocol (by broadcasting a message that is not legal), then his message is changed to some default legal value.

[3] Note that a player's legal actions include messages in the support of S_i^j for all types, not just the true one. This is so because the other players do not know i's true type.

Players may not legally base their messages in round j on the messages of other players in round j. However, since we can not guarantee simultaneity within a round, we allow the dishonest players to *rush*: they may base their messages on the messages of other players from the same round (but not from later rounds). A leader election protocol is *ε-resilient* to coalitions of size t if the following holds: If at most t players are playing a coordinated rushing strategy, then the probability that the elected leader is a cheating player is at most $1 - \varepsilon$. Often we will implicitly be referring to a family of protocols, one for each value of n. In this case, we say a protocol is resilient if there exists some $\varepsilon > 0$ such that all protocols in the family with enough players are ε-resilient.

A protocol is *oblivious* if players' messages are based only on their internal coin tosses. A protocol is *explicit* if players' messages and the function f are computable in probabilistic polynomial time (in the number of players and $\log(m)$).

2.2 Extensive-Form Games and Protocols

An n-player *extensive-form game* is specified by a game tree in which every node is owned by a player and outgoing edges are labelled by actions. The game begins at the root node and proceeds down the tree – at every node following the edge labelled by the action played by the node's owner. Payoffs for players are specified at the leaves.

Definition 2 (Nash equilibrium (NE)). *A* Nash equilibrium (NE) *in an extensive-form game is a mixed strategy for every player at every node that he owns, such that: if all players play their NE strategy, then no player obtains a higher expected payoff by deviating at any of his nodes.*

We note that in the games we consider, the NE will be completely mixed strategies (i.e. players will play every action with positive probability). Such Nash equilibria are in fact subgame perfect (see [30]).

Consider a selection protocol, where each player i derives some utility $u_i : T_i \times [m] \mapsto \mathbb{R}$ from outputs of the protocol. u_i is such that for $o \neq o' \in [m]$, we have that $u_i(t_i, o) > u_i(t_i, o')$ if and only if player i of type t_i strictly prefers o to o'. Then any protocol in which only one player sends a message in each round can be viewed as an extensive-form game[4]: if after $j - 1$ rounds and messages $M_1, \ldots, M_{j-1}$ player i plays in round j, i owns the node at level j in the game tree reached by the game path $M_1, \ldots, M_{j-1}$. Player i's payoff from an instance of play resulting in o is $u_i(t_i, o)$. Such a selection protocol is in a NE if, in the associated game, it is a NE for every player i to play according to strategy S_i^j if any of his nodes at level j is reached (we say that i follows strategy S_i).

2.3 Rationality in Selection Protocols – Definitions

We now define notions of what it means for a protocol to be faithful, i.e. in which it is in players' best interests to follow the protocol specification. Because

[4] While it is possible to model simultaneous play as an extensive-form game with *imperfect* information, the ability to rush and the lack of synchrony are more difficult to incorporate into this framework.

there is no synchrony within rounds of a selection protocol, we may view the possibility of rushing as a strategy for players. That is, a player may choose to wait until others have played, and only then submit his message. Thus, NE does not suffice as a solution concept for such games. However, if only one player plays in each round, then this does not matter (since rushing is only allowed *within* rounds), and so for such protocols NE is a reasonable solution concept. For general protocols, we would like the protocol to be optimal for players *regardless of the order of play within a round.* This motivates the following.

For any q-round protocol $\mathcal{P}$, we can construct protocol $\mathcal{P}'$ with at most qn rounds, and such that only one player sends a message in each round. We say that $\mathcal{P}'$ is a *linearization* of $\mathcal{P}$, and it is constructed as follows: Let $\pi : [n] \times [q] \mapsto [nq]$ be some bijective map. Then $\mathcal{P}'$ is such that in round ℓ, if $(i, k) = \pi^{-1}(\ell)$ then player i sends a message sampled from S_i^k. This is well-defined for oblivious protocols, and essentially means players play in an arbitrary order, but only one player per round. For non-oblivious protocols, we require π to be *round-respecting*: $\pi(i, k) = \ell$ if and only if for all $j \in [n]$ and $k' < k$ it holds that $\pi(j, k') < \ell$. That is, here the arbitrary ordering is only within rounds.

We note that the idea of considering all linearizations appears also in Ong et al. [29].

Our first solution concept for selection protocols in the full-information model is a full-information dominant strategy equilibrium, which essentially means that for any player i, regardless of the messages sent by others in all rounds, i can never strictly increase his utility by deviating from the protocol. The following generalizes the definition of [5] to multi-round protocols.

Definition 3 (full-information dominant strategy equilibrium). *An oblivious selection protocol $\mathcal{P}$ is in a* dominant strategy equilibrium *if for all type profiles, all linearizations of $\mathcal{P}$ are in a NE.*

An alternative, more direct but equivalent formulation is the following:

Definition 4 (full-information dominant strategy equilibrium – alternative formulation). *An oblivious n-player, q-round, m-item selection protocol is in a* full-information dominant strategy equilibrium *if for all $i \in [n]$ and messages $M_{-i} = \{M_k^l : k \in [n] \setminus \{i\}, l \in [q]\}$ sent by all other players in all rounds, it holds that $u_i(t_i, f(M_{-i}, M_i)) = u_i(t_i, f(M_{-i}, M_i'))$, where $M_i, M_i' \in supp(S_i^1) \times \ldots \times supp(S_i^q)$.*

Remark 1. The reason we have equality above, as opposed to an inequality, is that the actions in the support of S_i^j are *all* dominant. That is, all these actions are best-responses, even conditioned on the actions of others. It can thus not be that one such action is better than the other, for then the other would not be dominant.

The definition of a full-information dominant strategy equilibrium is rather strong, but still achievable (for example, Theorems 2, 3, and 4 below). We note that our impossibility result, Theorem 1, applies even to our weaker solution concepts.

In a full-information ex post NE the requirement is a bit relaxed: a player i can not strictly increase his expected utility in any round j by deviating, regardless of the messages of players in all rounds up to *and including* round j. That is, regardless of the order of play within the current round, i has no incentive to deviate (on expectation over play in future rounds). The following definition is new:

Definition 5 (full-information ex post Nash equilibrium). *A selection protocol $\mathcal{P}$ is in an* ex post NE *if for all type profiles, all round-respecting linearizations of $\mathcal{P}$ are in a NE.*

An alternative, more direct but equivalent formulation for this solution concept is a bit more involved, and appears in the full version of this paper [19].

2.4 Rationality in the Face of an Adversary – Definitions

A protocol that satisfies the definitions of Section 2.3 is an optimal strategy for players assuming all others also follow the protocol. If some of the players are adversarial, however, then this may not hold. In this case, we actually want a stronger guarantee. To this end, we need the following definition, first defined by [1] (for normal-form games):

Definition 6 (v-tolerant NE). *A* v-tolerant NE *in an extensive-form game is a mixed strategy for every player at every node that he owns, such that the following holds: for any $V \subset [n]$ of size at most v, if all players in $[n] \setminus \{V\}$ play their NE strategy, then none of them can obtain a higher expected payoff by deviating from the NE at any of their nodes* regardless of the actions of players in V.

The ideal faithfulness guarantee that we would like for selection protocols is roughly the following: no player should be able to strictly improve his expected payoff by deviating, assuming most players follow the protocol, some play arbitrarily, and the order within any round is also arbitrary.

Definition 7 (full-information v-tolerant ex post NE). *A leader election protocol $\mathcal{P}$ is in a* full-information v-tolerant ex post NE *if all round-respecting linearizations of $\mathcal{P}$ are in a v-tolerant NE.*

One possible weakening of this definition is to consider an adversary who does not act arbitrarily, but also has his own utility function u_A. Suppose an adversary corrupts a set V of players. Then we say he is playing a *coalition-optimal strategy* with respect to strategies $S = (S_1, \dots, S_n)$ if, when the players not in V follow strategies S_{-V}, the members of V play a coordinated strategy that maximizes the expectation of u_A. We say he is playing a *strictly coalition-optimal strategy* with respect to strategies S if the above holds, and if, at every node owned by some $i \in V$, i follows S_i if his part of the coordinated deviation does not strictly increase the expectation of u_A. (A more formal definition appears in the full version of this paper [19]).

Definition 8 (v-tolerant NE with (strictly) self-interested adversary). *A* v-tolerant NE with self-interested adversary *in an extensive-form game is a mixed strategy* S_j *for every player* j *for every node that he owns, such that the following holds: for any* $V \subset [n]$ *of size at most* v *and any player* $i \notin V$*, if the players in* V *play any coalition-optimal strategy and the others play their* S_j *strategy, then* i *can not increase his expected utility by deviating from* S_i*. If this holds only when the players* V *play a strictly coalition-optimal strategy, then the equilibrium is a* v-tolerant NE with strictly self-interested adversary.

Definition 9 (full-information v-tolerant ex post NE with (strictly) self-interested adversary). *A leader election protocol* $\mathcal{P}$ *is in a* full-information v-tolerant ex post NE with a (strictly) self-interested adversary *if all round-respecting linearizations of* $\mathcal{P}$ *are in a* v*-tolerant NE with a (strictly) self-interested adversary.*

2.5 Resilience to Rational Coalitions – Definitions

In Section 2.4 the adversarial coalition could act arbitrarily, or by maximizing some joint utility function u_A. In this section we define notions of rational coalitions – i.e. coalitions that rational players might reasonably want to form. In the following definitions, we assume there is some prescribed protocol $\mathcal{P}$ for the players. When we say players are "at least as well off" or "strictly gain", this is with respect to following the prescribed protocol.

Definition 10 (Pareto coalition). *A coalition* V *is a* Pareto coalition *if there exists a coordinated rushing strategy* S_V^* *for the players in* V *such that all players in* V *are at least as well off when playing* S_V^**, and one player strictly gains.*

Definition 11 (strong coalition). *A coalition* V *is* strong *if there exists a coordinated rushing strategy* S_V^* *for players* V *such that the expected utility of every* $i \in V$ *strictly increases when playing* S_V^**.*

Definition 12 (stable coalition). *A coalition* V *is* stable *if there exists a coordinated rushing strategy* S_V^* *for players* V *such that the expected utility of every* $i \in V$ *strictly increases when playing* S_V^**, and, in addition, for all sub-coalitions* $V' \subset V$ *and any coordinated rushing strategy* $S_{V'}^*$*, playing* $S_{V'}^*$ *does not increase the expected utility of all players in* V' *when players* $V \setminus V'$ *play* S_V^**.*

3 Impossibility with Complex Preferences

A player in a selection protocol has *complex preferences* if for any two outcomes $o \neq o'$ he strictly prefers one over the other. We now show that there are no faithful selection protocols for players with such preferences.

Theorem 1. *No selection protocol can be in a full-information ex post NE for players with complex preferences.*

Proof. Suppose there exists an *oblivious* selection protocol $\mathcal{P}$ in an ex post Nash equilibrium, and fix some round-respecting linearization of $\mathcal{P}$. Let T be the corresponding game tree, where some player i owns a node u (that is reached with positive probability) at the lowest non-leaf level ℓ. Suppose the protocol specification is for i to play mixed strategy S_i at level ℓ. Now, if different actions in $supp(S_i)$ result in leaves with different outcomes, then i prefers one outcome over the others (due to complex preferences). However, due to the full-information ex post NE this can not be the case: a player's different actions should not affect his expected utility, for otherwise he would have a beneficial deviation. We conclude that player i's actions do not influence the final choice of item. Hence, u can safely be omitted, resulting in a new, smaller tree. We continue shrinking the tree in this manner, yielding a deterministic selection protocol (a contradiction). The extension to non-oblivious protocols appears in the full version of this paper [19].

4 Rational Leader Election Protocols

4.1 Basic Faithful Leader Election Protocols

Because of Theorem 1, we must limit the preferences in order to obtain protocols. One natural setting for leader election is that of *self-interested* players: players care only about whether or not they are elected (they either want to win or want to not win), but are indifferent otherwise. Note that if a leader election protocol is in a full-information dominant strategy equilibrium for self-interested players, then the messages sent by others determine whether a player is elected or not (because the equilibrium holds for all type profiles). That player can only determine who is elected if he is not. The same holds for leader election protocols in a full-information ex post NE, but on expectation over messages in future rounds.

There are some basic protocols that we will use in our constructions. The first is a 1-round leader election protocol that is in a full-information dominant strategy equilibrium (but is not resilient). This protocol was given by Antonakopoulos [6] for the uniform distribution, and then generalized by Altman and Tennenholtz [5].

Theorem 2 ([5]). *For any $n \geq 4$ and any distribution $\mathcal{D}$ over $[n]$ there exists a 1-round, n-player leader election protocol $\mathcal{P}_{AT}$ in a full-information dominant strategy equilibrium, and in which each player i is elected with probability $\mathcal{D}(i)$.*

[5] also showed that there is no faithful 1-round leader election protocol for 3 players. The following protocol, which we will use in our constructions, does work for 3 players, albeit at the cost of having 2 rounds[5]. Fix any natural number $k \geq 3$, and denote $i_+ = (i \mod (k-1))+1$. Then for any positive $p_1, \ldots, p_k$ with $p_1 + \ldots + p_k = 1$ define

[5] Note that the case of 2 players is impossible by the lower bound of [32], regardless of the number of rounds.

Protocol $\mathcal{P}_k$

1. Player k chooses one player $i \neq k$, each with probability $\frac{p_{i_+}}{1-p_k}$.
2. For each $j \in \{1,\ldots,k-1\}$, if player j is chosen in round 1, he elects player k with probability p_k and player j_+ with probability $1-p_k$ as leader.

Proposition 1. *$\mathcal{P}_k$ is a leader election protocol in a full-information ex post NE that elects each player i with probability p_i.*

4.2 Combining Rationality and Resilience in Leader Election

Neither of the protocols of Section 4.1 is resilient for any $t > 1$. The following theorem can be combined with resilient leader election protocols to obtain protocols that are both resilient and in full-information dominant strategy equilibria.

Theorem 3. *For any $n \geq 4$, $k = \Omega(\sqrt{n})$, and any explicit, oblivious $r(n)$-round leader election protocol $\mathcal{P}$ there exists an explicit protocol $\mathcal{P}'$ in a dominant strategy equilibrium that has $r(\lceil n/4 \rceil)$ rounds. If $\mathcal{P}$ is resilient to $t(n)$ faults, then $\mathcal{P}'$ is resilient to $t(\lfloor n/4 \rfloor) - k$ faults.*

Proof. In the protocol below and the rest of the proof, indices are cyclical. We will prove the theorem for n a multiple of 4. The general case follows similar lines. The players are partitioned into 4 disjoint sets C_1, C_2, C_3, C_4, where $i \in C_j$ if $\lceil 4i/n \rceil = j$. The following is done in **parallel**:

1. Each set C_i runs protocol $\mathcal{P}$ to select a representative R_i.
2. For each i, R_i chooses a random player from C_{i+1}, say L_{i+1}, and outputs a random message b_i to $\mathcal{P}_{AT}$ (i.e. b_i is a random element of B_i, where $\mathcal{P}_{AT}$ takes inputs from $B_1 \times \ldots \times B_4$).
3. The winner is L_j, where j is the winner of $\mathcal{P}_{AT}$ with inputs b_1, b_2, b_3, b_4.

Since the 3 steps are done in parallel, all players choose a random player and a random input in step 2., but the output depends only on the choices of the R_i's.

Fix some player x, and suppose $x \in C_i$. x is chosen as the leader only if R_{i-1} chooses x. The probability that this occurs does not change regardless of the actions of x. Additionally, for x to win, i must be the winner of $\mathcal{P}_4$. However, since $\mathcal{P}_{AT}$ is in an ex post NE, no player in C_i can influence the probability that this occurs. Hence, from x's perspective, it does not matter who is chosen as L_i.

Now consider some cheating coalition of t players. In order for a member of the coalition to win, at least one member of the coalition must be chosen as L_i for some i. In order for this to occur, either R_{i-1} must be a member of the coalition (and then he can choose a fellow member in C_i), or R_{i-1} is an honest player who chooses a member of the coalition. Suppose there are c_1 faulty players in C_{i-1} and c_2 faulty players in C_i, where $c_1 + c_2 \leq t$. Then the probability that there are more than $c_2 + k$ honest players in C_{i-1} who choose a faulty player is at most a constant $e = \exp(-2k^2/(n/4 - c_1)) < \exp(-16k^2/n) < 1$ by a multiplicative Chernoff bound, and using the fact that $c_1 \leq t(n/4) < n/8$ (since no leader election protocol can be resilient to more than half the players). Thus, with probability at least $1 - e$, there are at most $c_1 + c_2 + k \leq t(n/4)$ players in C_{i-1} who choose a coalition member in C_i. The maximal probability that one of them wins and becomes R_{i-1} is at most a constant $\varepsilon < 1$ (since we can view the honest players who chose a coalition-member as additional faulty players). The probability that a coalition member becomes L_i for any i is thus at most $1 - (1 - \varepsilon)^4 \cdot (1 - e)^4$, which is some constant < 1.

In Theorem 3 is that the size of the coalition shrinks by about a factor of 4, and so we can not use it to get a faithful protocol with resilience close to the optimal $n/2$. The following protocol has optimal resilience, is in a full-information dominant strategy equilibrium, and has $\log^*(n) + O(1)$ rounds (same as in the state-of-the-art leader election protocols [31,14]). The proof is in the full version of this paper [19].

Theorem 4. *For every constant $\delta > 0$ and $n \geq 4$ there exists an explicit $(\log^* n + O(1))$-round leader election protocol resilient against $n(1/2 - \delta)$ faults that is in a full-information dominant strategy equilibrium.*

Extensions and Further Results. Theorem 3 can actually be generalized to obtain any distribution over the players. If we plug a 1-round leader election protocol into Theorem 3 with any distribution, we get a 1-round protocol that implements any distribution and is in a full-information dominant strategy equilibrium. This confirms a conjecture of Altman and Tennenholtz [4] about the existence of such protocols in which all players influence the outcome of the protocol in some instance. We can also construct protocols that satisfy a stronger notion of resilience against adversarial coalitions – namely, they have bounded cheaters' edge [6] – that are in a full-information ex post NE. Finally, our protocols can also be used to construct leader election protocols in which a player is elected at random *from the set of players who want to be elected.* All these extensions can be found in the full version of this paper [19].

4.3 Rationality in the Face of an Adversary

While the protocols of Section 4.2 are resilient against adversarial behavior, they are in equilibrium only if all players follow the protocol. What if this is not the case? Can an honest player's protocol specification be optimal even when some others play adversarially? The main difficulty here is that a player's actions may now also influence the actions of adversarial players in future rounds. Even if the protocol is oblivious, an adversary's strategy might not be. Definition 7 defines a the concept of an full-information v-tolerant ex post NE to deal precisely with this issue.

Unfortunately, Theorem 5 below implies that no leader election protocol can be in a v-tolerant ex post NE, and so we must look for some relaxation. For Definition 7 we make no assumptions about the adversary. If we assume that the adversary also has some preferences, then we may be able to weaken this restriction. We will assume here that the adversary's goal is to maximize the probability that some member of his coalition gets elected (the standard assumption for leader election) – that is, we consider Definition 9, where u_A is the probability that a member of the coalition gets elected. Theorem 5 also shows that this relaxation does not suffice:

Theorem 5. *There does not exist a leader election protocol in a v-tolerant ex post Nash equilibrium with self-interested adversary for any* $v \geq 1$.

Proof. Fix some protocol $\mathcal{P}$ in an ex post Nash equilibrium and a round-respecting linearization of $\mathcal{P}$. Suppose i is the first player who has a mixed strategy in the game, where two possible messages in i's support are I_1, I_2. Because i is eventually chosen by $\mathcal{P}$ with some probability that i can not influence himself (he wins with the same expected probability whether he plays I_1 or I_2), there must exist some other player whose choice of messages does influence this probability. In the subtree rooted at the node following i choosing I_1 there must exist some player j who has a strategy S_1 that increases the probability of i getting elected, and some other strategy S_2 that decreases this probability. Because $\mathcal{P}$ is in an ex post Nash equilibrium, these choices of player j do not harm his own chance of getting elected. A valid (adversarial) strategy for player j is to play S_1 whenever i plays I_1. Alternatively, j can play S_2 whenever i plays I_1. Because i does not know which strategy j is using (since j is adversarial), and in either case one of I_1 or I_2 is strictly better than the other, no single strategy of player i can be optimal in both cases.

If we limit the adversary even more by assuming that deviation is costly, we can get an explicit protocol. The following assumes that the adversary is strictly self-interested – he is self-interested, but also only deviates if he strictly gains from doing so (a formal definition appears in the full version of this paper [19]).

Theorem 6. *For any positive* k *and* $n = 3^k$ *there exists an explicit* n*-player* $2\log_3(n)$*-round leader election protocol* $\mathcal{P}$ *that is in a full-information* n*-tolerant ex post Nash equilibrium with a strictly self-interested adversary. Furthermore,* $\mathcal{P}$ *is resilient against* $n^{\log_3(2)}/2$ *faults.*

To get an idea for the proof, we show that $\mathcal{P}_k$ with $k = 3$ and the uniform distribution is in a full-information 3-tolerant ex post NE with a strictly self-interested adversary. If none or all of the players are adversarial, then all non-adversarial players should follow the protocol (since it is in a full-information ex post NE). If two players are adversarial, then they can always force a win, and so the third player may as well follow the protocol. Finally, if only one player is adversarial, then he can not increase his chance of winning (by the full-information ex post NE), and since the adversary is strictly self-interested he will not deviate. Hence, it is also a full-information ex post NE for the others to follow the protocol.

To generalize this to more players, we divide the players into sets of 3, each running $\mathcal{P}_3$. We then repeat this on the winners, until only one is left. The full version of this paper [19] contains further details and an analysis of the resilience of this protocol.

4.4 Resilience to Rational Coalitions

In Sections 4.2 and 4.3 the adversary corrupts some set of v players, and coordinates their actions. Here we let players form a "rational coalition" to benefit all members – namely, we consider the definitions of Section 2.5. For the following theorems (whose proofs appear in the full version [19]), we restrict ourselves to the case in which players are self-interested, and all *want* to be elected. First, we show that it is impossible to have resilience against our weakest notion of a rational coalition.

Theorem 7. *Every leader election protocol in a full-information ex post NE has a Pareto coalition of two players.*

For a stronger notion, however, we can get a protocol that side-steps the impossibility of leader election with adversarial coalitions of size $n/2$:

Theorem 8. *There exists an explicit 2-round leader election protocol in a full-information ex post Nash equilibrium with only 1 stable coalition. The coalition is of size $n-1$.*

The protocol that achieves this is $\mathcal{P}_k$ with $k = n$ and the uniform distribution (see Section 4.1). We also have the following theorem, as a weak illustration that we gained something by weakening our requirement from strong to stable coalitions.

Theorem 9. *For any n-player leader election protocol in a full-information ex post Nash equilibrium, all coalitions of size $n-1$ are strong.*

5 Rational Random Sampling Protocols

We consider some universe of m items, and will construct protocols that output each item with probability proportional to the number of players who (claim to) like that item most. In the full version [19] we discuss generalizations to other distributions. Due to Theorem 1, we restrict ourselves to *single-minded* players – each i's type $t_i \in [m]$ is the item he prefers, and he is indifferent about the others. Theorem 1 also implies that no random sampling is possible with $m = 2$. If $m > 2$ but players prefer only one of two items we are sampling from two items. So we must limit the type profiles. We do this by considering balanced profiles: a profile $(t_1, \ldots, t_n)$ for m items is z-balanced if each type occurs between $n/m+z$ and $n/m - z$ times.

Theorem 10. *For any $n \geq 66$ and explicit $r(n)$-round leader election protocol resilient to $t(n)$ faults in a full-information ex post NE, there exists an explicit $(r(n)+3)$-round random selection protocol for a universe of size $m \geq 66$ that is in a full-information ex post NE for all $(n(1/66 - 1/m))$-balanced type profiles. For such profiles, the random selection protocol is resilient to $t(\lfloor n/3 - n/66 \rfloor)$ faults.*

Proof. The protocol is the following:

1. Each player announces his preferred item. Players are split into 3 categories C_1, C_2, C_3 as follows: all players with the same announced type are in the same category, and the categories are "roughly" balanced: sets of players with the same declared type are greedily assigned to the smallest C_i. Fix $c_i = |C_i|$, and note that $|c_i - c_j| \leq d$ for $d = n/66$ (assuming at most one player lies about his preferred item).
2. For each i, players in C_i run the leader election protocol $\mathcal{P}$ to elect a representative R_i.
3. For each i, the players in $C_{i+1} \cup C_{i+2}$ run the leader election protocol $\mathcal{P}$, and the winner chooses a uniformly random player L_i from C_i.
4. R_1, R_2, and R_3 run $\mathcal{P}_3$. The protocol is run so that players are elected with probabilities $\frac{c_1}{n}$, $\frac{c_2}{n}$, and $\frac{c_3}{n}$ respectively.
5. The protocol's output is the announced item of player L_j, where j is the winner of $\mathcal{P}_3$ in the last round.

If a player $i \in C_j$ truthfully announces his type, then he can no longer change the probability of his type getting chosen: he only affects which of the other types are potential winners (via his choice of L_k for $k \neq j$) or which player from C_i participates in $\mathcal{P}_3$. However, since $\mathcal{P}_3$ is in a full-information ex post NE, this does not matter either.

It remains to show that it is optimal for i to truthfully reveal his type. Suppose i's preferred item is B, the fraction of other players who announce B is β, and they all get placed in C_j. Suppose i lies about his type and gets placed in $C_k \neq C_j$. How can i cheat? i wins the leader election protocol of step (2) with probability $1/c_k$ and the leader election protocol of step (3) (choosing L_j) with probability $1/(c_k + c_\ell)$ for $\ell \neq j, k$. If i is elected in both leader election protocols, he can force the winner to be a player who wants B with probability at most 1. If he wins only the leader election protocol of step (2), he can cause j to win in $\mathcal{P}_3$ with probability $1 - c_k/n$. If he wins only the leader election protocol of step (3), he can force L_j to be a player who wants B (but that player wins $\mathcal{P}_3$ with probability c_j/n. The probability that B is the chosen type given that i is cheating is

$$\begin{aligned}\Pr[B \text{ wins}] &< \left(1 - \frac{1}{c_k + c_\ell} - \frac{1}{c_k} + \frac{1}{c_k} \cdot \frac{1}{c_k + c_\ell}\right) \cdot \beta \\ &\quad + \frac{1}{c_k + c_\ell} \cdot \frac{c_j}{n} + \frac{1}{c_k} \cdot \frac{\beta n}{c_j}\left(1 - \frac{c_k}{n}\right) + \frac{1}{c_k} \cdot \frac{1}{c_k + c_\ell} \\ &= \beta - \frac{\beta}{c_k + c_\ell} - \frac{\beta}{c_k} + \frac{c_j}{(c_k + c_\ell)n} + \frac{\beta n}{c_k \cdot c_j} - \frac{\beta}{c_j} + \frac{1}{c_k} \cdot \frac{1}{c_k + c_\ell}.\end{aligned}$$

By our balancedness assumption, we know that $n/3 - d \leq c_1, c_2, c_3 \leq n/3 + d$, and that $\beta \leq d/n$. Plugging in these values (and performing some manipulations) yields

$$\Pr[B \text{ wins}] < \beta + \left(\frac{n + 3d}{2n - 6d}\right)\frac{1}{n} + \frac{9d + 18}{\left(\frac{n}{3} - d\right)^2}.$$

It can be verified that when $d \leq n/66$ and $n \geq 66$, we get that $\Pr[B \text{ wins}] < \beta + 1/n$. Now, if player i were to bid truthfully, then the probability that B wins would be $\beta + 1/n$ (since i's vote adds to B's chance of winning). Thus, it is an optimal strategy for i to bid truthfully.

What about resilience? Suppose there is an adversary of size at most $t(\lfloor n/3 - n/66 \rfloor)$ faults. In order to force an outcome in some predefined set, the adversary must win at least one of the 6 runs of the leader election protocol $\mathcal{P}$, and each runs on a set of at least $\lfloor n/3 - n/66 \rfloor$ players. Since $\mathcal{P}$ is resilient for this number of adversaries, the probability that the adversary loses all of them is at least ε^6 for some constant $\varepsilon > 0$.

The following works for smaller m, and is proved in the full version [19].

Theorem 11. *For $n \geq 3$, any explicit $r(n)$-round leader election protocol resilient up to $t(n)$ faults in a full-information ex post NE, and any constant natural number $m \geq 3$, there exists an explicit $(r(n)+4)$-round random selection protocol for a universe of size m that is in a full-information ex post NE for all z-balanced profiles, where $z = n/10m^2$. For such profiles, the random selection protocol is resilient up to $t(\lfloor n/m - z \rfloor)$ faults.*

6 Conclusion and Open Problems

Perhaps the main insight of this paper is that the full-information model is a setting that allows for a relatively clean examination of the interplay between rationality and adversarial behavior in the presence of asynchronous communication. While we have explored numerous aspects of this interplay, we are now faced with many more open questions.

The first set of questions consists of direct extensions of the results presented here. For example, can one generalize the types of preferences for which there are faithful and resilient protocols? For random selection protocols, for example, one might consider a setting in which each players likes some set of items, and dislikes the others. Are there random sampling protocols with weaker balancedness assumptions? How about such protocols that are rational in the face of an adversary, or resilient to rational coalitions? Also, are there protocols with few strong coalitions? Finally, one may consider approximate solution concepts: for example, one may desire all linearizations of a protocol to be in an ε-Nash equilibrium for a small but positive ε. Note that in this case our impossibility result of Theorem 1 no longer applies.

The second set of questions is more open-ended. What can one say about rationality for more general protocol problems in the full-information model? And are there other tractable models for the study of the interplay between rationality and adversarial behavior?

Acknowledgements. I would like to thank Ran Canetti, Moni Naor, and Omer Reingold for helpful conversations. I am also grateful to the anonymous referees for their comments.

References

1. Abraham, I., Dolev, D., Gonen, R., Halpern, J.: Distributed computing meets game theory: Robust mechanisms for rational secret sharing and multiparty computation. In: Proceedings of 25th Annual ACM Symposium on Principles of Distributed Computing, pp. 53–62 (2006)
2. Aiyer, A., Alvisi, L., Clement, A., Dahlin, M., Martin, J.-P., Porth, C.: Bar fault tolerance for cooperative services. In: Proceedings of 20th ACM Symposium on Operating Systems Principles, pp. 45–58 (2005)
3. Alon, N., Naor, M.: Coin-flipping games immune against linear-sized coalitions. SIAM Journal of Computing 22(2), 403–417 (1993)
4. Altman, A., Tennenholtz, M.: Selection games and deterministic lotteries (2008), http://iew3.technion.ac.il/~moshet/selection-lottery.pdf
5. Altman, A., Tennenholtz, M.: Strategyproof deterministic lotteries under broadcast communication. In: Proceedings of AAMAS (2008)
6. Antonakopoulos, S.: Fast leader-election protocols with bounded cheaters' edge. In: Proceedings of STOC, pp. 187–196 (2006)
7. Aumann, R.J.: Acceptable points in general cooperative n-person games. Contributions to the Theory of Games, Annals of Mathematical Studies IV, 287–324 (1959)
8. Ben-Or, M., Linial, N.: Collective coin fliping. Advances in Computing Research 5, 91–115 (1989)
9. Bernheim, B.D., Peleg, B., Whinston, M.: Coalition proof nash equilibrium: Concepts. Journal of Economic Theory 42(1), 1–12 (1989)
10. Brandt, F., Fischer, F., Shoham, Y.: On strictly competitive multi-player games. In: Proceedings of AAAI (2006)
11. Damgård, I.B.: Interactive hashing can simplify zero-knowledge protocol design without computational assumptions. In: Stinson, D.R. (ed.) CRYPTO 1993. LNCS, vol. 773, pp. 100–109. Springer, Heidelberg (1994)
12. Damgård, I.B., Goldreich, O., Wigderson, A.: Hashing functions can simplify zero-knowledge protocol design (too). Technical Report TR RS-94-39, BRICS (1994)
13. Ding, Y.Z., Harnik, D., Rosen, A., Shaltiel, R.: Constant-round oblivious transfer in the bounded storage model. In: Proceedings of STOC (2004)
14. Feige, U.: Noncryptographic selection protocols. In: Proceedings of 40th Annual Symposium on Foundations of Computer Science, pp. 142–152 (1999)
15. Feigenbaum, J., Shenker, S.: Distributed algorithmic mechanism design: Recent results and future directions. In: Proceedings of the 6th International Workshop on Discrete Algorithms and Methods for Mobile Computing and Communications, pp. 1–13 (2002)
16. Goldreich, O., Goldwasser, S., Linial, N.: Fault-tolerant computation in the full information model. SIAM J. Computing 27(2) (1998)
17. Goldreich, O., Sahai, A., Vadhan, S.: Honest-verifier statistical zero-knowledge equals general statistical zero-knowledge. In: Proceedings of 30th STOC (1998)
18. Gradwohl, R.: Fault tolerance in distributed mechanism design. In: Proceedings of the 4th Workshop on Internet and Network Economics (2008)
19. Gradwohl, R.: Rationality in the Full-Information Model, http://www.kellogg.northwestern.edu/faculty/Gradwohl/htm/papers/rle.pdf
20. Gradwohl, R., Vadhan, S., Zuckerman, D.: Random selection with an adversarial majority. In: Dwork, C. (ed.) CRYPTO 2006. LNCS, vol. 4117, pp. 409–426. Springer, Heidelberg (2006)

21. Halpern, J.: A computer scientist looks at game theory. Games and Economic Behavior 45(1), 114–131 (2003)
22. Halpern, J.: Computer science and game theory: A brief survey. The New Palgrave Dictionary of Economics (2008)
23. Kalai, E.: Large robust games. Econometrica 72(6), 1631–1665 (2004)
24. Katz, J.: Bridging game theory and cryptography: Recent results and future directions. In: Canetti, R. (ed.) TCC 2008. LNCS, vol. 4948, pp. 251–272. Springer, Heidelberg (2008)
25. Monderer, D., Tennenholtz, M.: Distributed games: from mechanisms to protocols. In: Proceedings of the 16th National Conference on Artificial Intelligence, pp. 32–37 (1999)
26. Moreno, D., Wooders, J.: Coalition-proof equilibrium. Games and Economic Behavior 17(1), 80–112 (1996)
27. Naor, M., Ostrovsky, R., Venkatesan, R., Yung, M.: Perfect zero-knowledge arguments for NP can be based on general complexity assumptions. J. Cryptology 11 (1998)
28. Ong, S.J., Parkes, D., Rosen, A., Vadhan, S.: Fairness with an honest minority and a rational majority. Preliminary version (October 2007)
29. Ong, S.J., Parkes, D., Rosen, A., Vadhan, S.: Fairness with an honest minority and a rational majority. In: Reingold, O. (ed.) TCC 2009. LNCS, vol. 5444, pp. 36–53. Springer, Heidelberg (2009)
30. Osborne, M.J., Rubinstein, A.: A Course in Game Theory. MIT Press, Cambridge (1994)
31. Russell, A., Zuckerman, D.: Perfect-information leader election in $\log^* n + O(1)$ rounds. Journal of Computer and System Sciences 63, 612–626 (2001)
32. Saks, M.: A robust noncryptographic protocol for collective coin flipping. SIAM Journal on Discrete Mathematics 2(2), 240–244 (1989)
33. Sanghvi, S., Vadhan, S.: The round complexity of two-party random selection. SIAM Journal of Computing 32(2), 523–550 (2008)
34. Shneidman, J., Parkes, D.C.: Specification faithfulness in networks with rational nodes. In: Proceedings of PODC (2004)

Efficient Rational Secret Sharing in Standard Communication Networks

Georg Fuchsbauer[1,*], Jonathan Katz[2,**], and David Naccache[1]

[1] École Normale Supérieure, LIENS - CNRS - INRIA, Paris, France
{georg.fuchsbauer,david.naccache}@ens.fr
[2] University of Maryland, USA
jkatz@cs.umd.edu

Abstract. We propose a new methodology for rational secret sharing leading to various instantiations (in both the two-party and multi-party settings) that are simple and efficient in terms of computation, share size, and round complexity. Our protocols do not require physical assumptions or simultaneous channels, and can even be run over asynchronous, point-to-point networks.

We also propose new equilibrium notions (namely, computational versions of *strict Nash equilibrium* and *stability with respect to trembles*) and prove that our protocols satisfy them. These notions guarantee, roughly speaking, that at each point in the protocol there is a *unique* legal message a party can send. This, in turn, ensures that protocol messages cannot be used as subliminal channels, something achieved in prior work only by making strong assumptions on the communication network.

1 Introduction

The classical problem of *t-out-of-n secret sharing* [28,5] involves a dealer D who distributes shares of a secret s to players $P_1, \ldots, P_n$ so that (1) any t or more players can reconstruct the secret without further involvement of the dealer, yet (2) any group of fewer than t players gets no information about s. For example, in Shamir's scheme [28] the secret s lies in a finite field $\mathbb{F}$, with $|\mathbb{F}| > n$. The dealer chooses a random polynomial $f(x)$ of degree at most $t-1$ with $f(0) = s$, and gives each player P_i the "share" $f(i)$. To reconstruct, t players broadcast their shares and interpolate the polynomial. Any set of fewer than t players has no information about s given their shares.

The implicit assumption in the original formulation of the problem is that each party is either honest or corrupt, and honest parties are all willing to cooperate

* Work supported by EADS, the French ANR-07-SESU-008-01 PAMPA Project, and the European Commission through the ICT Program under Contract ICT-2007-216646 ECRYPT II.

** Work done while visiting École Normale Supérieure and IBM, and supported by NSF CyberTrust grant #0830464, NSF CAREER award #0447075, and DARPA. The contents of this paper do not necessarily reflect the position or the policy of the US Government, and no official endorsement should be inferred.

D. Micciancio (Ed.): TCC 2010, LNCS 5978, pp. 419–436, 2010.

when reconstruction of the secret is desired. Beginning with the work of Halpern and Teague [13], protocols for secret sharing and other cryptographic tasks have begun to be re-evaluated in a game-theoretic light (see [7,16] for an overview of work in this direction). In this setting, parties are neither honest nor corrupt but are instead viewed as *rational* and are assumed (only) to act in their own self-interest.

Under natural assumptions regarding the utilities of the parties, standard secret-sharing schemes completely fail. For example, assume as in [13] that all players want to learn the secret above all else, but otherwise prefer that no other players learn the secret. (Later, we will treat the utilities of the players more precisely.) For t parties to reconstruct the secret in Shamir's scheme, each party is supposed to broadcast their share simultaneously. It is easy to see, however, that each player does no worse (and potentially does better) by withholding their share no matter what the other players do. Consider P_1: If fewer than $t-1$ players reveal their shares, P_1 does not learn the secret regardless of whether P_1 reveals his share or not. If more than $t-1$ other players reveal their shares, then everyone learns the secret and P_1's actions again have no effect. On the other hand, if *exactly* $t-1$ other players reveal their shares, then P_1 learns the secret (using his share) but prevents other players from learning the secret by *not* revealing his own share. The result is that if all players are rational then no one will broadcast their share and the secret will not be reconstructed.

Several works [13,11,24,1,18,19,27,26,4] have focused on designing *rational* secret-sharing protocols immune to the above problem. Protocols for rational secret sharing also follow from the more general results of Lepinski et al. [20,21, 15,14]. Each of these works has some or all of the following disadvantages:

On-line dealer or trusted/honest parties. Halpern and Teague [13] introduced a general approach to solving the problem that has been followed in most subsequent work. Their solution, however, requires the continual involvement of the dealer, even after the initial shares have been distributed. (The Halpern-Teague solution also applies only when $t, n \geq 3$.) Ong et al. [27] assume that sufficiently many parties behave honestly during the reconstruction phase.

Computational inefficiency. To eliminate the on-line dealer, several schemes rely on multiple invocations of protocols for generic secure multi-party computation [11,24,1,18,4]. It is unclear whether computationally *efficient* protocols with suitable functionality can be designed. The solutions of [20,21,15,14], though following a different high-level approach, also rely on generic secure multi-party computation.

Strong communication models. All prior schemes for $n > 2$ assume broadcast. The solutions in [13,11,24,1] assume *simultaneous broadcast* which means that parties must decide on what value (if any) to broadcast in a given round before observing the values broadcast by other parties. The solutions of [20,21, 15,14] rely on *physical* assumptions such as secure envelopes and ballot boxes.

Kol and Naor [18] show how to avoid simultaneous broadcast, at the cost of increasing the round complexity by a (multiplicative) factor linear in the size of the domain from which the secret is chosen; this approach cannot (efficiently) handle secrets of super-logarithmic length. Subsequent work by Kol and Naor [19] (see also [4]) shows how to avoid the assumption of simultaneous broadcast at the expense of increasing the round complexity by a (multiplicative) factor of t. We provide a detailed comparison of our results to those of [19] in Section 1.3.

1.1 Our Results

We show protocols for both 2-out-of-2 and t-out-of-n secret sharing (resilient to coalitions of size $t-1$) that do not suffer from any of the drawbacks mentioned above. We do not assume an on-line dealer or any trusted/honest parties, nor do we resort to generic secure multi-party computation. Our protocols are (arguably) simpler than previous solutions; they are also extremely efficient in terms of round complexity, share size, and required computation.

The primary advantage of our protocols, however, is that they do not require broadcast or simultaneous communication but can instead rely on synchronous (but *non*-simultaneous) point-to-point channels. Recall that all prior schemes for $n > 2$ assume broadcast; furthermore, the obvious approach of simulating broadcast by running a broadcast protocol over a point-to-point network will not, in general, work in the rational setting. Going further, we show that our protocol can be adapted for *asynchronous* point-to-point networks (with respect to a natural extension of the model for rational secret sharing), thus answering a question that had been open since the work of Halpern and Teague [13].

We also introduce two new equilibrium notions and prove that our protocols satisfy them. (A discussion of game-theoretic equilibrium notions used in this and prior work is given in Section 2.2.) The first notion we introduce is a computational version of *strict* Nash equilibrium. A similar notion was put forth by Kol and Naor [19], but they used an *information-theoretic* version of strict Nash and showed some inherent limitations of doing so. As in all of cryptography, we believe computational relaxations are meaningful and should be considered; doing so allows us to circumvent the limitations that hold in the information-theoretic case. We also formalize a notion of *stability with respect to trembles*, motivated by [16]; a different formalization of this notion, with somewhat different motivation, is given in [27].

Our definitions effectively rule out "signalling" via subliminal channels in the protocol. In fact, our protocols ensure that, at every point, *there is a* ***unique*** *legal message each party can send.* This prevents a party from outwardly appearing to follow the protocol while subliminally communicating (or trying to organize collusion) with other parties. Preventing subliminal communication is an explicit goal of some prior work (e.g., [15,21,3,2]), which achieved it only by relying on physical assumptions [15,21] or non-standard network models [3,2].

1.2 Overview of Our Approach

We follow the same high-level approach as in [13,11,24,1,18,19,4]. Our reconstruction protocol proceeds in a sequence of "fake" iterations followed by a single "real" iteration. Roughly speaking:

– In the real iteration, everyone learns the secret (assuming everyone follows the protocol).
– In a fake iteration, no information about the secret is revealed.
– No party can tell, in advance, whether the next iteration will be real or fake.

The iteration number i^* of the real iteration is chosen according to a geometric distribution with parameter $\beta \in (0,1)$ (where β depends on the players' utilities). To reconstruct the secret, parties run a sequence of iterations until the real iteration is identified, at which point all parties output the secret. If some party fails to follow the protocol, all parties abort. Intuitively, it is rational for P_i to follow the protocol as long as the expected gain of deviating, which is positive only if P_i aborts *exactly* in iteration i^*, is outweighed by the expected loss if P_i aborts before iteration i^*.

In most prior work [11,24,1,18], a secure multi-party computation is performed in each iteration to determine whether the given iteration should be real or fake. Instead we use the following approach, described in the 2-out-of-2 case (we omit some technical details in order to focus on the main idea): The dealer D chooses i^* from the appropriate distribution *in advance*, at the time of sharing. The dealer then generates two key-pairs (vk_1, sk_1), (vk_2, sk_2) for a *verifiable random function* (VRF) [25], where vk represents a verification key and sk represents a secret key, and we denote by $\mathsf{VRF}_{sk}(x)$ the evaluation of the VRF on input x using secret key sk. (See Appendix A for definitions of VRFs.) The dealer gives the verification keys to both parties, gives sk_1 to P_1, and gives sk_2 to P_2. It also gives $s_1 = s \oplus \mathsf{VRF}_{sk_2}(i^*)$ to P_1, and $s_2 = s \oplus \mathsf{VRF}_{sk_1}(i^*)$ to P_2. Each iteration consists of one message from each party: in iteration i, party P_1 sends $\mathsf{VRF}_{sk_1}(i)$ while P_2 sends $\mathsf{VRF}_{sk_2}(i)$. Observe that a fake iteration reveals nothing about the secret, in a computational sense. Furthermore, neither party can identify the real iteration in advance. (The description above relies on VRFs. We show that, in fact, trapdoor permutations suffice.)

To complete the protocol, we need to provide a way for parties to identify the real iteration. Previous work [11, 24, 1, 18] allows parties to identify the real iteration *as soon as it occurs*. We could use this approach for our protocol as well if we assumed simultaneous channels, since then each party must decide on its current-iteration message before it learns whether the current iteration is real or fake. When simultaneous channels are not available, however, this approach is vulnerable to an obvious rushing strategy.

Motivated by recent work on fairness (in the malicious setting) [10, 12], we suggest the following, new approach: delay the signal indicating whether a given iteration is real or fake until the *following* iteration. As before, a party cannot risk aborting until it is sure that the real iteration has occurred; the difference is that now, once a party learns that the real iteration occurred, the real iteration is over and all parties can reconstruct the secret. This eliminates the need for simultaneous channels, while adding only a single round. This approach can be adapted for t-out-of-n secret sharing and can be shown to work even when parties communicate over asynchronous, point-to-point channels.

Our protocol assumes parties have no auxiliary information about the secret s. (If simultaneous channels are assumed, then our protocol *does* tolerate auxiliary information about s.) We believe there are settings where this assumption is valid, and that understanding this case sheds light on the general question of rational computation. Prior work in the non-simultaneous model [18, 19] also fails in the presence of auxiliary information, and in fact this is inherent [4].

1.3 Comparison to the Kol-Naor Scheme

The only prior rational secret-sharing scheme that assumes no honest parties, is computationally efficient, and does not require simultaneous broadcast or physical assumptions is that of Kol and Naor [19] (an extension of this protocol is given in [4]). They also use the strict Nash solution concept and so their work provides an especially good point of comparison. Our protocols have the following advantages with respect to theirs:

Share size. In the Kol-Naor scheme, the shares of the parties have *unbounded* length. While not a significant problem in its own right, this *is* problematic when rational secret sharing is used as a sub-routine for rational computation of general functions. (See [18].) Moreover, the *expected* length of the parties' shares in their 2-out-of-2 scheme is $O(\beta^{-1} \cdot (|s| + k))$ (where k is a security parameter), whereas shares in our scheme have size $|s| + O(k)$.

Round complexity. The version of the Kol-Naor scheme that does not rely on simultaneous broadcast [19, Section 6] has expected round complexity $O(\beta^{-1} \cdot t)$, whereas our protocol has expected round complexity $O(\beta^{-1})$. (The value of β is roughly the same in both cases.)

Resistance to coalitions. For the case of t-out-of-n secret sharing, the Kol-Naor scheme is susceptible to coalitions of two or more players. We show t-out-of-n secret-sharing protocols resilient to coalitions of up to $(t-1)$ parties; see Section 4 for further details.

Avoiding broadcast. The Kol-Naor scheme for $n > 2$ assumes synchronous broadcast, whereas our protocols work even if parties communicate over an asynchronous, point-to-point network.

2 Model and Definitions

We denote the security parameter by k. A function $\epsilon : \mathbb{N} \to \mathbb{R}$ is *negligible* if for all $c > 0$ there is a $k_c > 0$ such that $\epsilon(k) < 1/k^c$ for all $k > k_c$; let negl denote a generic negligible function. We say ϵ is *noticeable* if there exist c, k_c such that $\epsilon(k) > 1/k^c$ for all $k > k_c$.

We define our model and then describe the game-theoretic concepts used. Even readers familiar with prior work in this area should skim the next few sections, since we formalize certain aspects of the problem slightly differently from prior work, and define new equilibrium notions.

2.1 Secret Sharing and Players' Utilities

A *t-out-of-n secret-sharing scheme for domain $\mathcal{S}$* (with $|\mathcal{S}| > 1$) is a two-phase protocol carried out by a dealer D and a set of n parties $P_1, \ldots, P_n$. In the first phase (the *sharing phase*), the dealer chooses a secret $s \in \mathcal{S}$. Based on this secret and a security parameter 1^k, the dealer generates shares $s_1, \ldots, s_n$ and gives s_i to player P_i. In the second phase (the *reconstruction phase*), some set I of $t^* \geq t$ active parties jointly reconstruct s. We impose the following requirements:

Secrecy. The shares of any $t-1$ parties reveal nothing about s, in an information-theoretic sense.

Correctness. For any set I of $t^* \geq t$ parties who run the reconstruction phase honestly, the correct secret s will be reconstructed, except possibly with probability negligible in k.

The above views parties as either malicious or honest. To model *rationality*, we consider players' utilities. Given a set I of $t^* \geq t$ parties active during the reconstruction phase, let the outcome o of the reconstruction phase be a vector of length t^* with $o_i = 1$ iff the output of P_i is equal to the initial secret s (i.e., P_i "learned the secret"). We consider a party to have learned the secret s if and only if it outputs s, and do not care whether that party "really knows" the secret or not. In particular, a party who outputs a random value in $\mathcal{S}$ without running the reconstruction phase at all "learns" the secret with probability $1/|\mathcal{S}|$. We model the problem this way for two reasons:

1. Our formulation lets us model a player learning *partial* information about the secret, something not reflected in prior work. In particular, partial information that increases the probability with which a party outputs the correct secret increases that party's expected utility.
2. It is difficult, in general, to formally model what it means for a party to "really" learn the secret, especially when considering arbitrary protocols and behaviors. Our notion is also better suited for a computational setting, where a party might "know" the secret from an information-theoretic point of view yet be unable to output it.

Let $\mu_i(o)$ be the utility of player P_i for the outcome o. Following [13] and most subsequent work (an exception is [4]), we make the following assumptions about the utility functions of the players:

- If $o_i > o'_i$, then $\mu_i(o) > \mu_i(o')$.
- If $o_i = o'_i$ and $\sum_i o_i < \sum_i o'_i$, then $\mu_i(o) > \mu_i(o')$.

That is, player P_i first prefers outcomes in which he learns the secret; otherwise, P_i prefers strategies in which the fewest number of other players learn the secret. For simplicity, in our analysis we distinguish three cases for the outcome o, described from the point of view of P_i (we could also work with utilities satisfying the more general constraints above, as long as utilities are known):

1. If P_i learns the secret and no other player does, then $\mu_i(o) \stackrel{\text{def}}{=} U^+$.
2. If P_i learns the secret and at least one other player does also, then $\mu_i(o) \stackrel{\text{def}}{=} U$.
3. If P_i does not learn the secret, then $\mu_i(o) \stackrel{\text{def}}{=} U^-$.

Our conditions impose $U^+ > U > U^-$. Define

$$U_{\text{random}} \stackrel{\text{def}}{=} \frac{1}{|S|} \cdot U^+ + \left(1 - \frac{1}{|S|}\right) \cdot U^- \ ; \tag{1}$$

this is the expected utility of a party who outputs a random guess for the secret (assuming other parties abort without any output, or with the wrong output). We will also assume that $U > U_{\text{random}}$; otherwise, players have (almost) no incentive to run the reconstruction phase at all.

In contrast to [4], we make no distinction between outputting the wrong secret and outputting a special "don't know" symbol; both are considered a failure to output the correct secret. By adapting techniques from their work, however, we can incorporate this distinction as well (as long as the relevant utilities are known). See Remark 2 in Section 3.

Strategies in our context refer to probabilistic polynomial-time interactive Turing machines. Given a vector of strategies $\boldsymbol{\sigma}$ for t^* parties active in the reconstruction phase, let $u_i(\boldsymbol{\sigma})$ denote the *expected* utility of P_i. (The expected utility is a function of the security parameter k.) This expectation is taken over the initial choice of s (which we will always assume to be uniform), the dealer's randomness, and the randomness of the players' strategies. Following the standard game-theoretic notation, we define $\boldsymbol{\sigma}_{-i} \stackrel{\text{def}}{=} (\sigma_1, \ldots, \sigma_{i-1}, \sigma_{i+1}, \ldots, \sigma_{t^*})$ and $(\sigma_i', \boldsymbol{\sigma}_{-i}) \stackrel{\text{def}}{=} (\sigma_1, \ldots, \sigma_{i-1}, \sigma_i', \sigma_{i+1}, \ldots, \sigma_{t^*})$; that is, $(\sigma_i', \boldsymbol{\sigma}_{-i})$ denotes the strategy vector $\boldsymbol{\sigma}$ with P_i's strategy changed to σ_i'.

2.2 Notions of Game-Theoretic Equilibria: A Discussion

The starting point for any discussion of game-theoretic equilibria is the *Nash equilibrium*. Roughly speaking, a protocol induces a Nash equilibrium if no party gains any advantage by deviating from the protocol, as long as all other parties follow the protocol. (In a *computational* Nash equilibrium, no *efficient* deviation confers any advantage.) As observed by Halpern and Teague [13], however, the Nash equilibrium concept is too weak for rational secret sharing. Halpern and Teague suggest, instead, to design protocols that induce a Nash equilibrium *surviving iterated deletion of weakly dominated strategies*; this notion was used in subsequent work of [11,24,1].

The notion of surviving iterated deletion, though, is also problematic in several respects. Kol and Naor [19] show a secret-sharing protocol that is "intuitively bad" yet technically satisfies the definition because no strategy weakly dominates any other. Also, a notion of surviving iterated deletion taking *computational* issues into account has not yet been defined (and doing so appears difficult). See [16,17] for other arguments against this notion.

Motivated by these drawbacks (and more), researchers have proposed other strengthenings of the Nash equilibrium concept [16,18,19]. Kol and Naor define *resistance to backward induction* [18], *everlasting equilibrium*, and *strict Nash equilibrium* [19]. The latter two notions are defined information-theoretically, and are overly conservative in that they rule out some "natural" protocols using cryptography. Nevertheless, the notion of strict Nash equilibrium is appealing. A protocol is in Nash equilibrium if no deviations are advantageous; it is in *strict* Nash equilibrium if all deviations are *dis*advantageous. Put differently, in the case of a Nash equilibrium there is no incentive to deviate whereas for a strict Nash equilibrium there is an incentive *not* to deviate.

Another advantage of strict Nash is that protocols satisfying this notion deter subliminal communication: since *any* (detectable) deviation from the protocol results in lower utility (when other parties follow the protocol), a party who tries to use protocol messages as a covert channel risks losing utility if there is any reasonable probability that other players are following the protocol.

We propose here a *computational* version of strict Nash equilibrium. Our definition retains the intuitive appeal of strict Nash, while meaningfully taking computational limitations into account. Moreover, our protocols satisfy the following, stronger condition: at every point in the protocol, there is a *unique* legal message that each party can send. Our protocols thus rule out subliminal communication in a strong sense, an explicit goal in prior work [20,22,21,3].

We also define a computational notion of *stability with respect to trembles*. Intuitively, stability with respect to trembles models players' uncertainty about other parties' behavior, and guarantees that even if a party P_i believes that other parties might play some arbitrary strategy with small probability δ (but follow the protocol with probability $1-\delta$), there is still no better strategy for P_i than to follow the protocol. Our formulation of this notion follows the general suggestion of Katz [16], but we flesh out the (non-trivial) technical details. Another formulation (*trembling-hand perfect equilibrium*), with somewhat different motivation, is discussed in [27].

As should be clear, determining the "right" game-theoretic notions for rational secret sharing is the subject of ongoing research. We do not suggest that the definitions proposed here are the *only* ones to consider, but we do believe they contribute to our understanding of the problem.

2.3 Definitions of Game-Theoretic Equilibria

We focus on the two-party case; the multi-party case is treated in the full version of this work [9]. Here, Π is a 2-out-of-2 secret-sharing scheme and σ_i is the prescribed action of P_i in the reconstruction phase.

Definition 1. *Π induces a* computational Nash equilibrium *if for any* PPT *strategy σ_1' of P_1 we have $u_1(\sigma_1', \sigma_2) \leq u_1(\sigma_1, \sigma_2) + \mathsf{negl}(k)$, and similarly for P_2.*

Our definitions of strict Nash and resistance to trembles require us to first define what it means to "follow a protocol". This is non-trivial since a *different* Turing machine ρ_1 might be "functionally identical" to the prescribed strategy σ_1 as far as the protocol is concerned: for example, ρ_1 may be the same as σ_1 except that it first performs some useless computation; the strategies may be identical except that ρ_1 uses pseudorandom coins instead of random coins; or, the two strategies may differ in the message(s) they send *after* the protocol ends. In any of these cases we would like to say that ρ_1 is essentially "the same" as σ_1. This motivates the following definition, stated for the case of a deviating P_1 (with an analogous definition for a deviating P_2):

Definition 2. *Define the random variable* view_2^{Π} *as follows:*

P_1 and P_2 interact, following σ_1 and σ_2, respectively. Let trans *denote the messages sent by P_1 but not including any messages sent by P_1 after it writes to its (write-once) output tape. Then view_2^{Π} includes the information given by the dealer to P_2, the random coins of P_2, and the (partial) transcript* trans.

Fix a strategy ρ_1 and an algorithm T. Define the random variable $\mathsf{view}_2^{T,\rho_1}$ as follows:

P_1 and P_2 interact, following ρ_1 and σ_2, respectively. Let trans *denote the messages sent by P_1. Algorithm T, given the entire view of P_1, outputs an arbitrary truncation* trans′ *of* trans. *(That is, it defines a cut-off point and deletes any messages sent after that point.) Then $\mathsf{view}_2^{T,\rho_1}$ includes the information given by the dealer to P_2, the random coins of P_2, and the (partial) transcript* trans′.

Strategy ρ_1 yields equivalent play with respect to Π, *denoted $\rho_1 \approx \Pi$, if there exists a* PPT *algorithm T such that for all* PPT *distinguishers D*

$$\left|\Pr[D(1^k, \mathsf{view}_2^{T,\rho_1}) = 1] - \Pr[D(1^k, \mathsf{view}_2^{\Pi}) = 1]\right| \leq \mathsf{negl}(k).$$

We write $\rho_1 \not\approx \Pi$ if ρ_1 does *not* yield equivalent play with respect to Π. Note that ρ_1 can yield equivalent play with respect to Π even if (1) it differs from the prescribed strategy when interacting with some *other* strategy σ_2' (we only care about the behavior of ρ_1 when the other party runs Π); (2) it differs from the prescribed strategy in its local computation or output; and (3) it differs from the prescribed strategy *after* P_1 computes its output. This last point models the fact that we cannot force P_1 to send "correct" messages once, as far as P_1 is concerned, the protocol is finished.

We now define the notion that *detectable* deviations from the protocol *decrease* a player's utility.

Definition 3. *Π induces a* computational strict Nash equilibrium *if*

1. *Π induces a computational Nash equilibrium;*
2. *For any* PPT *strategy σ_1' with $\sigma_1' \not\approx \Pi$, there is a $c > 0$ such that $u_1(\sigma_1, \sigma_2) \geq u_1(\sigma_1', \sigma_2) + 1/k^c$ for infinitely many values of k (with an analogous requirement for a deviating P_2).*

We next turn to defining stability with respect to trembles. We say that ρ_i *is δ-close to σ_i* if ρ_i takes the following form: with probability $1 - \delta$ party P_i plays

σ_i, while with probability δ it follows an arbitrary PPT strategy σ'_i. (In this case, we refer to σ'_i as the *residual strategy* of ρ_i.) The notion of δ-closeness is meant to model a situation in which P_i plays σ_i "most of the time," but with some (small) probability plays some other arbitrary strategy.

Intuitively, a pair of strategies (σ_1, σ_2) is *stable with respect to trembles* if σ_1 (resp., σ_2) remains a best response even if the other party plays a strategy other than σ_2 (resp., σ_1) with some small (but noticeable) probability δ. As in the case of strict Nash equilibrium, this notion is difficult to define formally because of the possibility that one party can do better (in case the other deviates) by performing some (undetectable) *local* computation.[1] Our definition essentially requires that this is the *only* way for either party to do better and so, in particular, each party will (at least outwardly) continue to follow the protocol until the other deviates. The fact that the prescribed strategies are in Nash equilibrium ensures that any (polynomial-time) local computation performed by either party is of no benefit as long as the other party follows the protocol.

Definition 4. *Π induces a* computational Nash equilibrium that is stable with respect to trembles *if*

1. *Π induces a computational Nash equilibrium;*
2. *There is a noticeable function δ such that for any* PPT *strategy ρ_2 that is δ-close to σ_2, and any* PPT *strategy ρ_1, there exists a* PPT *strategy $\sigma'_1 \approx \Pi$ such that $u_1(\rho_1, \rho_2) \leq u_1(\sigma'_1, \rho_2) + \mathsf{negl}(k)$ (with an analogous requirement for the case of deviations by P_2).*

Stated differently, even if a party P_i believes that the other party might play a different strategy with some small probability δ, there is still no better strategy for P_i than to outwardly follow the protocol[2] (while possibly performing some additional local computation).

3 Rational Secret Sharing: The 2-Out-of-2 Case

Let $\mathcal{S} = \{0,1\}^\ell$ be the domain of the secret. Let (Gen, Eval, Prove, Vrfy) be a VRF with range $\{0,1\}^\ell$, and let (Gen′, Eval′, Prove′, Vrfy′) be a VRF with range $\{0,1\}^k$. Protocol Π is defined as follows:

Sharing phase. Let s denote the secret. The dealer chooses an integer $i^* \in \mathbb{N}$ according to a geometric distribution with parameter β, where β is a constant that depends on the players' utilities but is independent of the security parameter; we discuss how to set β below. We assume $i^* < 2^k - 1$ since this occurs with all but negligible probability. (Technically, if $i^* \geq 2^k - 1$ the dealer can just send a special error message to each party.)

The dealer first computes the keys $(pk_1, sk_1), (pk_2, sk_2) \leftarrow \mathsf{Gen}(1^k)$ as well as $(pk'_1, sk'_1), (pk'_2, sk'_2) \leftarrow \mathsf{Gen}'(1^k)$. Then the dealer computes:

[1] As a trivial example, consider the case where with probability δ one party just sends its share to the other.

[2] Specifically, for any strategy ρ_i that does *not* yield equivalent play w.r.t. Π, there is a strategy σ'_i that *does* yield equivalent play w.r.t. Π and performs about as well.

– $\mathsf{share}_1 := \mathsf{Eval}_{sk_2}(i^*) \oplus s$ and $\mathsf{share}_2 := \mathsf{Eval}_{sk_1}(i^*) \oplus s$;
– $\mathsf{signal}_1 := \mathsf{Eval}'_{sk'_2}(i^*+1)$ and $\mathsf{signal}_2 := \mathsf{Eval}'_{sk'_1}(i^*+1)$.

Finally, the dealer gives to P_1 the values $(sk_1, sk'_1, pk_2, pk'_2, \mathsf{share}_1, \mathsf{signal}_1)$, and gives to P_2 the values $(sk_2, sk'_2, pk_1, pk'_1, \mathsf{share}_2, \mathsf{signal}_2)$.

As written, the share given to each party only hides s in a *computational* sense. Nevertheless, information-theoretic secrecy is easy to achieve; see Remark 1 at the end of this section.

Reconstruction phase

At the outset, P_1 chooses $s_1^{(0)}$ uniformly from $\mathcal{S} = \{0,1\}^\ell$ and P_2 chooses $s_2^{(0)}$ the same way. Then in each iteration $i = 1, \ldots$, the parties do the following:

(P_2 sends message to P_1:) P_2 computes

– $y_2^{(i)} := \mathsf{Eval}_{sk_2}(i)$, $\pi_2^{(i)} := \mathsf{Prove}_{sk_2}(i)$
– $z_2^{(i)} := \mathsf{Eval}'_{sk'_2}(i)$, $\bar{\pi}_2^{(i)} := \mathsf{Prove}'_{sk'_2}(i)$.

It then sends $(y_2^{(i)}, \pi_2^{(i)}, z_2^{(i)}, \bar{\pi}_2^{(i)})$ to P_1.

(P_1 receives message from P_2:) P_1 receives $(y_2^{(i)}, \pi_2^{(i)}, z_2^{(i)}, \bar{\pi}_2^{(i)})$ from P_2. If P_2 does not send anything, or if $\mathsf{Vrfy}_{pk_2}(i, y_2^{(i)}, \pi_2^{(i)}) = 0$ or $\mathsf{Vrfy}'_{pk'_2}(i, z_2^{(i)}, \bar{\pi}_2^{(i)}) = 0$, then P_1 outputs $s_1^{(i-1)}$ and halts.

If $\mathsf{signal}_1 \stackrel{?}{=} z_2^{(i)}$ then P_1 outputs $s_1^{(i-1)}$, sends its iteration-i message to P_2 (see below), and halts. Otherwise, it sets $s_1^{(i)} := \mathsf{share}_1 \oplus y_2^{(i)}$ and continues.

(P_1 sends message to P_2:) P_1 computes

– $y_1^{(i)} := \mathsf{Eval}_{sk_1}(i)$, $\pi_1^{(i)} := \mathsf{Prove}_{sk_1}(i)$
– $z_1^{(i)} := \mathsf{Eval}'_{sk'_1}(i)$, $\bar{\pi}_1^{(i)} := \mathsf{Prove}'_{sk'_1}(i)$.

It then sends $(y_1^{(i)}, \pi_1^{(i)}, z_1^{(i)}, \bar{\pi}_1^{(i)})$ to P_2.

(P_2 receives message from P_1:) P_2 receives $(y_1^{(i)}, \pi_1^{(i)}, z_1^{(i)}, \bar{\pi}_1^{(i)})$ from P_1. If P_1 does not send anything, or if $\mathsf{Vrfy}_{pk_1}(i, y_1^{(i)}, \pi_1^{(i)}) = 0$ or $\mathsf{Vrfy}'_{pk'_1}(i, z_1^{(i)}, \bar{\pi}_1^{(i)}) = 0$, then P_2 outputs $s_2^{(i-1)}$ and halts.

If $\mathsf{signal}_2 \stackrel{?}{=} z_1^{(i)}$ then P_2 outputs $s_2^{(i-1)}$ and halts. Otherwise, it sets $s_2^{(i)} := \mathsf{share}_2 \oplus y_1^{(i)}$ and continues.

Fig. 1. The reconstruction phase of secret-sharing protocol Π.

Reconstruction phase. A high-level overview of the protocol was given in Section 1.1, and we give the formal specification in Figure 1. The reconstruction phase proceeds in a series of iterations, where each iteration consists of one message sent by each party. Although these messages could be sent at the same time (since they do not depend on each other), we do not want to assume simultaneous communication and therefore simply require P_2 to communicate first in each iteration. (If one were willing to assume simultaneous channels then the protocol could be simplified by having P_2 send $\mathsf{Eval}'_{sk'_2}(i+1)$ at the same time as $\mathsf{Eval}_{sk_2}(i)$, and similarly for P_1.)

We give some intuition as to why the reconstruction phase of Π is a computational Nash equilibrium for an appropriate choice of β. Assume P_2 follows the protocol, and consider possible deviations by P_1. (Deviations by P_2 are even easier to analyze since P_2 goes first in every iteration.) P_1 can abort in iteration $i = i^* + 1$ (i.e., as soon as it receives $z_2^{(i)} = \mathsf{signal}_1$), or it can abort in some iteration $i < i^* + 1$. In the first case P_1 "knows" that it learned the dealer's secret in the preceding iteration (that is, in iteration i^*) and can thus output the correct secret; however, P_2 will output $s_2^{(i^*)} = s$ and so learns the secret as well. So P_1 does not increase its utility beyond what it would achieve by following the protocol. In the second case, when P_1 aborts in some iteration $i < i^* + 1$, the best strategy P_1 can adopt is to output $s_1^{(i)}$ and hope that $i = i^*$. The expected utility that P_1 obtains by following this strategy can be calculated as follows:

- P_1 aborts exactly in iteration $i = i^*$ with probability β. Then P_1 gets utility at most U^+.
- When $i < i^*$, player P_1 has "no information" about s and so the best it can do is guess. The expected utility of P_1 in this case is thus at most U_{random} (cf. Equation (1)).

Putting everything together, the expected utility of P_1 following this strategy is at most $\beta \times U^+ + (1-\beta) \times U_{\text{random}}$. Since $U_{\text{random}} < U$ by assumption, it is possible to set β so that the expected utility of this strategy is strictly less than U, in which case P_1 has no incentive to deviate.

That Π induces a *strict* computational Nash equilibrium (which is also stable with respect to trembles) follows since there is always a *unique* valid message a party can send; anything else is treated as an abort. A proof of the following theorem appears in the full version of this work [9].

Theorem 1. *If $\beta > 0$ and $U > \beta \cdot U^+ + (1-\beta) \cdot U_{\text{random}}$, then Π induces a computational strict Nash equilibrium that is stable with respect to trembles.*

Remark 1. The sharing phase, as described, guarantees computational secrecy only. A generic transformation from any such protocol (with bounded-size shares) to one that achieves information-theoretic secrecy follows: After D generates shares s_1, s_2 in the computationally secure scheme, it chooses random r_1, r_2 and random keys k_1, k_2, and gives to P_i the share $(s_i \oplus r_i,\ k_i,\ r_{3-i},\ \mathsf{MAC}_{k_{3-i}}(r_{3-i}))$, where MAC denotes an information-theoretically secure MAC. The reconstruction phase begins by having the parties exchange r_1 and r_2 along with the associated MAC tags, verifying the tags (and aborting if they are incorrect), and then recovering s_1, s_2. They then run the original protocol. It is easy to see that this maintains all the game-theoretic properties of the original protocol.

Remark 2. In the reconstruction phase, as described, one party can cause the other to output an incorrect secret (by aborting early). If the utilities of doing so are known, the protocol can be modified to rule out this behavior (in a game-theoretic sense) using the same techniques as in [4, Section 5.2]. Specifically, the dealer can — for each party — designate certain rounds as "completely fake", so that the party will know to output $\perp$ instead of an incorrect secret in case the other party aborts in that round. Using VRFs, this still can be achieved with bounded-size shares. Details will appear in the full version.

3.1 Using Trapdoor Permutations Instead of VRFs

The protocol from the previous section can be adapted easily to use trapdoor permutations instead of VRFs. The key observation is that the VRFs in the previous protocol are used only in a very specific way: they applied *sequentially* to values $1, 2, \ldots$. One can therefore use a trapdoor permutation f with associated hardcore bit h to instantiate the VRF in our scheme in the following way: The public key is a description of f along with a random element y in the domain of f; the secret key is the trapdoor enabling inversion of f. In iteration i, the "evaluation" of the VRF on input i is the ℓ-bit sequence

$$h\left(f^{-(i-1)\ell-1}(y)\right), h\left(f^{-(i-1)\ell-2}(y)\right), \ldots, h\left(f^{-(i-1)\ell-\ell}(y)\right),$$

and the "proof" is $\pi_i = f^{-(i-1)\ell-\ell}(y)$. Verification can be done using the original point y, and can also be done in time independent of i by using π_{i-1} (namely, by checking that $f^{\ell}(\pi_i) = \pi_{i-1}$), assuming π_{i-1} has already been verified.

The essential properties we need still hold: verifiability and uniqueness of proofs are easy to see, and pseudorandomness still holds with respect to a modified game where the adversary queries $\mathsf{Eval}_{sk}(1), \ldots, \mathsf{Eval}_{sk}(i)$ and then has to guess whether it is given $\mathsf{Eval}_{sk}(i+1)$ or a random string. We omit further details.

4 Rational Secret Sharing: The t-Out-of-n Case

In this section we describe extensions of our protocol to the t-out-of-n case, where we consider deviations by coalitions of up to $t-1$ parties. Formal definitions of game-theoretic notions in the multi-player setting, both for the case of single-player deviations as well as coalitions, are fairly straightforward adaptations of the definitions from Section 2.3 and are given in the full version of this work [9].

In describing our protocols we use VRFs for notational simplicity, but all the protocols given here can be instantiated using trapdoor permutations as described in Section 3.1.

A protocol for "exactly t-out-of-n" secret sharing. We begin by describing a protocol $\Pi_{t,n}$ for t-out-of n secret sharing that is resilient to coalitions of up to $t-1$ parties under the assumption that *exactly* t parties are active during the reconstruction phase. (We also require that the coalition be a subset of the active parties.) For now, we assume communication over a synchronous (but not simultaneous) point-to-point network.

Sharing Phase

To share a secret $s \in \{0,1\}^\ell$, the dealer does the following:

- Choose $r^* \in \mathbb{N}$ according to a geometric distribution with parameter β.
- Generate[a] VRF keys $(pk_1, sk_1), \ldots, (pk_n, sk_n) \leftarrow \mathsf{Gen}(1^k)$ followed by $(pk'_1, sk'_1), \ldots, (pk'_n, sk'_n) \leftarrow \mathsf{Gen}'(1^k)$.
- Choose random $(t-1)$-degree polynomials $G \in \mathbb{F}_{2^\ell}[x]$ and $H \in \mathbb{F}_{2^k}[x]$ such that $G(0) = s$ and $H(0) = 0$.
- Send sk_i, sk'_i to player P_i, and send to all parties the following values:
  1. $\{(pk_j, pk'_j)\}_{1 \le j \le n}$
  2. $\{g_j := G(j) \oplus \mathsf{Eval}_{sk_j}(r^*)\}_{1 \le j \le n}$
  3. $\{h_j := H(j) \oplus \mathsf{Eval}'_{sk'_j}(r^*+1)\}_{1 \le j \le n}$

Reconstruction Phase

Let I be the indices of the t active players. Each party P_i (for $i \in I$) chooses $s_i^{(0)}$ uniformly from $\{0,1\}^\ell$. In each iteration $r = 1, \ldots$, the parties do:

- For all $i \in I$ (in ascending order), P_i sends the following to all players:

$$\big(\, y_i^{(r)} := \mathsf{Eval}_{sk_i}(r),\ z_i^{(r)} := \mathsf{Eval}'_{sk'_i}(r),\ \mathsf{Prove}_{sk_i}(r),\ \mathsf{Prove}'_{sk'_i}(r) \,\big).$$

- If a party P_i receives an incorrect proof (or nothing) from any other party P_j, then P_i terminates and outputs $s_i^{(r-1)}$. Otherwise:
  - P_i sets $h_j^{(r)} := h_j \oplus z_j^{(r)}$ for all $j \in I$, and interpolates a degree-$(t-1)$ polynomial $H^{(r)}$ through the t points $\{h_j^{(r)}\}_{j \in I}$. If $H^{(r)}(0) \stackrel{?}{=} 0$ then P_i outputs $s_i^{(r-1)}$ immediately, and terminates after sending its current-iteration message.
  - Otherwise, P_i computes $s_i^{(r)}$ as follows: set $g_j^{(r)} := g_j \oplus y_j^{(r)}$ for all $j \in I$. Interpolate a degree-$(t-1)$ polynomial $G^{(r)}$ through the points $\{g_j^{(r)}\}_{j \in I}$, and set $s_i^{(r)} := G^{(r)}(0)$.

[a] Gen outputs VRF keys with range $\{0,1\}^\ell$, and Gen' outputs VRF keys with range $\{0,1\}^k$.

Fig. 2. Protocol $\Pi_{t,n}$ for "exactly t-out-of-n" secret sharing

As in the 2-out-of-2 case, every party is associated with two keys for a VRF. The dealer chooses an iteration r^* according to a geometric distribution, and also chooses two random $(t-1)$-degree polynomials G, H (over some finite field) such that $G(0) = s$ and $H(0) = 0$. Each party receives *blinded* versions of all n points $\{G(j), H(j)\}_{j=1}^n$: each $G(j)$ is blinded by the value of P_j's VRF on the input r^*, and each $H(j)$ is blinded by the value of P_j's VRF on the input r^*+1. In each iteration r, each party is supposed to send to all other parties the value of their VRFs evaluated on the current iteration number r; once this is done, every party can interpolate a polynomial to obtain candidate values for $G(0)$

and $H(0)$. When $H(0) = 0$ parties know the protocol is over, and output the $G(0)$ value reconstructed in the *previous* iteration. See Figure 2 for details.

Theorem 2. *If $\beta > 0$ and $U > \beta \cdot U^+ + (1-\beta) \cdot U_{\text{random}}$, then $\Pi_{t,n}$ induces a $(t-1)$-resilient computational strict Nash equilibrium that is stable with respect to trembles, as long as exactly t parties are active during the reconstruction phase.*

A proof is exactly analogous to the proof of Theorem 1.

Handling the general case. The prior solution assumes exactly t parties are active during reconstruction. If $t^* > t$ parties are active, the "natural" implementation of the protocol — where the lowest-indexed t parties run $\Pi_{t,n}$ and all other parties remain silent — is not a $(t-1)$-resilient computational Nash equilibrium. To see why, let the active parties be $I = \{1, \ldots t+1\}$ and let $\mathcal{C} = \{3, \ldots, t+1\}$ be a coalition of $t-1$ parties. In each iteration r, as soon as P_1 and P_2 send their values the parties in $\mathcal{C}$ can compute $t+1$ points $\{g_j^{(r)}\}_{j \in I}$. Because of the way these points are constructed, they are guaranteed to lie on a $(t-1)$-degree polynomial when $r = r^*$, but are unlikely to lie on a $(t-1)$-degree polynomial when $r < r^*$. This gives the parties in $\mathcal{C}$ a way to determine r^* as soon as that iteration is reached, at which point they can abort and output the secret while preventing P_1 and P_2 from doing the same.

Fortunately, a simple modification works: simply have the dealer run independent instances $\Pi_{t,n}, \Pi_{t+1,n}, \ldots, \Pi_{n,n}$; in the reconstruction phase, the parties run $\Pi_{t^*,n}$ where t^* denotes the number of active players. It follows as an easy corollary of Theorem 2 that this induces a $(t-1)$-resilient computational strict Nash equilibrium (that is also stable with respect to trembles) regardless of how many parties are active during the reconstruction phase. (As in previous work, we only consider coalitions that are subsets of the parties who are active during reconstruction. The protocol is no longer a computational Nash equilibrium if this is not the case.[3])

Asynchronous networks. Our protocol $\Pi_{t,n}$ can be adapted to work even when the parties communicate over an *asynchronous* point-to-point network. (In our model of asynchronous networks, messages can be delayed arbitrarily and delivered out of order, but any message that is sent is eventually delivered.) In this case parties cannot distinguish an abort from a delayed message and so we modify the protocol as follows: each party proceeds to the next iteration as soon as it receives $t-1$ valid messages from the previous iteration, and only halts if it receives an invalid message from someone. More formal treatment of the asynchronous case, including a discussion of definitions in this setting and a proof for the preceding protocol, is given in the full version of this work [9].

As before, we can handle the general case by having the dealer run the "exactly t^*-out-of-n" protocol just described for all values of $t^* \in \{t, \ldots, n\}$.

[3] This case can be addressed, however, by having the dealer run independent instances of $\Pi_{t,n}$ for all $\binom{n}{t}$ subsets of size t; to reconstruct, the t lowest-indexed active players run the instance corresponding to their subset while the remaining active players are silent. This is only efficient when t (or $n-t$) is small.

References

1. Abraham, I., Dolev, D., Gonen, R., Halpern, J.: Distributed computing meets game theory: robust mechanisms for rational secret sharing and multiparty computation. In: 25th ACM Symposium Annual on Principles of Distributed Computing, pp. 53–62. ACM Press, New York (2006)
2. Alwen, J., Katz, J., Lindell, Y., Persiano, G., Shelat, A., Visconti, I.: Collusion-free multiparty computation in the mediated model. In: Halevi, S. (ed.) CRYPTO 2009. LNCS, vol. 5677, pp. 524–540. Springer, Heidelberg (2009)
3. Alwen, J., Shelat, A., Visconti, I.: Collusion-free protocols in the mediated model. In: Wagner, D. (ed.) CRYPTO 2008. LNCS, vol. 5157, pp. 497–514. Springer, Heidelberg (2008)
4. Asharov, G., Lindell, Y.: Utility dependence in correct and fair rational secret sharing. In: Halevi, S. (ed.) CRYPTO 2009. LNCS, vol. 5677, pp. 559–576. Springer, Heidelberg (2009), A full version, containing additional results: http://eprint.iacr.org/209/373
5. Blakley, G.: Safeguarding cryptographic keys. In: Proc. AFIPS National Computer Conference, vol. 48, pp. 313–317 (1979)
6. Dodis, Y.: Efficient construction of (distributed) verifiable random functions. In: Desmedt, Y.G. (ed.) PKC 2003. LNCS, vol. 2567, pp. 1–17. Springer, Heidelberg (2002)
7. Dodis, Y., Rabin, T.: Cryptography and game theory. In: Nisan, N., Roughgarden, T., Tardos, E., Vazirani, V. (eds.) Algorithmic Game Theory, pp. 181–207. Cambridge University Press, Cambridge (2007)
8. Dodis, Y., Yampolskiy, A.: A verifiable random function with short proofs and keys. In: Vaudenay, S. (ed.) PKC 2005. LNCS, vol. 3386, pp. 416–431. Springer, Heidelberg (2005)
9. Fuchsbauer, G., Katz, J., Naccache, D.: Efficient rational secret sharing in standard communication networks, http://eprint.iacr.org/2008/488
10. Gordon, S.D., Hazay, C., Katz, J., Lindell, Y.: Complete fairness in secure two-party computation. In: 40th Annual ACM Symposium on Theory of Computing (STOC), pp. 413–422. ACM Press, New York (2008)
11. Gordon, S.D., Katz, J.: Rational secret sharing, revisited. In: De Prisco, R., Yung, M. (eds.) SCN 2006. LNCS, vol. 4116, pp. 229–241. Springer, Heidelberg (2006)
12. Gordon, S.D., Katz, J.: Partial fairness in secure two-party computation (2008), http://eprint.iacr.org/2008/206
13. Halpern, J., Teague, V.: Rational secret sharing and multiparty computation: Extended abstract. In: 36th Annual ACM Symposium on Theory of Computing (STOC), pp. 623–632. ACM Press, New York (2004)
14. Izmalkov, S., Lepinski, M., Micali, S.: Verifiably secure devices. In: Canetti, R. (ed.) TCC 2008. LNCS, vol. 4948, pp. 273–301. Springer, Heidelberg (2008)
15. Izmalkov, S., Micali, S., Lepinski, M.: Rational secure computation and ideal mechanism design. In: 46th Annual Symposium on Foundations of Computer Science (FOCS), pp. 585–595. IEEE, Los Alamitos (2005)
16. Katz, J.: Bridging game theory and cryptography: Recent results and future directions. In: Canetti, R. (ed.) TCC 2008. LNCS, vol. 4948, pp. 251–272. Springer, Heidelberg (2008)
17. Katz, J.: Ruminations on defining rational MPC. Talk given at SSoRC, Bertinoro, Italy (2008), http://www.daimi.au.dk/~jbn/SSoRC2008/program

18. Kol, G., Naor, M.: Cryptography and game theory: Designing protocols for exchanging information. In: Canetti, R. (ed.) TCC 2008. LNCS, vol. 4948, pp. 320–339. Springer, Heidelberg (2008)
19. Kol, G., Naor, M.: Games for exchanging information. In: 40th Annual ACM Symposium on Theory of Computing (STOC), pp. 423–432. ACM Press, New York (2008)
20. Lepinski, M., Micali, S., Peikert, C., Shelat, A.: Completely fair SFE and coalition-safe cheap talk. In: 23rd ACM Symposium Annual on Principles of Dis- tributed Computing, pp. 1–10. ACM Press, New York (2004)
21. Lepinski, M., Micali, S., Shelat, A.: Collusion-free protocols. In: 37th Annual ACM Symposium on Theory of Computing (STOC), pp. 543–552. ACM Press, New York (2005)
22. Lepinski, M., Micali, S., Shelat, A.: Fair-zero knowledge. In: Kilian, J. (ed.) TCC 2005. LNCS, vol. 3378, pp. 245–263. Springer, Heidelberg (2005)
23. Lysyanskaya, A.: Unique signatures and verifiable random functions from the DH-DDH separation. In: Yung, M. (ed.) CRYPTO 2002. LNCS, vol. 2442, pp. 597–612. Springer, Heidelberg (2002)
24. Lysyanskaya, A., Triandopoulos, N.: Rationality and adversarial behavior in multi-party computation. In: Dwork, C. (ed.) CRYPTO 2006. LNCS, vol. 4117, pp. 180–197. Springer, Heidelberg (2006)
25. Micali, S., Rabin, M.O., Vadhan, S.P.: Verifiable random functions. In: 40th Annual Symposium on Foundations of Computer Science (FOCS), pp. 120–130. IEEE, Los Alamitos (1999)
26. Miclai, S., Shelat, A.: Truly rational secret sharing. In: Reingold, O. (ed.) TCC 2009. LNCS, vol. 5444, pp. 54–71. Springer, Heidelberg (2009)
27. Ong, S.J., Parkes, D., Rosen, A., Vadhan, S.: Fairness with an honest minority and a rational majority. In: Reingold, O. (ed.) TCC 2009. LNCS, vol. 5444, pp. 36–53. Springer, Heidelberg (2009)
28. Shamir, A.: How to share a secret. Communications of the ACM 22(11), 612–613 (1979)

A Verifiable Random Functions (VRFs)

A VRF is a keyed function whose output is "random-looking" but can still be verified as correct, given an associated proof. The notion was introduced by Micali, Rabin, and Vadhan [25], and various constructions in the standard model are known [25, 6, 23, 8]. The definition we use is stronger than the "standard" one in that it includes a uniqueness requirement on the *proof* as well, but the constructions of [6, 8] achieve it. (Also, we use VRFs only as a stepping stone to our construction based on trapdoor permutations; see Section 3.1.)

Definition 5. *A* verifiable random function *(VRF) with range* $\mathcal{R} = \{\mathcal{R}_k\}$ *is a tuple of probabilistic polynomial-time algorithms* (Gen, Eval, Prove, Vrfy) *such that the following hold:*

Correctness: *For all k, any* (pk, sk) *output by* $\mathsf{Gen}(1^k)$, *the algorithm* Eval_{sk} *maps k-bit inputs to the set* $\mathcal{R}_k$. *Furthermore, for any* $x \in \{0,1\}^k$ *we have* $\mathsf{Vrfy}_{pk}(x, \mathsf{Eval}_{sk}(x), \mathsf{Prove}_{sk}(x)) = 1$.

Verifiability: *For all (pk, sk) output by $\mathsf{Gen}(1^k)$, there does not exist a tuple (x, y, y', π, π') with $y \neq y'$ and $\mathsf{Vrfy}_{pk}(x, y, \pi) = 1 = \mathsf{Vrfy}_{pk}(x, y', \pi')$.*

Unique proofs: *For all (pk, sk) output by $\mathsf{Gen}(1^k)$, there does not exist a tuple (x, y, π, π') with $\pi \neq \pi'$ and $\mathsf{Vrfy}_{pk}(x, y, \pi) = 1 = \mathsf{Vrfy}_{pk}(x, y, \pi')$.*

Pseudorandomness: *Consider the following experiment involving an adversary $\mathcal{A}$:*

1. *Generate $(pk, sk) \leftarrow \mathsf{Gen}(1^k)$ and give pk to $\mathcal{A}$. Then $\mathcal{A}$ adaptively queries a sequence of strings $x_1, \ldots, x_\ell \in \{0,1\}^k$ and is given $y_i = \mathsf{Eval}_{sk}(x_i)$ and $\pi_i = \mathsf{Prove}_{sk}(x_i)$ in response to each such query x_i.*
2. *$\mathcal{A}$ outputs a string $x \in \{0,1\}^k$ subject to the restriction $x \notin \{x_1, \ldots, x_\ell\}$.*
3. *A random bit $b \leftarrow \{0,1\}$ is chosen. If $b = 0$ then $\mathcal{A}$ is given $y = \mathsf{Eval}_{sk}(x)$; if $b = 1$ then $\mathcal{A}$ is given a random $y \leftarrow \mathcal{R}_k$.*
4. *$\mathcal{A}$ makes more queries as in step 2, as long as none of these queries is equal to x.*
5. *At the end of the experiment, $\mathcal{A}$ outputs a bit b' and* succeeds *if $b' = b$.*

We require that the success probability of any PPT *adversary $\mathcal{A}$ is $\frac{1}{2} + \mathsf{negl}(k)$.*

Bounds on the Sample Complexity for Private Learning and Private Data Release

Amos Beimel[1,*], Shiva Prasad Kasiviswanathan[2], and Kobbi Nissim[1,3,**]

[1] Dept. of Computer Science, Ben-Gurion University
[2] CCS-3, Los Alamos National Laboratory
[3] Microsoft Audience Intelligence

Abstract. Learning is a task that generalizes many of the analyses that are applied to collections of data, and in particular, collections of sensitive individual information. Hence, it is natural to ask what can be learned while preserving individual privacy. [Kasiviswanathan, Lee, Nissim, Raskhodnikova, and Smith; FOCS 2008] initiated such a discussion. They formalized the notion of *private learning*, as a combination of PAC learning and differential privacy, and investigated what concept classes can be learned privately. Somewhat surprisingly, they showed that, ignoring time complexity, every PAC learning task could be performed privately with polynomially many samples, and in many natural cases this could even be done in polynomial time.

While these results seem to equate non-private and private learning, there is still a significant gap: the sample complexity of (non-private) PAC learning is crisply characterized in terms of the VC-dimension of the concept class, whereas this relationship is lost in the constructions of private learners, which exhibit, generally, a higher sample complexity.

Looking into this gap, we examine several private learning tasks and give tight bounds on their sample complexity. In particular, we show strong separations between sample complexities of proper and improper private learners (such separation does not exist for non-private learners), and between sample complexities of efficient and inefficient proper private learners. Our results show that VC-dimension is not the right measure for characterizing the sample complexity of proper private learning.

We also examine the task of *private data release* (as initiated by [Blum, Ligett, and Roth; STOC 2008]), and give new lower bounds on the sample complexity. Our results show that the logarithmic dependence on size of the instance space is essential for private data release.

1 Introduction

Consider a scenario in which a survey is conducted among a sample of random individuals and datamining techniques are applied to learn information on the

* Research partially supported by the Israel Science Foundation (grant No. 938/09) and by the Frankel Center for Computer Science.
** Research partly supported by the Israel Science Foundation (grant No. 860/06).

D. Micciancio (Ed.): TCC 2010, LNCS 5978, pp. 437–454, 2010.

entire population. If such information will disclose information on the individuals participating in the survey, then they will be reluctant to participate in the survey. To address this question, Kasiviswanathan *et al.* [10] introduced the notion of *private learning*. Informally, a private learner is required to output a hypothesis that gives accurate classification while protecting the privacy of the individual samples from which the hypothesis was obtained. The formal notion of a private learner is a combination of two qualitatively different notions. One is that of PAC learning [17], the other of differential privacy [7]. In PAC (probably approximately correct) learning, a collection of samples (labeled examples) is generalized into a hypothesis. It is assumed that the examples are generated by sampling from some (unknown) distribution $\mathcal{D}$ and are labeled according to an (unknown) concept c taken from some concept class $\mathcal{C}$. The learned hypothesis h should predict with high accuracy the labeling of examples taken from the distribution $\mathcal{D}$, an *average-case* requirement. Differential privacy, on the other hand, is formulated as a *worst-case* requirement. It requires that the output of a learner should not be significantly affected if a particular example d is replaced with arbitrary d', for all d and d'. This strong notion provides rigorous privacy guarantees even against attackers empowered with arbitrary side information [11].

Recent research on privacy has shown, somewhat surprisingly, that it is possible to design differentially private variants of many analyses (see [6] for a recent survey). In this line, the work of [10] demonstrated that private learning is generally feasible – any concept class that is PAC learnable can be learned privately (but not necessarily efficiently), by a "Private Occam's Razor" algorithm, with sample complexity that is logarithmic in the size of the hypothesis class. Furthermore, taking into account the earlier result of [2] (that all concept classes that can be efficiently learned in the *statistical queries* model can be learned privately and efficiently) and the efficient private parity learner of [10], we get that most "natural" computational learning tasks can be performed privately and efficiently (i.e., with polynomial resources). This is important as learning problems generalize many of the computations performed by analysts over collections of sensitive data.

The results of [2, 10] show that private learning is feasible in an extremely broad sense, and hence one can essentially equate learning and private learning. However, the costs of the private learners constructed in [2, 10] are generally higher than those of non-private ones by factors that depend not only on the privacy, accuracy, and confidence parameters of the private learner. In particular, the well-known relationship between the sampling complexity of PAC learners and the VC-dimension of the concept class (ignoring computational efficiency) [5] does not hold for the above constructions of private learners – as their sample complexity is proportional to the logarithm of the size of the concept class. Recall that the VC-dimension of a concept class is bounded by the logarithm of its size, and is significantly lower for many interesting concept classes, hence there may exist learning tasks for which "very practical" non-private learner exists, but any private learner is "impractical".

The focus of this work is on a fine-grain examination of the differences in complexity between private and non-private learning. The hope is that such an examination will eventually lead to an understanding of which complexity measure is relevant for the sample complexity of private learning, similar to the well-understood relationship between the VC-dimension and sample complexity of PAC learning. Such an examination is interesting also for other tasks, and a second task we examine is that of releasing a *sanitization* of a data set that simultaneously protects privacy of individual contributors and offers utility to the data analyst. See the discussion in Section 1.1.

1.1 Our Contributions

We now give a brief account of our results. Throughout this rather informal discussion we will treat the accuracy, confidence, and privacy parameters as constants (a detailed analysis revealing the dependency on these parameters is presented in the technical sections). We use the term "efficient" for polynomial time computations.

Following standard computational learning terminology, we will call learners for a concept class $\mathcal{C}$ that only output hypotheses in $\mathcal{C}$ *proper*, and other learners *improper*. The original motivation in computational learning theory for this distinction is that there exist concept classes $\mathcal{C}$ for which proper learning is computationally intractable [16], whereas it is possible to efficiently learn $\mathcal{C}$ improperly [17]. As we will see below, the distinction between proper and improper learning is useful also when discussing private learning, and for more reasons than making intractable learning tasks tractable.

Proper and Improper Private Learning. It is instructive to look into the construction of the Private Occam's Razor algorithm of [10] and see why its sample complexity is proportional to the logarithm of the size of the hypothesis class used. The algorithm uses the exponential mechanism of McSherry and Talwar [14] to choose a hypothesis. The choice is probabilistic, where the probability mass that is assigned to each of the hypotheses decreases exponentially with the number of samples that are inconsistent with it. A union-bound argument is used in the claim that the construction actually yields a learner, and a sample size that is logarithmic in the size of the hypothesis class is needed for the argument to go through.

For our analyses in this paper, we consider a simple, but natural, class $POINT_d$ containing the concepts $c_j : \{0,1\}^d \rightarrow \{0,1\}$ where $c_j(x) = 1$ for $x = j$, and 0 otherwise. The VC-dimension of $POINT_d$ is one, and hence it can be learned (non-privately and efficiently, properly or improperly) with merely $O(1)$ samples.

In sharp contrast, (when used for properly learning $POINT_d$) the Private Occam's Razor algorithm requires $O(\log |POINT_d|) = O(d)$ samples – obtaining the largest possible gap in sample complexity when compared to non-private learners! Our first result is a matching lower bound. We prove that *any* proper private learner for $POINT_d$ must use $\Omega(d)$ samples, therefore, answering negatively the question (from [10]) of whether proper private learners should exhibit sample complexity that is approximately the VC-dimension (or even a function of the VC-dimension) of the concept class[1].

A natural way to improve on the sample complexity is to use the Private Occam's Razor to improperly learn $POINT_d$ with a smaller hypothesis class that is still expressive enough for $POINT_d$, reducing the sample complexity to the logarithm of the smaller hypothesis class. We show that this indeed is possible, as there exists a hypothesis class of size $O(d)$ that can be used for learning $POINT_d$ improperly, yielding an algorithm with sample complexity $O(\log d)$. Furthermore, this bound is tight, any hypothesis class for learning $POINT_d$ must contain $\Omega(d)$ hypotheses. These bounds are interesting as they give a separation between proper and improper private learning – proper private learning of $POINT_d$ requires $\Omega(d)$ samples, whereas $POINT_d$ can be improperly privately learned using $O(\log d)$ samples. Note that such a combinatorial separation does not exist for non-private learning, as a VC-dimension number of samples are needed and sufficient for both proper and improper non-private learners. Furthermore, the $\Omega(d)$ lower bound on the size of the hypothesis class maps a clear boundary to what can be achieved in terms of sample complexity using the Private Occam's Razor for $POINT_d$. It might even suggest that *any* private learner for $POINT_d$ should use $\Omega(\log d)$ samples.

It turns out, however, that the intuition expressed in the last sentence is at fault. We construct an efficient improper private learner for $POINT_d$ that uses merely $O(1)$ samples, hence establishing the strongest possible separation between proper and improper private learners. For the construction we extrapolate on a technique from the efficient private parity learner of [10]. The construction of [10] utilizes a natural non-private proper learner, and hence results in a proper private learner, whereas, due to the bounds mentioned above, we cannot use a proper learner for $POINT_d$, and hence we construct an improper (rather unnatural) learner to base our construction upon. Our construction utilizes a double-exponential hypothesis class, and hence is inefficient (even outputting a hypothesis requires super-polynomial time). We use a simple compression using pseudorandom functions (akin to [15]) to make the algorithm efficient.

Efficient and Inefficient Proper Private Learning. We apply the above lower bound on the number of samples for proper private learning $POINT_d$ to show a separation in the sample size between efficient and inefficient proper private learning. Assuming the existence of pseudorandom generators with exponential stretch, we present a concept class $\widehat{POINT}_d$ – a variant of $POINT_d$ – such that every efficient proper private learner for this class requires $\Omega(d)$ samples. In contrast, an inefficient proper private learner exists that uses only a super-logarithmic number of samples. This is the first example where requiring efficiency on top of privacy comes at a price of larger sample size.

The Sample Size of Non-Interactive Sanitization Mechanisms. Given a database containing a collection of individual information, a sanitization is

[1] Our proof technique yields lower bounds not only on private learning $POINT_d$ properly, but on private learning of any concept class $\mathcal{C}$ with various hypothesis classes that we call α-minimal for $\mathcal{C}$.

a release that protects the privacy of the individual contributors while offering utility to the analyst using the database. The setting is non-interactive if once the sanitization is released the original database and the curator play no further role. Blum *et al.* [3] presented a construction of such non-interactive sanitizers for count queries. Let $\mathcal{C}$ be a concept class consisting of efficiently computable predicates from a discretized domain X to $\{0,1\}$. Given a collection D of data items taken from X, Blum *et al.* employ the exponential mechanism [14] to (inefficiently) obtain another collection D' with data items from X such that D' maintains approximately correct count of $\sum_{d \in D} c(d)$ for all concepts $c \in \mathcal{C}$. Also, they show that it suffices for D to have a size that is $O(\log|X| \cdot VCDIM(\mathcal{C}))$. The database D' is referred to as a *synthetic* database as it contains data items drawn from the same universe (i.e., from X) as the original database D.

We provide new lower bounds for non-interactive sanitization mechanisms. We show that for $POINT_d$ every non-interactive sanitization mechanism that is useful[2] for $POINT_d$ requires a database of $\Omega(d)$ size. This lower bound is tight as the sanitization mechanism of Blum *et al.* for $POINT_d$ uses a database of $O(d \cdot VCDIM(POINT_d)) = O(d)$ size. Our lower bound holds even if the sanitized output is an arbitrary data structure and not a synthetic database.

1.2 Related Work

The notion of PAC learning was introduced by Valiant [17]. The notion of differential privacy was introduced by Dwork *et al.* [7]. Private learning was introduced in [10]. Beyond proving that (ignoring computation) every concept class can be PAC learned privately (see Theorem 2 below), they proved an equivalence between learning in the statistical queries model and private learning in the local communication model (aka randomized response). The general private data release mechanism we mentioned above was introduced in [3] along with a specific construction for halfspace queries. As we mentioned above, both [10] and [3] use the exponential mechanism of [14], a generic construction of differential private analyses, that (in general) does not yield efficient algorithms.

A recent work of Dwork *et al.* [8] considered the complexity of non-interactive sanitization under two settings: (a) sanitized output is a synthetic database, and (b) sanitized output is some arbitrary data structure. For the task of sanitizing with a synthetic database they show a separation between efficient and inefficient sanitization mechanisms based on whether the size of the instance space and the size of the concept class is polynomial in a (security) parameter or not. For the task of sanitizing with an arbitrary data structure they show a tight connection between the complexity of sanitization and traitor tracing schemes used in cryptography. They leave the problem of separating efficient private and inefficient private learning open.

It is well known that for all concept classes $\mathcal{C}$, every learner for $\mathcal{C}$ requires $\Omega(VCDIM(\mathcal{C}))$ samples [9]. This lower bound on the sample size also holds for

[2] Informally, a mechanism is useful for a concept class if for every input, the output of the mechanism maintains approximately correct counts for all concepts in the concept class.

private learning. Blum, Ligett, and Roth [4] have recently extended this result to the setting of private data release. They show that for all concept classes $\mathcal{C}$, every non-interactive sanitization mechanism that is useful for $\mathcal{C}$ requires $\Omega(VCDIM(\mathcal{C}))$ samples. We show in Section 4 that this bound is not tight – there exists a concept class $\mathcal{C}$ of constant VC-dimension such that every non-interactive sanitization mechanism that is useful for $\mathcal{C}$ requires a much larger sample size.

2 Preliminaries

Notation. We use $[n]$ to denote the set $\{1, 2, \ldots, n\}$. The notation $O_\gamma(g(n))$ is a shorthand for $O(h(\gamma) \cdot g(n))$ for some non-negative function h. Similarly, the notation $\Omega_\gamma(g(n))$. We use $\text{negl}(\cdot)$ to denote functions from $\mathbb{R}^+$ to $[0, 1]$ that decrease faster than any inverse polynomial.

2.1 Preliminaries from Privacy

A database is a vector $D = (d_1, \ldots, d_m)$ over a domain X, where each entry $d_i \in D$ represents information contributed by one individual. Databases D and D' are called *neighbors* if they differ in exactly one entry (i.e., the Hamming distance between D and D' is 1). An algorithm is private if neighboring databases induce nearby distributions on its outcomes. Formally:

Definition 1 (Differential Privacy [7]). *A randomized algorithm $\mathcal{A}$ is ϵ-differentially private if for all neighboring databases D, D', and for all sets $\mathcal{S}$ of outputs,*

$$\Pr[\mathcal{A}(D) \in \mathcal{S}] \leq \exp(\epsilon) \cdot \Pr[\mathcal{A}(D') \in \mathcal{S}]. \tag{1}$$

The probability is taken over the random coins of $\mathcal{A}$.

An immediate consequence of Equation (1) is that for any two databases D, D' (not necessarily neighbors) of size m, and for all sets $\mathcal{S}$ of outputs, $\Pr[\mathcal{A}(D) \in \mathcal{S}] \geq \exp(-\epsilon m) \cdot \Pr[\mathcal{A}(D') \in \mathcal{S}]$.

2.2 Preliminaries from Learning Theory

We consider Boolean classification problems. A concept is a function that labels *examples* taken from the domain X by the elements of the range $\{0, 1\}$. The domain X is understood to be an ensemble $X = \{X_d\}_{d \in \mathbb{N}}$. A *concept class* $\mathcal{C}$ is a set of concepts, considered as an ensemble $\mathcal{C} = \{\mathcal{C}_d\}_{d \in \mathbb{N}}$ where $\mathcal{C}_d$ is a class of concepts from $\{0, 1\}^d$ to $\{0, 1\}$.

A concept class comes implicitly with a way to represent concepts and size(c) is the size of the (smallest) representation of c under the given representation scheme. Let $\mathcal{D}$ be a distribution on X_d. PAC learning algorithms are designed assuming a promise that the examples are labeled consistently with some *target* concept c from a class $\mathcal{C}$. Define,

$$\underset{\mathcal{D}}{\text{error}}(c, h) = \Pr_{x \sim \mathcal{D}}[h(x) \neq c(x)].$$

Definition 2 (PAC Learning [17]). *An algorithm $\mathcal{A}$ is an (α, β)-*PAC learner *of a concept class $\mathcal{C}_d$ over X_d using hypothesis class $\mathcal{H}_d$ and sample size n if for all concepts $c \in \mathcal{C}_d$, all distributions $\mathcal{D}$ on X_d, given an input $D = (d_1, \cdots, d_n)$, where $d_i = (x_i, c(x_i))$ and x_i are drawn i.i.d. from $\mathcal{D}$ for $i \in [n]$, algorithm $\mathcal{A}$ outputs a hypothesis $h \in \mathcal{H}_d$ satisfying $\Pr[\text{error}_{\mathcal{D}}(c, h) \leq \alpha] \geq 1 - \beta$. The probability is taken over the random choice of the examples D and the coin tosses of the learner.*

An algorithm $\mathcal{A}$, whose inputs are d, α, β, and a set of samples (labeled examples) D, is a PAC learner *of a concept class $\mathcal{C} = \{\mathcal{C}_d\}_{d \in \mathbb{N}}$ over $X = \{X_d\}_{d \in \mathbb{N}}$ using hypothesis class $\mathcal{H}_d = \{\mathcal{H}_d\}_{d \in \mathbb{N}}$ if there exists a polynomial $p(\cdot, \cdot, \cdot, \cdot)$ such that for all $d \in \mathbb{N}$ and $0 < \alpha, \beta < 1$, the algorithm $\mathcal{A}(d, \alpha, \beta, \cdot)$ is an (α, β)-PAC learner of the concept class $\mathcal{C}_d$ over X_d using hypothesis class $\mathcal{H}_d$ and sample size $n = p(d, \text{size}(c), 1/\alpha, \log(1/\beta))$. If $\mathcal{A}$ runs in time polynomial in $d, \text{size}(c), 1/\alpha, \log(1/\beta)$, we say that it is an* efficient PAC learner. *Also, the learner is called a* proper *PAC learner if $\mathcal{H} = \mathcal{C}$, otherwise it is called an* improper *PAC learner.*

A concept class $\mathcal{C} = \{\mathcal{C}_d\}_{d \in \mathbb{N}}$ over $X = \{X_d\}_{d \in \mathbb{N}}$ is PAC learnable *using hypothesis class $\mathcal{H} = \{\mathcal{H}_d\}_{d \in \mathbb{N}}$ if there exists a PAC learner $\mathcal{A}$ learning $\mathcal{C}$ over X using hypothesis class $\mathcal{H}$. If $\mathcal{A}$ is an efficient PAC learner, we say that $\mathcal{C}$ is efficiently PAC learnable.*

It is well known that improper learning is more powerful than proper learning. For example, Pitt and Valiant [16] show that unless **RP=NP**, k-term DNF formulae are not learnable by k-term DNF, whereas it is possible to learn a k-term DNF using k-CNF [17]. For more background on learning theory, see, e.g., [13].

Definition 3 (VC-Dimension [18]). *Let $\mathcal{C} = \{\mathcal{C}_d\}$ be a class of concepts over $X = \{X_d\}$. We say that $\mathcal{C}_d$ shatters a point set $Y \subset X_d$ if $|\{c(Y) : c \in \mathcal{C}_d\}| = 2^{|Y|}$, i.e., the concepts in $\mathcal{C}_d$ when restricted to Y produce all the $2^{|Y|}$ possible assignments on Y. the VC-dimension of $\mathcal{C}$ is defined as the size of the maximum point set that is shattered by $\mathcal{C}_d$, as a function of d.*

Theorem 1 ([5]). *A concept class $\mathcal{C} = \{\mathcal{C}_d\}$ over $X = \{X_d\}$ is PAC learnable using $\mathcal{C}$ by a PAC learner $\mathcal{A}$ that uses $O((\mathit{VCDIM}(\mathcal{C}_d) \cdot \log \frac{1}{\alpha} + \log \frac{1}{\beta})/\alpha)$ samples.*

2.3 Private Learning

Definition 4 (Private PAC Learning [10]). *Let d, α, β be as in Definition 2 and $\epsilon > 0$. Concept class $\mathcal{C}$ is ϵ-differentially privately PAC learnable using $\mathcal{H}$ if there exists an algorithm $\mathcal{A}$ that takes inputs $\epsilon, d, \alpha, \beta, D$, where n, the number of samples (labeled examples) in D is polynomial in $1/\epsilon, d, \text{size}(c), 1/\alpha, \log(1/\beta)$, and satisfies*

PRIVACY. *For all d and $\epsilon, \alpha, \beta > 0$, algorithm $\mathcal{A}(\epsilon, d, \alpha, \beta, \cdot)$ is ϵ-differentially private (Definition 1);*

UTILITY. *For all $\epsilon > 0$, algorithm $\mathcal{A}(\epsilon, \cdot, \cdot, \cdot)$ PAC learns $\mathcal{C}$ using $\mathcal{H}$ (Definition 2).*

$\mathcal{A}$ is an efficient private PAC learner if it runs in time polynomial in $1/\epsilon$, d, size(c), $1/\alpha$, $\log(1/\beta)$. *Also, the private learner is called* proper *if $\mathcal{H} = \mathcal{C}$, otherwise it is called* improper.

Remark 1. The privacy requirement in Definition 4 is a worst-case requirement. That is, Equation (1) must hold for every pair of neighboring databases D, D' (even if these databases are not consistent with any concept in $\mathcal{C}$). In contrast, the utility requirement is an average-case requirement, where we only require the learner to succeed with high probability over the distribution of the databases. This qualitative difference between the utility and privacy of private learners is crucial. A wrong assumption on how samples are formed that leads to a meaningless outcome can usually be replaced with a better one with very little harm. No such amendment is possible once privacy is lost due to a wrong assumption.

Note also that each entry d_i in a database D is a labeled example. That is, we protect the privacy of both the example and its label.

Observation 1. The computational separation between proper and improper learning also holds when we add the privacy constraint. That is unless **RP=NP** no proper private learner can learn k-term DNF, whereas there exists an efficient improper private learner that can learn k-term DNF using a k-CNF. The efficient k-term DNF learner of [17] uses statistical queries (SQ) [12] which can be simulated efficiently and privately as shown by [2, 10].

More generally, such a gap can be shown for any concept class that cannot be properly PAC learned, but can be efficiently learned (improperly) in the statistical queries model.

3 Learning vs. Private Learning

We begin by recalling the upper bound on the sample (database) size for private learning from [10]. The bound in [10] is for agnostic learning, and we restate it for (non-agnostic) PAC learning using the following notion of α-representation:

Definition 5. *We say that a hypothesis class $\mathcal{H}_d$ α-represents a concept class $\mathcal{C}_d$ over the domain X_d if for every $c \in \mathcal{C}_d$ and every distribution $\mathcal{D}$ on X_d there exists a hypothesis $h \in \mathcal{H}_d$ such that* $\mathrm{error}_{\mathcal{D}}(c, h) \leq \alpha$.

Theorem 2 (Kasiviswanathan *et al.* [10], restated). *Assume that there is a hypothesis class $\mathcal{H}_d$ that α-represents a concept class $\mathcal{C}_d$. Then, there exists a private PAC learner for $\mathcal{C}_d$ using $\mathcal{H}_d$ that uses $O((\log |\mathcal{H}_d| + \log(1/\beta))/(\epsilon\alpha))$ labeled examples, where ϵ, α, and β are parameters of the private learner. The learner might not be efficient.*

In other words, using Theorem 2 the number of labeled examples required for learning a concept class $\mathcal{C}_d$ is logarithmic in the size of the smallest hypothesis class that α-represents $\mathcal{C}_d$. For comparison, the number of labeled examples required for learning $\mathcal{C}_d$ non-privately is proportional to the VC-dimension of $\mathcal{C}_d$ [5, 9].

3.1 Separation between Private and Non-private PAC Learning

Our first result shows that private learners may require many more samples than non-private ones. We consider a very simple concept class of VC-dimension one, and hence is (non-privately) properly learnable using $O_{\alpha,\beta}(1)$ labeled examples. We prove that for any proper learner for this class the required number of labeled examples is at least logarithmic in the size of the concept class, matching Theorem 2.

Proving the lower bound, we show that a large collection of m-record databases $D_1, \ldots, D_N$ exists, with the property that every PAC learner has to output a different hypothesis for each of these databases (recall that in our context a database is a collection of labeled examples, supposedly drawn from some distribution and labeled consistently with some target concept).

As any two databases D_a and D_b differ on at most m entries, a private learner must, because of the differential privacy requirement, output on input D_a the hypothesis that is accurate for D_b (and not accurate for D_a) with probability at least $(1-\beta)\cdot\exp(-\epsilon m)$. Since this holds for every pair of databases, unless m is large enough we get that the private learner's output on D_a is, with high probability, a hypothesis that is not accurate for D_a. We use the following notion of α-minimality:

Definition 6. *If $\mathcal{H}_d$ α-represents $\mathcal{C}_d$, and every $\mathcal{H}'_d \subsetneq \mathcal{H}_d$ does not α-represent $\mathcal{C}_d$, then we say that $\mathcal{H}_d$ is α-minimal for $\mathcal{C}_d$.*

Theorem 3. *Let $\mathcal{H}_d$ be an α-minimal class for $\mathcal{C}_d$. Then any private PAC learner that learns $\mathcal{C}_d$ using $\mathcal{H}_d$ requires $\Omega((\log|\mathcal{H}_d| + \log(1/\beta))/\epsilon)$ labeled examples.*

Proof. Let $\mathcal{C}_d$ be over the domain X_d and let $\mathcal{H}_d$ be α-minimal for $\mathcal{C}_d$. Since for every $h \in \mathcal{H}_d$, $\mathcal{H}_d \setminus \{h\}$ does not α-represent $\mathcal{C}_d$, we get that there exists a concept $c_h \in \mathcal{C}_d$ and a distribution $\mathcal{D}_h$ on X_d such that on inputs drawn from $\mathcal{D}_h$ labeled by c_h, every PAC learner (that learns $\mathcal{C}_d$ using $\mathcal{H}_d$) has to output h with probability at least $1-\beta$.

Let $\mathcal{A}$ be a private learner that learns $\mathcal{C}_d$ using $\mathcal{H}_d$, and suppose $\mathcal{A}$ uses m labeled examples. For every $h \in \mathcal{H}_d$, note that there exists a database $D_h \in X_d^m$ on which $\mathcal{A}$ has to output h with probability at least $1-\beta$. To see that, note that if $\mathcal{A}$ is run on m examples chosen i.i.d. from the distribution $\mathcal{D}_h$ and labeled according to c_h, then $\mathcal{A}$ outputs h with probability at least $1-\beta$ (where the probability is over the sampling from $\mathcal{D}_h$ and over the randomness of $\mathcal{A}$). Hence, a collection of m labeled examples over which $\mathcal{A}$ outputs h with probability $1-\beta$ exists, and D_h can be set to contain these m labeled examples.

Take $h, h' \in \mathcal{H}_d$ such that $h \neq h'$ and consider the two corresponding databases D_h and $D_{h'}$ with m entries each. Clearly, they differ in at most m entries, and hence we get by differential privacy of $\mathcal{A}$ that

$$\Pr[\mathcal{A}(D_h) = h'] \geq \exp(-\epsilon m)\cdot\Pr[\mathcal{A}(D_{h'}) = h'] \geq \exp(-\epsilon m)\cdot(1-\beta).$$

Since the above inequality holds for every pair of databases, we fix any h and get,

$$\Pr[\mathcal{A}(D_h) \neq h] = \Pr[\mathcal{A}(D_h) \in \mathcal{H}_d \setminus \{h\}] = \sum_{h' \in \mathcal{H}_d \setminus \{h\}} \Pr[\mathcal{A}(D_h) = h']$$
$$\geq (|\mathcal{H}_d| - 1) \cdot \exp(-\epsilon m) \cdot (1 - \beta).$$

On the other hand, we chose D_h such that $\Pr[\mathcal{A}(D_h) = h] \geq 1 - \beta$, equivalently, $\Pr[\mathcal{A}(D_h) \neq h] \leq \beta$. We hence get that $(|\mathcal{H}_d| - 1) \cdot \exp(-\epsilon m) \cdot (1 - \beta) \leq \beta$. Solving the last inequality for m, we get $m = \Omega((\log |\mathcal{H}_d| + \log(1/\beta))/\epsilon)$ as required. □

Using Theorem 3, we now prove a lower bound on the number of labeled examples needed for proper private learning a specific concept class. Let $T = 2^d$ and $X_d = \{1, \ldots, T\}$. Define the concept class $POINT_d$ to be the set of points over $\{1, \ldots, T\}$:

Definition 7 (Concept Class $POINT_d$). *For $j \in [T]$ define $c_j : [T] \to \{0, 1\}$ as $c_j(x) = 1$ if $x = j$, and $c_j(x) = 0$ otherwise. $POINT_d = \{c_j\}_{j \in [T]}$.*

We note that we use the set $\{1, \ldots, T\}$ for notational convenience only. We never use the fact that the set elements are integer numbers.

Proposition 1. *$POINT_d$ is α-minimal for itself.*

Proof. Clearly, $POINT_d$ α-represents itself. To show minimality, consider a subset $\mathcal{H}'_d \subsetneq POINT_d$, where $c_i \notin \mathcal{H}'_d$. Note that under the distribution $\mathcal{D}$ that chooses i with probability one, $\text{error}_{\mathcal{D}}(c_i, c_j) = 1$ for all $j \neq i$. Hence, $\mathcal{H}'_d$ does not α-represent $POINT_d$. □

The VC-dimension of $POINT_d$ is one[3]. It is well known that a standard (non-private) proper learner uses approximately VC-dimension number of labeled examples to learn a concept class [5]. In contrast, we get that far more labeled examples are needed for any proper private learner for $POINT_d$. The following corollary follows directly from Theorem 3 and Proposition 1:

Corollary 1. *Every proper private PAC learner for $POINT_d$ requires $\Omega((d + \log(1/\beta))/\epsilon)$ labeled examples.*

Remark 2. We note that the lower bound for $POINT_d$ can be improved to $\Omega((d + \log(1/\beta))/(\epsilon\alpha))$ labeled examples, matching the upper bound from Theorem 2. Also, the proper learner for $POINT_d$ from Theorem 2 can be made efficient. Details are deferred to the full version [1].

We conclude this section showing that every hypothesis class $\mathcal{H}$ that α-represents $POINT_d$ should have at least d hypotheses. Therefore, if we use Theorem 2 to

[3] Note that every singleton $\{j\}$ where $j \in [T]$ is shattered by $POINT_d$ as $c_j(j) = 1$ and $c_{j'}(j) = 0$ for all $j' \neq j$. No set of two points $\{j, j'\}$ is shattered by $POINT_d$ as $c_{j''}(j) = c_{j''}(j') = 1$ for no $j'' \in [T]$.

learn $POINT_d$ we need $\Omega(\log d)$ labeled examples. At first sight, it may seem that the relationship between $|\mathcal{H}|$ and the sample complexity is essential, and hence, the number of labeled examples needed for *every* private PAC learner for $POINT_d$ is super-constant. However, this turns out not to be the case. In Section 3.2, we present a private learner for $POINT_d$ that uses $O_{\alpha,\beta,\epsilon}(1)$ labeled examples. For this construction, we use techniques that are very different from those used in the proof of Theorem 2. In particular, our private learner uses a very large hypothesis class.

Lemma 1. *Let $\alpha < 1/2$. $|\mathcal{H}| \geq d$ for every hypothesis class $\mathcal{H}$ that α-represents $POINT_d$.*

Proof. Let $\mathcal{H}$ be a hypothesis class with $|\mathcal{H}| < d$. Consider a table whose $T = 2^d$ columns correspond to the possible 2^d inputs $1, \ldots, T$, and whose $|\mathcal{H}|$ rows correspond to the hypothesis in $\mathcal{H}$. The (i,j)th entry is 0 or 1 depending on whether the ith hypothesis gives 0 or 1 on input j. Since $|\mathcal{H}| < d = \log T$, at least two columns $j \neq j'$ are identical. That is, $h(j) = h(j')$ for every $h \in \mathcal{H}$. Consider the concept $c_j \in POINT_d$ (defined as $c_j(x) = 1$ if $x = j$, and 0 otherwise), and the distribution $\mathcal{D}$ with probability mass 1/2 on both j and j'. We get that $\mathrm{error}_{\mathcal{D}}(c_j, h) \geq 1/2 > \alpha$ for all $h \in \mathcal{H}$ (since any hypothesis either errs on j or on j'). Therefore, $\mathcal{H}$ does not α-represent $POINT_d$. □

3.2 Separation between Proper and Improper Private PAC Learning

We now use $POINT_d$ to show a separation between proper and improper private PAC learning. We show that $POINT_d$ can be privately (and efficiently) learned by an improper learner using $O_{\alpha,\beta,\epsilon}(1)$ labeled examples. We begin by presenting a non-private improper PAC learner $\mathcal{A}_1$ for $POINT_d$ that succeeds with only constant probability. Roughly, $\mathcal{A}_1$ applies a simple proper learner for $POINT_d$, and then modifies its outcome by adding random "noise". We then use sampling to convert $\mathcal{A}_1$ into a private learner $\mathcal{A}_2$, and like $\mathcal{A}_1$ the probability that $\mathcal{A}_2$ succeeds in learning $POINT_d$ is only a constant. Later we amplify the success probability of $\mathcal{A}_2$ to get a private PAC learner. Both $\mathcal{A}_1$ and $\mathcal{A}_2$ are inefficient as they output hypotheses with exponential description length. However, using a pseudorandom function it is possible to compress the outputs of $\mathcal{A}_1$ and $\mathcal{A}_2$, and hence achieve efficiency.

Algorithm $\mathcal{A}_1$. Given labeled examples $(x_1, y_1), \ldots, (x_m, y_m)$, algorithm $\mathcal{A}_1$ performs the following:

1. If $(x_1, y_1), \ldots, (x_m, y_m)$ are not consistent with any concept in $POINT_d$, return $\perp$ (this happens only if $x_i \neq x_j$ and $y_i = y_j = 1$ for some $i, j \in [m]$ or if $x_i = x_j$ and $y_i \neq y_j$).
2. If $y_i = 0$ for all $i \in [m]$, then let $c = \mathbf{0}$ (the all zero hypothesis); otherwise, let c be the (unique) hypothesis from $POINT_d$ that is consistent with the m input labeled examples.

3. Modify c at random to get a hypothesis h by letting $h(x) = c(x)$ with probability $1 - \alpha/8$, and $h(x) = 1 - c(x)$ otherwise for all $x \in [T]$. Return h.

Let $m = 2\ln(4)/\alpha$. We next argue that if m examples are drawn i.i.d. according to a distribution $\mathcal{D}$ on $[T]$, and the examples are labeled consistently according to some $c_j \in POINT_d$, then $\Pr[\mathrm{error}_{\mathcal{D}}(c_j, c) > \alpha/2] \leq 1/4$. If the examples are labeled consistently according to some $c_j \neq \mathbf{0}$, then $c \neq c_j$ only if $(j, 1)$ is not in the sample and in this case $c = \mathbf{0}$. If $\Pr_{x\sim\mathcal{D}}[x = j] \leq \alpha/2$ and $(j, 1)$ is not in the sample, then $c = \mathbf{0}$ and $\mathrm{error}_{\mathcal{D}}(c_j, \mathbf{0}) \leq \alpha/2$. If $\Pr_{x\sim\mathcal{D}}[x = j] \leq \alpha/2$ and $(j, 1)$ is in the sample, then $c = c_j$ and $\mathrm{error}_{\mathcal{D}}(c_j, c) = 0$. Otherwise if $\Pr_{x\sim\mathcal{D}}[x = j] > \alpha/2$, the probability that all m examples are not $(j, 1)$ is at most $(1 - \alpha/2)^m = ((1 - \alpha/2)^{2/\alpha})^{\ln 4} \leq 1/4$.

To see that $\mathcal{A}_1$ PAC learns $POINT_d$ (with accuracy α and confidence $1/4$) note that

$$\mathop{\mathbb{E}}_{h}[\mathrm{error}(c, h)] = \mathop{\mathbb{E}}_{\mathcal{D}} \mathop{\mathbb{E}}_{h\ x\sim\mathcal{D}}[|h(x) - c(x)|] = \mathop{\mathbb{E}}_{x\sim\mathcal{D}} \mathop{\mathbb{E}}_{h}[|h(x) - c(x)|] = \frac{\alpha}{8},$$

and hence, using Markov's Inequality, $\Pr_h[\mathrm{error}_{\mathcal{D}}(c, h) > \alpha/2] \leq 1/4$. Combining this with $\Pr[\mathrm{error}_{\mathcal{D}}(c_j, c) > \alpha/2] \leq 1/4$ and $\mathrm{error}_{\mathcal{D}}(c_j, h) \leq \mathrm{error}_{\mathcal{D}}(c_j, c) + \mathrm{error}_{\mathcal{D}}(c, h)$, implies that $\Pr[\mathrm{error}_{\mathcal{D}}(c_j, h) > \alpha] \leq 1/2$.

Algorithm $\mathcal{A}_2$. We now modify learner $\mathcal{A}_1$ to get a private learner $\mathcal{A}_2$ (a similar idea was used in [10] for learning parity functions). Given labeled examples $(x_1, y_1), \ldots, (x_{m'}, y_{m'})$, algorithm $\mathcal{A}_2$ performs the following:

1. With probability $\alpha/8$, return $\perp$.
2. Construct a set $S \subseteq [m']$ by picking each element of $[m']$ with probability $p = \alpha/4$. Run the non-private learner $\mathcal{A}_1$ on the examples indexed by S.

We first show that, given $m' = 8m/\alpha$ labeled examples, $\mathcal{A}_2$ PAC learns $POINT_d$ with confidence $\Theta(1)$. Note that, by Chernoff bound, $\Pr[|S| \leq m] \leq \exp(-m/4) = O_\alpha(1)$. Therefore, we get that $\mathcal{A}_2$ PAC learns $POINT_d$ with accuracy parameter $\alpha' = \alpha$ and confidence parameter $\beta' = 1/2 + \alpha/8 + \exp(-m/4) = \Theta(1)$. We now show that $\mathcal{A}_2$ is ϵ^*-differentially private with bounded ϵ^*.

Claim. Algorithm $\mathcal{A}_2$ is ϵ^*-differentially private, where $\epsilon^* = \ln(4)$.

Proof. Let D, D' be two neighboring databases, and assume that they differ on the ith entry. First let us analyze the probability of $\mathcal{A}_2$ outputting $\perp$:

$$\begin{aligned}\frac{\Pr[\mathcal{A}_2(D) = \perp]}{\Pr[\mathcal{A}_2(D') = \perp]} &= \frac{p \cdot \Pr[\mathcal{A}_2(D) = \perp \mid i \in S] + (1-p) \cdot \Pr[\mathcal{A}_2(D) = \perp \mid i \notin S]}{p \cdot \Pr[\mathcal{A}_2(D') = \perp \mid i \in S] + (1-p) \cdot \Pr[\mathcal{A}_2(D') = \perp \mid i \notin S]} \\ &\leq \frac{p \cdot 1 + (1-p) \cdot \Pr[\mathcal{A}_2(D) = \perp \mid i \notin S]}{p \cdot 0 + (1-p) \cdot \Pr[\mathcal{A}_2(D') = \perp \mid i \notin S]} \\ &= \frac{p}{(1-p) \cdot \Pr[\mathcal{A}_2(D') = \perp \mid i \notin S]} + 1 \leq \frac{8p}{\alpha(1-p)} + 1,\end{aligned}$$

where the last equality follows noting that if $i \notin S$ then $\mathcal{A}_2$ is equally likely to output $\perp$ on D and D', and the last inequality follows as $\perp$ is returned with probability $\alpha/8$ in Step 1 of Algorithm $\mathcal{A}_2$.

For the more interesting case, where $\mathcal{A}_2$ outputs a hypothesis h, we get:

$$\begin{aligned}\frac{\Pr[\mathcal{A}_2(D)=h]}{\Pr[\mathcal{A}_2(D')=h]} &= \frac{p\cdot\Pr[\mathcal{A}_2(D)=h\mid i\in S]+(1-p)\cdot\Pr[\mathcal{A}_2(D)=h\mid i\notin S]}{p\cdot\Pr[\mathcal{A}_2(D')=h\mid i\in S]+(1-p)\cdot\Pr[\mathcal{A}_2(D')=h\mid i\notin S]}\\ &\leq \frac{p\cdot\Pr[\mathcal{A}_2(D)=h\mid i\in S]+(1-p)\cdot\Pr[\mathcal{A}_2(D)=h\mid i\notin S]}{p\cdot 0+(1-p)\cdot\Pr[\mathcal{A}_2(D')=h\mid i\notin S]}\\ &= \frac{p}{1-p}\cdot\frac{\Pr[\mathcal{A}_2(D)=h\mid i\in S]}{\Pr[\mathcal{A}_2(D)=h\mid i\notin S]}+1,\end{aligned}$$

where the last equality uses the fact that if $i\notin S$ then $\mathcal{A}_2$ is equally likely to output h on D and D'. To conclude our proof, we need to bound the ratio of $\Pr[\mathcal{A}_2(D)=h\mid i\in S]$ to $\Pr[\mathcal{A}_2(D)=h\mid i\notin S]$.

$$\begin{aligned}&\frac{\Pr[\mathcal{A}_2(D)=h\mid i\in S]}{\Pr[\mathcal{A}_2(D)=h\mid i\notin S]}\\ &= \frac{\sum_{R\subseteq[m']\setminus\{i\}}\Pr[\mathcal{A}_2(D)=h\mid S=R\cup\{i\}]\cdot\Pr[\mathcal{A}_2 \text{ selects } R \text{ from } [m']\setminus\{i\}]}{\sum_{R\subseteq[m']\setminus\{i\}}\Pr[\mathcal{A}_2(D)=h\mid S=R]\cdot\Pr[\mathcal{A}_2 \text{ selects } R \text{ from } [m']\setminus\{i\}]}\\ &\leq \max_{R\subseteq[m']\setminus\{i\}}\frac{\Pr[\mathcal{A}_2(D)=h\mid S=R\cup\{i\}]}{\Pr[\mathcal{A}_2(D)=h\mid S=R]}.\end{aligned}$$

Now, having or not having access to (x_i,y_i) can only affect the choice of $h(x_i)$, and since, $\mathcal{A}_1$ flips the output with probability $\alpha/8$, we get

$$\max_{R\subseteq[m']\setminus\{i\}}\frac{\Pr[\mathcal{A}_2(D)=h\mid S=R\cup\{i\}]}{\Pr[\mathcal{A}_2(D)=h|S=R]}\leq\frac{1-\alpha/8}{\alpha/8}\leq\frac{8}{\alpha}.$$

Putting everything together, we get

$$\frac{\Pr[\mathcal{A}_2(D)=h]}{\Pr[\mathcal{A}_2(D')=h]}\leq\frac{8p}{\alpha(1-p)}+1=\frac{8}{(4-\alpha)}+1<3+1=e^{\epsilon^*}. \qquad \square$$

We can reduce ϵ^* to any desired ϵ using the following simple lemma (implicit in [10], see proof in [1]):

Lemma 2. *Let $\mathcal{A}$ be an ϵ^*-differentially private algorithm. Construct an algorithm $\mathcal{B}$ that on input a database $D=(d_1,\ldots,d_n)$ constructs a new database D_s whose ith entry is d_i with probability $f(\epsilon,\epsilon^*)=(\exp(\epsilon)-1)/(\exp(\epsilon^*)+\exp(\epsilon)-\exp(\epsilon-\epsilon^*)-1)$ and $\perp$ otherwise, and then runs $\mathcal{A}$ on D_s. Then, $\mathcal{B}$ is ϵ-differentially private.*

It is clearly possible to incorporate the sampling in the lemma directly in Step 2 of $\mathcal{A}_2$ (note that for small ϵ, $f(\epsilon,\epsilon^*)\approx\epsilon/(\exp(\epsilon^*)-1)$). We get that the number of labeled examples required to get a private learner with confidence parameter $\Theta(1)$ is $O_{\alpha,\epsilon}(1)$. The confidence parameter of the learner can be boosted privately from $\Theta(1)$ to any value $\beta>0$ as explained in [10]. In doing this boosting, the number of labeled examples required for the learner increases by a factor of $O(\log(1/\beta))$. Therefore, we get that a sample size that is polynomial in $1/\epsilon, 1/\alpha$, and $\log(1/\beta)$ is sufficient to learn $POINT_d$ improperly with privacy parameter ϵ, accuracy parameter α, and confidence parameter β.

Making the Learner Efficient. Recall that the outcome of $\mathcal{A}_1$ (hence $\mathcal{A}_2$) is an exponentially long description of a hypothesis. We now complete our construction by compressing this description using a pseudorandom function. We use a slightly non-standard definition of (non-uniform) pseudorandom functions from binary strings of size d to bits; these pseudorandom functions can be easily constructed given regular pseudorandom functions.

Definition 8. *Let $F = \{F_d\}_{d\in\mathbb{N}}$ be a function ensemble, where for every d, F_d is a set of functions from $\{0,1\}^d$ to $\{0,1\}$. We say that the function ensemble F is q-biased pseudorandom if for every family of polynomial-size circuits with oracle access $\{C_d\}_{d\in\mathbb{N}}$, every polynomial $p(\cdot)$, and all sufficiently large d's,*

$$|\Pr[C_d^f(1^d) = 1] - \Pr[C_d^{H_d^q}(1^d) = 1]| < \frac{1}{p(d)} ,$$

where f is chosen at random from F_d and $H_d^q : \{0,1\}^d \to \{0,1\}$ is a function and the value $H_d^q(x)$ for $x \in \{0,1\}^d$ are selected i.i.d. to be 1 with probability q and 0 otherwise. The probabilities are taken over the random choice of H_d^q, and f.

For convenience, for $d \in \mathbb{N}$, we consider F_d as a set of functions from $\{1,\ldots,T\}$ to $\{0,1\}$, where $T = 2^d$. We set $q = \alpha\beta/4$ in the above definition. Using an $\alpha\beta/4$-biased pseudorandom function ensemble F, we change Step 3 of algorithm $\mathcal{A}_1$ as follows:

3'. If $c = \mathbf{0}$, let h be a random function from F_d. Otherwise (i.e., $c = c_j$ for some $j \in [T]$), let h be a random function from F_d subject to $h(j) = 1$. Return h.

Call the resulting modified algorithm $\mathcal{A}_3$. We next show that $\mathcal{A}_3$ is a PAC learner. Note that the exists a negligible function negl such that for large enough d, $|\Pr[h(x) = 1|h(j) = 1] - \alpha\beta/4| \leq \text{negl}(d)$ for every $x \in \{1,\ldots,T\}$ (as otherwise, we get a non-uniform distinguisher for the ensemble F). Thus,

$$\begin{aligned}\mathop{\mathbb{E}}_{h\in F_d}[\mathop{\text{error}}_{\mathcal{D}}(c,h)] &= \mathop{\mathbb{E}}_{h\in F_d}\mathop{\mathbb{E}}_{x\sim\mathcal{D}}[|h(x) - c(x)|] \\ &\leq \mathop{\mathbb{E}}_{h\in F_d}\mathop{\mathbb{E}}_{x\sim\mathcal{D}}[h(x)] = \mathop{\mathbb{E}}_{x\sim\mathcal{D}}\mathop{\mathbb{E}}_{h\in F_d}[h(x)] \leq \frac{\alpha\beta}{4} + \text{negl}(d).\end{aligned}$$

The first inequality follows as for all $x \in [T]$, $h(x) \geq c(x)$ by our restriction on the choice of h. Thus, by the same arguments as for $\mathcal{A}_1$, Algorithm $\mathcal{A}_3$ is a PAC learner.

We next modify algorithm $\mathcal{A}_2$ by executing the learner $\mathcal{A}_3$ instead of the learner $\mathcal{A}_1$. Call the resulting modified algorithm $\mathcal{A}_4$. To see that algorithm $\mathcal{A}_4$ preserves differential privacy it suffices to give a bound on Equation (2). By comparing the case where $S = R$ with $S = R \cup \{i\}$, we get that the probability for a hypothesis h can increase only if $c = \mathbf{0}$ when $S = R$, and $c = c_{y_i}$ when $S = R \cup \{i\}$. Therefore,

$$\max_{R\subseteq[m']\setminus\{i\}} \frac{\Pr[\mathcal{A}_4(D) = h \mid S = R \cup \{i\}]}{\Pr[\mathcal{A}_4(D) = h \mid S = R]} \leq \frac{1}{(\alpha\beta/4) - \text{negl}(d)} \leq \frac{1}{(\alpha\beta/8)} = \frac{8}{\alpha\beta}.$$

Theorem 4. *There exists an efficient improper private PAC learner for $POINT_d$ that uses $O_{\alpha,\beta,\epsilon}(1)$ labeled examples, where ϵ, α, and β are parameters of the private learner.*

3.3 Separation between Efficient and Inefficient Proper Private PAC Learning

In this section, we use the sample size lower bound for proper private learning $POINT_d$ to obtain a separation between efficient and inefficient proper private PAC learning. Let U_r represent a uniformly random string from $\{0,1\}^r$. Let $\ell(d) : \mathbb{N} \to \mathbb{N}$ be a function and $G = \{G_d\}_{d\in\mathbb{N}}$ be a deterministic algorithm such that on input from $\{0,1\}^{\ell(d)}$ it returns an output from $\{0,1\}^d$. Informally, we say that G is pseudorandom generator if on $\ell(d)$ truly random bits it outputs d bits that are indistinguishable from d random bits. Formally, for every probabilistic polynomial time algorithm $\mathcal{B}$ there exists a negligible function negl(d) (i.e., a function that is asymptotically smaller than $1/d^c$ for all $c > 0$) such that

$$|\Pr[\mathcal{B}(G_d(U_{\ell(d)})) = 1] - \Pr[\mathcal{B}(U_d) = 1]| \leq \text{negl}(d).$$

Such exponential stretch pseudorandom generators G (i.e., with $\ell(d) = \omega(\log d)$) exist under various strong hardness assumptions.

Let $POINT_d = \{c_1, \ldots, c_{2^d}\}$. Now to a polynomially bounded private learner, $c_{G_d(U_{\ell(d)})}$ would appear with high probability as a uniformly random concept picked from $POINT_d$. We will show by using ideas similar to the proof of Theorem 3 that a polynomially bounded proper private learner would require $\Omega((d + \log(1/\beta))/\epsilon)$ labeled examples to learn $c_{G_d(U_{\ell(d)})}$. More precisely, define concept class

$$\widehat{POINT}_d = \bigcup_{r\in\{0,1\}^{\ell(d)}} \{c_{G_d(r)}\}.$$

Assume that there is an efficient proper private learner $\mathcal{A}$ for $\widehat{POINT}_d$ with sample size $m = o((d + \log(1/\beta))/\epsilon)$. We use $\mathcal{A}$ to construct a distinguisher for the pseudorandom generator: Given j we construct the database D with m entries $(j, 1)$. If $\mathcal{A}(D) = c_j$, then the distinguisher returns 1, otherwise it returns 0. If $j = G_d(r)$ for some r, then, by the utility of the private learner, $\mathcal{A}$ has to return c_j on this database with probability at least $1 - \beta$. Thus, the distinguisher returns 1 with probability at least $1 - \beta$ when j is chosen from $G_d(U_{\ell(d)})$. Assume that for (say) 1/4 of the values $j \in [2^d]$ algorithm $\mathcal{A}$, when applied to the database with m entries $(j, 1)$, returns c_j with probability at least 1/3. Then, we get a contradiction following the same argument as in the proof of Theorem 3 (as at least a fraction of 1/4 of the c_j's must have probability at least $(1/3)(1-\beta)\cdot\exp(-\epsilon m)$). Thus, the distinguisher returns 1 with probability at most $1/4 + 3/4 \cdot 1/3 = 1/2$ when j is chosen from U_d.

If the learner is not polynomially bounded, then it can use the algorithm from Theorem 2 to privately learn $\widehat{POINT}_d$. Since, $|\widehat{POINT}_d| = 2^{\ell(d)}$, the private learner from Theorem 2 uses $O((\ell(d) + \log(1/\beta))/(\epsilon\alpha))$ labeled examples. We get the following separation between efficient and inefficient proper private learning:

Theorem 5. *Let $\ell(d)$ be any function that grows as $\omega(\log d)$, and G be a be a pseudorandom generator with stretch $d - \ell(d)$. For the concept class $\widehat{POINT}_d$, every polynomial-time proper private PAC learner with probability at least $1 - \text{negl}(d)$ requires $\Omega((d + \log(1/\beta))/\epsilon)$ labeled examples, whereas there exists an inefficient proper private PAC learner that can learn $\widehat{POINT}_d$ using $O((\ell(d) + \log(1/\beta))/(\epsilon\alpha))$ labeled examples.*

Remark 3. In the non-private setting, there exists an efficient proper learner that can learn the concept class $\widehat{POINT}_d$ using $O((\log(1/\alpha) + \log(1/\beta))/\alpha)$ labeled examples (as $VCDIM(\widehat{POINT}_d) = 1$). In the non-private setting we also know that even inefficient learners require $\Omega(\log(1/\beta)/\alpha)$ labeled examples [9, 13]. Therefore, for $\widehat{POINT}_d$ the sample complexities of efficient non-private learners and inefficient non-private learners are almost the same.

4 Lower Bounds for Non-interactive Sanitization

We now prove a lower bound on the database size (or sample size) needed to privately release an output that is useful for all concepts in a concept class. We start by recalling a definition and a result of Blum *et al.* [3].

Let $X = \{X_d\}_{d \in \mathbb{N}}$ be some discretized domain and consider a class of predicates $\mathcal{C}$ over X. A database D contains points taken from X_d. A predicate query Q_c for $c : X_d \to \{0, 1\}$ in $\mathcal{C}$ is defined as

$$Q_c(D) = \frac{|\{d_i \in D : c(d_i) = 1\}|}{|D|}.$$

A sanitizer (or data release mechanism) is a differentially private algorithm $\mathcal{A}$ that on input a database D outputs another database $\widehat{D}$ with entries taken from X_d. An algorithm $\mathcal{A}$ is (α, β)-useful for predicates in class $\mathcal{C}$ if with probability at least $1 - \beta$ for every $c \in \mathcal{C}$, and every database D, for $\widehat{D} = \mathcal{A}(D)$,

$$|Q_c(D) - Q_c(\widehat{D})| < \alpha.$$

Theorem 6 (Blum *et al.* [3]). *For any class of predicates $\mathcal{C}$, and any database $D \in X_d^m$, such that*

$$m \geq O\left(\frac{\log |X_d| \cdot VCDIM(\mathcal{C}) \log(1/\alpha)}{\alpha^3 \epsilon} + \frac{\log(1/\beta)}{\epsilon\alpha}\right),$$

there exists an (α, β)-useful mechanism $\mathcal{A}$ that preserves ϵ-differential privacy. The algorithm might not be efficient.

We show that the dependency on $\log |X_d|$ in Theorem 6 is essential: there exists a class of predicates $\mathcal{C}$ with VC-dimension $O(1)$ that requires $|D| = \Omega_{\alpha,\beta,\epsilon}(\log |X_d|)$. For our lower bound, the sanitized output $\widehat{D}$ could be any arbitrary data structure (not necessarily a synthetic database). For simplicity, however, here we focus on the case where the output is a synthetic database. The proof of this lower bound uses ideas from Section 3.1.

Let $T = 2^d$ and $X_d = [T]$ be the domain. Consider the class $POINT_d$ (where $i \in [T]$). For every $i \in [T]$, construct a database $D_i \in X_d^m$ by setting $(1-3\alpha)m$ entries at 1 and the remaining $3\alpha m$ entries at i (for $i = 1$ all entries of D_1 are 1). For $i \in [T]\setminus\{1\}$ we say that a database $\widehat{D}$ is α-useful for D_i if $2\alpha < Q_{c_i}(\widehat{D}) < 4\alpha$ and $1 - 4\alpha < Q_{c_1}(\widehat{D}) < 1 - 2\alpha$. We say that $\widehat{D}$ is α-useful for D_1 if $1 - \alpha < Q_{c_1}(\widehat{D}) \leq 1$. It follows that for $i \neq j$ if $\widehat{D}$ is α-useful for D_i then it is not α-useful for D_j.

Let $\widehat{\mathbb{D}}_i$ be the set of all databases that are α-useful for D_i. Note that for all $i \neq 1$, D_1 and D_i differ on $3\alpha m$ entries, and by our previous observation, $\widehat{\mathbb{D}}_1 \cap \widehat{\mathbb{D}}_i = \emptyset$. Let $\mathcal{A}$ be an (α,β)-useful private release mechanism for $POINT_d$. For all i, on input D_i mechanism $\mathcal{A}$ should pick an output from $\widehat{\mathbb{D}}_i$ with probability at least $1-\beta$. We get by the differential privacy of $\mathcal{A}$ that

$$\Pr[\mathcal{A}(D_1) \in \widehat{\mathbb{D}}_i] \geq \exp(-3\epsilon\alpha m)\Pr[\mathcal{A}(D_i) \in \widehat{\mathbb{D}}_i] \geq \exp(-3\epsilon\alpha m)\cdot(1-\beta).$$

Hence,

$$\begin{aligned}\Pr[\mathcal{A}(D_1) \notin \widehat{\mathbb{D}}_1] &\geq \Pr[\mathcal{A}(D_1) \in \bigcup_{i\neq 1}\widehat{\mathbb{D}}_i] \\ &= \sum_{i\neq 1}\Pr[\mathcal{A}(D_1) \in \widehat{\mathbb{D}}_i] \qquad (\text{sets } \widehat{\mathbb{D}}_i \text{ are disjoint}) \\ &\geq (T-1)\exp(-3\epsilon\alpha m)\cdot(1-\beta).\end{aligned}$$

On the other hand, since $\mathcal{A}$ is (α,β)-useful, $\Pr[\mathcal{A}(D_1) \notin \widehat{\mathbb{D}}_1] < \beta$, and hence we get that $m = \Omega((d + \log(1/\beta))/(\epsilon\alpha))$.

Theorem 7. *Every ϵ-differentially private non-interactive mechanism that is (α,β)-useful for $POINT_d$ requires an input database of $\Omega((d+\log(1/\beta))/(\epsilon\alpha))$ size.*

Acknowledgments

We thank Benny Applebaum, Eyal Kushilevitz, and Adam Smith for helpful initial discussions.

References

[1] Beimel, A., Kasiviswanathan, S., Nissim, K.: Bounds on the Sample Complexity for Private Learning and Private Data Release (Full version) (2009)

[2] Blum, A., Dwork, C., McSherry, F., Nissim, K.: Practical privacy: The SuLQ framework. In: PODS, pp. 128–138. ACM, New York (2005)

[3] Blum, A., Ligett, K., Roth, A.: A learning theory approach to non-interactive database privacy. In: STOC, pp. 609–618. ACM, New York (2008)

[4] Blum, A., Ligett, K., Roth, A.: Private communication (2008)

[5] Blumer, A., Ehrenfeucht, A., Haussler, D., Warmuth, M.K.: Learnability and the Vapnik-Chervonenkis dimension. Journal of the Association for Computing Machinery 36(4), 929–965 (1989)
[6] Dwork, C.: The differential privacy frontier (extended abstract). In: Reingold, O. (ed.) TCC 2009. LNCS, vol. 5444, pp. 496–502. Springer, Heidelberg (2009)
[7] Dwork, C., McSherry, F., Nissim, K., Smith, A.: Calibrating noise to sensitivity in private data analysis. In: Halevi, S., Rabin, T. (eds.) TCC 2006. LNCS, vol. 3876, pp. 265–284. Springer, Heidelberg (2006)
[8] Dwork, C., Naor, M., Reingold, O., Rothblum, G., Vadhan, S.: On the complexity of differentially private data release. In: STOC, pp. 381–390. ACM, New York (2009)
[9] Ehrenfeucht, A., Haussler, D., Kearns, M.J., Valiant, L.G.: A general lower bound on the number of examples needed for learning. Inf. Comput. 82(3), 247–261 (1989)
[10] Kasiviswanathan, S.P., Lee, H.K., Nissim, K., Raskhodnikova, S., Smith, A.: What can we learn privately? In: FOCS, pp. 531–540. IEEE Computer Society, Los Alamitos (2008)
[11] Kasiviswanathan, S.P., Smith, A.: A note on differential privacy: Defining resistance to arbitrary side information. CoRR, arXiv:0803.39461 [cs.CR] (2008)
[12] Kearns, M.J.: Efficient noise-tolerant learning from statistical queries. Journal of the ACM 45(6), 983–1006 (1998); Preliminary version in Proceedings of STOC 1993
[13] Kearns, M.J., Vazirani, U.V.: An Introduction to Computational Learning Theory. MIT Press, Cambridge (1994)
[14] McSherry, F., Talwar, K.: Mechanism design via differential privacy. In: FOCS, pp. 94–103. IEEE, Los Alamitos (2007)
[15] Mishra, N., Sandler, M.: Privacy via pseudorandom sketches. In: PODS, pp. 143–152. ACM, New York (2006)
[16] Pitt, L., Valiant, L.G.: Computational limitations on learning from examples. Journal of the ACM 35(4), 965–984 (1988)
[17] Valiant, L.G.: A theory of the learnable. Communications of the ACM 27, 1134–1142 (1984)
[18] Vapnik, V.N., Chervonenkis, A.Y.: On the uniform convergence of relative frequencies of events to their probabilities. Theory of Probability and its Applications 16, 264 (1971)

New Techniques for Dual System Encryption and Fully Secure HIBE with Short Ciphertexts

Allison Lewko* and Brent Waters**

University of Texas, Austin
alewko@cs.utexas.edu, bwaters@cs.utexas.edu

Abstract. We construct a fully secure HIBE scheme with short ciphertexts. The previous construction of Boneh, Boyen, and Goh was only proven to be secure in the selective model, under a non-static assumption which depended on the depth of the hierarchy. To obtain full security, we apply the dual system encryption concept recently introduced by Waters. A straightforward application of this technique is insufficient to achieve short ciphertexts, since the original instantiation of the technique includes tags that do not compress. To overcome this challenge, we design a new method for realizing dual system encryption. We provide a system in composite order groups (of three primes) and prove the security of our scheme under three static assumptions.

1 Introduction

An IBE system is a public key system where an encryptor uses only the identity of the recipient and a set of global public parameters, so a separate public key for each entity is not required. A trusted authority holds a master secret key which allows it to create secret keys for identities and distribute them to authenticated users. A Hierarchical IBE system (HIBE) [1, 2] provides more functionality by forming levels of an organizational hierarchy. A user at level k can delegate secret keys to descendant identities at lower levels, but cannot decrypt messages intended for a recipient that is not among its descendants. For example, a user with the identity "University of Texas: computer science department" can delegate a key for the identity "University of Texas: computer science department: grad student", but cannot delegate keys for identities that do not begin with "University of Texas : computer science department". A more formal definition of an HIBE system is given in Section 2.

Most previous HIBE constructions were proven secure in the selective model of security (where an attacker must declare the identity he intends to attacker before seeing the public parameters of the system), with two recent exceptions. Gentry and Halevi [3] employ the techniques of [4] to obtain full security, but at

* Supported by National Defense Science and Engineering Graduate Fellowship.
** Supported by NSF CNS-0716199, CNS-0915361 and Air Force Office of Scientific Research (AFO SR) under the MURI award for "Collaborative policies and assured information sharing" (Project PRESIDIO).

D. Micciancio (Ed.): TCC 2010, LNCS 5978, pp. 455–479, 2010.

the cost of a strong assumption (the BDHE-Set assumption) and ciphertext size growing linearly in the depth of the hierarchy. Waters [5] obtained full security with his new dual system encryption methodology from the well-established d-BDH and decisional Linear assumptions, but also had ciphertexts with size growing linearly in the depth of the hierarchy. This fell short of the constant size ciphertexts achieved by Boneh, Boyen, and Goh [6], but their HIBE system was only proven to be selectively secure in the standard model (or fully secure in the random oracle model).

In this paper, we resolve the question of whether full security and short ciphertexts (like [6]) can be simultaneously achieved in a HIBE system. A natural approach is to combine the Waters realization of dual system encryption with the Boneh-Boyen-Goh construction. This direct combination presents two problems:

1. Tags for each level that do not compress
2. Keys that are not fully rerandomized at delegation.

In the Boneh-Boyen-Goh system, group elements corresponding to each level of an identity are compressed (multiplied together) into a constant number of ciphertext elements. The tags in the Waters system do not allow this. These tags also prevent a key from being fully rerandomized upon delegation, meaning that an attacker can tell the difference between a delegated key and one freshly generated by the key generation algorithm. This requires a security definition that keeps track of such subtleties, which substantially complicates the security proof. Removing the tags from the Waters realization of dual system encryption is a nontrivial task because the tags were used to avoid a potential paradox in the dual system proof strategy.

1.1 Our Approach

We develop a new realization of dual system encryption that does not use tags. This provides several benefits:

1. compression of ciphertext is now possible
2. negligible correctness error caused by the tags is removed
3. schemes appear very natural and closely related to prior schemes.

Before giving the details of our approach, we first review the concept of dual system encryption.

Dual System Encryption. In a dual system, ciphertexts and keys can take on two forms: normal or semi-functional. Semi-functional ciphertexts and keys are not used in the real system, they are only used in the security proof. A normal key can decrypt normal or semi-functional ciphertexts, and a normal ciphertext can be decrypted by normal or semi-functional keys. However, when a semi-functional key is used to decrypt a semi-functional ciphertext, decryption will fail. More specifically, the semi-functional components of the key and ciphertext will interact to mask the blinding factor by an additional random term. Security for dual systems is proved using a sequence of games which are shown to be indistinguishable. The first game is the real security game (with normal ciphertext and keys). In the next game, the ciphertext is semi-functional, while all the keys are normal. For an attacker that makes q key requests, games 1 through q follow. In game k, the first k keys are semi-functional while the remaining keys are normal. In game q, all the keys and the challenge ciphertext given to the attacker are semi-functional. Hence none of the given keys are useful for decrypting the challenge ciphertext. At this point, proving security becomes relatively easy.

The Waters Realization. When arguing that games k and $k-1$ are indistinguishable, we create a simulator who can use any legal identities for the challenge ciphertext and keys. This creates a potential problem. The simulator is prepared to make a semi-functional ciphertext for an identity ID and is also prepared to make the kth key for identity ID, so it may seem like the simulator can determine whether key k is semi-functional for itself by test decrypting with a semi-functional ciphertext for the same identity. To resolve this paradox, the Waters IBE scheme associates random tag values with each ciphertext and key. Decryption works only when the tag values of the ciphertext and decrypting key are unequal. If the simulator attempted to test semi-functionality of key k for itself by creating a semi-functional ciphertext for the same identity, it would only be able to create one with an equal tag, and hence decryption would *unconditionally fail.* This correlation of tags is hidden from an attacker who cannot request a key with the same identity as the challenge ciphertext, so the tags look randomly distributed from the attacker's point of view.

Tags are used similarly in the Waters HIBE scheme, but here they cause two additional problems. First, there is a separate tag value associated with each level of the identity in a ciphertext or key. All these tag values must be given out in a ciphertext, so this forces ciphertext size to grow linearly with the depth of the hierarchy. Secondly, there is no method for rerandomizing the tags in key delegation. This means that a key at level $d+1$ which is delegated from a key at level d will share its first d tag values, a property which links the distribution of a key to its lineage. Some previous security definitions for HIBE [1, 2] which did not keep track of delegation paths of keys are hence invalid for such a system. Security must be argued under a more complete definition introduced in [7].

Our Realization. The additional complications of the proof and the linear ciphertext size are undesirable artifacts of building the HIBE system with the same tag techniques as the IBE system. To remove the tags, we must find a different way to resolve the paradox. Instead of having decryption unconditionally fail when the simulator attempts to test semi functionality of the kth key, we design our system so that decryption will *unconditionally succeed.* We introduce a variant of semi-functional keys which we call *nominally* semi-functional keys. These keys are semi-functional in name only, meaning that they are distributed like semi-functional keys, but are actually correlated with semi-functional ciphertexts so that when a nominally semi-functional key is used to decrypt a semi-functional ciphertext, the interaction of the two semi-functional components results in cancelation and decryption is successful. If the simulator attempts to answer its own question by creating the kth key and challenge ciphertext for the same identity, the created key will be nominally semi-functional and hence test decrypting will not distinguish this from a normal key. This nominally semi-functional key will appear to be distributed like a regular semi-functional key to the attacker, who cannot request a key that can decrypt the challenge ciphertext.

With this technique, we are able to construct a fully secure IBE system with short parameters without tags, and also give a fully secure HIBE system with constant-size ciphertexts. Our proofs rely on simple (constant-size) assumptions which do not depend on the number of queries the attacker makes. Our proof for our HIBE system is considerably simplified by the fact that our keys can be fully rerandomized upon delegation, avoiding the corresponding difficulties of the Waters HIBE proof.

In our the main body we provide a construction under a group of composite order N where N is the product of three primes. In Appendix C, we provide an analog of this for prime order groups. Our analog takes advantage of asymmetric bilinear groups where there is no efficient isomorphism between G_1 and G_2.

An interesting observation arising from our work is that the existing Boneh-Boyen IBE [8] and Boneh-Boyen-Goh HIBE [6] schemes which were only proven to be selectively secure can be transformed into fully secure systems by embedding them in composite order groups. Our IBE and HIBE systems are *remarkably* similar to these schemes.

1.2 Related Work

Identity Based Encryption was introduced by Shamir [9] and first realized by Boneh and Franklin [10] and Cocks [11]. The Boneh-Franklin IBE construction [10] proved security in the random oracle model. Subsequent constructions by Canetti, Halevi, and Katz [12] and Boneh and Boyen [8] were proved secure in the standard model, but under the weaker notion of selective security. Later, Boneh and Boyen [13] and Waters [14] gave constructions which were fully secure in the standard model. The Waters system was efficient and fully secure in the standard model under the decisional Bilinear Diffie-Hellman assumption (d-BDH), but it had public parameters consisting of $\mathcal{O}(\lambda)$ group elements for security parameter λ. Gentry [4] constructed an IBE system with short public parameters and proved full security in the standard model, but used an assumption (q-ABHDE) which is substantially more complicated than d-BDH and depends on the number of queries made by the attacker. Gentry, Peikert, and Vaikuntanathan also gave an IBE construction based on lattice assumptions [15].

Hierarchical Identity Based Encryption was introduced by Horwitz and Lynn [2] and then constructed by Gentry and Silverberg [1] in the random oracle model. Boneh and Boyen [8] achieved security in the selective model without random oracles. Boneh, Boyen, and Goh [6] then gave an HIBE with constant size ciphertexts, also in the selective model under a q-based assumption. These short ciphertexts were particularly useful for applications, including forward secure encryption [12] and converting the NNL broadcast encryption system [16] into a public-key system [17]. Gentry and Halevi [3] constructed the first fully secure HIBE for polynomial depth, though also under a complex assumption. Waters [5] attained full security under the d-BDH and decisional Linear assumptions, but with ciphertext size growing linearly in the hierarchy depth. We note that Waters first instantiated this result in composite order groups. The complete definition of security for HIBE that we use in this paper was formulated by Shi and Waters [7].

1.3 Organization

In Section 2, we formally define an HIBE system and give the complete security definition, give background on bilinear groups, and state our assumptions. In Section 3, we present our IBE scheme and prove its security. In Section 4, we give our HIBE scheme and prove its security. In Section 6, we conclude and discuss open directions for further research.

2 Background

2.1 Hierarchical Identity Based Encryption

A Hierarchical Identity Based Encryption scheme has five algorithms: Setup, Encrypt, KeyGen, Decrypt, and Delegate.

Setup(λ) $\rightarrow PK, MSK$. The setup algorithm takes a security parameter λ as input and outputs the public parameters PK and a master secret key MSK.

KenGen($MSK, \boldsymbol{I}$) $\rightarrow SK_{\boldsymbol{I}}$. The key generation algorithm takes the master secret key and an identity vector $\boldsymbol{I}$ as input and outputs a private key $SK_{\boldsymbol{I}}$.

Delegate($PK, SK_{\boldsymbol{I}}, I$) $\rightarrow SK_{\boldsymbol{I}:I}$. The delegation algorithm takes a secret key for the identity vector $\boldsymbol{I}$ of depth d and an identity I as input and outputs a secret key for the depth $d+1$ identity vector $\boldsymbol{I} : I$ formed by concatenating I onto the end of $\boldsymbol{I}$.

Encrypt($PK, M, \boldsymbol{I}$) $\rightarrow CT$. The encryption algorithm takes the public parameters PK, a message M, and an identity vector $\boldsymbol{I}$ as input and outputs a ciphertext CT.

Decrypt(PK, CT, SK) $\rightarrow M$. The decryption algorithm takes the public parameters PK, a ciphertext CT, and a secret key SK as input and outputs the message M, if the ciphertext was an encryption to an identity vector $\boldsymbol{I}$ and the secret key is for the same identity vector.

Notice that the decryption algorithm is only required to work when the identity vector for the ciphertext matches the secret key exactly. However, someone who has a secret key for a prefix of this identity vector can delegate to themselves the required secret key and also decrypt.

Security definition. We give the complete form of the security definition [7] which keeps track of how keys are generated and delegated. Security is defined through the following game, played by a challenger and an attacker.

Setup. The challenger runs the Setup algorithm to generate public parameters PK which it gives to the adversary. We let S denote the set of private keys that the challenger has created but not yet given to the adversary. At this point, $S = \emptyset$.

Phase 1. The adversary makes Create, Delegate, and Reveal key queries. To make a Create query, the attacker specifies an identity vector $\boldsymbol{I}$. In response, the challenger creates a key for this vector by calling the key generation algorithm, and places this key in the set S. It only gives the attacker a reference to this key, not the key itself. To make a Delegate query, the attacker specifies a key $SK_{\boldsymbol{I}}$ in the set S and specifies an identity I'. In response, the challenger appends I' to $\boldsymbol{I}$ and makes a key for this new identity by running the delegation algorithm on $SK_{\boldsymbol{I}}$ and I'. It adds this key to the set S and again gives the attacker only a reference to it, not the actual key. To make a Reveal query, the attacker specifies an element of the set S. The challenger gives this key to the attacker and removes it from the set S. We note that the attacker need no longer make any delegation queries for this key because it can run the delegation algorithm on the revealed key for itself.

Challenge. The adversary gives the challenger two messages M_0 and M_1 and a challenge identity vector $\boldsymbol{I}^*$. This identity vector must satisfy the property that no revealed identity in Phase 1 was a prefix of it. The challenger sets $\beta \in \{0,1\}$ randomly, and encrypts M_β under $\boldsymbol{I}^*$. It sends the ciphertext to the adversary.

Phase 2. This is the same as Phase 1, with the added restriction that any revealed identity vector must not be a prefix of $\boldsymbol{I}^*$.

Guess. The adversary must output a guess β' for β.

The advantage of an adversary $\mathcal{A}$ is defined to be $Pr[\beta' = \beta] - \frac{1}{2}$.

Definition 1. *A Hierarchical Identity Based Encryption scheme is secure if all polynomial time adversaries achieve at most a negligible advantage in the security game.*

2.2 Composite Order Bilinear Groups

Composite order bilinear groups were first introduced in [18]. We define them by using a group generator $\mathcal{G}$, an algorithm which takes a security parameter λ as input and outputs a description of a bilinear group G. In our case, $\mathcal{G}$ outputs $(N = p_1p_2p_3, G, G_T, e)$ where p_1, p_2, p_3 are distinct primes, G and G_T are cyclic groups of order $N = p_1p_2p_3$, and $e : G^2 \to G_T$ is a map such that:

1. (Bilinear) $\forall g, h \in G, a, b \in \mathbb{Z}_N, e(g^a, h^b) = e(g,h)^{ab}$
2. (Non-degenerate) $\exists g \in G$ such that $e(g,g)$ has order N in G_T.

We further require that the group operations in G and G_T as well as the bilinear map e are computable in polynomial time with respect to λ. Also, we assume the group descriptions of G and G_T include generators of the respective cyclic groups. We let G_{p_1}, G_{p_2}, and G_{p_3} denote the subgroups of order p_1, p_2 and p_3 in G respectively. We note that when $h_i \in G_{p_i}$ and $h_j \in G_{p_j}$ for $i \neq j$, $e(h_i, h_j)$ is the identity element in G_T. To see this, suppose $h_1 \in G_{p_1}$ and $h_2 \in G_{p_2}$. We let g denote a generator of G. Then, $g^{p_1 p_2}$ generates G_{p_3}, $g^{p_1 p_3}$ generates G_{p_2}, and $g^{p_2 p_3}$ generates G_{p_1}. Hence, for some α_1, α_2, $h_1 = (g^{p_2 p_3})^{\alpha_1}$ and $h_2 = (g^{p_1 p_3})^{\alpha_2}$. We note:

$$e(h_1, h_2) = e(g^{p_2 p_3 \alpha_1}, g^{p_1 p_3 \alpha_2}) = e(g^{\alpha_1}, g^{p_3 \alpha_2})^{p_1 p_2 p_3} = 1.$$

This orthogonality property of $G_{p_1}, G_{p_2}, G_{p_3}$ will be a principal tool in our constructions.

We now give our complexity assumptions. These same assumptions will be used to prove the security of our IBE and HIBE systems. We note that they are static (not dependent on the depth of the hierarchy or the number of queries made by an attacker). The first assumption is just the subgroup decision problem in the case where the group order is a product of 3 primes. In Appendix A, we show that these assumptions hold in the generic group model if finding a nontrivial factor of the group order is hard. We prove this by applying the theorems of Katz, Sahai, and Waters [19]. Their work also used composite order bilinear groups and provided a general framework for proving generic security of assumptions in this setting.

In the assumptions below, we let $G_{p_1 p_2}$, e.g., denote the subgroup of order $p_1 p_2$ in G.

Assumption 1 (Subgroup decision problem for 3 primes). Given a group generator $\mathcal{G}$, we define the following distribution:

$$\mathbb{G} = (N = p_1 p_2 p_3, G, G_T, e) \xleftarrow{R} \mathcal{G},$$

$$g \xleftarrow{R} G_{p_1}, \; X_3 \xleftarrow{R} G_{p_3},$$

$$D = (\mathbb{G}, g, X_3),$$

$$T_1 \xleftarrow{R} G_{p_1 p_2}, \; T_2 \xleftarrow{R} G_{p_1}.$$

We define the advantage of an algorithm $\mathcal{A}$ in breaking Assumption 1 to be:

$$Adv1_{\mathcal{G},\mathcal{A}}(\lambda) := \left| Pr[\mathcal{A}(D, T_1) = 1] - Pr[\mathcal{A}(D, T_2) = 1] \right|.$$

We note that T_1 can be written (uniquely) as the product of an element of G_{p_1} and an element of G_{p_2}. We refer to these elements as the "G_{p_1} part of T_1" and the "G_{p_2} part of T_1" respectively. We will use this terminology in our proofs.

Definition 2. *We say that $\mathcal{G}$ satisfies Assumption 1 if $Adv1_{\mathcal{G},\mathcal{A}}(\lambda)$ is a negligible function of λ for any polynomial time algorithm $\mathcal{A}$.*

Assumption 2. Given a group generator $\mathcal{G}$, we define the following distribution:

$$\mathbb{G} = (N = p_1p_2p_3, G, G_T, e) \xleftarrow{R} \mathcal{G},$$

$$g, X_1 \xleftarrow{R} G_{p_1}, \; X_2, Y_2 \xleftarrow{R} G_{p_2}, \; X_3, Y_3 \xleftarrow{R} G_{p_3},$$

$$D = (\mathbb{G}, g, X_1X_2, X_3, Y_2Y_3),$$

$$T_1 \xleftarrow{R} G, \; T_2 \xleftarrow{R} G_{p_1p_3}.$$

We define the advantage of an algorithm $\mathcal{A}$ in breaking Assumption 2 to be:

$$Adv2_{\mathcal{G},\mathcal{A}}(\lambda) := \big|Pr[\mathcal{A}(D, T_1) = 1] - Pr[\mathcal{A}(D, T_2) = 1]\big|.$$

We use $G_{p_1p_3}$ to denote the subgroup of order p_1p_3 in G. We note that T_1 can be (uniquely) written as the product of an element of G_{p_1}, an element of G_{p_2}, and an element of G_{p_3}. We refer to these as the "G_{p_1} part of T_1", the "G_{p_2} part of T_1", and the "G_{p_3} part of T_1", respectively. T_2 can similarly be written as the product of an element of G_{p_1} and an element of G_{p_3}.

Definition 3. *We say that $\mathcal{G}$ satisfies Assumption 2 if $Adv2_{\mathcal{G},\mathcal{A}}(\lambda)$ is a negligible function of λ for any polynomial time algorithm $\mathcal{A}$.*

Assumption 3. Given a group generator $\mathcal{G}$, we define the following distribution:

$$\mathbb{G} = (N = p_1p_2p_3, G, G_T, e) \xleftarrow{R} \mathcal{G}, \alpha, s \xleftarrow{R} \mathbb{Z}_N,$$

$$g \xleftarrow{R} G_{p_1}, X_2, Y_2, Z_2 \xleftarrow{R} G_{p_2}, X_3 \xleftarrow{R} G_{p_3},$$

$$D = (\mathbb{G}, g, g^{\alpha}X_2, X_3, g^sY_2, Z_2),$$

$$T_1 = e(g, g)^{\alpha s}, T_2 \xleftarrow{R} G_T.$$

We define the advantage of an algorithm $\mathcal{A}$ in breaking Assumption 3 to be:

$$Adv3_{\mathcal{G},\mathcal{A}}(\lambda) := \big|Pr[\mathcal{A}(D, T_1) = 1] - Pr[\mathcal{A}(D, T_2) = 1]\big|.$$

Definition 4. *We say that $\mathcal{G}$ satisfies Assumption 3 if $Adv3_{\mathcal{G},\mathcal{A}}(\lambda)$ is a negligible function of λ for any polynomial time algorithm $\mathcal{A}$.*

3 Our IBE System

We begin by giving our new dual system encryption realization of IBE. Our construction will use composite order groups of order $N = p_1p_2p_3$ and identities in $\mathbb{Z}_N$. Remarkably, our construction looks almost exactly like the Boneh-Boyen IBE with keys additionally randomized in the subgroup G_{p_3}. This resemblance to preexisting selectively secure schemes will continue in our HIBE system as well. We regard this as a desirable feature of our approach.

We note that the subgroup G_{p_2} is not used in our actual scheme, instead it serves as our semi-functional space. Keys and ciphertexts will be semi-functional when they include terms in G_{p_2} and decryption will proceed by pairing key elements with ciphertext elements. This will give us the decryption functionality we need: when we pair a normal key with a semi-functional ciphertext or a normal ciphertext with a semi-functional key, the terms in G_{p_2} are orthogonal to terms in G_{p_1} and G_{p_3} under the pairing and will cancel out. When we pair a semi-functional key with a semi-functional ciphertext, we will get an additional term arising from the pairing of the terms in G_{p_2}.

3.1 Construction

Setup. The setup algorithm chooses a bilinear group G of order $N = p_1p_2p_3$ (where p_1, p_2, and p_3 are distinct primes). We let G_{p_i} denote the subgroup of order p_i in G. It then chooses $u, g, h \in G_{p_1}$ and $\alpha \in \mathbb{Z}_N$. The public parameters are published as:

$$PK = \{N, u, g, h, e(g, g)^{\alpha}\}.$$

The secret parameters are α and a generator of G_{p_3}.

Encrypt(M, ID). The encryption algorithm chooses $s \in \mathbb{Z}_N$ randomly and creates the ciphertext as:

$$C_0 = Me(g, g)^{\alpha s}, C_1 = (u^{ID}h)^s, C_2 = g^s.$$

KeyGen(ID, MSK). The key generation algorithm chooses $r \in \mathbb{Z}_N$ and R_3, $R_3' \in G_{p_3}$ randomly. (Random elements of G_{p_3} can be obtained by taking a generator of G_{p_3} and raising it to random exponents modulo N.) The key is formed as:

$$K_1 = g^r R_3, K_2 = g^{\alpha}(u^{ID}h)^r R_3'.$$

Decryption. If the ID's of the ciphertext and key are equal, the decryption algorithm computes the blinding factor as:

$$\frac{e(K_2, C_2)}{e(K_1, C_1)} = \frac{e(g, g)^{\alpha s} e(u^{ID}h, g)^{rs}}{e(u^{ID}h, g)^{rs}}.$$

3.2 Security

To prove security of our IBE system, we first define two additional structures: semi-functional keys and semi-functional ciphertexts. These will not be used in the real system, but they will be used in our proof.

Semi-functional Ciphertext. We let g_2 denote a generator of the subgroup G_{p_2}. A semi-functional ciphertext is created as follows: first, a normal ciphertext C_0', C_1', C_2' is generated by the encryption algorithm. Random exponents $x, z_c \in \mathbb{Z}_N$ are chosen. Then, C_0 is set to be C_0', C_1 is set to be $C_1' g_2^{x z_c}$, and C_2 is set to be $C_2' g_2^x$.

Semi-functional Key. A semi-functional key is created as follows: first, a normal key K_1', K_2' is generated by the key generation algorithm. Random exponents $\gamma, z_k \in \mathbb{Z}_N$ are chosen. K_1 is set to be $K_1' g_2^{\gamma}$ and K_2 is set to be $K_2' g_2^{\gamma z_k}$.

Notice that if a semi-functional key is used to decrypt a semi-functional ciphertext, the blinding factor will be obscured by an additional factor of $e(g_2, g_2)^{x\gamma(z_k - z_c)}$. If $z_c = z_k$, decryption will still work. In this case, we say that the key is *nominally* semi-functional: it has terms in G_{p_2}, but these do not hinder decryption.

Our proof of security relies on Assumptions 1, 2, 3 defined in Section 2. We will prove security by a hybrid argument using a sequence of games. The first game, Game_{Real}, will be the real security game. The next game, $\text{Game}_{Restricted}$, will be like the real security game except that the attacker cannot ask for keys for identities which are equal to the challenge identity modulo p_2. This is a stronger restriction than the real security game, where the identities must be unequal modulo N. We will retain this stronger restriction throughout the subsequent games. The reason for it will be explained in the proof. We let q denote the number of key queries the attacker makes. For k from 0 to q, we define Game_k as:

$Game_k$. This is like the restricted security game, except that the ciphertext given to the attacker is semi-functional and the first k keys are semi-functional. The rest of the keys are normal.

In Game_0, all the keys are normal and the ciphertext is semi-functional. In Game_q, the ciphertext and all of the keys are semi-functional. Our last game is Game_{Final}, which is the same as Game_q except that the ciphertext is a semi-functional encryption of a random message, not one of the two messages requested by the attacker. We will prove that each of these games is indistinguishable in the following four lemmas.

Lemma 1. *Suppose there exists an algorithm $\mathcal{A}$ such that $Game_{Real}Adv_{\mathcal{A}} - Game_{Restricted}Adv_{\mathcal{A}} = \epsilon$. Then we can build an algorithm $\mathcal{B}$ with advantage $\geq \frac{\epsilon}{2}$ in breaking either Assumption 1 or Assumption 2.*

Proof. Given g, X_3, $\mathcal{B}$ can simulate Game_{Real} with $\mathcal{A}$. With probability ϵ, $\mathcal{A}$ produces identities ID and ID^* such that $ID \neq ID^*$ modulo N and p_2 divides $ID - ID^*$. $\mathcal{B}$ uses these identities to produce a nontrivial factor of N by computing $a = \gcd(ID - ID^*, N)$. We set $b = \frac{N}{a}$. We note that p_2 divides a and $N = ab = p_1 p_2 p_3$. We consider two cases:

1. p_1 divides b
2. $a = p_1 p_2$ and $b = p_3$.

At least one of these cases must occur with probability $\geq \frac{\epsilon}{2}$. In case 1, $\mathcal{B}$ will break Assumption 1. Given g, X_3, T, $\mathcal{B}$ can determine that p_1 divides b by verifying that g^b is the identity and will then test whether T^b is the identity. If it is, then $T \in G_{p_1}$. If it is not, $T \in G_{p_1 p_2}$.

In case 2, $\mathcal{B}$ will break Assumption 2. Given $g, X_1 X_2, X_3, Y_2 Y_3$, $\mathcal{B}$ can determine that $a = p_1 p_2$ by verifying that $(X_1 X_2)^a$ is the identity and will then test whether $e((Y_2Y_3)^b, T)$ is the identity. If it is, then $T \in G_{p_1p_3}$. If it is not, then $T \in G$.

Lemma 2. *Suppose there exists an algorithm $\mathcal{A}$ such that $Game_{Restricted}Adv_{\mathcal{A}} - Game_0Adv_{\mathcal{A}} = \epsilon$. Then we can build an algorithm $\mathcal{B}$ with advantage ϵ in breaking Assumption 1.*

Proof. $\mathcal{B}$ first receives g, X_3, T. It simulates $\text{Game}_{Restricted}$ or Game_0 with $\mathcal{A}$. It sets the public parameters as follows. It chooses random exponents $\alpha, a, b \in \mathbb{Z}_N$ and sets $g = g, u = g^a$, $h = g^b$. It sends these public parameters $\{N, u, g, h, e(g,g)^\alpha\}$ to $\mathcal{A}$. Each time $\mathcal{B}$ is asked to provide a key for an identity ID_i, it chooses random exponents r_i, t_i, and $w_i \in \mathbb{Z}_N$ and sets:

$$K_1 = g^{r_i} X_3^{t_i}, K_2 = g^\alpha (u^{ID_i} h)^{r_i} X_3^{w_i}.$$

$\mathcal{A}$ sends $\mathcal{B}$ two messages, M_0 and M_1, and a challenge identity, ID. $\mathcal{B}$ chooses $\beta \in \{0,1\}$ randomly. The ciphertext is formed as follows:

$$C_0 = M_\beta e(T, g)^\alpha, C_1 = T^{aID+b}, C_2 = T.$$

(This implicitly sets g^s equal to the G_{p_1} part of T.) If $T \in G_{p_1p_2}$, then this is a semi-functional ciphertext with $z_c = aID + b$. We note that the value of z_c modulo p_2 is not correlated with the values of a and b modulo p_1, so this is properly distributed. If $T \in G_{p_1}$, this is a normal ciphertext. Hence, $\mathcal{B}$ can use the output of $\mathcal{A}$ to distinguish between these possibilities for T.

Lemma 3. *Suppose there exists an algorithm $\mathcal{A}$ such that $Game_{k-1}Adv_{\mathcal{A}} - Game_kAdv_{\mathcal{A}} = \epsilon$. Then we can build an algorithm $\mathcal{B}$ with advantage ϵ in breaking Assumption 2.*

Proof. $\mathcal{B}$ first receives $g, X_1X_2, X_3, Y_2Y_3, T$. $\mathcal{B}$ picks random exponents $a, b, \alpha \in \mathbb{Z}_N$ and sets the public parameters as: $g = g, u = g^a, h = g^b, e(g,g)^\alpha$. It sends these to $\mathcal{A}$. When $\mathcal{A}$ requests the ith key for ID_i when $i < k$, $\mathcal{B}$ creates a semi-functional key. It does this by choosing random exponents r_i, z_i, $t_i \in \mathbb{Z}_N$ and setting:

$$K_1 = g^{r_i} (Y_2Y_3)^{t_i}, K_2 = g^\alpha (u^{ID_i} h)^{r_i} (Y_2Y_3)^{z_i}.$$

This is a properly distributed semi-functional key with $g_2^\gamma = Y_2^{t_i}$. (We note that the values of t_i and z_i modulo p_2 and modulo p_3 are uncorrelated by the Chinese Remainder Theorem.)

For $i > k$, $\mathcal{B}$ generates normal keys by using random exponents $r_i, t_i, w_i \in \mathbb{Z}_N$ and setting:

$$K_1 = g^{r_i} X_3^{t_i}, K_2 = g^\alpha (u^{ID_i} h)^{r_i} X_3^{w_i}.$$

To create the kth requested key, $\mathcal{B}$ lets $z_k = aID_k + b$, chooses a random exponent $w_k \in \mathbb{Z}_N$, and sets:

$$K_1 = T, K_2 = g^\alpha T^{z_k} X_3^{w_k}.$$

At some point, $\mathcal{A}$ sends $\mathcal{B}$ two messages, M_0 and M_1, and a challenge identity, ID. $\mathcal{B}$ sets $\beta \in \{0,1\}$ randomly. The challenge ciphertext is formed as:

$$C_0 = M_\beta e(X_1X_2, g)^\alpha, C_1 = (X_1X_2)^{aID+b}, C_2 = X_1X_2.$$

We note that this sets $g^s = X_1$ and $z_c = aID + b$. Since $f(ID) = aID + b$ is a pairwise independent function modulo p_2, as long as $ID_k \neq ID (\bmod p_2)$, z_k and z_c will seem randomly distributed to $\mathcal{A}$ (again, we note that the values of a and b modulo p_2 are uncorrelated with their values modulo p_1). If $ID_k \equiv ID (\bmod p_2)$, then $\mathcal{A}$ has made an invalid key request. This is where we use our additional modular restriction.

Though it is hidden from $\mathcal{A}$, this relationship between z_c and z_k is crucial: if $\mathcal{B}$ attempts to test itself whether key k is semi-functional by creating a semi-functional ciphertext for ID_k and trying to decrypt, then decryption will work whether key k is semi-functional or not, because $z_c = z_k$. In other words, the simulator $\mathcal{B}$ can only make a nominally semi-functional key k.

If $T \in G_{p_1p_3}$, then $\mathcal{B}$ has properly simulated Game$_{k-1}$. If $T \in G$, then $\mathcal{B}$ has properly simulated Game$_k$. Hence, $\mathcal{B}$ can use the output of $\mathcal{A}$ to distinguish between these possibilities for T.

Lemma 4. *Suppose there exists an algorithm $\mathcal{A}$ such that $Game_q Adv_\mathcal{A} - Game_{Final} Adv_\mathcal{A} = \epsilon$. Then we can build an algorithm $\mathcal{B}$ with advantage ϵ in breaking Assumption 3.*

Proof. $\mathcal{B}$ first receives $g, g^\alpha X_2, X_3, g^s Y_2, Z_2, T$. $\mathcal{B}$ chooses random exponents $a, b \in \mathbb{Z}_N$ and sets the public parameters as $g = g, u = g^a, h = g^b, e(g,g)^\alpha = e(g^\alpha X_2, g)$. It sends these to $\mathcal{A}$. When $\mathcal{A}$ requests a key for identity ID_i, $\mathcal{B}$ generates a semi-functional key. It does this by choosing random exponents $c_i, r_i, t_i, w_i, \gamma_i \in \mathbb{Z}_N$ and setting:

$$K_1 = g^{r_i} Z_2^{\gamma_i} X_3^{t_i}, K_2 = g^\alpha X_2 (u^{ID_i} h)^{r_i} Z_2^{c_i} X_3^{w_i}.$$

$\mathcal{A}$ sends $\mathcal{B}$ two messages, M_0 and M_1, and a challenge identity, ID. $\mathcal{B}$ sets $\beta \in \{0,1\}$ randomly. It forms the challenge ciphertext as:

$$C_0 = M_\beta T, C_1 = (g^s Y_2)^{aID+b}, C_2 = g^s Y_2.$$

This sets $z_c = aID + b$. We note that the value of z_c only matters modulo p_2, whereas $u = g^a$ and $h = g^b$ are elements of G_{p_1}, so when a and b are chosen randomly modulo N, there is no correlation between the values of a and b modulo p_1 and the value $z_c = aID + b$ modulo p_2.

If $T = e(g,g)^{\alpha s}$, then this is a properly distributed semi-functional ciphertext with message M_β. If T is a random element of G_T, then this is a semi-functional ciphertext with a random message. Hence, $\mathcal{B}$ can use the output of $\mathcal{A}$ to distinguish between these possibilities for T.

We have now proven the following theorem:

Theorem 1. *If Assumptions 1, 2, and 3 hold, then our IBE system is secure.*

Proof. If Assumptions 1, 2, and 3 hold, then we have shown by the previous lemmas that the real security game is indistinguishable from Game_{Final}, in which the value of β is information-theoretically hidden from the attacker. Hence the attacker can attain no advantage in breaking the IBE system.

4 Our HIBE System

We build upon our IBE system and extend our techniques to give an HIBE system with short ciphertexts. The absence of tags allows us to compress the ciphertext into a constant number of group elements and also to rerandomize keys fully upon delegation. This dramatically simplifies our proof of security. Our construction again uses composite order groups of order $N = p_1p_2p_3$, and looks almost exactly like the Boneh-Boyen-Goh HIBE system with keys additionally randomized in subgroup G_{p_3}. G_{p_2} will be our semi-functional space, which is not used in the real system.

4.1 Construction

Setup. The setup algorithm chooses a bilinear group G or order $N = p_1p_2p_3$. We let ℓ denote the maximum depth of the HIBE. The setup algorithm chooses $g, h, u_1, \dots, u_\ell \in G_{p_1}$, $X_3 \in G_{p_3}$, and $\alpha \in \mathbb{Z}_N$. The public parameters are published as:

$$PK = \{N, g, h, u_1, \dots, u_\ell, X_3, e(g,g)^\alpha\}.$$

The secret parameter is α.

Encrypt(M, $(ID_1, \dots, ID_j)$). The encryption algorithm chooses $s \in Z_N$ randomly. It sets:

$$C_0 = Me(g,g)^{\alpha s}, C_1 = \left(u_1^{ID_1} \cdots u_j^{ID_j} h\right)^s, C_2 = g^s.$$

KeyGen(MSK, $(ID_1, \dots, ID_j)$). The key generation algorithm chooses $r \in \mathbb{Z}_N$ randomly and also chooses random elements $R_3, R_3', R_{j+1}, \dots, R_\ell$ of G_{p_3}. It sets:

$$K_1 = g^r R_3, K_2 = g^\alpha \left(u_1^{ID_1} \cdots u_j^{ID_j} h\right)^r R_3', E_{j+1} = u_{j+1}^r R_{j+1}, \dots, E_\ell = u_\ell^r R_\ell.$$

Delegate. Given a key $K_1', K_2', E_{j+1}', \dots, E_\ell'$ for $(ID_1, \dots, ID_j)$, the delegation algorithm creates a key for $(ID_1, \dots, ID_{j+1})$ as follows. It chooses a random $r' \in \mathbb{Z}_N$ and random elements of G_{p_3} denoted, e.g., by $\tilde{R}_3$. The new key is set as:

$$K_1 = K_1' g^{r'} \tilde{R}_3,$$

$$K_2 = K_2' \left(u_1^{ID_1} \cdots u_j^{ID_j} h\right)^{r'} (E_{j+1}')^{ID_{j+1}} u_{j+1}^{r' ID_{j+1}} \tilde{R}_3',$$

$$E_{j+2} = E_{j+2}' u_{j+2}^{r'} \tilde{R}_{j+2}, \dots, E_\ell = E_\ell' u_\ell^{r'} \tilde{R}_\ell.$$

We note that this new key is fully rerandomized: its only tie to the previous key is in the values $ID_1, \dots, ID_j$.

Decrypt. The decryption algorithm assumes that the key and ciphertext both correspond to the same identity $(ID_1, \ldots, ID_j)$. If the key identity is a prefix of this instead, then the decryption algorithm starts by running the key delegation algorithm to create a key with identity matching the ciphertext identity exactly. The decryption algorithm then computes the blinding factor as:

$$\frac{e(K_2, C_2)}{e(K_1, C_1)} = \frac{e(g,g)^{\alpha s} e(u_1^{ID_1} \cdots u_j^{ID_j} h, g)^{rs}}{e(g, u_1^{ID_1} \cdots u_j^{ID_j} h)^{rs}} = e(g,g)^{\alpha s}.$$

4.2 Security

To prove security of our HIBE system, we again rely on the static Assumptions 1, 2, and 3. We first define two additional structures: semi-functional ciphertexts and semi-functional keys. These will not be used in the real system, but will be used in our proof.

Semi-functional Ciphertext. We let g_2 denote a generator of G_{p_2}. A semi-functional ciphertext is created as follows: first, we use the encryption algorithm to form a normal ciphertext C'_0, C'_1, C'_2. We choose random exponents $x, z_c \in \mathbb{Z}_N$. We set:

$$C_0 = C'_0, C_1 = C'_1 g_2^{x z_c}, C_2 = C'_2 g_2^{x}.$$

Semi-functional Keys. To create a semi-functional key, we first create a normal key $K'_1, K'_2, E'_{j+1}, \ldots, E'_\ell$ using the key generation algorithm. We choose random exponents $\gamma, z_k, z_{j+1}, \ldots, z_\ell \in \mathbb{Z}_N$. We set:

$$K_1 = K'_1 g_2^{\gamma}, K_2 = K'_2 g_2^{\gamma z_k}, E_{j+1} = E'_{j+1} g_2^{\gamma z_{j+1}}, \ldots, E_\ell = E'_\ell g_2^{\gamma z_\ell}.$$

We note that when a semi-functional key is used to decrypt a semi-functional ciphertext, the decryption algorithm will compute the blinding factor multiplied by the additional term $e(g_2, g_2)^{x\gamma(z_k - z_c)}$. If $z_c = z_k$, decryption will still work. In this case, the key is nominally semi-functional.

Our proof of security will again be structured as a hybrid argument over a sequence of games. The first game, Game_{Real}, is the real HIBE security game. The next game, $\text{Game}_{Real'}$, is the same as the real game except that all key queries will be answered by fresh calls to the key generation algorithm (the challenger will not be asked to delegate keys in a particular way). The next game, $\text{Game}_{Restricted}$ is the same as $\text{Game}_{Real'}$ except that the attacker cannot ask for keys for identities which are prefixes of the challenge identity modulo p_2. We will retain this restriction in all subsequent games. We let q denote the number of key queries the attacker makes. For k from 0 to q, we define Game_k as:

Game_k. This is like $\text{Game}_{Restricted}$, except that the ciphertext given to the attacker is semi-functional and the first k keys are semi-functional. The rest of the keys are normal.

In Game_0, only the challenge ciphertext is semi-functional. In Game_q, the challenge ciphertext and all of the keys are semi-functional. We define Game_{Final} to be like Game_q, except that the challenge ciphertext is a semi-functional encryption of a random message, not one of the messages provided by the attacker. We will show these games are indistinguishable in five lemmas. The proofs are very similar to the proofs for our IBE system, and can be found in Appendix B.

5 Moving to Prime Order Groups

In Appendix C we show an analog of our previous construction in prime order groups. The prime order group construction we give takes advantage of asymmetric groups where there is a pairing function $e : G_1 \times G_2 \rightarrow G_T$, but there is not believed to be an efficient isomorphism from either G_1 to G_2 or G_2 to G_1.

Our prime construction can be viewed as an analog of the composite order one where we "emulate" the three subgroups with multiple group elements to create three subspaces. Our "emulation" technique uses some ideas from the Waters [5] prime order group realization; however, we are able to "squeeze" things down by using asymmetric groups.

A potential future direction is be to realize our methods in prime order groups without relying on the lack of isomorphism for security. A natural approach would be to use an "unsqueezed" version of our techniques. It is possible that this approach might give a reduction with more cancelations that in turn provides security from even simpler assumptions.

6 Conclusions and Open Directions

We have given the first HIBE system with constant size ciphertext that is fully secure in the standard model from simple assumptions. In doing so, we discovered that instantiations of the selectively secure Boneh-Boyen IBE and Boneh-Boyen-Goh HIBE schemes in composite order bilinear groups can be proved to be fully secure using the dual encryption technique of Waters. We overcame the initial challenges introduced by the use of tags in the original Waters IBE and HIBE systems by introducing the concept of nominally semi-functional keys. Our work further demonstrates the power and versatility of the dual system encryption technique, which we believe will have many future applications.

We leave it as an open problem to transfer our IBE and HIBE systems into prime order groups with security proven from standard assumptions such as the decisional Linear assumption and d-BDH. This kind of translation was previously achieved by Waters [5] for his IBE and HIBE systems, which were originally constructed in composite order groups.

References

1. Gentry, C., Silverberg, A.: Hierarchical id-based cryptography. In: Zheng, Y. (ed.) ASIACRYPT 2002. LNCS, vol. 2501, pp. 548–566. Springer, Heidelberg (2002)
2. Horwitz, J., Lynn, B.: Toward hierarchical identity-based encryption. In: Knudsen, L.R. (ed.) EUROCRYPT 2002. LNCS, vol. 2332, pp. 466–481. Springer, Heidelberg (2002)

3. Gentry, C., Halevi, S.: Hierarchical identity based encryption with polynomially many levels. In: Reingold, O. (ed.) TCC 2009. LNCS, vol. 5444, pp. 437–456. Springer, Heidelberg (2009)
4. Gentry, C.: Practical identity-based encryption without random oracles. In: Vaudenay, S. (ed.) EUROCRYPT 2006. LNCS, vol. 4004, pp. 445–464. Springer, Heidelberg (2006)
5. Waters, B.: Dual system encryption: realizing fully secure ibe and hibe under simple assumptions. In: Halevi, S. (ed.) CRYPTO 2009. LNCS, vol. 5677, pp. 619–636. Springer, Heidelberg (2009)
6. Boneh, D., Boyen, X., Goh, E.: Hierarchical identity based encryption with constant size ciphertext. In: Cramer, R. (ed.) EUROCRYPT 2005. LNCS, vol. 3494, pp. 440–456. Springer, Heidelberg (2005)
7. Shi, E., Waters, B.: Delegating capabilities in predicate encryption systems. In: Aceto, L., Damgård, I., Goldberg, L.A., Halldórsson, M.M., Ingólfsdóttir, A., Walukiewicz, I. (eds.) ICALP 2008, Part II. LNCS, vol. 5126, pp. 560–578. Springer, Heidelberg (2008)
8. Boneh, D., Boyen, X.: Efficient selective-ID secure identity-based encryption without random oracles. In: Cachin, C., Camenisch, J.L. (eds.) EUROCRYPT 2004. LNCS, vol. 3027, pp. 223–238. Springer, Heidelberg (2004)
9. Shamir, A.: Identity-based cryptosystems and signature schemes. In: Blakely, G.R., Chaum, D. (eds.) CRYPTO 1984. LNCS, vol. 196, pp. 47–53. Springer, Heidelberg (1985)
10. Boneh, D., Franklin, M.: Identity based encryption from the weil pairing. In: Kilian, J. (ed.) CRYPTO 2001. LNCS, vol. 2139, pp. 213–229. Springer, Heidelberg (2001)
11. Cocks, C.: An identity based encryption scheme based on quadratic residues. In: Honary, B. (ed.) Cryptography and Coding 2001. LNCS, vol. 2260, p. 360. Springer, Heidelberg (2001)
12. Canetti, R., Halevi, S., Katz, J.: A forward-secure public-key encryption scheme. In: Biham, E. (ed.) EUROCRYPT 2003. LNCS, vol. 2656, pp. 255–271. Springer, Heidelberg (2003)
13. Boneh, D., Boyen, X.: Secure identity based encryption without random oracles. In: Franklin, M. (ed.) CRYPTO 2004. LNCS, vol. 3152, pp. 443–459. Springer, Heidelberg (2004)
14. Waters, B.: Efficient identity-based encryption without random oracles. In: Cramer, R. (ed.) EUROCRYPT 2005. LNCS, vol. 3494, pp. 114–127. Springer, Heidelberg (2005)
15. Gentry, C., Peikert, C., Vaikuntanathan, V.: Trapdoors for hard lattices and new cryptographic constructions. In: Proceedings of the 40th annual ACM Symposium on Theory of Computing, pp. 197–206. ACM, New York (2008)
16. Naor, D., Naor, M., Lotspiech, J.: Revocation and tracing schemes for stateless receivers. In: Kilian, J. (ed.) CRYPTO 2001. LNCS, vol. 2139, pp. 41–62. Springer, Heidelberg (2001)
17. Dodis, Y., Fazio, N.: Public key broadcast encryption for stateless receivers. In: Feigenbaum, J. (ed.) DRM 2002. LNCS, vol. 2696, pp. 61–80. Springer, Heidelberg (2003)
18. Boneh, D., Goh, E., Nissim, K.: Evaluating 2-DNF formulas on ciphertexts. In: Kilian, J. (ed.) TCC 2005. LNCS, vol. 3378, pp. 325–341. Springer, Heidelberg (2005)
19. Katz, J., Sahai, A., Waters, B.: Predicate encryption supporting disjunctions, polynomial equations, and inner products. In: Smart, N.P. (ed.) EUROCRYPT 2008. LNCS, vol. 4965, pp. 146–162. Springer, Heidelberg (2008)

A Generic Security of Our Complexity Assumptions

We now prove our three complexity assumptions hold in the generic group model, as long as it is hard to find a nontrivial factor of the group order, N. We adopt the notation of [19] to express our assumptions. We fix generators $g_{p_1}, g_{p_2}, g_{p_3}$ of the subgroups $G_{p_1}, G_{p_2}, G_{p_3}$ respectively. Every element of G can then be expressed as $g_{p_1}^{a_1} g_{p_2}^{a_2} g_{p_3}^{a_3}$ for some values of a_1, a_2, a_3. We denote an element of G by (a_1, a_2, a_3). The element $e(g_{p_1}, g_{p_1})^{a_1} e(g_{p_2}, g_{p_2})^{a_2} e(g_{p_3}, g_{p_3})^{a_3}$ in G_T will be denoted by $[a_1, a_2, a_3]$. We use capital letters to denote random variables, and we reuse random variables to denote relationships between elements. For example, $X = (X_1, Y_1, Z_1)$ is a random element of G, and $Y = (X_1, Y_2, Z_2)$ is another random element that shares the same component in the G_{p_1} subgroup.

Given random variables $X, \{A_i\}$ expressed in this form, we say that X is *dependent* on $\{A_i\}$ if there exists values $\lambda_i \in \mathbb{Z}_n$ such that $X = \sum_i \lambda_i A_i$ as formal random variables. Otherwise, we say that X is *independent* of $\{A_i\}$. We note the following two theorems from [19]:

Theorem 2. *(Theorem A.1 of [19]) Let $N = \prod_{i=1}^m p_i$ be a product of distinct primes, each greater than 2^λ. Let $\{A_i\}$ be random variables over G, and let $\{B_i\}, T_0, T_1$ be random variables over G_T, where all random variables have degree at most t. Consider the following experiment in the generic group model:*

An algorithm is given $N, \{A_i\}$, and $\{B_i\}$. A random bit b is chosen, and the adversary is given T_b. The algorithm outputs a bits b', and succeeds if $b' = b$. The algorithm's advantage is the absolute value of the difference between its success probability and $\frac{1}{2}$.

Say each of T_0 and T_1 is independent of $\{B_i\} \cup \{e(A_i, A_j)\}$. Then given any algorithm $\mathcal{A}$ issuing at most q instructions and having advantage δ in the above experiment, $\mathcal{A}$ can be used to find a nontrivial factor of N (in time polynomial in λ and the running time of $\mathcal{A}$) with probability at least $\delta - \mathcal{O}(q^2 t/2^\lambda)$.

Theorem 3. *(Theorem A.2 of [19]) Let $N = \prod_{i=1}^m p_i$ be a product of distinct primes, each greater than 2^λ. Let $\{A_i\}, T_0, T_1$ be random variables over G, and let $\{B_i\}$ be random variables over G_T, where all random variables have degree at most t. Consider the same experiment as in the theorem above.*

Let $S := \{i | e(T_0, A_i) \neq e(T_1, A_i)\}$ (where inequality refers to inequality as formal polynomials). Say each of T_0 and T_1 is independent of $\{A_i\}$, and furthermore that for all $k \in S$ it holds that $e(T_0, A_k)$ is independent of $\{B_i\} \cup \{e(A_i, A_j)\} \cup \{e(T_0, A_i)\}_{i \neq k}$, and $e(T_1, A_k)$ is independent of $\{B_i\} \cup \{e(A_i, A_j)\} \cup \{e(T_1, A_i)\}_{i \neq k}$. Then given any algorithm $\mathcal{A}$ issuing at most q instructions and having advantage δ, the algorithm can be used to find a nontrivial factor of N (in time polynomial in λ and the running time of $\mathcal{A}$) with probability at least $\delta - \mathcal{O}(q^2 t/2^\lambda)$.

We apply these theorems to prove the security of our assumptions in the generic group model.

Assumption 1. We apply Theorem 3. We can express this assumption as:

$$A_1 = (1, 0, 0), A_2 = (0, 0, 1),$$

$$T_0 = (X_1, X_2, 0), T_1 = (X_1, 0, 0).$$

We note that $S = \emptyset$ in this case. It is clear that T_0 and T_1 are both independent of $\{A_1, A_2\}$ because X_1 does not appear in A_1 or A_2. Thus, Assumption 1 is generically secure, assuming it is hard to find a nontrivial factor of N.

Assumption 2. We apply Theorem 3. We can express this assumption as:

$$A_1 = (1, 0, 0), A_2 = (X_1, 1, 0), A_3 = (Y_1, 0, 0), A_4 = (0, X_2, 1),$$

$$T_0 = (Z_1, Z_2, Z_3), T_1 = (Z_1, 0, Z_3).$$

We note that $S = \{2, 4\}$ in this case. It is clear that T_0 and T_1 are both independent of $\{A_i\}$ since Z_1 does not appear in the A_i's, for example. We see that $e(T_0, A_2)$ is independent of $\{e(A_i, A_j)\} \cup \{e(T_0, A_i)\}_{i \neq 2}$ because it is impossible to obtain $X_1 Z_1$ in the first coordinate of a combination of elements of $\{e(A_i, A_j)\} \cup \{e(T_0, A_i)\}_{i \neq 2}$. This also allows us to conclude that $e(T_1, A_2)$ is independent of $\{e(A_i, A_j)\} \cup \{e(T_1, A_i)\}_{i \neq 2}$. We similarly note that $e(T_0, A_4)$ is independent of $\{e(A_i, A_j)\} \cup \{e(T_0, A_i)\}_{i \neq 4}$ and $e(T_1, A_4)$ is independent of $\{e(A_i, A_j)\} \cup \{e(T_1, A_i)\}_{i \neq 4}$ because we cannot obtain Z_3 in the third coordinate. Thus, Assumption 2 is generically secure, assuming it is hard to find a nontrivial factor of N.

Assumption 3. We apply Theorem 2. We can express this assumption as:

$$A_1 = (1, 0, 0), A_2 = (B, 1, 0), A_3 = (0, 0, 1), A_4 = (S, X_2, 0), A_5 = (0, Y_2, 0),$$

$$T_0 = [BS, 0, 0], T_2 = [Z_1, Z_2, Z_3].$$

T_1 is independent of $\{e(A_i, A_j)\}$ because Z_1, Z_2, Z_3 do not appear in $\{A_i\}$. T_0 is independent of $\{e(A_i, A_j)\}$ because the only way to obtain BS in the first coordinate is to take $e(A_2, A_4)$, but then we are left with an X_2 in the second coordinate that cannot be canceled. Thus, Assumption 3 is generically secure, assuming it is hard to find a nontrivial factor of N.

B HIBE Security Proof

Lemma 5. *For any algorithm $\mathcal{A}$, $Game_{Real} Adv_{\mathcal{A}} = Game_{Real'} Adv_{\mathcal{A}}$.*

Proof. We note that keys are identically distributed whether they are produced by the key delegation algorithm from a previous key or from a fresh call to the key generation algorithm. Thus, in the attacker's view, there is no difference between these games.

Lemma 6. *Suppose there exists an algorithm $\mathcal{A}$ such that $Game_{Real'}Adv_{\mathcal{A}} - Game_{Restricted}Adv_{\mathcal{A}} = \epsilon$. Then we can build an algorithm $\mathcal{B}$ with advantage $\geq \frac{\epsilon}{2}$ in breaking either Assumption 1 or Assumption 2.*

Proof. This proof is identical to the proof of Lemma 5.

Lemma 7. *Suppose there exists an algorithm $\mathcal{A}$ such that $Game_{Restricted}Adv_{\mathcal{A}} - Game_0 Adv_{\mathcal{A}} = \epsilon$. Then we can build an algorithm $\mathcal{B}$ with advantage ϵ in breaking Assumption 1.*

Proof. $\mathcal{B}$ first receives g, X_3, T. It simulates Game_{Real} or Game_0 with A. It sets the public parameters as follows. It chooses random exponents $\alpha, a_1, \ldots, a_\ell, b \in \mathbb{Z}_N$ and sets $g = g, u_i = g^{a_i}$ for i from 1 to ℓ and $h = g^b$. It sends these public parameters $\{N, g, u_1, \ldots, u_\ell, h, e(g,g)^\alpha\}$ to $\mathcal{A}$. Each time $\mathcal{B}$ is asked to provide a key for an identity $(ID_1, \ldots, ID_j)$, it chooses random exponents r, t, w, $v_{j_1}, \ldots, v_\ell \in \mathbb{Z}_N$ and sets:

$$K_1 = g^r X_3^t, K_2 = g^\alpha (u_1^{ID_1} \cdot u_j^{ID_j} h)^r X_3^w, E_{j+1} = u_{j+1}^r X_3^{v_{j+1}}, \ldots, E_\ell = u_\ell^r X_3^{v_\ell}.$$

$\mathcal{A}$ sends $\mathcal{B}$ two messages, M_0 and M_1, and a challenge identity, $(ID_1^*, \ldots, ID_j^*)$. $\mathcal{B}$ chooses $\beta \in \{0,1\}$ randomly. The ciphertext is formed as follows:

$$C_0 = M_\beta e(T,g)^\alpha, C_1 = T^{a_1 ID_1^* + \cdots a_j ID_j^* + b}, C_2 = T.$$

(This implicitly sets g^s equal to the G_{p_1} part of T.) If $T =\in G_{p_1 p_2}$, then this is a semi-functional ciphertext with $z_c = a_1 ID_1^* + \cdots + a_j ID_j^* + b$. If $T \in G_{p_1}$, this is a normal ciphertext. Hence, $\mathcal{B}$ can use the output of $\mathcal{A}$ to distinguish between these possibilities for T.

Lemma 8. *Suppose there exists an algorithm $\mathcal{A}$ such that $Game_{k-1}Adv_{\mathcal{A}} - Game_k Adv_{\mathcal{A}} = \epsilon$. Then we can build an algorithm $\mathcal{B}$ with advantage ϵ in breaking Assumption 2.*

Proof. $\mathcal{B}$ first receives $g, X_1X_2, X_3, Y_2Y_3, T$. $\mathcal{B}$ picks random exponents $a_1, \ldots, a_\ell$, $b \in \mathbb{Z}_N$ and sets the public parameters as: $g = g, u_i = g^{a_i}, h = g^b, e(g,g)^\alpha$. It sends these to $\mathcal{A}$. When $\mathcal{A}$ requests the ith key for $(ID_1, \ldots, ID_j)$ when $i < k$, $\mathcal{B}$ creates a semi-functional key. It does this by choosing random exponents r, z, t, $z_{j+1}, \ldots, z_\ell \in \mathbb{Z}_N$ and setting:

$$K_1 = g^r (Y_2Y_3)^t, K_2 = g^\alpha (u_1^{ID_1} \cdots u_j^{ID_j} h)^r (Y_2Y_3)^z,$$

$$E_{j+1} = u_{j+1}^r (Y_2Y_3)^{z_{j+1}}, \ldots, E_\ell = u_\ell^r (Y_2Y_3)^{z_\ell}.$$

This is a properly distributed semi-functional key with $g_2^\gamma = Y_2^t$.

For $i > k$, $\mathcal{B}$ generates normal keys by calling the usual key generation algorithm.

To create the kth requested key for $(ID_1, \ldots, ID_j)$, $\mathcal{B}$ lets $z_k = a_1 ID_1 + \cdots a_j ID_j + b$, chooses random exponents $w_k, w_{j+1}, \cdots, w_\ell \in \mathbb{Z}_N$, and sets:

$$K_1 = T, K_2 = g^\alpha T^{z_k} X_3^{w_k}, E_{j+1} = T^{a_{j+1}} X_3^{w_{j+1}}, \ldots, E_\ell = T^{a_\ell} X_3^{w_\ell}.$$

If $T \in G_{p_1 p_3}$, this is a normal key with g^r equal to the G_{p_1} part of T. If $T \in G$, this is a semi-functional key.

At some point, $\mathcal{A}$ sends $\mathcal{B}$ two messages, M_0 and M_1, and a challenge identity, $(ID_1^*, \ldots, ID_j^*)$. $\mathcal{B}$ sets $\beta \in \{0,1\}$ randomly. The challenge ciphertext is formed as:

$$C_0 = M_\beta e(X_1 X_2, g)^\alpha, C_1 = (X_1 X_2)^{a_1 ID_1^* + \cdots + a_j ID_j^* + b}, C_2 = X_1 X_2.$$

We note that this sets $g^s = X_1$ and $z_c = a_1 ID_1^* + \cdots a_j ID_j^* + b$. Since the kth key is not a prefix of the challenge key modulo p_2, z_k and z_c will seem randomly distributed to $\mathcal{A}$. Though it is hidden from $\mathcal{A}$, this relationship between z_c and z_k is crucial: if $\mathcal{B}$ attempts to test itself whether key k is semi-functional by creating a semi-functional ciphertext for this identity and trying to decrypt, then decryption will work whether key k is semi-functional or not, because $z_c = z_k$. In other words, the simulator can only create a nominally semi-functional key k.

If $T \in G_{p_1 p_3}$, then $\mathcal{B}$ has properly simulated Game$_{k-1}$. If $T \in G$, then $\mathcal{B}$ has properly simulated Game$_k$. Hence, $\mathcal{B}$ can use the output of $\mathcal{A}$ to distinguish between these possibilities for T.

Lemma 9. *Suppose there exists an algorithm $\mathcal{A}$ such that $Game_q Adv_{\mathcal{A}} - Game_{Final} Adv_{\mathcal{A}} = \epsilon$. Then we can build an algorithm $\mathcal{B}$ with advantage ϵ in breaking Assumption 3.*

Proof. $\mathcal{B}$ first receives $g, g^\alpha X_2, X_3, g^s Y_2, Z_2, T$. $\mathcal{B}$ chooses random exponents $a_1, \ldots, a_\ell, b \in \mathbb{Z}_N$ and sets the public parameters as $g = g, u_1 = g^{a_1}, \ldots, u_\ell = g^{a_\ell}, h = g^b, e(g,g)^\alpha = e(g^\alpha X_2, g)$. It sends these to $\mathcal{A}$. When $\mathcal{A}$ requests a key for identity $(ID_1, \ldots, ID_j)$, $\mathcal{B}$ generates a semi-functional key. It does this by choosing random exponents c, r, t, w, z, z_{j+1}, $\ldots$, z_ℓ, $w_{j+1}, \ldots, w_\ell \in \mathbb{Z}_N$ and setting:

$$K_1 = g^r Z_2^z X_3^t, K_2 = g^\alpha X_2 Z_2^c (u_1^{ID_1} \cdots u_j^{ID_j} h)^r X_3^w,$$

$$E_{j+1} = u_{j+1}^r Z_2^{z_{j+1}} X_3^{w_{j+1}}, \ldots, E_\ell = u_\ell^r Z_2^{z_\ell} X_3^{w_\ell}.$$

$\mathcal{A}$ sends $\mathcal{B}$ two messages, M_0 and M_1, and a challenge identity, $(ID_1^*, \ldots, ID_j^*)$. $\mathcal{B}$ sets $\beta \in \{0,1\}$ randomly. It forms the challenge ciphertext as:

$$C_0 = M_\beta T, C_1 = (g^s Y_2)^{a_1 ID_1^* + \cdots a_j ID_j^* + b}, C_2 = g^s Y_2.$$

This sets $z_c = a_1 ID_1^* + \cdots + a_j ID_j^* + b$. We note that the value of z_c only matters modulo p_2, whereas $u_1 = g^{a_1}, \ldots, u_\ell = g^{a_\ell}$, and $h = g^b$ are elements of G_{p_1}, so when $a_1, \ldots, a_\ell$ and b are chosen randomly modulo N, there is no correlation between the values of $a_1, \ldots, a_\ell, b$ modulo p_1 and the value $z_c = a_1 ID_1^* + \cdots + a_j ID_j^* + b$ modulo p_2.

If $T = e(g,g)^{\alpha s}$, then this is a properly distributed semi-functional ciphertext with message M_β. If T is a random element of G_T, then this is a semi-functional ciphertext with a random message. Hence, $\mathcal{B}$ can use the output of $\mathcal{A}$ to distinguish between these possibilities for T.

We have now proven the following theorem:

Theorem 4. *If Assumptions 1, 2, and 3 hold, then our HIBE system is secure.*

Proof. If Assumptions 1, 2, and 3 hold, then we have shown by the previous lemmas that the real security game is indistinguishable from Game_{Final}, in which the value of β is information-theoretically hidden from the attacker. Hence the attacker can attain no advantage in breaking the HIBE system.

C IBE in Prime Order Groups

Our construction essentially replaces each single group element in our composite order construction with a 3-tuple of group elements. This 3-tuple is inspired by a simplification of the Waters dual encryption IBE system [5]. We prove security under 3 new static assumptions. We leave it as an open problem to obtain security from the decisional Linear and d-BDH assumptions. One approach would be to use more of the Waters system (with less simplification).

C.1 Construction

For our construction, we employ prime order groups G_1, G_2, G_T of order p such that there is an efficient bilinear map $e : G_1 \times G_2 \rightarrow G_T$ but no efficient isomorphism between G_1 and G_2. We use subscripts to clarify which elements are in G_1 and which are in G_2, for example, $g_1 \in G_1$.

Setup. Our setup algorithm chooses groups G_1, G_2, G_T of order p as above. It chooses $g_1, u_1, h_1 \in G_1, g_2 \in G_2$ randomly. It sets u_2 and h_2 so that the discrete log of u_2, h_2 base g_2 is equal to the discrete log of u_1, h_1 base g_1 respectively. It chooses $a, \alpha \in Z_p$ randomly. It chooses $v_2, v_2', f_2 \in G_2$ randomly and sets $\tau \in Z_p$ to satisfy $f_2^{\tau} = v_2(v_2')^a$. It publishes the public parameters as:

$$\{g_1, u_1, h_1, g_1^a, u_1^a, h_1^a, g_1^{\tau}, u_1^{\tau}, h_1^{\tau}, e(g_1, g_2)^{\alpha}\}.$$

The master secret key is $g_2, \alpha, v_2, v_2', u_2, h_2, f_2$.

Encrypt(M, ID). The encryption algorithm randomly chooses $s \in Z_p$ and creates the ciphertext as:

$$C_0 = Me(g_1, g_2)^{\alpha s}, C_{1,1} = (u_1^{ID} h_1)^s, C_{1,2} = (u_1^{ID} h)^{as}, C_{1,3} = (u_1^{ID} h_1)^{-s\tau},$$

$$C_{2,1} = g_1^s, C_{2,2} = g_1^{as}, C_{2,3} = g_1^{\tau s}.$$

KeyGen(ID, MSK). The key generation algorithm chooses random values $y, c_1, c_2 \in \mathbb{Z}_p$. It creates the key as:

$$K_{1,1} = g_2^y v_2^{c_1}, K_{1,2} = (v_2')^{c_1}, K_{1,3} = f_2^{c_1},$$

$$K_{2,1} = g_2^{\alpha}(u_2^{ID} h_2)^y v_2^{c_2}, K_{2,2} = (v_2')^{c_2}, K_{2,3} = f_2^{c_2}.$$

Decryption. If the ID's of the ciphertext and key are equal, the decryption algorithm computes the blinding factor as:

$$\frac{e(C_{2,1}, K_{2,1})e(C_{2,2}, K_{2,2})e(C_{2,3}, K_{2,3})}{e(C_{1,1}, K_{1,1})e(C_{1,2}, K_{1,2})e(C_{1,3}K_{1,3})}.$$

C.2 Complexity Assumptions

We state the assumptions we will rely on in our security proof. These are non-standard assumptions, but we emphasize that they are static.

Assumption 1. Let $f_1 \in G_1$ and $f_2 \in G_2$ be chosen randomly. Let $a, b, s \in \mathbb{Z}_p$ be chosen randomly. Given

$$\{f_1, f_1^{bs}, f_1^s, f_1^a, f_1^{ab^2}, f_1^b, f_1^{b^2}, f_1^{as}, f_1^{b^2 s}, f_1^{b^3}, f_1^{b^3 s}, T \in G_1, f_2, f_2^b \in G_2\},$$

it should be hard to distinguish $T = f_1^{asb^2}$ from random.

Assumption 2. Let $f_1 \in G_1$ and $f_2 \in G_2$ be chosen randomly. Let $d, b, c, x \in \mathbb{Z}_p$ be chosen randomly. Given

$$\{f_1, f_1^d, f_1^{d^2}, f_1^{bx}, f_1^{dbx}, f_1^{d^2 x} \in G_1, f_2, f_2^d, f_2^b, f_2^c \in G_2, T \in G_2\},$$

it should be hard to distinguish $T = f_2^{bc}$ from random.

Assumption 3. Let $f_1 \in G_1$ and $f_2 \in G_2$ be chosen randomly. Let $d, b, c \in \mathbb{Z}_p$ be chosen randomly. Given

$$\{f_1, f_1^a, f_1^b, f_1^c \in G_1, f_2, f_2^a, f_2^b, f_2^c \in G_2, T \in G_T\},$$

it should be hard to distinguish $T = e(f_1, f_2)^{abc}$ from random.

C.3 Security

We first define semi-functional keys and ciphertexts.

Semi-functional Ciphertext. We let f_1, v_1' denote elements of G_1 such that the discrete log of v_1' base f_1 is the same as the discrete log of v_2' base f_2. We let t, z_c denote random exponents in $\mathbb{Z}_p$. A semi-functional ciphertext is created as follows: first, a normal ciphertext $C_0', C_{1,1}', C_{1,2}', C_{1,3}', C_{2,1}', C_{2,2}', C_{2,3}'$ is created. Then, C_0 is set to be C_0', $C_{1,1} = C_{1,1}'$, $C_{1,2} = C_{1,2}' f_1^{tz_c}$, $C_{1,3} = C_{1,3}'(v_1')^{-tz_c}$, $C_{2,1} = C_{2,1}'$, $C_{2,2} = C_{2,2}' f_1^t$, $C_{2,3} = C_{2,3}'(v_1')^{-t}$.

Semi-functional Key. A semi-functional key is created as follows: a normal key $K_{1,1}'$, $K_{1,2}'$, $K_{1,3}'$, $K_{2,1}', K_{2,2}', K_{2,3}'$ is generated. Random exponents $w, z_k \in \mathbb{Z}_p$ are chosen. Then we set: $K_{1,1} = K_{1,1}' f_2^{-aw}$, $K_{1,2} = K_{1,2}' f_2^w$, $K_{1,3} = K_{1,3}'$, $K_{2,1} = K_{2,1}' f_2^{-awz_k}$, $K_{2,2} = K_{2,2}' f_2^{wz_k}$, $K_{2,3} = K_{2,3}'$.

We note that when a semi-functional key is paired with a normal ciphertext or a normal key is paired with a semi-functional ciphertext, decryption still works. When a semi-functional key is paired with a semi-functional ciphertext, the blinding factor is obscured by an additional term: $e(f_1, f_2)^{tw(z_k - z_c)}$. (When $z_k = z_c$, decryption will still work.)

We will prove security through a hybrid argument over a sequence of games. Game_{Real} is the real security game. Game_0 is like the real security game, except with a semi-functional ciphertext. $Game_k$ for k from 1 to q (where q is the number of queries by the attacker) is like Game_0, except that the first k requested keys are semi-functional and the rest are normal. In Game_{Final}, the semi-functional encryption is of a random message instead of one of the requested messages. We will rely on Assumptions 1, 2, 3 as defined in the subsection above. We prove security through the following 3 lemmas.

Lemma 10. *Suppose there exists an algorithm $\mathcal{A}$ such that $Game_{Real}Adv_{\mathcal{A}} - Game_0 Adv_{\mathcal{A}} = \epsilon$. Then we can build an algorithm $\mathcal{B}$ with advantage ϵ in breaking Assumption 1.*

Proof. $\mathcal{B}$ is given

$$\{f_1, f_1^{bs}, f_1^s, f_1^a, f_1^{ab^2}, f_1^b, f_1^{b^2}, f_1^{as}, f_1^{b^2 s}, f_1^{b^3}, f_1^{b^3 s}, T \in G_1, f_2, f_2^b \in G_2\}.$$

It chooses random exponents $\alpha, A, B, y_g, y_u, y_h, y'_v \in \mathbb{Z}_p$ and sets the parameters as:

$$g_1 = f_1^{b^2} f_1^{y_g}, u_1 = (f_1^{b^2})^A f_1^{y_u}, h_1 = (f_1^{b^2})^B f_1^{y_h}, g_1^a, u_1^a, h_1^a,$$

$$f_2 = f_2, v_2 = f_2^b, v'_2 = f_2^{y'_v}, \tau = b + a y'_v.$$

Here, a is from the assumption and g_1^a, u_1^a, h_1^a can be computed from f_1^a and $f_1^{ab^2}$. We note that $\mathcal{B}$ can also compute g_1^τ, u_1^τ, and h_1^τ using $f_1^{b^3}$, $f_1^{b_2 a}$, f_1^a, and f_1^b. It can also compute $e(g_1, g_2)^\alpha$ using $f_1^{b^2}, f_2^b$, and $f_1^{b^3}$.

To construct a normal key for ID, $\mathcal{B}$ chooses random exponents $c'_1, c'_2, y \in \mathbb{Z}_p$ and sets $f_2^{c_1} = f_2^{c'_1}(f_2^b)^{-y}$ and $f_2^{c_2} = f_2^{c'_2}(f_2^b)^{-yAID+B-\alpha}$. Then the key can be formed as:

$$K_{1,1} = f_2^{y_g y}(f_2^b)^{c'_1}, K_{1,2} = (f_2^{c_1})^{y'_v}, K_{1,3} = f_2^{c_1},$$

$$K_{2,1} = f_2^{\alpha y_g}(f_2^b)^{c'_2} f_2^{y(y_u ID + y_h)}, K_{2,2} = (f_2^{c_2})^{y'_v}, K_{2,3} = f_2^{c_2}.$$

To construct the challenge ciphertext for M_β and ID^*, $\mathcal{B}$ sets $s = s$ from the assumption. Then $C_{1,1}$ and $C_{2,1}$ can be computed from f_1^s and $f_1^{b^2 s}$. Next,

$$C_{1,2} = T^{AID+B}(f_1^{as})^{y_I D + y_h}, C_{2,2} = T(f_1^{as})^{y_g}.$$

We can create $C_{2,3}$ as:

$$C_{2,3} = f_1^{b^3 s}(f_1^{bs})^{y_g} T^{y'_v}(f_1^{as})^{y_g y'_v}.$$

$C_{1,3}$ can similarly be constructed using $T^{y'_v(AID+B)}$. We note that A, B are information-theoretically hidden from the attacker.

Lemma 11. *Suppose there exists an algorithm $\mathcal{A}$ such that $Game_{k-1}Adv_{\mathcal{A}} - Game_k Adv_{\mathcal{A}} = \epsilon$. Then we can build an algorithm $\mathcal{B}$ with advantage ϵ in breaking Assumption 2.*

Proof. $\mathcal{B}$ is given $\{f_1, f_1^d, f_1^{d^2}, f_1^{bx}, f_1^{dbx}, f_1^{d^2x} \in G_1, f_2, f_2^d, f_2^b, f_2^c \in G_2, T \in G_2\}$. It chooses random exponents $\alpha, a, A, B, y_u, y_h, y_v \in \mathbb{Z}_p$. It sets the parameters as follows:

$$g_1 = f_1^d, u_1 = (f_1^d)^A f_1^{y_u}, h_1 = (f_1^d)^B f_1^{y_h}, g_1^a, u_1^a, h_1^a, g_2 = f_2^d, u_2 = (f_2^d)^A f_2^{y_u},$$

$$h_2 = (f_2^d)^B f_2^{y_h}, v_2' = f_2^b, v_2 = f_2^d f_2^{-ba} f_2^{y_v}, f_2 = f_2.$$

This sets $\tau = d - ba + y_v + ab = d + y_v$, so the simulator $\mathcal{B}$ can also compute $g_1^\tau, u_1^\tau, h_1^\tau$ and send all of the public parameters to $\mathcal{A}$.

To make normal keys for key queries $< k$, $\mathcal{B}$ can choose $y, c_1, c_2 \in \mathbb{Z}_p$ randomly and generate the keys from the MSK. To make semi-functional keys for key queries $> k$, $\mathcal{B}$ can choose $y, c_1, c_2, w, z_k \in \mathbb{Z}_p$ randomly to generate the semi-functional key.

To make the challenge key k for ID, $\mathcal{B}$ chooses $y', c_2' \in \mathbb{Z}_p$ randomly and implicitly sets $y = -c + y'$, $c_1 = c$, $c_2 = c(AID + B) + c_2'$. The key can then be formed as:

$$K_{1,1} = (f_2^d)^{y'} T^{-a} (f_2^c)^{y_v}, K_{1,2} = T, K_{1,3} = f_2^c,$$

$$K_{2,1} = g_2^\alpha T^{-a(AID+B)} (f_2^b)^{-ac_2'} (f_2^d)^{y'(AID+B)+c_2'} (f_2^c)^{y_u ID + y_h + y_v(AID+B)} f_2^{y'(y_u ID + y_h) + c_2' y_v},$$

$$K_{2,2} = T^{AID+B} (f_2^b)^{c_2'}, K_{2,3} = (f_2^c)^{AID+B} f_2^{c_2'}.$$

We note that this sets $z_k = AID + B$.

At some point, $\mathcal{A}$ sends two messages, M_0, M_1, to $\mathcal{B}$ along with a challenge identity ID^*. $\mathcal{B}$ chooses $\beta \in \{0,1\}$ randomly and generates a semi-functional ciphertext for M_β and ID^* as follows. $\mathcal{B}$ chooses a random exponent $s' \in \mathbb{Z}_p$ and implicitly sets $s = bx + s'$, $t = -d^2x$. The ciphertext is formed as follows:

$$C_0 = M_\beta e(f_1^{dbx}, f_2^d) e(g_1, g_2)^{\alpha s'}, C_{1,1} = (f_1^{dbx})^{AID^*+B} (f_1^{bx})^{y_u ID^* + B} (u_1^{ID^*} h_1)^{s'},$$

$$C_{1,2} = (f_1^{dbx})^{a(AID^*+B)} (f_1^{d^2x})^{-(AID^*+B)} (u_1^{ID^*} h_1)^{as'},$$

$$C_{1,3} = (f_1^{dbx})^{-y_v(AID^*+B)} (f_1^{d^2})^{-s'(AID^*+B)} (f_1^d)^{-y_v s'(AID^*+B)},$$

$$C_{2,1} = f_1^{dbx} (f_1^d)^{s'}, C_{2,2} = (f_1^{dbx})^a (f_1^d)^{s'a} (f_1^{d^2x})^{-1},$$

$$C_{2,3} = (f_1^{dbx})^{-y_v} (f_1^{d^2})^{-s'} (f_1^d)^{-y_v s'}.$$

We note that $z_c = AID^* + B$. Since A and B are information-theoretically hidden from the attacker, this will seem properly distributed to the attacker. If $T = f_2^{bc}$, then $\mathcal{B}$ has properly simulated Game$_{k-1}$, and if T is random, then $\mathcal{B}$ has properly simulated Game$_k$.

Lemma 12. *Suppose there exists an algorithm $\mathcal{A}$ such that $Game_q Adv_{\mathcal{A}} - Game_{Final} Adv_{\mathcal{A}} = \epsilon$. Then we can build an algorithm $\mathcal{B}$ with advantage ϵ in breaking Assumption 3.*

Proof. $\mathcal{B}$ is given

$$\{f_1, f_1^d, f_1^{d^2}, f_1^{bx}, f_1^{dbx}, f_1^{d^2x} \in G_1, f_2, f_2^d, f_2^b, f_2^c \in G_2, T \in G_2\}.$$

It will implicitly set $\alpha = ab$, $s = c$, and $a = a$. $\mathcal{B}$ chooses random exponents $y_g, y_u, y_v, y_v, y'_v \in \mathbb{Z}_p$. It sets the parameters as:

$$g_1 = f_1^{y_g}, u_1 = f_1^{y_u}, h_1 = f_1^{y_h}, v_2 = f_2^{y_v}, v'_2 = f_2^{y'_v}, g_2 = f_2^{y_g}$$

and sets $\tau = y_v + a y'_v$. From this, it can calculate the rest of the public parameters as:

$$g_1^a = (f_1^a)^{y_g}, u_1^a = (f_1^a)^{y_u}, h_1^a = (f_1^a)^{y_h}, g_1^\tau = (g_1^a)^{y'_v} g_1^{y_v},$$

$$u_1^\tau = (u_1^a)^{y'_v} u_1^{y_v}, h_1^\tau = (h_1^a)^{y'_v} h_1^{y_v}, e(g_1, g_2)^\alpha = e(f_1^a, f_2^b)^{y_g^2}.$$

To make semi-functional keys, $\mathcal{B}$ must cancel the term g_2^α in $K_{2,1}$ since this is unknown. To do this, the simulator randomly chooses $w, c_1, c_2, y, \gamma \in \mathbb{Z}_p$ and implicitly sets $wz_k = b + \gamma$.

To make the challenge ciphertext for M_β and ID^*, $\mathcal{B}$ sets $s = c$ and chooses random values $\delta, \delta' \in \mathbb{Z}_p$. It implicitly sets $ca+t = \delta$ and $ca(y_u ID^* + y_h) + tz_c = \delta'$ and $acy_g + t = \delta$.

$$C_0 = M_\beta T, C_{1,1} = (f_1^c)^{y_u ID^* + y_h}, C_{1,2} = f_1^{\delta'}, C_{1,3} = (f_1^c)^{-y_v(y_u ID^* + y_h)} f_1^{-y'_v \delta'},$$

$$C_{2,1} = (f_1^c)^{y_g}, C_{2,2} = f_1^\delta, C_{2,3} = (f_1^c)^{-y_v y_g} (f_1^\delta)^{-y'_v}.$$

Robust Encryption

Michel Abdalla[1], Mihir Bellare[2], and Gregory Neven[3,4]

[1] Departement d'Informatique, École normale supérieure, Paris, France
Michel.Abdalla@ens.fr
http://www.di.ens.fr/users/mabdalla
[2] Department of Computer Science & Engineering,
University of California San Diego, USA
mihir@cs.ucsd.edu
http://www.cs.ucsd.edu/users/mihir
[3] Department of Electrical Engineering, Katholieke Universiteit Leuven, Belgium
[4] IBM Research – Zurich, Switzerland
nev@zurich.ibm.com
http://www.neven.org

Abstract. We provide a provable-security treatment of "robust" encryption. Robustness means it is hard to produce a ciphertext that is valid for two different users. Robustness makes explicit a property that has been implicitly assumed in the past. We argue that it is an essential conjunct of anonymous encryption. We show that natural anonymity-preserving ways to achieve it, such as adding recipient identification information before encrypting, fail. We provide transforms that do achieve it, efficiently and provably. We assess the robustness of specific encryption schemes in the literature, providing simple patches for some that lack the property. We present various applications. Our work enables safer and simpler use of encryption.

1 Introduction

This paper provides a provable-security treatment of encryption "robustness." Robustness reflects the difficulty of producing a ciphertext valid under two different encryption keys. The value of robustness is conceptual, "naming" something that has been undefined yet at times implicitly (and incorrectly) assumed. Robustness helps make encryption more mis-use resistant. We provide formal definitions of several variants of the goal; consider and dismiss natural approaches to achieve it; provide two general robustness-adding transforms; test robustness of existing schemes and patch the ones that fail; and discuss some applications.

THE DEFINITIONS. Both the PKE and the IBE settings are of interest and the explication is simplified by unifying them as follows. Associate to each identity an *encryption key*, defined as the identity itself in the IBE case and its (honestly generated) public key in the PKE case. The adversary outputs a pair id_0, id_1 of distinct identities. For strong robustness it also outputs a ciphertext C^*; for weak, it outputs a message M^*, and C^* is defined as the encryption of M^* under the encryption key ek_1 of id_1. The adversary wins if the decryptions of

D. Micciancio (Ed.): TCC 2010, LNCS 5978, pp. 480–497, 2010.

C^* under the decryption keys dk_0, dk_1 corresponding to ek_0, ek_1 are *both* non-$\perp$. Both weak and strong robustness can be considered under chosen plaintext or chosen ciphertext attacks, resulting in four notions (for each of PKE and IBE) that we denote WROB-CPA, WROB-CCA, SROB-CPA, SROB-CCA.

WHY ROBUSTNESS? The primary security requirement for encryption is data-privacy, as captured by notions IND-CPA or IND-CCA [18,21,16,5,11]. Increasingly, we are also seeing a market for *anonymity*, as captured by notions ANO-CPA and ANO-CCA [4,1]. Anonymity asks that a ciphertext does not reveal the encryption key under which it was created.

Where you need anonymity, there is a good chance you need robustness too. Indeed, we would go so far as to say that robustness is an essential companion of anonymous encryption. The reason is that without it we would have security without basic communication correctness, likely upsetting our application. This is best illustrated by the following canonical application of anonymous encryption, but shows up also, in less direct but no less important ways, in other applications. A sender wants to send a message to a *particular* target recipient, but, to hide the identity of this target recipient, anonymously encrypts it under her key and broadcasts the ciphertext to a larger group. But as a member of this group I need, upon receiving a ciphertext, to know whether or not I am the target recipient. (The latter typically needs to act on the message.) Of course I can't tell whether the ciphertext is for me just by looking at it since the encryption is anonymous, but decryption should divulge this information. It does, unambiguously, if the encryption is robust (the ciphertext is for me iff my decryption of it is not $\perp$) but otherwise I might accept a ciphertext (and some resulting message) of which I am not the target, creating mis-communication. Natural "solutions," such as including the encryption key or identity of the target recipient in the plaintext before encryption and checking it upon decryption, are, in hindsight, just attempts to add robustness without violating anonymity and, as we will see, don't work.

We were lead to formulate robustness upon revisiting Public key Encryption with Keyword Search (PEKS) [9]. In a clever usage of anonymity, Boneh, Di Crescenzo, Ostrovsky and Persiano (BDOP) [9] showed how this property in an IBE scheme allowed it to be turned into a privacy-respecting communications filter. But Abdalla et. al [1] noted that the BDOP filter could lack *consistency*, meaning turn up false positives. Their solution was to modify the construction. What we observed instead was that consistency would in fact be provided by the *original* construct if the IBE scheme was robust. PEKS consistency turns out to correspond exactly to communication correctness of the anonymous IBE scheme in the sense discussed above. (Because the PEKS messages in the BDOP scheme are the recipients identities from the IBE perspective.) Besides resurrecting the BDOP construct, the robustness approach allows us to obtain the first consistent IND-CCA secure PEKS without random oracles.

Sako's auction protocol [23] is important because it was the first truly practical one to hide the bids of losers. It makes clever use of anonymous encryption for privacy. But we present an attack on fairness whose cause is ultimately a lack of robustness in the anonymous encryption scheme (cf. [2]).

All this underscores a number of the claims we are making about robustness: that it is of conceptual value; that it makes encryption more resistant to mis-use; that it has been implicitly (and incorrectly) assumed; and that there is value to making it explicit, formally defining and provably achieving it.

Weak versus strong. The above-mentioned auction protocol fails because an adversary can create a ciphertext that decrypts correctly under any decryption key. Strong robustness is needed to prevent this. Weak robustness (of the underlying IBE) will yield PEKS consistency for honestly-encrypted messages but may allow spammers to bypass all filters with a single ciphertext, something prevented by strong robustness. Strong robustness trumps weak for applications and goes farther towards making encryption mis-use resistant. We have defined and considered the weaker version because it can be more efficiently achieved, because some existing schemes achieve it and because attaining it is a crucial first step in our method for attaining strong robustness.

Achieving robustness. As the reader has surely already noted, robustness (even strong) is trivially achieved by appending the encryption key to the ciphertext and checking for it upon decryption. The problem is that the resulting scheme is not anonymous and, as we have seen above, it is exactly for anonymous schemes that robustness is important. Of course, data privacy is important too. Letting AI-ATK = ANO-ATK + IND-ATK for ATK $\in$ {CPA, CCA}, our goal is to achieve AI-ATK + XROB-ATK, ideally for both ATK $\in$ {CPA, CCA} and X $\in$ {W, S}. This is harder.

Transforms. It is natural to begin by seeking a general transform that takes an arbitrary AI-ATK scheme and returns a AI-ATK + XROB-ATK one. This allows us to exploit known constructions of AI-ATK schemes, supports modular protocol design and also helps understand robustness divorced from the algebra of specific schemes. Furthermore, there is a natural and promising transform to consider. Namely, before encrypting, append to the message some redundancy, such as the recipient encryption key, a constant, or even a hash of the message, and check for its presence upon decryption. (Adding the redundancy before encrypting rather than after preserves AI-ATK.) Intuitively this should provide robustness because decryption with the "wrong" key will result, if not in rejection, then in recovery of a garbled plaintext, unlikely to possess the correct redundancy.

The truth is more complex. We consider two versions of the paradigm and summarize our findings in Fig. 1. In encryption with *unkeyed redundancy*, the redundancy is a function RC of the message and encryption key alone. In this case we show that the method fails spectacularly, not providing even *weak* robustness *regardless of the choice of the function* RC. In encryption with *keyed redundancy*, we allow RC to depend on a key K that is placed in the public parameters of the transformed scheme, out of direct reach of the algorithms of the original scheme.

In this form, the method can easily provide weak robustness, and that too with a very simple redundancy function, namely the one that simply returns K.

But we show that even encryption with keyed redundancy fails to provide *strong* robustness. To achieve the latter we have to step outside the encryption with redundancy paradigm. We present a strong robustness conferring transform that uses a (non-interactive) commitment scheme. For subtle reasons, for this transform to work the starting scheme needs to already be weakly robust. If it isn't already, we can make it so via our weak robustness transform.

In summary, on the positive side we provide a transform conferring weak robustness and another conferring strong robustness. Given any AI-ATK scheme the first transform returns a WROB-ATK + AI-ATK one. Given any AI-ATK + WROB-ATK scheme the second transform returns a SROB-ATK + AI-ATK one. In both cases it is for both ATK = CPA and ATK = CCA and in both cases the transform applies to what we call general encryption schemes, of which both PKE and IBE are special cases, so both are covered.

ROBUSTNESS OF SPECIFIC SCHEMES. The robustness of existing schemes is important because they might be in use. We ask which specific existing schemes are robust, and, for those that are not, whether they can be made so at a cost lower than that of applying one of our general transforms. There is no reason to expect schemes that are only AI-CPA to be robust since the decryption algorithm may never reject, so we focus on schemes that are known to be AI-CCA. This narrows the field quite a bit. Our findings and results are summarized in Fig. 1.

Canonical AI-CCA schemes in the PKE setting are Cramer-Shoup ($\mathcal{CS}$) in the standard model [15,4] and $\mathcal{DHIES}$ in the random oracle (RO) model [3,4]. We show that both are WROB-CCA but neither is SROB-CCA, the latter because encryption with 0 randomness yields a ciphertext valid under any encryption key. We present modified versions $\mathcal{CS}^*$, $\mathcal{DHIES}^*$ of the schemes that we show are SROB-CCA. Our proof that $\mathcal{CS}^*$ is SROB-CCA builds on the information-theoretic part of the proof of [15]. The result does not need to assume hardness of DDH. It relies instead on pre-image security of the underlying hash function for random range points, something not implied by collision-resistance but seemingly possessed by candidate functions.

In the IBE setting, the CCA version $\mathcal{BF}$ of the RO model Boneh-Franklin scheme is AI-CCA [10,1], and we show it is SROB-CCA. The standard model Boyen-Waters scheme $\mathcal{BW}$ is AI-CCA [13], and we show it is neither WROB-CCA nor SROB-CCA. It can be made either via our transforms but we don't know of any more direct way to do this.

$\mathcal{BF}$ is obtained via the Fujisaki-Okamoto (FO) transform [17] and $\mathcal{BW}$ via the Canetti-Halevi-Katz (CHK) transform [14,8]. We can show that neither transform *generically* provides strong robustness. This doesn't say whether they do or not when applied to specific schemes, and indeed the first does for $\mathcal{BF}$ and the second does not for $\mathcal{BW}$.

SUMMARY. Protocol design suggests that designers have the intuition that robustness is naturally present. This seems to be more often right than wrong

Transform	WROB-ATK	SROB-ATK
Encryption with unkeyed redundancy (EuR)	No	No
Encryption with keyed redundancy (EkR)	Yes	No

Scheme	setting	AI-CCA	WROB-CCA	SROB-CCA	RO model
$\mathcal{CS}$	PKE	Yes [15,4]	Yes	No	No
$\mathcal{CS}^*$	PKE	Yes	Yes	Yes	No
$\mathcal{DHIES}$	PKE	Yes [3]	Yes	No	Yes
$\mathcal{DHIES}^*$	PKE	Yes	Yes	Yes	Yes
$\mathcal{BF}$	IBE	Yes [10,1]	Yes	Yes	Yes
$\mathcal{BW}$	IBE	Yes [13]	No	No	No

Fig. 1. Achieving Robustness. The first table summarizes our findings on the encryption with redundancy transform. "No" means the method fails to achieve the indicated robustness for *all* redundancy functions, while "yes" means there exists a redundancy function for which it works. The second table summarizes robustness results about some specific AI-CCA schemes.

when considering *weak* robustness of *specific* AI-CCA schemes. Prevailing intuition about *generic* ways to add even weak robustness is wrong, yet we show it can be done by an appropriate tweak of these ideas. Strong robustness is more likely to be absent than present in specific schemes, but important schemes can be patched. Strong robustness can also be added generically, but with more work.

RELATED WORK. There is growing recognition that robustness is important in applications and worth defining explicitly, supporting our own claims to this end. In particular the correctness requirement for predicate encryption [20] includes a form of weak robustness and, in recent work concurrent to, and independent of, ours, Hofheinz and Weinreb [19] introduced a notion of *well-addressedness* of IBE schemes that is just like weak robustness except that the adversary gets the IBE master secret key. Neither work considers or achieves strong robustness, and neither treats PKE.

2 Definitions

NOTATION AND CONVENTIONS. If x is a string then $|x|$ denotes its length, and if S is a set then $|S|$ denotes its size. The empty string is denoted ε. By $a_1 \| \ldots \| a_n$, we denote a string encoding of $a_1, \ldots, a_n$ from which $a_1, \ldots, a_n$ are uniquely recoverable. (Usually, concatenation suffices.) By $a_1 \| \ldots \| a_n \leftarrow a$, we mean that a is parsed into its constituents $a_1, \ldots, a_n$. Similarly, if $a = (a_1, \ldots, a_n)$ then $(a_1, \ldots, a_n) \leftarrow a$ means we parse a as shown. Unless otherwise indicated, an algorithm may be randomized. By $y \stackrel{\$}{\leftarrow} A(x_1, x_2, \ldots)$ we denote the operation of running A on inputs $x_1, x_2, \ldots$ and fresh coins and letting y denote the output. We denote by $[A(x_1, x_2, \ldots)]$ the set of all possible outputs of A on inputs $x_1, x_2, \ldots$. We assume that an algorithm returns $\perp$ if any of its inputs is $\perp$.

proc Initialize	proc Dec(C, id)
$(pars, msk) \xleftarrow{\$} \mathsf{PG}$; $b \xleftarrow{\$} \{0,1\}$	If $id \notin U$ then return $\perp$
$S, T, U, V \leftarrow \emptyset$	If $(id, C) \in T$ then return $\perp$
Return $pars$	$M \leftarrow \mathsf{Dec}(pars, \mathsf{EK}[id], \mathsf{DK}[id], C)$
proc GetEK(id)	Return M
$U \leftarrow U \cup \{id\}$	**proc LR**$(id_0^*, id_1^*, M_0^*, M_1^*)$
$(\mathsf{EK}[id], \mathsf{DK}[id]) \xleftarrow{\$} \mathsf{KG}(pars, msk, id)$	If $(id_0^* \notin U) \vee (id_1^* \notin U)$ then return $\perp$
Return $\mathsf{EK}[id]$	If $(id_0^* \in V) \vee (id_1^* \in V)$ then return $\perp$
proc GetDK(id)	If $\lvert M_0^* \rvert \neq \lvert M_1^* \rvert$ then return $\perp$
If $id \notin U$ then return $\perp$	$C^* \xleftarrow{\$} \mathsf{Enc}(pars, \mathsf{EK}[id_b], M_b^*)$
If $id \in S$ then return $\perp$	$S \leftarrow S \cup \{id_0^*, id_1^*\}$
$V \leftarrow V \cup \{id\}$	$T \leftarrow T \cup \{(id_0^*, C^*), (id_1^*, C^*)\}$
Return $\mathsf{DK}[id]$	Return C^*
	proc Finalize(b')
	Return $(b' = b)$

Fig. 2. Game $\mathrm{AI}_{\mathcal{GE}}$ defining AI-ATK security of general encryption scheme $\mathcal{GE} = (\mathsf{PG}, \mathsf{KG}, \mathsf{Enc}, \mathsf{Dec})$

GAMES. Our definitions and proofs use code-based game-playing [6]. Recall that a game —look at Fig. 2 for an example— has an **Initialize** procedure, procedures to respond to adversary oracle queries, and a **Finalize** procedure. A game G is executed with an adversary A as follows. First, **Initialize** executes and its outputs are the inputs to A. Then A executes, its oracle queries being answered by the corresponding procedures of G. When A terminates, its output becomes the input to the **Finalize** procedure. The output of the latter, denoted G^A, is called the output of the game, and we let "G^A" denote the event that this game output takes value true. Boolean flags are assumed initialized to false. Games $\mathrm{G}_i, \mathrm{G}_j$ are *identical until* bad if their code differs only in statements that follow the setting of bad to true. Our proofs will use the following.

Lemma 1 [6] *Let* $\mathrm{G}_i, \mathrm{G}_j$ *be identical until* bad *games, and* A *an adversary. Then*

$$\left|\Pr\left[\mathrm{G}_i^A\right] - \Pr\left[\mathrm{G}_j^A\right]\right| \leq \Pr\left[\mathrm{G}_j^A \text{ sets bad}\right] .$$

The running time of an adversary is the worst case time of the execution of the adversary with the game defining its security, so that the execution time of the called game procedures is included.

GENERAL ENCRYPTION. We introduce and use general encryption schemes, of which both PKE and IBE are special cases. This allows us to avoid repeating similar definitions and proofs. A *general encryption* (GE) scheme is a tuple $\mathcal{GE} = (\mathsf{PG}, \mathsf{KG}, \mathsf{Enc}, \mathsf{Dec})$ of algorithms. The parameter generation algorithm PG takes no input and returns common parameter *pars* and a master secret key *msk*. On input *pars*, *msk*, *id*, the key generation algorithm KG produces an encryption key *ek* and decryption key *dk*. On inputs *pars*, *ek*, M, the encryption algorithm Enc produces a ciphertext C encrypting plaintext M. On input *pars*, *ek*, *dk*, C,

proc Initialize
$(pars, msk) \xleftarrow{\$} \mathsf{PG}$; $U, V \leftarrow \emptyset$
Return *pars*

proc GetEK(id)
$U \leftarrow U \cup \{id\}$
$(\mathsf{EK}[id], \mathsf{DK}[id]) \xleftarrow{\$} \mathsf{KG}(pars, msk, id)$
Return $\mathsf{EK}[id]$

proc GetDK(id)
If $id \notin U$ then return $\perp$
$V \leftarrow V \cup \{id\}$
Return $\mathsf{DK}[id]$

proc Dec(C, id)
If $id \notin U$ then return $\perp$
$M \leftarrow \mathsf{Dec}(pars, \mathsf{EK}[id], \mathsf{DK}[id], C)$
Return M

proc Finalize(M, id_0, id_1) // $\mathrm{WROB}_{\mathcal{GE}}$
If $(id_0 \notin U) \vee (id_1 \notin U)$ then return false
If $(id_0 \in V) \vee (id_1 \in V)$ then return false
If $(id_0 = id_1)$ then return false
$M_0 \leftarrow M$; $C \xleftarrow{\$} \mathsf{Enc}(pars, \mathsf{EK}[id_0], M_0)$
$M_1 \leftarrow \mathsf{Dec}(pars, \mathsf{EK}[id_1], \mathsf{DK}[id_1], C)$
Return $(M_0 \neq \perp) \wedge (M_1 \neq \perp)$

proc Finalize(C, id_0, id_1) // $\mathrm{SROB}_{\mathcal{GE}}$
If $(id_0 \notin U) \vee (id_1 \notin U)$ then return false
If $(id_0 \in V) \vee (id_1 \in V)$ then return false
If $(id_0 = id_1)$ then return false
$M_0 \leftarrow \mathsf{Dec}(pars, \mathsf{EK}[id_0], \mathsf{DK}[id_0], C)$
$M_1 \leftarrow \mathsf{Dec}(pars, \mathsf{EK}[id_1], \mathsf{DK}[id_1], C)$
Return $(M_0 \neq \perp) \wedge (M_1 \neq \perp)$

Fig. 3. Games $\mathrm{WROB}_{\mathcal{GE}}$ and $\mathrm{SROB}_{\mathcal{GE}}$ defining WROB-ATK and SROB-ATK security (respectively) of general encryption scheme $\mathcal{GE} = (\mathsf{PG}, \mathsf{KG}, \mathsf{Enc}, \mathsf{Dec})$. The procedures on the left are common to both games, which differ only in their **Finalize** procedures.

the deterministic decryption algorithm Dec returns either a plaintext message M or $\perp$ to indicate that it rejects. We say that $\mathcal{GE}$ is a public-key encryption (PKE) scheme if $msk = \varepsilon$ and KG ignores its *id* input. To recover the usual syntax we may in this case write the output of PG as *pars* rather than $(pars, msk)$ and omit *msk*, *id* as inputs to KG. We say that $\mathcal{GE}$ is an identity-based encryption (IBE) scheme if $ek = id$, meaning the encryption key created by KG on inputs *pars*, *msk*, *id* always equals *id*. To recover the usual syntax we may in this case write the output of KG as *dk* rather than (ek, dk). It is easy to see that in this way we have recovered the usual primitives. But there are general encryption schemes that are neither PKE nor IBE schemes, meaning the primitive is indeed more general.

CORRECTNESS. Correctness of a general encryption scheme $\mathcal{GE} = (\mathsf{PG}, \mathsf{KG}, \mathsf{Enc}, \mathsf{Dec})$ requires that, for all $(pars, msk) \in [\mathsf{PG}]$, all plaintexts M in the underlying message space associated to *pars*, all identities *id*, and all $(ek, dk) \in [\mathsf{KG}(pars, msk, id)]$, we have $\mathsf{Dec}(pars, ek, dk, \mathsf{Enc}(pars, ek, M)) = M$ with probability one, where the probability is taken over the coins of Enc.

AI-ATK SECURITY. Historically, definitions of data privacy (IND) [18,21,16,5,11] and anonymity (ANON) [4,1] have been separate. We are interested in schemes that achieve both, so rather than use separate definitions we follow [12] and capture both simultaneously via game $\mathrm{AI}_{\mathcal{GE}}$ of Fig. 2. A cpa adversary is one that makes no **Dec** queries, and a cca adversary is one that might make such queries. The ai-advantage of such an adversary, in either case, is

$$\mathbf{Adv}^{\mathrm{ai}}_{\mathcal{GE}}(A) = 2 \cdot \Pr\left[\,\mathrm{AI}^{A}_{\mathcal{GE}}\,\right] - 1 .$$

We will assume an ai-adversary makes only one **LR** query, since a hybrid argument shows that making q of them can increase its ai-advantage by a factor of at most q.

Oracle **GetDK** represents the IBE key-extraction oracle [11]. In the PKE case it is superfluous in the sense that removing it results in a definition that is equivalent up to a factor depending on the number of **GetDK** queries. That's probably why the usual definition has no such oracle. But conceptually, if it is there for IBE, it ought to be there for PKE, and it does impact concrete security.

ROBUSTNESS. Associated to general encryption scheme $\mathcal{GE} = (\mathsf{PG}, \mathsf{KG}, \mathsf{Enc}, \mathsf{Dec})$ are games WROB, SROB of Fig. 3. As before, a cpa adversary is one that makes no **Dec** queries, and a cca adversary is one that might make such queries. The wrob and srob advantages of an adversary, in either case, are

$$\mathbf{Adv}^{\mathrm{wrob}}_{\mathcal{GE}}(A) = \Pr\left[\mathrm{WROB}^{A}_{\mathcal{GE}}\right] \quad \text{and} \quad \mathbf{Adv}^{\mathrm{srob}}_{\mathcal{GE}}(A) = \Pr\left[\mathrm{SROB}^{A}_{\mathcal{GE}}\right].$$

The difference between WROB and SROB is that in the former the adversary produces a message M, and C is its encryption under the encryption key of one of the given identities, while in the latter it produces C directly, and may not obtain it as an honest encryption. It is worth clarifying that in the PKE case the adversary does *not* get to choose the encryption (public) keys of the identities it is targeting. These are honestly and independently chosen, in real life by the identities themselves and in our formalization by the games.

3 Robustness Failures of Encryption with Redundancy

A natural privacy-and-anonymity-preserving approach to add robustness to an encryption scheme is to add redundancy before encrypting, and upon decryption reject if the redundancy is absent. Here we investigate the effectiveness of this encryption with redundancy approach, justifying the negative results discussed in Section 1 and summarized in the first table of Fig. 1.

REDUNDANCY CODES AND THE TRANSFORM. A redundancy code $\mathcal{RED} = (\mathsf{RKG}, \mathsf{RC}, \mathsf{RV})$ is a triple of algorithms. The redundancy key generation algorithm RKG generates a key K. On input K and data x the redundancy computation algorithm RC returns redundancy r. Given K, x, and claimed redundancy r, the deterministic redundancy verification algorithm RV returns 0 or 1. We say that $\mathcal{RED}$ is unkeyed if the key K output by RKG is always equal to ε, and keyed otherwise. The correctness condition is that for all x we have $\mathsf{RV}(K, x, \mathsf{RC}(K, x)) = 1$ with probability one, where the probability is taken over the coins of RKG and RC. (We stress that the latter is allowed to be randomized.)

Given a general encryption scheme $\mathcal{GE} = (\mathsf{PG}, \mathsf{KG}, \mathsf{Enc}, \mathsf{Dec})$ and a redundancy code $\mathcal{RED} = (\mathsf{RKG}, \mathsf{RC}, \mathsf{RV})$, the *encryption with redundancy transform* associates to them the general encryption scheme $\overline{\mathcal{GE}} = (\overline{\mathsf{PG}}, \overline{\mathsf{KG}}, \overline{\mathsf{Enc}}, \overline{\mathsf{Dec}})$ whose algorithms are shown on the left side of Fig. 5. Note that the transform has the first of our desired properties, namely that it preserves AI-ATK.

RKG	$\mathsf{RC}(K, ek\|M)$	$\mathsf{RV}(K, ek\|M, r)$
Return $K \leftarrow \varepsilon$	Return ε	Return 1
Return $K \leftarrow \varepsilon$	Return 0^k	Return $(r = 0^k)$
Return $K \leftarrow \varepsilon$	Return ek	Return $(r = ek)$
Return $K \leftarrow \varepsilon$	$L \stackrel{\$}{\leftarrow} \{0,1\}^k$; Return $L\|H(L, ek\|M)$	$L\|h \leftarrow r$; Return $(h = H(L, ek\|M))$
Return $K \stackrel{\$}{\leftarrow} \{0,1\}^k$	Return K	Return $(r = K)$
Return $K \stackrel{\$}{\leftarrow} \{0,1\}^k$	Return $H(K, ek\|M)$	Return $(r = H(K, ek\|M))$

Fig. 4. Examples of redundancy codes, where the data x is of the form $ek\|M$. The first four are unkeyed and the last two are keyed.

Also if $\mathcal{GE}$ is a PKE scheme then so is $\overline{\mathcal{GE}}$, and if $\mathcal{GE}$ is an IBE scheme then so is $\overline{\mathcal{GE}}$, which means the results we obtain here apply to both settings.

Fig. 4 shows example redundancy codes for the transform. With the first, $\overline{\mathcal{GE}}$ is identical to $\mathcal{GE}$, so that the counterexample below shows that AI-CCA does not imply WROB-CPA. The second and third rows show redundancy equal to a constant or the encryption key as examples of (unkeyed) redundancy codes. The fourth row shows a code that is randomized but still unkeyed. The hash function H could be a MAC or a collision resistant function. The last two are keyed redundancy codes, the first the simple one that just always returns the key, and the second using a hash function. Obviously, there are many other examples.

SROB FAILURE. We show that encryption with redundancy fails to provide strong robustness for *all* redundancy codes, whether keyed or not. More precisely, we show that for any redundancy code $\mathcal{RED}$ and both ATK $\in$ {CPA, CCA}, there is an AI-ATK encryption scheme $\mathcal{GE}$ such that the scheme $\overline{\mathcal{GE}}$ resulting from the encryption-with-redundancy transform applied to $\mathcal{GE}, \mathcal{RED}$ is not SROB-CPA. We build $\mathcal{GE}$ by modifying a given AI-ATK encryption scheme $\mathcal{GE}^* = (\mathsf{PG}, \mathsf{KG}, \mathsf{Enc}^*, \mathsf{Dec}^*)$. Let l be the number of coins used by RC, and let $\mathsf{RC}(x; \omega)$ denote the result of executing RC on input x with coins $\omega \in \{0,1\}^l$. Let M^* be a function that given *pars* returns a point in the message space associated to *pars* in $\mathcal{GE}^*$. Then $\mathcal{GE} = (\mathsf{PG}, \mathsf{KG}, \mathsf{Enc}, \mathsf{Dec})$ where the new algorithms are shown on the bottom right side of Fig. 5. The reason we used 0^l as coins for RC here is that Dec is required to be deterministic.

Our first claim is that the assumption that $\mathcal{GE}^*$ is AI-ATK implies that $\mathcal{GE}$ is too. Our second claim, that $\overline{\mathcal{GE}}$ is not SROB-CPA, is demonstrated by the following attack. For a pair id_0, id_1 of distinct identities of its choice, the adversary A, on input $(pars, K)$, begins with queries $ek_0 \stackrel{\$}{\leftarrow} \mathbf{GetEK}(id_0)$ and $ek_1 \stackrel{\$}{\leftarrow} \mathbf{GetEK}(id_1)$. It then creates ciphertext $C \leftarrow 0 \,\|\, K$ and returns (id_0, id_1, C). We claim that $\mathbf{Adv}^{\mathrm{srob}}_{\overline{\mathcal{GE}}}(A) = 1$. Letting dk_0, dk_1 denote the decryption keys corresponding to ek_0, ek_1 respectively, the reason is the following. For both $b \in \{0,1\}$, the output of $\mathsf{Dec}(pars, ek_b, dk_b, C)$ is $M^*(pars)\|r_b(pars)$ where $r_b(pars) = \mathsf{RC}(K, ek_b\|M^*(pars); 0^l)$. But the correctness of $\mathcal{RED}$ implies

Algorithm $\overline{\mathsf{PG}}$
$(pars, msk) \xleftarrow{\$} \mathsf{PG}$; $K \xleftarrow{\$} \mathsf{RKG}$
Return $((pars, K), msk)$

Algorithm $\overline{\mathsf{KG}}((pars, K), msk, id)$
$(ek, dk) \xleftarrow{\$} \mathsf{KG}(pars, msk, id)$
Return ek

Algorithm $\overline{\mathsf{Enc}}((pars, K), ek, M)$
$r \xleftarrow{\$} \mathsf{RC}(K, ek \| M)$
$C \xleftarrow{\$} \mathsf{Enc}(pars, ek, M \| r)$
Return C

Algorithm $\overline{\mathsf{Dec}}((pars, K), ek, dk, C)$
$M \| r \leftarrow \mathsf{Dec}(pars, ek, dk, C)$
If $\mathsf{RV}(K, ek \| M, r) = 1$ then return M
Else return $\bot$

Algorithm $\mathsf{Enc}(pars, ek, M)$
$C \xleftarrow{\$} \mathsf{Enc}^*(pars, ek, M)$
Return C

Algorithm $\mathsf{Dec}(pars, ek, dk, C)$
$M \leftarrow \mathsf{Dec}^*(pars, ek, dk, C)$
If $M = \bot$ then
$M \leftarrow M^*(pars) \| \mathsf{RC}(\varepsilon, ek \| M^*(pars); 0^l)$
Return M

Algorithm $\mathsf{Enc}(pars, ek, M)$
$C^* \xleftarrow{\$} \mathsf{Enc}^*(pars, ek, M)$
Return $1 \| C^*$

Algorithm $\mathsf{Dec}(pars, ek, dk, C)$
$b \| C^* \leftarrow C$
If $b = 1$ then return $\mathsf{Dec}^*(pars, ek, dk, C^*)$
Else return $M^*(pars) \| \mathsf{RC}(C^*, ek \| M^*(pars); 0^l)$

Fig. 5. Left: Transformed scheme for the encryption with redundancy paradigm. **Top Right**: Counterexample for WROB. **Bottom Right**: Counterexample for SROB.

that $\mathsf{RV}(K, ek_b \| M^*(pars), r_b(pars)) = 1$ and hence $\overline{\mathsf{Dec}}((pars, K), ek_b, dk_b, C)$ returns $M^*(pars)$ rather than $\bot$.

WROB FAILURE. We show that encryption with redundancy fails to provide even *weak* robustness for all *unkeyed* redundancy codes. This is still a powerful negative result because many forms of redundancy that might intuitively work, such the first four of Fig. 4, are included. More precisely, we claim that for any unkeyed redundancy code $\mathcal{RED}$ and both ATK $\in$ {CPA, CCA}, there is an AI-ATK encryption scheme $\mathcal{GE}$ such that the scheme $\overline{\mathcal{GE}}$ resulting from the encryption-with-redundancy transform applied to $\mathcal{GE}, \mathcal{RED}$ is not WROB-CPA. We build $\mathcal{GE}$ by modifying a given AI-ATK + WROB-CPA encryption scheme $\mathcal{GE}^* = (\mathsf{PG}, \mathsf{KG}, \mathsf{Enc}^*, \mathsf{Dec}^*)$. With notation as above, the new algorithms for the scheme $\mathcal{GE} = (\mathsf{PG}, \mathsf{KG}, \mathsf{Enc}, \mathsf{Dec})$ are shown on the top right side of Fig. 5.

Our first claim is that the assumption that $\mathcal{GE}^*$ is AI-ATK implies that $\mathcal{GE}$ is too. Our second claim, that $\overline{\mathcal{GE}}$ is not WROB-CPA, is demonstrated by the following attack. For a pair id_0, id_1 of distinct identities of its choice, the adversary A, on input $(pars, \varepsilon)$, makes queries $ek_0 \xleftarrow{\$}$ **GetEK**(id_0) and $ek_1 \xleftarrow{\$}$ **GetEK**(id_1) and returns $(id_0, id_1, M^*(pars))$. We claim that $\mathbf{Adv}^{\mathrm{wrob}}_{\overline{\mathcal{GE}}}(A)$ is high. Letting dk_1 denote the decryption key for ek_1, the reason is the following. Let $r_0 \xleftarrow{\$} \mathsf{RC}(\varepsilon, ek_0 \| M^*(pars))$ and $C \xleftarrow{\$} \mathsf{Enc}(pars, ek_0, M^*(pars) \| r_0)$. The assumed WROB-CPA security of $\mathcal{GE}^*$ implies that $\mathsf{Dec}(pars, ek_1, dk_1, C)$ is most probably $M^*(pars) \| r_1(pars)$ where $r_1(pars) = \mathsf{RC}(\varepsilon, ek_1 \| M^*(pars); 0^l)$. But the correctness of $\mathcal{RED}$ implies that $\mathsf{RV}(\varepsilon, ek_1 \| M^*(pars), r_1(pars)) = 1$ and hence $\overline{\mathsf{Dec}}((pars, \varepsilon), ek_1, dk_1, C)$ returns $M^*(pars)$ rather than $\bot$.

4 Transforms That Work

We present a transform that confers weak robustness and another that confers strong robustness. They preserve privacy and anonymity, work for PKE as well as IBE, and for CPA as well as CCA. In both cases the security proofs surface some delicate issues. Besides being useful in its own right, the weak robustness transform is a crucial step in obtaining strong robustness, so we begin there.

WEAK ROBUSTNESS TRANSFORM. We saw that encryption-with-redundancy fails to provide even weak robustness if the redundancy code is unkeyed. Here we show that if the redundancy code is keyed, even in the simplest possible way where the redundancy is just the key itself, the transform does provide weak robustness, turning any AI-ATK secure general encryption scheme into an AI-ATK + WROB-ATK one, for both ATK $\in \{\text{CPA}, \text{CCA}\}$.

The transformed scheme encrypts with the message a key K placed in the public parameters. In more detail, the *weak robustness transform* associates to a given general encryption scheme $\mathcal{GE} = (\mathsf{PG}, \mathsf{KG}, \mathsf{Enc}, \mathsf{Dec})$ and integer parameter k, representing the length of K, the general encryption scheme $\overline{\mathcal{GE}} = (\overline{\mathsf{PG}}, \overline{\mathsf{KG}}, \overline{\mathsf{Enc}}, \overline{\mathsf{Dec}})$ whose algorithms are depicted in Fig. 6. Note that if $\mathcal{GE}$ is a PKE scheme then so is $\overline{\mathcal{GE}}$ and if $\mathcal{GE}$ is an IBE scheme then so is $\overline{\mathcal{GE}}$, so that our results, captured by Theorem 2 below, cover both settings.

The intuition for the weak robustness of $\overline{\mathcal{GE}}$ is that the $\mathcal{GE}$ decryption under one key, of an encryption of $\overline{M}\|K$ created under another key, cannot, by the assumed AI-ATK security of $\mathcal{GE}$, reveal K, and hence the check will fail. This is pretty much right for PKE, but the delicate issue is that for IBE, information about K can enter via the identities, which in this case are the encryption keys and are chosen by the adversary as a function of K. The AI-ATK security of $\mathcal{GE}$ is no protection against this. We show however that this can be dealt with by making K sufficiently longer than the identities.

Theorem 2. *Let $\mathcal{GE} = (\mathsf{PG}, \mathsf{KG}, \mathsf{Enc}, \mathsf{Dec})$ be a general encryption scheme with identity space $\{0,1\}^n$, and let $\overline{\mathcal{GE}} = (\overline{\mathsf{PG}}, \overline{\mathsf{KG}}, \overline{\mathsf{Enc}}, \overline{\mathsf{Dec}})$ be the general encryption scheme resulting from applying the weak robustness transform to $\mathcal{GE}$ and integer parameter k. Then*

1. AI-ATK*: Let A be an ai-adversary against $\overline{\mathcal{GE}}$. Then there is an ai-adversary B against $\mathcal{GE}$ such that $\mathbf{Adv}^{\mathrm{ai}}_{\overline{\mathcal{GE}}}(A) = \mathbf{Adv}^{\mathrm{ai}}_{\mathcal{GE}}(B)$. Adversary B inherits the query profile of A and has the same running time as A. If A is a cpa adversary then so is B.*
2. WROB-ATK*: Let A be a wrob adversary against $\overline{\mathcal{GE}}$ with running time t, and let $\ell = 2n + \lceil \log_2(t) \rceil$. Then there is an ai-adversary B against $\mathcal{GE}$ such that $\mathbf{Adv}^{\mathrm{wrob}}_{\overline{\mathcal{GE}}}(A) \leq \mathbf{Adv}^{\mathrm{ai}}_{\mathcal{GE}}(B) + 2^{\ell-k}$. Adversary B inherits the query profile of A and has the same running time as A. If A is a cpa adversary then so is B.* ∎

The first part of the theorem implies that if $\mathcal{GE}$ is AI-ATK then $\overline{\mathcal{GE}}$ is AI-ATK as well. The second part of the theorem implies that if $\mathcal{GE}$ is AI-ATK and k is

Algorithm $\overline{\mathsf{PG}}$	**Algorithm** $\overline{\mathsf{KG}}((pars, K), msk, id)$
$(pars, msk) \xleftarrow{\$} \mathsf{PG}$	$(ek, dk) \xleftarrow{\$} \mathsf{KG}(pars, msk, id)$
$K \xleftarrow{\$} \{0,1\}^k$	Return (ek, dk)
Return $((pars, K), msk)$	**Algorithm** $\overline{\mathsf{Dec}}((pars, K), ek, dk, C)$
Algorithm $\overline{\mathsf{Enc}}((pars, K), ek, \overline{M})$	$M \leftarrow \mathsf{Dec}(pars, ek, dk, C)$
$C \xleftarrow{\$} \mathsf{Enc}(pars, ek, \overline{M} \| K))$	If $M = \perp$ then return $\perp$
Return C	$\overline{M} \| K^* \leftarrow M$
	If $(K = K^*)$ then return $\overline{M}$
	Else Return $\perp$

Fig. 6. General encryption scheme $\overline{\mathcal{GE}} = (\overline{\mathsf{PG}}, \overline{\mathsf{KG}}, \overline{\mathsf{Enc}}, \overline{\mathsf{Dec}})$ resulting from applying our weak-robustness transform to general encryption scheme $\mathcal{GE} = (\mathsf{PG}, \mathsf{KG}, \mathsf{Enc}, \mathsf{Dec})$ and integer parameter k

chosen sufficiently larger than $2n + \lceil \log_2(t) \rceil$ then $\overline{\mathcal{GE}}$ is WROB-ATK. In both cases this is for both ATK $\in$ {CPA, CCA}. The theorem says it directly for CCA, and for CPA by the fact that if A is a cpa adversary then so is B. When we say that B inherits the query profile of A we mean that for every oracle that B has, if A has an oracle of the same name and makes q queries to it, then this is also the number B makes. The proof of the first part of the theorem is straightforward and is omitted. The proof of the second part is given in [2]. It is well known that collision-resistant hashing of identities preserves AI-ATK and serves to make them of fixed length [7] so the assumption that the identity space is $\{0,1\}^n$ rather than $\{0,1\}^*$ is not really a restriction. In practice we might hash with SHA256 so that $n = 256$, and, assuming $t \leq 2^{128}$, setting $k = 768$ would make $2^{\ell-k} = 2^{-128}$.

Commitment schemes. Our strong robustness transform will use commitments. A commitment scheme is a 3-tuple $\mathcal{CMT} = (\mathsf{CPG}, \mathsf{Com}, \mathsf{Ver})$. The parameter generation algorithm CPG returns public parameters *cpars*. The committal algorithm Com takes *cpars* and data x as input and returns a commitment *com* to x along with a decommittal key *dec*. The deterministic verification algorithm Ver takes *cpars*, x, *com*, *dec* as input and returns 1 to indicate that accepts or 0 to indicate that it rejects. Correctness requires that, for any $x \in \{0,1\}^*$, any $cpars \in [\mathsf{CPG}]$, and any $(com, dec) \in [\mathsf{Com}(cpars, x)]$, we have that $\mathsf{Ver}(cpars, x, com, dec) = 1$ with probability one, where the probability is taken over the coins of Com. We require the scheme to have the *uniqueness* property, which means that for any $x \in \{0,1\}^*$, any $cpars \in [\mathsf{CPG}]$, and any $(com, dec) \in [\mathsf{Com}(cpars, x)]$ it is the case that $\mathsf{Ver}(cpars, x, com^*, dec) = 0$ for all $com^* \neq com$. In most schemes the decommittal key is the randomness used by the committal algorithm and verification is by re-applying the committal function, which ensures uniqueness. The advantage measures $\mathbf{Adv}^{\mathrm{hide}}_{\mathcal{CMT}}(A)$ and $\mathbf{Adv}^{\mathrm{bind}}_{\mathcal{CMT}}(A)$, referring to the standard hiding and binding properties, are recalled in [2]. We refer to the corresponding notions as HIDE and BIND.

Algorithm $\overline{\mathsf{PG}}$	Algorithm $\overline{\mathsf{KG}}((pars, cpars), msk, id)$
$(pars, msk) \xleftarrow{\$} \mathsf{PG}$	$(ek, dk) \xleftarrow{\$} \mathsf{KG}(pars, msk, id)$
$cpars \xleftarrow{\$} \mathsf{CPG}$	Return (ek, dk)
Return $((pars, cpars), msk)$	Algorithm $\overline{\mathsf{Dec}}((pars, cpars), ek, dk, (C, com))$
Algorithm $\overline{\mathsf{Enc}}((pars, cpars), ek, \overline{M})$	$M \leftarrow \mathsf{Dec}(pars, ek, dk, C)$
$(com, dec) \xleftarrow{\$} \mathsf{Com}(cpars, ek)$	If $M = \bot$ then return $\bot$
$C \xleftarrow{\$} \mathsf{Enc}(pars, ek, \overline{M} \| dec))$	$\overline{M} \| dec \leftarrow M$
Return (C, com)	If $(\mathsf{Ver}(cpars, ek, com, dec) = 1)$ then return $\overline{M}$
	Else Return $\bot$

Fig. 7. General encryption scheme $\overline{\mathcal{GE}} = (\overline{\mathsf{PG}}, \overline{\mathsf{KG}}, \overline{\mathsf{Enc}}, \overline{\mathsf{Dec}})$ resulting from applying our strong robustness transform to general encryption scheme $\mathcal{GE} = (\mathsf{PG}, \mathsf{KG}, \mathsf{Enc}, \mathsf{Dec})$ and commitment scheme $\mathcal{CMT} = (\mathsf{CPG}, \mathsf{Com}, \mathsf{Ver})$

THE STRONG ROBUSTNESS TRANSFORM. The idea is for the ciphertext to include a commitment to the encryption key. The commitment is *not* encrypted, but the decommittal key is. In detail, given a general encryption scheme $\mathcal{GE} = (\mathsf{PG}, \mathsf{KG}, \mathsf{Enc}, \mathsf{Dec})$ and a commitment scheme $\mathcal{CMT} = (\mathsf{CPG}, \mathsf{Com}, \mathsf{Ver})$ the *strong robustness transform* associates to them the general encryption scheme $\overline{\mathcal{GE}} = (\overline{\mathsf{PG}}, \overline{\mathsf{KG}}, \overline{\mathsf{Enc}}, \overline{\mathsf{Dec}})$ whose algorithms are depicted in Fig. 7. Note that if $\mathcal{GE}$ is a PKE scheme then so is $\overline{\mathcal{GE}}$ and if $\mathcal{GE}$ is an IBE scheme then so is $\overline{\mathcal{GE}}$, so that our results, captured by the Theorem 3, cover both settings.

In this case the delicate issue is not the robustness but the AI-ATK security of $\overline{\mathcal{GE}}$ in the CCA case. Intuitively, the hiding security of the commitment scheme means that a $\overline{\mathcal{GE}}$ ciphertext does not reveal the encryption key. As a result, we would expect AI-ATK security of $\overline{\mathcal{GE}}$ to follow from the commitment hiding security and the assumed AI-ATK security of $\mathcal{GE}$. This turns out not to be true, and demonstrably so, meaning there is a counterexample to this claim. (See below.) What we show is that the claim is true if $\mathcal{GE}$ is additionally WROB-ATK. This property, if not already present, can be conferred by first applying our weak robustness transform.

Theorem 3. *Let* $\mathcal{GE} = (\mathsf{PG}, \mathsf{KG}, \mathsf{Enc}, \mathsf{Dec})$ *be a general encryption scheme, and let* $\overline{\mathcal{GE}} = (\overline{\mathsf{PG}}, \overline{\mathsf{KG}}, \overline{\mathsf{Enc}}, \overline{\mathsf{Dec}})$ *be the general encryption scheme resulting from applying the strong robustness transform to* $\mathcal{GE}$ *and commitment scheme* $\mathcal{CMT} = (\mathsf{CPG}, \mathsf{Com}, \mathsf{Ver})$*. Then*

1. <u>AI-ATK:</u> *Let A be an ai-adversary against $\overline{\mathcal{GE}}$. Then there is a wrob adversary W against $\mathcal{GE}$, a hiding adversary H against $\mathcal{CMT}$ and an ai-adversary B against $\mathcal{GE}$ such that*

$$\mathbf{Adv}^{\mathrm{ai}}_{\overline{\mathcal{GE}}}(A) \leq 2 \cdot \mathbf{Adv}^{\mathrm{wrob}}_{\mathcal{GE}}(W) + 2 \cdot \mathbf{Adv}^{\mathrm{hide}}_{\mathcal{CMT}}(H) + 3 \cdot \mathbf{Adv}^{\mathrm{ai}}_{\mathcal{GE}}(B) .$$

Adversaries W, B inherit the query profile of A, and adversaries W, H, B have the same running time as A. If A is a cpa adversary then so are W, B.

2. <u>SROB-ATK</u>: *Let A be a srob adversary against $\overline{\mathcal{GE}}$ making q* ***GetEK*** *queries. Then there is a binding adversary B against $\mathcal{CMT}$ such that*

$$\mathbf{Adv}^{\mathrm{srob}}_{\overline{\mathcal{GE}}}(A) \leq \mathbf{Adv}^{\mathrm{bind}}_{\mathcal{CMT}}(B) + \binom{q}{2} \cdot \mathbf{Coll}_{\mathcal{GE}} .$$

Adversary B has the same running time as A. ∎

The first part of the theorem implies that if $\mathcal{GE}$ is AI-ATK and WROB-ATK and $\mathcal{CMT}$ is HIDE then $\overline{\mathcal{GE}}$ is AI-ATK, and the second part of the theorem implies that if $\mathcal{CMT}$ is BIND secure and $\mathcal{GE}$ has low encryption key collision probability then $\overline{\mathcal{GE}}$ is SROB-ATK. In both cases this is for both ATK $\in$ {CPA, CCA}. We remark that the proof shows that in the CPA case the WROB-ATK assumption on $\mathcal{GE}$ in the first part is actually not needed. The encryption key collision probability $\mathbf{Coll}_{\mathcal{GE}}$ of $\mathcal{GE}$ is defined as the maximum probability that $ek_0 = ek_1$ in the experiment where we let $(pars, msk) \stackrel{\$}{\leftarrow} \mathsf{PG}$ and then let $(ek_0, dk_0) \stackrel{\$}{\leftarrow} \mathsf{KG}(pars, msk, id_0)$ and $(ek_1, dk_1) \stackrel{\$}{\leftarrow} \mathsf{KG}(pars, msk, id_1)$, where the maximum is over all distinct identities id_0, id_1. The collision probability is zero in the IBE case since $ek_0 = id_0 \neq id_1 = ek_1$. It is easy to see that $\mathcal{GE}$ being AI implies $\mathbf{Coll}_{\mathcal{GE}}$ is negligible, so asking for low encryption key collision probability is in fact not an extra assumption. (For a general encryption scheme the adversary needs to have hardwired the identities that achieve the maximum, but this is not necessary for PKE because here the probability being maximized is the same for all pairs of distinct identities.) The reason we made the encryption key collision probability explicit is that for most schemes it is unconditionally low. For example, when $\mathcal{GE}$ is the ElGamal PKE scheme, it is $1/|\mathbb{G}|$ where $\mathbb{G}$ is the group being used. Proofs of both parts of the theorem are in [2].

THE NEED FOR WEAK-ROBUSTNESS. As we said above, the AI-ATK security of $\overline{\mathcal{GE}}$ won't be implied merely by that of $\mathcal{GE}$. (We had to additionally assume that $\mathcal{GE}$ is WROB-ATK.) Here we justify this somewhat counter-intuitive claim. This discussion is informal but can be turned into a formal counterexample. Imagine that the decryption algorithm of $\mathcal{GE}$ returns a fixed string of the form $(\hat{M}, \hat{dec})$ whenever the wrong key is used to decrypt. Moreover, imagine $\mathcal{CMT}$ is such that it is easy, given $cpars, x, dec$, to find com so that $\mathsf{Ver}(cpars, x, com, dec) = 1$. (This is true for any commitment scheme where dec is the coins used by the Com algorithm.) Consider then the AI-ATK adversary A against the transformed scheme that that receives a challenge ciphertext (C^*, com^*) where $C^* \leftarrow \mathsf{Enc}(pars, \mathsf{EK}[id_b], M^* \| dec^*)$ for hidden bit $b \in \{0,1\}$. It then creates a commitment $\hat{com}$ of $\mathsf{EK}[id_1]$ with opening information $\hat{dec}$, and queries $(C^*, \hat{com})$ to be decrypted under $\mathsf{DK}[id_0]$. If $b = 0$ this query will probably return $\perp$ because $\mathsf{Ver}(cpars, \mathsf{EK}[id_0], \hat{com}, dec^*)$ is unlikely to be 1, but if $b = 1$ it returns $\hat{M}$, allowing A to determine the value of b. The weak robustness of $\mathcal{GE}$ rules out such anomalies.

Algorithm PG
$K \xleftarrow{\$} \mathsf{Keys}(H)$; $g_1 \xleftarrow{\$} \mathbb{G}^*$; $w \xleftarrow{\$} \mathbb{Z}_p^*$; $g_2 \leftarrow g_1^w$; Return (g_1, g_2, K)

Algorithm $\mathsf{KG}(g_1, g_2, K)$
$x_1, x_2, y_1, y_2, z_1, z_2 \xleftarrow{\$} \mathbb{Z}_p$; $e \leftarrow g_1^{x_1} g_2^{x_2}$; $f \leftarrow g_1^{y_1} g_2^{y_2}$; $h \leftarrow g_1^{z_1} g_2^{z_2}$
Return $((e, f, h), (x_1, x_2, y_1, y_2, z_1, z_2))$

Algorithm $\mathsf{Enc}((g_1, g_2, K), (e, f, h), M)$
$u \xleftarrow{\$} \mathbb{Z}_p^{\boxed{*}}$; $a_1 \leftarrow g_1^u$; $a_2 \leftarrow g_2^u$; $b \leftarrow h^u$; $c \leftarrow b \cdot M$; $v \leftarrow H(K, (a_1, a_2, c))$; $d \leftarrow e^u f^{uv}$
Return (a_1, a_2, c, d)

Algorithm $\mathsf{Dec}((g_1, g_2, K), (e, f, h), (x_1, x_2, y_1, y_2, z_1, z_2), C)$
$(a_1, a_2, c, d) \leftarrow C$; $v \leftarrow H(K, (a_1, a_2, c))$; $M \leftarrow c \cdot a_1^{-z_1} a_2^{-z_2}$
If $d \neq a_1^{x_1+y_1 v} a_2^{x_2+y_2 v}$ Then $M \leftarrow \perp$
$\boxed{\text{If } a_1 = \mathbf{1} \text{ Then } M \leftarrow \perp}$
Return M

Fig. 8. The original CS scheme [15] does not contain the boxed code while the variant $\mathcal{CS}^*$ does. Although not shown above, the decryption algorithm in both versions always checks to ensure that the ciphertext $C \in \mathbb{G}^4$. The message space is $\mathbb{G}$.

5 A SROB-CCA Version of Cramer-Shoup

Let $\mathbb{G}$ be a group of prime order p, and H: $\mathsf{Keys}(H) \times \mathbb{G}^3 \to \mathbb{G}$ a family of functions. We assume $\mathbb{G}, p, H$ are fixed and known to all parties. Fig. 8 shows the Cramer-Shoup (CS) scheme and the variant $\mathcal{CS}^*$ scheme where $\mathbf{1}$ denotes the identity element of $\mathbb{G}$. The differences are boxed. Recall that the CS scheme was shown to be IND-CCA in [15] and ANO-CCA in [4]. However, for any message $M \in \mathbb{G}$ the ciphertext $(\mathbf{1}, \mathbf{1}, M, \mathbf{1})$ in the CS scheme decrypts to M under *any* *pars*, *pk*, and *sk*, meaning in particular that the scheme is not even SROB-CPA. The modified scheme $\mathcal{CS}^*$ —which continues to be IND-CCA and ANO-CCA— removes this pathological case by having Enc choose the randomness u to be non-zero —Enc draws u from $\mathbb{Z}_p^*$ while the CS scheme draws it from $\mathbb{Z}_p$— and then having Dec reject (a_1, a_2, c, d) if $a_1 = \mathbf{1}$. This thwarts the attack, but is there any other attack? We show that there is not by proving that $\mathcal{CS}^*$ is actually SROB-CCA. Our proof of robustness relies only on the security — specifically, pre-image resistance— of the hash family H: it does not make the DDH assumption. Our proof uses ideas from the information-theoretic part of the proof of [15].

We say that a family H: $\mathsf{Keys}(H) \times \mathsf{Dom}(H) \to \mathsf{Rng}(H)$ of functions is *pre-image resistant* if, given a key K and a *random* range element v^*, it is computationally infeasible to find a pre-image of v^* under $H(K, \cdot)$. The notion is captured formally by the following advantage measure for an adversary I:

$$\mathbf{Adv}_H^{\text{pre-img}}(I) = \Pr\Big[\, H(K, x) = v^* \;:\; K \xleftarrow{\$} \mathsf{Keys}(H) \,;\, v^* \xleftarrow{\$} \mathsf{Rng}(H) \,;\, x \xleftarrow{\$} I(K, v^*) \,\Big] .$$

Pre-image resistance is not implied by the standard notion of one-wayness, since in the latter the target v^* is the image under $H(K,\cdot)$ of a random domain point, which may not be a random range point. However, it seems like a fairly mild assumption on a practical cryptographic hash function and is implied by the notion of "everywhere pre-image resistance" of [22], the difference being that, for the latter, the advantage is the maximum probability over all $v^* \in \mathsf{Rng}(H)$. We now claim the following.

Theorem 4. *Let B be an adversary making two* **GetEK** *queries, no* **GetDK** *queries and at most $q-1$* **Dec** *queries, and having running time t. Then we can construct an adversary I such that*

$$\mathbf{Adv}^{\mathrm{srob}}_{\mathcal{CS}^*}(A) \;\leq\; \mathbf{Adv}^{\mathrm{pre\text{-}img}}_{H}(I) + \frac{2q+1}{p}\,. \tag{1}$$

Furthermore, the running time of I is $t + q\cdot O(t_{\exp})$ where $t_{\exp}$ denotes the time for one exponentiation in $\mathbb{G}$.

Since $\mathcal{CS}^*$ is a PKE scheme, the above automatically implies security even in the presence of multiple **GetEK** and **GetDK** queries as required by game $\mathrm{SROB}_{\mathcal{CS}^*}$. Thus the theorem implies that $\mathcal{CS}^*$ is SROB-CCA if H is pre-image resistant. A detailed proof of Theorem 4 is in [2]. Here we sketch some intuition.

We begin by conveniently modifying the game interface. We replace B with an adversary A that gets input $(g_1, g_2, K), (e_0, f_0, h_0), (e_1, f_1, h_1)$ representing the parameters that would be input to B and the public keys returned in response to B's two **GetEK** queries. Let $(x_{01}, x_{02}, y_{01}, y_{02}, z_{01}, z_{02})$ and $(x_{11}, x_{12}, y_{11}, y_{12}, z_{11}, z_{12})$ be the corresponding secret keys. The decryption oracle takes (only) a ciphertext and returns its decryption under *both* secret keys, setting a WIN flag if these are both non-$\perp$. Adversary A no longer needs an output, since it can win via a **Dec** query.

Suppose A makes a **Dec** query (a_1, a_2, c, d). Then the code of the decryption algorithm Dec from Fig. 8 tells us that, for this to be a winning query, it must be that

$$d \;=\; a_1^{x_{01}+y_{01}v} a_2^{x_{02}+y_{02}v} \;=\; a_1^{x_{11}+y_{11}v} a_2^{x_{12}+y_{12}v}$$

where $v = H(K,(a_1,a_2,c))$. Letting $u_1 = \log_{g_1}(a_1), u_2 = \log_{g_2}(a_2)$ and $s = \log_{g_1}(d)$, we have

$$s \;=\; u_1(x_{01}+y_{01}v) + wu_2(x_{02}+y_{02}v) \;=\; u_1(x_{11}+y_{11}v) + wu_2(x_{12}+y_{12}v) \tag{2}$$

However, even acknowledging that A knows little about $x_{b1}, x_{b2}, y_{b1}, y_{b2}$ ($b \in \{0,1\}$) through its **Dec** queries, it is unclear why Equation (2) is prevented by pre-image resistance —or in fact any property short of being a random oracle— of the hash function H. In particular, there seems no way to "plant" a target v^* as the value v of Equation (2) since the adversary controls u_1 and u_2. However, suppose now that $a_2 = a_1^w$. (We will discuss later why we can assume this.) This implies $wu_2 = wu_1$ or $u_2 = u_1$ since $w \neq 0$. Now from Equation (2) we have

$$u_1(x_{01}+y_{01}v) + wu_1(x_{02}+y_{02}v) - u_1(x_{11}+y_{11}v) - wu_1(x_{12}+y_{12}v) \;=\; 0\,.$$

We now see the value of enforcing $a_1 \neq 1$, since this implies $u_1 \neq 0$. After canceling u_1 and re-arranging terms, we have

$$v(y_{01} + wy_{02} - y_{11} - wy_{12}) + (x_{01} + wx_{02} - x_{11} - wx_{12}) = 0 . \quad (3)$$

Given that $x_{b1}, x_{b2}, y_{b1}, y_{b2}$ ($b \in \{0,1\}$) and w are chosen by the game, there is at most one solution v (modulo p) to Equation (3). We would like now to design I so that on input K, v^* it chooses $x_{b1}, x_{b2}, y_{b1}, y_{b2}$ ($b \in \{0,1\}$) so that the solution v to Equation (3) is v^*. Then (a_1, a_2, c) will be a pre-image of v^* which I can output.

To make all this work, we need to resolve two problems. The first is why we may assume $a_2 = a_1^w$ —which is what enables Equation (3)— given that a_1, a_2 are chosen by A. The second is to properly design I and show that it can simulate A correctly with high probability. To solve these problems, we consider, as in [15], a modified check under which decryption, rather than rejecting when $d \neq a_1^{x_1+y_1v} a_2^{x_2+y_2v}$, rejects when $a_2 \neq a_1^w$ or $d \neq a_1^{x+yv}$, where $x = x_1 + wx_2$, $y = y_1 + wy_2$, $v = H(K, (a_1, a_2, c))$ and (a_1, a_2, c, d) is the ciphertext being decrypted. See [2].

Acknowledgments

First and third authors were supported in part by the European Commission through the ICT Program under Contract ICT-2007-216646 ECRYPT II. First author was supported in part by the French ANR-07-SESU-008-01 PAMPA Project. Second author was supported in part by NSF grants CNS-0627779 and CCF-0915675. Third author was supported in part by a Postdoctoral Fellowship from the Research Foundation – Flanders (FWO – Vlaanderen) and by the European Community's Seventh Framework Programme project PrimeLife (grant agreement no. 216483).

We thank Chanathip Namprempre, who declined our invitation to be a co-author, for her participation and contributions in the early stage of this work.

References

1. Abdalla, M., Bellare, M., Catalano, D., Kiltz, E., Kohno, T., Lange, T., Malone-Lee, J., Neven, G., Paillier, P., Shi, H.: Searchable encryption revisited: Consistency properties, relation to anonymous IBE, and extensions. Journal of Cryptology 21(3), 350–391 (2008)
2. Abdalla, M., Bellare, M., Neven, G.: Robust encryption. Cryptology ePrint Archive (2009), Full version of this paper: `http://eprint.iacr.org/`
3. Abdalla, M., Bellare, M., Rogaway, P.: The oracle Diffie-Hellman assumptions and an analysis of DHIES. In: Naccache, D. (ed.) CT-RSA 2001. LNCS, vol. 2020, pp. 143–158. Springer, Heidelberg (2001)
4. Bellare, M., Boldyreva, A., Desai, A., Pointcheval, D.: Key-privacy in public-key encryption. In: Boyd, C. (ed.) ASIACRYPT 2001. LNCS, vol. 2248, pp. 566–582. Springer, Heidelberg (2001)
5. Bellare, M., Desai, A., Pointcheval, D., Rogaway, P.: Relations among notions of security for public-key encryption schemes. In: Krawczyk, H. (ed.) CRYPTO 1998. LNCS, vol. 1462, pp. 26–45. Springer, Heidelberg (1998)

6. Bellare, M., Rogaway, P.: The security of triple encryption and a framework for code-based game-playing proofs. In: Vaudenay, S. (ed.) EUROCRYPT 2006. LNCS, vol. 4004, pp. 409–426. Springer, Heidelberg (2006)
7. Boneh, D., Boyen, X.: Efficient selective-ID secure identity based encryption without random oracles. In: Cachin, C., Camenisch, J.L. (eds.) EUROCRYPT 2004. LNCS, vol. 3027, pp. 223–238. Springer, Heidelberg (2004)
8. Boneh, D., Canetti, R., Halevi, S., Katz, J.: Chosen-ciphertext security from identity-based encryption. SIAM Journal on Computing 36(5), 915–942 (2006)
9. Boneh, D., Di Crescenzo, G., Ostrovsky, R., Persiano, G.: Public key encryption with keyword search. In: Cachin, C., Camenisch, J.L. (eds.) EUROCRYPT 2004. LNCS, vol. 3027, pp. 506–522. Springer, Heidelberg (2004)
10. Boneh, D., Franklin, M.K.: Identity-based encryption from the Weil pairing. In: Kilian, J. (ed.) CRYPTO 2001. LNCS, vol. 2139, pp. 213–229. Springer, Heidelberg (2001)
11. Boneh, D., Franklin, M.K.: Identity based encryption from the Weil pairing. SIAM Journal on Computing 32(3), 586–615 (2003)
12. Boneh, D., Gentry, C., Hamburg, M.: Space-efficient identity based encryption without pairings. In: 48th FOCS, pp. 647–657. IEEE Computer Society Press, Los Alamitos (2007)
13. Boyen, X., Waters, B.: Anonymous hierarchical identity-based encryption (without random oracles). In: Dwork, C. (ed.) CRYPTO 2006. LNCS, vol. 4117, pp. 290–307. Springer, Heidelberg (2006)
14. Canetti, R., Halevi, S., Katz, J.: Chosen-ciphertext security from identity-based encryption. In: Cachin, C., Camenisch, J.L. (eds.) EUROCRYPT 2004. LNCS, vol. 3027, pp. 207–222. Springer, Heidelberg (2004)
15. Cramer, R., Shoup, V.: Design and analysis of practical public-key encryption schemes secure against adaptive chosen ciphertext attack. SIAM Journal on Computing 33(1), 167–226 (2003)
16. Dolev, D., Dwork, C., Naor, M.: Nonmalleable cryptography. SIAM Journal on Computing 30(2), 391–437 (2000)
17. Fujisaki, E., Okamoto, T.: Secure integration of asymmetric and symmetric encryption schemes. In: Wiener, M. (ed.) CRYPTO 1999. LNCS, vol. 1666, pp. 537–554. Springer, Heidelberg (1999)
18. Goldwasser, S., Micali, S.: Probabilistic encryption. Journal of Computer and System Sciences 28(2), 270–299 (1984)
19. Hofheinz, D., Weinreb, E.: Searchable encryption with decryption in the standard model. Cryptology ePrint Archive, Report 2008/423 (2008), `http://eprint.iacr.org/`
20. Katz, J., Sahai, A., Waters, B.: Predicate encryption supporting disjunctions, polynomial equations, and inner products. In: Smart, N.P. (ed.) EUROCRYPT 2008. LNCS, vol. 4965, pp. 146–162. Springer, Heidelberg (2008)
21. Rackoff, C., Simon, D.R.: Non-interactive zero-knowledge proof of knowledge and chosen ciphertext attack. In: Feigenbaum, J. (ed.) CRYPTO 1991. LNCS, vol. 576, pp. 433–444. Springer, Heidelberg (1992)
22. Rogaway, P., Shrimpton, T.: Cryptographic hash-function basics: Definitions, implications, and separations for preimage resistance, second-preimage resistance, and collision resistance. In: Roy, B., Meier, W. (eds.) FSE 2004. LNCS, vol. 3017, pp. 371–388. Springer, Heidelberg (2004)
23. Sako, K.: An auction protocol which hides bids of losers. In: Imai, H., Zheng, Y. (eds.) PKC 2000. LNCS, vol. 1751, pp. 422–432. Springer, Heidelberg (2000)

Privacy-Enhancing Cryptography: From Theory into Practice

Jan Camenisch

IBM Research – Zurich, Rüschlikon, Switzerland
jca@zurich.ibm.com

Abstract. We conduct an increasing part of our daily transactions electronically and thereby we leave an eternal electronic trail of personal data. We are almost never able to see what data about us we imprint, where it is processed or where it is stored. Indeed, controlling the dispersal of our data and protecting our privacy has become virtually impossible.

In this talk we will investigate the extent to which tools from cryptography and other technical means can help us to regain control of our data and to save our privacy. To this end, we will review the most important of the practical cryptographic mechanisms and discuss how they could be applied. In a second part, we will report on the readiness of the industry to indeed employ such technologies and on how governments address the current erosion of privacy.

D. Micciancio (Ed.): TCC 2010, LNCS 5978, p. 498, 2010.

Concise Mercurial Vector Commitments and Independent Zero-Knowledge Sets with Short Proofs

Benoît Libert[1,⋆] and Moti Yung[2]

[1] Université catholique de Louvain, Crypto Group, Belgium
[2] Google Inc. and Columbia University, USA

Abstract. Introduced by Micali, Rabin and Kilian (MRK), the basic primitive of zero-knowledge sets (ZKS) allows a prover to commit to a secret set S so as to be able to prove statements such as $x \in S$ or $x \notin S$. Chase *et al.* showed that ZKS protocols are underlain by a cryptographic primitive termed *mercurial commitment*. A (trapdoor) mercurial commitment has two commitment procedures. At committing time, the committer can choose not to commit to a specific message and rather generate a dummy value which it will be able to softly open to any message without being able to completely open it. Hard commitments, on the other hand, can be hardly or softly opened to only one specific message. At Eurocrypt 2008, Catalano, Fiore and Messina (CFM) introduced an extension called trapdoor q-mercurial commitment (qTMC), which allows committing to a vector of q messages. These qTMC schemes are interesting since their openings w.r.t. specific vector positions can be short (ideally, the opening length should not depend on q), which provides zero-knowledge sets with much shorter proofs when such a commitment is combined with a Merkle tree of arity q. The CFM construction notably features short proofs of *non-membership* as it makes use of a qTMC scheme with short soft openings. A problem left open is that hard openings still have size $O(q)$, which prevents proofs of membership from being as compact as those of non-membership. In this paper, we solve this open problem and describe a new qTMC scheme where hard and short position-wise openings, both, have *constant size*. We then show how our scheme is amenable to constructing independent zero-knowledge sets (i.e., ZKS's that prevent adversaries from correlating their set to the sets of honest provers, as defined by Gennaro and Micali). Our solution retains the short proof property for this important primitive as well.

Keywords: Zero-knowledge databases, mercurial commitments, efficiency, independence.

⋆ This author acknowledges the Belgian National Fund for Scientific Research (F.R.S.-F.N.R.S.) for their support and the BCRYPT Interuniversity Attraction Pole.

D. Micciancio (Ed.): TCC 2010, LNCS 5978, pp. 499–517, 2010.

1 Introduction

Introduced by Micali, Rabin and Kilian [21], zero-knowledge sets (ZKS) are fundamental secure data structures which allow a prover P to commit to a finite set S in such a way that, later on, he will be able to efficiently (and non-interactively) prove statements of the form $x \in S$ or $x \notin S$ without revealing anything else on S, not even its size. Of course, the prover should not be able to cheat and prove different statements about an element x. The more general notion of zero-knowledge elementary databases (ZK-EDB) generalizes zero-knowledge sets in that each element x has an associated value $D(x)$ in the committed database.

In [21], Micali *et al.* described a beautiful construction of ZK-EDB based on the discrete logarithm assumption. The MRK scheme relies on the shared random string model (where a random string chosen by some trusted entity is made available to all parties) and suitably uses an extension of Pedersen's trapdoor commitment [23]. In 2005, Chase *et al.* [10] gave general constructions of zero-knowledge databases and formalized a primitive named *mercurial commitment* which they proved to give rise to ZK-EDB protocols. The MRK construction turned out to be a particular instance of a general design combining mercurial commitments with a Merkle tree [20], where each internal node contains a mercurial commitment to its two children.

Informally speaking, mercurial commitments are commitments where the binding property is slightly relaxed in that the committer is allowed to *softly* open a commitment and say "if the commitment can be opened at all, then it opens to that message". Upon committing, the sender has to decide whether the commitment will be a hard commitment, that can be hard/soft-opened to only one message, or a soft one that can be soft-opened to any arbitrary message without committing the sender to a specific one. Unlike soft commitments that cannot be hard-opened, hard commitments can be opened either in the soft or the hard manner but soft openings can never contradict hard ones. In addition, hard and soft commitments should be computationally indistinguishable.

Related Work. Promptly after the work of Micali, Rabin and Kilian, Ostrovsky, Rackoff and Smith [22] described protocols for generalized queries (beyond membership/non-membership) for committed databases and also show how to add privacy to their schemes. Liskov [18] also extended the construction of Chase *et al.* [10] to obtain updatable zero-knowledge databases in the random oracle model. Subsequently, Catalano, Dodis and Visconti [8] gave simplified security definitions for (trapdoor) mercurial commitments and notably showed how to construct them out of one-way functions in the shared random string model.

In order to extend the properties of non-malleable commitments to zero-knowledge databases, Gennaro and Micali [15] formalized the notion of *independent* ZK-EDBs. Informally, this notion prevents adversaries from correlating their committed databases to those produced by honest provers.

More recently, Prabhakaran and Xue [24] defined the related notion of statistically hiding sets that requires the hiding property of zero-knowledge sets to be preserved against unbounded verifiers. At the same time, their notion of zero-knowledge was relaxed to permit unbounded simulators.

At Eurocrypt 2008, Catalano, Fiore and Messina [9] addressed the problem of compressing proofs in ZK-EDB schemes and gave significant improvements.

Our Contribution. The original construction of zero-knowledge database [21,10] considers a binary Merkle tree of height $O(\lambda)$, where λ is the security parameter (in such a way that the upper bound on the database size is exponential in λ and leaks no information on its actual size). Each internal node contains a mercurial commitment to (a hash value of) its two children whereas each leaf node is a mercurial commitment to a database entry. The crucial idea is that internal childless nodes contain soft commitments, which keeps the commitment generation phase efficient (*i.e.*, polynomial in λ). A proof of membership for the entry x consists of a sequence of hard openings for commitments appearing in nodes on the path from leaf x to the root. Proofs of non-membership proceed similarly but rather use soft openings along the path.

As noted in [9], the above approach often results in long proofs, which may be problematic in applications, like mobile Internet connections, where users are charged depending on the number of blocks that they send/receive. To address this issue, Catalano, Fiore and Messina (CFM) suggested to increase the branching factor q of the tree and to use a primitive called *trapdoor q-mercurial commitment* (qTMC). The latter is like an ordinary mercurial commitment with the difference that it allows committing to a vector of q messages at once. With regular mercurial commitments, increasing the arity of the tree is not appropriate as generating proofs entails to reveal q values (instead of 2) at each level of the tree. However, it becomes interesting with qTMC schemes that can be opened with respect to specific vector positions without having to disclose each one of the q committed messages. The CFM construction makes use of an elegant qTMC scheme where soft commitment openings consist of a single group element, which yields dramatically shorter proofs of non-membership. On the other hand, hard openings unfortunately comprise $O(q)$ elements in the qTMC scheme described in [9]. For this reason, proofs of membership remain significantly longer than proofs of non-membership.

In this paper, we solve a problem left open in [9] and consider a primitive called *concise* mercurial vector commitment, which is a qTMC scheme allowing to commit to a q-vector in such a way that (1) hard and soft position-wise openings *both* have constant (*i.e.*, independent of q) size; (2) the committer can hard-open the commitment at position $i \in \{1, \ldots, q\}$ without revealing anything on messages at other positions in the vector. We describe a simple and natural example of such scheme. Like the CFM q-mercurial commitment, our realization relies on a specific number theoretic assumption in bilinear groups. Implementing the CFM flat-tree system with our scheme immediately yields very short proofs of membership and while retaining short proofs of non-membership. Assuming that 2^{λ} is a theoretical bound on the database size, we obtain proofs comprising $O(\lambda/\log(q))$ group elements for membership and non-membership. In the CFM system, proofs of membership grow as $O(\lambda \cdot q/\log(q))$, which prevents one from compressing proofs of non-membership without incurring a blow-up in the length of proofs of membership. Using our commitment scheme, both kinds of proof can be shortened by increasing q as long as the common reference string (which has size $O(q)$ as in [9]) is not too large. With $q = 128$ for instance, proofs do not exceed 2 kB in instantiations using suitable parameters.

In addition, we also show that our qTMC scheme easily lends itself to the construction of independent zero-knowledge databases. To construct such protocols satisfying a strong definition of independence, Gennaro and Micali [15] used *multi-trapdoor* mercurial commitments that can be seen as families of mercurial commitments (in the same way as multi-trapdoor commitments [14] are families of trapdoor commitments). Modulo appropriate slight modifications, our scheme can be turned into a concise multi-trapdoor qTMC scheme. It thus gives rise to the first ZK-EDB realization that simultaneously provides independence and short proofs.

ORGANIZATION. Section 2 recalls the definitions of qTMC schemes and zero-knowledge databases. We describe the new q-mercurial commitment scheme and discuss its efficiency impact in sections 3 and 4. Section 5 finally explains how the resulting ZK-EDB scheme can be made independent.

2 Background

2.1 Complexity Assumptions

We use groups $(\mathbb{G}, \mathbb{G}_T)$ of prime order p with an efficiently computable map $e : \mathbb{G} \times \mathbb{G} \rightarrow \mathbb{G}_T$ such that $e(g^a, h^b) = e(g, h)^{ab}$ for any $(g, h) \in \mathbb{G} \times \mathbb{G}$, $a, b \in \mathbb{Z}$ and $e(g, h) \neq 1_{\mathbb{G}_T}$ whenever $g, h \neq 1_{\mathbb{G}}$. In this mathematical setting, we rely on a computational assumption previously used in [5,6].

Definition 1 ([5]). *Let $\mathbb{G}$ be a group of prime order p and $g \in \mathbb{G}$. The* **q-Diffie-Hellman Exponent** *(q-DHE) problem is, given a tuple of elements $(g, g_1, \ldots, g_q, g_{q+2}, \ldots, g_{2q})$ such that $g_i = g^{(\alpha^i)}$, for $i = 1, \ldots, q, q+2, \ldots, 2q$ and where $\alpha \stackrel{R}{\leftarrow} \mathbb{Z}_p^*$, to compute the missing group element $g_{q+1} = g^{(\alpha^{q+1})}$.*

As noted in [6], this problem is not easier than the one used in [5], which is to compute $e(g, h)^{(\alpha^{q+1})}$ on input of the same values and the additional element $h \in \mathbb{G}$. The generic hardness of q-DHE is thus implied by the generic security of the family of assumptions described in [4].

2.2 Trapdoor q-Mercurial Commitments

A trapdoor q-mercurial commitment (qTMC) consists of a set of efficient algorithms (qKeygen, qHCom, qHOpen, qHVer, qSCom, qSOpen, qSVer, qFake, qHEquiv, qSEquiv) with the following specifications.

qKeygen(λ, q): takes as input a security parameter λ and the number q of messages that can be committed to in a single commitment. The output is a pair of public/private keys (pk, tk).

$\mathsf{qHCom}_{pk}(m_1, \ldots, m_q)$: takes as input an ordered tuple of messages. It outputs a hard commitment C to $(m_1, \ldots, m_q)$ under the public key pk and some auxiliary state information aux.

$\mathsf{qHOpen}_{pk}(m, i, \mathsf{aux})$: is a hard opening algorithm. Given a pair $(C, \mathsf{aux}) = \mathsf{qHCom}_{pk}(m_1, \ldots, m_q)$, it outputs a hard de-commitment π of C w.r.t. position i if $m = m_i$. If $m \neq m_i$, it returns $\perp$.

$\mathsf{qHVer}_{pk}(m, i, C, \pi)$: is the hard verification algorithm. It outputs 1 if π gives evidence that C is a commitment to a sequence $(m_1, \ldots, m_q)$ such that $m_i = m$. Otherwise, it outputs 0.

$\mathsf{qSCom}_{pk}()$: is a probabilistic algorithm that generates a soft commitment and some auxiliary information aux. Such a commitment is not associated with a specific sequence of messages.

$\mathsf{qSOpen}_{pk}(m, i, \mathsf{flag}, \mathsf{aux})$: generates a soft de-commitment (a.k.a. "tease") τ of C to the message m at position i. The variable $\mathsf{flag} \in \{\mathbb{H}, \mathbb{S}\}$ indicates whether the state information aux corresponds to a hard commitment $(C, \mathsf{aux}) = \mathsf{qHCom}_{pk}(m_1, \ldots, m_q)$ or a soft one $(C, \mathsf{aux}) = \mathsf{qSCom}_{pk}()$. If $\mathsf{flag} = \mathbb{H}$ and $m \neq m_i$, the algorithm returns the error message $\perp$.

$\mathsf{qSVer}_{pk}(m, i, C, \tau)$: returns 1 if τ is a valid soft de-commitment of C to m at position i and 0 otherwise. If τ is valid and C is a hard commitment, its hard opening must be to m at index i.

$\mathsf{qFake}_{pk,tk}()$: is a randomized algorithm that takes as input the trapdoor tk and generates a q-fake commitment C and some auxiliary information aux. The commitment C is not bound to any sequence of messages. The q-fake commitment C is similar to a soft de-commitment with the difference that it can be hard-opened using the trapdoor tk.

$\mathsf{qHEquiv}_{pk,tk}(m_1, \ldots, m_q, i, \mathsf{aux})$: is a non-adaptive hard equivocation algorithm. Namely, given $(C, \mathsf{aux}) = \mathsf{qFake}_{pk,tk}()$, it generates a hard de-commitment π for C at the ith position of the sequence $(m_1, \ldots, m_q)$. The algorithm is non-adaptive in that the sequence of messages has to be determined once-and-for-all before the execution of $\mathsf{qHEquiv}$.

$\mathsf{qSEquiv}_{pk,tk}(m, i, \mathsf{aux})$: is a soft equivocation algorithm. Given the auxiliary information aux returned by $(C, \mathsf{aux}) = \mathsf{qFake}_{pk,tk}()$, it creates a soft de-commitment τ to m at position i.

Standard trapdoor mercurial commitments can be seen as a special case of qTMC schemes where $q = 1$.

CORRECTNESS. The correctness requirements are similar to those of standard mercurial commitments. For any sequence $(m_1, \ldots, m_q)$, these statements must hold with overwhelming probability.

- Given a hard commitment $(C, \mathsf{aux}) = \mathsf{qHCom}_{pk}(m_1, \ldots, m_q)$, for all indices $i \in \{1, \ldots, q\}$, it must hold that $\mathsf{qHVer}_{pk}(m_i, i, C, \mathsf{qHOpen}_{pk}(m_i, i, \mathsf{aux})) = 1$ and $\mathsf{qSVer}_{pk}(m_i, i, C, \mathsf{qSOpen}_{pk}(m_i, i, \mathbb{H}, \mathsf{aux})) = 1$.
- If $(C, \mathsf{aux}) = \mathsf{qSCom}_{pk}()$, then $\mathsf{qSVer}_{pk}(m_i, i, C, \mathsf{qSOpen}_{pk}(m_i, i, \mathbb{S}, \mathsf{aux})) = 1$ for $i = 1, \ldots, q$.

- Given a fake commitment $(C, \mathsf{aux}) = \mathsf{qFake}_{pk,tk}()$, for each $i \in \{1, \ldots, q\}$, we must have $\mathsf{qHVer}_{pk}(m_i, i, C, \mathsf{qHEquiv}_{pk,tk}(m_1, \ldots, m_q, i, \mathsf{aux})) = 1$ and $\mathsf{qSVer}_{pk}(m_i, i, C, \mathsf{qSEquiv}_{pk,tk}(m_i, i, \mathsf{aux})) = 1$.

SECURITY. The security properties of a trapdoor q-mercurial commitment are stated as follows:

- **q-Mercurial binding:** given the public key pk, it should be computationally infeasible to output a commitment C, an index $i \in \{1, \ldots, q\}$ and pairs (m, π), (m', π') that satisfy either of these two conditions which are respectively termed "hard collision" and "soft collision":
  - $\mathsf{qHVer}_{pk}(m, i, C, \pi) = 1$, $\mathsf{qHVer}_{pk}(m', i, C, \pi') = 1$ and $m \neq m'$.
  - $\mathsf{qHVer}_{pk}(m, i, C, \pi) = 1$, $\mathsf{qSVer}_{pk}(m', i, C, \pi') = 1$ and $m \neq m'$.
- **q-Mercurial hiding:** on input of pk, no PPT adversary can find a tuple $(m_1, \ldots, m_q)$ and an index $i \in \{1, \ldots, q\}$ for which it is able to distinguish $(C, \mathsf{qSOpen}_{pk}(m_i, i, \mathbb{H}, \mathsf{aux}))$ from $(C', \mathsf{qSOpen}_{pk}(m_i, i, \mathbb{S}, \mathsf{aux}'))$, where $(C, \mathsf{aux}) = \mathsf{qHCom}_{pk}(m_1, \ldots, m_q)$, $(C', \mathsf{aux}') = \mathsf{qSCom}_{pk}()$.
- **Equivocations:** given the public key pk *and the trapdoor* tk, no PPT adversary $\mathcal{A}$ should be able to win the following games with non-negligible probability. In these games, $\mathcal{A}$ aims to distinguish the "real" world from the corresponding "ideal" one. The kind of world that $\mathcal{A}$ is faced with depends on a random $b \stackrel{R}{\leftarrow} \{0, 1\}$ flipped by the challenger. If $b = 0$, the challenger plays the "real" game and provides $\mathcal{A}$ with a real commitment/de-commitment tuple. If $b = 1$, the adversary $\mathcal{A}$ rather receives a fake commitment and equivocations. More precisely, $\mathcal{A}$ is required to guess the bit $b \in \{0, 1\}$ with no better advantage than 1/2 in the following games:
  - **q-HHEquivocation:** when $\mathcal{A}$ chooses a message sequence $(m_1, \ldots, m_q)$, the challenger computes $(C, \mathsf{aux}) = \mathsf{qHCom}_{pk}(m_1, \ldots, m_q)$ if $b = 0$ and $(C, \mathsf{aux}) = \mathsf{qFake}_{pk,tk}()$ if $b = 1$. In either case, $\mathcal{A}$ receives C. When $\mathcal{A}$ chooses $i \in \{1, \ldots, q\}$, the challenger returns $\pi = \mathsf{qHOpen}_{pk}(m_i, i, \mathsf{aux})$ if $b = 0$ and $\pi = \mathsf{qHEquiv}_{pk,tk}(m_1, \ldots, m_q, i, \mathsf{aux})$ if $b = 1$.
  - **q-HSEquivocation:** when $\mathcal{A}$ chooses a message sequence $(m_1, \ldots, m_q)$, the challenger computes $(C, \mathsf{aux}) = \mathsf{qHCom}_{pk}(m_1, \ldots, m_q)$ if $b = 0$ and $(C, \mathsf{aux}) = \mathsf{qFake}_{pk,tk}()$ if $b = 1$. In either case, C is given to $\mathcal{A}$ who then chooses $i \in \{1, \ldots, q\}$. If $b = 0$, the challenger replies with $\tau = \mathsf{qSOpen}_{pk}(m_i, i, \mathbb{H}, \mathsf{aux})$. If $b = 1$, $\mathcal{A}$ receives $\tau = \mathsf{qSEquiv}_{pk,tk}(m_i, i, \mathsf{aux})$.
  - **q-SSEquivocation:** if $b = 0$, the challenger creates a soft commitment $(C, \mathsf{aux}) = \mathsf{qSCom}_{pk}()$ and hands C to $\mathcal{A}$. If $b = 1$, $\mathcal{A}$ rather obtains a fake commitment C, which is obtained as $(C, \mathsf{aux}) = \mathsf{qFake}_{pk,tk}()$. Then, $\mathcal{A}$ chooses $m \in \mathcal{M}$ and $i \in \{1, \ldots, q\}$ and gets $\tau = \mathsf{qSOpen}_{pk}(m, i, \mathbb{S}, \mathsf{aux})$ if $b = 0$ and $\tau = \mathsf{qSEquiv}_{pk,tk}(m, i, \mathsf{aux})$ if $b = 1$.

As pointed out in [8] in the case of ordinary trapdoor mercurial commitments, any qTMC scheme satisfying the q-HSEquivocation and q-SSEquivocation properties also satisfies the q-mercurial hiding requirement.

In the following, we say that a qTMC scheme is a *concise* mercurial vector commitment if the output sizes of qHOpen and qSOpen do not depend on q and if, when invoked on the index $i \in \{1, \ldots, q\}$, qHOpen does not reveal any information on messages m_j with $j \neq i$.

2.3 Zero-Knowledge Sets and Databases

An elementary database D (EDB) is a set of pairs $(x, y) \subset \{0,1\}^* \times \{0,1\}^*$, where x is called *key* and y is termed *value*. The support $[D]$ of D is the set of $x \in \{0,1\}^*$ for which there exists $y \in \{0,1\}^*$ such that $(x,y) \in D$. When $x \notin [D]$, one usually writes $D(x) = \perp$. When $x \in [D]$, the associated value $y = D(x)$ must be unique: if $(x,y) \in D$ and $(x,y') \in D$, then $y = y'$. A zero-knowledge EDB allows a prover to commit to such a database D while being able to non-interactively prove statements of the form "$x \in [D]$ and $y = D(x)$ is the associated value" or "$x \notin [D]$" without revealing any further information on D (not even the cardinality of $[D]$). Zero-knowledge sets are specific ZK-EDBs where each key is assigned the value 1.

The prover and the verifier both take as input a string σ that can be a random string (in which case, the protocol stands in the common random string model) or have a specific structure (in which case we are in the trusted parameters model). An EDB scheme is formally defined by a tuple (CRS-Gen, P1, P2, V) such that:

- CRS-Gen generates a common reference string σ on input of a security parameter λ.
- P1 is the commitment algorithm that takes as input the database D and σ. It outputs commitment and de-commitment strings (Com, Dec).
- P2 is the proving algorithm that, given σ, the commitment/de-commitment pair (Com, Dec) and a key $x \in \{0,1\}^*$, outputs a proof π_x.
- V is the verification algorithm that, on input of σ, Com, x and π_x, outputs either y (which must be $\perp$ if $x \notin [D]$) if it is convinced that $D(x) = y$ or *bad* if it believes that the prover is cheating.

The security requirements are formally defined in appendix A. In a nutshell, they are as follows. Correctness mandates that honestly generated proofs always satisfy the verification test. Soundness requires that provers be unable to come up with a key x and convincing proofs π_x, π_x' such that $y = \mathsf{V}(\sigma, Com, x, \pi_x) \neq \mathsf{V}(\sigma, Com, x, \pi_x) = y'$. Finally, zero-knowledge means that each proof π_x only reveals the value $D(x)$ and nothing else: for any computable database D, there must exist a simulator that outputs a simulated reference string σ' and a simulated commitment Com' that does not depend on D. For any key $x \in \{0,1\}^*$ and with oracle access to D, the simulator should be able to simulate proofs π_x that are indistinguishable from real proofs.

3 A Construction of Concise qTMC Scheme

Our idea is to build on the accumulator of Camenisch, Kohlweiss and Soriente [6], which is itself inspired by the Boneh-Gentry-Waters broadcast encryption system [5]. In the former, the public key comprises a sequence of group elements $(g, g_1, \ldots, g_q, g_{q+2}, \ldots, g_{2q})$, where q is the maximal number of accumulated values and $g_i = g^{(\alpha^i)}$ for each i. Elements of $\mathcal{V} \subseteq \{1, \ldots, q\}$ are accumulated by computing $V = \prod_{j \in \mathcal{V}} g_{q+1-j}$ and the witness for the accumulation of $i \in \mathcal{V}$ consists of $W_i = \prod_{j \in \mathcal{V} \setminus \{i\}} g_{q+1-j+i}$, which always satisfies $e(g_i, V) = e(g, W_i) \cdot e(g_1, g_q)$.

To obtain a commitment scheme, we modify this construction in order to accumulate messages $m_i \in \mathbb{Z}_p^*$ in a position-sensitive manner and we also add some randomness $\gamma \in \mathbb{Z}_p$ to have a hiding commitment. More precisely, we commit to $(m_1, \ldots, m_q)$ by computing $V = g^\gamma \cdot \prod_{j=1}^q g_{q+1-j}^{m_j}$ and obtain a kind of generalized Pedersen commitment [23]. Thanks to the specific choice of base elements however, $W_i = g_i^\gamma \cdot \prod_{j=1, j \neq i}^q g_{q+1-j+i}^{m_j}$ can serve as evidence that m_i was the ith committed message as it satisfies the relation $e(g_i, V) = e(g, W_i) \cdot e(g_1, g_q)^{m_i}$. Moreover, the opening W_i at position i does not reveal anything about other components of the committed vector, which is a property that can be useful in other applications. This commitment can be proved binding under the q-DHE assumption, which would be broken if the adversary was able to produce two distinct openings of V at position i. It is also a trapdoor commitment since anyone holding $g_{q+1} = g^{(\alpha^{q+1})}$ can trapdoor open a commitment as he likes.

The scheme can be made mercurial by observing that its binding property disappears if the verification equation becomes $e(g_i, V) = e(g_1, W_i) \cdot e(g_1, g_q)^{m_i}$. The key idea is then to use commitments of the form (C, V) where $C = g^\theta$, for some $\theta \in \mathbb{Z}_p$, in hard commitments and $C = g_1^\theta$ in soft commitments. The verification equation thus becomes $e(g_i, V) = e(C, W_i) \cdot e(g_1, g_q)^{m_i}$.

Description. We assume that committed messages are elements of $\mathbb{Z}_p^*$. In practice, arbitrary messages can be committed to by first applying a collision-resistant hash function with range $\mathbb{Z}_p^*$.

qKeygen(λ, q): chooses bilinear groups $(\mathbb{G}, \mathbb{G}_T)$ of prime order $p > 2^\lambda$ and $g \xleftarrow{R} \mathbb{G}$. It picks $\alpha \xleftarrow{R} \mathbb{Z}_p^*$ and computes $g_1, \ldots, g_q, g_{q+2}, \ldots, g_{2q}$, where $g_i = g^{(\alpha^i)}$ for $i = 1, \ldots, q, q+2, \ldots, 2q$. The public key is defined to be $pk = \{g, g_1, \ldots, g_q, g_{q+2}, \ldots, g_{2q}\}$ and the trapdoor is $tk = g_{q+1} = g^{(\alpha^{q+1})}$.

qHCom$_{pk}(m_1, \ldots, m_q)$: to hard-commit to a sequence $(m_1, \ldots, m_q) \in (\mathbb{Z}_p^*)^q$, this algorithm chooses $\gamma, \theta \xleftarrow{R} \mathbb{Z}_p$ and computes the commitment as the pair

$$C = g^\theta \qquad V = g^\gamma \cdot \prod_{j=1}^{q} g_{q+1-j}^{m_j} = g^\gamma \cdot g_q^{m_1} \cdots g_1^{m_q}.$$

The output is (C, V) and the auxiliary information is $\mathsf{aux} = (m_1, \ldots, m_q, \gamma, \theta)$.

qHOpen$_{pk}(m_i, i, \mathsf{aux})$: parses aux as $(m_1, \ldots, m_q, \gamma, \theta)$ and calculates

$$W_i = \Big(g_i^\gamma \cdot \prod_{j=1, j \neq i}^{q} g_{q+1-j+i}^{m_j} \Big)^{1/\theta}. \tag{1}$$

The hard opening of (C, V) consists of $\pi = (\theta, W_i) \in \mathbb{Z}_p \times \mathbb{G}$.

qHVer$_{pk}(m_i, i, (C, V), \pi)$: parses π as $(\theta, W_i) \in \mathbb{Z}_p \times \mathbb{G}$ and returns 1 if $C, V \in \mathbb{G}$ and it holds that

$$e(g_i, V) = e(C, W_i) \cdot e(g_1, g_q)^{m_i} \quad \text{and} \quad C = g^{\theta}. \tag{2}$$

Otherwise, it returns 0.

qSCom$_{pk}()$: chooses $\theta, \gamma \xleftarrow{R} \mathbb{Z}_p$ and computes $C = g_1^{\theta}$, $V = g_1^{\gamma}$. The output is (C, V) and the auxiliary information is $\mathsf{aux} = (\theta, \gamma)$.

qSOpen$_{pk}(m, i, \mathsf{flag}, \mathsf{aux})$: if $\mathsf{flag} = \mathbb{H}$, aux is parsed as $(m_1, \ldots, m_q, \gamma, \theta)$. The algorithm returns $\perp$ if $m \neq m_i$. Otherwise, it computes the soft opening as $W_i = \left(g_i^{\gamma} \cdot \prod_{j=1, j\neq i}^{q} g_{q+1-j+i}^{m_j}\right)^{1/\theta}$. If $\mathsf{flag} = \mathbb{S}$, the algorithm parses aux as (θ, γ) and soft-de-commits to m using $W_i = \left(g_i^{\gamma} \cdot g_q^{-m}\right)^{1/\theta}$. In either case, the algorithm returns $\tau = W_i \in \mathbb{G}$.

qSVer$_{pk}(m, i, (C, V), \tau)$: parses τ as $W_i \in \mathbb{G}$ and returns 1 if and only if it holds that $C, V \in \mathbb{G}$ and the first verification equation of (2) is satisfied.

qFake$_{pk,tk}()$: the fake commitment algorithm chooses $\theta, \gamma \xleftarrow{R} \mathbb{Z}_p$ and returns $(C, V) = (g^{\theta}, g^{\gamma})$. The auxiliary information is $\mathsf{aux} = (\theta, \gamma)$.

qHEquiv$_{pk,tk}(m_1, \ldots, m_q, i, \mathsf{aux})$: parses aux as $(\theta, \gamma) \in (\mathbb{Z}_p)^2$. Using the trapdoor $tk = g_{q+1} \in \mathbb{G}$, it computes $W_i = \left(g_i^{\gamma} \cdot g_{q+1}^{-m_i}\right)^{1/\theta}$. The de-commitment consists of $\pi = (\theta, W_i)$.

qSEquiv$_{pk,tk}(m, i, \mathsf{aux})$: parse aux as (θ, γ) and returns $W_i = \left(g_i^{\gamma} \cdot g_{q+1}^{-m}\right)^{1/\theta}$.

Correctness. In hard commitments, we can check that properly generated hard de-commitments always satisfy the verification test (2) since

$$\begin{aligned}
\frac{e(g_i, V)}{e(C, W_i)} &= e\big(g^{(\alpha^i)}, g^{\gamma + \sum_{j=1}^{q} m_j(\alpha^{q+1-j})}\big) / e\big(g^{\theta}, g^{(\gamma(\alpha^i) + \sum_{j=1, j\neq i}^{q} m_j(\alpha^{q+1-j+i}))/\theta}\big) \\
&= e\big(g, g^{\gamma(\alpha^i) + \sum_{j=1}^{q} m_j(\alpha^{q+1-j+i})}\big) / e\big(g, g^{\gamma(\alpha^i) + \sum_{j=1, j\neq i}^{q} m_j(\alpha^{q+1-j+i})}\big) \\
&= e(g, g)^{m_i(\alpha^{q+1})} = e(g_1, g_q)^{m_i}.
\end{aligned}$$

As for soft commitments, soft de-commitments always satisfy the first relation of (2) since

$$\begin{aligned}
e(C, W_i) \cdot e(g_1, g_q)^{m_i} &= e\big(g_1^{\theta}, (g_i^{\gamma} \cdot g_q^{-m_i})^{1/\theta}\big) \cdot e(g_1, g_q)^{m_i} \\
&= e\big(g_1, g_i^{\gamma} \cdot g_q^{-m_i}\big) \cdot e(g_1, g_q)^{m_i} = e(g_1^{\gamma}, g_i) = e(g_i, V).
\end{aligned}$$

We finally observe that, in any fake commitment $(C, V) = (g^{\theta}, g^{\gamma})$, the hard de-commitment (θ, W_i) successfully passes the verification test as

$$\begin{aligned}
e(C, W_i) \cdot e(g_1, g_q)^{m_i} &= e\big(g^{\theta}, (g_i^{\gamma} \cdot g_{q+1}^{-m_i})^{1/\theta}\big) \cdot e(g_1, g_q)^{m_i} \\
&= e\big(g, g_i^{\gamma} \cdot g_{q+1}^{-m_i}\big) \cdot e(g_1, g_q)^{m_i} = e(g_i, g^{\gamma}) = e(g_i, V).
\end{aligned}$$

Security. To prove the security of the scheme, we first notice that it is a "proper" qTMC [8] since, in hard commitments, the soft de-commitment is a proper subset of the hard de-commitment.

Theorem 1. *The above scheme is a secure concise qTMC if the q-DHE assumption holds in* $\mathbb{G}$.

Proof. We first show the q-mercurial binding property. Let us assume that, given the public key, an adversary $\mathcal{A}$ is able to generate soft collisions (since the scheme is "proper", the case of hard collisions immediately follows). That is, $\mathcal{A}$ comes up with a commitment $(C, V) \in \mathbb{G}^2$, an index $i \in \{1, \ldots, q\}$, a valid hard decommitment $\pi = (\theta, W_i) \in \mathbb{Z}_p \times \mathbb{G}$ to m_i at position i and a valid soft decommitment $\tau = W_i' \in \mathbb{G}$ to m_i' such that $m_i \neq m_i'$. We must have

$$e(g_i, V) = e(g^\theta, W_i) \cdot e(g_1, g_q)^{m_i} \qquad e(g_i, V) = e(g^\theta, W_i') \cdot e(g_1, g_q)^{m_i'},$$

so that $e(g^\theta, W_i/W_i') = e(g_1, g_q)^{m_i' - m_i}$ and $e\big(g, (W_i/W_i')^{\theta/(m_i'-m_i)}\big) = e(g_1, g_q)$. Since $m_i \neq m_i'$, the latter relation implies that $g_{q+1} = (W_i/W_i')^{\theta/(m_i'-m_i)}$ is revealed by the soft collision, which contradicts the q-DHE assumption.

We now turn to the q-HHE, q-HSE and q-SSE equivocation properties (which imply q-mercurial hiding). A fake commitment has the form $(C, V) = (g^\theta, g^\gamma)$ and its hard equivocation to (m_i, i) is the pair $\big(\theta, W_i = (g_i^\gamma \cdot g_{q+1}^{-m_i})^{1/\theta}\big)$. For any sequence of messages $(m_1, \ldots, m_q) \in (\mathbb{Z}_p^*)^q$, there exists $\gamma' \in \mathbb{Z}_p$ such that

$$V = g^{\gamma'} \cdot \prod_{j=1}^{q} g_{q+1-j}^{m_j}. \tag{3}$$

Then, the corresponding hard opening of (C, V) w.r.t. m_i at position i should be obtained as $W_i' = (g_i^{\gamma'} \cdot \prod_{j=1, j\neq i}^{q} g_{q+1-j+i}^{m_j})^{1/\theta}$. Since V also equals g^γ, if we raise both members of (3) to the power α^i, we find that

$$g_i^\gamma = g_i^{\gamma'} \cdot \prod_{j=1}^{q} g_{q+1-j+i}^{m_j}.$$

Therefore, the element $W_i = (g_i^\gamma \cdot g_{q+1}^{-m_i})^{1/\theta}$ returned by the hard equivocation algorithm can also be written $W_i = (g_i^{\gamma'} \cdot \prod_{j=1, j\neq i}^{q} g_{q+1-j+i}^{m_j})^{1/\theta}$. It comes that fake commitments and hard equivocations have exactly the same distribution as hard commitments and their hard openings.

The q-HSEquivocation property follows from the above arguments (since the scheme is "proper"). To prove the indistinguishability in the q-SSEquivocation game, we note that fake commitments $(C, V) = (g^\theta, g^\gamma)$ have the same distribution as soft ones as they can be written $(C, V) = (g_1^{\tilde{\theta}}, g_1^{\tilde{\gamma}})$ where $\tilde{\theta} = \theta/\alpha$ and $\tilde{\gamma} = \gamma/\alpha$. Their soft equivocation $W_i = (g_i^\gamma \cdot g_{q+1}^{-m_i})^{1/\theta}$ can be written $(g_i^{\alpha\tilde{\gamma}} \cdot g_{q+1}^{-m_i})^{1/(\alpha\tilde{\theta})} = (g_i^{\tilde{\gamma}} \cdot g_q^{-m_i})^{1/\tilde{\theta}}$ and has the distribution of a soft opening. □

Instantiation with Asymmetric Pairings. It is simple[1] to describe the construction in terms of asymmetric pairings $e : \mathbb{G} \times \hat{\mathbb{G}} \rightarrow \mathbb{G}_T$, where $\mathbb{G} \neq \hat{\mathbb{G}}$ and

[1] The security then relies on the hardness of computing $\psi(\hat{g})^{(\alpha^{q+1})}$ on input of $(\hat{g}, \hat{g}_1, \ldots, \hat{g}_q, \hat{g}_{q+2}, \ldots, \hat{g}_{2q}) \in \hat{\mathbb{G}}^{2q}$, where $\hat{g}_i = \hat{g}^{(\alpha^i)}$ for each i.

an isomorphism $\psi : \hat{\mathbb{G}} \rightarrow \mathbb{G}$ is efficiently computable. The public key comprises generators $\hat{g} \in \hat{\mathbb{G}}$ and $\hat{g}_i$ for $i = 1, \ldots, q, q+2, \ldots, 2q$. Then, hard (resp. soft) commitments $(C, V) \in \hat{\mathbb{G}} \times \mathbb{G}$ are pairs of group elements obtained as $C = \hat{g}^{\theta}$ and $V = \psi(\hat{g})^{\gamma} \cdot \prod_{j=1}^{q} \psi(\hat{g}_{q+1-j})^{m_j}$ (resp. $C = \hat{g}_1^{\theta}$ and $V = \psi(\hat{g}_1)^{\gamma}$). Hard openings are pairs $(\theta, W_i) \in \mathbb{Z}_p^* \times \mathbb{G}$, where $W_i = \psi(\hat{g}_i)^{\gamma/\theta} \cdot \prod_{j=1, j \neq i}^{q} \psi(\hat{g}_{q+1-j+i})^{m_j/\theta}$ and they are verified by checking that $C = \hat{g}^{\theta}$ and $e(V, \hat{g}_i) = e(W_i, C) \cdot e(\psi(\hat{g}_1), \hat{g}_q)^{m_i}$. Using the trapdoor $\hat{g}_{q+1}$, fake commitments $(C, V) = (\hat{g}^{\theta}, \psi(\hat{g})^{\gamma})$ can be equivocated by outputting θ and $W_i = \psi(\hat{g}_i)^{\gamma/\theta} \cdot \psi(\hat{g}_{q+1})^{-m_i/\theta}$.

4 Implications on the Efficiency of ZK-EDBs

The construction [9] of ZK-EDB from qTMC schemes goes as follows. Each key x is assigned to a leaf of a q-ary tree of height h (and can be seen as the label of the leaf, expressed in q-ary encoding), so that q^h is the theoretical bound on the size of the EDB.

The committing phase is made efficient by pruning subtrees where all leaves correspond to keys that are *not* in the database. Only the roots (called "frontier nodes" and at least one sibling of which is an ancestor of a leaf in the EDB) of these subtrees are kept in the tree and contain soft q-commitments. For each key x such that $D(x) \neq \perp$, the corresponding leaf contains a standard hard mercurial commitment to a hash value of $D(x)$. As for remaining nodes, each internal one contains a hard q-commitment to messages obtained by hashing its children. The q-commitment at the root then serves as a commitment to the entire EDB.

To convince a verifier that $D(x) = v \neq \perp$ for some key x, the prover generates a proof of membership consisting of hard openings for commitments in nodes on the path connecting leaf x to the root. At each level of the tree, the q-commitment is hard-opened with respect to the position determined by the q-ary encoding of x at that level.

To provide evidence that some key x does not belong to the database (in other words, $D(x) = \perp$), the prover first generates the missing portion of the subtree where x lies. Then, it reveals soft openings for all (hard or soft) commitments contained in nodes appearing in the path from x to the root.

As in the original zero-knowledge EDB construction [21], only storing commitments in subtrees containing leaves x for which $D(x) \neq \perp$ (and soft commitments at nodes that have no descendants) is what allows committing with complexity $O(h \cdot |D|)$ instead of $O(q^h)$.

The advantage of using qTMC schemes and q-ary (with $q > 2$) trees lies in that proofs can be made much shorter if, at each level, commitments can be opened w.r.t. the required position $i \in \{1, \ldots, q\}$ without having to reveal q values. The qTMC scheme of [9] features soft openings consisting of a single group element and, for an appropriate branching factor q, allows reducing proofs of non-membership by 73% in comparison with [21]. On the other hand, hard openings still have length $O(q)$ and proofs of membership thus remain significantly longer than proofs of non-membership. If h denotes the height of the tree,

q	h	Membership	Non-Membership	Membership in [9]
8	43	220	176	521
16	32	165	132	643
32	26	135	108	941
64	22	115	92	1501
128	19	100	80	2513

Fig. 1. Required number of group elements per proof

the former consist of $h(q+4)+5$ elements of $\mathbb{G}$ (in an implementation with asymmetric pairings) while the latter only demand $4h+4$ such elements.

If we plug our qTMC scheme into the above construction, proofs of membership become essentially as short as proofs of non-membership. At each internal node, each hard opening only requires to reveal $(C, V) \in \hat{\mathbb{G}} \times \mathbb{G}$ and $(\theta, W_i) \in \mathbb{Z}_p \times \mathbb{G}$. At the same time, proofs of non-membership remain as short as in [9] since, at each internal node, the prover only discloses (C, V) and W_i.

To concretely assess proof sizes, we assume (as in [9]) that elements of $\hat{\mathbb{G}}$ count as two elements of $\mathbb{G}$ (since their representation is usually twice as large using suitable parameters and optimizations such as those of [2]), each one of which costs $|p|$ bits to represent. Then, we find that proofs of membership and non-membership eventually amount to $5h+5$ and $4h+4$ elements of $\mathbb{G}$, respectively. These short hard openings allow us to increase the branching factor of the tree as long as the length of the common reference string is deemed acceptable.

The table of figure 1 summarizes the proof lengths (expressed in numbers of $\mathbb{G}$ elements and in comparison with [9]) for various branching factors and assuming that $q^h \approx 2^{128}$ theoretically bounds the EDB's size. In the MRK construction, membership (resp. non-membership) can be proved using 773 (resp. 644) group elements. The best tradeoff achieved in [9] was for $q = 8$, where proofs of non-membership could be reduced to 176 elements but proofs of membership still took 521 elements. With $q = 8$, we have equally short proofs of non-membership and only need 220 elements to prove membership, which improves CFM [9] by about 57% and MRK [21] by 71%.

Moreover, we can shorten both kinds of proof by increasing q: with $q = 128$ for instance, no more than 100 group elements (or 13% of the original length achieved in [21]) are needed to prove membership whereas 2513 elements are necessary in [9]. Instantiating our scheme with curves of [2] yields proofs of less than 2 kB when $q = 128$. For such relatively small values of q, Cheon's attack [12] does not require to increase the security parameter λ and it is reasonable to use groups $(\mathbb{G}, \hat{\mathbb{G}})$ where elements of $\mathbb{G}$ have a 161-bit representation.

5 Achieving Strong Independence

In [15], Gennaro and Micali formalized the notion of *independent zero-knowledge* EDBs which requires that adversaries be unable to correlate their database to those created by honest provers.

The strongest flavor of independence considers two-stage adversaries $\mathcal{A} = (\mathcal{A}_1, \mathcal{A}_2)$. First, $\mathcal{A}_1$ observes ℓ honest provers' commitments $(Com_1, \ldots, Com_\ell)$ and queries proofs for keys of her choice in underlying databases $D_1, \ldots, D_\ell$ before outputting her own commitment Com. Then, two copies of $\mathcal{A}_2$ are executed: in the first one, $\mathcal{A}_2$ is given oracle access to provers that "open" Com_i w.r.t D_i whereas, in the second run, $\mathcal{A}_2$ has access to provers for different[2] databases D'_i that agree with D_i for the set Q_i of queries made by $\mathcal{A}_1$. Eventually, both executions of $\mathcal{A}_2$ end with $\mathcal{A}_2$ outputting a key x, which is identical in both runs, and a proof π_x. The resulting database value $D(x)$ is required to be the same in the two copies, meaning that it was fixed at the end of the committing stage.

In the strongest definition of [15], $\mathcal{A}_1$ is allowed to copy one of the honest provers' commitment (say Com_i) as long as the key x returned by $\mathcal{A}_2$ is never queried to $\mathsf{Sim}_2(St_i, Com_i)$ by $\mathcal{A}_1$ or $\mathcal{A}_2$: in other words, $\mathcal{A}_2$'s answer must be fixed on all values x that were not queried to the ith prover.

Definition 2. *[15] A ZK-EDB protocol is strongly independent if, for any polynomial ℓ, any PPT adversary $\mathcal{A} = (\mathcal{A}_1, \mathcal{A}_2)$ and any databases $D_1, \ldots, D_\ell$, $D'_1, \ldots, D'_\ell$, the following probability is negligible.*

$$\begin{aligned}
\Pr\Big[&(\sigma, St_0) \leftarrow \mathsf{Sim}_0(\lambda);\ (Com_i, St_i) \leftarrow \mathsf{Sim}_1(St_0)\ \forall i = 1, \ldots, \ell;\\
&(Com, \omega) \leftarrow \mathcal{A}_1^{\mathsf{Sim}_2^{D_i(\cdot)}(St_i, Com_i)}(\sigma, Com_1, \ldots, Com_\ell);\\
&(x, \pi_x) \leftarrow \mathcal{A}_2^{\mathsf{Sim}_2^{D_i(\cdot)}(St_i, Com_i)}(\sigma, \omega);\ (x, \pi'_x) \leftarrow \mathcal{A}_2^{\mathsf{Sim}_2^{D'_i \dashv_{Q_i} D_i(\cdot)}(St_i, Com_i)}(\sigma, \omega);\\
&(bad \neq \mathsf{V}(\sigma, Com, x, \pi_x) \neq \mathsf{V}(\sigma, Com, x, \pi'_x) \neq bad) \wedge \Big((\forall i : Com \neq Com_i)\\
&\vee (\exists i : (Com = Com_i) \wedge (x \notin Q_i \cup Q'_i))\Big)\Big],
\end{aligned}$$

where Q_i (resp. Q'_i) stands for the list of queries made by $\mathcal{A}_1$ (resp. $\mathcal{A}_2$) to $\mathsf{Sim}_2^{D_i(\cdot)}(St_i, Com_i)$ (resp. $\mathsf{Sim}_2^{D_i(\cdot)}(St_i, Com_i)$ and $\mathsf{Sim}_2^{D'_i \dashv_{Q_i} D_i(\cdot)}(St_i, Com_i)$) and $D'_i \dashv_{Q_i} D_i$ denotes a database that agrees with D'_i on all keys but those in Q_i where it agrees with D_i.

An efficient construction of independent ZK-EDB was proved in [15] to satisfy the above definition under the strong RSA assumption. It was obtained by extending Gennaro's multi-trapdoor commitment scheme [14] and making it mercurial.

We show how to turn our qTMC scheme into a multi-trapdoor q-mercurial commitment scheme that yields strongly independent EDBs with short proofs.

MULTI-TRAPDOOR Q-MERCURIAL COMMITMENTS. A multi-trapdoor qTMC can be seen as extending qTMC schemes in the same way as multi-trapdoor commitments generalize ordinary trapdoor commitments. It can be defined as a family of trapdoor q-mercurial commitments, each member of which is identified

[2] For this reason, commitments $(Com_1, \ldots, Com_\ell)$ are produced using the ZK-EDB simulator, whose definition is recalled in appendix A, as the two executions of $\mathcal{A}_2$ proceed as if underlying databases were different.

by a string tag and has its own trapdoor tk_{tag}. The latter is generated from tag using a master trapdoor TK that matches the master public key PK.

$\mathsf{qKeygen}(\lambda, q)$: has the same specification as in section 2.2 but, in addition to the master key pair (PK, TK), it outputs the description of a tag space $\mathcal{T}$.

$\mathsf{qHCom}_{PK}(m_1, \ldots, m_q, tag)$: given an ordered tuple $(m_1, \ldots, m_q)$ and $tag \in \mathcal{T}$, this algorithm outputs a hard commitment C under (PK, tag) and some auxiliary state information aux.

$\mathsf{qHOpen}_{PK}(m, i, tag, \mathsf{aux})$: given a pair $(C, \mathsf{aux}) = \mathsf{qHCom}_{PK}(m_1, \ldots, m_q, tag)$, this algorithm outputs a hard de-commitment π of C w.r.t. position i if $m = m_i$. If $m \neq m_i$, it returns $\perp$.

$\mathsf{qHVer}_{PK}(m, i, C, tag, \pi)$: outputs 1 if and only if π gives evidence that, under the tag tag, C is bound to a sequence $(m_1, \ldots, m_q)$ such that $m_i = m$.

$\mathsf{qSCom}_{PK}()$: generates a soft commitment and some auxiliary information aux. Such a commitment is not associated with any specific messages or tag.

$\mathsf{qSOpen}_{PK}(m, i, \mathsf{flag}, tag, \mathsf{aux})$: generates a soft de-commitment τ of C to m at position i and w.r.t. tag. The variable $\mathsf{flag} \in \{\mathbb{H}, \mathbb{S}\}$ indicates whether τ pertains to a hard commitment $(C, \mathsf{aux}) = \mathsf{qHCom}_{PK}(m_1, \ldots, m_q, tag)$ or a soft commitment $(C, \mathsf{aux}) = \mathsf{qSCom}_{PK}()$. If $\mathsf{flag} = \mathbb{H}$ and $m \neq m_i$, the algorithm returns $\perp$.

$\mathsf{qSVer}_{PK}(m, i, C, \tau, tag)$ returns 1 if, under $tag \in \mathcal{T}$, τ is deemed as a valid soft de-commitment of C to m at position i and 0 otherwise.

$\mathsf{qTrapGen}_{PK,TK}(tag)$: given a string $tag \in \mathcal{T}$, this algorithm generates a tag-specific trapdoor tk_{tag} using the master trapdoor TK.

$\mathsf{qFake}_{PK,tk_{tag}}()$: outputs a q-fake commitment C and some auxiliary state information aux.

$\mathsf{qHEquiv}_{PK,tk_{tag}}(m_1, \ldots, m_q, i, tag, \mathsf{aux})$: given $(C, \mathsf{aux}) = \mathsf{qFake}_{PK,tk_{tag}}()$, this algorithm generates a hard de-commitment π for C and $tag \in \mathcal{T}$ at the ith position of the sequence $(m_1, \ldots, m_q)$. The sequence of messages has to be determined once-and-for-all before the execution of $\mathsf{qHEquiv}$.

$\mathsf{qSEquiv}_{PK,tk_{tag}}(m, i, tag, \mathsf{aux})$: using the trapdoor tk_{tag} and the state information aux returned by $(C, \mathsf{aux}) = \mathsf{qFake}_{PK,tk_{tag}}()$, this algorithm creates a soft de-commitment τ to m at position i and w.r.t. $tag \in \mathcal{T}$.

Again, we call such a scheme *concise* if it satisfies the same conditions as those mentioned at the end of section 2.2.

The security properties are expressed by naturally requiring the q-mercurial hiding and equivocation properties to hold for each $tag \in \mathcal{T}$. In equivocation games, the adversary should be unable to distinguish the two games even knowing the master trapdoor TK. As for the q-mercurial binding property, it states that no PPT adversary $\mathcal{A}$ should have non-negligible advantage in this game:

q-Mercurial binding game: $\mathcal{A}$ chooses strings $tag_1, \ldots, tag_\ell \in \mathcal{T}$. Then, the challenger generates a master key pair $(TK, PK) \leftarrow \mathsf{qKeygen}(\lambda, q)$ and gives PK to $\mathcal{A}$ who starts invoking a trapdoor oracle $\mathcal{TG}$: the latter receives $tag \in \{tag_1, \ldots, tag_\ell\}$ and returns $tk_{tag} \leftarrow \mathsf{qTrapGen}_{PK,TK}(tag)$. Eventually, $\mathcal{A}$ chooses a family $tag^\star \in \mathcal{T} \backslash \{tag_1, \ldots, tag_\ell\}$ for which she aims to generate a collision: she wins if she outputs C, an index $i \in \{1, \ldots, q\}$ and pairs (m, π), (m', π') (resp. (m, π) and (m', τ)) such that $\mathsf{qHVer}_{PK}(m, i, C, tag^\star, \pi) = 1$ and $\mathsf{qHVer}_{PK}(m', i, C, tag^\star, \pi') = 1$ (resp. $\mathsf{qHVer}_{PK}(m, i, C, tag^\star, \pi) = 1$ and $\mathsf{qSVer}_{PK}(m', i, C, tag^\star, \tau) = 1$) but $m \neq m'$.

As in [14], the latter definition captures security in a non-adaptive sense in that the adversary chooses $tag_1, \ldots, tag_\ell$ before seeing the public key PK. As noted in [13,19] in the case of ordinary multi-trapdoor commitments, some applications might require to consider a notion of adaptive security where, much in the fashion of identity-based trapdoor commitments [1,7], the adversary can query $\mathcal{TG}$ in an adaptive fashion. In the present context, non-adaptive security suffices.

A Construction of Multi-Trapdoor qTMC. The construction combines the qTMC scheme of section 3 with a programmable hash function $H_{\mathbb{G}} : \mathcal{T} \to \mathbb{G}$ and techniques that were introduced in [3]. Programmable hash functions, as formalized by Hofheinz and Kiltz [17], are designed in such a way that a trapdoor information makes it possible to relate the output $H_{\mathbb{G}}(M)$, which lies in a group $\mathbb{G}$, to computable values $a_M, b_M \in \mathbb{Z}_p$ satisfying $H_{\mathbb{G}}(M) = g^{a_M} \cdot h^{b_M}$. Informally (see [17] for a formal definition), a (m, n)-programmable hash function is such that, for any $M_1, \ldots, M_m$, $M'_1, \ldots, M'_n$ such that $M_i \neq M'_j$, there is a non-negligible probability that $b_{M_i} = 0$ and $b_{M'_j} \neq 0$ for $i = 1, \ldots, m$ and $j = 1, \ldots, n$. The number theoretic hash function used in [11,25] is an example of such a $(1, \ell)$-programmable hash function, for some polynomial ℓ.

qKeygen(λ, q): is as in section 3 but the algorithm also chooses a tag space $\mathcal{T} = \{0,1\}^L$ and a $(1, \ell)$-programmable hash function $H_{\mathbb{G}} : \mathcal{T} \to \mathbb{G}$ for some polynomials ℓ, L. The public key is $PK = \{\mathcal{T}, g, g_1, \ldots, g_q, g_{q+2}, \ldots, g_{2q}, H_{\mathbb{G}}\}$ and the master trapdoor is $TK = g_{q+1} = g^{(\alpha^{q+1})}$.

qHCom$_{PK}(m_1, \ldots, m_q, tag)$: to hard-commit to a sequence $(m_1, \ldots, m_q) \in (\mathbb{Z}_p^*)^q$, this algorithm chooses $\gamma, \theta \stackrel{R}{\leftarrow} \mathbb{Z}_p$ and computes $(C, V) = (g^\theta, g^\gamma \cdot \prod_{j=1}^{q} g_{q+1-j}^{m_j})$. The output is (C, V) and the auxiliary information is $\mathsf{aux} = (m_1, \ldots, m_q, \gamma, \theta)$.

qHOpen$_{PK}(m_i, i, tag, \mathsf{aux})$: parses aux as $(m_1, \ldots, m_q, \gamma, \theta)$, chooses $r \stackrel{R}{\leftarrow} \mathbb{Z}_p^*$ and computes

$$(W_i, Z_i) = \Big(\big(g_i^\gamma \cdot \prod_{j=1, j\neq i}^{q} g_{q+1-j+i}^{m_j} \cdot H_{\mathbb{G}}(tag)^r\big)^{1/\theta}, g^{-r}\Big), \tag{4}$$

The hard opening of (C, V) with respect to $tag \in \mathcal{T}$ consists of the triple $\pi = (\theta, W_i, Z_i) \in \mathbb{Z}_p \times \mathbb{G}^2$.

qHVer$_{PK}(m_i, i, (C, V), tag, \pi)$: parses π as $(\theta, W_i, Z_i) \in \mathbb{Z}_p \times \mathbb{G}^2$ and returns 1 if $C, V \in \mathbb{G}$ and relations (5) are both satisfied. Otherwise, it returns 0.

$$e(g_i, V) = e(C, W_i) \cdot e(g_1, g_q)^{m_i} \cdot e(H_{\mathbb{G}}(tag), Z_i) \qquad C = g^\theta. \tag{5}$$

qSCom$_{PK}()$: chooses $\theta, \gamma \stackrel{R}{\leftarrow} \mathbb{Z}_p$ and computes $C = g_1^\theta$, $V = g_1^\gamma$. The output is (C, V) and the auxiliary information is $\mathsf{aux} = (\theta, \gamma)$.

$\mathsf{qSOpen}_{PK}(m, i, \mathsf{flag}, tag, \mathsf{aux})$: if $\mathsf{flag} = \mathbb{H}$, aux is parsed as $(m_1, \ldots, m_q, \gamma, \theta)$. The algorithm returns $\perp$ if $m \neq m_i$. Otherwise, the soft opening $\tau = (W_i, Z_i)$ is generated as per (4). If $\mathsf{flag} = \mathbb{S}$, the algorithm parses aux as (θ, γ) and soft-decommits to m using

$$(W_i, Z_i) = \Big(\big(g_i^{\gamma} \cdot g_q^{-m} \cdot H_{\mathbb{G}}(tag)^r\big)^{1/\theta}, g_1^{-r} \Big), \qquad (6)$$

where $r \xleftarrow{R} \mathbb{Z}_p^*$. In either case, the algorithm returns $\tau = (W_i, Z_i) \in \mathbb{G}^2$.

$\mathsf{qSVer}_{pk}(m, i, (C, V), \tau, tag)$: parses τ as $(W_i, Z_i) \in \mathbb{G}$ and returns 1 if and only if $C, V \in \mathbb{G}$ and the first verification equation of (5) is satisfied.

$\mathsf{qTrapGen}_{PK,TK}(tag)$: given $TK = g_{q+1}$, a trapdoor for $tag \in \mathcal{T}$ is computed $tk_{tag} = (t_{tag,1}, t_{tag,2}) = (g_{q+1} \cdot H_{\mathbb{G}}(tag)^s, g^{-s})$ for a random $s \xleftarrow{R} \mathbb{Z}_p^*$.

$\mathsf{qFake}_{PK,tk_{tag}}()$: outputs a pair $(C, V) = (g^{\theta}, g^{\gamma})$, where $\theta, \gamma \xleftarrow{R} \mathbb{Z}_p^*$, and retains the state information $\mathsf{aux} = (\theta, \gamma)$.

$\mathsf{qHEquiv}_{PK,tk_{tag}}(m_1, \ldots, m_q, i, tag, \mathsf{aux})$: parses aux as $(\theta, \gamma) \in (\mathbb{Z}_p^*)^2$ and the trapdoor tk_{tag} as $(t_{tag,1}, t_{tag,2}) \in \mathbb{G}^2$. It randomly picks $r \xleftarrow{R} \mathbb{Z}_p^*$ and computes $(W_i, Z_i) = \Big(\big(g_i^{\gamma} \cdot t_{tag,1}^{-m_i} \cdot H_{\mathbb{G}}(tag)^r\big)^{1/\theta}, t_{tag,2}^{-m_i} \cdot g^{-r} \Big)$. The de-commitment is $\pi = (\theta, W_i, Z_i) = \Big(\theta, \big(g_i^{\gamma} \cdot g_{q+1}^{-m_i} \cdot H_{\mathbb{G}}(tag)^{r'}\big)^{1/\theta}, g^{-r'}\Big)$, where $r' = -sm_i + r$.

$\mathsf{qSEquiv}_{PK,tk_{tag}}(m, i, tag, \mathsf{aux})$: parse aux as (θ, γ) and computes (W_i, Z_i) as in $\mathsf{qHEquiv}_{PK,tk_{tag}}$.

Theorem 2. *The scheme is a concise multi-trapdoor qTMC if the q-DHE assumption holds.*

Proof. Given in the full version of the paper. □

STRONGLY INDEPENDENT ZK-EDBS FROM MULTI-TRAPDOOR QTMC. Following [15], a multi-trapdoor qTMC can be combined with a digital signature and a collision-resistant hash function $H : \{0,1\}^* \rightarrow \mathcal{T}$ to give a strongly independent ZK-EDB. To commit to a database D, the prover first generates a key pair (SK, VK) for an existentially unforgeable (in the sense of [16]) signature scheme $\Sigma = (\mathcal{G}, \mathcal{S}, \mathcal{V})$ [16]. The commitment string is (Com, VK), where all commitments are produced using the qTMC family (with $q = 1$ at the leaves and $q > 1$ at internal nodes) indexed by the tag $H(\mathsf{VK})$. To generate a proof for some key x, the prover generates a proof π_x (by opening the appropriate commitments using Dec) and outputs π_x and $\mathsf{sig}_x = \mathcal{S}(\mathsf{SK}, (Com, x))$. Verification entails to check π_x and that $\mathcal{V}(\mathsf{sig}_x, \mathsf{VK}, (Com, x)) = 1$. The security proof of this scheme (detailed in the full version of the paper) is similar to that of theorem 3 in [15].

References

1. Ateniese, G., de Medeiros, B.: Identity-Based Chameleon Hash and Applications. In: Juels, A. (ed.) FC 2004. LNCS, vol. 3110, pp. 164–180. Springer, Heidelberg (2004)
2. Barreto, P., Naehrig, M.: Pairing-Friendly Elliptic Curves of Prime Order. In: Preneel, B., Tavares, S. (eds.) SAC 2005. LNCS, vol. 3897, pp. 319–331. Springer, Heidelberg (2006)

3. Boneh, D., Boyen, X.: Efficient Selective-ID Secure Identity-Based Encryption Without Random Oracles. In: Cachin, C., Camenisch, J.L. (eds.) EUROCRYPT 2004. LNCS, vol. 3027, pp. 223–238. Springer, Heidelberg (2004)
4. Boneh, D., Boyen, X., Goh, E.-J.: Hierarchical Identity-Based encryption with Constant Size Ciphertext. In: Cramer, R. (ed.) EUROCRYPT 2005. LNCS, vol. 3494, pp. 440–456. Springer, Heidelberg (2005)
5. Boneh, D., Gentry, C., Waters, B.: Collusion Resistant Broadcast Encryption with Short Ciphertexts and Private Keys. In: Shoup, V. (ed.) CRYPTO 2005. LNCS, vol. 3621, pp. 258–275. Springer, Heidelberg (2005)
6. Camenisch, J., Kohlweiss, M., Soriente, C.: An Accumulator Based on Bilinear Maps and Efficient Revocation for Anonymous Credentials. In: Jarecki, S., Tsudik, G. (eds.) PKC 2009. LNCS, vol. 5443, pp. 481–500. Springer, Heidelberg (2009)
7. Canetti, R., Dodis, Y., Pass, R., Walfish, S.: Universally Composable Security with Global Setup. In: Vadhan, S.P. (ed.) TCC 2007. LNCS, vol. 4392, pp. 61–85. Springer, Heidelberg (2007)
8. Catalano, D., Dodis, Y., Visconti, I.: Mercurial Commitments: Minimal Assumptions and Efficient Constructions. In: Halevi, S., Rabin, T. (eds.) TCC 2006. LNCS, vol. 3876, pp. 120–144. Springer, Heidelberg (2006)
9. Catalano, D., Fiore, D., Messina, M.: Zero-Knowledge Sets with Short Proofs. In: Smart, N.P. (ed.) EUROCRYPT 2008. LNCS, vol. 4965, pp. 433–450. Springer, Heidelberg (2008)
10. Chase, M., Healy, A., Lysyanskaya, A., Malkin, T., Reyzin, L.: Mercurial Commitments with Applications to Zero-Knowledge Sets. In: Cramer, R. (ed.) EUROCRYPT 2005. LNCS, vol. 3494, pp. 422–439. Springer, Heidelberg (2005)
11. Chaum, D., Evertse, J.-H., van de Graaf, J.: An Improved Protocol for Demonstrating Possession of Discrete Logarithms and Some Generalizations. In: Price, W.L., Chaum, D. (eds.) EUROCRYPT 1987. LNCS, vol. 304, pp. 127–141. Springer, Heidelberg (1988)
12. Cheon, J.H.: Security Analysis of the Strong Diffie-Hellman Problem. In: Vaudenay, S. (ed.) EUROCRYPT 2006. LNCS, vol. 4004, pp. 1–11. Springer, Heidelberg (2006)
13. Di Raimondo, M., Gennaro, R.: New Approaches for Deniable Authentication. In: ACM-CCS 2005, pp. 112–121 (2005)
14. Gennaro, R.: Multi-trapdoor Commitments and Their Applications to Proofs of Knowledge Secure Under Concurrent Man-in-the-Middle Attacks. In: Franklin, M. (ed.) CRYPTO 2004. LNCS, vol. 3152, pp. 220–236. Springer, Heidelberg (2004)
15. Gennaro, R., Micali, S.: Independent Zero-Knowledge Sets. In: Bugliesi, M., Preneel, B., Sassone, V., Wegener, I. (eds.) ICALP 2006. LNCS, vol. 4052, pp. 34–45. Springer, Heidelberg (2006)
16. Goldwasser, S., Micali, S., Rivest, R.: A digital signature scheme secure against adaptive chosen message attacks. SIAM J. of Computing 17(2), 281–308 (1988)
17. Hofheinz, D., Kiltz, E.: Programmable Hash Functions and Their Applications. In: Wagner, D. (ed.) CRYPTO 2008. LNCS, vol. 5157, pp. 21–38. Springer, Heidelberg (2008)
18. Liskov, M.: Updatable Zero-Knowledge Databases. In: Roy, B. (ed.) ASIACRYPT 2005. LNCS, vol. 3788, pp. 174–198. Springer, Heidelberg (2005)
19. MacKenzie, P., Yang, K.: On Simulation-Sound Trapdoor Commitments. In: Cachin, C., Camenisch, J.L. (eds.) EUROCRYPT 2004. LNCS, vol. 3027, pp. 382–400. Springer, Heidelberg (2004)

20. Merkle, R.: A Digital Signature Based on a Conventional Encryption Function. In: Pomerance, C. (ed.) CRYPTO 1987. LNCS, vol. 293, pp. 369–378. Springer, Heidelberg (1988)
21. Micali, S., Rabin, M.-O., Kilian, J.: Zero-Knowledge Sets. In: FOCS 2003, pp. 80–91 (2003)
22. Ostrovsky, R., Rackoff, C., Smith, A.: Efficient Consistency Proofs for Generalized Queries on a Committed Database. In: Díaz, J., Karhumäki, J., Lepistö, A., Sannella, D. (eds.) ICALP 2004. LNCS, vol. 3142, pp. 1041–1053. Springer, Heidelberg (2004)
23. Pedersen, T.: Non-Interactive and Information-Theoretic Secure Verifiable Secret Sharing. In: Feigenbaum, J. (ed.) CRYPTO 1991. LNCS, vol. 576, pp. 129–140. Springer, Heidelberg (1992)
24. Prabhakaran, M., Xue, R.: Statistically Hiding Sets. In: Fischlin, M. (ed.) CT-RSA 2009. LNCS, vol. 5473, pp. 100–116. Springer, Heidelberg (2009)
25. Waters, B.: Efficient Identity-Based Encryption Without Random Oracles. In: Cramer, R. (ed.) EUROCRYPT 2005. LNCS, vol. 3494, pp. 114–127. Springer, Heidelberg (2005)

A Security Properties of Zero-Knowledge Databases

The completeness, soundness and zero-knowledge properties of ZK-EDBs are formally stated as follows.

Completeness: For all databases D and for all keys x, it must hold that

$$\Pr\big[\sigma \leftarrow \mathsf{CRS\text{-}Gen}(\lambda); (Com, Dec) \leftarrow \mathsf{P1}(\sigma, D); \\ \pi_x \leftarrow \mathsf{P2}(\sigma, D, Com, Dec, x) : \mathsf{V}(\sigma, Com, x, \pi_x) = D(x)\ \big] = 1 - \nu.$$

for some negligible function ν.

Soundness: For all keys x and for any probabilistic poly-time algorithm P', the following probability is negligible:

$$\Pr\big[\sigma \leftarrow \mathsf{CRS\text{-}Gen}(\lambda); (Com, x, \pi_x, \pi'_x) \leftarrow \mathsf{P}'(\sigma, D); \\ \mathsf{V}(\sigma, Com, x, \pi_x) = y \neq bad \wedge \mathsf{V}(\sigma, Com, x, \pi'_x) = y' \neq bad \wedge (y \neq y')\ \big].$$

Zero-knowledge: for any PPT adversary $\mathcal{A}$ and any efficiently computable database D, there must exist an efficient simulator $(\mathsf{Sim}_0, \mathsf{Sim}_1, \mathsf{Sim}_2^D)$ such that the outputs of the following experiments are indistinguishable:

Real experiment:

1. Set $\sigma \leftarrow \mathsf{CRS\text{-}Gen}(\lambda)$, $(Com, Dec) \leftarrow \mathsf{P1}(\sigma, D)$ and $s_0 = \varepsilon$, $\pi_0 = \varepsilon$.
2. For $i = 1, \ldots, n$, $\mathcal{A}$ outputs $(x_i, s_i) \leftarrow \mathcal{A}(\sigma, Com, \pi_0, \ldots, \pi_{i-1}, s_{i-1})$ and obtains a real proof $\pi_i = \mathsf{P2}(\sigma, D, Com, Dec, x_i)$.

The output is $(\sigma, x_1, \pi_1, \ldots, x_n, \pi_n)$.

Ideal experiment:

1. Set $(\sigma', St_0) \leftarrow \mathsf{Sim}_0(\lambda)$, $(Com', St_1) \leftarrow \mathsf{Sim}_1(St_0)$ as well as $s_0 = \varepsilon$, $\pi'_0 = \varepsilon$.
2. For $i = 1, \ldots, n$, $\mathcal{A}$ outputs $(x_i, s_i) \leftarrow \mathcal{A}(\sigma', Com', \pi'_0, \ldots, \pi'_{i-1}, s_{i-1})$ and gets a simulated proof $\pi'_i \leftarrow \mathsf{Sim}_2^D(\sigma', St_1, x_i)$.

The output of the experiment is $(\sigma', x_1, \pi'_1, \ldots, x_n, \pi'_n)$.

In the above, Sim_2^D is an oracle that is permitted to invoke a database oracle $D(.)$ and obtain values $D(x)$ for the keys x chosen by $\mathcal{A}$.

Eye for an Eye: Efficient Concurrent Zero-Knowledge in the Timing Model

Rafael Pass*, Wei-Lung Dustin Tseng**,
and Muthuramakrishnan Venkitasubramaniam

Cornell University, NY, USA

Abstract. We present new and efficient concurrent zero-knowledge protocols in the timing model. In contrast to earlier works—which through artificially-imposed delays require *every* protocol execution to run at the speed of the *slowest* link in the network—our protocols essentially only delay messages based on the *actual* response time of each verifier (which can be significantly smaller).

1 Introduction

Zero-knowledge (ZK) interactive proofs [GMR89] are paradoxical constructs that allow one player (called the prover) to convince another player (called the verifier) of the validity of a mathematical statement $x \in L$, while providing *zero additional knowledge* to the verifier. This is formalized by requiring that for every PPT adversary verifier V^*, there is a PPT *simulator* S that can simulate the view of V^* interacting with the honest prover P. The idea behind this definition is that whatever V^* might have learned from interacting with P could have been learned by simply running the simulator S. The notion of concurrent ZK (cZK), first introduced and achieved by Dwork, Naor and Sahai [DNS04] extends the notion of ZK protocols to a concurrent and asynchronous setting. More precisely, we consider a single adversary mounting a coordinated attack by acting as a verifier in many concurrent sessions, possibly with many independent provers. cZK protocols are significantly harder to construct and analyze, and are often less efficient than the "standalone" ZK protocols.

The original constant-round cZK protocol of [DNS04] is constructed in the timing model (also explored in [Gol02]). Informally speaking, the timing model assumes that every party (in our case every honest prover) has a local clock, and that all these local clocks are roughly synchronized (1 second is roughly the same on every clock). Also, all parties know a (pessimistic) upper-bound,

* Supported in part by a Microsoft New Faculty Fellowship, NSF CAREER Award CCF-0746990, AFOSR Award FA9550-08-1-0197 and BSF Grant 2006317.

** Supported in part by a NSF Graduate Research Fellowship.

D. Micciancio (Ed.): TCC 2010, LNCS 5978, pp. 518–534, 2010.

Δ, on the time it takes to deliver a message on the network. As argued by Goldreich [Gol02], this assumption seems to be most reasonable for systems today. The problem, however, is that known constructions of cZK protocols in the timing model [DNS04, Gol02] are not very efficient in terms of execution time: Despite having a constant number of rounds (4 or 5 messages), the prover in these protocols delays the response of certain messages by time Δ. In other words, every instance of the protocol must take time longer than the pessimistic bound on the max latency of the network (rather than being based on the actual message-delivery time).

Leaving the timing model, Richardson and Kilian [RK99] (and subsequent improvments by Kilian and Petrank [KP01] and Prabhakaran, Rosen and Sahai [PRS02]) show how to construct cZK protocols in the standard model (without clocks). Here the protocols are "message-delivery" driven, but there is a significant increase in round-complexity: Whereas constant-round ZK protocols exists in the standalone setting, $\tilde{O}(\log n)$-rounds are both necessary and sufficient for (black-box) cZK protocols [PRS02, KPR98, Ros00, CKPR01]. Another related work of Pass and Venkitasubramaniam [PV08] gives a constant-round cZK protocol without clocks, but at the expense of having quasi-polynomial time simulators (against quasi-polynomial time adversaries).

In this work we revisit the timing model. Ideally, we want to construct cZK protocols that are efficient in all three manners mentioned so far: Small (constant) round-complexity, low imposed delays, and fast simulation. As communicated by Goldreich [Gol02], Barak and Micciancio suggested the following possible improvement to cZK protocols in the timing model: The prover may only need to impose a delay δ that is a linear fraction of Δ (say $\delta = \Delta/d$), at the expense of increasing the running time of the ZK simulator exponentially (around $n^{O(d)}$). In other words, there could be a compromise between protocol efficiency and knowledge security [Gol01, MP06] (i.e., simulator running-time). However, as discussed in [Gol02], this suggestion has not been proven secure. We show that such a trade-off is not only possible, but can be significantly improved.[1]

Trading rounds for minimum delays. The original work of Richardson and Kilian [RK99] shows that increasing the number of communication rounds can decrease the running-time of the simulator. Our first result shows that by only slightly increasing the number of rounds, but still keeping it constant (e.g., 10 messages), the prover may reduce the imposed delay to $\delta = \Delta/2^d$, while keeping the simulator running time at $n^{O(d)}$. This is accomplished by combining simulation techniques from both the timing model [DNS04, Gol02] (polynomial time simulation but high timing constraints) and the standard model [RK99, PV08] (quasi-polynomial time simulation but no timing constraints). As far as we know, this yeilds the first formal proof that constant-round concurrent zero-knowledge protocols are possible using a delay δ that is smaller than Δ.

[1] It seems that traditional techniques can be used to demonstrate the Barak-Micciancio trade-off when the adversary employs a *static* scheduling of messages. However, complications arise in the case of *adaptive* schedules. See Section 3.1 for more details.

"Eye-for-an-eye" delays. The traditional approach for constructing cZK protocols is to "penalize" all parties equally, whether it is in the form of added round complexity or imposed timing delays. One may instead consider the notion of punishing only adversarial behaviour, similar to the well-known "tit-for-tat" or "eye-for-an-eye" technique of game theory (see e.g., [Axe84]). The work of Cohen, Kilian and Petrank [CKP01] first implemented such a strategy (with respect to cZK) using an iterated protocol where in each iteration, the verifier is given a time constraint under which it must produce all of its messages; should a verifier exceed this constraint, the protocol is restarted with doubled the allowed time constraint (the punishment here is the resetting); their protocol had $\tilde{O}(log^2 n)$ rounds and $\tilde{O}(\log n)$ "responsive complexity"—namely, the protocol takes time $\tilde{O}(\log n)T$ to complete if each verifier message is sent within time T. The work of Persiano and Visconti [PV05] and Rosen and shelat [Rs09] takes a different approach and punish adversaries that perform "bad" schedulings of messages by adaptively adding more rounds to the protocol; their approaches, however, only work under the assumption that there is a *single* prover, or alternatively that all messages on the network are exposed on a broadcast channel (so that the provers can check if a problematic scheduling of messages has occurred).

In our work, we instead suggest the following simple approach: Should a verifier provide its messages with delay t, the prover will delay its message accordingly so that the protocol completes in time $p(t) + \delta$, where p is some *penalty* function and δ is some small minimal delay. We note that, at a high-level, this approach is somewhat reminiscent of how message delivery is performed in TCP/IP.

As we show, such penalty-based *adaptive delays* may significantly improve the compromise between protocol efficiency and knowledge security. For example, setting $p(t) = 2t$ (i.e., against a verifier that responds in time $t < \Delta$, the prover responds in time $t + \delta$) has a similar effect as increasing the number of rounds: The prover may reduce the minimal imposed delay to $\delta = \Delta/2^d$, while keeping the simulator running time at $n^{O(d)}$. Moreover, if we are willing to use more aggressive penalty functions, such as $p(t) = t^2$, the minimal delay may be drastically reduced to $\delta = \Delta^{1/2^d}$, greatly benefiting "honest" parties that respond quickly, while keeping the same simulator running time. Note that, perhaps surprisingly, we show that such a "tit-for-tat" technique, which is usually employed in the setting of rational players, provides significant efficiency improvements even with respect to fully adversarial players.

Combining it all. Finally, we combine our techniques by both slightly increasing the round complexity and implementing penalty-based delays. We state our main theorem below for $p(t) = t$ (no penalty), ct (linear penalty), and t^c (polynomial penalty) (in the main text we provide an expression for a generic $p(t)$):

Theorem 1. *Let Δ be an upper-bound on the time it takes to deliver a message on the network. Let r and d be integer parameters, and $p(t)$ be a (penalty) function. Then, assuming the existence of claw-free permutations, there is a $(2r+6)$-message black-box perfect cZK argument for all of NP with the following properties:*

– *The simulator has running time* $(rn)^{O(d)}$.
– *For any verifier that cumulatively delays its message by time at most* T*, the prover will provide its last message in time at most* $p(T) + \delta$*, where*

$$\delta = \begin{cases} 2\Delta/r^d & \text{if } p(t) = t \text{ (no penalty)} \\ 2\Delta/(cr)^d & \text{if } p(t) = ct \text{ (linear penalty)} \\ \frac{(2\Delta)^{1/c^d}}{r^{1+1/c+\cdots+1/c^{d-1}}} \leq \frac{(2\Delta)^{1/c^d}}{r} & \text{if } p(t) = t^c \text{ (polynomial penalty)} \end{cases}$$

Remark 1 (On the number of rounds). Even without penalty-based delays, if $r = 2$, we achieve an exponential improvement in the imposed delay ($\delta = \Delta/2^d$), compared to the suggestion by Barak and Micciancio (which required a delay of $\delta = \Delta/d$). Larger r (i.e., more rounds) allows us to further improve the delay.

Remark 2 (On adversarially controlled networks). If an adversary controls the whole network, it may also delay messages from the honest players. In this case, honest players (that answer as fast as they can) are also penalized. However, the adversary can anyway delay message delivery to honest players, so this problem is unavoidable. What we guarantee is that, if a pair of honest players are communicating over a channel that is not delayed (or only slightly delayed) by the adversary, then the protocol will complete fast.

Remark 3 (On networks with failure). Note that even if the network is not under adversarial control, messages from honest parties might be delayed due to network failures. We leave it as an open question to (experimentally or otherwise) determine the "right" amount of penalty to employ in real-life networks: Aggressive delays allow us to minimize the imposed delay δ, but can raise the expected protocol running time if network failures are common.

Remark 4 (On concurrent multi-part computation). [KLP05] and [LPV09] show that concurrent multi-party computation (MPC) is possible in the timing model using delays of length $O(\Delta)$. Additionally, [LPV09] shows that at least $\Delta/2$ delays are necessary to achieve concurrent MPC in the timing model. In retrospect, this separation between concurrent ZK and MPC should not be surprising since cZK can be constructed in the plain model [RK99, KP01, PRS02], but concurrent MPC cannot [CF01, Lin04].

1.1 Organization

In Sect. 2 we give definitions regarding the timing model and primitives used in our constructions. An overview of our protocol and zero-knowledge simulator, followed by their formal descriptions, is given in Sect. 3. Actual formal analyses are given in Sect. 4 and the appendix.

2 Preliminaries

We assume familiarity with indistinguishability, interactive proofs and arguments, and stand-alone (black-box) zero-knowledge. Let N denote the set of natural numbers. Given a function $g : \mathsf{N} \to \mathsf{N}$, let $g^k(n)$ be the function computed by composing g together k times, i.e., $g^k(n) = g(g^{k-1}(n))$ and $g^0(n) = n$.

2.1 Timing Model

In the timing model, originally introduced by Dwork, Naor and Sahai [DNS04], we consider a model that incorporates a "timed" network. Informally, in such a network, a (known) maximum network latency Δ—the time it takes for a message to be computed and delivered over the network—is assumed. Moreover, each party (in our case the honest provers) possesses a local clock that is somewhat synchronized with the others (in the sense that a second takes about the same time on each clock).

As in [DNS04, Gol02, KLP05], we model all the parties in the timing model as interactive Turing machines that have an extra input tape, called the **clock tape**. In an adversarial model, the adversary has full control of the content of everyone's clock tape (it can initialize and update the tape value at will), while each machine only has read access to its own clock tape. More precisely, when a party P_i is invoked, the adversary initializes the local clock of P_i to some time t of its choice. Thereafter the adversary may, at any time, overwrite the all existing clock tapes with new time values. To model that in reality most clocks are reasonably but not perfectly synchronized, we consider adversaries that are **ϵ-drift preserving**, as defined below:

Let $\sigma_1, \sigma_2, \ldots$ be a series of global states of all machines in play; these states are recorded whenever the adversary initiates a new clock or updates the existing clocks. Denote by $\text{CLK}_P(\sigma)$ the value of the local clock tape of machine P at state σ. We say that an adversary is **ϵ-drift preserving** if for every pair of parties P and P' and every pair of states σ and σ', it holds that

$$\frac{1}{\epsilon}(\text{CLK}_P(\sigma) - \text{CLK}_P(\sigma')) \leq \text{CLK}_{P'}(\sigma) - \text{CLK}_{P'}(\sigma') \leq \epsilon(\text{CLK}_P(\sigma) - \text{CLK}_P(\sigma'))$$

As in [DNS04, Gol02, KLP05], we use the following constructs that utilize the clock tapes. Below, by local time we mean the value of the local clock tape.

Delays: When a party is instructed to delay sending a message m by δ time, it records the present local time t, checks its local clock every time it is updated, and sends the message when the local time reaches $t + \delta$.

Time-out: When a party is instructed to time-out if a response from some other party P_i does not arrive in δ time, it records the present time t. When the message from P_i does arrive, it aborts if the local time is greater than $t + \delta$.

Measure: When a party is instructed to measure the time elapsed between two messages, it simply reads the local time t when the first message is sent/received, and reads the local time t' again when the second message is sent/received. The party then outputs the elapsed time $t' - t$.

Although the measure operator is not present in previous works, it is essentially the quantitative version of the time-out operation, and can be implemented without additional extensions of the timing model. For simplicity, we focus on the model where the adversary is 1-drift preserving, i.e. all clocks are synchronized, but our results easily extend to ϵ-drift preserving adversaries.

2.2 Black-Box Concurrent Zero-Knowledge in the Timing Model

The standard notion of concurrent zero-knowledge extends straightforwardly to the timing model; all machines involved are simply augmented with the aforementioned clock tape. The view of a party still consists of all incoming messages as well as the parties random tape. In particular, the view of the adversary determines the value of all the clocks. We repeat the standard definition of black-box concurrent zero-knowledge below.

Let $\langle P, V \rangle$ be an interactive proof for a language L, and let V^* be a concurrent adversarial verifier that may interact with multiple independent copies of P concurrently, without any restrictions over the scheduling of the messages in the different interactions with P. Let $\{\mathsf{VIEW}_2[P(x) \leftrightarrow V^*(x,z)]\}$ denote the random variable describing the view of the adversary V^* in an interaction with P on common input x and auxiliary input z.

Definition 1. *Let $\langle P, V \rangle$ be an interactive proof system for a language L. We say that $\langle P, V \rangle$ is* black-box concurrent zero-knowledge *if for every polynomials q and m, there exists a probabilistic polynomial time algorithm $S_{q,m}$, such that for every concurrent adversary V^* that on common input x and auxiliary input z opens up $m(|x|)$ sessions and has a running-time bounded by $q(|x|)$, $S_{q,m}(x,z)$ runs in time polynomial in $|x|$. Furthermore, it holds that the ensembles $\{\mathsf{VIEW}_2[P(x) \leftrightarrow V^*(x,z)]\}_{x \in L, z \in \{0,1\}^*}$ and $\{S_{q,m}(x,z)\}_{x \in L, z \in \{0,1\}^*}$ are computationally indistinguishable over $x \in L$. We say $\langle P, V \rangle$ is* black-box perfect concurrent zero-knowledge *if the above ensembles are identical.*

Remark: [Gol02] defines concurrent ZK in the timing model with the assumption (WLOG) that the adversary never trigger a time-out from any prover. [Gol02] also made the assumption that the adversary always delays the verifier messages as much as permitted, but is assumption is no longer WLOG for protocols with penalty-based delays. Therefore in our model, the adversary is given total control over all the clocks (subject to ϵ-drift preserving), similar to the definition of [KLP05] for the setting of concurrent multi-party computation.

2.3 Other Primitives

We informally define other primitives used in the construction of our protocols.

Special-sound proofs: A 3-round public-coin interactive proof for the language $L \in \mathcal{NP}$ with witness relation R_L is special-sound with respect to R_L, if for any two transcripts (α, β, γ) and $(\alpha', \beta', \gamma')$ such that the initial messages α, α' are the same but the challenges β, β' are different, there is a deterministic procedure to extract the witness from the two transcripts that runs in polynomial time. Special-sound WI proofs for languages in NP can be based on the existence of non-interactive commitment schemes, which in turn can be based on one-way permutations. Assuming only one-way functions, 4-round special-sound WI proofs for NP exists[2]. For simplicity, we use 3-round special-sound proofs in our protocol though our proof works also with 4-round proofs.

Proofs of knowledge: Informally an interactive proof is a proof of knowledge if the prover convinces the verifier not only of the validity of a statement, but also that it possesses a witness for the statement. If we consider computationally bounded provers, we only get a "computationally convincing" notion of a proof of knowledge (aka *arguments of knowledge*).

3 Our Protocol and Simulator

3.1 Protocol Overview

Following the works of [FS90, GK96], later extended to the concurrent setting by [RK99, KP01, PRS02, PV08], we consider ZK protocols with two stages:

Stage 1: First the verifier V "commits to a trapdoor" (the start message). This is followed by one or multiple **slots**; each slot consists of a prover challenge (the opening of the slot) followed by a verifier response (the closing of the slot). A rewinding black-box ZK simulator can rewind any one of these slots to extract the verifier trapdoor.

Stage 2: The protocol ends with a modified proof of the original statement that can be simulated given the verifier trapdoor.

To generate the view of an adversarial verifier V^* in the standalone setting, a black-box simulator simply rewinds a slot to learn the trapdoor, and use it to simulate the final modified proof.

In the concurrent setting, however, V^* may *fully nest* another session inside a slot (i.e., after the prover sends the opening message, V^* schedules a full session before replying with closing message). In order for the simulator to rewind this slot, it would need to simulate the view of the nested session twice. Therefore, repeated nesting may cause a naive simulator to have super-polynomial running time [DNS04]. Different techniques were employed in different models to circumvent this difficulty caused by nesting. In the timing model, [DNS04, Gol02] shows that by delaying the Stage 2 proof and limiting the time allowed between the opening and closing of any slot, we can avoid the nesting situation all together. On the other hand, [RK99] showed that if the protocol has enough slots, the simulator can always find a slot that isn't "too nested" to rewind.

The work of Pass and Venkitasubramaniam describes a simulator (based on the work of [RK99]) that works also for constant-round protocols. Its running

[2] A 4-round protocol is special sound if a witness can be extracted from any two transcripts $(\tau, \alpha, \beta, \gamma)$ and $(\tau', \alpha', \beta', \gamma')$ such that $\tau = \tau'$, $\alpha = \alpha'$ and $\beta \neq \beta'$.

time (implicitly) depends on the maximum nesting level/depth of the least nested slot. Specifically, the running time of the simulator is $n^{O(d)}$ when this maximum depth of nesting is d. Building upon this, we now focus on reducing the maximum depth of nesting in the timing model.

In the following overview of our techniques, we assume that V^* interleaves different sessions in a *static* schedule; the full generality of dynamic scheduling is left for our formal analysis. Additionally, we keep track of the running time of our protocols as a function of T—the total amount of accumulated delay caused by the verifier in all the messages.

Imposing traditional timing delays with one slot. We first review the works of [DNS04, Gol02]. Recall that Δ is the maximum network latency—the time it takes for a message to be computed and delivered over the network. We require that the time between the opening and closing of each slot be bounded by 2Δ (otherwise the prover aborts); this is the smallest time-out value that we may ask of the honest verifier. At the same time, the prover delays the Stage 2 proof by δ time (after receiving the closing message of the last slot), where δ is a parameter (Fig. 1(a)). It is easy to see that if $\delta = 2\Delta$, then no nesting can occur (Fig. 1(b)). In this case the running time of the protocol is $T + \Delta$.

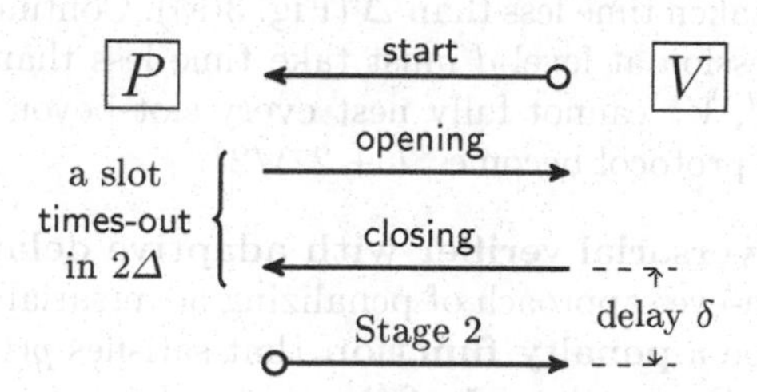

(a) 1 slot protocol with traditional timing constraints

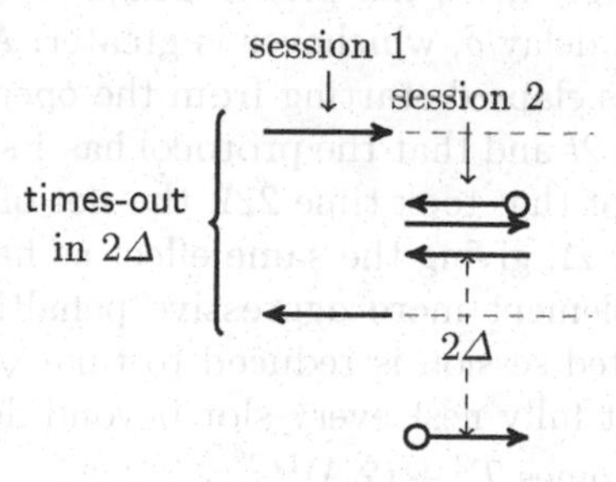

(b) $\delta = 2\Delta$ prevents nesting.

Fig. 1. Traditional timing delays with 1 slot

If we consider the suggestion of Barak and Micciancio and set $\delta = 2\Delta/d$, then up to d levels of nesting can occur (Fig. 2). In this case, the running time of the protocol is $T + 2\Delta$ and $T + 2\Delta/d$, respectively.

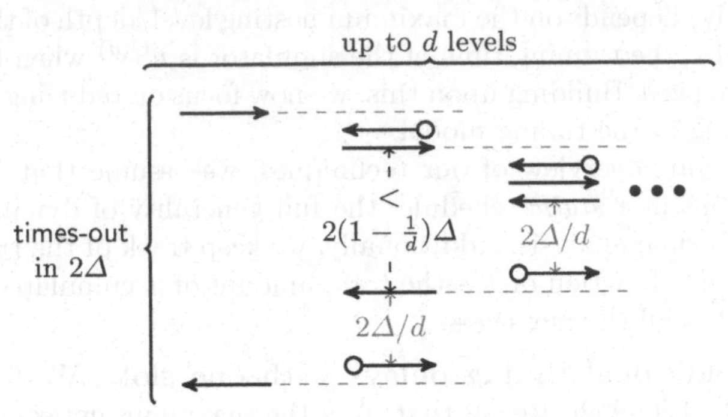

Fig. 2. $\delta = 2\Delta/d$ gives at most d levels of nesting

Increasing the number of slots. This idea was first explored by [RK99] in the standard model where intuitively, more slots translates to more rewinding opportunities for the simulator. In the timing model, the effect of multiple slots is much more direct. Let us look at the case of 2 slots. Suppose in some session, V^* delays the closing of a slot by the maximum allowed time, 2Δ. Further suppose that V^* nests an entire session inside this slot. Then in this nested session, one of the slots must have taken time less than Δ (Fig. 3(a)). Continuing this argument, some fully nested session at level d must take time less than $2\Delta/2^d$. Therefore if we set $\delta = 2\Delta/2^d$, V^* cannot fully nest every slot beyond depth d, and the running time of the protocol becomes $T + 2\Delta/2^d$.

Penalizing the adversarial verifier with adaptive delays. Here we implement our "eye-for-an-eye" approach of penalizing adversarial verifiers that delay messages. Let $p(t)$ be a **penalty function** that satisfies $p(t) > t$ and is monotonically increasing. During Stage 1 of the protocol, the prover measures t, the total time elapsed from the opening of the first slot to the closing of the last slot. Based on this measurement, the prover delays Stage 2 by time $p(t) - t$ or by the minimal imposed delay δ, whichever is greater. As a result, Stage 2 only starts after $p(t)$ time has elapsed starting from the opening of the first slot. For example, suppose $p(t) = 2t$ and that the protocol has 1 slot. Then for V^* to fully nest a session inside a slot that took time 2Δ, the slot of the nested session must have taken time at most Δ, giving the same effect as having 2 slots (Fig. 3(b)). Furthermore, if we implement more aggressive penalties, such as $p(t) = t^2$,[3] then the slot of the nested session is reduced to time $\sqrt{2\Delta}$. Therefore if we set $\delta = (2\Delta)^{1/2^d}$, V^* cannot fully nest every slot beyond depth d, and the running time of the protocol becomes $T^2 + (2\Delta)^{1/2^d}$.

Combining the techniques. In general, we can consider concurrent ZK protocols that both contain multiple slots and impose penalty-based delays (e.g., Fig. 4). If we have r slots and impose $p(t)$ penalty on delays, and define $g(t) = p(rt)$, then δ can be decreased to

[3] Formally we may use $p(t) = t^2 + 1$ to ensure that $p(t) > t$.

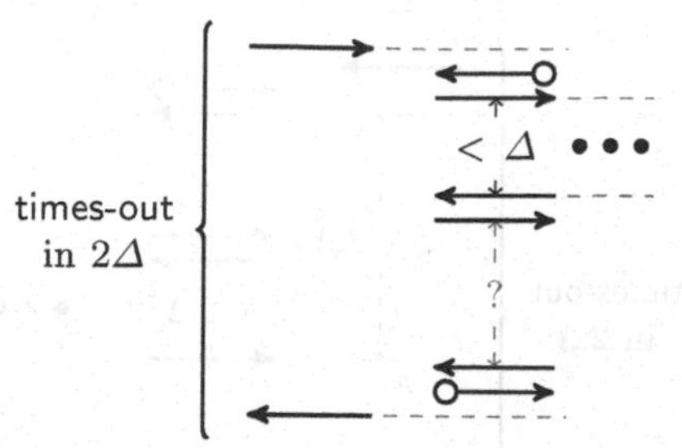

(a) **2 slots, no penalty.** One of the nested slot must have half the delay.

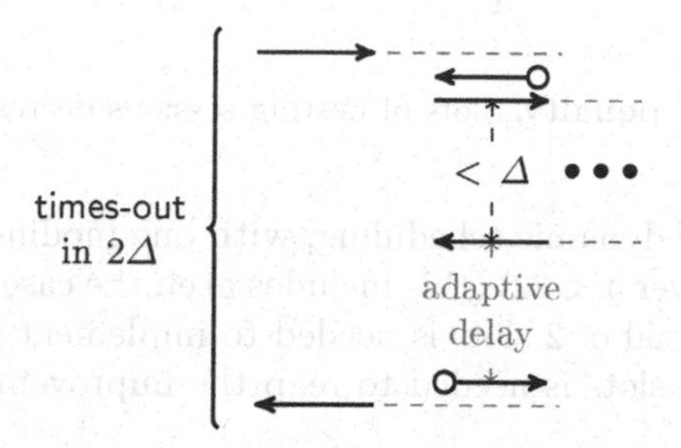

(b) **1 slot, $2t$ penalty**. The nested slot must have half the delay as well.

Fig. 3. Our main techniques of restricting the nesting depth of V^*

$$d \text{ times} \left\{ \frac{p^{-1}\left(\cdots \dfrac{p^{-1}\left(\dfrac{p^{-1}(2\Delta)}{r}\right)}{r} \right)}{r} \right. = (g^{-1})^d(2\Delta)$$

$$= \begin{cases} 2\Delta/r^d & \text{if } p(t) = t \quad \text{(no penalty)} \\ 2\Delta/(cr)^d & \text{if } p(t) = ct \text{ (linear penalty)} \\ \frac{(2\Delta)^{1/c^d}}{r^{1+1/c+\cdots+1/c^{d-1}}} \leq \frac{(2\Delta)^{1/c^d}}{r} & \text{if } p(t) = t^c \text{ (polynomial penalty)} \end{cases}$$

while keeping the simulator running time at $(rn)^{O(d)}$. The running time of the protocol is then $p(T) + \delta$.

Handling dynamic scheduling. So far we have discussed our analysis (and have drawn our diagrams) assuming that V^* follows a static schedule when interleaving multiple sessions. In general though, V^* may change the scheduling dynamically based on the content of the prover messages. As a result, the schedule (and nesting) of messages may change drastically when a black-box simulator rewinds V^*. This phenomenon introduces many technical difficulties into the analysis, but fortunately the same difficulties were also present and resolved [PV08]. By adapting the analysis in [PV08], we give essentially the same

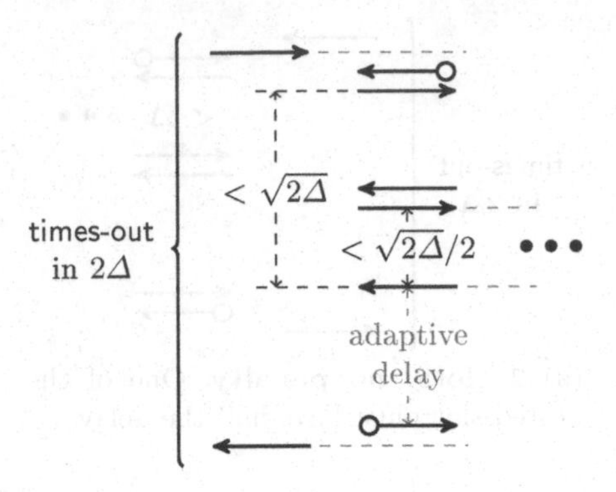

Fig. 4. 2 slots and t^2 penalty. Slots of nesting sessions decrease in size very quickly.

results in the case of dynamic scheduling, with one modification: An additional slot is needed whenever $\delta < 2\Delta$ (this includes even the case illustrated in Fig. 2). For example, a minimal of 2 slots is needed to implement penalty-based delays, and a minimum of 3 slots is needed to reap the improvements that result from multiple slots.

Handling ϵ-drifts in clock tapes. As in the work of [DNS04, Gol02] we merely need to scale the time-out values in our protocols when the local clocks are not perfectly synchronized. Specifically, if the adversary is ϵ-drift preserving for some $\epsilon \geq 1$, then our protocol will impose a minimal delay of $\epsilon\delta$ and an adaptive delay of $\epsilon p(t)$ (when applicable) between the closing of the last slot and Stage 2.

3.2 Description of the Protocol

Our concurrent ZK protocol is a slight variant of the precise ZK protocol of [MP06], which in turn is a modification of the Feige-Shamir protocol [FS90]. Given a one-way function f, a parameter r, a penalty function $p(t)$, and a minimal delay δ, our protocol for language $L \in \mathsf{NP}$ proceeds in the following two stages on common input $x \in \{0,1\}^*$ and security parameter n:

Stage 1: The verifier picks two random strings $s_1, s_2 \in \{0,1\}^n$ and sends $c_1 = f(s_1), c_2 = f(s_2)$ to the prover. The verifier also sends $\alpha_1, \ldots, \alpha_{r+1}$, the first messages of $r+1$ invocations of a WI special-sound proof of the statement "c_1 and c_2 are in the image set of f". These proofs are then completed sequentially in $r+1$ iterations.

In the jth iteration, the prover first sends $\beta_j \leftarrow \{0,1\}^{n^2}$, a random second message for the jth proof (opening of the jth slot), then the verifier replies with the third message γ_j of the jth proof (closing of the jth slot). The prover times-out the closing of each slot with time 2Δ, and measures the time that elapsed between the opening of the first slot and the closing of the $r+1^{\text{st}}$ slot as t.

Stage 2: The prover delays by time $\max\{p(t) - t, \delta\}$, and then provides a WI proof of knowledge of the statement "either $x \in L$, or that (at least) one of c_1 and c_2 are in the image set of f".

More precisely, let L' be the language characterized by the witness relation $R_{L'}(c_1, c_2) = \{(s_1, s_2) \mid f(s_1) = c_1 \text{ or } f(s_2) = c_2\}$. Let f be a one-way function, r and δ be integers, $p(t) : \mathsf{N} \rightarrow \mathsf{N}$ be a monotonically increasing function satisfying $p(t) > t$, and L be a language in NP. Our ZK argument for L, ConcZKArg, is depicted in Figure 5.

Protocol ConcZKArg

Common Input: an instance x of a language L with witness relation R_L.

Auxiliary Input for Prover: a witness w, such that $(x, w) \in R_L(x)$.

Parameters: r (round complexity), p (penalty function), Δ (max delay), δ (min delay)

Stage 1:

V uniformly chooses $s_1, s_2 \in \{0, 1\}^n$.

V $\rightarrow$ P: $c_1 = f(s_1)$, $c_2 = f(s_2)$, and $r + 1$ first messages $\alpha_1, \ldots, \alpha_{r+1}$ of WI special-sound proofs of knowledge of the statement $(c_1, c_2) \in L'$ (called the start message). The proof of knowledge is with respect to the witness relation $R_{L'}$.

For $j = 1$ to $r + 1$ do

P $\rightarrow$ V [opening of slot j]: Select a second message $\beta_j \leftarrow \{0, 1\}^{n^2}$ for the jth WI special-sound proof. P times-out if the next verifier message is not received in time 2Δ.

V $\rightarrow$ P [closing of slot j]: Third message γ_j for the jth WI special-sound proof.

P measures the time elapsed between the opening of the first slot and the closing of the $r + 1^{\text{st}}$ slot as t.

Stage 2:

P delays the next message by time $\max\{p(t) - t, \delta\}$.

P $\leftrightarrow$ V: A perfect-WI argument of knowledge of the statement $(x, c_1, c_2) \in L \vee L'$, where $L \vee L'$ is characterized by the witness relation

$$R_{L \vee L'}(x, c_1, c_2) = \{(w, s_1', s_2') \mid w \in R_L(x) \vee (r_1', r_2') \in R_{L'}(c_1, c_2)$$

The argument of knowledge is with respect $R_{L \vee L'}$.

Fig. 5. Concurrent Perfect ZK argument for NP

The soundness and the completeness of the protocol follows directly from the proof of Feige and Shamir [FS90]; in fact, the protocol is an instantiation of theirs. Intuitively, to cheat in the protocol a prover must "know" an inverse to either c_1 or c_2, which requires inverting the one-way function f.

3.3 Simulator Overview

At a very high-level our simulator follows that of Feige and Shamir [FS90]. The simulator will attempt to rewind one of the special-sound proofs (i.e., the slots), because whenever the simulator obtains two accepting proof transcripts, the special-soundness property allows the simulator to extract a "fake witness" r_i such that $c_i = f(r_i)$. This witness can later be used in the second phase of the protocol. At any point in the simulation, we call a session of the protocol **solved** if such a witness has been extracted. On the other hand, if the simulation reaches Stage 2 of a session without extracting any "fake witnesses", we say the simulation is **stuck**.

In more detail, our simulator is essentially identical to that of [PV08], which in turn is based on the simulator of [RK99]. The general strategy of the simulator is to find and rewind the "easiest" slot for each session; during a rewind, the simulator recursively invokes itself on any nested sessions when necessary. The main difference between our work and that of [RK99, PV08] lies in determining which slot to rewind. In [RK99, PV08], a slot that contains a "small" amount of start messages (freshly started sessions) is chosen, whereas in our simulation, a slot with *"little" elapsed time* (between the opening and the closing) is rewound. As we will see, part of the analysis from [PV08] applies directly to our simulator modulo some changes in parameters; we only need to ensure that our definition of "little" elapsed time allows the simulator to always find a slot to rewind (formally argued in Claim 2).

3.4 Description of the Simulator

Our simulator is defined recursively. Intuitively on recursive level 0, the simulator's goal is to generate a view of V^*, while on all other recursive levels, the simulator's goal is to rewind a particular slot (from a previous recursion level). On recursive level ℓ, the simulator starts by feeding random Stage 1 messages to V^*. Whenever a slot s closes, S decides whether or not to rewind s depending on the time elapsed between the opening and the closing of s. If the elapsed time is "small" (where the definition of small depends on the level ℓ), S begins to rewind the slot. That is, S recursively invokes itself on level $\ell + 1$ starting from the opening of slot s with a new (random) message β, with the goal of reaching the closing message of slot s. While in level $\ell + 1$, S continues the simulation until one of the following happens:

1. *The* closing *message* γ *for slot* s *occurs*: S extracts a "fake" witness using the special-sound property and continues its simulation (on level ℓ).
2. V^* *aborts or delays "too much" in the rewinding:* S restarts its rewinding using a new challenge β for s. We show in expectation, S only restarts $O(1)$ times (intuitively, this follows since during the execution at level ℓ, S only starts rewinding a slot if V^* did not abort and only took "little time").
3. S *is "stuck" at Stage 2 of an unsolved session that started at level* $\ell + 1$*:* S halts and outputs fail (we later show that this never happens).

4. *S is "stuck" at Stage 2 of an unsolved session that started at level ℓ:* Again, S restarts its rewinding. We show that this case can happen at most $m-1$ times, where m is the total number of sessions.
5. *S is "stuck" at Stage 2 of an unsolved session that started at level $\ell' < \ell$:* S returns the view to level ℓ' (intuitively, this is just case 4 for the recursion at level ℓ').

In the unlikely event that S asks the same challenge β twice, S performs a brute-force search for the witness. Furthermore, to simplify the analysis of the running-time, the simulation is cut-off if it runs "too long" and S extracts witnesses for each session using brute-force search.

$\text{SOLVE}_d^{V^*}(x, \ell, h_{initial}, s, \mathsf{W}, \mathsf{R})$:

Let $h \leftarrow h_{initial}$. Note that $h_{initial}$ contains all sessions that are started on previous recursion levels.

Repeat forever:

1. If v is a Stage 2 verifier message of some session, continue.
2. If V^* aborts in the sessions of slot s, or the time elapsed since $h_{initial}$ exceeds $g^{\mathsf{d}+1-\ell}(\delta)$, restart SOLVE from $h_{initial}$.
3. If the next scheduled message is a Stage 2 prover message for session i and $\mathsf{W}(i) \neq \bot$, then use $\mathsf{W}(i)$ to complete the WI proof of knowledge; if $\mathsf{W}(i) = \bot$ and start message of session i is in $h_{initial}$ return h, otherwise halt with output **fail**.
4. If the next scheduled message is a Stage 1 prover message for slot s', pick a random message $\beta \leftarrow \{0,1\}^{n^2}$. Append β to h. Let $v \leftarrow V^*(h)$.
5. Otherwise, if v is the closing message for $s' = \mathsf{slot}\ (i', j')$, then update W with v (using R) and proceed as follows.
   (a) If $s = s'$, then return h.
   (b) Otherwise, if session i' starts in $h_{initial}$, then return h.
   (c) Otherwise, if $\mathsf{W}(i') \neq \bot$ or the time elapsed since the opening of $\mathsf{slot}\ (i', j')$ exceeds $g^{\mathsf{d}-\ell}$, then continue.
   (d) Otherwise, let h' be the prefix of the history h where the prover message for s' is generated. Set $\mathsf{R}' \leftarrow \phi$. Repeat the following m times:
      i. $h^* \leftarrow \text{SOLVE}_d^{V^*}(x, \ell+1, h', s', \mathsf{W}, \mathsf{R}')$
      ii. If h^* contains an accepting proof transcript for slot s', extract witness for session i' from h and h^* and update W.
      iii. Otherwise, if the last message in h^* is the closing message for the last slot of an session that started in $h_{initial}$ return h^*.
      iv. Otherwise, add h^* to R'.

$S^{V^*}(x, z)$:

Let $\mathsf{d} \leftarrow \min_d\{g^d(\delta) > 2\Delta\}$. Run $\text{SOLVE}_d^{V^*}(x, 0, , , ,)$ and output whatever SOLVE outputs with one exception. If an execution of $\text{SOLVE}_d^{V^*}(x, 0, , , ,)$ queries V^* more that 2^n times, proceed as follows:

Let h denote the view reached in the "main-line" simulation (i.e., in the top-level of the recursion). Continue the simulation in a "straight-line" fashion from h by using a brute-force search to find a "fake" witness each time Stage 2 of an session i is reached.

Fig. 6. Description of our black-box ZK simulator

The basic idea behind the simulation is similar to [PV08]: We wish to define "little time" appropriately, so that some slot of every session is rewound and that expected running time is bounded. For a technical reason (used later in Claim 2), we actually want the simulator to rewind one of the first r (out of $r+1$) slots of each session.

Take for example $p(t) = 2t$ and $r = 2$ (3 slots). Based on our intuition from Sect. 3.1, a good approach would be to ensure that the simulation at recursive level ℓ finishes within time $2\Delta/4^{\ell}$, and define "little time" on level ℓ to be $2\Delta/4^{\ell+1}$. Then, we know that any session that is fully executed at recursive level ℓ must have taken time less than $2\Delta/(4^{\ell} \cdot 2)$ in Stage 1 (due to penalty-based delays), and therefore one of the first two slot must have taken time less than $2\Delta/4^{\ell+1}$, making it eligible for rewind. To show that the expected running time is bounded, we simply set δ appropriately (as a function of d, Δ and r) as in Sect. 3.1, and this would guarantee that the recursion depth of the simulator is bounded.

A formal description of our simulator can be found in Figure 6. We rely on the following notation.

- Define the function $g : \mathsf{N} \to \mathsf{N}$ by $g(n) = p(rn)$. Recall that $g^k(n)$ be the function computed by composing g together k times, i.e., $g^k(n) = g(g^{k-1}(n))$ and $g^0(n) = n$. Let d (the maximum depth of recursion) be $\min_d\{g^d(\delta) > 2\Delta\}$. Note that if $\delta = (g^{-1})^k(2\Delta)$, then $\mathsf{d} = k$.
- slot (i, j) will denote slot j of session i.
- W is a repository that stores the witness for each session. The update W command extracts a witness from two transcripts of a slot (using the special-sound property). If the two transcripts are identical (i.e. the openings of the slot are the same), the simulator performs a brute-force search to extract a "fake" witness s_i s.t. $c_i = f(s_i)$ for $i \in \{1, 2\}$.
- R is a repository that stores the transcripts of slots of unsolved sessions. Transcripts are stored in R when the simulator gets stuck in a rewinding (cases 4 and 5 mentioned in the high-level description).

4 Analysis of the Simulator

To prove correctness of the simulator, we show that the output of the simulator is correctly distributed and its expected running-time is bounded. We first prove in Claim 2 that the simulator never outputs fail. Using Claim 2, we show that the output distribution of the simulator is correct in Prop. 3, and that the expected running time of the simulator is at most $poly(m^{\mathsf{d}} r^{\mathsf{d}})$ in Prop. 4. Theorem 1 then follows from Prop. 3 and 4, together with the fact that if $\delta = (g^{-1})^k(2\Delta)$ then $\mathsf{d} = k$.

Claim 2. *For every $x \in L$, $S^{V^*}(x, z)$ never outputs* fail.

Proposition 3. *The ensembles* $\{\mathsf{VIEW}_2[P(x,w) \leftrightarrow V^*(x,z)]\}$ *and* $\{S^{V^*}(x,z)\}$ *are identical over* $x \in L, w \in R_L(x), z \in \{0,1\}^*$.

Proposition 4. *For all $x \in L, z \in \{0,1\}^*$, and all V^* such that $V^*(x,z)$ opens up at most m sessions,* $E[\mathsf{time}_{\tilde{S}^{V^*(x,z)}}] \leq poly(m^{\mathsf{d}} r^{\mathsf{d}})$

The proof of Claim 2 is given below, while the proofs of Prop. 3 and 4 are given in the full version of the paper; in any case, the proofs of Prop. 3 and 4 are essentially identical to [PV08], modulo a change of parameters. Throughout the analysis we assume without loss of generality that the adversary verifier V^* is deterministic (as it can always get its random coins as part of the auxiliary input).

Proof: (Claim 2) Recall that $S^{V^*}(x,z)$ outputs fail only if $\text{SOLVE}_d^{V^*}(x,0,,,)$ outputs fail. Furthermore, SOLVE outputs fail at recursive level ℓ only if it reaches Stage 2 of an unsolved session that started at level ℓ (see Step 3 of SOLVE). We complete the proof in two parts. First we show $\text{SOLVE}_d^{V^*}$ will rewind at least one of the first r slots of every session at level ℓ. Then, we show that SOLVE always extracts a witness when it rewinds a slot.

In order for SOLVE to be stuck at a session i that starts at recursive level ℓ, session i must reach Stage 2 within $g^{(\mathsf{d}-\ell)}(\delta)$ time-steps (otherwise SOLVE would have rewound as per Step 2). This implies that t, the time between the opening of the first slot and the closing of th last slot of session i, must satisfy $p(t) \leq g^{(\mathsf{d}-\ell)}(\delta)$ (due to penalty-based delays). This in turn implies that one of the first r slots of session i must have taking time at most

$$\frac{t}{r} \leq \frac{p^{-1}(g^{(\mathsf{d}-\ell)}(\delta))}{r} \leq g^{(\mathsf{d}-\ell-1)}(\delta)$$

(here we use the monotonicity of p). By construction, SOLVE would have rewound this slot (i.e., execute Step 5.(d)).

Next we show that whenever SOLVE rewinds a slot, a witness for that session is extracted. Assume for contradiction that SOLVE fails to extract a witness after rewinding a particular slot. Let level ℓ and slot j of session i be the first time this happens. This means at the end of Step 5.(d), m views are obtained, yet none of them contained a second transcript for slot j. Observe that in such a view, SOLVE most have encountered Stage 2 of some unsolved session i' (i.e., stuck). Yet, we can show that the $m-1$ other sessions can each cause SOLVE to be stuck at most once; this contradicts the fact that SOLVE is stuck on all m good views.

For every session i' that SOLVE gets stuck on, both the opening and the closing of the last slot occurs inside the rewinding of slot (i,j); otherwise, SOLVE would have rewound one of the r slots that occurred before the opening of slot (i,j) successfully and extracted a witness for session i' (l,i,j was the first "failed" slot). Furthermore, the transcript of this slot enables SOLVE to never get stuck on session i' again, since the next time that the last slot of session i' closes will allow SOLVE to extract a witness for session i'. □

Acknowledgments

We would like to thank the anonymous TCC reviewers for their helpful comments.

References

[Axe84] Axelrod, R.: The evolution of cooperation. Basic Books, New York (1984)

[CF01] Canetti, R., Fischlin, M.: Universally composable commitments. In: Kilian, J. (ed.) CRYPTO 2001. LNCS, vol. 2139, pp. 19–40. Springer, Heidelberg (2001)

[CKP01] Cohen, T., Kilian, J., Petrank, E.: Responsive round complexity and concurrent zero-knowledge. In: Boyd, C. (ed.) ASIACRYPT 2001. LNCS, vol. 2248, pp. 422–441. Springer, Heidelberg (2001)

[CKPR01] Canetti, R., Kilian, J., Petrank, E., Rosen, A.: Black-box concurrent zero-knowledge requires $\tilde{\omega}(\log n)$ rounds. In: STOC 2001, pp. 570–579 (2001)

[DNS04] Dwork, C., Naor, M., Sahai, A.: Concurrent zero-knowledge. J. ACM 51(6), 851–898 (2004)

[FS90] Feige, U., Shamir, A.: Witness indistinguishable and witness hiding protocols. In: STOC 1990, pp. 416–426 (1990)

[GK96] Goldreich, O., Kahan, A.: How to construct constant-round zero-knowledge proof systems for NP. Journal of Cryptology 9(3), 167–190 (1996)

[GMR89] Goldwasser, S., Micali, S., Rackoff, C.: The knowledge complexity of interactive proof systems. SIAM J. Comput. 18(1), 186–208 (1989)

[Gol01] Goldreich, O.: Foundations of Cryptography — Basic Tools. Cambridge University Press, Cambridge (2001)

[Gol02] Goldreich, O.: Concurrent zero-knowledge with timing, revisited. In: STOC 2002, pp. 332–340 (2002)

[KLP05] Kalai, Y.T., Lindell, Y., Prabhakaran, M.: Concurrent general composition of secure protocols in the timing model. In: STOC 2005, pp. 644–653 (2005)

[KP01] Kilian, J., Petrank, E.: Concurrent and resettable zero-knowledge in polylogarithmic rounds. In: STOC 2001, pp. 560–569 (2001)

[KPR98] Kilian, J., Petrank, E., Rackoff, C.: Lower bounds for zero knowledge on the internet. In: FOCS 1998, pp. 484–492 (1998)

[Lin04] Lindell, Y.: Lower bounds for concurrent self composition. In: Naor, M. (ed.) TCC 2004. LNCS, vol. 2951, pp. 203–222. Springer, Heidelberg (2004)

[LPV09] Lin, H., Pass, R., Venkitasubramaniam, M.: A unified framework for concurrent security: universal composability from stand-alone non-malleability. In: STOC 2009, pp. 179–188 (2009)

[MP06] Micali, S., Pass, R.: Local zero knowledge. In: STOC 2006, pp. 306–315 (2006)

[PRS02] Prabhakaran, M., Rosen, A., Sahai, A.: Concurrent zero knowledge with logarithmic round-complexity. In: FOCS 2002, pp. 366–375 (2002)

[PV05] Persiano, G., Visconti, I.: Single-prover concurrent zero knowledge in almost constant rounds. In: Caires, L., Italiano, G.F., Monteiro, L., Palamidessi, C., Yung, M. (eds.) ICALP 2005. LNCS, vol. 3580, pp. 228–240. Springer, Heidelberg (2005)

[PV08] Pass, R., Venkitasubramaniam, M.: On constant-round concurrent zero-knowledge. In: Canetti, R. (ed.) TCC 2008. LNCS, vol. 4948, pp. 553–570. Springer, Heidelberg (2008)

[RK99] Richardson, R., Kilian, J.: On the concurrent composition of zero-knowledge proofs. In: Stern, J. (ed.) EUROCRYPT 1999. LNCS, vol. 1592, pp. 415–432. Springer, Heidelberg (1999)

[Ros00] Rosen, A.: A note on the round-complexity of concurrent zero-knowledge. In: Bellare, M. (ed.) CRYPTO 2000. LNCS, vol. 1880, pp. 451–468. Springer, Heidelberg (2000)

[Rs09] Rosen, A., shelat, a.: A rational defense against concurrent attacks (2009) (manuscript)

Efficiency Preserving Transformations for Concurrent Non-malleable Zero Knowledge

Rafail Ostrovsky[1,*], Omkant Pandey[1,**], and Ivan Visconti[2,***]

[1] University of California, Los Angeles, USA
{rafail,omkant}@cs.ucla.edu
[2] University of Salerno, Italy
visconti@dia.unisa.it

Abstract. Ever since the invention of Zero-Knowledge by Goldwasser, Micali, and Rackoff [1], Zero-Knowledge has become a central building block in cryptography - with numerous applications, ranging from electronic cash to digital signatures. The properties of Zero-Knowledge range from the most simple (and not particularly useful in practice) requirements, such as honest-verifier zero-knowledge to the most demanding (and most useful in applications) such as non-malleable and concurrent zero-knowledge. In this paper, we study the complexity of *efficient* zero-knowledge reductions, from the first type to the second type. More precisely, under a standard complexity assumption (DDH), on input a public-coin honest-verifier statistical zero knowledge argument of knowledge π' for a language L we show a compiler that produces an argument system π for L that is concurrent non-malleable zero-knowledge (under non-adaptive inputs – which is the best one can hope to achieve [2,3]). If κ is the security parameter, the overhead of our compiler is as follows:

- The round complexity of π is $r + \tilde{O}(\log \kappa)$ rounds, where r is the round complexity of π'.
- The new prover $\mathcal{P}$ (resp., the new verifier $\mathcal{V}$) incurs an additional overhead of (at most) $r + \kappa \cdot \tilde{O}(\log^2 \kappa)$ modular exponentiations. If tags of length $\tilde{O}(\log \kappa)$ are provided, the overhead is only $r + \tilde{O}(\log^2 \kappa)$ modular exponentiations.

The only previous concurrent non-malleable zero-knowledge (under non-adaptive inputs) was achieved by Barak, Prabhakaran and Sahai [4]. Their construction, however, mainly focuses on a *feasibility* result rather than efficiency, and requires expensive $\mathcal{NP}$-reductions.

* Supported in part by IBM Faculty Award, Xerox Innovation Group Award, the Okawa Foundation Award, Intel, Teradata, NSF grants 0716835, 0716389, 0830803, 0916574 and U.C. MICRO grant.

** Supported in part by NSF grants 0716835, 0716389, 0830803, 0916574, and the European Commission grants of the third author.

*** Supported in part by the European Commission through the EU IST program under Contract IST-2002-507932 ECRYPT, and through the EU ICT program under Contract ICT-2007-216646 ECRYPT II.

D. Micciancio (Ed.): TCC 2010, LNCS 5978, pp. 535–552, 2010.

1 Introduction

In this paper, we consider Zero-Knowledge argument systems that are non-malleable and secure against concurrent man-in-the-middle attacks. In such systems, the adversary has complete control over the communication channel and can behave as honest prover and honest verifier in any polynomial number of protocols, therefore controlling the scheduling of the messages. We aim at designing efficient argument systems secure against these attacks, namely *efficient concurrent non-malleable zero knowledge argument systems.* Despite the extreme importance of these proof systems, no efficient and secure (plain model) protocol for such settings is known until today. Feasibility results have been given originally by Dolev, Dwork, and Naor (DDN) [5], restricting the adversary to two simultaneous proofs. In recent work, Barak, Prabhakaran and Sahai [4] have obtained concurrent and non-malleable zero-knowledge without restricting the adversary to a bounded number of proofs, however the solutions proposed there can be viewed as constructing feasibility results only as their methods require NP reductions and are highly inefficient.

The need of *efficient* instantiations of concurrent NMZK for useful languages and its applicability as sub-protocols motivated the introduction of several strong set-up assumptions [6,7,8]. In this paper, we focus on achieving efficient transformations in the *plain* model which does not rely on any setup assumptions. We show a transformation that on input a public-coin honest-verifier statistical zero knowledge argument of knowledge π' for a language L produces a *concurrent non-malleable* zero-knowledge argument system π for L. Further, our transformation is an *efficiency preserving transformation* that does not require any $\mathcal{NP}$-reduction and works assuming standard number-theoretic assumptions (see theorem 1 for a precise statement).

It should be noted that CNMZK arguments are significantly harder to construct and analyze. In fact, Lindell proved that in the most general form of the attack, (non-trivial) CNMZK arguments do not even exist [2,3]. However, assuming that the honest parties' inputs are fixed in advance (i.e., are not chosen *adaptively* based on the protocol execution), CNMZK was shown to be achievable by Barak, Prabhakaran, and Sahai (BPS). The impossibility results discussed in [2,3,9] and the plausibility results of [4] suggest that CNMZK (under the non-adaptive input notion) is the best notion of security for proof systems that one can hope to achieve in the plain model. Our results are the *first* efficiency preserving transformation for the concurrent man-in-the-middle setting in the plain model, gaining dramatic efficiency improvements over [4] (see further discussion on efficiency immediately after the statement of our main result).

OUR RESULTS. Assuming the hardness of (standard) decisional Diffie-Hellman assumption, we show CNMZK argument-of-knowledge (see theorem 1). Our results require that the HVZK argument system admit *statistical* simulation and be an "argument of knowledge"[1]. We remark that the statistical simulation requirement for the given HVZK argument, is easy to achieve as most HVZK protocols that we know of already admit statistical simulation (by using statistically hiding commitments such as [10] – which exist under the DDH assumption).

Theorem 1 (Main Result). *Let* $\pi' : \langle \mathcal{P}', \mathcal{V}' \rangle$ *be a* public coin *honest verifier* statistical *zero-knowledge argument of knowledge, for some language* $L \in \mathcal{NP}$. *Let* κ *be a security parameter, and* q *be a prime number whose length is determined by* κ. *Then, assuming that the Decisional Diffie-Hellman Assumption holds, it is possible to transform* π' *into a new argument system* $\pi : \langle \mathcal{P}, \mathcal{V} \rangle$ *such that,*

– *Protocol* π *is a* computational concurrent non-malleable zero-knowledge *argument of knowledge for* L.
– *Protocol* π *has* $r + \tilde{O}(\log \kappa)$ *rounds of interaction, where* r *is the round complexity of* π'.
– *The new prover* $\mathcal{P}$ *(resp., the new verifier* $\mathcal{V}$*) incurs an additional overhead of* $r + \kappa \cdot \tilde{O}(\log^2 \kappa)$ *exponentiations in* $\mathbb{Z}_q$. *For tag-based non-malleability, the overhead is only* $r + \tilde{O}(\log^2 \kappa)$ *additional exponentiations in* $\mathbb{Z}_q$*, assuming tags of length* $\tilde{O}(\log \kappa)$.

Although our main focus is the plain model, our results about tag-based non-malleability, lead to more efficient constructions in the Bare-Public-Key (BPK) model [11]. The BPK model, assumes an *untrusted* setup which brings it very close to the plain model. Like the plain model, our results in the BPK model are the first efficient transformations (see section 5 for more details).

Our starting point to avoid $\mathcal{NP}$-reductions is "Simulatable Commitments" as defined by Micciancio and Petrank [12] (though our construction and proof requires development of several new techniques and ideas on top of this work). Using simulatable commitments, Micciancio and Petrank demonstrate how to efficiently transform any HVZK argument system into a *concurrent* ZK argument system which is secure against a cheating verifier V^* mounting a *concurrent* attack. Their transformation increases the round complexity of the original argument system by $\tilde{O}(\log \kappa)$ and incurs an additional overhead of $r + \tilde{O}(\log \kappa)$ exponentiations in $\mathbb{Z}_q$.

TECHNICAL OVERVIEW AND MAIN DIFFICULTIES. We design a new protocol to make the given protocol π_{HV} secure in the CNMZK model without much compromise in its efficiency. As we explain below, our transformation is conceptually different from the only known CNMZK protocol of BPS. Due to this conceptual difference in the construction, our proof of security is entirely new.

[1] The "argument of knowledge" requirement is actually due to the particular definition of security we aim to achieve, namely *simulation extractability* (see Definition 1). If the given protocol is not an argument of knowledge, our transformation still delivers *simulation soundness*.

To explain the main conceptual ideas/differences, we now sketch our transformation.[2] At a very high level, our transformation has following structure: (1) Our verifier, $\mathcal{V}$, first executes a KP/PRS preamble for a secret v; (2) Our prover, $\mathcal{P}$, then commits to 0^κ using a (properly instantiated) DDN-commitment; (3) $\mathcal{V}$ then reveals v; (4) and finally, $\mathcal{P}$ proves to $\mathcal{V}$ that "$x \in L$ OR $\mathcal{P}$ committed to v".

Note that in phase-(3) we need an *efficient* version of DDN-commitments.[3] Also, phase 4 needs typically required NP reductions to apply "FLS"-trick which requires an NP reduction. Instead, we design a new protocol by extending and applying in a non-trivial way the Micciancio-Petrank (MP) [12] transformation to the input protocol π_{HV}. We now explain the main conceptual differences from BPS and their proof.

Note that our protocol has only four phases whereas BPS has five: we do not require a separate phase involving a statistically hiding commitment to 0^κ, followed by a SZKAOK for the knowledge of randomness to the commitment. This phase is *crucial* for BPS-proof to go through. This changes the proof significantly – we directly rely on phase-(3) and phase-(4). Next, it is clear that simulation will proceed by $\mathcal{S}$ extracting v (from KP/PRS-preamble) and then committing to v (instead of 0^κ) in the left sessions of DDN. Protocol of BPS commits to the *witness* (of the statement) instead and relies on this stage for extraction. Clearly, this completely changes how our extractor would work. Instead, we must rely on the last phase to perform extraction. This is more involved than it seems: when simulator uses commitment to v as witness for succeeding in the last phase. Thus, to be able to argue correctness of extraction, we need *statistical simulation*.[4] Unfortunately, because of MP-transformation, transformed π_{HV} loses its statistical simulation – making the proof stuck. However, we identify a new property: MP-transformation admits "statistical simulation with respect to *lucky* provers" (see section 3), and this suffices to argue the correctness of extraction. Briefly. a "lucky" prover is one who can *guess* the PRS-secret correctly, in advance. Our extractor also differs from "standard" methods: we first test whether man-in-the-middle has succeeded in setting up a trapdoor by doing a preliminary DDN-extraction before performing actual extraction from the last phase (otherwise the extractor may not be expected-PPT).

Other Related Work. Achieving practical constructions/instantiations of advanced cryptographic tasks has become an increasingly popular research direction in recent years. To gain efficiency, $\mathcal{NP}$-reductions has been a common bottleneck that most of these research works also aim at avoiding. Among these,

[2] Unfortunately, here reader's familiarity with the BPS-protocol and their proof structure is required.

[3] Interestingly, this is not immediately clear. Before this work, to the best of our knowledge, the only hope for achieving an efficient non-malleable commitment was from a *recent* protocol of Lin, Pass, and Venkitasubramaniam [13]. Here, we show a new and simple technique which provides an efficient instantiation of DDN-commitments (see section 3.2).

[4] This is a somewhat common issue in non-malleability proofs when going from one hybrid to another (e.g., the non-malleable commitments of Pass and Rosen [14]).

the most relevant works are those of Garay, MacKenzie, and Yang [6], and De Santis, Di Crescenzo, Ostrovsky, Persiano, and Sahai [15] (CRS model), and Micciancio and Petrank [12] (plain model). In the area of secure two-party computation, see the works of Mohassel and Franklin [16], Woodruff [17], Lindell and Pinkas [18], and Goyal, Mohassel, and Smith [19]. For non-interactive zero-knowledge see Chase and Lysyanskaya [20], and Groth, Ostrovsky, and Sahai [21].

2 Definitions

In this section we present relevant definitions. We assume familiarity with (standard) cryptographic concepts such as computational and statistical indistinguishability, $\mathcal{NP}$-relations, interactive proof and argument systems, simulation paradigm, etcetera (see [22]). In the following, L is an $\mathcal{NP}$-language with witness relation R_L. That is, a statement $x \in L$ iff there exists a y of length poly($|x|$) such that $R_L(x, y) = 1$.

Concurrent Man-in-the-Middle Attack. The concurrent man-in-the-middle setting proceeds as follows. First, the inputs to the honest provers, i.e., statements $x_1, \ldots, x_{m_L} \in L \cap \{0,1\}^n$ are chosen; thereafter, m_L honest provers, $P_i \stackrel{\text{def}}{=} P(x_i, y_i; \omega_i)$, are constructed (for $i \in [m_L]$) such that $R_L(x_i, y_i) = 1$, and ω_i is a uniformly chosen random tape of sufficient (polynomial in κ) length. Adversary M may now start interacting with these provers while playing the role of a verifier of π with each one of them. These interactions are called "left" interactions. At any point, M, may adaptively output a new statement $\tilde{x}_i \in L \cap \{0,1\}^n$. Whenever it does so, an honest verifier $V_i \stackrel{\text{def}}{=} V(\tilde{x}_i; \tilde{\omega}_i)$, is created with input $\tilde{x}_i$ and uniformly chosen randomness $\tilde{\omega}_i$. Such verifiers are created to the "right" of M who may try to convince V_i of the validity of statement $\tilde{x}_i$ by playing the role of the prover in a session of π. These interactions are called the "right" interactions, and M may simultaneously continue its left interactions. Let m_R denote the number of right hand side sessions before M halts.

A *concurrent non-malleable attack*, a man-in-the-middle adversary M interacts with provers $P_1, \ldots, P_{m_L}$ in m_L "left sessions" and verifiers $V_1, \ldots, V_{m_R}$ in m_R "right sessions" of the protocol with M controlling the scheduling of all the sessions. "Left inputs" $x_1, \ldots, x_{m_L}$ are fixed in advance, whereas "right inputs" $\tilde{x}_1, \ldots, \tilde{x}_{m_R}$ can be decided by M adaptively. We consider only non-uniform PPT adversaries M, and so both m_L, m_R are polynomial in κ.

Following the work of Pass and Rosen [14], when dealing with non-malleability it is sometimes easier to work with a somewhat stronger notion called the *simulation-extractability*. They demonstrate that simulation-extractability implies non-malleable zero-knowledge argument (proof) of knowledge property. This approach was also followed by RPS, and we stick to their definition.

Definition 1. *A protocol* $\pi \stackrel{def}{=} \langle P, V \rangle$ *is said to be a Concurrent Non-Malleable Zero Knowledge (*CNMZK*) argument of knowledge for membership in an* $\mathcal{NP}$ *language* L *with witness relation* R_L*, if it is an interactive argument system between a prover and verifier (both PPT) such that the following conditions hold.*

Completeness. *For every x, y such that $R_L(x,y) = 1$, $P(x,y)$ makes V accept with probability 1.*

Soundness, Zero Knowledge, and Non-malleability. *For every PPT adversary M launching a concurrent non-malleable attack as above (i.e., M interacts with $P_1, \ldots, P_{m_L}$ in "left sessions" and $V_1, \ldots, V_{m_R}$ in right sessions as defined above), there exists an expected polynomial time simulator-extractor $\mathcal{S}$ such that for every set of "left inputs" $x_1, \ldots, x_{m_L}$ we have $\mathcal{S}(x_1, \ldots, x_{m_L}) = (\nu, \tilde{y}_1, \ldots, \tilde{y}_{m_R})$ such that,*

- *ν is the simulated joint view of M and $V_1, \ldots, V_{m_R}$. Further, for any set of witnesses $(y_1, \ldots, y_{m_L})$ defining the provers $P_1, \ldots, P_{m_L}$, the view ν is distributed computationally indistinguishably from the view of M in a real execution.*
- *In the view ν, let TRANS_h denote the transcript of hth left execution, and $\tilde{\text{TRANS}}_\ell$ that of ℓth right execution, $h \in [m_L], \ell \in [m_R]$. If $\tilde{x}_\ell$ is the common input in $\tilde{\text{TRANS}}_\ell$, $\tilde{\text{TRANS}}_\ell \neq \text{TRANS}_h$ (for all h) and V_ℓ accepts, then $R_L(\tilde{x}_\ell, \tilde{y}_\ell) = 1$ except with probability negligible in κ.*

The probability is taken over the random coins of $\mathcal{S}$. Further, the protocol is black-box CNMZK, *if $\mathcal{S}$ is an universal simulator that uses M only as an oracle, i.e., $\mathcal{S} = \mathcal{S}^M$.*

The second condition in the definition of soundness above, says that if some right session is not an *exact copy* of any of the left sessions, then $\mathcal{S}$ should output a valid witness for the statement of that right session.

The DDH Assumption. Let q be a sufficiently large randomly chosen prime such that there exists another sufficiently large prime p that divides $q-1$. Let G_p be an order p (multiplicative cyclic) subgroup of $\mathbb{Z}_q$, with some generator g. Then, the DDH assumption states that for randomly and independently chosen $a, b, c \in \mathbb{Z}_p$, the following two distributions are computationally indistinguishable: (g^a, g^b, g^{ab}) and (g^a, g^b, g^c).

Strong Signatures. A signature scheme $(\mathcal{K}, \text{SIGN}, \text{VERIFY})$ is said to be *strongly unforgeable* if no efficient adversary, with access to a signing oracle with respect to verification key VK, can output a pair (m, σ) with non-negligible probability, such that: $\text{VERIFY}(m, \sigma, \text{VK}) = 1$ and the pair (m, σ) does not correspond to the input-output pair of a performed oracle query. A *strong signature scheme* is a signature scheme that is strongly unforgeable.

Notation. Throughout the paper, $\mu : \mathbb{N} \to \mathbb{R}$ denotes a negligible function in κ (the security parameter). If a message u appears in the "left" session, then its counterpart in the "right" session will be denoted by $\tilde{u}$.

3 Building Blocks: Efficient Instantiations

We discuss two of our main building blocks: **(a)** simulatable commitments, and **(b)** DDN-commitments. We assume here familiarity with *commitment schemes* and their computational/statistical/perfect binding and hiding properties (see [22]). Simulatable commitments were used by [12] to compile HVZK arguments into concurrent ZK arguments (for us just stand-alone ZK suffices) – which we discuss briefly. Thereafter, we discuss an efficient implementation of the DDN-commitment scheme. Our descriptions are brief, and we refer the reader to the respective works for more details.

3.1 Simulatable Commitments

A simulatable commitment [12] scheme is a tuple (COM, DCOM, $P_{\text{COM}}, V_{\text{COM}}, S_{\text{COM}}$) such that (COM, DCOM) specifies an usual (*non-interactive, perfectly binding, computationally hiding*) commitment scheme. Additionally, it comes with a *3-round* HVZK proof system ($P_{\text{COM}}, V_{\text{COM}}$) to show, given two strings (c, v), c is a commitment to v (i.e., $\exists r$ s.t. $c \leftarrow \text{COM}(v; r)$). The proof system has perfect completeness, optimal soundness[5], and efficient prover (given input r); S_{COM} is the simulator for the HVZK property of the system. A construction of a simulatable commitment scheme, based on the DDH assumption, is given in the full version of this paper (the construction is due to [12], and admits *statistical* simulation). Note that because of the (computational) hiding property, it follows that the output of S_{COM} on input a true statement (c, v), is computationally indistinguishable from its output on input a false statement (c, v').

HVZK to Stand-alone ZK. Using simulatable commitments, [12] show how to transform any public coin HVZK argument system $\pi_{\text{HV}} : \langle P_{\text{HV}}, V_{\text{HV}} \rangle$ to a new system, which is zero-knowledge with respect to *any* (PPT) verifier (i.e., the new system is stand-alone ZK). We call this transformation the Micciancio-Petrank transformation, and denote the new system by $\pi_{\text{MP}} : \langle P_{\text{MP}}, V_{\text{MP}} \rangle$.[6]

To pinpoint a crucial property we need, we briefly explain how the transformed protocol π_{MP} proceeds. First, parties P_{MP} and V_{MP} execute a preamble phase, in which V_{MP} commits to a value $v \in \mathbb{Z}_p$, using a statistically hiding commitment; P_{MP} then commits to 0^κ using *simulatable commitments* (let the commitment be denoted by c). Finally V_{MP} opens the value v to P_{MP}. The transcript of conversation is thus (c, v). Now the second phase of the proof starts, in which V_{MP} acts like V_{HV}, but each challenge of V_{MP} is decided using "coin-tossing"-type style (see the full version of this paper for concrete details). The proof system, that comes with simulatable commitments, is used for this purpose with input statement (c, v). Statement (c, v) is false in a real execution with high probability (which results in uniform output for V_{HV}'s challenges), but the simulator can setup a *true* (c, v) via rewinding (and hence bias the output of coin-tossing to any value). The protocol is thus both: ZK and sound. The crucial property that we need, is described next.

---

[5] Informally, it means that for a false statement (c, v), given the first prover-message and the verifier-query, there is exactly one convincing answer.

[6] As mentioned earlier, the main result of [12] gives *concurrent* ZK; stand-alone ZK is a special case and adds only *four* more rounds to π_{HV}.

Statistical *Simulation with respect to "lucky" Provers.* In general, π_{MP} is only computational ZK, since the prover commits to 0^κ while the simulator commits to v (the message opened by V_{MP}). However, consider a prover who can always guess the value v correctly and commits to it instead of 0^κ (but uses its witness in the rest of the execution of P_{MP}). Call such a prover "lucky".[7] Then, for such provers, the statement (c, v) (from first phase) is always *true*. Thus, if π_{HV} admits *statistical* simulation, the protocol π_{MP} also admits statistical simulation *with respect to the "lucky" provers.* Formally, there exists a simulator S_{MP} for every verifier V^*_{MP} (of the protocol π_{MP}) such that the output of S_{MP} is *statistically indistinguishable* from the view of V^*_{MP} in a real execution with a "lucky" prover (say $P^{(\text{lucky})}_{\text{MP}}$).

3.2 The DDN Commitment

Our construction needs an efficient instantiation of the DDN-commitment protocol. The DDN-commitment protocol is *non-malleable* which means – intuitively – given a commitment c on some message v, knowledge of c does not help a man-in-the-middle adversary in constructing a new commitment c' of a *related* message v'. The formal definition that we shall stick to appears in the full version of this paper.(This definition is satisfied by the variant of DDN-commitment protocol given in [4].) An efficient instantiation appears in Fig. 1.

In step 2, we mention the use of an efficient SZKAOK. An appropriate SZKAOK would be the one obtained by *sequentially* repeating the Schnorr protocol [23] $\omega(1)$ times. The size of verifier's challenge in each execution of Schnorr protocol, however, would only be $\log \kappa$.

In step 3 of the BCK protocol, we need an efficient proof system for statements of type: "c, c_{1-r} are commitments to v, x_{1-r} resp., s.t. $\alpha = x_{1-r} + v \mod p$". Informally, it can be achieved as follows. Commitment c is a pair of values in $\mathbb{Z}_q$: (a, b). Similarly, $c_{1-r} = (a', b')$. Compute $A = aa' \mod q, B = bb' \mod q$. Now use the proof system of simulatable commitments, to prove that (A, B) is a commitment to α. (Note that the proof system is only HVZK, but it can be first converted to (general) ZK by using the Micciancio-Petrank transformation once again before it is used in step 3 of the BCK protocol). The details are an easy exercise, which we defer to the full version of the paper.

4 An Efficiency Preserving Transformation

4.1 The Extraction Preamble

The extraction preamble is just the the "KP/PRS-preamble". This is a protocol between two players: a sender, A, and a receiver, B. The sender holds a value $v \in \mathbb{Z}_p$.[8] Let $a_i \stackrel{\text{def}}{=} \{(v_0^{i,j}, v_1^{i,j})\}_{j=1}^{\beta}$ be the list of pairs such that $v_0^{i,j} + v_1^{i,j} = v$

[7] Note that in real executions provers will not be "lucky" w.h.p.; the simulator will, however, setup the situation of the "lucky" prover to succeed.

[8] Here, and everywhere else in this paper, when we mention $\mathbb{Z}_p$, it should be assumed that $\mathbb{Z}_p$ is an appropriately chosen order p subgroup of $\mathbb{Z}_q$ in which DDH is hard, where p, q are as defined in the DDH assumption.

The DDN-commitment protocol.

1. $\mathcal{S}_{\text{DDN}}$ sends VK – the verification key of a strong signature scheme, computed using $\mathcal{K}(1^\kappa)$. Let $|\text{VK}| = \kappa$.
2. $\mathcal{S}_{\text{DDN}}$ commits to v using the simulatable commitment scheme (COM, DCOM), and sends $c \leftarrow \text{COM}(v; \omega)$ to $\mathcal{R}_{\text{DDN}}$. $\mathcal{S}_{\text{DDN}}$ then proves to $\mathcal{R}_{\text{DDN}}$ the knowledge of (v, ω) using an efficient $\omega(1)$-round public-coin statistical ZK argument of knowledge (SZKAOK). The last message of this SZKAOK is called the "Knowledge Determining Message" (KDM).
3. For $i = 1, \ldots, \kappa$, define $t^{(i)} = i \circ \text{VK}_i$. Thus, $|t^{(i)}| = 1 + \log \kappa$. Let $\text{BCK}^{\|}$ denote the protocol obtained by composing β parallel executions of the BCK protocol (described below), here $\beta \in \omega(\log \kappa)$. Recall that DDN defines two types of scheduling for $\text{BCK}^{\|}$: type-0 and type-1 (see [5]).
4. In *parallel*, for $i = 1, \ldots, |\text{VK}|$, execute the following protocol
   – For $j = 1, \ldots, (1 + \log \kappa)$ do *sequentially* –
   Execute $\text{BCK}^{\|}$ with type-$t_j^{(i)}$ scheduling.
   Execute $\text{BCK}^{\|}$ with type-$(1 - t_j^{(i)})$ scheduling.
5. $\mathcal{S}_{\text{DDN}}$ signs the full transcript of execution, and sends the signature σ to $\mathcal{R}_{\text{DDN}}$. $\mathcal{R}_{\text{DDN}}$ verifies the signature.

The BCK protocol mentioned in step 3 above.

1. $\mathcal{S}_{\text{DDN}}$ chooses $x_0, x_1 \in \mathbb{Z}_p$, and commits to each one of them using simulatable commitments; $c_b \leftarrow \text{COM}(x_b; \omega_b), b \in \{0, 1\}$. (Step BCK1)
2. $\mathcal{R}_{\text{DDN}}$ sends a bit r to $\mathcal{S}_{\text{DDN}}$. (Step BCK2)
3. $\mathcal{S}_{\text{DDN}}$ opens x_r and sends $\alpha = x_{1-r} + v \mod p$. $\mathcal{S}_{\text{DDN}}$ then proves to $\mathcal{R}_{\text{DDN}}$ using an efficient ZK protocol that: "c, c_{1-r} are commitments to v, x_{1-r} resp., s.t. $\alpha = x_{1-r} + v \mod p$". This protocol is discussed in section 3.2. (Step BCK3)

Fig. 1. The $O(\log \kappa)$-round DDN commitment scheme. $\mathcal{S}_{\text{DDN}}$ holds a value $v \in \mathbb{Z}_p$.

mod p where values $v_b^{i,j} \in \mathbb{Z}_p$ for all $b \in \{0, 1\}$ and $i, j \in [\beta]$. Here $\beta = \beta(\kappa)$ is any function in $\omega(\log \kappa)$. So there are β such lists, each consisting of β pairs.

The preamble consists of three steps. First step is the *commitment* step. Sender A chooses the parameters for the (perfectly binding) simulatable commitment scheme[9], and sends commitments to value v and to each share $v_b^{i,j} \in \mathbb{Z}_p$ (defined as above), using COM. The second step, (called the *challenge-response* step), is an interactive protocol consisting of β rounds, where in round i, player B sends a *challenge* $r_i \in \{0, 1\}^\beta$, and A sends a *response* as follows. The response of A consists of an opening of the commitments to one of the elements of each pair in a_i. That is, if $r_i^j = b$ (the jth-bit of r_i), then A includes in its response the value $v_b^{i,j}$, and the randomness it used to commit to $v_b^{i,j}$. At the end of this step,

[9] These commitments will sometimes be referred to as Micciancio-Petrank commitments.

we say that the preamble has *concluded.* The final step is the *opening* step. This step consists of A sending to B, the decommitment information corresponding to *all* the commitments of the *commitment* step. That is, A sends to B the values $v, v_b^{i,j}$, and the randomness it used to commit to them.

There can be other messages in the protocol between the prover concluding the preamble and the verifier opening the commitments. It is easy to see that if COM is a commitment scheme[10], the extraction-preamble is an interactive commitment scheme. We now state a result from PRS [25].

Lemma 1. *(Adapted from [25]) Consider provers $P_1, \dots, P_m$ and an adversarial verifier A_{PRS} running m sessions of a protocol with the extraction-preamble as described above, where m is polynomial in κ. Then except with negligible probability in κ, in every thread of execution output by the* KP/PRS*simulator, if the simulation reaches a point where P_i accepts the extraction-preamble with v as the secret of the sender (in that particular thread), then at the point when the preamble was concluded, the simulator would have already recorded the value v.*

In fact, we will also need a refinement of this lemma. However, both the lemma and the refinement are not needed until the analysis of hybrid simulators (which appears in the full version of this paper). Thus, the refinement and a more detailed discussion is provided in the full version of this paper.

4.2 The Transformation

Overview. We provide an overview of our transformation here in order to present the basic ideas in the construction (issues originating in the proof due to these ideas, were discussed in the introduction). The transformed protocol has the following structure. In the first phase, the verifier $\mathcal{V}$ executes the extraction preamble (of $\tilde{O}(\log \kappa)$ rounds with a value $v \in \mathbb{Z}_p$ chosen uniformly. In the second phase, the prover commits to 0^κ using our efficient DDN-commitment scheme. Note that the first message of this DDN-commitment phase includes a perfectly binding commitment to 0^κ using a simulatable commitment scheme – which we denote by c^*. $\mathcal{V}$ now opens the value v in the preamble (along with opening all other commitments of the preamble). This defines the pair (c^*, v).

Let the input protocol be $\langle \pi_{\mathrm{HV}} \rangle$. Recall that the Micciancio-Petrank transformation goes in two steps. In the first step a preamble is run, to obtain a pair (c_1, v_1) and then the second step uses this pair to enforce random challenges from the verifier of π_{HV}. In our protocol also, both $\mathcal{P}, \mathcal{V}$ now proceed exactly like this transformation, except that the first step of the transformation is not executed. Instead, (c^*, v) is used in place of (c_1, v_1).(We also use the standard trick of sending a verification key VK of a strong signature scheme to be used as the identity for the DDN-commitment, and in the end sign the whole transcript).

[10] In our description, COM is chosen to be a simulatable commitment which is perfectly binding. For the extraction-preamble, however, a perfectly hiding commitment scheme (such as [10]) may be used as well. Also, for simplicity, we have chosen to use the extraction preamble in the PRS-style, but the original style of Richardson-Kilian [24] will be more efficient.

The Compiler (from HVZK to CNMZK): The given HVZK argument is $\pi_{\text{HV}} : \langle P_{\text{HV}}, V_{\text{HV}} \rangle$.
$\mathcal{P} \to \mathcal{V}$: Run the key generation algorithm, $(\text{VK}, \text{SK}) \leftarrow \mathcal{K}(1^\kappa)$. Send VK to $\mathcal{V}$.

$\mathcal{V} \leftrightarrow \mathcal{P}$: $\mathcal{V}$ chooses a value $v \in_r \mathbb{Z}_p$. $\mathcal{V}$ and $\mathcal{P}$ then execute the "extraction preamble" where $\mathcal{V}$ plays the role of the sender, with input v; $\mathcal{P}$ plays the role of the receiver. Let $c \leftarrow \text{COM}(v; \omega), c_b^{i,j} \leftarrow \text{COM}(v_b^{i,j}; \omega_b^{i,j})$ denote the corresponding commitments. Recall that: $b \in \{0,1\}, i, j \in [\beta], v_0^{i,j} + v_1^{i,j} = v$. Here, $\omega, \omega_b^{i,j}$ denote the randomness used by the commitment scheme.

$\mathcal{P} \leftrightarrow \mathcal{V}$: $\mathcal{P}$ and $\mathcal{V}$ execute a DDN-commitment protocol in which $\mathcal{P}$ plays the role of $\mathcal{S}_{\text{DDN}}$ with input 0^κ, and $\mathcal{V}$ plays the role of $\mathcal{R}_{\text{DDN}}$. Let $\text{FM}^* = (p^*, q^*, g^*, h^*, \text{VK}^*, c^* \leftarrow \text{COM}(0^\kappa; \omega^*))$, denote the first message of the DDN-commitment protocol.

$\mathcal{V} \to \mathcal{P}$: $\mathcal{V}$ executes the *opening* step of the "extraction preamble", by sending the opening of all commitments sent in phase 2. That is, $\mathcal{V}$ sends to $\mathcal{P}$ the values: $v, v_b^{i,j}$ and randomness $\omega, \omega_b^{i,j}$, where b, i, j are defined as above.

$\mathcal{P} \leftrightarrow \mathcal{V}$: $\mathcal{P}$ (resp., $\mathcal{V}$) applies the Micciancio-Petrank transformation to P_{HV} (resp., V_{HV}) to obtain the algorithm P_{MP} (resp., V_{MP}). Now, $\mathcal{P}$ and $\mathcal{V}$ execute the $(P_{\text{MP}}, V_{\text{MP}})$ protocol with common input (x, c^*, v) in which $\mathcal{P}$ (using P_{MP}) proves to $\mathcal{V}$ (using V_{MP}) that $x \in L \cap \{0,1\}^n$.

$\mathcal{P} \to \mathcal{V}$: Let trans denote the transcript of communication so far. $\mathcal{P}$ computes $\sigma \leftarrow \text{SIGN}(\mathsf{trans}, \text{SK}, \text{VK})$, and sends σ to $\mathcal{V}$.

Fig. 2. The Transformed Argument System $\pi : \langle \mathcal{P}, \mathcal{V} \rangle$

The formal description of our transformation (sometimes also referred to as the *compiler*) is given in Figure 2. The compiler transforms any given public coin statistical HVZK argument of knowledge in a CNMZK argument of knowledge.

In all steps above, whenever a message is not according to the protocol specifications, an honest party aborts the protocol.[11] We will frequently refer to above steps as *phases*. Thus, our transformation has six phases, where in phase-1 $\mathcal{P}$ sends a verification key to $\mathcal{V}$, in phase-2 $\mathcal{V}$ and $\mathcal{P}$ execute an extraction preamble, and so on.

4.3 Proving Concurrent Non-malleability

We now proceed to the actual proof that $\pi : \langle \mathcal{P}, \mathcal{V} \rangle$ is indeed a CNMZK argument of knowledge, given that $\pi_{\text{HV}} : \langle P_{\text{HV}}, V_{\text{HV}} \rangle$ is a public coin honest verifier statistical zero-knowledge argument of knowledge. Using a series of hybrid simulators,

[11] In particular, this means that in second phase (extraction preamble phase), all commitments, challenges, and responses (i.e., openings) are valid; and during the fourth (i.e., opening) phase, $\mathcal{P}$ confirms that all openings are valid and that $v_0^{i,j} + v_1^{i,j} = v \mod p$ for all values of i, j.

we will show how to simulate the joint view of M and $V_1, \ldots, V_{m_R}$, while simultaneously extracting a witness for each $\tilde{x}_\ell$ whenever V_ℓ's view is accepting and $\tilde{\text{TRANS}}_\ell \neq \text{TRANS}_h$ (for all h). Assume M to be deterministic, without loss of generality. It is easy to see that if $\tilde{\text{VK}}_\ell = \text{VK}_h$ for some ℓ, h (i.e., M copies the tag), then due to the *strong unforgeability* of the signature scheme, it holds that $\tilde{\text{TRANS}}_\ell = \text{TRANS}_h$ except with negligible probability. Thus, in the proof we will not attempt to extract a witness for $\tilde{x}_\ell$ whenever $\tilde{\text{VK}}_\ell = \text{VK}_h$. We now define some random variables.

Let ν be a random variable denoting the joint view of M and $V_1, \ldots, V_{m_R}$ in a real execution of π. Similarly, $\nu^{(i)}$ will be the random variable denoting the output of hybrid simulator $\mathcal{H}_i$, $i = 1, 2, \ldots$. For every "left" session $h \in [m_L]$, let $v_h^{(i)}$ denote the value committed to by M in phase-2 (i.e., extraction-preamble) of session h; and let $\mathbf{v}_h^{(i)}$ denote the value committed to by prover P_h in phase-3 (i.e., the DDN-commitment phase) of that session. Of course, $\mathbf{v}_h^{(i)} = 0$ for an honest prover. Define random variables $\tilde{v}_\ell^{(i)}, \tilde{\mathbf{v}}_\ell^{(i)}$ for right sessions $\ell \in [m_R]$, analogously. Thus, $\tilde{v}_\ell^{(i)}$ denotes the value committed to by V_ℓ in phase-2 of ℓth right session; and $\tilde{\mathbf{v}}_\ell^{(i)}$ denotes the value committed to by M to V_ℓ in phase-3 of the same session on right, here $\ell \in [m_R]$. Finally, define $b_\ell^{(i)}$ to be a random boolean variable denoting whether in right-session ℓ, V_ℓ rejects ($b_\ell^{(i)} = 0$ and 1 otherwise) at the end of phase-3 (i.e., the DDN-commitment phase) in a simulation by $\mathcal{H}_i$.

Overall strategy of the proof. In our proof the key-idea is to ensure that $\forall \ell, \tilde{\mathbf{v}}_\ell^{(i)} \neq \tilde{v}_\ell^{(i)}$ while at the same time $\mathbf{v}_h^{(i)} = v_h^{(i)}$ $(\forall h)$ with high probability.

We do this by designing a series of hybrid experiments $\mathcal{H}_i$ setting up $\mathbf{v}_h^{(i)} = v_h^{(i)}$ one-by-one for all left sessions h; it would be done while maintaining $\tilde{\mathbf{v}}_\ell^{(i)} \neq \tilde{v}_\ell^{(i)}$ for every right session in all the hybrid experiments with high probability. This would result in our final simulator using the Micciancio-Petrank method to succeed on left; whereas the adversary M will be forced to use the real witness due to the aforementioned condition on right. We start by presenting our first hybrid.

Simulator $\mathcal{H}_0$. This simulator is provided with auxiliary inputs $y_\ell \in R_L(x_\ell)$ for all left statements x_ℓ for $\ell = 1, \ldots, m_L$. Let γ denote the uniformly chosen random tape of $\mathcal{H}_0$. The simulator starts interacting with $M(\boldsymbol{x}, z)$, where z is M's auxiliary input and $\boldsymbol{x} \stackrel{\text{def}}{=} (x_1, \ldots, x_{m_L})$. On left, $\mathcal{H}_0$ acts as honest provers $P_1, \ldots, P_{m_L}$ (with independent and uniform random tapes) using inputs $y_1, \ldots, y_{m_L}$. On right, $\mathcal{H}_0$ acts as honest verifiers $V_1, \ldots, V_{m_R}$ (with independent and uniform random tapes). When M halts, $\mathcal{H}_0$ outputs the (joint) view of M and all $V_\ell, \ell \in [m_R]$, and halts. Recall that ν denotes the joint view in a real execution of π, and $\nu^{(0)}$ is the output of $\mathcal{H}_0$. The simulation is perfect, and so $\nu \equiv \nu^{(0)}$. Because we use a perfectly binding commitment scheme, values $\tilde{v}_\ell^{(0)}, \tilde{\mathbf{v}}_\ell^{(0)}$ are well defined. Let p_0 be the probability that there exists a right session ℓ such that $\left(\tilde{v}_\ell^{(0)} = \tilde{\mathbf{v}}_\ell^{(0)}\right)$ conditioned on the occurrence of the event "V_ℓ accepts".

Claim. $p_0 \leq \mu(\kappa)$

Proof. Contrary to the claim, suppose that $p_0 \geq 1/s(\kappa)$ for some polynomial $s(\cdot)$. Hence, for a non-negligible fraction of random tapes γ, it holds that for one of the right-sessions (say ℓ^{th}) M succeeds in setting $\tilde{v}_\ell^{(0)} = \tilde{\mathbf{v}}_\ell^{(0)}$, *and* V_ℓ accepts at the end of DDN-commitment phase of right-session ℓ (i.e., $b_\ell^{(0)} = 1$). We construct two machines M^*, M^*_{DDN}, and use them to break the semantic security of the commitment scheme denoted by the extraction preamble.

Machine M^* incorporates $M(\boldsymbol{x}, z)$ and interacts with it exactly as $\mathcal{H}_0$ except for the following two differences. First, in the ℓth-right-session, V_ℓ does not execute the extraction-preamble internally; instead it receives the commitment from an outside party A. That is, it chooses two values $v_0', v_1' \in \mathbb{Z}_p$ uniformly at random, and sends them to the outside sender A (of the extraction-preamble[12]). A then commits to v_b', where $b \in_R \{0,1\}$ which M^* forwards to $M(\boldsymbol{x}, z)$ as part of V_ℓ. Second, as soon as the preamble (of ℓth-right-session) concludes, M^* outputs its complete internal state, denoted ST_{M^*}, and halts.

Next, we use M^* to construct a DDN-sender M^*_{DDN} as follows. M^*_{DDN} starts with state ST_{M^*} and continues the rest of the execution internally exactly as $\mathcal{H}_0$, except for the following difference. In the DDN-commitment phase of ℓth-right-session, instead of internally emulating the actions of a DDN-receiver, V_ℓ (which is an internal part of M^*_{DDN}) interacts with an external DDN-receiver $\mathcal{R}_{\text{DDN}}$. M^*_{DDN} halts as soon as this phase finishes.

Finally, to break the semantic security of the extraction-preamble, our adversary (say $\mathcal{A}_{\text{COM}}$) proceeds as follows. Given $M(\boldsymbol{x}, z)$, $\mathcal{A}_{\text{COM}}$ first acts as M^* to receive a commitment from external A. Once, this interaction is over, we have the state ST_{M^*} and hence the adversary M^*_{DDN}. (By construction, the execution of M^*_{DDN} is identical to that of $\mathcal{H}_0$ up to the point where DDN-phase completes). Now $\mathcal{A}_{\text{COM}}$ interacts with M^*_{DDN} while acting as $\mathcal{R}_{\text{DDN}}$, and if the interaction is accepting, it applies the DDN-extractor, E_{DDN}, to M^*_{DDN} and outputs whatever E_{DDN} outputs.

M^*_{DDN} is a machine that succeeds in committing to v_b' with probability $\geq p_0$ over the randomness of whole experiment. It follows that for at least $p_0/2$ fraction of views ST_{M^*}, M^*_{DDN} (using $M(\boldsymbol{x}, z)$) successfully commits the value v_b' to $\mathcal{R}_{\text{DDN}}$ with probability at least $p_0/2$. Thus, from the properties of the DDN-commitment scheme, we conclude that E_{DDN} extracts v_b' with probability $p_0/2 - \mu(\kappa)$ by running in expected polynomial time. Hence, $\mathcal{A}_{\text{COM}}$ can guess b with probability $p_0/2(p_0/2 - \mu(\kappa)) \geq p_0^2/8$ contradicting the semantic security of the extraction-preamble. □

Before proceeding further with the proof, imagine the following hybrid experiment $\mathcal{H}_1'$: it is the same as $\mathcal{H}_0$ except that it also performs the extraction of KP/PRS-secrets v_h on left by running both main as well as look-ahead threads just like the KP/PRS-simulator.[13]

[12] Recall that we can look at the extraction-preamble as an interactive commitment scheme.

[13] Values $v_h, \mathbf{v}_h, \tilde{v}_\ell, \tilde{\mathbf{v}}_\ell$ are defined for session h of the *main* thread. For look-ahead threads, we'll introduce a new variable when needed.

Note that $\mathcal{H}_1'$ simulates honest verifiers $V_1, \ldots, V_{m_R}$ on right, and runs *real* provers $P_1, \ldots, P_{m_L}$ on left of $M(x, z)$ in executing all the threads. If extraction of KP/PRS-secrets fails, (i.e., KP/PRS-simulator gets "stuck") than $\mathcal{H}_1'$ aborts.

Recall that the "threads" of a KP/PRS-simulator are classified into three types: a *main* thread, look-ahead threads that share a prefix with the main thread, and look-ahead threads that do not share any prefix with the main thread. Furthermore, all these threads can be *ordered* by their finishing time: thread 1 is the one that finishes first, thread 2 is the one that finishes second, and so on.

Threads contain several left and right sessions. In each left session belonging to a thread, if the execution of that session reaches the DDN-commitment phase, that session will contain the first message FM^*. Each thread can contain at most m_L such first messages, and there are at most $N \in O((\beta m_L)^2)$ FM^*s that ever appear in an execution of $\mathcal{H}_1'$. Further, in any given thread, these FM^*s can be ordered by their order of appearance, and since each thread can be ordered as explained above, we have an implicit ordering on these first messages which we denote by $\text{FM}_1^*, \ldots, \text{FM}_N^*$.

Observe that instead of executing all look-ahead threads at once, it is possible to only execute look-ahead threads of $\mathcal{H}_1'$ up to a specific point (e.g., up to the point where a specific first message FM_i^* appears) and then from thereon stop running any look-ahead threads and just complete the main-thread from where it was left.

We are now ready to explain our next $3N + 1$ hybrid simulators: $\mathcal{H}_{i:0}, \mathcal{H}_{i:1}$, and $\mathcal{H}_{i:2}$ for $i = 1, \ldots, N$. Define $\mathcal{H}_{0:2}$ to be the same as $\mathcal{H}_0$.

Simulator $\mathcal{H}_{i:0}$. This experiment is the same as $\mathcal{H}_{i-1:2}$ except that it runs look-ahead threads up to the point where FM_i^* gets generated. After this point, the experiment continues the execution of main-thread directly *without running any look-ahead threads* at all. Note that up to this point, KP/PRS-secret v_j must have been extracted for all left sessions j for which the extraction preamble concludes successfully (in *any* thread), with high probability; and the experiment aborts if this is not the case.

Simulator $\mathcal{H}_{i:1}$. Consider an execution of our previous simulator $\mathcal{H}_{i:0}$. Since the execution reaches to the point FM_i^*, $\mathcal{H}_{i:0}$ must have extracted the value committed to in the extraction preamble of the session to which FM_i^* belongs. Denote this value by e_i.

Simulator $\mathcal{H}_{i:1}$ is the same as $\mathcal{H}_{i:0}$ except that when creating FM_i^*, it commits to e^i instead of committing to 0^κ using uniform randomness λ_i. It uses (e_i, λ_i) to complete the DDN-commitment phase of this session when needed.

Note that e_i is extracted "correctly" (i.e., equals the value opened by M later on in this session) with high probability. $\mathcal{H}_{i:1}$ aborts if this is not the case.

Simulator $\mathcal{H}_{i:2}$. This simulator is the same as $\mathcal{H}_{i:1}$ except that in all sessions j (across all threads) that belong to FM_i^*, if phase-5 is ever reached, it uses the Micciancio-Petrank simulator along with "trapdoor" (e_i, λ_i) defined in previous hybrid-simulator to succeed in this phase.

Our final simulator-extractor will use $\mathcal{H}_{N:2}$, to construct the final view. Due to space constraints, an analysis of above hybrid simulators is provided in the full version of this paper. We move on to present our final simulator.

The Final Simulator-extractor $\mathcal{S}$. For succinctness, let us denote $\mathcal{H}_{N:2}$ by $\mathcal{H}_2$. Our simulator-extractor $\mathcal{S}$ works as follows. It first runs the hybrid simulator $\mathcal{H}_2$ to produce a joint view $\nu^{(2)}$. The statements in the right executions (in $\nu^{(2)}$) are $\tilde{x}_1, \ldots, \tilde{x}_{m_R}$. For each right session $\ell \in [m_R]$, if V_ℓ accepts the proof and $\forall h \in [m_L]$ $\tilde{\text{VK}}_\ell \neq \text{VK}_h$, the simulator $\mathcal{S}$ extracts (a witness) $\tilde{y}_\ell \in R_L(\tilde{x}_L)$. The extraction for each such ℓ is performed *one by one*, as follows.

1. First, $\mathcal{S}$ defines an adversarial machine $\mathcal{A}_{\text{DDN}}^{(\ell)}$ as follows. $\mathcal{A}_{\text{DDN}}^{(\ell)}$ incorporates $M(\boldsymbol{x}, z)$ and proceeds exactly as $\mathcal{H}_2$ by internally simulating all the honest parties, except for the part of V_ℓ in the *main* thread which receives the phase-3 DDN-commitment. $\mathcal{A}_{\text{DDN}}^{(\ell)}$ terminates the execution after sending the knowledge-determining-message (KDM) to the external receiver. Now, $\mathcal{S}$ uses the (guaranteed) extractor which can work on this prefix (up to the KDM) to extract the value committed to by $\mathcal{A}_{\text{DDN}}^{(\ell)}$ in view $\nu^{(2)}$. Let the extracted value be u ($\mathcal{S}$ aborts if extraction fails).
2. If $u = \tilde{v}_\ell^{(2)}$, $\mathcal{S}$ aborts the extraction and halts. Otherwise, it defines a new machine $\mathcal{A}_{\text{MP}}^{(\ell)}$ as follows. $\mathcal{A}_{\text{MP}}^{(\ell)}$ is exactly as $\mathcal{H}_2$ that incorporates $M(\boldsymbol{x}, z)$ and all the simulated honest parties internally, except for the part of V_ℓ in the main thread which receives the phase-5 (i.e., Micciancio-Petrank) proof. $\mathcal{A}_{\text{MP}}^{(\ell)}$ is then a Micciancio-Petrank prover. It then applies the extractor guaranteed for such a prover, to extract a value $\tilde{y}_\ell$ – supposedly a witness for $\tilde{x}_\ell$ (repeat this procedure to obtain $\tilde{y} = \{\tilde{y}_1, \ldots, \tilde{y}_{m_R}\}$). It then outputs $\tilde{y}$ and halts.

A few remarks are in order. First, let us mention why we need to execute the first step involving $\mathcal{A}_{\text{DDN}}^{(\ell)}$, and why not directly execute the second step and extract $\tilde{y}_\ell$ using $\mathcal{A}_{\text{MP}}^{(\ell)}$. This is done in order to ensure that the extraction procedure has expected polynomial running time. Because otherwise, if $\tilde{\mathbf{v}}_\ell^{(2)} = \tilde{v}_\ell^{(2)}$ (even if with only negligible probability – equation ??), the extraction procedure would never halt. As a result, the running time of $\mathcal{S}$ will not be bounded by any polynomial. Extracting u ($=\tilde{\mathbf{v}}_\ell^{(2)}$) using $\mathcal{A}_{\text{DDN}}^{(\ell)}$ allows $\mathcal{S}$ to abort whenever it is in this case.

Second, a subtlety in constructing $\mathcal{A}_{\text{DDN}}^{(\ell)}$ (and $\mathcal{A}_{\text{MP}}^{(\ell)}$ as well) is worth mentioning here. $\mathcal{A}_{\text{DDN}}^{(\ell)}$ acts as $\mathcal{H}_2$ internally and hence executes various "threads of execution" which may share a prefix with the main thread. When $\mathcal{A}_{\text{DDN}}^{(\ell)}$ interacts with the external receiver, it may define parts of some look-ahead threads. If the KDM did not appear in the shared prefix, $\mathcal{H}_2$ will have to *internally continue* the execution of these look-ahead threads who share a prefix with the main-thread (defined by the external receiver). The fact that the protocol is *public coin* up to the KDM, allows $\mathcal{H}_2$ to do that if required. Thus $\mathcal{A}_{\text{DDN}}^{(\ell)}$ (and for the same reason, $\mathcal{A}_{\text{MP}}^{(\ell)}$) is indeed well defined. From here, deriving our main theorem (Theorem 1) is not hard. Due to space constraints, this proof appears in the full version of this paper.

5 Efficiency

The Actual Cost. It is easy to see that the additional overhead incurred by the new prover and verifier, is dominated by three steps (overhead from all other steps is a *small* additive constant). First overhead is β^2 exponentiations (in $\mathbb{Z}_q$) due to the extraction-preamble.[14] The second overhead is due to the DDN-commitment phase, which as we discuss shortly, is $\kappa \cdot \tilde{O}(\log^2 \kappa)$ exponentiations. Finally, the last overhead is due to the Micciancio-Petrank transformation, which is r exponentiations, where r is the round complexity of π_{HV}. As $\beta \in \tilde{O}(\log \kappa)$, it follows that the additional overhead incurred by each party is (at most) $r + \kappa \cdot \tilde{O}(\log^2 \kappa)$ exponentiations in $\mathbb{Z}_q$.

The overhead in DDN-commitments is as follows. The cost is dominated by the following steps (overhead from all other steps is a *small* additive constant). First costly operation is the execution of SZKAOK, which requires $\omega(1)$ exponentiations. The next (in fact, the main) costly operation is the execution of step 3. This involves performing $\kappa \cdot (1 + \log \kappa) \cdot 2$ executions of $\mathrm{BCK}^{\parallel}$. As $\mathrm{BCK}^{\parallel}$ repeats BCK, in parallel, β times, and each BCK has an overhead of constant (less than 10) exponentiations in $\mathbb{Z}_q$, it follows that the overall overhead is $\kappa \cdot \tilde{O}(\log^2 \kappa)$ exponentiations in $\mathbb{Z}_q$.

Cost for Tag-based Non-malleability. Historically, the verification key VK used in DDN-commitment protocol, is also called an *identity* or *tag*. Currently, the size of this tag is κ. If identities (or tags) are given to exist, then the first and the last steps of the protocol are unnecessary (and hence are not executed). Non-malleability in such cases requires the extraction of witness only when the adversary does not copy the tag entirely, and is called "tag-based" non-malleability. If tags of shorter length are possible, it results in more efficient protocols. The two mainly cited reasons for justifying this notion are the following ones. First, for some applications, it may be reasonable to assume that all parties have unique identities. As there are only polynomially many parties in real world protocols, they can all be represented by using tags of length at most $\omega(\log \kappa)$. Second, non-malleable protocols are typically used as building blocks in larger protocols. The execution of these larger protocols, may somehow, result in establishing tags for this building block.

For tag-based non-malleability, assuming the tag-length, $|\mathrm{VK}|$, is $\tilde{O}(\log \kappa)$ – which we believe is reasonable – the overhead in the DDN-commitment phase would only be $\tilde{O}(\log^2 \kappa)$. And thus, the overhead incurred by each party in our transformation would be at most $r + \tilde{O}(\log^2 \kappa)$.

We would like to mention here that our transformed protocol is very suitable for the employment of preprocessing and batching techniques.

Efficient CNMZK *in the* BPK *Model.* In the full version of this paper we show that our tag-based non-malleable protocols lead to first truly efficient constructions in the BPK model [11]. This model has been used in sequence of papers [26,27,28] to

[14] This overhead is only β exponentiations, if one chooses RK/KP-type preamble.

initially achieve round and computationally efficient concurrent zero knowledge and later constant-round concurrent non-malleable zero-knowledge [29,30].

We give an efficiency preserving compiler for obtaining CNMZK arguments from any HVSZK argument π' in the (true) BPK model. We obtain these results by applying our efficient tag-based constructions of CNMZK arguments in the plain model. (When coupled with a proper π', this gives efficient constructions of comparable efficiency.) By efficient we mean that the round complexity of the new protocol is $r + \tilde{O}(\log \kappa)$ while the additional computational overhead incurred by each party would be at most $r + \tilde{O}(\log^2 \kappa)$.

References

1. Goldwasser, S., Micali, S., Rackoff, C.: The knowledge complexity of interactive proof-systems. In: Proc. 17th STOC, pp. 291–304 (1985)
2. Lindell, Y.: General composition and universal composability in secure multi-party computation. In: Proc. 44th FOCS, pp. 394–403 (2003)
3. Lindell, Y.: Lower bounds for concurrent self composition. In: Naor, M. (ed.) TCC 2004. LNCS, vol. 2951, pp. 203–222. Springer, Heidelberg (2004)
4. Barak, B., Prabhakaran, M., Sahai, A.: Concurrent non-malleable zero knowledge. In: FOCS 2006 (2006); Full version on Cryptology ePrint Archive report, `http://eprint.iacr.org/`
5. Dolev, D., Dwork, C., Naor, M.: Nonmalleable cryptography. SIAM Journal on Computing 30(2), 391–437 (2000); (electronic) Preliminary version in STOC 1991 (1991)
6. Garay, J.A., MacKenzie, P.D., Yang, K.: Strengthening zero-knowledge protocols using signatures. In: Biham, E. (ed.) EUROCRYPT 2003. LNCS, vol. 2656, pp. 177–194. Springer, Heidelberg (2003)
7. MacKenzie, P., Yang, K.: On Simulation-Sound Trapdoor Commitments. In: Cachin, C., Camenisch, J.L. (eds.) EUROCRYPT 2004. LNCS, vol. 3027, pp. 382–400. Springer, Heidelberg (2004)
8. Gennaro, R.: Multi-trapdoor Commitments and Their Applications to Proof s of Knowledge Secure Under Concurrent Man-in-the-Middle Attacks. In: Franklin, M. (ed.) CRYPTO 2004. LNCS, vol. 3152, pp. 220–236. Springer, Heidelberg (2004)
9. Damgård, I., Nielsen, J.B., Orlandi, C.: On the necessary and sufficient assumptions for uc computation. In: Micciancio, D. (ed.) TCC 2010. LNCS, vol. 5978. Springer, Heidelberg (2010)
10. Pedersen, T.P.: Non-interactive and information-theoretic secure verifiable secret sharing. In: Feigenbaum, J. (ed.) CRYPTO 1991. LNCS, vol. 576, pp. 129–140. Springer, Heidelberg (1992)
11. Canetti, R., Goldreich, O., Goldwasser, S., Micali, S.: Resettable zero-knowledge. In: Proc. 32th STOC, pp. 235–244 (2000)
12. Micciancio, D., Petrank, E.: Simulatable commitments and efficient concurrent zoro-knowledge. In: Biham, E. (ed.) EUROCRYPT 2003. LNCS, vol. 2656, pp. 140–159. Springer, Heidelberg (2003)
13. Lin, H., Pass, R., Venkitasubramaniam, M.: Concurrent non-malleable commitments from any one-way function. In: Canetti, R. (ed.) TCC 2008. LNCS, vol. 4948, pp. 571–588. Springer, Heidelberg (2008)
14. Pass, R., Rosen, A.: New and improved constructions of non-malleable cryptographic protocols. In: Proc. 37th STOC (2005)

15. De Santis, A., Di Crescenzo, G., Ostrovsky, R., Persiano, G., Sahai, A.: Robust non-interactive zero knowledge. In: Kilian, J. (ed.) CRYPTO 2001. LNCS, vol. 2139, pp. 566–598. Springer, Heidelberg (2001)
16. Mohassel, P., Franklin, M.K.: Efficiency tradeoffs for malicious two-party computation. In: Yung, M., Dodis, Y., Kiayias, A., Malkin, T.G. (eds.) PKC 2006. LNCS, vol. 3958, pp. 458–473. Springer, Heidelberg (2006)
17. Woodruff, D.P.: Revisiting the efficiency of malicious two-party computation. In: Naor, M. (ed.) EUROCRYPT 2007. LNCS, vol. 4515, pp. 79–96. Springer, Heidelberg (2007)
18. Lindell, Y., Pinkas, B.: An efficient protocol for secure two-party computation in the presence of malicious adversaries. In: Naor, M. (ed.) EUROCRYPT 2007. LNCS, vol. 4515, pp. 52–78. Springer, Heidelberg (2007)
19. Goyal, V., Mohassel, P., Smith, A.: Efficient two party and multi party computation against covert adversaries. In: Smart, N.P. (ed.) EUROCRYPT 2008. LNCS, vol. 4965, pp. 289–306. Springer, Heidelberg (2008)
20. Chase, M., Lysyanskaya, A.: Simulatable vrfs with applications to multi-theorem nizk. In: Menezes, A. (ed.) CRYPTO 2007. LNCS, vol. 4622, pp. 303–322. Springer, Heidelberg (2007)
21. Groth, J., Ostrovsky, R., Sahai, A.: Perfect non-interactive zero knowledge for NP. In: Vaudenay, S. (ed.) EUROCRYPT 2006. LNCS, vol. 4004, pp. 339–358. Springer, Heidelberg (2006)
22. Goldreich, O.: Foundations of Cryptography: Basic Tools. Cambridge University Press, Cambridge (2001)
23. Schnorr, C.P.: Efficient identification and signatures for smart cards (abstract). In: Quisquater, J.-J., Vandewalle, J. (eds.) EUROCRYPT 1989. LNCS, vol. 434, pp. 688–689. Springer, Heidelberg (1990)
24. Richardson, R., Kilian, J.: On the concurrent composition of zero-knowledge proofs. In: Stern, J. (ed.) EUROCRYPT 1999. LNCS, vol. 1592, pp. 415–432. Springer, Heidelberg (1999)
25. Prabhakaran, M., Rosen, A., Sahai, A.: Concurrent zero knowledge with logarithmic round-complexity. In: FOCS, pp. 366–375 (2002)
26. Di Crescenzo, G., Persiano, G., Visconti, I.: Constant-round resettable zero knowledge with concurrent soundness in the bare public-key model. In: Franklin, M. (ed.) CRYPTO 2004. LNCS, vol. 3152, pp. 237–253. Springer, Heidelberg (2004)
27. Di Crescenzo, G., Visconti, I.: Concurrent zero knowledge in the public-key model. In: Caires, L., Italiano, G.F., Monteiro, L., Palamidessi, C., Yung, M. (eds.).ICALP 2005. LNCS, vol. 3580, pp. 816–827. Springer, Heidelberg (2005)
28. Visconti, I.: Efficient zero knowledge on the internet. In: Bugliesi, M., Preneel, B., Sassone, V., Wegener, I. (eds.) ICALP 2006. LNCS, vol. 4052, pp. 22–33. Springer, Heidelberg (2006)
29. Ostrovsky, R., Persiano, G., Visconti, I.: Constant-round concurrent nmwi and its relation to nmzk. Technical Report ECCC Report TR06-095, ECCC (2006)
30. Ostrovsky, R., Persiano, G., Visconti, I.: Constant-round concurrent non-malleable zero knowledge in the bare public-key model. In: Aceto, L., Damgård, I., Goldberg, L.A., Halldórsson, M.M., Ingólfsdóttir, A., Walukiewicz, I. (eds.) ICALP 2008, Part II. LNCS, vol. 5126, pp. 548–559. Springer, Heidelberg (2008)

Efficiency Limitations for Σ-Protocols for Group Homomorphisms*

Endre Bangerter[1], Jan Camenisch[2], and Stephan Krenn[3]

[1] Bern University of Applied Sciences, Biel-Bienne, Switzerland
endre.bangerter@bfh.ch
[2] IBM Research — Zurich, Rüschlikon, Switzerland
jca@zurich.ibm.com
[3] Bern University of Applied Sciences, Biel-Bienne, Switzerland, and
University of Fribourg, Fribourg, Switzerland
stephan.krenn@bfh.ch

Abstract. Efficient zero-knowledge proofs of knowledge for group homomorphisms are essential for numerous systems in applied cryptography. Especially, Σ-protocols for proving knowledge of discrete logarithms in known and hidden order groups are of prime importance. Yet, while these proofs can be performed very efficiently within groups of known order, for hidden order groups the respective proofs are far less efficient.

This paper shows strong evidence that this efficiency gap cannot be bridged. Namely, while there are efficient protocols allowing a prover to cheat only with negligibly small probability in the case of known order groups, we provide strong evidence that for hidden order groups this probability is bounded below by 1/2 for all efficient Σ-protocols not using common reference strings or the like.

We prove our results for a comprehensive class of Σ-protocols in the generic group model, and further strengthen them by investigating certain instantiations in the plain model.

Keywords: Generic Group Model, Σ-Protocols, Proofs of Knowledge, Error Bounds.

1 Introduction

A *Zero-Knowledge Proof of Knowledge (ZK-PoK)* is a two party protocol between a prover and a verifier enabling the prover to convince the verifier that he knows some secret value, without the verifier being able to learn anything about it. More precisely, in a ZK-PoK an honest prover can always convince the verifier, while no malicious prover (not knowing the secret) can do so with a probability larger than some threshold value (the *knowledge error*).

Fundamental results show that there are ZK-PoK for all languages in $\mathcal{NP}$ [2]. Yet, the respective protocols are of theoretical interest only, because executing

* This work was in part funded by the European Community's Seventh Framework Programme (FP7) under grant agreements no. 216499 and 216483. We also refer to the full version of this paper [1].

D. Micciancio (Ed.): TCC 2010, LNCS 5978, pp. 553–571, 2010.

them once is either computationally and communicationally too expensive for real world use, or enables the prover to cheat with a high probability. In the latter case, the protocols have to be repeated numerous times to reduce the knowledge error (remember that r repetitions of a ZK-PoK with knowledge error κ result in a protocol with knowledge error κ^r), and thus they become inefficient again.

A (group) homomorphism is a mapping between two groups $\mathcal{G}$ and $\mathcal{H}$ satisfying $\phi(a+b) = \phi(a) \cdot \phi(b)$ for all $a, b \in \mathcal{G}$. Proving knowledge of a preimage under a homomorphism (i.e., of w satisfying $x = \phi(w)$) can often be done very efficiently by using the so-called *Σ^ϕ-protocol* (i.e., the Schnorr [3] or Guillou/Quisquater [4] protocol generalized to arbitrary homomorphisms [5,6,7]). This protocol consists of three messages being exchanged: the prover chooses r at random from the domain of the homomorphism, and sends the *commitment* $t := \phi(r)$ to the verifier. The verifier then chooses a random *challenge* c from a predefined challenge set $\mathcal{C}$, and sends it to the prover, who computes its response $s := r + c \cdot w$. The verifier now accepts the proof, if and only if $\phi(s) = x^c \cdot t$. Standard techniques [8] allow one to transform this protocol into non-interactive versions or so called signatures of knowledge.

The Σ^ϕ-protocol is a very efficient proof of knowledge for many proof goals existing in cryptography (e.g., knowledge of a discrete logarithm in a known order group, or of the plaintext encrypted in a Paillier ciphertext). The reason is that for the respective homomorphisms, a negligibly small knowledge error can be obtained in a *single run* of the Σ^ϕ-protocol. Yet, the situation is different for the important class of exponentiation homomorphisms with hidden order co-domain (e.g., $\phi(\cdot) : \mathbb{Z} \to \mathbb{Z}_n^* : a \mapsto g^a$, where g is a generator of the quadratic residues modulo n). Such homomorphisms play an important role for many cryptographic applications, e.g., [9,10,11,12,13,14,15,16], including *Direct Anonymous Attestation (DAA)* [17], and the *identity mixer (idemix)* anonymous credential system [18]. In this case, the Σ^ϕ-protocol is only known to be a PoK with knowledge error 1/2, and hence must be repeated sequentially to get a sufficiently small knowledge error (e.g., 80 sequential repetitions are required to obtain a knowledge error of $1/2^{80}$). The resulting computational and communicational costs are much too high for many practical applications.

A number of authors have tried to overcome the above problem by proposing alternative protocols for exponentiation homomorphisms with hidden order co-domain [5,19,20,21,22,23]. All these protocols build on a basic idea put forth by Fujisaki and Okamoto [22], and we thus call them *FO-based* henceforth. Unfortunately, none of these FO-based protocols is fully satisfactory, neither from a practical nor from a theoretical point of view:

- One run of any FO-based protocol is much more expensive than running the Σ^ϕ-protocol once. Moreover, if only standard complexity assumptions (i.e. the Strong RSA Assumption [22]) are made, a recent analysis has revealed that in many cases FO-based protocols are even more expensive than the sequential repetition of the Σ^ϕ-protocol with knowledge error 1/2 [20].
- The FO-based protocols in [5,19,20,21,22] make use of a common reference string, which is either issued by a trusted third party or generated in an

expensive interactive setup phase. Yet, the presence of common reference strings reduces the modularity, and thus increases the complexity of the security analysis of larger applications (as discussed, e.g., in [23,24,25]). The security proofs for the protocols in [5,19] additionally assume the existence of ideal hash functions, and thus only hold true in the random oracle model[1].

Because of these disadvantages, the natural question arises *whether it is necessary to use FO-based protocols at all*? After all, the possibilities of Σ-protocols have not yet been explored thoroughly, and it could be possible that a novel, cleverly designed Σ-protocol or even the existing Σ^{ϕ}-protocol could be used to overcome the current efficiency limitations. (We note that the latter could be quite possible, if one could find a new knowledge extractor working for the Σ^{ϕ}-protocol with a suitably chosen challenge set that allows one to obtain a small knowledge error in a single execution of the protocol.).

Contribution and Results. In this paper we are aiming at answering this question. We provide ample evidence suggesting that the known minimal knowledge error of the Σ^{ϕ}-protocol cannot be underrun, neither by a better knowledge extractor for the Σ^{ϕ}-protocol nor by any other Σ-protocol. In particular, our results indicate that using Σ-protocols the knowledge error of 1/2 cannot be decreased for exponentiation homomorphisms with hidden order co-domain.

More precisely, we first consider PoK based on Σ-protocols in the generic group model. That is, Σ-protocols where prover, verifier, and knowledge extractor are generic algorithms that can only access the homomorphism and its domain and co-domain through an oracle. We then show that there are lower bounds on the knowledge error for (almost) arbitrary Σ-protocols. These lower bounds on the knowledge error in turn imply efficiency limitations for most possible protocol instances. Roughly, these follow by the fact that a PoK with a large knowledge error needs to be repeated sequentially to reduce the knowledge error, which results in a high computational and communicational overhead. Within the generic group model our efficiency analysis shows that the existing Σ^{ϕ}-protocol is *optimal* and there cannot be another, more efficient Σ-protocol.

We further complement our results by proving lower bounds on the knowledge error of the Σ^{ϕ}-protocol in the plain model. First, for homomorphisms of the form $w \mapsto w^e$ in RSA groups we show that $1/d$ is a lower bound on the knowledge error, where d is the smallest divisor of e. Then, we show that for exponentiation homomorphisms with hidden order co-domain, 1/2 is a lower bound on the knowledge error for all knowledge extractors structurally related to the only one currently known. These results are in accord with those in the generic model and again suggest that the knowledge error that is currently known to be achievable and the associated efficiency limitations cannot be underrun.

Finally, we note that our results do not rule out entirely the possibility to obtain efficient PoK using Σ-protocols. On the one hand, we describe a large number of cases (i.e., instances of Σ-protocols) where this is indeed impossible,

[1] For completeness, we note that while the protocol in [23] yields ZK-PoK in the plain model, it is by far too inefficient for practical usage.

indicating that there are inherent efficiency limitations for Σ-protocols. On the other hand, the cases that are not covered by our results also seem to be valuable, since they provide cues for protocol designers on how it could be possible to conceive novel Σ-protocols that overcome current efficiency limitations.

Related Work. Given the abundant usage of Σ-protocols, very little work on their theoretical foundations has been done. Shoup [26] shows that the knowledge error of 1/2 for homomorphisms of the form $\phi(w) = w^{2^t}$ in RSA groups cannot be improved. One of our results in the plain model extends this to arbitrary exponents. Further, parts of our results are based on unpublished results of one of the authors [5]. Apart from this we are not aware of any other work on efficiency limitations of Σ-protocols. Yet, technically we make use of generic group proof techniques devised by Shoup [26] as well as the extension of these techniques to groups of hidden order by Damgård/Koprowski [27].

The generic group model goes back to Nechaev and Shoup [28,29]. It has been extensively used since then to provide evidence for the security of various cryptographic systems, e.g., [27,28,29,30,31,32,33,34,35,36,37,38]. The model is often criticized, because of the risk of lulling a user in a false sense of security. Indeed, there are cases where information only available in the plain model (i.e., obtained from encoding specific properties of the group) can be used to break a system which was proved secure in the generic model [39,40]. Yet, the implications of these observations are different for all the systems cited above than for our results. All the proofs in the former case are used to give evidence for the security of a cryptographic system. Thus, if any of them does not hold true in the plain model, the security of the according system can be flawed, resulting in dire consequences for all applications using the respective scheme. In contrast to this, we use the generic group model in a more conservative way. Namely, we show efficiency limitations on the efficiency of a cryptographic primitive. Thus, if our results do not hold true in the plain model this means that the efficiency of the scheme can be increased, but the security of the scheme is not affected by any means.

We finally remark that our results do not conflict with those in [41]. The authors there show how to build efficient Σ-protocols for certain exponentiation homomorphisms with hidden order co-domain. Yet, their approach is not generic, but rather uses certain properties of the homomorphism at hand. Further, only very few proofs of practical interest can be performed with their technique.

Structure of this Document. In §2 we recap the basic definitions, and introduce the notion of lower bounds and the class of Σ-protocols for which our results hold true. In §3 we then formulate our main result in the generic group model. This result is strengthened in §4, where we give results in the plain model. We finally conclude and point out some open problems in §5.

2 Preliminaries

In §2.1 we give a short introduction to ZK-PoK and briefly discuss the Σ^ϕ-protocol in §2.2. Then, in §2.3 we introduce the notion of lower bounds on the knowledge error of a protocol. In §2.4 we recap the generic group model we are working in, and finally describe the class of protocols for which our results in the generic group model hold true in §2.5.

2.1 Zero-Knowledge Proofs of Knowledge

After having defined ZK-PoK We recall the widely accepted definition of zero-knowledge proofs of knowledge (ZK-PoK) [42,43]. We by $(\mathsf{P}(w), \mathsf{V})(x)$ denote a two party protocol between a prover P and a verifier V with common input x and private input w to P.

Definition 1 (Computational Proof of Knowledge [42,43]). *A* computational proof of knowledge *for a binary relation $\mathcal{R}$ with knowledge error $\kappa(\cdot) : \mathbb{N} \to [0,1]$ is a two party protocol $(\mathsf{P}(w), \mathsf{V})(x)$, satisfying the following two conditions:*

Completeness: *The verifier always accepts the proof, if $(x, w) \in \mathcal{R}$.*
Soundness: *There exists a polynomial* poly$(\cdot)$, *and a probabilistic algorithm* M *(the* knowledge extractor*) with input x and rewindable black-box access to the prover, such that the following holds true. For every probabilistic polynomial-time (PPT) prover P^* that can make V accept the proof with probability $\varepsilon(x) > \kappa(x)$, M outputs w' satisfying $(x, w') \in R$ in expected time at most*

$$t^+(\varepsilon, \kappa, x) := \frac{\text{poly}(\|x\|)}{\varepsilon(x) - \kappa(\|x\|)},$$

where access to P^* counts as one step only.

The computational aspect of this definition, i.e., the restriction of P^* to be a PPT algorithm, is of importance for our results, as it (almost) allows us to stay in the standard complexity class of PPT algorithms. This issuc will also be discussed in §2.3.

A proof of knowledge (PoK) is called *honest verifier zero knowledge (HVZK)*, if no verifier following the protocol is able to gain any information about the secret value w except that it satisfies the stated relation. For a formal description we refer to [43]. There are well known techniques to transform HVZK protocols into protocols which are zero-knowledge also against maliciously behaving verifiers [8].

2.2 The Σ^ϕ-Protocol in Hidden-Order Groups

Most practical applications using ZK-PoK make use of the Σ^ϕ-protocol explained in §1. This allows one to prove knowledge of a preimage w of a public value x under some group homomorphism $\phi(\cdot) : \mathcal{G} \to \mathcal{H}$. If $\phi(\cdot)$ is an exponentiation homomorphism with hidden order co-domain, e.g., $\phi(\cdot) : \mathbb{Z} \to \mathbb{Z}_n^* : a \mapsto g^a$ for some RSA modulus n, the domain of the homomorphism is infinite. To circumvent the problem of drawing random values from an infinite set in P's first step, the random choice $r \in_R \mathcal{G} = \mathbb{Z}$ is substituted by $r \in_R \mathcal{G}' = \{-\Delta w, \dots, \Delta w\}$

with $(\Delta w - \operatorname{ord} \mathcal{G}) / \operatorname{ord} \mathcal{G}$ being negligibly small. The rest of the protocol remains unchanged. This approach can be generalized also to the case $\mathcal{G} = \mathbb{Z}^u$ for some integer u. For more details see, e.g., [5,23].

It is well known that the Σ^ϕ-protocol is a PoK with knowledge error 1/2 for exponentiation homomorphisms with hidden order co-domain. For homomorphisms with a co-domain of known order v, and power homomorphisms $(w_1, w_2) \mapsto \psi(w_1) \cdot w_2^e$, the protocol is known to have a knowledge error of $1/d$, where d is the smallest prime dividing v in the former, respectively e in the latter case [6].

2.3 Lower Bounds of the Knowledge Error

Let us now introduce the notion of *lower bounds*, which is a key to our results stated in the following. Intuitively, β is a lower bound of the knowledge error of a protocol, if for this protocol it is not possible to achieve any knowledge error *smaller than or equal to* β:

Definition 2 (Lower Bound). *A function $\beta(\cdot) : \mathbb{N} \to [0,1]$ is called a* lower bound *on the knowledge error of the protocol* (P, V) *for a binary relation $\mathcal{R}$, if* (P, V) *is not a computational proof of knowledge for $\mathcal{R}$ for any $\kappa'(\cdot) : \mathbb{N} \to [0,1]$ with $\kappa'(\cdot) \leq \beta(\cdot)$.*

An alternative but equivalent characterization is that of $\beta(\cdot)$ being a lower bound if and only if (P, V) is not a computational PoK with knowledge error $\beta(\cdot)$ for the given relation.

All our results on lower bounds are proven by showing that the conditions of the following theorem are satisfied.

Theorem 3 (Sufficient Conditions for Lower Bounds). *Let* (P, V) *be a two-party protocol, let $\mathcal{R}$ be a binary relation, and let $\beta(\cdot) : \mathbb{N} \to [0,1]$ be a function. Then $\beta(\cdot)$ is a lower bound on the knowledge error of* (P, V) *for $\mathcal{R}$, if the following two conditions are satisfied:*

Uniformity: *There are a polynomial* $\operatorname{poly}(\cdot)$ *and PPT algorithms* P^* *and* D *such that $\varepsilon(x) - \beta(\|x\|) \geq 1/\operatorname{poly}(\|x\|)$ holds for all sufficiently long x generated by* D*, where $\varepsilon(x)$ is the probability that* P^* *makes* V *accept on common input x.*

Hardness: *For all expected PPT algorithms* M *having rewindable black-box access to* P^**, the probability that* M *outputs a w' with $(x, w') \in \mathcal{R}$ is negligible.*

From the uniformity condition and Definition 1 it follows that any hypothetical knowledge extractor must be an expected PPT algorithm. This is important, as in our results we show that the hardness condition has to be satisfied by showing that otherwise the respective knowledge extractor could be used to break a cryptographic standard assumption, which is typically defined against PPT attackers. Still, we will have to adopt these assumptions in a natural way. As the standard definition of PoK allows the knowledge extractor to be an *expected* time algorithm [42,43], we have to generalize the class of attackers the cryptographic assumption holds against to *expected* PPT algorithms as well. Yet, we believe that this generalization is reasonable as by Markov's inequality we see that an expected PPT algorithm may only run super-polynomially long for a small fraction of its executions.

2.4 The Generic Group Model and Groups of Hidden Order

Our main result holds in the generic group model, which we briefly recap next.

The *generic group model* is used to analyze the complexity of problems by considering algorithms in groups whose representation does not reveal any information to the algorithm. That is, such an algorithm must not exploit encoding dependent properties of the group, but is restricted to only use group operations. The hardness of a problem in the generic model is a necessary but not sufficient condition for a problem to be hard in the plain model [39,40].

Various formalizations of this model have been proposed [28,29,35,44]. They all have in common that an algorithm does not get the concrete group description, but only handles to group elements (e.g., via random encodings [29] or indices to elements [35]). Further, the algorithm gets access to an oracle. To evaluate a group operation, the algorithm inputs the handles of elements and the operation to perform to the oracle, which then returns the handle of the result. Similarly, a homomorphism $\phi(\cdot) : \mathcal{G} \to \mathcal{H}$ has to be evaluated through an oracle.

We call an algorithm a *generic homomorphism algorithm* for $\phi(\cdot) : \mathcal{G} \to \mathcal{H}$, if, through an oracle $\mathcal{O}^{\phi(\cdot)}$, it might perform the following operations.

- $+$: Evaluation of the group operation within $\mathcal{G}$ or $\mathcal{H}$,
- $-$: inverting an element within $\mathcal{G}$ or $\mathcal{H}$,
- $\stackrel{?}{=}$: testing the equality of two elements from the same group,
- $\in_\mathbf{R}$: choosing a group element uniformly at random within $\mathcal{G}$ and $\mathcal{H}$, and
- $\phi(\cdot)$: evaluating the homomorphism on arbitrary elements $a \in \mathcal{G}$.

When proving our results, we show that any generic algorithm, acting as hypothetical knowledge extractor for a knowledge error smaller than the stated lower bounds, must fail with overwhelming probability. We therefore describe next which operations such an algorithm may perform.

Definition 4 (Generic Black-Box Algorithm). *A* generic black-box algorithm *is a generic homomorphism algorithm for $\phi(\cdot)$ with oracle $\mathcal{O}^{\phi(\cdot)}$, which additionally has rewindable black-box access to* P^**. That is, it can (i) execute* P^**, (ii) choose the random inputs of* P^**, and (iii) repeatedly reset* P^**. Resetting* P^* *does not reset* $\mathcal{O}^{\phi(\cdot)}$.

We remark that the black-box property of such an algorithm is exactly the same as for a knowledge extractor according to Definition 1.

Groups of Hidden Order. In the following we will be interested in group homomorphisms with hidden order co-domain (resp., image). Intuitively this means that the order of the co-domain (image of $\phi(\cdot)$, denoted by Im $\phi(\cdot)$) cannot be computed with non-negligible probability. More precisely, using the formalization of Damgård/Koprowski [27], we let π be the largest prime dividing the order of the co-domain (Im $\phi(\cdot)$), and let $\alpha(\pi)$ denote the maximal probability that π occurs when $\phi(\cdot)$ is chosen randomly from a predefined finite set of homomorphisms. Then $\phi(\cdot)$ is said to have a *hidden order co-domain (image)*, if $\alpha(\pi)$ is negligibly small.

2.5 Generic Σ-Protocols

We call the class of protocols for which our results hold true generic Σ-protocols. Informally, this class consists of almost all HVZK Σ-protocols of the following form. The prover is allowed to compute and send arbitrary elements obtained from generic homomorphism algorithms in both moves. The verifier may send multiple randomly chosen challenges in its first move, and use an arbitrary generic algorithm to decide whether to accept or to reject the proof.

Definition 5 (Generic (Group) Σ-Protocols). *Let $a_{ij}, b_{ij}, d_i, e_i, f_i, g_i$ be integer coefficients, let $\{(b_{11}, \dots, b_{1l}), \dots, (b_{n1}, \dots, b_{nl})\}$ be linearly independent over the integers, and let $\mathcal{C}_1, \dots, \mathcal{C}_p \subseteq \mathbb{Z}$ be arbitrary finite sets. Let further* $\mathsf{Verify}(\cdot, \dots, \cdot)$ *be a generic homomorphism algorithm, and let the verifier always accept for an honest prover. We then call an HVZK two party protocol a* generic (group) Σ-protocol *for a homomorphism $\phi(\cdot) : \mathcal{G} \to \mathcal{H}$, if it has the form depicted in Fig. 1.*

It can easily be seen that this class covers the existing Σ^ϕ-protocol as well as the parallel execution of multiple instantiations thereof. Yet, a much broader set of protocols is covered by the class of generic Σ-protocols.

We make two minor remarks on this definition. First, the required linear independence can often be inferred from the HVZK property. Namely, if the

$$
\begin{array}{lll}
\mathsf{P}(x, w) & & \mathsf{V}(x) \\
\hline
r_i \in_R \mathcal{G} & \forall\, 1 \le i \le l^- & \\
t_j := \phi(\sum a_{ji} r_i + f_j w) & \forall\, 1 \le j \le m^- & \\
s_k := \sum b_{ki} r_i + d_k w & \forall\, 1 \le k \le n^- & \\
& \xrightarrow{t_1, \dots, t_{m^-}, s_1, \dots, s_{n^-}} & \\
& & (c_1, \dots, c_p) \in_R \mathcal{C} := \mathcal{C}_1 \times \dots \times \mathcal{C}_p \\
& \xleftarrow{c_1, \dots, c_p} & \\
r_i \in_R \mathcal{G} & \forall\, l^- < i \le l & \\
t_j := \phi(\sum a_{ji} r_i + (f_j + \sum g_{ji} c_i) w) & \forall\, m^- < j \le m & \\
s_k := \sum b_{ki} r_i + (d_k + \sum e_{ki} c_i) w & \forall\, n^- < k \le n & \\
& \xrightarrow{t_{m^-+1}, \dots, t_m, s_{n^-+1}, \dots, s_n} & \\
& & \mathsf{Verify}(x, c_1, \dots, c_p, s_1, \dots, s_n, t_1, \dots, t_m)
\end{array}
$$

Fig. 1. Structure of a generic Σ-protocol for a homomorphism $\phi : \mathcal{G} \to \mathcal{H}$

vectors were not linearly independent the verifier could compute a multiple of w, and using Shamir's trick [5] could thus often compute the secret. Second, the definition of generic homomorphism algorithms also allows to draw random choices in the co-domain of the homomorphism. The above definition allows to draw random choices in the image by drawing $r \in_R \mathcal{G}$ and computing $\phi(r)$.

3 Efficiency Limitations in the Generic Group Model

In this section we describe lower bounds on the knowledge error for generic Σ-protocols with generic black-box algorithms as knowledge extractors. From these lower bounds we infer efficiency limitations for ZK-PoK using Σ-protocols.

In the statement of our results we refer to the notion of *expected PPT pseudo random functions*. Such functions are defined just as pseudo random functions (cf., e.g., [43]), except for one minor modification. Namely, we require that no *expected* PPT algorithm can distinguish such a function from a truly random one (usually one considers only *strict* PPT distinguishers). See §2.3 for a brief discussion why we resort to expected PPT time assumptions.

We are now ready to formulate our main result in the generic group model.

Theorem 6 (Lower Bounds in the Generic Group Model). *Let be given an arbitrary but fixed polynomial* poly$(\cdot)$, *a homomorphism* $\phi(\cdot) : \mathcal{G} \to \mathcal{H}$ *with hidden order image*[2], *and* $x \in \mathcal{H}$, *for which knowledge of a preimage under* $\phi(\cdot)$ *shall be proven. Consider a generic Σ-protocol as in Definition 5, and let* q *be the number of responses sent by the prover in its second step, i.e.,* $q := n - n^{-} + m - m^{-}$. *Assuming that expected PPT pseudo random functions exist, the knowledge error of this protocol in the generic group model is lower bounded by*

$$\frac{1}{2^{\min(p,q)}} - \frac{1}{\text{poly}(\|x\|)}.$$

Let us briefly discuss the relevance and implications of this result.

- Our results indicates that a knowledge error of $1/2$ is an inherent limitation of the Σ^{ϕ}-protocol for homomorphisms with hidden order co-domain, which especially cover exponentiation homomorphisms in RSA groups.
- The best known technique to decrease the knowledge error is to repeat the Σ^{ϕ}-protocol, sequentially or in parallel. In either case, the number of elements sent by the prover and the verifier increases by the number of repetitions. Our results show that at least for the second and third move, i.e., the challenges sent by V and the responses sent by P, this growth cannot be avoided.

  Put differently, Theorem 6 shows that the number p of challenges, and the number q of responses are the key parameters determining the size of the knowledge error. This implies that the strategy of repeating the Σ^{ϕ}-protocol parallely is optimal concerning the second and third move of the protocol.

[2] Note that this is a stronger requirement than the requirement that the co-domain has hidden order. Yet, typically these two properties accompany each other.

– Finally, a protocol designer can deduce from Theorem 6 how an alternative for the Σ^{ϕ}-protocol must not look like. Namely, it must either not be a generic Σ-protocol, or the protocol must have a non-generic knowledge extractor, which uses particulars of the homomorphism.

3.1 Generalization to Other Classes of Homomorphisms

So far we have considered homomorphisms with hidden order co-domain. Yet, in practice this information is sometimes available and could potentially be used to decrease the lower bounds on the knowledge error.

More generally, we thus consider the class of *special* homomorphisms next. A homomorphisms $\phi(\cdot) : \mathcal{G} \to \mathcal{H}$ is called special, if for every $x \in \mathcal{H}$ a pair $(u, v) \in \mathcal{G} \times \mathbb{Z} \setminus \{0\}$ satisfying $\phi(u) = x^v$ can be computed efficiently. The pair (u, v) is called *pseudo-preimage* of x under $\phi(\cdot)$. Besides homomorphisms with known order co-domain, also power homomorphisms are known to be special.

We model this property by adding one more query to the oracle $\mathcal{O}^{\phi(\cdot)}$, i.e., we allow a generic homomorphism algorithm to request a pseudo-preimage under $\phi(\cdot)$ for arbitrary elements from the co-domain of the homomorphism. We then obtain the following lemma:

Lemma 7 (Lower Bounds for Special Homomorphisms)

(i) For power homomorphisms $(w_1, w_2) \mapsto \psi(w_1) \cdot w_2^e$ with hidden order co-domain, Theorem 6 can be generalized to a lower bound of $\frac{1}{d^{\min(p,q)}} - \frac{1}{\text{poly}(\|x\|)}$, where d is the smallest prime dividing e.

(ii) For arbitrary homomorphisms with a co-domain of known order v, Theorem 6 generalizes literally with a lower bound of $\frac{1}{d^{\min(p,q)}} - \frac{1}{\text{poly}(\|x\|)}$, where d is the smallest prime dividing v, if v has a super-polynomially large prime factor.

Note that no such generalization is suitable for exponentiation homomorphisms with hidden order co-domain, as they are not known to be special. Analogue observations as for Theorem 6 on the implications of this lemma hold. Especially, in the generic group model the known knowledge error of the Σ^{ϕ}-protocol cannot be underrun for special homomorphisms.

Examples for homomorphisms of Case (i) are those used in the RSA, Paillier, and Damgård/Jurik encryption schemes [45,46,47,48]. The ZK-PoK for these homomorphisms was introduced by Guillou/Quisquater [4]. Case (ii) covers the homomorphisms underlying the ElGamal encryption scheme [49] and the ZK-PoK for it was proposed by Schnorr [3].

3.2 Proof of Theorem 6

The remainder of this section is now dedicated to proving the theorem. We therefore recap the following lemma introduced by Damgård/Koprowski.

Lemma 8 (Lemma 3 of [27]). *Let* $E := a_1X_1 + \cdots + a_uX_u \in \mathbb{Z}[X_1, \ldots, X_u]$ *be a non-zero polynomial, and let* $z \geq |a_i|$ *for all* i*. Let further* $\mathcal{G}$ *be a group of hidden order, and* $x_1, \ldots, x_u \in_R \mathcal{G}$*. For any positive* A*, we then have*

$$Pr[a_1x_1 + \cdots + a_ux_u = 0] \leq \frac{1}{A} + (\log_2 z + A)\alpha(\pi).$$

Proof (of Theorem 6 – Sketch). The proof is structured as follows. We describe a prover P^* for which we show that it satisfies the conditions of Theorem 3. We will see that the uniformity condition holds true by definition. For the hardness condition we simulate the behavior of P^* in the additive subgroup of a suitable polynomial ring. We then estimate the success probability of this simulated game and the error made when making this simulation.

We start with describing a malicious prover P^*. This cheating prover essentially behaves like the honest prover, except that it does not answer all challenges but only certain ones. Depending on whether $p \leq q$ or not, the set $\mathcal{C}'$ of answered challenges is defined as follows:

$p \leq q$: For $i = 1, \ldots, p$, let $\bar{c}_i \in \{0, 1\}$ such that at least half of the elements of $\mathcal{C}_i$ have the same parity as $\bar{c}_i$. Then $\mathcal{C}' := \{(c_1, \ldots, c_p) \in \mathcal{C} | c_i \equiv \bar{c}_i \mod 2\}$.
$q < p$: We define $\mathcal{C}'$ as a subset of $\mathcal{C}$, which has a cardinality of at least $\#\mathcal{C}/2^q$, and all $(c_1, \ldots, c_p), (c'_1, \ldots, c'_p) \in \mathcal{C}'$ satisfy the following q equations for all $j = m^- + 1, \ldots, m$ and all $k = n^- + 1, \ldots, n$:

$$\sum g_{ji}c_i \equiv \sum g_{ji}c'_i \mod 2 \quad \text{and} \quad \sum e_{ki}c_i \equiv \sum e_{ki}c'_i \mod 2$$

We next describe P^*. We therefore make the random input $\zeta = (\zeta_1, \ldots, \zeta_l)$ to the prover explicit, and let $\rho(\cdot)$ be a pseudo random function.

(i) It sets $r'_i := \rho(\zeta_i)$ for $i = 1, \ldots, l$, and using these random elements, it behaves just as an honest prover.
(ii) If $c_i \in \mathcal{C}'$, P^* behaves like an honest prover, using $(r'_{l^-+1}, \ldots, r'_l)$ as random elements. Otherwise it halts.

The **uniformity property** of Theorem 3 is obviously satisfied, as the prover answers a fraction of at least $1/2^{\min(p,q)}$ of all challenges, and makes the verifier accept (because the verifier would accept for an honest prover).

Let us now turn towards the **hardness property**. We say that a generic black-box algorithm succeeds, if after v steps it outputs the handle corresponding to a preimage of x under $\phi(\cdot)$. Now, instead of letting the knowledge extractor interact with P^* and the oracle $\mathcal{O}^{\phi(\cdot)}$, we play the following game. We substitute $\mathcal{G}$ and $\mathcal{H}$ by the following subgroups of the polynomial rings over the indeterminantes $W, O_{ij}, R_{ij}, T_{ij}$:

$$\mathcal{G}' := \langle W, O_{11}, \ldots, O_{1l}, \ldots, O_{v1}, \ldots, O_{vl}, R_{11}, \ldots, R_{1m}, \ldots, R_{v1}, \ldots, R_{vm} \rangle$$
$$\mathcal{H}' := \langle \mathcal{G}', T_{11}, \ldots, T_{1n}, \ldots, T_{v1}, \ldots, T_{vn} \rangle.$$

Accordingly, the oracle $\mathcal{O}'^{\phi(\cdot)}$ now performs its computations within $\mathcal{G}'$ and $\mathcal{H}'$.

The prover P^* is adopted as described next. It maintains a list L, which is initially empty, and sets $u := 0$. On random input ζ, it performs the following steps:

(i) For each ζ_i, it checks whether there is a pair $(\zeta_{ji}, \bar{R_{ji}})$ with $\zeta_i = \zeta_{ji}$ in L. If so, it sets $\hat{R}_i := \bar{R_{ji}}$. Otherwise, it increases u by 1 (but at most once in each run), sets $\hat{R}_i := R_{ui}$, and adds $(\zeta_i, \hat{R}_i)$ to L. Then it sends

$$\left((\sum a_{ji} \cdot \hat{R}_i + f_j \cdot W)_{j=1}^{\bar{m}}, (\sum b_{ki} \cdot \hat{R}_i + d_k \cdot W)_{k=1}^{\bar{n}}\right)$$

to V. Former are marked as elements of $\mathcal{G}'$, latter as elements of $\mathcal{H}'$.

(ii) If $c_i \in \mathcal{C}'$, P^* analogously computes its response according to the protocol. Otherwise, if $c \notin \mathcal{C}'$, P^* halts.

By $\mathbf{r}$ we denote an element from the set of from which the oracle and by the generator of the input to the protocol draw their random choices, i.e.,

$$\mathbf{r} \in \Big\{ (\phi(\cdot), x, w, \rho, o, t) \,\Big|\, \phi(\cdot) : \mathcal{G} \to \mathcal{H} \text{ has hidden order co-domain,} \\ x = \phi(w), \rho(\cdot) \text{ pseudo random}, o \in \mathcal{G}^{v \times l}, t \in \mathcal{H}^{v \times m} \Big\}$$

We then define the following two mappings. By $\iota^{\mathbf{r}}_{\mathcal{G}'}(\cdot)$ we denote the evaluation homomorphism from $\mathcal{G}'$ into $\mathcal{G}$. That is, by $\iota^{\mathbf{r}}_{\mathcal{G}'}(E)$ we denote the element in $\mathcal{G}$ which results when all indeterminantes in E are substituted in the following way:

$$W \mapsto w \quad O_{ij} \mapsto o_{ij} \quad R_{ij} \mapsto r'_{ij}.$$

In absolute analogy we let $\iota^{\mathbf{r}}_{\mathcal{H}'}(\cdot)$ be the evaluation homomorphism from $\mathcal{H}'$ into $\mathcal{H}$. That is, the substitution is given by:

$$W \mapsto \phi(w) \quad O_{ij} \mapsto \phi(o_{ij}) \quad R_{ij} \mapsto \phi(r'_{ij}) \quad T_{ij} \mapsto t_{ij}.$$

We observe that for all $E \in \mathcal{G}'$ we have $\phi(\iota^{\mathbf{r}}_{\mathcal{G}'}(E)) = \iota^{\mathbf{r}}_{\mathcal{H}'}(E)$.

During its computation the generic black-box algorithm maintains a list of elements $E_i \in \mathcal{G}'$ respectively $F_i \in \mathcal{H}'$. We say that the algorithm wins this modified game, if one of the following to cases occurs. In case (a), the algorithm finds a preimage of x under $\phi(\cdot)$, while in case (b) there is a pair $i \neq j$ satisfying the following. For a randomly chosen $\mathbf{r}$, we either have $E_i \neq E_j$ and $\iota^{\mathbf{r}}_{\mathcal{G}'}(E_i - E_j) = 0$, or $F_i \neq F_j$ and $\iota^{\mathbf{r}}_{\mathcal{H}'}(F_i - F_j) = 0$.

Observing that the behavior of this game and the actual interaction between the algorithm and the real oracle are indistinguishable as long as the above game is not won, we get that the success probability of the generic black-box algorithm is upper bounded by the probability that the algorithm wins the game [50].

Case (a). Finding a preimage means to compute E_i such that $\phi(\iota^{\mathbf{r}}_{\mathcal{G}'}(E_i)) = x$. Using the observation that we always have $\iota^{\mathbf{r}}_{\mathcal{H}'}(W) = x$ this means to find an E_i such that $\iota^{\mathbf{r}}_{\mathcal{H}'}(E_i - X) = 0$. By introspection of how the E_i's are computed, and by using the linear independency of the vectors $\{(b_{11}, \ldots, b_{1l}), \ldots, (b_{n1}, \ldots, b_{nl})\}$, one can show that $W \neq E_i$ for all i.

Let $K := K(\mathcal{C}, a_{ji}, b_{ki}, g_{ji}, e_{ji}, f_j, d_k)$ be an integer such that K is larger than the absolute values of all coefficients occurring in the definition of the examined generic Σ-protocol. Using that $E_i \neq W$ and Lemma 8, and noting that after v oracle queries for E_i's and F_j's each, all coefficients are smaller than $2^v \cdot K$, we get

$$Pr[(a)] \leq \frac{1}{A} + (v + \log_2 K + A)\alpha(\pi) \qquad \text{for all} \qquad A \in \mathbb{Z}.$$

Case (b). Using K as before, and observing that there are at most v different E_i's and F_j's each, we get by a similar argument that the probability for (b) is bounded by

$$Pr[(b)] \leq v^2 \left(\frac{1}{A} + (v + \log_2 K + A)\alpha(\pi)\right) \qquad \text{for all} \qquad A \in \mathbb{Z}.$$

We here assumed that $\phi(\cdot)$ is surjective, and that $\rho(\cdot)$ is a truly random function. The former can easily be seen to be just a technical issue to ease presentation, and the latter yields only a negligible error as $\rho(\cdot)$ is pseudo random by definition.

Demonstration of Hardness Condition. The overall probability that the algorithm wins the game described above is hence limited by

$$Pr[(a)] + Pr[(b)] \leq (v^2 + 1)\left(\frac{1}{A} + (v + \log_2 K + A)\alpha(\pi)\right) \qquad \text{for all} \qquad A \in \mathbb{Z}.$$

for a fixed choice of $\mathbf{r}$. We now set the so far arbitrary value of A to $A := \sqrt{1/\alpha(\pi)}$ such that both $1/A$ and $A \cdot \alpha(\pi)$ are negligible, and observe that K and $\alpha(\pi)$ are independent from $\mathbf{r}$. Using now that for the hardness condition to be satisfied we only need to consider generic black-box algorithms the expected number v of steps of which is polynomially bounded, and computing the expectation value over all choices of $\mathbf{r}$, we get that the success probability of the generic black-box algorithm is negligible. □

4 Lower Bounds for the Σ^ϕ-Protocol in the Plain Model

As pointed out by Dent and Fischlin [39,40], restrictions proven in the generic model do not necessarily hold true in the plain model as well. In this section we thus confirm our results obtained in the generic model by showing the existence of lower bounds in the plain model. That is, we provide evidence that for exponentiation homomorphisms with hidden order co-domain, and for power homomorphisms of the form $\phi(\cdot) : \mathcal{H} \to \mathcal{H} : w \mapsto w^e$, no smaller knowledge error than in the generic model can be reached in the plain model. The results only hold for the Σ^ϕ-protocol, and not for the entire class of generic Σ-protocols.

The following results are based on a generalization of the Root Assumption [47], which we call the *Expected Root Assumption.* We say that the Expected

Root Assumption holds for a group $\mathcal{H}$ if there exists no expected PPT algorithm that on input a random element $h \in_R \mathcal{H}$ and $e \geq 2$ outputs an eth root of h with non-negligible probability. In contrast to the standard Root Assumption, we here also require that no *expected* PPT algorithm has a noticeable success probability. This requirement naturally arises from the fact that the definition of PoK only restricts the *expected* running time of the knowledge extractor, cf. §2.3.

4.1 Lower Bounds for Power Homomorphisms

We first consider the Σ^ϕ-protocol for power homomorphisms of the form $\phi_P(\cdot) : \mathcal{H} \to \mathcal{H} : w \mapsto w^e$. This is a generalization of the protocol proposed Guillou/Quisquater [4]. We generalize the result from Shoup [26] from exponents of the form $e = 2^t$ to arbitrary values of e.

In the following we use the following notation. For a set $\mathcal{S}$ and $r \in \mathbb{Z}$, we define $\mathrm{Div}(\mathcal{S}, r)$ to be all multiples of r within $\mathcal{S}$, i.e., $\mathrm{Div}(\mathcal{S}, r) := \{s : s \in \mathcal{S}, r|s\}$.

Theorem 9 (Bounds for Power Homomorphisms). *Let* $\mathrm{poly}(\cdot)$ *be an arbitrary but fixed polynomial. Then for every power homomorphism* $\phi_P(\cdot) : \mathcal{H} \to \mathcal{H} : w \mapsto w^e$ *with* $e \geq 2$*, the knowledge error of the* Σ^ϕ*-protocol for* $\phi_P(\cdot)$ *is lower bounded by*

$$\max_{2 \leq r \leq e, r|e} \frac{\#\,\mathrm{Div}(\mathcal{C}, r)}{\#\mathcal{C}} - \frac{1}{\mathrm{poly}(\|x\|)},$$

if the Expected Root Assumption is satisfied for $\mathcal{H}$ *and* $\gcd(e, \mathrm{ord}\,\mathcal{H}) = 1$.

Note here that, if $\mathcal{H}$ is an RSA group, i.e., $\mathcal{H} = \mathbb{Z}_n^*$ for a composite modulus n of unknown factorization, the condition $\gcd(e, \mathrm{ord}\,\mathcal{H}) = 1$ is always satisfied.

We stress that, if the challenge set $\mathcal{C}$ is an integer interval, the theorem implies a lower bound which is equal to the smallest knowledge error that is currently known to be achievable:

Corollary 10. *Let the conditions of Theorem 9 be satisfied, and let the challenge set be an integer interval (i.e.,* $\mathcal{C} = \{a, \ldots, b\}$ *for some* $a, b \in \mathbb{Z}$*). Let* d *denote the smallest divisor of* e*. Then knowledge error of the* Σ^ϕ*-protocol is bounded from below by*

$$\frac{1}{d} - \frac{1}{\mathrm{poly}(\|x\|)}.$$

Theorem 9 becomes meaningless if all elements of $\mathcal{C}$ are co-prime (e.g., if all elements of $\mathcal{C}$ are primes), as it then implies a lower bound of 0. Though, the result is still relevant when seen in connection with Theorem 6. Namely, while the latter states that any hypothetical knowledge extractor has to use encoding specific properties of the homomorphism $\phi_P(\cdot)$, the former further restricts the situations where the generic result could potentially be violated in the plain model. In summary, the existence of an extractor underrunning the limitation of $1/d$ seems unlikely.

4.2 Lower Bounds for Exponentiation Homomorphisms

For exponentiation homomorphisms $\phi_E(\cdot) : \mathbb{Z} \to \mathcal{H} : w \mapsto h^w$ with hidden order co-domain $\mathcal{H}$, the Σ^ϕ-protocol is only known to be a PoK with knowledge error 1/2. In this section we show that (if existing at all) any knowledge error achieving a smaller knowledge error in this case would require fundamentally new insights to the Σ^ϕ-protocol.

Although being used for numerous different homomorphisms, essentially only one knowledge extractor is known for the Σ^ϕ-protocol. This standard knowledge extractor works as described next. In a first phase, it is given rewindable black-box access to the prover, and extracts a *pseudo* preimage (u, v), i.e., a pair satisfying $v \neq 0$ and $x^v = \phi_E(u)$, cf. §3.1. Then, in a second phase in which the extractor does not have access to the prover any more, it computes a preimage of x from this pseudo preimage. We call knowledge extractors working this way *pseudo preimage based.* We show that no such knowledge extractor can underrun a knowledge error of 1/2 for the Σ^ϕ-protocol and exponentiation homomorphisms with hidden order co-domain.

Let us introduce some notation: for a set $\mathcal{S}$ of integers, we write Diff$(\mathcal{S})$ for the set of all possible absolute values of differences between different elements of $\mathcal{S}$, i.e., Diff$(\mathcal{S}) := \{|s_1 - s_2| : s_1 \neq s_2 \in \mathcal{S}\}$. We further say that an integer d and a set $\mathcal{S}$ are co-prime, if $\gcd(d, s) = 1$ for all $s \in \mathcal{S}$.

Theorem 11 (Bounds for Exponentiation Homomorphisms). *Let* poly$(\cdot)$ *be an arbitrary but fixed polynomial. Then for every exponentiation homomorphisms $\phi_E(\cdot) : \mathbb{Z} \to \mathcal{H}' : w \mapsto h^w$, with $h \in \mathcal{H}'$, the knowledge error of the Σ^ϕ-protocol for $\phi_E(\cdot)$ is lower bounded by*

$$\frac{1}{2} - \frac{1}{\text{poly}(\|x\|)},$$

against pseudo preimage based knowledge extractors, if the following conditions are satisfied. The co-domain $\mathcal{H}'$ is a large subgroup of $\mathcal{H}$ (i.e., $\#\mathcal{H}'/\#\mathcal{H}$ is not negligible), the Expected Root Assumption is satisfied for $\mathcal{H}$, and ord $\mathcal{H}'$ *and* Diff$(\mathcal{C})$ *are co-prime.*

We remark that this result can straightforwardly be generalized to homomorphisms of the form $\phi_M(\cdot) : \mathcal{G}^r \to \mathcal{H} : (w_1, \ldots, w_r) \mapsto h_1^{w_1} \ldots h_r^{w_r}$.

In practice the conditions of this theorem are most often satisfied. For instance consider the case where $\mathcal{H} = \mathbb{Z}_n^*$ for a safe RSA modulus n, i.e., $n = (2p + 1) \cdot (2q+1)$, where p, q, $(2p+1)$, and $(2q+1)$ are primes. Then $\mathcal{H}'$ is usually given by the set of quadratic residues modulo n, and we have $\#\mathcal{H}'/\#\mathcal{H} = 1/4$. Further, ord $\mathcal{H}' = p \cdot q$, and hence any challenge set $\mathcal{C}$ only containing elements smaller than p, q will satisfy the condition of Diff$(\mathcal{C})$ and ord $\mathcal{H}'$ being co-prime.

Although this result only considers pseudo preimage based knowledge extractors, it is still relevant for the following reason. Together with the results in the generic group model in §3, Theorem 11 implies that a knowledge extractor for exponentiation homomorphisms with hidden order co-domain must neither be generic nor pseudo preimage based. Thus, if possible at all, substantially new insights were required to underrun the restriction of 1/2 in this case. According to current knowledge, we doubt the existence of such an extractor. We thus believe that for reaching a small knowledge error in the case of exponentiation homomorphisms with hidden order co-domain, either running the Σ^ϕ-protocol repeatedly or employing an FO-based protocol cannot be avoided.

5 Conclusion

We have introduced the class of generic Σ-protocols, and have shown that in the generic group model a knowledge error of $1/2^n$ (where n is the minimum of the number of challenges and responses sent in the protocol) is inherent to any of these protocols for homomorphisms with hidden order co-domain. We further generalized this result to special homomorphisms as well, covering essentially all homomorphisms being used in cryptography. Especially, those underlying various crypto systems fall into this class [45,46,47,48,49]. We then confirmed our results for the Σ^ϕ-protocol and certain homomorphisms in the plain model as well.

Besides pointing out these limitations, our results also give insights in how these restrictions could be overcome. Namely, any Σ-protocol overcoming these bounds must either be substantially different from the Σ^ϕ-protocol (i.e., it must not be a generic Σ-protocol), or it must have a non-generic knowledge extractor.

The former seems to be hard to achieve without using auxiliary constructions resulting from a common reference string as done in [5,20,21], because the class of generic Σ-protocols does not leave much design options for other Σ-protocols to look like. Yet, the latter also is unlikely, because of our results in the plain model. Thus, although being riddled with various limitations from a theoretical point of view, FO-based protocols [5,19,20,21,22,23] using common reference strings seem to be inevitable for many real systems.

References

1. Bangerter, E., Camenisch, J., Krenn, S.: Efficiency limitations for Σ-protocols for group homomorphisms. Cryptology ePrint Archive, Report 2009/595 (2009)
2. Goldreich, O., Micali, S., Wigderson, A.: Proofs that yield nothing but their validity or all languages in NP have zero-knowledge proof systems. Journal of the ACM 38, 691–729 (1991); Preliminary version in 27th FOCS (1986)
3. Schnorr, C.: Efficient signature generation by smart cards. Journal of Cryptology 4, 161–174 (1991)
4. Guillou, L., Quisquater, J.J.: A "paradoxical" identity-based signature scheme resulting from zero-knowledge. In: Goldwasser, S. (ed.) CRYPTO 1988. LNCS, vol. 403, pp. 216–231. Springer, Heidelberg (1990)
5. Bangerter, E.: Efficient Zero-Knowledge Proofs of Knowledge for Homomorphisms. PhD thesis, Ruhr-University Bochum (2005)
6. Cramer, R.: Modular Design of Secure yet Practical Cryptographic Protocols. PhD thesis, CWI and University of Amsterdam (1997)

7. Maurer, U.: Unifying zero-knowledge proofs of knowledge. In: Preneel, B. (ed.) AFRICACRYPT 2009. LNCS, vol. 5580, pp. 272–286. Springer, Heidelberg (2009)
8. Fiat, A., Shamir, A.: How to prove yourself: practical solutions to identification and signature problems. In: Odlyzko, A.M. (ed.) CRYPTO 1986. LNCS, vol. 263, pp. 186–194. Springer, Heidelberg (1987)
9. Boudot, F.: Efficient proofs that a committed number lies in an interval. In: Preneel, B. (ed.) EUROCRYPT 2000. LNCS, vol. 1807, pp. 431–444. Springer, Heidelberg (2000)
10. Camenisch, J.: Better privacy for trusted computing platforms (extended abstract). In: Samarati, P., Ryan, P.Y.A., Gollmann, D., Molva, R. (eds.) ESORICS 2004. LNCS, vol. 3193, pp. 73–88. Springer, Heidelberg (2004)
11. Camenisch, J., Michels, M.: Proving in zero-knowledge that a number is the product of two safe primes. In: Stern, J. (ed.) EUROCRYPT 1999. LNCS, vol. 1592, pp. 107–122. Springer, Heidelberg (1999)
12. Camenisch, J., Shoup, V.: Practical verifiable encryption and decryption of discrete logarithms. In: Boneh, D. (ed.) CRYPTO 2003. LNCS, vol. 2729, pp. 126–144. Springer, Heidelberg (2003)
13. Lipmaa, H.: On diophantine complexity and statistical zero-knowledge arguments. In: Laih, C.-S. (ed.) ASIACRYPT 2003. LNCS, vol. 2894, pp. 398–415. Springer, Heidelberg (2003)
14. Song, D.X.: Practical forward secure group signature schemes. In: ACM CCS 2001, pp. 225–234. ACM, New York (2001)
15. Tang, C., Liu, Z., Wang, M.: A verifiable secret sharing scheme with statistical zero-knowledge. Cryptology ePrint Archive, Report 2003/222 (2003)
16. Tsang, P.P., Wei, V.K., Chan, T.K., Au, M.H., Liu, J.K., Wong, D.S.: Separable linkable threshold ring signatures. In: Canteaut, A., Viswanathan, K. (eds.) INDOCRYPT 2004. LNCS, vol. 3348, pp. 384–398. Springer, Heidelberg (2004)
17. Brickell, E., Camenisch, J., Chen, L.: Direct anonymous attestation. In: ACM CCS 2004, pp. 132–145. ACM, New York (2004)
18. Camenisch, J., Herreweghen, E.V.: Design and implementation of the idemix anonymous credential system. In: ACM CCS 2002, pp. 21–30. ACM, New York (2002)
19. Bangerter, E., Camenisch, J., Maurer, U.: Efficient proofs of knowledge of discrete logarithms and representations in groups with hidden order. In: Vaudenay, S. (ed.) PKC 2005. LNCS, vol. 3386, pp. 154–171. Springer, Heidelberg (2005)
20. Bangerter, E., Krenn, S., Sadeghi, A.R., Schneider, T., Tsay, J.K.: Design and implementation of efficient zero-knowledge proofs of knowledge. In: SPEED-CC 2009 (2009)
21. Damgård, I., Fujisaki, E.: A statistically-hiding integer commitment scheme based on groups with hidden order. In: Zheng, Y. (ed.) ASIACRYPT 2002. LNCS, vol. 2501, pp. 125–142. Springer, Heidelberg (2002)
22. Fujisaki, E., Okamoto, T.: Statistical zero knowledge protocols to prove modular polynomial relations. In: Kaliski Jr., B.S. (ed.) CRYPTO 1997. LNCS, vol. 1294, pp. 16–30. Springer, Heidelberg (1997)
23. Camenisch, J., Kiayias, A., Yung, M.: On the portability of Generalized Schnorr Proofs. In: Joux, A. (ed.) EUROCRYPT 2009. LNCS, vol. 5479, pp. 425–442. Springer, Heidelberg (2009)
24. Canetti, R., Goldreich, O., Halevi, S.: The random oracle methodology, revisited. Journal of the ACM 51, 557–594 (2004); Preliminary version in STOC (1998)

25. Pass, R.: On deniability in the common reference string and random oracle model. In: Boneh, D. (ed.) CRYPTO 2003. LNCS, vol. 2729, pp. 316–337. Springer, Heidelberg (2003)
26. Shoup, V.: On the security of a practical identification scheme. In: Maurer, U.M. (ed.) EUROCRYPT 1996. LNCS, vol. 1070, pp. 344–353. Springer, Heidelberg (1996)
27. Damgård, I., Koprowski, M.: Generic lower bounds for root extraction and signature schemes in general groups. In: Knudsen, L.R. (ed.) EUROCRYPT 2002. LNCS, vol. 2332, pp. 256–271. Springer, Heidelberg (2002)
28. Nechaev, V.I.: Complexity of a determinate algorithm for the discrete logarithm. Mathematical Notes 55, 165–172 (1994); Translated from Matematicheskie Zametki 55(2), 91–101 (1994)
29. Shoup, V.: Lower bounds for discrete logarithms and related problems. In: Fumy, W. (ed.) EUROCRYPT 1997. LNCS, vol. 1233, pp. 256–266. Springer, Heidelberg (1997)
30. Abe, M., Fehr, S.: Perfect NIZK with adaptive soundness. In: Vadhan, S.P. (ed.) TCC 2007. LNCS, vol. 4392, pp. 118–136. Springer, Heidelberg (2007)
31. Aggarwal, D., Maurer, U.: Breaking RSA generically is equivalent to factoring. In: Joux, A. (ed.) EUROCRYPT 2009. LNCS, vol. 5479, pp. 36–53. Springer, Heidelberg (2009)
32. Boneh, D., Boyen, X., Goh, E.J.: Hierarchical identity based encryption with constant size ciphertext. In: Cramer, R. (ed.) EUROCRYPT 2005. LNCS, vol. 3494, pp. 440–456. Springer, Heidelberg (2005)
33. Brown, D.: Generic groups, collision resistance, and ECDSA. Cryptology ePrint Archive, Report 2002/026 (2002)
34. Dent, A.W.: The hardness of the DHK problem in the generic group model. Cryptology ePrint Archive, Report 2006/156 (2006)
35. Maurer, U.: Index search, discrete logarithms, and Diffie-Hellman. In: Number-theoretic cryptography workshop. Mathematical Sciences Research Institute, Berkeley (2000)
36. Maurer, U., Wolf, S.: Lower bounds on generic algorithms in groups. In: Nyberg, K. (ed.) EUROCRYPT 1998. LNCS, vol. 1403, pp. 72–84. Springer, Heidelberg (1998)
37. Schnorr, C., Jakobsson, M.: Security of signed elgamal encryption. In: Okamoto, T. (ed.) ASIACRYPT 2000. LNCS, vol. 1976, pp. 73–89. Springer, Heidelberg (2000)
38. Smart, N.P.: The exact security of ECIES in the generic group model. In: Honary, B. (ed.) Cryptography and Coding 2001. LNCS, vol. 2260, pp. 73–84. Springer, Heidelberg (2001)
39. Dent, A.W.: Adapting the weaknesses of the random oracle model to the generic group model. In: Zheng, Y. (ed.) ASIACRYPT 2002. LNCS, vol. 2501, pp. 100–109. Springer, Heidelberg (2002)
40. Fischlin, M.: A note on security proofs in the generic model. In: Okamoto, T. (ed.) ASIACRYPT 2000. LNCS, vol. 1976, pp. 458–469. Springer, Heidelberg (2000)
41. Cramer, R., Damgård, I.: On the amortized complexity of zero-knowledge protocols. In: Halevi, S. (ed.) CRYPTO 2009. LNCS, vol. 5677, pp. 177–191. Springer, Heidelberg (2009)
42. Bellare, M., Goldreich, O.: On defining proofs of knowledge. In: Brickell, E.F. (ed.) CRYPTO 1992. LNCS, vol. 740, pp. 390–420. Springer, Heidelberg (1993)
43. Goldreich, O.: Foundations of Cryptography – Basic Tools. Cambridge University Press, Cambridge (2001)

44. Babai, L., Szemerédi, E.: On the complexity of matrix group problems I. In: IEEE FOCS 1984, pp. 229–240 (1984)
45. Damgård, I., Jurik, M.: A generalisation, a simplification and some applications of Paillier's probabilistic public-key system. In: Kim, K.-c. (ed.) PKC 2001. LNCS, vol. 1992, pp. 119–136. Springer, Heidelberg (2001)
46. Paillier, P.: Public-key cryptosystems based on composite degree residuosity classes. In: Stern, J. (ed.) EUROCRYPT 1999. LNCS, vol. 1592, pp. 223–238. Springer, Heidelberg (1999)
47. Rivest, R., Shamir, A., Adleman, L.: A method for obtaining digital signatures and public-key cryptosystems. Communications of the ACM 21, 120–126 (1978)
48. Takagi, T.: Fast RSA-type cryptosystem modulo $p^k q$. In: Krawczyk, H. (ed.) CRYPTO 1998. LNCS, vol. 1462, pp. 318–326. Springer, Heidelberg (1998)
49. Gamal, T.E.: A public key cryptosystem and a signature scheme based on discrete logarithms. In: Blakely, G.R., Chaum, D. (eds.) CRYPTO 1984. LNCS, vol. 196, pp. 10–18. Springer, Heidelberg (1985)
50. Shoup, V.: OAEP reconsidered. In: Kilian, J. (ed.) CRYPTO 2001. LNCS, vol. 2139, pp. 239–259. Springer, Heidelberg (2001)

Composition of Zero-Knowledge Proofs with Efficient Provers*

Eleanor Birrell[1] and Salil Vadhan[2]

[1] Department of Computer Science, Cornell University
eleanor@cs.cornell.edu

[2] School of Engineering and Applied Sciences and Center for Research on Computation and Society, Harvard University**
salil@seas.harvard.edu

Abstract. We revisit the composability of different forms of zero-knowledge proofs when the honest prover strategy is restricted to be polynomial time (given an appropriate auxiliary input). Our results are:

1. When restricted to efficient provers, the original Goldwasser–Micali–Rackoff (GMR) definition of zero knowledge (STOC '85), here called *plain zero knowledge*, is closed under a constant number of sequential compositions (on the same input). This contrasts with the case of unbounded provers, where Goldreich and Krawczyk (ICALP '90, SICOMP '96) exhibited a protocol that is zero knowledge under the GMR definition, but for which the sequential composition of 2 copies is not zero knowledge.
2. If we relax the GMR definition to only require that the simulation is indistinguishable from the verifier's view by uniform polynomial-time distinguishers, with no auxiliary input beyond the statement being proven, then again zero knowledge is not closed under sequential composition of 2 copies.
3. We show that auxiliary-input zero knowledge with efficient provers is not closed under *parallel* composition of 2 copies under the assumption that there is a secure key agreement protocol (in which it is easy to recognize valid transcripts). Feige and Shamir (STOC '90) gave similar results under the seemingly incomparable assumptions that (a) the discrete logarithm problem is hard, or (b) $\mathcal{UP} \not\subseteq \mathcal{BPP}$ and one-way functions exist.

1 Introduction

Composition has been one of the most active subjects of research on zero-knowledge proofs. The goal is to understand whether the zero-knowledge property is preserved when a zero-knowledge proof is repeated many times. The

* These results first appeared in the first author's undergraduate thesis [5] and in the full version of the paper is available on the *Cryptology ePrint Archive* [6].

** 33 Oxford Street, Cambridge, MA 02138. http://seas.harvard.edu/~salil/.
Supported by NSF grant CNS-0831289.

D. Micciancio (Ed.): TCC 2010, LNCS 5978, pp. 572–587, 2010.

answers vary depending on the variant of zero knowledge in consideration and the form of composition (e.g. sequential, parallel, or concurrent). The study of composition was first aimed at reducing the soundness error of basic constructions of zero-knowledge proofs (via sequential or parallel composition), but was later also motivated by considering networked environments in which an adversary might be able to open several instances of a protocol (even concurrently).

Soon after Goldwasser, Micali, and Rackoff introduced the concept of zero-knowledge proofs [20], it was realized that composability is a subtle issue. In particular, this motivated a strengthening of the GMR definition, known as *auxiliary-input zero knowledge* [21,19,9], which was shown to be closed under sequential composition [19]. The need for this stronger definition was subsequently justified by a result of Goldreich and Krawczyk [16], who showed that the original GMR definition is not closed under sequential composition. Specifically, they exhibited a protocol that is *plain zero knowledge* when executed once, but fails to be zero knowledge when executed twice sequentially.

The starting point for our work is the realization that the Goldreich–Krawczyk protocol is not an entirely satisfactory counterexample, because the prover strategy is inefficient (i.e. super-polynomial time). Most cryptographic applications of zero-knowledge proofs require a prover strategy that can be implemented efficiently given an appropriate auxiliary input (e.g. NP witness). Prover efficiency can intuitively have an impact on the composability of zero-knowledge proofs, because an adversarial verifier may be able to use the extra computational power of one prover copy to "break" the zero-knowledge property of another copy. Indeed, known positive results on the parallel and concurrent composability of witness-indistinguishable proofs (a weaker variant of zero-knowledge proofs) rely on prover efficiency [9].

Thus, we revisit the sequential composability of plain zero knowledge, but restricted to efficient provers. Our first result is positive, and shows that such proofs *are* closed under any constant number of sequential compositions (in contrast to the Goldreich–Krawczyk result with unbounded provers). The case of a superconstant or polynomial number of compositions remains an interesting open question. This positive result refers to the standard formulation of plain zero knowledge, where the simulation and the verifier's view are required to be indistinguishable by nonuniform polynomial-time distinguishers (or distinguishers that are given the prover's auxiliary input in addition to the statement being proven).

We then consider the case where the distinguishers are uniform probabilistic polynomial-time algorithms, whose only additional input is the statement being proven. In this case, we obtain a negative result analogous to the one of Goldreich and Krawczyk, showing that zero knowledge is not closed under sequential composition of even 2 copies (assuming that $\mathcal{NP} \not\subseteq \mathcal{BPP}$). Informally, these two results say that plain zero knowledge is closed under a constant number of sequential compositions if and only if the distinguishers are at least a powerful as the prover.

We also examine the *parallel* composability of *auxiliary-input* zero knowledge. Here, too, Goldreich and Krawczyk [16] gave a negative result that utilizes an inefficient prover. Feige and Shamir [9], however, gave a negative result with an efficient prover, under the assumption that the discrete logarithm is hard, or more generally under the assumptions that $\mathcal{UP} \not\subseteq \mathcal{BPP}$ and one-way functions exist. We are interested in whether the complexity assumption used by Feige and Shamir can be weakened. To this end, we provide a negative result under a seemingly incomparable assumption, namely that there exists a key agreement protocol (in which it is easy to recognize valid transcripts).

2 Definitions and Preliminaries

2.1 Interactive Proofs

Given two interactive Turing machines – a prover P and a verifier V – we consider two types of interactive protocols: proofs of language membership (interactive proofs) and proofs of knowledge. In each case, both parties receive a common input x, and P is trying to convince V that $x \in L$ for some language L. We will allow P to have an extra "auxiliary input" or "witness" y. We use the notation (P, V) to denote an interactive protocol and the notation $\langle P(x,y), V(x)\rangle$ to denote the verifier V's view of that protocol with inputs (x, y) and x respectively. The choices for y will be given by a relation of the following kind:

Definition 2.1 (Poly-balanced Relation). *A binary relation R is* poly-balanced *if there exists a polynomial p such that for all $(x,y) \in R$, $|y| \leq p(|x|)$. The language generated by such a relation is denoted $L_R = \{x : (x,y) \in R\}$.*

Observe that we don't require R to be polynomial-time verifiable, so *every* language L is generated by such a relation, for example the relation $R = \{(x,y) : |y| = |x| \text{ and } x \in L\}$.

Definition 2.2 (Interactive Proof). *We say that an interactive protocol (P, V) is an* interactive proof *system for a language L if there exists a poly-balanced relation R such that $L = L_R$ and the following properties hold:*

- *(Verifier Efficiency): The verifier V runs in time at most poly$(|x|)$ on input x.*
- *(Completeness): If $(x,y) \in R$ then the verifier $V(x)$ accepts with probability 1 after interacting with the prover $P(x,y)$ on common input x and prover auxiliary input y.*
- *(Soundness): There exists a function $s(n) \leq 1 - 1/\text{poly}(n)$ (called the* soundness error*) for which it holds that for all $x \notin L$ and for all prover strategies P^*, the verifier $V(x)$ accepts with probability at most $s(|x|)$ after interacting with P^* on common input x and prover auxiliary input y.*

Definition 2.3 (Proof of Knowledge). *Let R be a poly-balanced relation. Given an interactive protocol (P, V), we let $p(x,y,r)$ be the probability that V accepts on common input x when y is P's auxiliary input and r is the random input generated by P's random coin flips. Let $P_{x,y,r}$ be the function such that $P_{x,y,r}(\overline{m})$ is the message sent by P after receiving messages $\overline{m}$. An interactive protocol $(P(x,y), V(x))$ is an* interactive proof of knowledge *for the relation R if the following three properties hold:*

- *(Verifier Efficiency): The verifier V runs in time at most poly($|x|$) on input x.*
- *(Completeness): If $(x, y) \in R$, then V accepts after interacting with P on common input x.*
- *(Extraction): There exists a function $s(n) \leq 1 - 1/poly(n)$ (called the* soundness error*), a polynomial q, and a probabilistic oracle machine K such that for every $x, y, r \in \{0,1\}^*$, K satisfies the following condition: if $p(x, y, r) > s(|x|)$ then on input x and with access to oracle $P_{x,y,r}$ machine K outputs w such that $(x, w) \in R$ within an expected number of steps bounded by $q(|x|)/(p(x, y, r) - s(|x|))$.*

Observe that extraction implies soundness, so a proof of knowledge for R is also an interactive proof for L_R.

Although the above definitions require a polynomial-time verifier, neither places any restriction on the computational power of the prover P. In keeping with the standard model of "realistic" computation, we sometimes prefer to limit the computational resources of both parties to polynomial time. Specifically, we add the additional requirement that there exists a polynomial p such that the prover $P(x, y)$ runs in time $p(|x|, |y|)$ where x is the common input and y is the prover's auxiliary input. We refer to such protocols as *efficient* or *efficient-prover* proofs.

2.2 Zero Knowledge

In keeping with the literature, we define zero knowledge in terms of the indistinguishability of the output distributions.

Definition 2.4 (Uniform/Nonuniform Indistinguishability). *Two ensembles of probability distributions $\{\Pi_1(x)\}_{x \in S}$ and $\{\Pi_2(x)\}_{x \in S}$ are* uniformly (resp. nonuniformly) indistinguishable *if for every uniform (resp. nonuniform) probabilistic polynomial-time algorithm D, there exists a negligible function μ such that for every $x \in S$,*

$$\left| \Pr[D(1^{|x|}, \Pi_1(x)) = 1] - \Pr[D(1^{|x|}, \Pi_2(x)) = 1] \right| \leq \mu(|x|),$$

where the probability is taken over the samples of $\Pi_1(x)$ and $\Pi_2(x)$ and the coin tosses of D.

Often, definitions of computational indistinguishability give the distinguisher the index x (not just its length). This makes no difference for nonuniform distinguishers – since they can have x hardwired in – but it does matter for uniform distinguishers. Indeed, we will see that zero-knowledge proofs demonstrate different properties under composition depending on how much information the distinguisher is given about the inputs.

Also, uniform indistinguishability is usually not defined with a universal quantifier over $x \in S$, but instead with respect to all polynomial-time samplable distributions on $x \in S$ (e.g. [2][12]). We use the above definition for simplicity, but our results also extend to the usual definition.

For the purposes of this paper, we consider two different definitions of zero knowledge. The first, which has primarily been of interest for historical reasons, is the one originally introduced by Goldwasser, Micali, and Rackoff [20]:

Definition 2.5 (Plain Zero Knowledge). *An interactive proof system (P, V) for a language $L = L_R$ is* plain zero knowledge *(with respect to nonuniform distinguishers) if for all probabilistic polynomial-time machines V^*, there exists a probabilistic polynomial-time algorithm M_{V^*} that on input x produces an output probability distribution $\{M_{V^*}(x)\}$ such that $\{M_{V^*}(x)\}_{(x,y)\in R}$ and $\{\langle P(x,y), V^*(x)\rangle\}_{(x,y)\in R}$ are nonuniformly indistinguishable.*

As is standard, the above definition refers to *nonuniform* distinguishers (which can have x, y and any additional information depending on x, y hardwired in as nonuniform advice). However, it is also natural to consider *uniform* distinguishers. In this setting, it is important to differentiate between the case where the distinguisher is only given the single verifier input x and the case where the distinguisher is given both x and the prover's auxiliary input y.

Definition 2.6. *An interactive proof system (P, V) for a language $L = L_R$ is* plain zero knowledge with respect to V-uniform distinguishers *if for all probabilistic polynomial-time machines V^*, there exists a probabilistic polynomial-time algorithm M_{V^*} that on input x produces an output probability distribution $\{M_{V^*}(x)\}$ such that $\{(x, M_{V^*}(x))\}_{(x,y)\in R}$ and $\{(x, \langle P(x,y), V^*(x)\rangle)\}_{(x,y)\in R}$ are uniformly indistinguishable.*

Definition 2.7. *An interactive proof system (P, V) for a language $L = L_R$ is* plain zero knowledge with respect to P-uniform distinguishers *if for all probabilistic polynomial-time machines V^*, there exists a probabilistic polynomial-time algorithm M_{V^*} that on input x produces an output probability distribution $\{M_{V^*}(x)\}$ such that $\{(x, y, M_{V^*}(x))\}_{(x,y)\in R}$ and $\{(x, y, \langle P(x,y), V^*(x)\rangle)\}_{(x,y)\in R}$ are uniformly indistinguishable.*

The next definition of zero knowledge that we will consider is the more standard definition which incorporates an auxiliary input for the verifier.

Definition 2.8 (Auxiliary-Input Zero Knowledge). *An interactive proof system (P, V) for a language L is* auxiliary-input zero knowledge *if for every probabilistic polynomial-time machine V^* and every polynomial p there exists a probabilistic polynomial-time machine M_{V^*} such that the probability ensembles $\{\langle P(x,y), V^*(x,z)\rangle\}_{(x,y)\in R, z\in\{0,1\}^{p(|x|)}}$ and $\{M_{V^*}(x,z)\}_{(x,y)\in R, z\in\{0,1\}^{p(|x|)}}$ are nonuniformly indistinguishable.*

Observe that although this last definition is given only in terms of nonuniform indistinguishability, this is actually equivalent to requiring only uniform indistinguishability; any nonuniform advice used by the distinguisher can instead be incorporated into the verifier's auxiliary input z.

2.3 Composition

In this section, we explicitly state the definitions of sequential and parallel composition that will be used throughout this paper. These definitions can be applied to any of the definitions of zero knowledge given in the previous section.

Definition 2.9. *Given an interactive proof system (P, V) and a polynomial $t(n)$, we consider the* t(n)-fold sequential composition *of this system to be the interactive system consisting of $t(n)$ copies of the proof executed in sequence. The ith copy of the protocol is initialized after the $(i-1)^{th}$ copy has concluded. All copies of the protocol are initialized with the same inputs.*

We can extend our notion of zero knowledge to this setting in the natural way.

Definition 2.10. *An interactive proof (P, V) for the language L is* sequential zero knowledge *if for all polynomials $t(n)$, the $t(n)$-fold sequential composition of (P, V) is a zero knowledge proof for L.*

Note that although the verifiers in the different proof copies may be distinct entities and may in fact be honest, this definition implicitly assumes the worst case in which a single adversary controls all verifier copies. That is, it considers a sequential adversary (verifier) to be an interactive Turing machine V^* that is allowed to interact with $t(n)$ independent copies of P (all on common input x) in sequence.

Our definition of parallel composition is analogous to the above definition:

Definition 2.11. *Given an interactive proof system (P, V) and a polynomial $t(n)$, we consider the* t(n)-fold parallel composition *of this system to be the interactive system consisting of $t(n)$ copies of the proof executed in parallel. Each message in the ith round of a copy of the protocol must be sent before any message from the $(i+1)^{th}$ round. All copies of the protocol are initialized with the same inputs.*

We can again extend our notion of zero knowledge to this setting:

Definition 2.12. *An interactive proof (P, V) for the language L is* parallel zero knowledge *if for all polynomials $t(n)$ the $t(n)$-fold parallel composition of (P, V) is a zero-knowledge proof for L.*

Thus a parallel adversary (verifier) is an interactive Turing machine V^* that is allowed to interact with $t(n)$ independent copies of P (all on common input x) in parallel. That is the ith message in each copy is sent before the $(i+1)^{th}$ message of any copy of the protocol.

3 Sequential Zero Knowledge

3.1 Previous Results

In the area of sequential zero knowledge, there are two major results. The first is a negative result concerning the composition of plain zero-knowledge proofs.

Theorem 3.1 (Goldreich and Krawczyk [16]). *There exists a plain zero-knowledge proof (with respect to nonuniform distinguishers) whose 2-fold sequential composition is not plain zero-knowledge.*

The second significant result to emerge from the area concerns the composition of auxiliary-input zero-knowledge proofs. In this case it is possible to show that the zero-knowledge property is retained under sequential composition.

Theorem 3.2 (Goldreich and Oren [19]). *If* (P, V) *is auxiliary-input zero knowledge, then* (P, V) *is auxiliary-input sequential zero knowledge.*

These two results provide a context for our new results on sequential composition.

3.2 New Results

While Theorem 3.1 demonstrates that the original definition of zero knowledge is not closed under sequential composition, it relies on the fact that the prover can be computationally unbounded. In this section, we address the question: what happens when you compose *efficient-prover* plain zero-knowledge proofs? We obtain two results that partially characterize this behavior.

First we show that the Goldreich and Krawczyk result (Theorem 3.1) cannot be extended to efficient-prover plain zero-knowledge proofs. Indeed, we show that such proofs are closed under a *constant* number of compositions.

Theorem 3.3. *If* (P, V) *is an efficient-prover plain zero-knowledge proof system with respect to nonuniform (resp., P-uniform) distinguishers then for every constant* k*, the* k*-fold sequential composition of* (P, V) *is also plain zero knowledge w.r.t. nonuniform (resp., P-uniform) distinguishers.*

We leave the case of a super-constant number of compositions as an intriguing open problem.

Next we consider the case of V-uniform distinguishers, and we show that such protocols are *not* closed under 2-fold sequential composition with efficient provers.

Theorem 3.4. *If* $\mathcal{NP} \not\subseteq \mathcal{BPP}$ *then there exists an efficient-prover plain zero-knowledge proof with respect to* V*-uniform distinguishers whose 2-fold composition is not plain zero knowledge with respect to* V*-uniform distinguishers.*

Informally, Theorems 3.3 and 3.4 say that plain zero knowledge is closed under a constant number of sequential compositions if and only if the distinguishers are at least as powerful as P.

Proof of Theorem 3.3. We now prove that efficient-prover plain zero-knowledge is closed under $O(1)$-fold sequential composition.

Proof. Let (P_k, V_k) denote the sequential composition of k copies of (P, V). We prove by induction on k that (P_k, V_k) is plain zero knowledge with respect to nonuniform (resp., P-uniform) distinguishers.

(P_1, V_1) is zero knowledge by assumption.

Assume for induction that (P_{k-1}, V_{k-1}) is zero knowledge, and consider the interactive protocol (P_k, V_k). Let V_k^* be some sequential verifier strategy for interacting with P_k, and let V_{k-1}^* denote the sequential verifier that emulates V_k^*'s interactions with the first $k-1$ copies of the the proof system (P, V) and then halts. Since (P_{k-1}, V_{k-1}) is zero knowledge, there exists a simulator M_{k-1} that successfully simulates V_{k-1}^*.

Define H_k^* to be the "hybrid" verifier strategy (for interaction with P) that consists of running the simulator M_{k-1} to obtain a simulated view v of the first $k-1$ interactions, and then emulates V_k^* (starting from the simulated view v) in the kth interaction. Since (P, V) is plain zero knowledge, there exists a polynomial-time simulator M_k for this verifier strategy.

We now show that M_k is also a valid simulator for (P_k, V_k^*). Since by induction (P_{k-1}, V_{k-1}) is plain zero knowledge versus nonuniform (resp., P-uniform) distinguishers, the ensembles $\Pi_1(x, y) = (x, y, \langle P_{k-1}(x, y), V_{k-1}^*(x)\rangle)$ and $\Pi_2(x, y) = (x, y, M_{k-1}(x))$ are nonuniformly (resp., uniformly) indistinguishable when $(x, y) \in R$. Consider the function $f(x, y, v) = (x, y, v')$ that emulates V_k^* starting from view v in one more interaction with $P(y)$ to obtain view v'. Since f is polynomial-time computable, we have that $f(\Pi_1(x, y))$ and $f(\Pi_2(x, y))$ are also nonuniformly (resp., uniformly) indistinguishable. Observe that $f(\Pi_1(x, y)) = (x, y, \langle P_k(x, y), V_k^*(x)\rangle)$ and $f(\Pi_2(x, y)) = (x, y, M_k(x))$ therefore M_k is a valid simulator for (P_k, V_k^*) and hence (P_k, V_k) is plain zero knowledge with respect to nonuniform (resp., P-uniform) distinguishers. □

In this proof, we implicitly rely on the fact that the number of copies k is a constant. It is possible that the running time of the simulation is $\Theta(n^{g(k)})$ for some growing function g, and hence super-polynomial for nonconstant k.

Note that this result doesn't conflict with either Theorem 3.1 (in which the prover was allowed to use exponential time and was therefore able to distinguish between a simulated interaction and a real interaction) or Theorem 3.4 (in which the prover is polynomial time but the distributions are only indistinguishable to a V-uniform distinguisher, so the prover was still able to distinguish between a simulated interaction and a real interaction). Instead, it demonstrates that when neither party has more computational resources than the distinguisher, it is possible to prove a sequential closure result for plain zero knowledge, albeit restricted to a constant number of compositions.

Proof of Theorem 3.4. We now prove Theorem 3.4, showing that plain zero knowledge with respect to V-uniform distinguishers is *not* closed under sequential composition. Our proof of Theorem 3.4 is a variant of the Goldreich-Krawczyk [16] proof of Theorem 3.1, so we be begin by reviewing their construction.

Overview of the Goldreich-Krawczyk Construction [16]. In the proof of Theorem 3.1, the key to constructing a zero-knowledge protocol that breaks under sequential composition lies in taking advantage of the difference in computational power between the unbounded prover and the polynomial-time verifier. The proof requires the notion of an *evasive pseudorandom ensemble.* This is simply a collection of sets $S_i \subseteq \{0,1\}^{p(i)}$ such that each set is pseudorandom and no polynomial-time algorithm can generate an element of S_i with non-negligible probability. The existence of such ensembles was proven by Goldreich and Krawczyk in [17]. Using this, Goldreich and Krawczyk [16] construct a protocol such that in the first sequential copy, the verifier learns some element $s \in S_{|x|}$. In the second iteration, the verifier uses this s (whose membership in $S_{|x|}$ can be confirmed by the prover) to extract information from P. A polynomial-time prover would be unable to generate or verify $s \in S_{|x|}$, therefore the result inherently relies on the super-polynomial time allotted to the prover.

Overview of our Construction. As in the Goldreich-Krawczyk construction, we take advantage of the difference in computational power between the two parties. However, since both are required to be polynomial-time machines, the only advantage that the prover has over the verifier is in the amount of nonuniform input each machine receives. The prover is allowed poly($|x|$) bits of auxiliary input y whereas the verifier receives only the $|x|$ bits from the common input x. In order to take advantage of this difference, we define *efficient bounded-nonuniform* evasive pseudorandom ensembles. Using the newly defined ensembles, we construct an analogous protocol; in the first iteration, the verifier learns some element of an efficient bounded-nonuniform evasive pseudorandom ensemble, and in the second it uses this information to extract otherwise unobtainable information from P.

Definition 3.5. *Let q be a polynomial and let $S = \{S_1, S_2, \dots\}$ be a sequence of (non-empty) sets such that each $S_n \subseteq \{0,1\}^n$. We say that S is a* efficient $q(n)$-nonuniform evasive pseudorandom ensemble *if the following three properties hold:*

(1) For all probabilistic polynomial-time machines A with at most $q(n)$ bits of nonuniformity, S_n is indistinguishable from the uniform distribution on strings of length n. That is, there exists a negligible function ϵ such that for all sufficiently large n,

$$\left| \Pr_{x \in S_n}[A(x) = 1] - \Pr_{x \in U_n}[A(x) = 1] \right| \leq \epsilon(n).$$

(2) For all probabilistic polynomial-time machines B with at most $q(n)$ bits of nonuniformity, it is infeasible for B to generate any element of S_n except with negligible probability. That is, there exists a negligible function ϵ such that for all sufficiently large n,

$$\Pr_{r \in \{0,1\}^{q(n)}}[B(x, r) \in S_n] \leq \epsilon(n).$$

(3) *There exists a polynomial $p(n)$ and a sequence of strings $\{\pi_n\}_{n\in\mathbb{N}}$ of length $|\pi_n| = p(n)$ such that:*
   (a) *There exists a probabilistic polynomial-time machine D such that for all $n \in \mathbb{N}$ and $x \in \{0,1\}^n$, $D(\pi_n, x) = 1$ if $x \in S_n$ and $D(\pi_n, x) = 0$ else.*
   (b) *There exists an expected probabilistic polynomial-time machine E such that for all n $E(\pi_n)$ is a uniformly random element of S_n.*

   That is there exist efficient algorithms with polynomial-length advice for checking membership in the ensemble and for choosing an element uniformly at random.

This definition is similar in spirit to the notion of an evasive pseudorandom ensemble used by Goldreich and Krawczyk in the proof of Theorem 3.1. However, we add the additional requirement that a polynomial-time machine with an appropriate advice string π_n can identify and generate elements of the ensemble. In order for this to be possible, we relax the pseudorandomness and evasiveness requirements to only hold with respect to distinguishers with bounded nonuniformity rather than with respect to nonuniform distinguishers.

The introduction of this definition begs the question of whether or not such ensembles exist. Fortunately it turns out that they do.

Theorem 3.6. *There exists an efficient $n/4$-nonuniform evasive pseudorandom ensemble.*

The proof of this theorem appears in the full version [6]. It shows that if we select a hash function $h_n : \{0,1\}^n \rightarrow \{0,1\}^{5n/16}$ from an appropriate pairwise independent family then with high probability $S_n = h_n^{-1}(0^{5n/16})$ is an $n/4$-nonuniform evasive pseudorandom set. The pseudorandomness and evasiveness conditions (items (1) and (2)) are obtained by using pairwise independence and taking a union bound over all algorithms with $n/4$ bits of nonuniformity. The efficiency condition (item (3)) is obtained by taking h_n to be from a standard family (e.g., $h_n(x) =$ the first $5n/16$ bits of $a \cdot x + b$) and taking π_n to be the descriptor of h_n (e.g., (a, b)).

We use this result to demonstrate that efficient-prover plain zero-knowledge proofs with respect to V-uniform distinguishers are not closed under sequential composition. The construction is analogous to the one by Goldreich and Krawczyk, and can be found in the full version of the paper [6].

4 Parallel Zero Knowledge

4.1 Previous Results

There are two classic results that provide context for our new result concerning the parallel composition of efficient-prover zero-knowledge proof systems. In both cases, the result applies to auxiliary-input (as well as plain) zero knowledge, and both results are negative.

The first result establishes the existence of non-parallelizable zero-knowledge proofs independent of any complexity assumptions.

Theorem 4.1 (Goldreich and Krawczyk [16]). *There exists an auxiliary-input zero knowledge proof whose 2-fold parallel composition is not auxiliary-input zero knowledge (or even plain zero knowledge with respect to nonuniform distinguishers).*

While this result demonstrates that zero knowledge is not closed under parallel composition in general, the proof (like that of Theorem 3.1) inherently relies on the unbounded computational power of the provers. Without the additional computational resources necessary to generate a string and test membership in an evasive pseudorandom ensemble, the prover would be unable to execute the defined protocol.

The second such result constructs an *efficient-prover* non-parallelizable zero-knowledge proof based on a zero-knowledge proof of knowledge of the discrete-logarithm relation.

Theorem 4.2 (Feige and Shamir [9]). *If the discrete logarithm assumption holds then there exists an efficient-prover auxiliary-input zero-knowledge proof whose 2-fold parallel composition is not auxiliary-input zero knowledge (or even plain zero knowledge with respect to V-uniform distinguishers).*

This proof relies on the very specific assumption that the discrete logarithm problem is intractable. However as Feige and Shamir observed [9], the only properties of this problem which are actually necessary are the fact that discrete logarithms are unique and that they have a zero-knowledge proof of knowledge. It is therefore natural to consider generalizing the result to proofs of language membership for any language $L \in \mathcal{NP}$ with exactly one witness for each element $x \in L$. The class of such languages is known as $\mathcal{UP}$. Moreover, if one-way functions exist, then every problem in $\mathcal{NP}$ (and hence in $\mathcal{UP}$) has a zero-knowledge proof of knowledge [18]. Thus:

Theorem 4.3 (Feige and Shamir [9]). *If $\mathcal{UP} \nsubseteq \mathcal{BPP}$ and one-way functions exist then there exists an efficient-prover auxiliary-input zero-knowledge proof whose 2-fold parallel composition is not auxiliary-input zero knowledge (or even plain zero knowledge with respect to V-uniform distinguishers).*

4.2 New Results

In this work, we broaden the complexity assumptions under which we have *efficient-prover* non-parallelizable zero-knowledge proofs under more general complexity assumptions. Specifically, we show that such protocols can be constructed from any key agreement protocol (satisfying an additional technical condition). Following the standard notion of key agreement, we introduce the following definition.

Definition 4.4. *A* key agreement protocol *is an efficient protocol between two parties P_1, P_2 with the following four properties:*

- *Input: Both parties have common input 1^ℓ which is a security parameter written in unary.*

- *Output: The outputs of both parties are k-bit strings (for some* $k = \text{poly}(\ell)$*).*
- *Correctness: The parties have the same output with probability 1 (when they follow the protocol). This common output is called the* key.
- *Secrecy: No probabilistic polynomial time Turing machine E given* 1^ℓ *and the transcript of the protocol (messages between* P_1, P_2*) can distinguish with non-negligible advantage the key from a uniformly distributed k-bit string. That is,* $\{(1^\ell, transcript(P_1, P_2), output(P_1, P_2))\}_{1^\ell:\ell\in\mathbb{N}}$ *is nonuniformly indistinguishable from* $\{(1^\ell, transcript(P_1, P_2), U_k)\}_{1^\ell:\ell\in\mathbb{N}}$.

For technical reasons, we impose an additional technical condition.

Definition 4.5. *Let* (P_1, P_2) *be a key agreement protocol. We say that a pair* $(i, r) \in \{1, 2\} \times \{0, 1\}^*$ *is* consistent *with a transcript t of messages if the messages from* P_i *in t are what* P_i *would have sent had its coin tosses been r and had it received the prior messages specified by t. We say that t is* valid *if there exist* r_1, r_2 *such that t is consistent with both* $(1, r_1)$ *and* $(2, r_2)$*; that is, t occurs with nonzero probability when the honest parties* P_1 *and* P_2 *interact. We say that* (P_1, P_2) *has* verifiable transcripts *if there is a polynomial-time algorithm that can decide whether a transcript t is valid given t and any pair* (i, r) *consistent with t.*

We note that many existing key agreement protocols have verifiable transcripts, including the Diffie-Hellman key exchange and the protocols constructed from any public-key encryption scheme with verifiable public keys.

Our main result on non-parallelizable zero knowledge proofs follows:

Theorem 4.6. *If key agreement protocols with verifiable transcripts exist then there exists an efficient-prover auxiliary-input zero-knowledge proof whose 2-fold parallel composition is not auxiliary-input zero knowledge (or even plain zero knowledge with respect to V-uniform distinguishers).*

The existence of secure key agreement protocols with verifiable transcripts seems incomparable to the assumption that $\mathcal{UP} \nsubseteq \mathcal{BPP}$ which was used in Theorem 4.3.

Proof of Theorem 4.6

Proof. By assumption, key agreement protocols with verifiable transcripts exist. We consider an occurrence of a key agreement protocol to consist of the coin tosses of the two parties (r_1, r_2 respectively) together with the transcript t of messages exchanged between the parties during the protocol.

Define a language $L = \{t : \exists(i, r_i) \text{ consistent with } t\}$. $L = L_R$ for the relation $R = \{(t, (i, r_i)) : (i, r_i) \text{ is consistent with } t\}$; we do not claim or require that $L \notin \mathcal{BPP}$. Observe that $L \in \mathcal{NP}$, so there exists an efficient-prover zero-knowledge proof of knowledge (ZKPOK) of a pair (i, r_i) that is consistent with t with error $s(n) \leq 2^{-m}$ where m is the maximum length of a witness (i, r_i)[18]. If necessary, the required error can be achieved by sequential composition of any initial ZKPOK.

We can use this proof as a subprotocol for constructing the following interactive proof for the language L. V begins by sending the message $c = 0$ to P. If

$c = 0$, then P uses the ZKPOK to demonstrate that he knows (i, r_i) consistent with the transcript t. If $c \neq 0$, V demonstrates knowledge of (j, r_j) using the same ZKPOK. If the proof is successful and the transcript is valid (which can be checked by P by our assumption of verifiable transcripts), then P shows in zero knowledge that he too knows a witness (i, r_i) and then sends the common key k to V.

The protocol is summarized below.

Step	$P(t, (i, r_i))$		$V(t)$
1			$c = 0$
		$\leftarrow$	c
2	if $c = 0$: ZKPOK of (i, r_i) consistent with t	$\rightarrow$	
		$\leftarrow$	if $c \neq 0$: ZKPOK of (j, r_j) consistent with t
3	if $c \neq 0$: ZKPOK of (i, r_i) consistent with t	$\rightarrow$	
4	if $c \neq 0$, V's ZKPOK is successful, and t is valid: send k	$\rightarrow$	

Fig. 1. A efficient-prover non-parallelizable zero-knowledge proof for L

The described protocol is a zero-knowledge proof for the language L.

Efficient-Prover Interactive Proof. The fact that this protocol is an interactive proof follows directly from the fact that the subprotocol is (by assumption) a proof of knowledge. Completeness and soundness follow from completeness and extraction properties of the ZKPOK that P conducts in Step 2 or Step 3 respectively. Prover and verifier efficiency likewise follow from the respective properties of the ZKPOK subprotocol.

Zero Knowledge. Given any verifier strategy V^* we can construct a simulator M_{V^*}. M_{V^*} begins by randomly choosing and fixing the coin tosses of the verifier V^*, and then runs the verifier V^* in order to obtain its first message c. If $c = 0$, M_{V^*} then emulates the simulator for the ZKPOK to simulate Step 2. It then does nothing for Step 3. If $c \neq 0$, then M_{V^*} simulates the ZKPOK in Step 2 by following the correct "verifier" protocol and running V^* in order to simulate the "prover" half of the protocol. M_{V^*} then simulates Step 3 using the simulator for the subprotocol. The expected time of all of these steps is polynomial; this follows directly from the running time of the simulators provided by the various subprotocols.

Finally, the simulator proceeds to Step 4. If $c = 0$ then there is no message sent in Step 4. If $c \neq 0$ and the ZKPOK in Step 2 was unsuccessful, then there is again no message sent in Step 4. If $c \neq 0$ and the proof in Step 2 was successful, then M_{V^*} runs the following two extraction techniques in parallel, halting when one succeeds: First, it attempts to extract some (j, r_j) consistent with t by employing the extractor K using V^*'s strategy from Step 2 as an "oracle." Second it attempts to learn some witness (j, r_j) by trying each of the 2^m possible witnesses in sequence. If M_{V^*} has successfully found a witness, it uses (j, r_j) together with the transcript t to determine whether t is valid and then to determine the common key k by emulating the actions of one party and responding to the "messages" from the other party as described in the transcript t. This key k is then used to simulate Step 4.

The indistinguishability and expected polynomial running time of the simulation follow from those of the ZKPOK simulator, except for the simulation of Step 4 in the case $c \neq 0$. To analyze this, let p be the probability that V^* succeeds in the ZKPOK in Step 2. If $p > 2 \cdot 2^{-m}$, then there exists such an extractor K that extracts a witness (j, r_j) in expected time $q(|x|)/(p - s(|x|)$. Since this occurs with probability p, the expected time for this case is bounded by $(p \cdot q(|x|))/(p - s(|x|)) \leq (p \cdot q(|x|))/(p - 2^{-m}) \leq (p \cdot q(|x|))/(p/2) \leq 2q(|x|) = \text{poly}(|x|)$. If $p \leq 2 \cdot 2^m$ then the brute force technique will find a witness in expected time $p \cdot 2^m \leq 2 = \text{poly}(|x|)$. Checking t's validity takes polynomial time by assumption, and determining k takes time $\Theta(|x|)$, therefore the entire simulation runs in expected polynomial time.

The indistinguishability of the final step of this simulation relies on the fact that the transcript t is valid. Therefore, by the correctness of the key agreement protocol, the same key will be computed using the extracted witness (j, r_j) as with the prover's witness (i, r_i) even if they are not the same, so the simulation is polynomially indistinguishable from V^*'s view of the interactive protocol.

Parallel Execution. Consider now two executions, $(\widetilde{P}_1, \widetilde{V})$ and $(\widetilde{P}_2, \widetilde{V})$ in parallel. A cheating verifier V^* can always extract some witness $w \in \{(1, r_1), (2, r_2)\}$ from $\widetilde{P}_1$ and $\widetilde{P}_2$ using the following strategy: in Step 1, V^* sends $c = 0$ to $\widetilde{P}_1$ and $c = 1$ to $\widetilde{P}_2$. Now V^* has to execute the protocol (P, V) twice: once as a verifier talking to the prover $\widetilde{P}_1$, and once as a prover talking to the verifier $\widetilde{P}_2$. This he does by serving as an intermediary between $\widetilde{P}_1$ and $\widetilde{P}_2$, sending $\widetilde{P}_1$'s messages to $\widetilde{P}_2$, and $\widetilde{P}_2$'s messages to $\widetilde{P}_1$. Now $\widetilde{P}_2$ willfully sends k to $\widetilde{V}$ (which, by the secrecy property of the key agreement protocol, $\widetilde{V}$ is incapable of computing on his own). □

5 Conclusions and Open Problems

We view our results as pointing out the significance of prover efficiency, as well as the power of the distinguishers, in the composability of zero-knowledge proofs. Indeed, we have shown that with prover efficiency, the original GMR definition enjoys a greater level of composability than without. Nevertheless, the now-standard notion of auxiliary input zero knowledge still seems to be the appropriate one for most purposes. In particular, we still do not know whether plain zero knowledge is closed under a super-constant number of compositions. We also have not considered the case that different statements are being proven in each of the copies, much less (sequential) composition with arbitrary protocols. For these, it seems likely that auxiliary input zero knowledge, or something similar, is necessary.

One way in which our negative result on sequential composition (of plain zero knowledge with respect to V-uniform distinguishers, Theorem 3.4) can be improved is to provide an example where the prover's auxiliary inputs are defined by a relation that can be decided in polynomial time (in contrast to our construction, where the prover's auxiliary input contains the advice string π_{4n}, which may be hard to recognize).

For the parallel composition of auxiliary-input zero knowledge with efficient provers, it remains open to determine whether a negative result can be proven under a more general assumption such as the existence of one-way functions. The methods of Feige and Shamir [9] (Theorem 4.3) can be generalized to replace the assumption $\mathcal{UP} \not\subseteq \mathcal{BPP}$ with the assumption that there is a a problem in $\mathcal{NP}$ for which the witnesses have a "uniquely determined feature" [22] that is hard to compute. That is, there is a poly-balanced, poly-time relation R, an efficiently computable f, and a function g such that (a) if $(x, w) \in R$, then $f(x, w) = g(x)$, and (b) there is no probabilistic polynomial-time algorithm that computes $g(x)$ correctly for all $x \in L_R$. (The assumption that $\mathcal{UP} \not\subseteq \mathcal{BPP}$ corresponds to the case that $f(x, w) = w$. In general, we allow the witnesses for x to have a "unique part," namely $g(x)$, which is still hard to compute.) Our result (Theorem 4.6) can be viewed as constructing such an R, f, and g from a key agreement protocol.

Our construction complements that of Haitner, Rosen, and Shaltiel [22] — they consider the parallel repetition of natural zero-knowledge proofs (such as 3-Coloring [18] or Hamiltonicity [7]), and argue that "certain black-box techniques" cannot prove that a feature $g(x)$ will remain hard to compute by the verifier (on average). In contrast, we consider the parallel repetition of a contrived zero-knowledge proof and show that a cheating verifier can always learn a certain hard-to-compute feature $g(x)$.

Acknowledgments

We thank the TCC 2010 reviewers for helpful comments.

References

1. Barak, B.: How to go beyond the Black-Box Simulation Barrier. In: 42nd IEEE Symposium on Foundations of Computer Science, pp. 106–115 (2001)
2. Barak, B., Lindell, Y., Vadhan, S.: Lower Bounds for Non-Black-Box Zero Knowledge. In: Proc. of the 44th IEEE Symposium on the Foundation of Computer Science, pp. 384–393 (2003)
3. Bellare, M., Goldreich, O.: On defining proofs of knowledge. In: Brickell, E.F. (ed.) CRYPTO 1992. LNCS, vol. 740, pp. 390–420. Springer, Heidelberg (1993)
4. Ben-Or, M., Goldreich, O., Goldwasser, S., Hastad, J., Kilian, J., Micali, S., Rogaway, P.: Everything provable is provable in zero-knowledge. In: Goldwasser, S. (ed.) CRYPTO 1988. LNCS, vol. 403, pp. 37–56. Springer, Heidelberg (1990)

5. Birrell, E.: Composition of Zero-Knowledge Proofs. Undergraduate Thesis. Harvard University (2009)
6. Birrell, E., Vadhan, S.: Composition of Zero Knowledge Proofs with Efficient Provers. Cryptology eprint archive (2009)
7. Blum, M.: How to prove a theorem so no one else can claim it. In: Proceedings of the International Congress of Mathematicians, pp. 1444–1451 (1987)
8. Diffie, W., Hellman, M.: New Directions in Cryptography. IEEE Trans. on Info. Theory IT-22, 644–654 (1976)
9. Feige, U., Shamir, A.: Witness Indistinguishability and Witness Hiding Protocols. In: 22nd ACM Symposium on the Theory of Computing, pp. 416–426 (1990)
10. Feige, U., Shamir, A.: Zero-Knowledge Proofs of Knowledge in Two Rounds. In: Brassard, G. (ed.) CRYPTO 1989. LNCS, vol. 435, pp. 526–544. Springer, Heidelberg (1990)
11. Goldreich, O.: Foundations of Cryptography - Basic Tools. Cambridge University Press, Cambridge (2001)
12. Goldreich, O.: A Uniform Complexity Treatment of Encryption and Zero Knowledge. Journal of Cryptology 6(1), 21–53 (1993)
13. Goldreich, O.: Zero-Knowledge twenty years after its invention. Cryptology ePrint Archive, Report 2002/186 (2002), http://eprint.iacr.org/
14. Goldreich, O., Goldwasser, S., Micali, S.: How to Construct Random Functions. Journal of the Association for Computing Machinery 33(4), 792–807 (1986)
15. Goldreich, O., Kahan, A.: How to Construct Constant-Round Zero-Knowledge Proof Systems for NP. Journal of Cryptology 9(2), 167–189 (1996)
16. Goldreich, O., Krawczyk, H.: On the Composition of Zero-Knowledge Proof Systems. SIAM Journal on Computing 25(1), 169–192 (1996); Preliminary version in ICALP 1990
17. Goldreich, O., Krawczyk, H.: Sparse Pseudorandom Distributions. Random Structures & Algorithms 3(2), 163–174 (1992)
18. Goldreich, O., Micali, S., Wigderson, A.: Proofs that Yield Nothing but their Validity or All Languages in NP have Zero-Knowledge Proof Systems. Journal of the ACM 38(1), 691–729 (1991)
19. Goldreich, O., Oren, Y.: Definitions and Properties of Zero-Knowledge Proof Systems. Journal of Cryptology 7(1), 1–32 (1994)
20. Goldwasser, S., Micali, S., Rackoff, C.: Knowledge Complexity of Interactive Proofs. In: Proc. 17th STOC, pp. 291–304 (1985)
21. Goldwasser, S., Micali, S., Rackoff, C.: The Knowledge Complexity of Interactive Proof Systems. SIAM Journal on Computing 18, 186–208 (1989)
22. Haitner, I., Rosen, A., Shaltiel, R.: On the (Im)possibility of Arthur-Merlin Witness Hiding Protocols. In: Reingold, O. (ed.) TCC 2009. LNCS, vol. 5444, pp. 220–237. Springer, Heidelberg (2009)
23. Vadhan, S.: Pseudorandomness. Foundations and Trends in Theoretical Computer Science (to appear, 2010)

Private Coins versus Public Coins in Zero-Knowledge Proof Systems

Rafael Pass* and Muthuramakrishnan Venkitasubramaniam

Cornell University
{rafael,vmuthu}@cs.cornell.edu

Abstract. Goldreich-Krawczyk (Siam J of Comp'96) showed that only languages in BPP have constant-round *public-coin* black-box zero-knowledge protocols. We extend their lower bound to "fully black-box" *private-coin* protocols based on one-way functions. More precisely, we show that only languages in $\mathsf{BPP}^{\mathsf{Sam}}$—where Sam is a "collision-finding" oracle in analogy with Simon (Eurocrypt'98) and Haitner et. al (FOCS'07)—can have constant-round fully black-box zero-knowledge proofs; the same holds for constant-round fully black-box zero-knowledge *arguments* with sublinear verifier communication complexity. We also establish near-linear lower bounds on the round complexity of fully black-box concurrent zero-knowledge proofs (or arguments with sublinear verifier communication) for languages outside $\mathsf{BPP}^{\mathsf{Sam}}$.

The technique used to establish these results is a transformation from private-coin protocols into Sam-relativized public-coin protocols; for the case of fully black-box protocols based on one-way functions, this transformation preserves zero knowledge, round complexity and communication complexity.

1 Introduction

Roughly speaking, interactive proofs, introduced by Goldwasser, Micali and Rackoff [9] and Babai and Moran [1]), are protocols that allow one party P—called the *Prover* (or Merlin)—to convince a computationally-bounded party V—called the *Verifier* (or Arthur)—of the validity of some statement $x \in L$. While, the notion of interactive proofs introduced by Goldwasser, Micali and Rackoff considers arbitrary probability polynomial time verifiers, the notion introduced by Babai and Moran, called *Arthur-Merlin games* considers verifiers that only send truly random messages; such proof systems are also called *public coin.* Soon after their introduction, a surprisingly result by Goldwasser and Sipser [11] showed that the two notions in fact are equivalent in their expressive power: Any private coin protocol $\langle P, V\rangle$ for a language L can be transformed into a public-coin $\langle \widehat{P}, \widehat{V}\rangle$ for L with the same round-complexity. Their result has played an important role in subsequent complexity-theoretic work. However,

* Supported in part by a Microsoft New Faculty Fellowship, NSF CAREER Award CCF-0746990, AFOSR Award FA9550-08-1-0197 and BSF Grant 2006317.

D. Micciancio (Ed.): TCC 2010, LNCS 5978, pp. 588–605, 2010.

from a cryptographic perspective, the transformation is somewhat unsatisfactory as it does not preserve the efficiency of the prover—and can thus not be applied to "computationally-sound" protocols (a.k.a. *arguments*)—or properties such as *zero-knowledge*—the principal notion introduced in [9]. By a result of Vadhan [26], any transformation that uses the original private-coin protocol $\langle P, V\rangle$ as a black-box, in fact, must require the prover to run in super-polynomial time.

In this work, we provide different and "robust" transformations from private-coin protocols to public-coin protocols. Our transformations preserve zero-knowledge, computational and communication complexity, but instead require the prover and the verifier to have oracle access to a certain "collision-finding" oracle [25,13], denoted Sam. Our transformation is black-box and thus by Vadhan's results we are required to use a super-polynomial time oracle. Nevertheless, the Sam oracle is not "too" powerful; in particular, as shown by Haitner, Hoch, Reingold and Segev [13] it cannot be used to invert *one-way functions*. Therefore, if the security properties (namely, zero-knowledge and computational soundness) of the private-coin protocol are based on the hardness of inverting one-way functions (or even *trapdoor permutations*), we can use our transformation to extended lower bounds for public-coin protocols to private-coin protocols.

More precisely, Goldreich and Krawczyk [8] showed that only languages in BPP can have constant-round public-coin *black-box* zero-knowledge protocols. Recently, Pass, Tseng and Wikström [21] extended this results to include all (even super-constant round) black-box zero-knowledge protocols that remain secure under concurrent (or even parallel) composition (a.k.a *concurrent zero-knowledge* protocols). Combining our transformation with these results, we obtain new lower bounds for *fully black-box constructions* of general, potentially private-coin, black-box zero-knowledge protocols based on the existence of one-way permutations.

Theorem 1 (Lower Bounds for Fully Black-Box Zero Knowledge—Informally stated). *Let $\langle P, V\rangle$ be a fully black-box construction of a zero-knowledge proof (or argument) for the language L from one-way permutations. Then, $L \in \mathsf{BPP}^{\mathsf{Sam}}$ if any of the following hold:*

1. *$\langle P, V\rangle$ is an $O(1)$-round proof.*
2. *$\langle P, V\rangle$ is an $O(1)$-round argument with $o(n)$ verifier communication complexity.*
3. *$\langle P, V\rangle$ is an $o(\frac{n}{\log n})$-round concurrent zero-knowledge proof.*
4. *$\langle P, V\rangle$ is an $o(\frac{n}{\log n})$-round concurrent zero-knowledge argument with $o(n)$ verifier communication complexity.*

We remark that all the above type of protocols can be achieved for languages in NP, assuming the existence of *collision-resistant hash-functions* [7,17,20]. Assuming only one-way permutations, however, the best zero-knowledge proofs require a super-constant number of rounds [10], and $O(n/\log n)$-rounds for concurrent zero-knowledge [22]. As such, assuming $\mathsf{NP} \not\subseteq \mathsf{BPP}^{\mathsf{Sam}}$, Theorem 1 is tight.

In Section 3, we discuss the complexity of $\mathsf{BPP}^{\mathsf{Sam}}$. We observe that the class SZK, of languages having *statistical* zero-knowledge proofs, is contained in $\mathsf{BPP}^{\mathsf{Sam}}$. This should not be surprising as Ong and Vadhan provide unconditional constructions of constant-round black-box zero-knowledge proofs for languages in SZK [19]. By extending the result of [13] we also observe that $\mathsf{BPP}^{\mathsf{Sam}}$ does not "generically" decide all NP languages, and seems thus like an interesting and natural complexity class in its own right.

We finally mention that the techniques used in our transformation are interesting in their own right. First, it directly follows that that there is no fully black-box construction of a one-way function, that compresses its input by more than a constant factor, from one-way permutations. Next, as pointed out to us by Haitner, it would seem that by our techniques, the black-box lower bounds from [14] can be extended also to honest-but-curious protocols; see the proof of Lemma 2 for more details.

2 Preliminaries and Definitions

We assume familiarity with the basic notions of an Interactive Turing Machine (ITM for brevity) and a protocol (in essence a pair of ITMs). We denote by $\mathcal{PPT}$ the class of probabilistic polynomial time Turing machines and $n.u.\mathcal{PPT}$, the class of non-uniform $\mathcal{PPT}$ machines. We denote by $M^\bullet$ an oracle machine; we sometimes drop $\bullet$ when it is clear from the context. As usual, if $M^\bullet$ is an oracle machine, M^O denotes the joint execution of M with oracle access to O. Let $\mathcal{O}$ be a random variable over functions from $\{0,1\}^* \to \{0,1\}^*$. Then, $M^{\mathcal{O}}$ denotes the execution of M^O, where O is sampled according to $\mathcal{O}$. Let Π_n denote the set of all permutations on $\{0,1\}^n$ and Π denote the set of all permutations $\{0,1\}^* \to \{0,1\}^*$ (obtained by choosing a π_n from Π_n for every n).

2.1 Fully Black-Box Constructions

A construction of a cryptographic primitive p from a primitive q is said to be *fully black-box* if both the implementation and the proof of correctness are black-box. (See [24] for more details on black-box constructions and reductions.) Here, we focus on fully black-box constructions from one-way permutations. For simplicity, we show our results only for one-way permutations, but analogous to [13], our results extend to trapdoor permutations as well. We proceed to define fully black-box constructions of arguments and zero-knowledge.

Definition 1 (Fully black-box interactive arguments). *Let $\langle P^\bullet, V^\bullet\rangle$ be an interactive argument for a language $L \subseteq \{0,1\}^*$. We say that $\langle P^\bullet, V^\bullet\rangle$ is a fully black-box construction from one-way permutations, if there exists a $\mathcal{PPT}$ machine $A^\bullet$, and a polynomial $q(\cdot)$ such that for every permutation $\pi = \{\pi_n\}_{n=1}^\infty$, malicious prover $P^{*\bullet}$, sequence $\{x_n\}_{n=1}^\infty$ where $x_n \in \bar{L} \cap \{0,1\}^n$ and polynomial $p(\cdot)$, if $\Pr[\langle P^\pi, V^\pi\rangle(x_n) = 1] \geq \frac{1}{p(n)}$ for infinitely many n, then*

$$Pr[A^{\pi, P^{*\pi}(x_n)}(1^n, y) = \pi_n^{-1}(y)] > \frac{1}{q((p(n))}$$

for infinitely many n, where the probability is taken uniformly over $y \in \{0,1\}^n$ (and over all the internal coin tosses of A).

Definition 2 (Fully black-box computational zero-knowledge). *Let $\langle P^\bullet, V^\bullet \rangle$ be an interactive proof (or argument) system for a language L. We say that $\langle P^\bullet, V^\bullet \rangle$ is a* fully black-box *construction of a computational zero-knowledge proof (or argument) from one-way permutations, if there exists an expected $\mathcal{PPT}$ simulator $S^\bullet$, a $\mathcal{PPT}$ machine $A^\bullet$, and a polynomial $q(\cdot)$ such that for every permutation $\pi = \{\pi_n\}_{n=1}^\infty$, a distinguisher D, malicious verifier $V^{*\bullet}$, sequence $\{(x_n, z_n)\}_{n=1}^\infty$ where $x_n \in L \cap \{0,1\}^n$, $z_n \in \{0,1\}^*$ and polynomial $p(\cdot)$, if for infinitely many n, D distinguishes $\left\{S^{\pi, V_r^{*\pi}(x_n,z_n)}(x_n)\right\}$ and $\left\{\langle P^\pi, V_r^{*\pi}(z_n)\rangle(x_n)\right\}$ with probability at least $\frac{1}{p(n)}$ where $\langle P^\pi, V_r^{*\pi}(z)\rangle(x)$ denotes the output of $V^{*\pi}$ in an interaction between P^π and $V_r^{*\pi}(z)$ on common input x, then*

$$Pr[A^{\pi, V_r^{*\pi}(x_n,z_n)}(1^n, y) = \pi_n^{-1}(y)] > \frac{1}{q(p((n))}$$

for infinitely many values of n, where the probability is taken uniformly over $y \in \{0,1\}^n$ (and over all the internal coin tosses of A).

Remark 1. Note that in Definition 2, the simulator S unconditionally runs in expected polynomial time. One can consider a weaker definition where A is required to invert π when the expected running time of S exceeds polynomial time. For simplicity (and due to the fact that all known black-box zero-knowledge proofs satisfy this property), we consider the stronger definition, but our results extend also to the weaker definition.

A fully black-box construction of a computational zero-knowledge arguments refers to a construction that is a fully black-box construction in the argument sense and the zero-knowledge sense.

3 The Collision Finding Class

Our transformation makes use of a "collision finding" oracle. Such an oracle was introduced by Simon [25]. In this work, we require a slightly stronger oracle that finds "collisions" in interactive protocols. Such an oracle—referred to as Sam—was recently introduced by Haitner, Hoch, Reingold and Segev [13]. The oracle comes with a permutation π and a parameter d; the depth parameter d denotes the number of rounds in the protocol on which it finds collisions. We denote the oracle by Sam_d^π. Below, we recall the Sam oracle from [13].

3.1 The Oracle Sam_d^π

Informally, Sam_d^π is an oracle, that takes as input a probabilistic interactive turing machine (ITM) $M^\bullet$ and a partial transcript *trans* of an interaction with M of d or fewer rounds, and

– If *trans* was an output of a previous query, Sam samples a random tape τ for M^π among all random tapes that are consistent with *trans*, and generates M^π's next message m using τ and outputs $trans :: m$.
– Otherwise, outputs $\perp$.

Description of $\mathsf{Sam}^\pi_{d(n)}$. Let $\pi = \{\pi_n\}_{n=1}^\infty$ be a permutation and $M^\bullet$ be a probabilistic oracle ITM that runs a d-round protocol and has access to π. Let $trans_i = (a_1, b_1, \ldots, a_i, b_i)$ be a partial transcript of the messages exchange with M^π in an execution; Define $R_{trans_i}(M^\pi)$ to be the set of all random tapes τ for which $M^\pi_\tau(a_1, b_1, \ldots, b_{j-1}) = a_j$ for all $j < i$; we say that such a τ is *consistent* w.r.t $trans_i$. Without loss of generality, we assume that M^π sends the first message (i.e. outputs a message on initiation). An input query for $\mathsf{Sam}^\pi_{d(n)}$ is of the form $Q = (M^\pi, trans_i, r)$ where $trans_{i-1} = (a_1, b_1, \ldots, b_{i-1})$ and $r \in \{0,1\}^*$. It outputs $(\tau', trans_{i-1} :: a_i)$ such that $\tau' \in R_{trans_j}(M^\pi)$ and $M^\pi_{\tau'}(trans_i) = a_i$, with the following restrictions:

1. If $i > 1$, then $(a_1, b_1, \ldots, a_{i-1})$ was the result of a previous query of the form $(M^\pi, (a_1, b_1, \ldots, b_{i-2}), r')$ for some $r' \in \{0,1\}^*$.
2. τ' is uniformly distributed in $R_{trans_{i-1}}(M^\pi)$ over the randomness of $\mathsf{Sam}^\pi_{d(n)}$, independent of all other queries.
3. $\mathsf{Sam}^\pi_{d(n)}$ answers queries only up to a depth $d(n)$, i.e. $i \leq d(n)$.

Otherwise, it outputs $\perp$. We remark that the role of r in the query is to obtain new and independent samples for each r and allow a verifier to obtain the same sample query by querying on the same r.

Our above description of the $\mathsf{Sam}^\pi_{d(n)}$-oracle is a stateful instantiation of the oracle defined in [13]. Just as in [13], for our results, we need the oracle to be stateless; [13] specify how to modify the oracle to achieve this (using "signatures"); we omit the details. It was shown by Haitner et. al that random permutations are hard to invert for polynomial time machines that query Sam oracle upto depth $o(\frac{n}{\log n})$.

Theorem 2 ([13]). *For every $\mathcal{PPT}$ machine $A^\bullet$, there exists a negligible function $\nu(\cdot)$, such that, for all n,* $\Pr[A^{\pi, \mathsf{Sam}^\pi_{o(\frac{n}{\log n})}}(y) = \pi_n^{-1}(y)] \leq \nu(n)$ *where the probability is taken uniformly over the randomness of $\mathsf{Sam}^\pi_{o(\frac{n}{\log n})}$, random permutation $\pi = \{\pi_n\}_{n=1}^\infty$ and $y \in \{0,1\}^n$.*

Looking ahead, in Section 4.2, we show that this result is optimal w.r.t the depth: $\mathsf{Sam}^\pi_{\frac{n}{\log n}}$ can be used to invert π.

3.2 The Complexity Class CF_d

We introduce a new complexity classes CF_d, which we call the "collision-finding class", that we use as part of our characterization of zero-knowledge protocols.

Definition 3. *A language $L \in \mathsf{CF}_d = \mathsf{BPP}^{\mathsf{Sam}_d}$, if there exists a $\mathcal{PPT}$ machine $M^\bullet$ such that:*

Completeness: *If* $x \in L$, $M^{\pi,\mathsf{Sam}_d^{\pi}}$ *outputs 1 with probability at least* $\frac{2}{3}$.
Soundness: *If* $x \notin L$, $M^{\pi,\mathsf{Sam}_d^{\pi}}$ *outputs 1 with probability at most* $\frac{1}{3}$.

where both the probabilities are taken uniformly over the random coins of M, *the randomness of* Sam *and a random permutation* π.

The complexity class CF_d seems to be interesting classes that lies between P and NP. Below we state some properties about this class. The formal proofs of these statements are postponed to the full version.

1. For every d, CF_d is closed under complement, (follows from the definition).
2. $\mathsf{SZK} \subseteq \mathsf{CF}_1$.
3. $\mathsf{CF}_{o(\frac{n}{\log n})}$ does not "generically" solve NP.
4. $\mathsf{CF}_{\frac{l(n)}{\log n}}$ can invert any one-way function with output length $l(n)$ on length n inputs (Theorem 3 in Section 4.2)

We leave a fuller exploration of the collision-finding class for future work. Note that, by property (4), if $\mathsf{NP} \not\subseteq \mathsf{CF}_{poly(n)}$, we have a "natural" complexity class that can inverts all one-way functions but not decide NP.

We mention that a somewhat weaker (and perhaps even more natural) definition of the collision-finding class—let us denote it $\mathsf{CF'}_d$—is defined identically, but without giving M, or Sam, access to a random permutation π. That is, in our notation $\mathsf{CF'}_d = \mathsf{BPP}^{\mathsf{Sam}_d^{\perp}}$, where $\perp$ is the all zero oracle. Clearly $\mathsf{CF'}_d \subseteq \mathsf{CF}_d$, but all the properties above continue to hold also for CF'.

A very recent work by Haitner, Mahmoody-Ghidary and Xiao [16] takes a step towards showing that $\mathsf{CF'}_{O(1)}$ does not contain NP; they show that if the deciding machine M only makes a constant number of adaptive queries to Sam, then the language it decides is in coAM.

4 From Private Coins to Public Coins

In this section, we provide our transformation from private-coin to public-coin protocols. We provide two transformations: The first transformation—or *weak duality*—converts any private coin zero-knowledge proof into a public-coin zero-knowledge proof in the Sam-hybrid model, where the prover, verifier and simulator have oracle access to Sam. The second transformation—or *strong duality*—converts any private coin zero-knowledge argument with sublinear verifier communication complexity into a public coin zero-knowledge argument in the Sam-hybrid model. While the first transformation is *oracle efficient* (the maximum depth it queries Sam is "small"), the second transformation is *computationally efficient* (the soundness reduction is polynomial-time) and thus can be applied to arguments.

Our transformations consider zero-knowledge proofs and arguments in an oracle world. Let $\mathcal{O}$ be a set of oracles $O : \{0,1\}^* \to \{0,1\}^*$.

Definition 4 ($\mathcal{O}$-relativized Interactive Proofs). *A pair of interactive machines* $\langle P^{\bullet}, V^{\bullet}\rangle$ *is called an* $\mathcal{O}$*-relativized* interactive proof system *for a language* L *if machine* $V^{\bullet}$ *is polynomial-time and the following two conditions hold :*

– Completeness: *There is a negligible function* $\nu(\cdot)$, *such that for every* n, $x \in L \cap \{0,1\}^n$,

$$\Pr\left[\langle P^{\mathcal{O}}, V^{\mathcal{O}}\rangle(x) = 1\right] \geq 1 - \nu(n)$$

where the probability is taken over all the internal coin tosses of P, V *and uniformly chosen* $\mathcal{O} \in \mathcal{O}$.
– Soundness: *For every machine* $B^{\bullet}$, *there exists a negligible function* $\nu(\cdot)$, *such that, for every* $x \in \overline{L} \cap \{0,1\}^n$,

$$\Pr\left[\langle B^{\mathcal{O}}, V^{\mathcal{O}}\rangle(x) = 1\right] \leq \nu(n)$$

where the probability is taken over all the internal coin tosses of V *and uniformly chosen* $\mathcal{O} \in \mathcal{O}$.

If the soundness holds only against n.u.PPT B, *then* $\langle P, V\rangle$ *is called an* $\mathcal{O}$-*relativized* interactive argument system.

Definition 5 ($\mathcal{O}$-relativized black-box $\mathcal{ZK}$). *Let* $\langle P^{\bullet}, V^{\bullet}\rangle$ *be an* $\mathcal{O}$*-relativized interactive proof (argument) system for the language* $L \in \mathcal{NP}$ *with the witness relation* R_L. *We say that* $\langle P^{\bullet}, V^{*\bullet}\rangle$ *is* $\mathcal{O}$-relativized computational black-box $\mathcal{ZK}$, *if there exists a probabilistic expected polynomial time oracle machine* $S^{\bullet}$ *such that for every* $\mathcal{PPT}$ *machine* $V^{*\bullet}$, *and* $\mathcal{PPT}$ *distinguisher* $D^{\bullet}$, *there exists a negligible function* $\nu(\cdot)$, *such that for all* n, $x \in L \cap \{0,1\}^n$, $z \in \{0,1\}^*$,

$$\left|\Pr[D^{\mathcal{O}}(S^{\mathcal{O},V^{\mathcal{O}}(x)}(x)) = 1] - \Pr[D^{\mathcal{O}}(\langle P^{\mathcal{O}}, V^{*\mathcal{O}}(z)\rangle(x)) = 1]\right| < \nu(n)$$

4.1 Weak Duality Lemma

Lemma 1 (Weak Duality). *Let* $\langle P^{\bullet}, V^{\bullet}\rangle$ *be a* d*-round fully black-box zero-knowledge proof for a language* L *from one-way permutations with verifier communication complexity* $c(n)$ *and prover communication complexity* $p(n)$. *Then, there exists a* d*-round public-coin protocol* $\langle \widehat{P}^{\mathsf{Sam}_d^{\bullet}}, \widehat{V}^{\mathsf{Sam}_d^{\bullet}}\rangle$ *with the verifier communication complexity* $O(dc(n))$ *and prover communication complexity* $p(n)$ *that is* $(\pi, \mathsf{Sam}_d^{\pi})$*-relativized black-box zero-knowledge proof.*

Proof. $\widehat{V}^{\mathsf{Sam}_d^{\pi}}$ is a d-round public-coin verifier that sends random coins in each round. On a high-level, $\widehat{P}^{\mathsf{Sam}_d^{\pi}}$ is a machine that internally incorporates the code of P^{π} and emulates an interaction with P^{π} by supplying verifier messages according to the $\langle P^{\pi}, V^{\pi}\rangle$ protocol. For every verifier round in the internal emulation, $\widehat{P}^{\mathsf{Sam}_d^{\pi}}$ first receives random coins externally from $\widehat{V}^{\mathsf{Sam}_d^{\pi}}$. Using that, it samples a random message q for V^{π} that is "consistent" with the interaction with P^{π}; this is made possible using Sam_d^{π}. Next, it q feeds internally to P^{π}. Upon receiving a message a from P^{π}, $\widehat{P}^{\mathsf{Sam}_d^{\pi}}$ forwards a to $\widehat{V}^{\mathsf{Sam}_d^{\pi}}$ and proceeds to the next round. Finally, $\widehat{V}^{\mathsf{Sam}_d^{\pi}}$ reconstructs the interaction emulated by P^{π} (again made possible using Sam_d^{π}) and outputs the verdict of V^{π} on that transcript. We remark that since V^{π} is a verifier for a d-round protocol, the maximum depth of a Sam_d^{π} query made by $\widehat{P}^{\mathsf{Sam}_d^{\pi}}$ and $\widehat{V}^{\mathsf{Sam}_d^{\pi}}$ is d.

PROTOCOL $\langle\widehat{P},\widehat{V}\rangle$

Let $\langle P^{\pi}, V^{\pi}\rangle$ be a d-round protocol with oracle access to the a permutations π. Each communication round consists of a message sent from the verifier to the prover followed by a message sent from the prover to the verifier. Without loss of generality, we assume that the verifier sends the first message and the prover sends the last message. Also, the verifier outputs its view at the end of the protocol.

1. **Common Input:** Statement $x \in L$, security parameter n.
2. **Private Input:** The statement x, for P^{π} and auxiliary input $z \in \{0,1\}^*$ for V^{π}.
3. $\widehat{P}^{\mathsf{Sam}_d^{\pi}}$ internally incorporates the code for P^{π}. Set $trans_0 = \perp$.
4. **for** $i = 1$ to d
   (a) $\widehat{V}^{\mathsf{Sam}_d^{\pi}}$ uniformly chooses $s_i \in \{0,1\}^{12(l_i+\log d)}$ where l_i is the length of V^{π}'s i^{th} message.
   (b) $\widehat{V} \to \widehat{P} : s_i$
   (c) $\widehat{P}$ queries Sam_d^{π} on input $(V^{\bullet}, trans_{i-1}, s_i)$ and obtains as response $(trans_{i-1} :: q_i, r_i)$. $\widehat{P}$ runs $P^{\pi}(trans_{i-1} :: q_i)$ and obtains its response a_i. Set $trans_i = trans_{i-1} :: q_i :: a_i$.
   (d) $\widehat{P} \to \widehat{V} : a_i$.
5. $\widehat{V}^{\mathsf{Sam}_d^{\pi}}$ computes $trans_i$ for all i, by querying Sam_d^{π} on $(V^{\bullet}, trans_{i-1}, s_i)$. $\widehat{V}^{\mathsf{Sam}_d^{\pi}}$ chooses $s \in \{0,1\}^n$, queries Sam_d^{π} on $(V^{\pi}, trans_d, s)$ and obtains as response (b, r_{d+1}), ($b = 1$ means V^{π} accepts). It **outputs** b.

Fig. 1. Weak Duality Protocol

Informally, the completeness of the protocol follows from the fact that, the internal emulation carried out by $\widehat{P}^{\mathsf{Sam}_d^{\pi}}$ proceeds exactly as an execution between P^{π} and the honest verifier V^{π}. The soundness and zero-knowledge of $\langle\widehat{P},\widehat{V}\rangle$, on the other hand, holds as the transformation essentially ensures that the messages from $\widehat{V}^{\mathsf{Sam}_d^{\pi}}$ carry the same amount of "knowledge" as messages from V^{π}. This is because, in each round, $\widehat{P}^{\mathsf{Sam}_d^{\pi}}$ samples a fresh random tape for V^{π} that is consistent with the partial conversation and obtains V^{π}'s next message by running V^{π} on that tape. Thus, the only extra knowledge that $\widehat{P}$ possesses in each round is the random tape sampled and (an unbounded) P^{π} can obtain these samples too. A formal description of the transformation is provided in Figure 1. We now proceed to prove correctness.

Claim 1 (Completeness). *For all $x \in L$, $\widehat{P}^{\mathsf{Sam}_d^{\pi}}$ convinces $\widehat{V}^{\mathsf{Sam}_d^{\pi}}$ w.p. $1 - \nu(|x|)$ where the randomness is taken over π, Sam_d^{π} and the internal coin tosses of $\widehat{P}$ and $\widehat{V}$, for some negligible function $\nu(\cdot)$.*

Proof. We show that, for every permutation π, the probability that $\widehat{V}^{\mathsf{Sam}_d^{\pi}}$ accepts is identical to the probability V^{π} accepts in an interaction with P (where the probability is over Sam_d^{π}). The completeness of $\langle\widehat{P},\widehat{V}\rangle$ then follows from the completeness of $\langle P^{\bullet}, V^{\bullet}\rangle$.

Towards this, fix a permutation π. Consider an intermediate verifier $V'^{\bullet}$ that uses Sam_d^{π} and interacts with P^{π}. Informally, this verifier $V'^{\mathsf{Sam}_d^{\pi}}$ generates messages exactly as $\widehat{P}^{\mathsf{Sam}_d^\pi}$ does in the internal emulation with P^π. More precisely, for a partial transcript $trans_{i-1}$ at the end of round $i-1$, $V'^{\mathsf{Sam}_d^\pi}$ samples a consistent random tape r for V^π (using Sam_d^π) and runs V_r^π on $trans_{i-1}$ to generate the next verifier message q_i. At the end of the protocol, $V'^{\mathsf{Sam}_d^\pi}$, samples a random tape r' consistent on the entire transcript and outputs $V_{r'}^\pi$'s verdict on the transcript. It follows from the construction that the probability that $\widehat{V}^{\mathsf{Sam}_d^\pi}$ accepts is equal to the probability that $V'^{\mathsf{Sam}_d^\pi}$ accepts in an interaction with P^π. In Claim 1 below, we prove that the probability that P^π convinces $V'^{\mathsf{Sam}_d^\pi}$ accepts is equal to the probability that P^π convinces V^π. Therefore, combining the two facts, we get that the probability $\widehat{V}^{\mathsf{Sam}_d^\pi}$ accepts is equal to the probability V^π accepts.

Sub-Claim 1. *For every $x, z \in \{0,1\}^*$, $\pi \in \Pi$ the following distributions are identical: $D_1 = \left\{\langle P^\pi, V^\pi(z)\rangle(x)\right\}$ and $D_2 = \left\{\langle P^\pi, V'^{\mathsf{Sam}_d^\pi}(z)\rangle(x)\right\}$ where the distributions are generated by the internal coin tosses of P, V, V' and Sam_d^π.*

Proof. Recall that the only difference between $V'^{\mathsf{Sam}_d^\pi}$ and V^π in an interaction with P^π is that V^π selects a uniform random tape at the beginning of the execution and uses that for the entire execution, while $V'^{\mathsf{Sam}_d^\pi}$ selects a (uniformly chosen) random tape consistent with the partial transcript in each round and executes V^π on that tape. First, we observe that every verifier message in D_1 and D_2 are generated by running V^π on a particular random tape. For a transcript $trans$, let $R(trans)$ denote the set of all random tapes of V^π consistent with $trans$. We show for D_1 and D_2, separately, that for every $trans$, conditioned on the history being $trans$, the random tape used to generate the next message is uniformly distributed in $R(trans)$. This shows that the process for generating verifier messages in D_1 and D_2 are identical and that concludes the proof of the claim. For D_2, this holds directly from the definition of Sam_d^π. For D_1, we prove this fact by induction on the number of verifier messages. The base case requires that the random tape is uniformly distributed over all possible tapes; this clearly holds. Suppose that, conditioned on the transcript $trans_{i-1}$, every random tape in $R_{trans_{i-1}}$ is equally likely. Let m be a possible message for V^π, given the history is $trans_{i-1}$. Since, $R(trans_{i-1} :: m)$ are disjoint sets for different m, we have that conditioned on the transcript $trans_{i-1} :: m$, every tape in $R(trans_{i-1} :: m)$ is equally likely to be chosen. This concludes the induction step. ∎

∎

Claim 2 (Soundness). *Let $x \in \{0,1\}^*$. If $\widehat{P}^{*\mathsf{Sam}_d^\pi}$ convinces $\widehat{V}^{\mathsf{Sam}_d^\pi}$ on x with probability p, then there exists a prover $P^{*\mathsf{Sam}_d^\pi}$ that convinces V^π on input x with probability at least $\frac{p}{2}$. (As usual, the probability are taken over Sam_d^π and the internal coin tosses of $\widehat{V}$ and V.)*

Proof. We prove the statement of the claim for every permutation π and over the randomness of Sam_d^π. We construct a machine $P^{*\mathsf{Sam}_d^\pi}$ that internally incorporates $\widehat{P}^{*\mathsf{Sam}_d^\pi}$ and emulates an interaction with it, while externally interacting

THE ALGORITHM FOR $P^{*\mathsf{Sam}_d^{\pi}}$

Let $\widehat{P}^{*\mathsf{Sam}_d^{\pi}}$ be the cheating prover for $\langle \widehat{P}, \widehat{V} \rangle$. Denote the length of ith verifier message in $\langle \widehat{P}, \widehat{V} \rangle$ by L_i. Then, $L_i = 12(l_i + \log d)$.

1. Internally incorporate $\widehat{P}^{\mathsf{Sam}_d^{\pi}}$. Let $trans_0 = \perp$.
2. **for** $i = 1$ to d
   (a) Receive q_i from V^{π}.
   (b) $ctr \leftarrow 0$, $found \leftarrow false$.
   (c) **while** $ctr < 2^{\frac{L_i}{3}}$ and $not(found)$.
      – Choose s_i uniformly from $\{0,1\}^{L_i}$. Let $(r_i, trans_i :: q) \leftarrow \mathsf{Sam}_d^{\pi}(V^{\bullet}, trans_{i-1}, s_i)$. If $q = q_i$, $found = true$.
      – $ctr \leftarrow ctr + 1$
   (d) **if** $found = false$, **abort**. Otherwise, compute $\widehat{P}^{*}$'s next message on transcript $trans_{i-1} :: q_i$. Let it be a_i. Set $trans_i = trans_{i-1} :: q_i :: a_i$.

Fig. 2. Proof of Soundness

with V^{π}. The high-level idea is to make $P^{*\mathsf{Sam}_d^{\pi}}$ convince V^{π} whenever $\widehat{P}^{*\mathsf{Sam}_d^{\pi}}$ succeeds in the internal execution. To ensure this, for every private-coin message q_i that $P^{*\mathsf{Sam}_d^{\pi}}$ receives externally from V^{π}, it needs to find a corresponding public-coin message s_i and feed it to $\widehat{P}^{*\mathsf{Sam}_d^{\pi}}$. Let $trans_{i-1}$ be the transcript of messages exchanged with V^{π} externally. Then, the message that $P^{*\mathsf{Sam}_d^{\pi}}$ needs to find, is a string s_i such that Sam_d^{π} on input $(V^{\bullet}, trans_{i-1}, s_i)$ outputs (q_i, r_i). We let $P^{*\mathsf{Sam}_d^{\pi}}$ sample s_i until it hits the "right" one; it cuts itself off, if it runs "too long". It then feeds it to $\widehat{P}^{*\mathsf{Sam}_d^{\pi}}$ internally and obtains a response a_i, which it forwards outside to V^{π}. A formal description is provided in Figure 2.

In Claim 2, we show that $P^{*\mathsf{Sam}_d^{\pi}}$ aborts with probability at most $\frac{1}{2}$. In Claim 3, we show that conditioned on $P^{*\mathsf{Sam}_d^{\pi}}$ not aborting, each verifier message fed internally to $\widehat{P}^{*\mathsf{Sam}_d^{\pi}}$ is uniformly distributed and thus identical to distribution of the messages received by $\widehat{P}^{*\mathsf{Sam}_d^{\pi}}$ in a real interaction with $\widehat{V}^{\mathsf{Sam}_d^{\pi}}$. Combining the two claims, we have that the probability that $P^{*\mathsf{Sam}_d^{\pi}}$ succeeds is at least $\frac{1}{2}\Pr[\widehat{P}^{*\mathsf{Sam}_d^{\pi}}$ succeeds$]$.

Sub-Claim 2. *$P^{*\mathsf{Sam}_d^{\pi}}$ aborts with probability at most $\frac{1}{2}$ (with probability taken over Sam_d^{π}).*

Proof. We analyze the abort probability by identifying three bad events for each round and bound their probabilities separately. Then, using an union bound over the bad events for each round, we conclude that the probability of aborting is at most $\frac{1}{2}$. Let $trans_{i-1}$ be the partial transcript at the end of $i-1$ rounds. Consider the following events:

1. **$P^{*\mathsf{Sam}_d^{\pi}}$ picks the same sample s_i twice:** The probability that two strings in $2^{\frac{L_i}{3}}$ trials are the same is at most $\frac{2^{\frac{2L_i}{3}}}{2^{L_i}} \leq \frac{1}{2^{4(l_i+\log d)}}$ using the birthday bound.

2. **V^π sends an "unlikely" message m:** Let p_m be the probability that V^π sends m conditioned on $trans_{i-1}$ being the transcript at the end of $i-1$ rounds. We say that m is unlikely if $p_m \leq \frac{1}{2^{2l_i+2\log d}}$. Using a union bound over all m we obtain that the probability of an unlikely m being sent is at most $2^{l_i}\frac{1}{2^{2l_i+2\log d}} = \frac{1}{2^{l_i+2\log d}}$
3. **For a "likely" message m, all trials fail:** The probability of a "likely" message m occurring is at least $> \frac{1}{2^{2(l_i+\log d)}}$. Therefore, the probability that all $2^{\frac{L_i}{3}}$ trials fails is at most

$$\left(1 - \frac{1}{2^{2(l_i+\log d)}}\right)^{2^{\frac{L_i}{3}}} \leq e^{-2^{2(l_i+\log d)}}$$

If the bad events do not occur in round i, then the message m is a "likely" message and some trial succeeds, which implies that $P^{*\mathsf{Sam}_d^\pi}$ does not abort in round i. Using the union bound, we obtain that $P^{*\mathsf{Sam}_d^\pi}$ aborts with probability at most $\frac{1}{d2^{l_i}}$ in round i. Using the union bound again over all the d rounds, the probability that $P^{*\mathsf{Sam}_d^\pi}$ aborts is at most $\sum_{i=1}^{d} \frac{1}{d2^{l_i}} \leq \frac{1}{2^L}$ where L is the length of the shortest message. Thus, $P^{*\mathsf{Sam}_d^\pi}$ aborts with probability at most $\frac{1}{2}$. ∎

Sub-Claim 3. *Conditioned on $P^{*\mathsf{Sam}_d^\pi}$ not aborting, the probability that $P^{*\mathsf{Sam}_d^\pi}$ succeeds in convincing V is identical to the probability that $\widehat{P}^{*\mathsf{Sam}_d^\pi}$ succeeds in convincing $\widehat{V}^{\mathsf{Sam}_d^\pi}$.*

Proof. Recall that, in every round $P^{*\mathsf{Sam}_d^\pi}$ samples public-coin messages a fixed number of times and aborts if none of them correspond to the private-coin message received externally from V^π. We observe that, the process that decides whether the random coins sampled by $P^{*\mathsf{Sam}_d^\pi}$ are the "right" ones depends only on the randomness of Sam_d^π and, in particular, is independent of the actual public-coin message sampled by $P^{*\mathsf{Sam}_d^\pi}$. Therefore, conditioned on $P^{*\mathsf{Sam}_d^\pi}$ not aborting, the messages fed internally to $\widehat{P}^{*\mathsf{Sam}_d^\pi}$ are uniformly distributed. Since, $\langle\widehat{P},\widehat{V}\rangle$ is a public-coin protocol, we have that the distribution of messages internally fed to $\widehat{P}^{*\mathsf{Sam}_d^\pi}$ is identically distributed to the messages generated in a real interaction with $\widehat{V}^{\mathsf{Sam}_d^\pi}$, and hence the probability that the internal emulation leads to a successful interaction is identical to the probability that $\widehat{P}^{*\mathsf{Sam}_d^\pi}$ succeeds in a real interaction. Recall that the acceptance condition in the internal emulation is decided by reconstructing a $\langle P^\pi, V^\pi\rangle$ transcript, sampling a fresh random tape consistent with the entire transcript and running V^π on that tape to obtain the verdict. By our construction, the transcript of the internal emulation with $P^{*\mathsf{Sam}_d^\pi}$ is identical to the transcript between $P^{*\mathsf{Sam}_d^\pi}$ and the external V^π. However, the random coins of the external V^π might not be the same as the ones sampled internally. Nevertheless, using the same proof as in Sub-Claim 1, it follows that conditioned on any complete transcript, the probability that the external verifier V^π and the internally emulated $\widehat{V}^{\mathsf{Sam}_d^\pi}$ accept are identical. ∎

∎

Remark 2. Note that in proof of Claim 2 we provide an algorithmic description of the cheating prover $P^{*\mathsf{Sam}_d^\pi}$ although we only need to contradict "unconditional soundness". This algortihm will be useful in proving the strong duality lemma (see Lemma 2) where consider also computationally-sound protocols.

Remark 3. The expected running time of $P^{*\mathsf{Sam}_d^\pi}$ in round i for a partial transcript $trans_{i-1}$ of the first $i-1$ rounds, is bounded by $\sum_m p_m \frac{1}{p_m} = 2^{l_i}$ where p_m is the conditional probability that V^π sent m in round i given $trans_{i-1}$. Therefore, the total expected running time of $P^{*\mathsf{Sam}_d^\pi}$ is at most $d \cdot 2^L = d2^L$ where L is the length of the longest message that V sends. If either the length of a message or the number of rounds is super-logarithmic, then the cheating prover P^* does not run in polynomial time. In the strong duality lemma, we show how to overcome this problem, as long as the verifier communication complexity is sublinear; this, however, requires querying Sam on larger depths.

Simulation. Let S^π be the simulator for $\langle P^\pi, V^\pi \rangle$. We construct a simulator $\widehat{S}^{\mathsf{Sam}_d^\pi}$ for $\langle \widehat{P}, \widehat{V} \rangle$ using S^π that has oracle access to Sam_d^π. Let $\widehat{V}^{*\mathsf{Sam}_d^\pi}$ be a malicious verifier for $\langle \widehat{P}, \widehat{V} \rangle$. On a high-level, $\widehat{S}^{\mathsf{Sam}_d^\pi}$ transforms $\widehat{V}^{*\mathsf{Sam}_d^\pi}$ to a verifier $V^{*\mathsf{Sam}_d^\pi}$ for $\langle P, V \rangle$ and simulates $V^{*\mathsf{Sam}_d^\pi}$ using S^π. The verifier $V^{*\mathsf{Sam}_d^\pi}$ with oracle access to $\widehat{V}^{*\mathsf{Sam}_d^\pi}$ and Sam_d^π proceeds as follows. In each round, on receiving a message from P^π, $V^{*\mathsf{Sam}_d^\pi}$ feeds that message to $\widehat{V}^{*\mathsf{Sam}_d^\pi}$. It obtains $\widehat{V}^{*\mathsf{Sam}_d^\pi}$'s next public-coin message r. $V^{*\mathsf{Sam}_d^\pi}$ queries Sam_d^π using r and generates the next message of V^π (i.e. generates a message following $\widehat{P}^{\mathsf{Sam}_d^\pi}$'s procedure) and forwards that to P^π. Finally, $V^{*\mathsf{Sam}_d^\pi}$ outputs what $\widehat{V}^{*\mathsf{Sam}_d^\pi}$ outputs. The simulator for $\langle \widehat{P}, \widehat{V} \rangle$, $\widehat{S}^{\mathsf{Sam}_d^\pi}$, internally incorporates S^π and verifier $V^{*\mathsf{Sam}_d^\pi}$, emulates an execution of S^π with $V^{*\mathsf{Sam}_d^\pi}$ and outputs what S^π outputs.

To show correctness of simulation, we need to show that $\langle \widehat{P}, \widehat{V} \rangle$ is $(\pi, \mathsf{Sam}_d^\pi)$-relativized zero-knowledge. Assume for contradiction, that there is a distinguisher $D^\bullet$ that can distinguish the simulation of $\widehat{V}^{*\mathsf{Sam}_d^\pi}$ by $\widehat{S}^{\mathsf{Sam}_d^\pi}$ from the real interaction for a random $(\pi, \mathsf{Sam}_d^\pi)$. More precisely, there exists a $\mathcal{PPT}$ distinguisher $D^\bullet$, polynomial $p(n)$, sequence $\{x_n, z_n\}_{n=1}^\infty$, $x_n \in L \cap \{0,1\}^n$, $z_n \in \{0,1\}^*$ such that for infinitely many n, $D^{\mathsf{Sam}_d^\pi}$ distinguishes the output of $\widehat{S}^{\mathsf{Sam}_d^\pi, \widehat{V}^{*\mathsf{Sam}_d^\pi}(x_n,z_n)}(x_n, 1^n)$ for a random $(\pi, \mathsf{Sam}_d^\pi)$ and the output of $\widehat{V}^{*\mathsf{Sam}_d^\pi}(x_n, z_n)$ in a real interaction with probability $\frac{1}{p(n)}$ (with probability taken over a random π, Sam_d^π). Using the Borel-Cantelli lemma, it follows that for measure 1 over permutations π, $D^{\mathsf{Sam}_d^\pi}$ distinguishes $\widehat{S}^{\mathsf{Sam}_d^\pi, \widehat{V}^{*\mathsf{Sam}_d^\pi}(x_n,y_n)}(x_n, 1^n)$ and the output of $\widehat{V}^{*\mathsf{Sam}_d^\pi}(x_n, y_n)$ in a real interaction with probability $\frac{1}{n^2 p(n)}$ for infinitely many n (with probability over Sam_d^π). Fix a $\pi = \{\pi_n\}_{n=1}^\infty$ for which this happens. It follows by the construction of $V^{*\mathsf{Sam}_d^\pi}$ that the output of $\widehat{S}^{\mathsf{Sam}_d^\pi}$ on $\widehat{V}^{*\mathsf{Sam}_d^\pi}$ is identically distributed to the output of S^π on $V^{*\mathsf{Sam}_d^\pi}$. We further claim that the output of $V^{*\mathsf{Sam}_d^\pi}$ in a real interaction with P^π is identically distributed to the output of $\widehat{V}^{*\mathsf{Sam}_d^\pi}$ with $\widehat{P}^{\mathsf{Sam}_d^\pi}$ (over a random Sam_d^π). The proof of this identically follows from the proof of Claim 1. Hence, $D^{\mathsf{Sam}_d^\pi}$ distinguishes the output of S^π with $V^{*\mathsf{Sam}_d^\pi}$ from the output of $V^{*\mathsf{Sam}_d^\pi}$ in a real interaction with

P^π with probability $\frac{1}{p(n)}$. Recall that, $\langle P, V\rangle$ is a fully black-box zero-knowledge based on one-way permutations, there exists a $\mathcal{PPT}$ machine $B^\bullet$, that with oracle access to $D^{\mathsf{Sam}_d^\pi}$ and $V^{*\mathsf{Sam}_d^\pi}$ inverts π (over a random Sam_d^π) for infinitely many lengths, for measure 1 over permutations π (and hence for a random π). From Theorem 2, we know that if $d \in o(\frac{n}{\log n})$, then no $\mathcal{PPT}$ machine with oracle access to Sam_d^π can invert a random one-way permutation π with more than negligible probability. Therefore, we arrive at a contradiction. This establishes that $\langle \widehat{P}, \widehat{V}\rangle$ is a $(\pi, \mathsf{Sam}_d^\pi)$-relativized black-box zero-knowledge proof. ■

4.2 Strong Duality Lemma

Lemma 2 (Strong Duality). *Let $\langle P^\bullet, V^\bullet\rangle$ be a d-round fully black-box zero-knowledge argument for a language L from one-way permutations with verifier communication complexity $c(n)$ and prover communication complexity $p(n)$. Then, there exists a d-round public-coin protocol $\langle \widehat{P}^{\mathsf{Sam}_D^\bullet}, \widehat{V}^{\mathsf{Sam}_D^\bullet}\rangle$ with the verifier communication complexity $O(Dc(n))$ (where $D = \frac{c(n)}{\log n}$) and prover communication complexity $p(n)$ that is $(\pi, \mathsf{Sam}_D^\pi)$-relativized black-box zero-knowledge argument.*

Proof. We modify the construction and proof from the previous lemma to obtain this lemma. From Remark 3, we know that the running time of P^*, is $d2^L$ where L is the length of the longest V-message. In order to use the previous construction and obtain an efficient P^*, we need the length of every verifier message to be logarithmic. Alternatively, if we split every message into segments of length $\log n$ bits and use the random tape sampled by Sam to generate one segment of the verifier message at a time, this also makes the running time of P^* polynomial. However, now, we need only to ensure the verifier's communication complexity is $o(n)$ (as this guarantees that the maximum depth is $o(\frac{n}{\log n})$). We note that the idea of splitting messages into segments of $\log n$ bits was used in [14] but their use of this technique is not sufficient for our application. More precisely, in [14] it is only shown how Sam can be used to generate a new random tape assuming that the original random tape was also generated using the random oracle. In our application, we need to be able to find a random tape for the "external" verifier. (As observed by Haitner in a personal communication, it would seem that by using our techniques (from Lemma 1) the results of [14] could be extended to rule-out also constructions that are secure with respect to only *honest-but-curious* players.)

We describe the procedure for generating a verifier message using Sam and the rest of the proof follows identically by plugging in this procedure wherever V^π's message is required to be generated. Without loss of generality, we assume that V^π's message in the ith round is a multiple of $\log n$, say $k \log n$. We describe how to sample the ith round message given $trans_{i-1}$, the partial transcript for the first $i-1$ rounds and a random string s_i. We first split s_i into k equal parts, $s_i^1, \ldots, s_i^k$. Using s_i^1, we samples a random tape r_1 for V^π consistent with $trans_{i-1}$ using Sam. We then run $V_{r_1}^\pi$ to generate only the first $\log n$ bits of

V^π's message in round i, say q_i^1. Next, we sample another random tape r_2 using s_i^2, but now r_2 is consistent with the extended transcript $trans_{i-1} :: q_i^1$. We run $V_{r_2}^\pi$ and obtain the next $\log n$ bits, q_i^2. In this manner we generate every $\log n$ segment up to q_i^k, each time ensuring that it is consistent with all previous segments. The depth of the maximum Sam_D^π query is total number segments counted over all verifier messages, i.e. $D = \frac{c(n)}{\log n}$ where $c(n)$ is the total verifier communication complexity. Since, $c(n) \in o(n)$, the maximum depth of a Sam query is $o(\frac{n}{\log n})$.

Completeness, Soundness and Zero-Knowledge. The proof of completeness follows exactly as before. As show for the weak-duality, we prove that if there exists a cheating prover $\widehat{P}^*$ for $\langle \widehat{P}, \widehat{V} \rangle$ that succeeds with probability p, there is a prover P^* with oracle access to Sam_D^π succeeds in $\langle P, V \rangle$ with probability $\frac{p}{2}$. The running time of P^*, computed as before, is $\frac{c(n)}{\log n} 2^{\log n} = (\frac{nc(n)}{\log n})$ which is polynomial since $c(n) \in o(n)$. Therefore, there exists a $\mathcal{PPT}$ prover P^* with oracle access to $\mathsf{Sam}_{o(\frac{n}{\log n})}^\pi$ that cheats with probability $\frac{p}{2}$. If $\langle \widehat{P}, \widehat{V} \rangle$ is not a $(\pi, \mathsf{Sam}_D^\pi)$-relativized argument for a random permutation π, then there is a sequence $\{x_n\}_{n=1}^\infty$, $x_n \in \overline{L} \cap \{0,1\}^n$, polynomial p such that P^* succeeds in convincing on V on x_n with probability $\frac{1}{p(n)}$ over a random permutation π for infinitely many n. Applying the Borel-Cantelli lemma, we again have that with measure 1 over permutations π, P^* cheats for infinitely many n. Using the fully black-box property, we have that for measure 1 over permutations π, there exists an adversary A that inverts π (and hence, for a random π); this violates Theorem 2 and we arrive at a contradiction. This completes the proof of soundness. To prove zero-knowledge, we use the same simulator from the weak duality, with the exception that it treats the verifier messages in $\log n$-bit segments. The rest of the proof follows as before. ■

We mention that the proof of the strong duality transformation shows that Theorem 2 (due to [13]) is optimal.

Theorem 3. *Let $f^\bullet$ be a function that on inputs of length n has output length $l(n)$. Then, for any π, there exists an oracle $\mathcal{PPT}$ machine $A^\bullet$, such that $A^{\mathsf{Sam}^\pi_{\frac{l(n)}{\log n}}}$ inverts f^π.*

Proof. First, we construct a 1-round protocol $\langle P^\bullet, V^\bullet \rangle$ for the empty language as follows: On input 1^n, V^π computes $y = f^\pi(r)$, where r is its random tape and sends y to P^π. P^π sends a string x' to V^π. V^π accepts if $f^\pi(x') = y$. Next, we apply the strong-duality transformation to the protocol $\langle P^\bullet, V^\bullet \rangle$ and obtain $\langle \widehat{P}, \widehat{V} \rangle$. In $\langle \widehat{P}, \widehat{V} \rangle$, $\widehat{P}^{\mathsf{Sam}^\pi_{l(n)/\log n}}$ on receiving a random string s from $\widehat{V}^{\mathsf{Sam}^\pi_{l(n)/\log n}}$, queries $\mathsf{Sam}^\pi_{l(n)/\log n}$ with input s and obtains a random tape r for V^π. $\widehat{P}^{\mathsf{Sam}^\pi_{l(n)/\log n}}$ runs V_r^π and obtains $f^\pi(r)$. $\widehat{V}^{\mathsf{Sam}^\pi_{l(n)/\log n}}$ accepts at the end, if $\widehat{P}^{\mathsf{Sam}^\pi_{l(n)/\log n}}$ can send r' such that $f^\pi(r') = f^\pi(r)$. Notice that $\widehat{P}^{\mathsf{Sam}^\pi_{l(n)/\log n}}$ knows r from $\mathsf{Sam}^\pi_{l(n)/\log n}$'s response, and can just forward r directly to $\widehat{V}^{\mathsf{Sam}^\pi_{l(n)/\log n}}$. Therefore, there is a cheating prover for $\langle \widehat{P}, \widehat{V} \rangle$, that succeeds with probability 1. From the proof of soundness of the strong duality lemma, we know how to construct a cheating prover $P^{*\mathsf{Sam}^{\pi}_{l(n)/\log n}}$ that convinces V^{π} with probability at least $\frac{1}{2}$. This means that $P^{*\mathsf{Sam}^{\pi}_{l(n)/\log n}}$ inverts f^{π} with probability at least $\frac{1}{2}$. The maximum depth of a query by $P^{*\mathsf{Sam}^{\pi}_{l(n)/\log n}}$ is $\frac{l(n)}{\log n}$. ■

We call a function *compressing* if it on inputs of length n has output length $o(n)$.

Corollary 1. *There exists no fully black-box construction of a compressing one-way function from one-way permutations.*

Proof. From Theorem 3, we have an adversary that inverts f^{π} with oracle access to $\mathsf{Sam}^{\pi}_{o(\frac{n}{\log n})}$. By the fully black-box property, we have an adversary A with oracle access to $\mathsf{Sam}^{\pi}_{o(\frac{n}{\log n})}$ that inverts π. Since this holds for every π, we arrive at a contradiction to Theorem 2. ■

5 Black-Box Lower Bounds for Zero Knowledge

All our black-box lower bounds follow by combining the weak or strong duality lemma with known lower bounds for public-coin protocols.

5.1 Lower Bounds Zero-Knowledge Proofs and Arguments

Goldreich-Krawczyk [8] show that only languages L in BPP have black-box constant-round public-coin zero-knowledge proofs. We remark that the proof of GK uses the simulator as a black-box to decide the language L, and relativizes. We therefore have:

Theorem 4 (Implicit in [8]). *Let $\langle P^{\bullet}, V^{\bullet}\rangle$ be a $\mathcal{O}$-relativized constant-round public-coin zero-knowledge proof for a language L with a black-box simulator S. Then, there exists a $\mathcal{PPT}$ machine $M^{\bullet}$, such that $M^{V^{O},S^{O}}$ decides L with probability $\frac{2}{3}$ when the probability is taken over a uniformly chosen $O \in \mathcal{O}$.*

Combining this theorem with the weak-duality lemma, we obtain the following corollary.

Corollary 2 (Constant-round Zero-Knowledge Proofs). *For any constant d, only languages L in CF_d have d-round fully black-box zero-knowledge proofs from one-way permutations.*

Proof. Let $\langle P, V\rangle$ be fully black-box zero-knowledge proof based on one-way permutations. Applying the weak duality lemma, we obtain a protocol $\langle \widehat{P}, \widehat{V}\rangle$ that is public-coin protocol where the prover, verifier and the simulator have access to Sam^{π}_{d}, that is $(\pi, \mathsf{Sam}^{\pi}_{d})$-relativized black-box zero-knowledge proof. Using Theorem 4, we have that $L \in \mathsf{CF}_d$. ■

We remark that if $\mathsf{NP} \not\subseteq \mathsf{CF}_{O(1)}$, then the corollary is tight; Goldreich, Micali and Wigderson [10] present a fully black-box construction of an $\omega(1)$-round protocol for NP based on one-way functions. On the other hand, Goldreich and Kahan [7], present a fully black-box $O(1)$-round zero-knowledge proofs for all of NP using claw-free permutations.[1]

Remark 4. A very recent work by Gordon, Wee, Xiao and Yerukhimovich [12] strengthens Corollary 2 by removing the usage of the random oracle π, and thus placing the class of languages having $O(1)$-round fully black-box zero-knowledge proofs from one-way permutations in $\mathsf{CF}'_{O(1)}$ (see Section 3.2). By relying on the recent work of [16], they obtain as a corollary that only languages in coAM have constant-round fully black-box zero-knowledge proofs from one-way permutations where the black-box simulator only makes a "constant number of adaptive queries" (where adaptive queries are defined in an appropriate way).

Using the strong-duality transformation, we obtain an analogous result for zero-knowledge arguments as well.

Corollary 3 (Constant-round Zero-Knowledge Arguments). *For any constant d, only languages L in $\mathsf{CF}_{o(\frac{n}{\log n})}$ have d-round fully black-box computational zero-knowledge argument based on one-way permutations where the total verifier communication complexity c(n) is sub-linear (i.e. o(n)).*

Proof. Applying the strong-duality lemma, there exists a protocol $\langle \widehat{P}, \widehat{V} \rangle$, that is public-coin protocol where the prover, verifier and the simulator have access to $\mathsf{Sam}^{\pi}_{o(\frac{n}{\log n})}$ that is $(\pi, \mathsf{Sam}^{\pi}_{o(\frac{n}{\log n})})$-relativized black-box zero-knowledge argument. Thus, using Theorem 4, we have that $L \in \mathsf{CF}_{o(\frac{n}{\log n})}$. ∎

If $\mathsf{NP} \not\subseteq \mathsf{CF}_{o(\frac{n}{\log n})}$, then the corollary is essentially tight. Feige and Shamir [5] and Pass and Wee [22] present an $O(1)$-round zero-knowledge arguments based on one-way functions. While, the former construction relies on one-way functions in a non black-box way, the latter is a fully black-box construction. Nevertheless, both the constructions require superlinear verifier communication complexity. On the other hand, efficient zero-knowledge arguments due to Kilian [17] have poly-logarithmic communication complexity, but are fully black-box based only on collision-resistant hash functions.

5.2 Lower Bounds for Concurrent Zero Knowledge

The notion of concurrent zero-knowledge introduced by Dwork, Naor and Sahai [4], considers the execution of zero-knowledge in a concurrent setting. That is, a single adversary participates as a verifier in many concurrent executions (see [23] for a formal definition and discussion). Analogous strong and weak duality transformation for concurrent zero-knowledge proofs and arguments follow directly by the proof of Lemma 1 and 2. We now turn to prove our lower bounds.

[1] Goldreich-Kahan use claw-free permutations to construct constant-round statistically-hiding commitments. However, these can be constructed under the potentially weaker assumption of collision-resistant hash functions [3,15]. Therefore, there also exists constant-round black-box zero-knowledge proofs for all of NP based on collision-resistant hash functions.

Recently, Pass, Tseng and Wikström in [21] prove that only languages in BPP have public-coin black-box concurrent zero-knowledge proofs or arguments. As the result of Goldreich-Krawczyk [8], this proof uses the simulator as a black-box to decide the language L, and relativizes. We therefore have:

Theorem 5 (Implicit in [21]). *Let $\langle P^\bullet, V^\bullet \rangle$ be a $\mathcal{O}$-relativized public-coin concurrent zero-knowledge proof (or argument) for a language L with a black-box simulator S (and negligible soundness error). Then, there exists a $\mathcal{PPT}$ machine $M^\bullet$, such that M^{V^O, S^O} decides L with probability $\frac{2}{3}$ when the probability is taken over a uniformly chosen $O \in \mathcal{O}$ and the internal coin tosses of M.*

As corollary of the strong and weak duality transformation for concurrent zero-knowledge, we obtain the following.

Corollary 4. *A language L has a $o(\frac{n}{\log n})$-round fully black-box concurrent zero-knowledge proof (or argument with $o(n)$ verifier communication complexity) based on one-way permutations, then $L \in \mathsf{CF}_{o(\frac{n}{\log n})}$.*

This result is tight if $\mathsf{NP} \not\subseteq \mathsf{CF}_{o(\frac{n}{\log n})}$; Prabhakaran, Rosen and Sahai [20] provide a fully black-box constructions of $\omega(\log n)$-round concurrent zero-knowledge proofs, or arguments with polylogarithmic communication complexity, based on collision-resistant hash functions; Pass and Wee [22] provide an $O(n)$-round fully black-box argument based on one-way functions.

Acknowledgements

We would like to thank the anonymous TCC reviewers for their helpful comments, and Iftach Haitner for pointing out the connection to the work of [14].

References

1. Babai, L., Moran, S.: Arthur-Merlin games: A randomized proof system, and a hierarchy of complexity classes. JCSS 36, 254–276 (1988)
2. Barak, B.: How to go Beyond the Black-Box Simulation Barrier. In: 42nd FOCS, pp. 106–115 (2001)
3. Damgård, I.B., Pedersen, T.P., Pfitzmann, B.: On the Existence of Statistically Hiding Bit Commitment Schemes and Fail-Stop Signatures. In: Stinson, D.R. (ed.) CRYPTO 1993. LNCS, vol. 773, pp. 250–265. Springer, Heidelberg (1994)
4. Dwork, C., Naor, M., Sahai, A.: Concurrent Zero-Knowledge. In: 30th STOC, pp. 409–418 (1998)
5. Feige, U., Shamir, A.: Zero Knowledge Proofs of Knowledge in Two Rounds. In: Brassard, G. (ed.) CRYPTO 1989. LNCS, vol. 435, pp. 526–544. Springer, Heidelberg (1990)
6. Goldreich, O.: Foundation of Cryptography – Basic Tools. Cambridge University Press, Cambridge (2001)
7. Goldreich, O., Kahan, A.: How to Construct Constant-Round Zero-Knowledge Proof Systems for NP. Journal of Cryptology 9(2), 167–189 (1996)

8. Goldreich, O., Krawczyk, H.: On the Composition of Zero-Knowledge Proof Systems. SIAM Jour. on Computing 25(1), 169–192 (1996)
9. Goldwasser, S., Micali, S., Rackoff, C.: The Knowledge Complexity of Interactive Proof Systems. SIAM Jour. on Computing 18(1), 186–208 (1989)
10. Goldreich, O., Micali, S., Wigderson, A.: Proofs that Yield Nothing But Their Validity or All Languages in NP Have Zero-Knowledge Proof Systems. J. ACM 38(1), 691–729 (1991)
11. Goldwasser, S., Sipser, M.: Private Coins versus Public Coins in Interactive Proof Systems. In: 18th STOC, pp. 59–68 (1986)
12. Gordon, S.D., Wee, H., Xiao, D., Yerukhimovich, A.: On the Round Complexity of Zero-Knowledge Proofs Based on One-Way Permutations (2009) (manuscript)
13. Haitner, I., Hoch, J., Reingold, O., Segev, G.: Finding Collisions in Interactive Protocols - A Tight Lower Bound on the Round Complexity of Statistically-Hiding Commitments. In: 48th FOCS, pp. 669–679 (2007)
14. Haitner, I., Hoch, J., Segev, G.: A Linear Lower Bound on the Communication Complexity of Single-Server Private Information Retrieval. In: Canetti, R. (ed.) TCC 2008. LNCS, vol. 4948, pp. 445–464. Springer, Heidelberg (2008)
15. Halevi, S., Micali, S.: Practical and Provably-Secure Commitment Schemes from Collision-Free Hashing. In: Koblitz, N. (ed.) CRYPTO 1996. LNCS, vol. 1109, pp. 201–215. Springer, Heidelberg (1996)
16. Haitner, I., Mahmoody-Ghidary, M., Xiao, D.: A constant-round public-coin protocol for sampling with size, and applications. Technical Report TR-867-09, Princeton University (2009)
17. Kilian, J.: A Note on Efficient Zero-Knowledge Proofs and Arguments. In: 24th STOC, pp. 723–732 (1992)
18. Kilian, J., Petrank, E.: Concurrent and resettable zero-knowledge in polylogarithmic rounds. In: 33rd STOC, pp. 560–569 (2001)
19. Ong, S.J., Vadhan, S.: An Equivalence between Zero Knowledge and Commitments. In: Canetti, R. (ed.) TCC 2008. LNCS, vol. 4948, pp. 482–500. Springer, Heidelberg (2008)
20. Prabhakaran, M., Rosen, A., Sahai, A.: Concurrent zero-Knowledge with logarithmic round complexity. In: 43rd FOCS, pp. 366–375 (2002)
21. Pass, R., Tseng, W., Wikström, D.: On the Composition of Public-Coin Zero-Knowledge Protocols. In: Halevi, S. (ed.) CRYPTO 2009. LNCS, vol. 5677, pp. 160–176. Springer, Heidelberg (2009)
22. Pass, R., Wee, H.: Black-box constructions of two-party primitives from one-way functions. In: Reingold, O. (ed.) TCC 2009. LNCS, vol. 5444, pp. 403–418. Springer, Heidelberg (2009)
23. Rosen, A.: Concurrent Zero-Knowledge. Springer, Heidelberg (2006)
24. Reingold, O., Trevisan, L., Vadhan, S.: Notions of reducibility between cryptographic primitives. In: Naor, M. (ed.) TCC 2004. LNCS, vol. 2951, pp. 1–20. Springer, Heidelberg (2004)
25. Simon, D.: Finding collisions on a one-way street: Can secure hash functions be based on general assumptions? In: Nyberg, K. (ed.) EUROCRYPT 1998. LNCS, vol. 1403, pp. 334–345. Springer, Heidelberg (1998)
26. Vadhan, S.: On Transformations of Interactive Proofs that Preserve Prover's Complexity. In: 32nd STOC, pp. 200–207 (2000)

Author Index